Statistical Data Analysis

Statistical Data Analysis

A Practical Guide

Complete with 1250 exercises and

answer key on CD

Milan Meloun
and
Jiří Militký

WOODHEAD PUBLISHING INDIA PVT LTD

New Delhi • Cambridge • Oxford • Philadelphia

Published by Woodhead Publishing India Pvt. Ltd.
Woodhead Publishing India Pvt. Ltd., G-2, Vardaan House, 7/28, Ansari Road
Daryaganj, New Delhi – 110002, India
www.woodheadpublishingindia.com

Woodhead Publishing Limited, 80 High Street Sawston Cambridge
CB22 3HJ UK
www.woodheadpublishing.com

Woodhead Publishing USA 1518 Walnut Street, Suite 1100
Philadelphia PA 19102-3406 USA

First published 2011, Woodhead Publishing India Pvt. Ltd.
© Woodhead Publishing India Pvt. Ltd., 2011

Woodhead Publishing India Pvt. Ltd. ISBN 13: 978-93-80308-11-1
Woodhead Publishing India Pvt. Ltd. EAN: 9789380308111

Woodhead Publishing Ltd. ISBN 13: 978-0-85709-109-3

Typeset by 3rdEyeQ, New Delhi
Printed and bound by Replika Press, New Delhi

Contents

The application of computer oriented statistical methods in scientific, medical, technical, and gnoseological fields enables not only the use of information hidden in data and the generalization of combined results from different sources, but also the creation of models, optimizations, and possible solutions. It is a multidisciplinary movement on the frontier of the scientific disciplines of statistics and informatics, which have led to the rise of new fields such as chemometrics, biometrics, psychometrics, econometrics, technometrics and others.

Focus of the book

The statistical analysis of data continues to gain in importance as an essential approach in numerous natural, medical, technical and social sciences. Contrary to classical mathematical statistics, the emphasis here is placed not on specific methods, but on their suitable combination, allowing the assessment of data quality and the selection of a suitable statistical model, including its verification. The goal of data processing and the level of expertise of the problems solved are always determinant factors, which affect the analysis approach and selection of methods used. This is particularly evident in the statistical analysis of data samples in practice. On one hand the diagnostics itself of hidden trends in data is an effective manner of study. Solving interesting and uncommon practical exercises allows one to better understand the limits and possibilities of various methods and to select through analogy the manner for processing one's own exercises. On the other hand, in practice many data samples lack a processing goal, which tends to be defined only vaguely, and no information is given about the current state of knowledge, known to only a limited circle of specialists. Practical exercises of data analysis also contain a number of unexpected obstacles:

- the extent of processed data is generally not large and due to reasons: of registration is full of variables which are insignificant for and the actual solution (redundant information),
- the data exhibits significant non-linearity, multicolinearity, non-additivity, and hidden ties, anomalies and influential points, which complicate data analysis,
- the distribution of experimental data rarely corresponds to the norm,
- the data exhibits deviant measurements, i.e. high heterogeneity,
- the parameters of the models sought have defined physical significance and their numeric estimates must therefore be suitable in terms of both size and sign.

In solving exercises in this book, we selected standard processing approaches aimed primarily at the statistical properties of data. This is exploratory data analysis to verify the basic assumptions about data with regard to the basic data model statistical method. The construction of statistical models here is more for illustration and the interpretation of results is merely general. Experts in the given scientific fields will clearly be able in a number of cases to recommend alternative solution goals or alternative interpretations of the results.

Intended audience

The submitted Guide represents a collection of methods with 1,100 exercises on computer processing of experimental data stored on CD. It is intended primarily for university students and graduates in the fields of technology, natural science, as well as humanities and economics. It is also intended for general reading and self study. It gives instructions on how any person can assess his or her own data. It does not contain definitions, principles or proofs of the mathematical theories of the materials practiced. The Guide has been written so that it can also be comprehended by readers with a high school education. The goal is to provide instructions for the logical assessment of data in everyday use. Along with the recommended statistical software, i.e. the use of certain programs such as ADSTAT, QC-Expert, NCSS2004, STATGRAPHICS, STATISTICA, MINITAB, SCAN, S-Plus and others, it provides an effective aid for the analysis of experimental data in laboratory research institutions, corporate and state inspection laboratories, and quality control centres. The Guide is also a welcome aid for managers of medical, veterinary, and water management laboratories, food and agricultural inspection agencies, chemical and pharmaceutical manufacturers. Suitable application may also be found in the field of environmental protection for all branches of industry, power generation and agriculture, for technology specialists, quality control employees, and especially management employees. With the selected appendixes examples, exercises, and data on the accompanying CD, the Guide may be directly applied in chemical laboratories or in chemometrics, biometrics and also in economics in econometrics, sociology, medicine, and in monitoring the environment. It is intended for people who are interested in revealing information contained in experimental data with the support of computers.

Chapter structure

The Guide is organized according to individual schematics: in the first part of each chapter there is a brief elucidation of the technique of data analysis. Where classic and well known statistical approaches are used there is only a brief introduction to the given problem. For less well known approaches however, there is a more detailed account to help comprehend the outputs of the example exercises. The solution to the example exercises including computer program details is also accompanied by interpretation. Individual chapters may be studied separately, even if the order of the chapters is intended to gradually acquaint the reader with all statistically related techniques.

On the accompanying CD there are practical exercises from various scientific fields along with interim results, indicating the correctness of the approach primarily from a numerical perspective. The exercises are partially taken from literature and the internet, although the vast majority come from the laboratory and industrial practice if participants in a post-graduate study of statistical data processing compiled over the past 18 years at the University of Pardubice.

Acknowledgement

It is not possible to thank all co-workers, students studying for their statistical data processing certification and post-graduate students who have helped us or contributed practical exercises, advice, and constructive criticism. We would like to give special thanks to the copyreaders for their general revision of the manuscript and a number of insightful practical comments and calculations. The production of this book was made possible by Ministry of Education, Youth and Sports financial grant for scientific projects no. MSM0021627502, and Ministry of Education, Youth and Sports project "Textile Center" 1M4674788501, for which the authors express their thanks.

December 2009

Milan Meloun and

Jiří Militký

Errors in instrumental measurements

Observation and experiment constitute the basis of all natural science. Observations and experiments that yield numbers – (the results of measurements – measurements) are of special significance. The correct analysis of such observations leads to a theoretical interpretation of the results, and to the final goal of natural science: the establishment of laws that make possible the prediction of the future behavior of the phenomena studied. The analysis of observations refers to operations on numbers that are obtained directly from said observations. To develop a theory based on the basis of the computation of quantities that are not directly observed, but which are derived from the analysis of observations, however, it is necessary to use various mathematical devices.

Chemistry, physics and other sciences deal with many quantities, some of which are purely physical (time, mass, volume, temperature and electrical parameters), some of which are physico-chemical (pH, potential, viscosity), and some of which are connected with the composition of the system. A description of composition requires the determination of a number of quantitative characteristics, and of the compositions of the components. Due to limited precision of measurement instruments, variation of experimental conditions and variability of measurement materials are results of measurements subjected to errors, i.e. are random variables.

A specialized subsystem is necessary in every branch of research to obtain data on compositions. This subsystem consists of the following operations:

(1) Sample preparation and treatment, including operations that are mostly carried out outside the laboratory, to ensure that the sample is representative.

(2) Sample preparation for measurement, including sample decomposition, separation operations, procedures defining the substance, and also the actual instrumental measurement. In instrumental methods, there is an attempt to combine the operations necessary for forming a measurable quantity with the actual measurement operation.

(3) Measurement, corresponding to the monitoring of the response signal (or often only the signal or measurement).

The signal usually corresponds to a physical quantity and is measured instrumentally. Two characteristics of the signal can be distinguished: signal magnitude (e.g. radiation intensity at a given wavelength) and signal position (e.g. wavelength). In identification analysis, the signal position is decisive, and its determination assumes a certain minimal size; in quantitative analysis, on the other hand, the signal magnitude is decisive. The magnitude of the signal, S, is in general a function of the quantities characterizing the test component, A, M, L, H, ... etc., and of the variable x_i which can be the reagent volume, temperature, etc.; $S = f(A, M, L, H; x_i)$. The measurement method is characterised by its sensitivity, defined as the derivative of the signal with respect to the quantities characterizing the test component

$$\left[\frac{\partial S}{\partial A}\right]_{M,L,H,x_i}$$ where the other quantities of M, L, H, and variables x_i are kept

constant so that the signal can be considered to be a function of a single variable A. The corresponding dependence $S = f(A)$ is then termed the *single calibration function*.

Since all measurements contain errors of diverse origin, the results of calculations made with numbers corresponding to these also contain errors. It is very important to be able to estimate both the errors incurred in making the measurements, and the errors resulting from operations on those measurements. Both the measurements and the calculations must be organized in such a way as to minimise the errors in the results.

Among errors in measurements a prominent place is occupied by random errors, i.e. errors with values that cannot be precisely predicted before the signal measurement. We might also note that they have random nature and can be described by proper probability density function, cannot be evaluated even after observation, since the presence of random errors makes it impossible for us to determine the exact value of the signal measured. The analysis of measurements containing random errors utilises the theory of probability, which is also necessary in statistical work. The measurements are treated according to the following rules:

(1) There is always a limit to the precision of measurements. Even results obtained with the best instrumental precision and experimental care are not precise, but have approximate values.

(2) Only some physical quantities can be measured directly, i.e. length, mass, and temporal intervals. Most other quantities are measured indirectly. For example, the relationship between the analytical signal and the analyte concentration is derived from a mathematical relationship and yields a mathematical model. The measurements results are approximate values, so the calculated indirect results must also be an approximate value.

(3) The arithmetic mean of repeated measurements corresponds to an estimate of the true value of the measured quantity; the true value, however, remains unknown.

(4) Finding empirical model functions is the first step in finding more basic relationships. In addition to the test quantities, the model also contains unknown parameters that are estimated from the experimental data by regression methods.

This chapter begins with the basic principles of statistical data treatment and theory of errors [1], so that the student may become familiar with the system of notation and terminology.

1.1 Types of Measurement Error

The results of laboratory measurements are always approximate; therefore, all measurements contain errors of diverse origin. It is customary to classify these errors according to their source in the measurement process into four types.

(i) *Instrumental errors* are caused by the construction of the instrument used and are usually known and specified by the instrument manufacturer.

(ii) *Methodology errors* are caused by use of an inappropriate method. Examples include inappropriate data acquisition, interference by certain external effects, faulty experimentation strategy, etc.

(iii) *Theoretical errors* are caused by the use of a false principle of measurement or inappropriate physical model, etc.

(iv) *Data treatment errors* are caused by the use of inappropriate numerical methods of data evaluation or statistical data treatment.

Errors may also be classified according to their effect on the evaluation of the results. Again, there are four types.

(1) *Systematic errors.* The most important systematic errors are instrumental errors, which result from incorrect instrumental settings, constant distortions, insufficient chemical purity, and imperfect standardisation and calibration.

Additive systematic errors (fixed bias errors) arise from simple instrumental errors, such as taking an imperfect or wrong zero point reading. In consequence because of such an error, all signal measurements are distorted by the same amount, which may be positive or negative.

A multiplicative systematic error (relative bias errors) depends in some definite way on other quantities, in particular, on the measured signal quantity itself. Systematic instrumental errors must be investigated and eliminated from the results of measurements.

(2) *Random errors.* Experiment has shown that successive measurements of a single fixed quantity, made with the greatest possible care, give different numerical values even after all the known systematic errors are allowed for. This shows that many causes have an effect on the results of measurements – causes for which we cannot make allowance. A whole series of similar random causes may produce deviations from an exact value. In each case, the deviation is slight; otherwise, it would be noticed and investigated. The total effect of all these causes can however yield significant deviations. The theory of errors usually refers to the theory of random errors; for the construction of such a theory, the nature of random errors suggests the application of probability theory.

(3) *Personal errors.* The results of measurements depend to some degree on the physical peculiarities of the observer (under otherwise equal conditions). For example, in recording the instant of a phenomenon, one observer may regularly notice a phenomenon somewhat sooner than another. Repeated study of the personal errors of different observations has shown that these errors can be both systematic and random. Some personal error is associated with the observer, and this error should be considered to be systematic and taken into consideration in the analysis of observations. Often observations can be made to determine the personal error, and the results of these observations are analyzed in much the same way as for random errors, in order to obtain their average value.

(4) *Gross errors (outliers).* In the analysis of observations, it is necessary to allow for the possibility of blunders or external influences that cause completely inaccurate results. One of the simplest of these would be for an observer to read 20.0 and write down 200, for example. The presence of gross errors is detected by the fact that in a succession of comparatively close results only one or a few values will differ appreciably from the general level of the values, i.e. will stand out. If the discrepancy is great enough for us to be sure that it is result of an error, the signal measurement can be disregarded.

Let us suppose that a certain measured quantity has a definite theoretical (numerical) value μ that remains unchanged during the entire process of measurement. Let us also suppose that the repeated measurement made of this quantity yields the values x_i, $i = 1, ..., n$. The difference between the exact and each actual value of measurement

$$\Delta_i = x_i - \mu \qquad (1.1)$$

is called the *absolute error.* This definition is convenient in that the concept of the absolute error coincides with the concept of the criterion additive errors since $x_i = \mu + \Delta_i$; that is, the absolute error is the number that must be added to the exact value μ in order to obtain the approximate number x_i.

When no gross errors are present in a set of n repeated observations, the average value

$$\bar{\Delta} = \frac{1}{n} \sum_{i=1}^{n} \Delta_i \tag{1.2}$$

represents the systematic error found in observations, and the difference $\bar{\Delta} - \Delta_i$ is the estimate of random error.

1.2 The Precision and Accuracy of Instrumental Measurements

In instrumental metrology [3] the terms accuracy and precision refer to characteristics of the measuring process. *Accuracy* refers to the typical closeness of n measurement results x_i, (or their average $\bar{x}$) to the true value μ while *precision* refers to the typical closeness of n measurement results x_i, together for the conceptually large population size n of results that might have been, or could be, obtained. When measurements of the same quantity are repeated, the dispersion of the results can be seen by examining the data. A set of data that shows little variation may be said to have greater precision than a set of data showing larger variation.

It is instructive to distinguish between random errors and systematic errors, and between precision and accuracy, in some cases typical of different measurement processes. Figure 1.1 shows four drawings, each of which depicts a distribution of the measurements x_i.

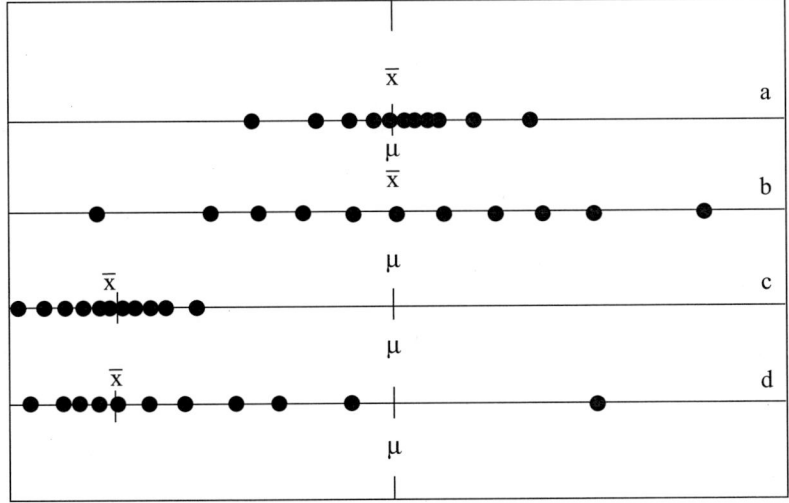

Figure 1.1   Classification of repeated measurements: (a) accurate and precise, (b) accurate but not precise, (c) not accurate but precise, and (d) not accurate and not precise

In case (a) each of the observations x_i is relatively close to the true value μ (accurate) and to the others (precise). In case (b) the individual observations do not have good accuracy, but their mean $\bar{x}$ would be reasonably accurate. The precision is relatively poor, as indicated by the wide spread of the data. In case (c), none of the observations is relatively close to the true value μ. The observations are close together and it can be said that the measurement process is precise but not accurate. Here the distinction between accuracy and precision is quite clear. In case (d), although one of the observations happens to be near the true value, the accuracy exhibited by most of the individual measurements is relatively poor, the accuracy exhibited by the mean of the set is relatively poor, and the precision of the set is relatively poor.

Only in case (a) can the measurement process be called accurate. In case (b) the accuracy can be improved by improving the precision, in case (c) the accuracy of the measurement process can be improved by correcting the systematic error, and in case (d) both factors need to be improved.

1.2.1 Absolute and Relative Errors

The calibration of an instrument requires that for known values of the input quantity x, (e.g., the concentration of hydrogen ions [H^+], in solution), the corresponding values of the measured output analytical signal y_i (e.g., the EMF of a glass electrode cell) be at disposal measured. Repeated measurements are made for each of several values of x_i so that the dependence $y = f(x)$ can be found. An approximate graphical interpretation shows the uncertainty band in the plot (Figure 1.2).

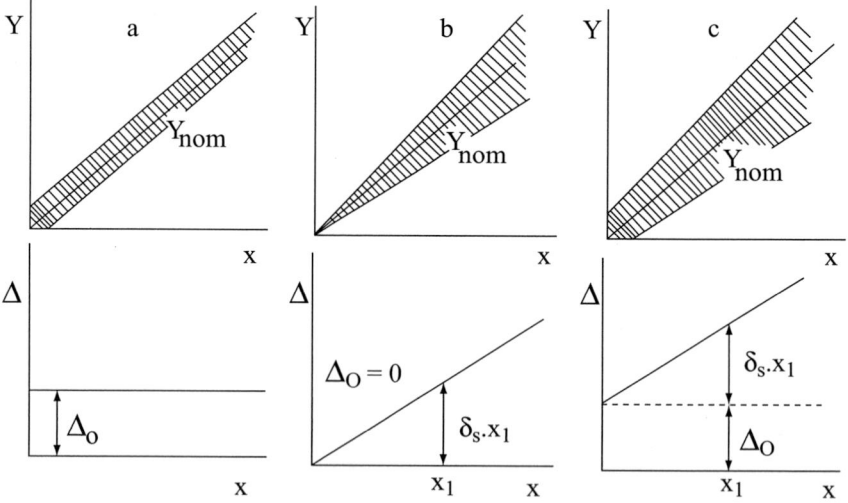

Figure 1.2 The uncertainty band and three types of instrumental errors: (a) additive, (b) multiplicative, and (c) combined error

The middle curve in the uncertainty band is called the nominal characteristic y_{nom} (or x_{nom}) and it is usually declared by the instrument manufacturer. The nominal characteristic y_{nom} (or x_{nom}) differs from the real characteristic y_{real} (or x_{real}) by the error of the instrument, $\Delta^* = y_{real} - y_{nom}$. For any selected y the error due to the instrument is $\Delta = x_{real} - x_{nom}$. [We will use errors Δ, which are in units of measurement quantities (e.g. pH)].

The *absolute error* of the signal measurement Δ is not convenient for expressing an instrument's precision because it is given in the specific units of the instrument used; more convenient is the *relative error* defined by

$$\delta = 100\, \Delta/x \qquad [\%] \qquad (1.3)$$

or the *reduced relative error* defined by

$$\delta_R = 100\, \Delta/(x_{max} - x_{min}) = 100\, \Delta/R \qquad [\%] \qquad (1.4)$$

where R is the range of measurements.

The *limiting error* of an instrument Δ_0 (absolute) or δ_0 (relative) is, under given experimental conditions, the highest possible error which is not obscured by any other random errors.

The *reduced limiting error* of an instrument $\delta_{0,R}$ (relative), for the actual value of the measured variable x_i and the given experimental conditions, is defined by the ratio of the limiting error Δ_0 to the highest value of the instrumental range R, $\delta_{0,R} = \Delta_0/R$. Often, the reduced limiting error is given as a percentage of the instrument range, [see Eq. (1.5)].

From the shape of the uncertainty band, various types of measurement errors can be identified, hence some corrections for their elimination may be suggested.

(a) Absolute measurement errors of an instrument are limited for the whole signal range by the constant limiting error Δ_0 that corresponds to the *additive errors model*. The systematic additive error is a result of the incorrect setting of an instrument's zero point. Today, however, instruments contain an automatic correction for zero, and systematic additive errors thus appear rarely.

(b) The magnitude of absolute measurement errors grows with the value of the input quantity x and for $x = 0$, it is $\Delta = 0$. This is a case of the *multiplicative errors model*. Such errors are called errors of instrument sensitivity. Systematic multiplicative errors are caused by defects of the instrument.

Real instruments have errors that include both types of effects, and are expressed by the nonlinear function $y = f(x)$.

Problem 1.1 *The absolute and relative errors of a pH-meter*

A glass electrode for pH measurement has a resistance of 500 MΩ at 25°C, and the input impedance of the pH meter is 2×10^{11} Ω. Estimate the absolute (Δ_0) and relative (δ_0) error of EMF measurement when the voltage measured is $U = 0.624$ V.

Solution: $U_{corr} = 0.624 \ (2 \cdot 10^{11} + 5 \cdot 10^8)/(2 \cdot 10^{11}) = 0.6254$ V; $\Delta_0 = 0.6254 - 0.624 = 0.0016$ V; $\delta_0 = 100 \cdot 0.0016/0.624 = 0.25\%$.

Conclusion: The absolute error is 1.6mV and the relative error 0.25%.

1.2.2 The Classification of Instrumental Precision

Limiting errors of measurement are used to express some metrological properties of instruments. These upper bounds of error will seldom be exceeded, and they also express the class of instrumental precision.

Class is an important instrumental precision parameter. It describes the highest absolute value of the reduced limiting errors found under given experimental conditions for the whole range of the instrument. To express the class of instrumental precision for additive, multiplicative and combined errors, the following limiting errors are used:

(1) For the additive errors model, the class of instrumental precision δ_0 is equal to the *limiting reduced relative error* $\delta_{0,R}$ defined by

$$\delta_{0,R} = 100 \ \Delta_0/(x_{max} - x_{min}) = 100 \ \Delta_0/R \qquad [\%] \qquad (1.5)$$

where R is the instrument range. The relative error δ decreases hyperbolically with an increasing value of x. The sensitivity limit x_c is the input value x for which the limiting absolute error Δ_0 is equal to x_c, i.e. $\Delta_0 = x_c$ or $\delta(x_c) = 100\%$. When the class of instrument precision δ_0 and the range R are known, the sensitivity limit can instead be calculated from

$$x_c = \delta_0 R/100 \qquad (1.6)$$

To ensure that the relative error of the instrument is sufficiently low, the lower limit of the working interval x_s is defined such that the relative error is kept equal to p (%), usually 4% or 10%. The lower limit of the working interval is defined as

$$x_s = 100 \ \Delta_0/p = 100 \ x_c/p \qquad (1.7)$$

When instrumental errors are additive, the range of use is limited to the region of low values of the input quantity x.

(2) In the case of the multiplicative errors model, the class of instrumental precision δ_s is expressed by the relative error of sensitivity calculated by

$$\delta_s = 100\,\Delta_0/x \qquad [\%] \tag{1.8}$$

and reaches a constant value in a limited range of the instrument scale; this will be declared by the instrument manufacturer.

(3) In the case of the combined errors model, the absolute error Δ may be written as the sum of the additive Δ_0 and multiplicative $\delta_s\,x$ parts by the expression

$$\Delta = \Delta_0 + \delta_s\,x \tag{1.9}$$

The combined uncertainty band is then formed by the addition of the two separate uncertainty bands. The limiting reduced relative error expressed by

$$\Delta_{0,R} = \delta_0 + \delta_s\,x/R \tag{1.10}$$

grows monotonically with increasing x. The growth of $\delta_{0,R}$ starts later for larger values of the ratio δ_s/δ_0. To express the level of instrumental precision in this case, two quantities are used:

(a) the reduced relative error δ_0,

(b) the relative error at the upper limit of the measurement scale δ_K expressed by

$$\delta_K = \delta_0 + \delta_s \tag{1.11}$$

Instruments can be classified according to their level of precision in the series: 6%, 4%, 2.5%, 1.5%, 1.0%, 0.5%, 0.2%, 0.1%, 0.05%, 0.02%, 0.01%, 0.005%, 0.002%, 0.001%, with the error type indicated by δ_S (for a multiplicative error), δ_0 (for an additive error) or δ_K/δ_0 (for a combined error), as follows.

(1) For pure multiplicative errors the level of instrumental precision is expressed by the relative error of the sensitivity δ_S, and is usually written as the number in a circle, e.g. $\delta_s = 1.5\%$ is written as ⓵.⑤.

(2) For pure additive errors the level of instrumental precision is expressed by the reduced relative error δ_0 and $R = x_{max}$ is the upper end of the scale range x_{max} when $x_{min} = 0$, e. g., $\delta_0 = 1.5\%$ is written 1.5.

(3) When the instrument has a strongly nonlinear scale the level of instrumental precision is expressed by the relative error and associated scale range $R = x_{max} - x_{min}$, e.g. 1.5, 10 and means $\delta_0 = 1.5\%$ and $R = 10$.

(4) For combined multiplicative and additive errors the level of instrumental precision is expressed by the ratio δ_K/δ_0, e.g., 1.5/1 and means $\delta_K = 1.5\%$ and $\delta_0 = 1\%$.

Based on the instrumental precision, it is possible to compute the maximum deviation likely to be caused by instrumental error (Table 1.1).

Table 1.1 Relative and absolute errors of an instrument expressed for the actual measured quantity x and the level of instrumental precision p or p_1/p_2

Type of error	Class of precision	Scale range	Relative error δ [%]	Absolute error Δ
Additive	p	x_{max}, $(x_{min} = 0)$	$p\,(x_{max}/x)$	$p\,x_{max}/100$
	p	$x_{max} - x_{min}$	$p\,(x_{max} - x_{min})/x$	$p\,(x_{max} - x_{min})/100$
Multiplicative	p	x_{max}, $(x_{min} = 0)$	p	$p\,x/100$
Combined	p_1/p_2	x_{max}, $(x_{min} = 0)$	$p_1 + p_2$ $(x_{max}/x - 1)$	$(p_1 x + p_2$ $(x_{max} - x))/100$

Problem 1.2   Determination of ammeter precision

If an ammeter with range $R = 60$ mA gives a mean reading $\bar{x} = 49.6$ mA when the true value of the electric current is 50 mA, what is the level of precision? Use the limiting absolute error and the limiting reduced relative error. Also, calculate the sensitivity limit.

Solution: $\Delta_0 = 50.0 - 49.6 = 0.4$ mA, $\delta_0 = 100 \times 0.4/60 = 0.67\%$ rounded off to the nearest larger value in the series of allowed values of instrument precision, i.e. 1%; [(Eq. (1.5)], $x_c = 0.67 \times (60/100) = 0.402$ mA [(Eq. (1.6)].

Conclusion: The class of instrumental precision is 1% and the sensitivity limit 0.402 mA.

Problem 1.3   Limiting and reduced values of errors for an ammeter

The manufacturer claims the following data for an ammeter: $p = 2$, $R = 60$ mA meaning that $\delta_{0,R} = 2\%$, $x_{min} = 0$ and $x_{max} = 60$ mA. Estimate the limiting absolute error Δ_0, the reduced relative error at x_{max}, i.e. $\delta_0(x_{max})$ and the reduced relative error at x_{min}, i.e. $\delta_0(x_{min})$ for this instrument.

Solution: $\Delta_0 = 2\,(60/100) = 1.2$mA; $\delta_0(x_{min}) = \infty$ for x $= 0$ mA; $\delta_0(x_{max}) = 2\,(60/60) = 2\%$ for $x_{max} = 60$ mA.

Conclusion: The relative error is not useful for expressing the error in a value close to zero. The minimum value for the reduced relative error is equal to the instrumental precision, i.e. 2% here.

Problem 1.4   Determination of the level of voltmeter precision

Determine the precision p for a voltmeter with a range from $x_{min} = 0$ to $x_{max} = 40$ mV, for which it is known that the error is a combination of additive and multiplicative errors. It was found that for $x = 10$ mV, $\delta_{0,R}(10) = 2\%$ and for 40mV, $\delta_{0,R}(40) = 5.2\%$ so that $\delta_K = 5.2\%$.

Solution: Since $\delta_K = 5.2\%$, δ_0 can be calculated from Eqs (1.10) and (1.11), i.e. $\delta_0 = (\delta_{0,R}(x) - \delta_K x/R)/(1 - x/R)$ where $R = x_{max} - x_{min} = 40$. Therefore $\delta_0 = 0.93\%$. The value δ_K is rounded off to the nearest higher value in the series of instrument precision, i.e. 6% and then δ_0 to 1%.

Conclusion: The voltmeter precision, expressed by the ratio δ_K/δ_0, is numerically equal to 6/1.

1.2.3 Decomposition of Measurement Error

The precision of instruments is expressed by the absolute error of the instrument Δ_{inst}, which represents the first part of the decomposed absolute error of the measurements signal, Δ_V. The second part of the measurements signal error is the variability of measured material Δ_M, the square of which is proportional to the variance σ^2. When the two parts of the error are uncorrelated, the decomposition of measurements signal error Δ_V may be expressed by the equation

$$\Delta_V = \sqrt{\Delta_{inst}^2 + \Delta_M^2} \qquad (1.12)$$

(1) With the most precise instruments, the smallest error of signal measurements Δ_V will be controlled by the material error Δ_M alone, so that $\Delta_{inst} \ll \Delta_M$. The precision of the measurements signal can be increased by making a greater number of repeated signal measurements n.

(2) For an instrument with error $\Delta_{inst} \approx \Delta_M/3$, the measurements signal error Δ_V will be only slightly higher than for a very precise instrument.

(3) For an instrument with $\Delta_{inst} \ll \Delta_M$, the error of the measurements signal will be $\Delta_V \approx 1.4\,\Delta_M$. For n repeated signal measurements, the error of measurements signal Δ_V will be decreased by $\sqrt{n}$, and consequently the random part of the instrument error Δ_{inst} will also decrease.

(4) For an instrument with $\Delta_{inst} \gg \Delta_M$, the error of signal measurement Δ_V will be proportional to the instrument error Δ_{inst}, i.e. $\Delta_V \approx \Delta_{inst}$. Repeated signal measurement cannot bring any improvement in the precision of the signal. An improvement of signal measurement is possible only with the use of a more precise instrument.

It may be concluded that a suitable choice of instrument is one with an error Δ_{inst} equal to $\Delta_M/3$ or less.

1.3 Models of Measurements

The statistical analysis of errors involves the statistical treatment of repeated measurements of the signal S. One of the following models of measurement is assumed:

(1) The additive model of measurements is the simplest: the ith measurement or observation is expressed by the equation

$$x_i = \mu + \varepsilon_i \qquad (1.13)$$

where ε_i represents the ith random error. It is obviously assumed that random errors have a mean of zero, constant variance and are not correlated. This model corresponds to the non-random signal μ, measured by an instrument that causes only random errors of measurement, ε_i; and the variability due to the material is equal to zero, $\Delta_M = 0$.

The additive model describes the measurement of a random signal variable ξ by its realisations x_i, with an ideally precise instrument for which $\Delta_{ins} = 0$. Here μ corresponds to the mean value.

A realistic approach involves a random variable ξ with a probability density function f_ξ, measured by an instrument which introduces errors ε with a probability density function f_ε. For an additive model of measurements [Eq. (1.13)] the probability density function f_x of the measured variable x may be expressed as

$$f_x(x) = \int_{-\infty}^{\infty} f_\xi(x-\varepsilon) f_\varepsilon(\varepsilon) d\varepsilon \qquad (1.14)$$

It may be concluded that

(a) if f_ξ is a probability density function with normal distribution $N(\mu, \sigma^2)$, and f_ε is a probability density function with normal distribution $N(0, \tau^2)$, the probability density function f_x will then also have a normal distribution, given by $N(\mu, \sigma^2 + \tau^2)$;

(b) the bimodal distribution may be formed by convolution of some unimodal symmetric distributions;

(c) if the variance of an instrument τ^2 is not constant, even for a normal distribution of f_ξ, the distribution of results is rather complex.

(2) The multiplicative model of signal measurements supposes that the errors have the following effect:

$$x_i = \mu \exp(\varepsilon_i) \qquad (1.15)$$

where the error ε_i has the same properties as in the previous model. The variable $\ln x_i$ has a distribution with mean $\ln \mu$ and variance σ^2. The variance of the measured quantities x_i is given by

$$\sigma^2(x) = x^2 \, \sigma^2 \qquad (1.16a)$$

and the corresponding relative error by

$$\delta(x) = \sigma(x) / x \qquad (1.16b)$$

For the multiplicative model, the relative error is constant. When random errors ε_i have a normal distribution, the measured variables x_i have a log-normal distribution.

(3) In a model with a systematic error, the measurement includes the systematic error of the instrument. The simplest model of this type is expressed by

$$x_i = \mu + \varepsilon_i + a \qquad (1.17)$$

where a is the constant systematic error of the instrument. If measurements are made at only one level (i.e. the level of μ), it is not possible to determine the systematic error a. If n replicate of measurements x are made at each of several different levels $\mu_j, j = 1, ..., m$, the resulting model

$$x_{ij} = \mu_j + \varepsilon_{ij} + a \qquad (1.18)$$

may be examined by one-way analysis of variance (Chapter 5). This model assumes that

(1) the mean error $\bar{\varepsilon}_j$ of repeated measurements is equal to zero;
(2) the variance of errors is constant;
(3) errors of repeated measurements are not correlated,
(4) errors at different levels μ_j are also not correlated [5].

To analyze a set of measurements, it is necessary to know the error distribution. Distribution determination needs a fairly large number of signal measurements. We will consider the four main types of error distribution which are illustrated in Fig. 1.3.

(1) A *rectangular distribution* occurs when the measurements have errors that are formed by rounding off the numbers.
(2) A *normal distribution* is observed for signal measurements when the errors ε result from the sum of the partial errors, and the measurements were performed at constant variance.
(3) A *Laplace (two-tailed exponential) distribution* is observed for signal measurements when the variance of the measurements oscillates around a mean value.
(4) A *log-normal distribution* is observed for signal measurements when the errors ε are a product of elemental errors. Measurements must

be positive and performed at constant relative error (coefficient of variation).

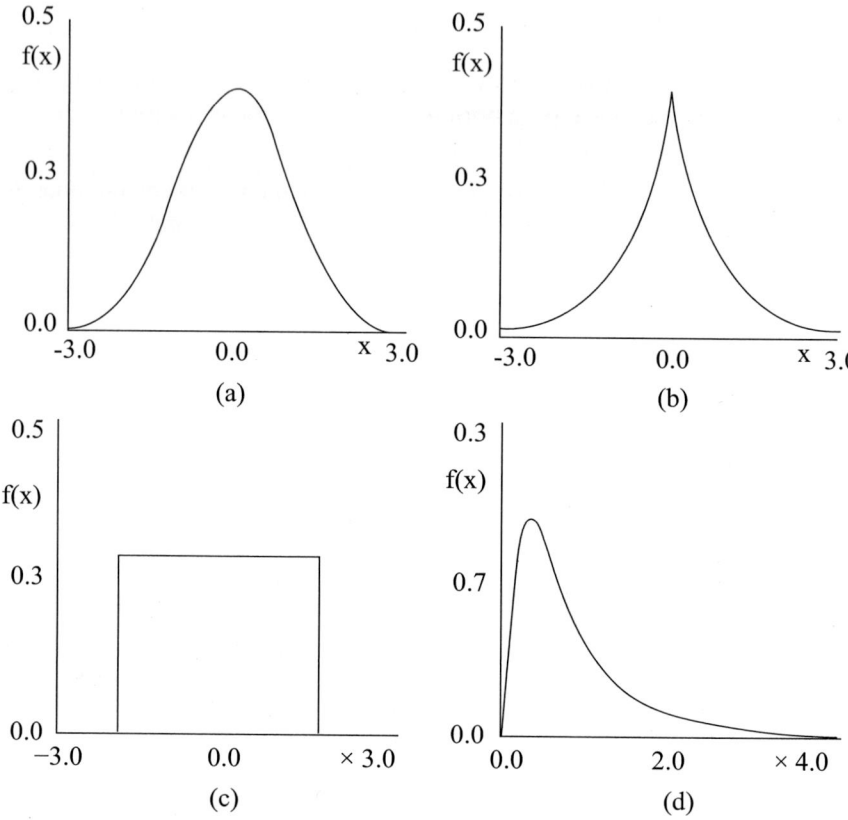

Figure 1.3   Selected distributions of random errors: (a) normal, (b) Laplace, (c) rectangular, and (d) log-normal

When the error distribution is bounded by some finite interval (e.g., as for a rectangular distribution) the limiting error of measurement may be calculated as half of the bounding interval. Because there is a need for exact expressions of error ε, various quantile and moment estimates of random errors are used.

Supplemented subchapters 1.4 through 1.10 are on enclosed CD

1.4 Error Propagation

The direct results x_j of measurements are always approximate, mainly because of the limited accuracy of measuring instruments. The results of analysis y are often calculated from several measured quantities $x_1, ..., x_m$ by the function $y = G(x_1, ..., x_j)$. We will formulate the following:

(1) the estimation of the results of the measurements, i.e. the mean value $\bar{y}$ and variance $s^2(y)$;

(2) the estimation of the total error of measurements $s(y)$ from the known errors of several measured quantities, $s(x_i)$;

(3) the reverse estimation of the limiting errors of the measured quantities $s(x_i)$ from the allowed error of the results of the measurements, $s(y)$.

The standard deviation $s(x_i)$ is convenient for the expression of the absolute error of the ith variable x_i; for the relative error of x_i the relative standard deviation (or coefficient of variation) is used

$$\delta(x_i) = s(x_i) / x_i \qquad (1.35)$$

For the first task the mean value $\bar{y}$ and variance $s^2(y)$ are evaluated from estimates of mean $\bar{x}_i$, variance $s^2(x_i)$, skewness $\hat{g}_{1,i}$, and kurtosis $\hat{g}_{2,i}$ calculated by the methods described in Chapter 3 are used.

For the second task the expressions for variance $s^2(y)$ as a function of the individual variances $s^2(x_i)$ are used. A simplification can be achieved by the use of relative errors.

For the third task the expressions for variance $s^2(y)$ or coefficient of variation $\delta(y) = s(y) / \bar{y}$ are used, with an assumption that the individual measured quantities x_i have the same relative effects.

To solve all three tasks the mean y and its variance $s^2(y)$ of a function $y = G(x_1, ..., x_m)$ must be evaluated. The estimates $\bar{y}$ and $s^2(y)$ may be obtained by any of the following methods:

(1) Taylor series expansion of the function $y = G(x_1, ..., x_m)$;

(2) two-points estimates;

(3) Monte-Carlo simulation.

While the Taylor expansion requires a knowledge of at least the first and second derivative of the function $y = G(x_1, ..., x_m)$, the other two methods may be computer-assisted and can often give more reliable results.

1.4.1 The Taylor Series Expansion Method

When a function of random variables is analysed, it should be noted that each non-linear transformation of the random variable distorts its distribution, and therefore changes the dependence of variance on the mean value. Where the measured variable x has a constant variance $\sigma^2(x)$, the results of analysis $y = G(x)$ have a non-constant variance

$$\sigma^2(y) = \left(\frac{dG(x)}{dx} \right)^2 \sigma^2(x) \qquad (1.36)$$

Moreover, the mean $\bar{y}$ cannot be estimated by direct substitution of the mean $\bar{x}$ into the function $G(x)$, i.e.

$$\bar{y} \neq G(\bar{x}) \tag{1.37}$$

To estimate the mean $\bar{y}$, the variance $s^2(y)$, and higher moments, the Taylor series expansion of function $G(x)$ can be used.

Suppose that the function $y = G(x_1,...,x_m)$ is known. Let $G(x)$ is a differentiable function. On writing the Taylor series expansion in the neighborhood of the vector of means $\bar{x} = (\bar{x}_1,...,\bar{x}_m)^T$ we obtain

$$G(x) \approx G(\bar{x}) + \sum_{i=1}^{m} \frac{\partial G(x)}{\partial x_i}(x_i - \bar{x}_i) + \frac{1}{2}\sum_{i=1}^{m} \frac{\partial^2 G(x)}{\partial x_i^2}(x_i - \bar{x}_i)^2$$

$$+ \sum_{i=1}^{m-1} \sum_{i>j}^{m} \frac{\partial^2 G(x)}{\partial x_i \, \delta x_j}(x_i - \bar{x}_i)(x_j - \bar{x}_j) + ... \tag{1.38}$$

where all first and second derivatives are calculated for the vector of mean values $\bar{x}$. By using a mean value operator E(.) at both sides of Eq. (1.38) the expression for the estimate of mean $\bar{y}$ may be written as

$$\bar{y} \approx G(\bar{x}) + \frac{1}{2}\sum_{i=1}^{m} \frac{\partial^2 G(x)}{\partial x_i^2}s^2(x_i) + \sum_{i=1}^{m-1}\sum_{i>j}^{m} \frac{\partial^2 G(x)}{\partial x_i \, \delta x_j}\mathrm{cov}(x_i,x_j) \tag{1.39}$$

where $\bar{y} = E(y) = E(G(x))$, $s^2(x_i) = E[(x_i - \bar{x}_i)^2]$ and where $E[(x_i - \bar{x}_i)] = 0$. The symbol $\mathrm{cov}(x_i, x_j)$ stands for the covariance which gives "a measure of linear dependence" between the two variables x_i and x_j.

Where the variance $s^2(y)$ is determined by an approximation [Eq. (1.38)], higher moments (i.e., the skewness and kurtosis) are neglected. The resulting approximate relation for variance is termed the rule of propagation of absolute errors and expressed by

$$s^2(y) \approx \sum_{i=1}^{m} \left[\frac{\partial G(x)}{\partial x_i}\right]^2 s^2(x_i) + 2\sum_{i=1}^{m-1}\sum_{i>j}^{m} \frac{\partial^2 G(x)}{\partial x_i \delta x_j}\mathrm{cov}(x_i,x_j)$$

$$+ \sum_{i=1}^{m-1}\sum_{i>j}^{m} \frac{\partial^2 G(x)}{\partial x_i \partial x_j}s^2(x_i)s^2(x_j) \tag{1.40}$$

The third term of Eq. (1.40) is usually neglected. When the resulting error $s(y)$ is formed from m sources of errors and each source has its own variance $s^2(x_i)$, the following expression for the error estimate can be used [8]

$$s^2(y) = \sum_{i=1}^{m} s^2(x_i) + 2\sum_{i=1}^{m-1}\sum_{i>j}^{m}\mathrm{cov}(x_i,x_j) \tag{1.41}$$

where $\text{cov}(x_i, x_j)$ is again a measure of the linear dependence between the two variables x_i and x_j. There are two limiting cases of estimation of the total error of measurements, $s(y)$:

(1) the sources of errors are independent, so that the covariance $\text{cov}(x_i, x_j)$ is equal to zero. The resulting estimate of the error will be proportional only to the quadratic mean of errors $s(x_i)$ coming from m sources,

$$s(y) = \sqrt{\sum_{i=1}^{m} s^2(x_i)} \qquad (1.42)$$

(2) the sources of errors are linearly dependent and the covariance cov (x_i, x_j) is given by $\text{cov}(x_i, x_j) = \sqrt{s^2(x_i) s^2(x_j)}$. The resulting estimates of the total error will be proportional to the arithmetic mean of errors $s(x_i)$ coming from all m sources

$$s(y) = \frac{1}{m} \sum_{i=1}^{m} s(x_i) \qquad (1.43)$$

For various experimental operations and measurements in a practice, the function $G(x)$ can be expressed by a power-type relationship

$$y = G(x) = x_1^{a_1} . x_2^{a_2} .... x_m^{a_m} = \prod_{i=1}^{m} x_i^{a_i} \qquad (1.44)$$

where a_i are known coefficients usually equal to ± 1. The estimation of the absolute error $s(y)$ or $s^2(y)$ by Eq. (1.40) is then rather complicated. Simplification results from the logarithmic transformation

$$\ln G(x) = \sum_{i=1}^{m} a_i \ln x_i \qquad (1.45)$$

Then

$$\frac{\partial \ln G(x)}{\partial x} = \frac{1}{G(x)} \frac{\partial G(x)}{\partial x} \qquad (1.46)$$

Substitution from Eq. (1.46) into Eq. (1.40) and rearrangement leads to a simplified form for the relative error (variation coefficient)

$$\delta^2(y) \approx \sum_{i=1}^{m} a_i^2 \delta^2(x_i) + 2 \sum_{i=1}^{m-1} \sum_{i>j}^{m} a_i a_j r_{ij} \delta(x_i) \delta(x_j) \qquad (1.47)$$

where r_{ij} represents the correlation coefficient expressing the closeness of linear dependence between variables x_i and x_j. Eq. (1.47) is called *the rule of propagation of relative errors*. The quality of the estimates $\bar{y}$, $s^2(y)$ and $\delta^2(y)$ is dependent on the quality of the approximation of the function $G(x)$ by the quadratic or linear function.

Although the estimate $\bar{y}$ is normally sufficiently accurate, some inaccuracy may be found in the estimation $s^2(y)$ [9].

Equation (1.47) may be used to estimate relative errors $\delta(x_i)$ such that the relative error of results $\delta(y)$ will not be greater than the selected value for H in %, i.e. $100\,\delta(y) \le H$. In solving this inversion problem, the independence of the measured variables x_i and the principle of the same relative influence $|a_1|\delta(x_1) \approx |a_2|\delta(x_2) \approx ... \approx |a_m|\delta(x_m) \approx H\,/\,m$ are assumed. Here a_p, $i = 1,..., m$, are coefficients of the function $G(x)$, [Eq. (1.44)]. For other types of function $G(x)$ the expression $a_i \approx |d\ln G(x)/dx_i|$ is used. For estimating the mean $\bar{y}$ the second derivatives of $G(x)$ play an important role.

For the case of a ratio x_1/x_2, an estimate of the mean $\bar{y}$ is controlled only by the variance $\sigma^2(x_2)$ and not by the variance $\sigma^2(x_1)$.

Problem 1.5 *Error in the isotope dilution method by Taylor's expansion*

Arsenic was determined by the method of isotope dilution. The initial specific activity was $a_2 = 3.7 \times 10^4$ s^{-1}. After addition of the standard, $m_1 = 5 \times 10^{-7}$g of arsenic, the specific activity was $a_1 = 5.3 \times 10^6$s^{-1}. Estimate the relative error of the arsenic content in the sample when the relative error of weighing is $\delta(m) = 0.03\%$, and the relative error of activity measurement $\delta(a_1) = \delta(a_2) = 1\%$.

Program: ADSTAT or QC-Expert: Basic Statistics: Error propagation.

Solution: The content of arsenic m_x in the samples is calculated by the relationship $m_x = m_1 (a_1 - a_2)/a_2$. Because this expression is not in the form of Eq. (1.44), Eq. (1.47) cannot be used. The relative error will be estimated by Eq. (1.35). Assuming that the quantities m_1, a_1, and a_2 are not correlated, substitution into Eq. (1.39) gives.

$$\bar{m}_x \approx m_1 (a_1 - a_2)\,/\,a_2 + m_1\,a_1\,s^2(a_2)\,/\,a_2^3 = 7.112 \times 10^{-5} + 7.162 \times 10^{-9}$$
$$= 7.113 \times 10^{-5}\text{g}$$

The variance is expressed by Eq. (1.40), after omitting the third term, as

$$s^2(m_x) = (a_1\,/\,a_2 - 1)^2\,s^2(m) + (m_1\,/\,a_2)^2\,s^2(a_1) - (m_1 a_1\,/\,a_2^2)\,s^2(a_2)$$

$$s^2(m_x) = (a_1\,/\,a_2 - 1)^2\,m_1^2\,\delta^2(m) + (m_1 a_1\,/\,a_2)^2\,(\delta^2(a_1) + \delta^2(a_2))$$

$$s^2(m_x) = 3.2 \times 10^{-18} + 1.0259 \times 10^{-12} = 1.0259 \times 10^{-12}.\ \text{The relative}$$
error is $\delta(m_x) = 100\,s(m_x)/m_x = 1.424\%$.

Conclusion: When an expression for the determination of results of the measurements analyte content or concentration is not in the form of Eq. (1.44), Eq. (1.35) should be used.

1.4.2 Method of Two-point Approximation

Manly's procedure [7] of two-point approximation is based on the replacement of the probability distribution of the function $G(x)$ by a two-point distribution with the same mean and variance. The estimate of the mean is then expressed by

$$\bar{y} \approx (G(\bar{x} + s(x)) + G(\bar{x} - s(x))) / 2 \qquad (1.48)$$

and the estimate of variance by

$$s^2(y) \approx (G(\bar{x} + s(x)) - G(\bar{x} - s(x)))^2 / 4 \qquad (1.49)$$

Both simple relationships give better results than Taylor's formula for a function of type (1.44). When the $G(x)$ is a function of m independent random variables $x_1, ..., x_m$, the summation of Eqs. (1.48) and (1.49) can be used

$$\bar{y} \approx \sum_{i=1}^{m} (G(\bar{x}_i + s(x_i)) + G(\bar{x}_i - s(x_i))) / 2m \qquad (1.50)$$

and

$$s^2(y) \approx \sum_{i=1}^{m} (G(\bar{x}_i + s(x_i)) - G(\bar{x}_i - s(x_i)))^2 / 4m \qquad (1.51)$$

1.4.3 Monte-Carlo Simulation Method

The mean $\bar{y}$ and variance $s^2(y)$, as a function of random variables x, may be determined by computer-assisted Monte-Carlo simulation methods. Schwartz [9] showed that this general procedure is well suited to the simulation of the statistical behavior of even rather complex systems. The following steps can be formulated:

(1) Selection of the function $G(x)$: for many chemical problems the function $G(x)$ is usually known. The great advantage of the Monte-Carlo simulation method is that the function $G(x)$ need not necessarily be expressed in explicit form.

(2) Distribution of the measured variables: in sciences it is usually assumed that measured variables are independent and have a normal distribution. The Monte-Carlo simulation method requires numerical values of the quantities $\bar{x}_i$, $s(x_i)$, $i = 1, ..., m$ only.

When these values are not available, two limiting values of the interval $[A, B]$ in which the variables x_i are expected should be supplied. The approximate probability density function is then expressed by the parabolic distribution

$$f(x_i) = 6(x_i - A)(B - x_i) / (B - A)^2 \qquad (1.52)$$

for $A \leq x_i \leq B$. The situation is more complicated when some correlation among the input variables exists. Here, the simultaneous distribution of all the variables x_i, $i = 1, ..., m$, should be specified; this will be simple only in cases of normal distribution.

(3) Generation of random numbers: most computer languages contain a function that will generate pseudo-random numbers from a rectangular distribution $R(0,1)$. For two independent random numbers, R_j, R_{j+1}, the Box-Müller transformation is used to generate two independent random numbers N_j, $N_j +_1$

$$N_j = \sqrt{(-2 \ln R_j)} \ \sin(2\pi R_{j+1}) \tag{1.53}$$

$$N_{j+1} = \sqrt{(-2 \ln R_j)} \ \cos(2\pi R_{j+1}) \tag{1.54}$$

which have a standardised normal distribution. The jth simulated values of the ith variable x_i will be expressed by

$$x^*_{i,j} = N_j \ s(x_i) + \overline{x}_i \tag{1.55}$$

For the parabolic distribution (1.52), the simulated quantity x^*_{ij} is the solution of the cubic equation

$$x^{*2}_{i,j} / 2 - x^*_{i,j} - x^{*3}_{i,j} / 3 + \alpha = R_i \ \beta \tag{1.56}$$

where $\alpha = A^3 - A^2 + A$ and $\beta = 6/(B - A)^2$.

(4) Selection of the number of simulations: the rules for the determination of the necessary number of simulations are the same as for the determination of sample size. The minimum number of simulations for the requested $100(1 - \alpha)\%$ confidence interval length D of the population mean is expressed by

$$n_{min} = [4u^2_{1-\alpha/2} \ s^2(y)] / D^2 + 1 \tag{1.57}$$

where $u_1 - \alpha/2$ is the quantile of standardised normal distribution and $s^2(y)$ is the estimate of variance from the first 50 simulations.

(5) Display of results: this includes a listing of an empirical probability density function for the distribution of the simulated data (y_j^*), $j = 1, ..., n_{min}$, and a subsequent calculation of the estimates of location and spread, $\overline{y}^*$ and $s(\overline{y}^*)$.

Problem 1.6 *Determination of the error of measured viscosity*

Calculate the viscosity of glycerol by the Stokes method, from the following experimental data: radius of the ball $r = 0.0112 \pm 0.0001$ m; density of the ball $d_0 = 1335$ kg m^{-3}, density of glycerol $d = 1280$ kg m^{-3}, trajectory

$l = 31.23 \pm 0.05$ cm, time $t = 62.1 \pm 0.2$ s, and acceleration due to gravity $g = 9.801$ m.s^{-1}.

Program: ADSTAT or QC-Expert: Basic statistics: Error propagation.

Solution: Viscosity η determined by the Stokes method is calculated from the expression $\eta = 2\, g\, r^2\, (d_0 - d)\, t\, /(9\, l)$. Because this expression is not of type (1.44), the relative error cannot be calculated from a simple relationship. Using the two-points approximation, the following values are calculated: $\overline{\eta} = 0.0299$ Pa s, $s(\eta) = 5.422\ 10^{-4}$ Pa s and the relative error $\delta(\eta) = 1.82\%$. By the Monte-Carlo simulation method $\overline{\eta} = 0.0299$ Pa s, $s(\eta) = 5.387\ 10^{-4}$ Pa sec, and $\hat{g}_1 = 0.038$ and $\hat{g}_2 = 2.77$.

Conclusion: The two methods, two-point and Monte-Carlo simulation, give the same results. The viscosity distribution is approximately symmetrical and flatter than a normal distribution.

1.5 Summary: The Determination of Measurement Errors

(1) The relative δ and absolute Δ_0 errors of signal measurement are calculated by using the expressions in Table 1.1. The sensitivity limit x_c (1.6) and the lower limit of the working interval x_s (1.7) are also calculated.

(2) The absolute error of measurements instrument Δ consists of an instrumental part Δ_V and a contribution from the variability of the measured material Δ_M, expressed by Eq. (1.12). Measurement errors models of the signal may include the additive model of errors (1.13), multiplicative model (1.15), or the model with a systematic error (1.17). Errors usually arise out of the rectangular, normal, log-normal or Laplace distributions.

(3) The measurement error may be estimated with the use of the interquantile range (1.19) for a given value of statistical certainty P. For $P = 0.5$ the resulting error $\sigma_{\Delta_{0.5}}$ is termed the mean error (1.20), and for $P = 0.683$, the probable error $\sigma_{\Delta_{0.689}}$. For $P = 0.9$ the limiting quantile error $\sigma_{\Delta_{0.9,i}}$ (1.22) can be added even if the distribution of partial errors $\sigma_{\Delta_{0.9,i}}$ is not known (1.23), or (1.24–1.31).

(4) The moment estimate of error with the use of standard deviation enables calculation of the probable error interval (1.32) which contains all the random errors with a probability $P = 1 - 2\alpha$ or of the tolerance interval of error (1.33–1.34) which uses the estimate of standard deviation.

(5) The final total error of an indirectly measured quantity (the concentration, the content, etc.) is often a result of the law of propagation of all kinds of errors concerning various experimental and instrumental operations. In addition to the classic method of Taylor series expansion (1.40−1.47), two computer-assisted methods may be applied, i.e. the two-point approximation (1.48−1.51) and the Monte-Carlo simulation method (1.52−1.57).

1.6 References

[1]   Taylor J. R.: *An Introduction to Error Analysis.* University Science Books, Mill Valey, California 1982.

[2]   Lyon A. J.: *Dealing with Data*, Pergamon Press, London 1970.

[3]   Zelený F.: *Základní vlastnosti měřicích přístrojů*, SNTL Praha 1976.

[4]   Novickij P. V., Zograf I. A.: *Oceňka pogrešnostej rezultatov izmerenij.* Atomizdat, Moskva 1985.

[5]   Hahn G. J., Nelson W.: *Technometrics* **12**, 95 (1970).

[6]   Mandel J.: *The Statistical Analysis of Experimental Data*, Interscience, New York 1964.

[7]   Manly B. F. J.: *Biom. J.* **28**, 949 (1986).

[8]   Müller J. W.: *Nucl. Instr. Meth.* **163**, 241 (1979).

[9]   Schwartz L. M.: *Anal. Chem.* **47**, 963 (1975).

[10] Shapiro S. S., Gross A. J.: *Statistical Modelling Techniques.* Marcel Dekker Inc., New York 1981.

[11] *Quantifying Uncertainty in Analytical Measurement,* EURACHEM 1995.

[12] Taylor B., Kuyatt CH. E. : *Guidelines for Evaluation and Expressing the Uncertainty of NIST Measurement Results,* NIST Tech. Note 1297, 1994.

[13] Agostini D. G.: *Probability and Measurement Uncertainty in Physic,* Rept. DESY 95−242, Roma December 1995.

[14] Phillips S. D., Eberhart K. R., Parry B.: *Guidelines for Expressing the Uncertainty of Measurement Results Containing Uncorrected Bias,* J. Res. Natl. Inst. of Standards **102**, 577 (1997).

[15] Meloun M., Militký J., Forina M.: *Chemometrics for Analytical Chemistry, Volume 1,* Ellis Horwood, Chichester, 1992.

[16] Meloun M., Militký J.: *Statistické zpracování experimentálních dat,* Plus Praha 1994 (1. vydání), EAST PUBLISHING, Praha 1998 (2. vydání), Academia Praha 2004 (3. vydání).

[17] Elishakoff I.: *Convex Modeling – a Generalization of Interval Analysis for Non-probabilistic Treatment of Uncertainty,* Proc. Int. Conf. APIC 95, El Paso, 1995 (a supplement to the international Journal of Reliable Computing).

[18] Ratschek, H. : *SIAM J. Numer. Anal.* **17,** 656 (1980).

Supplemented material (Review Questions, Exercises, Results of Exercises) to Chapter 1 is on enclosed CD.

2

The exploratory and confirmatory analysis of univariate data

The main aim of the exploratory analysis of univariate data is to isolate certain basic statistical features and patterns of data. *Exploratory data analysis* (*EDA*) often provides the first contact with the data and serves to uncover unexpected departures from familiar models [1]. An important element of the exploratory approach is flexibility in responding to the patterns uncovered by the successive steps of the analysis. Essentially, exploratory data analysis emphasises a flexible search for clues and evidence, whereas *confirmatory data analysis* (*CDA*) stresses evaluation of the available evidence. Four major facets of exploratory data analysis stand out:

(a) *Revelation* through visual display, which meets the need to look at the behaviour of data, diagnostic tests, fits and residuals, and thus to highlight unexpected features as well as the familiar regularities.

(b) *Resistance*, which provides insensitivity to localised misbehaviour in the data. Resistant methods are influenced mostly by the main body of the data, and less by outliers. Resistance ensures that a few extraordinary data values do not unduly influence the results of the analysis. We can distinguish between resistance and the related notion of robustness. *Robustness* generally implies insensitivity to departures from assumptions about an underlying model.

(c) *Residuals*, which focus attention on what remains of the data after some analysis, after a fitted model has been subtracted from the data, i.e. residual = measured data − calculated data.

(d) *Transformation* with subsequent re-expression of data, which involves finding a scale (e.g., logarithmic or square root) that can clarify the analysis of the data, or simplify the behaviour of the data. A transformation into another scale may help to promote symmetry, constancy of variability, linearity, or additivity of effect, depending on the structure of the data.

Procedure for univariate data analysis

(A) Exploratory data analysis (EDA):

Examination of the symmetry and shape of sample distribution;
Indication of local concentrations of the sample elements;
Detection of outliers and suspicious points in the sample;
Estimation of sample distribution; and
Power and Box-Cox transformations of the data.

(B) Confirmatory analysis of assumptions about the data:

Examination for minimum sample size;
Examination for independence of sample elements;
Testing for normality of sample distribution;
Testing for sample homogeneity.

Problem 2.1 *Analysis of data with normal and log.-normal distribution*

The exploratory data analysis of simulated sample data, from (a) normal distribution N(10, 0.1) denoted by *norm* and (b) log.-normal distribution ln (5, 2) denoted by *log.*

Data: the entire data set is included on the accompanying CD.

(a) Sample *norm*

10.0005	10.185	10.05	10.042	10.197	10.021	10.033	9.99985	9.826	10.076
10.053	10.079	9.9998	10.026	9.9969	9.98995	10.035	10.064	9.9985	10.093
10.132	10.047	9.877	9.931	10.002	9.929	9.959	9.846	10.029	10.029
9.994	10.113	10.158	9.999	10.1414	10.004	10.067	9.995	10.091	10.088
10.06	9.9998	10.017	9.865	9.907	10.037	10.081	10.018	9.987	10.115
10.037	10.063	9.928	9.975	9.937	9.933	9.942	10.106	10.039	9.989
9.906	9.894	9.946	9.955	9.98	10.108	10.05	9.948	9.974	9.986
9.986	10.105	10.037	9.955	10.025	9.949	9.879	10.042	10.052	9.92
10.064	10.075	10.028	9.955	9.987	9.957	9.969	9.9999	9.9995	10.021
10.069	9.975	10.109	10.024	9.984	10.122	9.885	10.011	10.013	10.011

(b) Sample *log*

2.408	5.389	2.259	2.439	2.173	1.157	0.892	0.498	0.351	1.229
1.356	4.719	1.445	1.023	1.723	0.572	2.012	0.212	0.305	0.993
11.993	2.247	0.973	0.418	2.27	12.03	1.321	3.076	1.355	4.54
0.216	10.159	0.346	1.078	0.206	0.116	1.733	0.55	0.762	2.689
1.798	1.522	2.763	0.536	0.21	2.462	0.516	0.421	1.588	2.54
7.48	0.881	0.841	1.039	0.966	0.49	1.476	1.185	0.875	0.557
1.464	0.308	0.097	1.137	2.247	0.084	0.217	1.885	0.204	2.786
2.341	0.466	0.712	0.401	0.404	1.027	0.623	0.139	2.905	0.111
0.958	0.188	0.611	0.243	5.331	0.745	0.367	0.919	1.236	1.912
2.816	0.666	4.972	0.451	1.316	3.241	0.316	2.2	8.291	0.815

Solution: Every individual diagnostic diagram or plot presents both samples, i.e. those from both the symmetric normal $N(10, 0.1)$ and asymmetric log.-normal $LN(5, 2)$ distributions. On the pages that follow, the reader can compare how all of the diagnostics monitor symmetric and strongly asymmetric distributions.

2.1 Sampling, Sorting and Ranking

In an ideal case, the known conditions of a chemical experiment fully determine the outcome. In practice, however, some factors are usually not fully controlled, and others are random in nature. The observations (responses) resulting from experiments are then *random quantities*. The complete collection of all possible outcomes from a chemical experiment, if the experiment is repeated an infinite number of times, is called the *population space*. Observations represent points in this population space. The population is *discrete* when there are a finite number of possible outcomes, and *continuous* when all real values are possible within a certain interval (finite or infinite), or series of intervals. When only one variable is recorded in an experiment, then the actual observations form a *univariate sample*. If more than one variable is obtained from a single experiment, a *multivariate sample* is obtained; e.g. if two values are obtained, the sample is *bivariate*. The aim of data analysis is to make *inferences* about population characteristics because of a *representative random sample* of items from the population. There are several reasons why it is usual to analyse a representative sample from the population rather than the whole population:

(a) The population, although finite, may be so large as to make all possible inspections too costly, or take too long a time.

(b) The experiment may involve a destructive process or the consumption of expensive chemicals.

(c) The whole population may not be available for analysis.

(d) The population may be infinite.

In practice the observation could result from an experiment under conditions which, for reasons outside the experimenter's control, may vary each time the experiment is repeated. The population in this case represents the set of observations that would be obtained if the experiments were repeated an infinite number of times.

A *sample* is said to be representative if it gives a sufficiently complete view of the population involved. All sample members have the same probability of being selected from the population, equal to $1/n$. If the experimenter has no prior information about the population, the only way to ensure representativeness is by random sampling or by impartial selection, which is given the statistical term *randomisation*. From the randomness of samples, it then immediately follows that any judgment passed on the population based on a sample is random as well.

The process of putting a set of numbers into order is known as *sorting*. Because an ordered sample batch makes it easy to pick out the letter values, as well as to detect possible stray values at either end, sorting is important in exploratory data analysis. The sample values $x_1, ..., x_n$ can be sorted such that $x_{(1)} < x_{(2)} < ... < x_{(n)}$. More formally, $x_{(1)}, x_{(2)}, ..., x_{(n)}$ are called the *order statistics* of the sample $x_1, x_2, ..., x_n$, and $x_{(i)}$ is the *i*th *order statistic* (see Fig. 2.1). Because of the sorting, we can define the *rank* of an observation in two ways: we can either count from the smallest value, or count down from the largest. The first of these yields the observation's *upward rank* that $R_{P_i} = i$ i.e. $x_{(2)}$ has upward rank 2 and, in general, $x_{(i)}$ has upward rank *i*. Counting down from the largest yields an observation's *downward rank* $K_{P_i} = n+1-i$ i.e. $x_{(n-1)}$ has downward rank 2, and generally, $x_{(i)}$ has downward rank $K_{P_i} = n+1-i$. Considering both of these rankings together, we see that for any data value $R_{P_i} + K_{P_i} = n+1$.

Sometimes it is useful to think in terms of the original observations. For example, if, through the sorting process, the raw observation x_i becomes the order statistic $x_{(j)}$, then the upward rank of x_i is *j*.

Often we want to give equal attention to both ends of a sample batch. A convenient way of handling this is to use the two ranks, upward and downward, in defining *depth*. The depth of the *i*th element in a sample is the smaller of its upward rank and its downward rank $H_i = \min(R_{P_i}, K_{P_i})$. The depth of each data value expresses how far it is from the low end or high end of the sample.

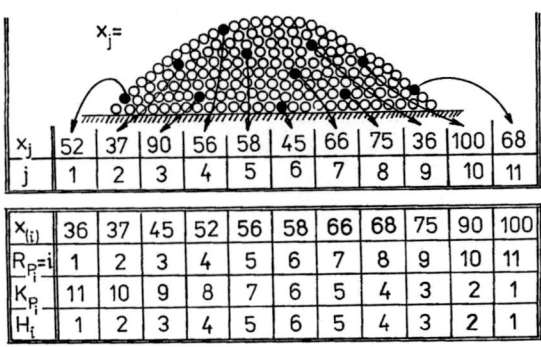

$x_j=$											
x_j	52	37	90	56	58	45	66	75	36	100	68
j	1	2	3	4	5	6	7	8	9	10	11

$x_{(i)}$	36	37	45	52	56	58	66	68	75	90	100
$R_{P_i}=i$	1	2	3	4	5	6	7	8	9	10	11
K_{P_i}	11	10	9	8	7	6	5	4	3	2	1
H_i	1	2	3	4	5	6	5	4	3	2	1

Figure 2.1   The sampling, sorting, ranking and depth of a sample. From the population of $N = 59$ values of the melting point of wax (°C), m.p. $= 63.00 + x/100$, the random sample of $n = 11$ values is taken by the selection of every fifth value. The order statistic $x_{(i)}$, the upward rank R_{P_i}, the downward rank K_{P_i}, and the depth H_i of the ith statistic, are shown.

2.2 Order Statistics, Quantiles and Letter Values

The method of exploratory data analysis (EDA) examines certain basic features of the statistical properties of the observations (experimental data). Graphical treatment of the data is used to identify the type of sample distribution, and to analyse and sometimes to re-express it. EDA is detective work that has a firm probability base and uses quantile descriptive statistics, whereas confirmatory data analysis is judicial or quasi-judicial in character.

Initially, the sample values $x_1, ..., x_n$ are sorted into ascending order to yield $x_{(1)} < x_{(2)} <.... < x_{(n)}$, the order statistics.

The P_ith *sample quantile* (or *percentile*) is defined to be the value of x below or at which $100 \times P_i\%$ of the sample values lie. The P_ith quantile is $\tilde{x}_{P_i} = x_{(j)}$ where $j = (n + 1)P_i$. Parameter P_i is usually termed as the *cumulative* or *rank probability* and is given by

$$P_i = \frac{i}{n+1} \tag{2.1}$$

If the index j is not an integer but lies between two integers m and $m + 1$, the P_ith quantile $\tilde{x}_{P_i}$ may be calculated by interpolating between $x_{(m)}$ and $x_{(m+1)}$ according to the formula

$$\tilde{x}_{P_i} = x_{(m)} + (j - m)(x_{(m+1)} - x_{(m)}) \tag{2.2}$$

For $P = 25\%$, 50%, 75% the 25th, 50th, and 75th quantiles (or percentiles) are called the *first* (or *lower*) *quartile*, the *second quartile* (or *median*) and the *third* (or *upper*) *quartile* of the sample.

The method of evaluating P_i depends on the nature of the sample distribution. The order statistics $x_{(i)}$ divide the real x-axis into $(n + 1)$ intervals, and any observation x will have the same $1/(n + 1)$ probability of appearing in any one of these. The cumulative probability is then given by $P_i = \dfrac{i}{n+1}$. For a normal distribution the expression $P_i = \dfrac{i - 3/8}{n + 1/4}$ is often used, but EDA uses

$$P_i = \frac{i - 1/3}{n + 1/3} \tag{2.3}$$

The plot of the order statistics $x_{(i)}$ against the cumulative probability P_i when $0 \leq P_i \leq 1$, for $i = 1,..., n$ is called the *quantile function* $Q(P)$. This is, in fact, an inverse function of the sample distribution function. For any value α from the interval $[0, 1]$, the 100αth quantile $\tilde{x}_\alpha$ may be calculated by linear interpolation

$$\tilde{x}_\alpha = x_{(i)} + (n+1)\left[\alpha - \frac{i}{n+1}\right](x_{(i+1)} - x_{(i)}) \tag{2.4}$$

where

$$\frac{i}{n+1} \leq \alpha \leq \frac{i+1}{n+1} \tag{2.5}$$

The variance of the sample quantile $\tilde{x}_\alpha$ for a sample size n is given by

$$D(\tilde{x}_\alpha) = \frac{\alpha(1 - \alpha)}{n\,[f(\tilde{x}_\alpha)]^2} \tag{2.6}$$

where $f(\tilde{x}_\alpha)$ is the value of the sample probability density function at point $\tilde{x}_\alpha$. An example of a quantile function is given in Fig. 2.2.

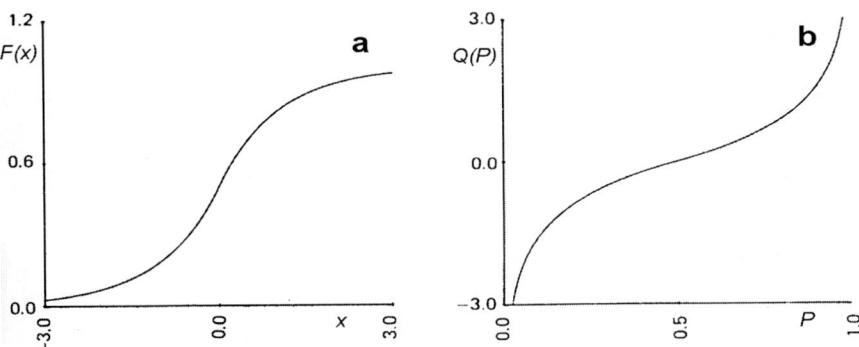

Figure 2.2 (a) The distribution function $F(x)$, and (b) the quantile function $Q(P)$ for a Laplace distribution with a mean of zero and variance of 2.

Some methods of EDA are based on certain selected quantiles Q being calculated for selected cumulative probabilities $P_{(i)} = 2^{-i}$, $i = 1, 2,...$ These quantiles are termed *letter values* (Table 2.1).

The symbol u_{P_i} is used to denote the quantiles of the standard normal distribution $N(0, 1)$ (Section 3.3.2). The median corresponds to $i = 1$, and for each $i > 1$ there is a pair of quantiles, the lower (Q_L) and upper (Q_U) letter values. The lower letter value is calculated for a cumulative probability $P_i = 2^{-i}$ and the upper for $P_i = 1 - 2^{-i}$.

Table 2.1 A survey of selected letter values

i	ith quantile	Cumulative probability	Letter value	Normal quantile u_{P_i}
1	Median	$2^{-1} = 0.500$	M	0
2	Quartiles	$2^{-2} = 0.250$	F or H	-0.674
3	Octiles	$2^{-3} = 0.125$	E	-1.15
4	Sedeciles	$2^{-4} = 0.0625$	D	-1.53

Letter values are estimated by the *rank-and-depth* method. The *rank* of an observation is defined by counting from the smallest value (*upward rank*), or by counting down from the largest (*downward rank*). The order statistic $x_{(i)}$ has an upward rank $R_{P_i} = i$ and a downward rank $K_{P_i} = (n + 1 - i)$. The depth H_i of the ith observation is defined as the lower value of the two ranks, R_{P_i} and K_{P_i}, i. e., $H_i = \min(R_{P_i}, K_{P_i})$, and the depth of the median is given by

$$H_M = \frac{n+1}{2} \tag{2.7}$$

If H_M is an integer (i.e. n is an odd number), then the median is equal to $\tilde{x}_{0.5} = M = x_{(H)}$; otherwise it is halfway between, $x_{(n/2)}$ and $x_{(n/2+1)}$. The depth of lower letter values is calculated from

$$H_Q = \frac{1 + \mathrm{int}(H_{Q-1})}{2} \tag{2.8}$$

where Q stands for the letters $F, E, D, \ldots$ and $\mathrm{int}(x)$ means the integer part of a number x. If Q is F, then we would say that $Q - 1 = M$, etc. When H_Q is an integer, the lower quantile Q_L is $x_{(H_Q)}$ while the upper quantile Q_U is $x_{(n+1-H_Q)}$. When H_Q is not an integer, the following linear interpolation is carried out

$$Q_L = \frac{x_{(\mathrm{int}(H_Q))} + x_{(\mathrm{int}(H_Q)+1)}}{2} \tag{2.9}$$

$$Q_U = \frac{x_{(n+1-\mathrm{int}(H_Q))} + x_{(n+2-\mathrm{int}(H_Q))}}{2} \tag{2.10}$$

For lower values of H_Q and quantiles near to $x_{(1)}$ and $x_{(n)}$, the procedure based on Eqs. (2.9) and (2.10) is more robust than that based on Eq. (2.4). The number of letter values for a sample depends on the sample size. For a given sample size n, this number, which includes the median, is given by

$$n_Q \approx 1.44 \ \ln(n+1) \tag{2.11}$$

The letters used as tags for the letter values start with M for median and F for fourths (quartiles), E for eighths (octiles), etc. The extremes have no tag other than 1, their depth. Letter values are used to provide a convenient summary of data, and the *5-number summary* (*1FMF1*) or the *7-number summary* (*1EFMFE1*) provide about the right amount of detail. More information is available in larger batches and we might use a fuller set of seven or more letter values if necessary (Fig. 2.3).

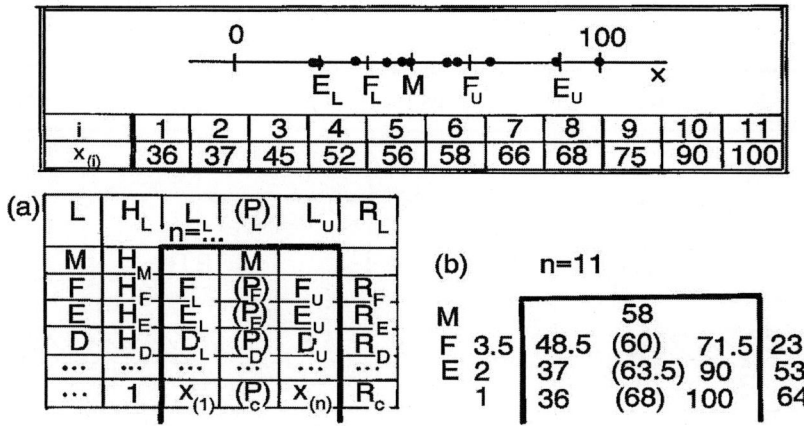

Figure 2.3 Data summarisation: (a) General construction of the skeleton of the letter-value display, (b) the letter-value display for a melting-point sample from Fig. 2.1.

Problem 2.2 *Use of the rank-and-depth method*

For the first 9 digits (1, 2,..., 9) determine letter values and both ranks, with depth.

Solution: The first row of Table 2.2 shows the order statistics $x_{(i)}$, the second row, the upward rank R, the third row the downward rank K, and the fourth row the depth H, calculated by Eq. (2.6). From Eq. (2.7), the depth of the median $H_M = (9+1)/2 = 5$ and the median is equal to $M = \tilde{x}_{(H_M)} = 5$. From Eq. (2.8), the depth of both quartiles is $H_F = 3$ and of the octiles $H_E = 2$. The letter values corresponding to the quartiles are $F_L = 3$ and $F_U = 7$, and to the octiles $E_L = 2$ and $E_U = 8$. The letter values in Table 2.2 are in a square. The corresponding diagram is shown in Fig. 2.4.

Table 2.2 The rank-and-depth method

$x_{(i)}$	1	2	3	4	5	6	7	8	9
R	1	2	3	4	5	6	7	8	9
K	9	8	7	6	5	4	3	2	1
H_i	1	2	3	4	5	4	3	2	1

Conclusion: The rank-and-depth method allows easy determination of letter values with pencil and paper.

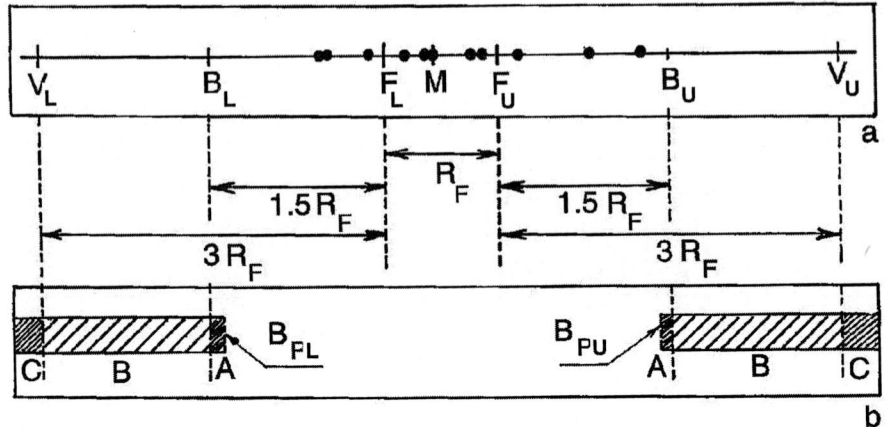

Figure 2.4 Dot diagram G2 (*x-axis* x values and on the *y-axis* a selected level, usually
$y = 0$) showing the letter values: (a) the dot diagram with median *M*, F_L
(lower) and F_U (upper) quartiles, inner B_L (lower) and B_U (upper) bounds,
and outer V_L (lower) and V_U (upper) bounds, (b) the area of outliers: A
close outliers, B near far outliers, C far outliers.

2.3 Plots and Displays in Exploratory Data Analysis

The basic features and statistical properties of experimental data are described
by the symmetry and kurtosis of the sample distribution, the variability of
the data, and the presence or absence of outliers. The various exploratory
diagnostic plots (EDA plots Gl–G23) offer information about these statistical
features of the data.

G1: Quantile plot

(*x-axis*: the cumulative (order) probability P_i; *y-axis*: the order statistic $x_{(i)}$)

The quantile plot permits identification of any peculiarities of the shape of
the sample distribution, which might be symmetrical or skewed to higher or
lower values. A real sample distribution can readily be compared with a normal
one if the quantile functions for the normal distribution $Q(u_p) = \mu + \sigma\, u_p$ for
$0 \leq P \leq 1$ are plotted on the same graph, with (1) the classical estimators of μ
and σ^2 ($\hat{\mu} = \bar{x}$ and $\hat{\sigma}^2 = s^2$) and (2) the robust estimators of μ and σ^2 ($\hat{\mu} = \tilde{x}_{0.5}$
and $\hat{\sigma}^2 = (R_F/1.349)^2$).

G2: Dot diagram

(*x-axis*: x values; *y-axis*: selected level, usually $y = 0$)

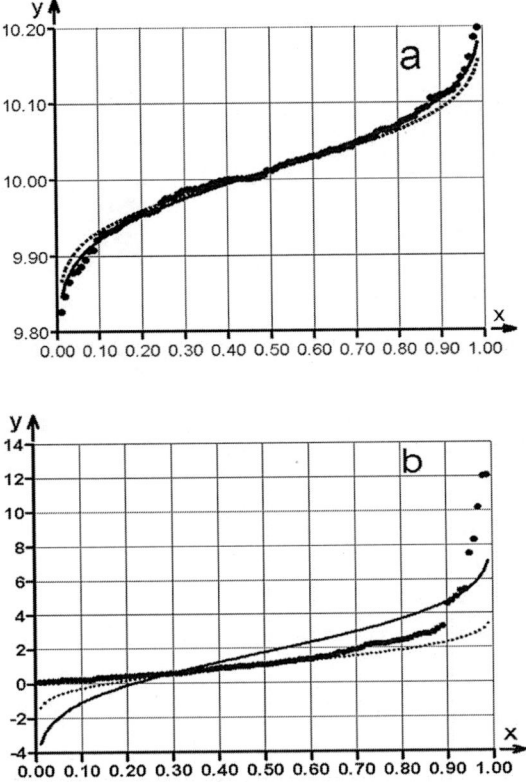

Figure 2.5 Quantile plot G1 (*x-axis*: the cumulative order probability P_i; *y-axis*: the order statistic $x_{(i)}$) for samples with (a) *norm*, symmetric (Gaussian, normal), and (b) *log*, asymmetric (log.-normal) distributions, (quantile functions for the normal distribution : classical – solid line, robust dotted line) *QC-EXPERT*.

The dot diagram is a univariate projection of the quantile plot onto the *x*-axis. It is a one-dimensional scatter plot of the data. The dot diagram indicates local concentrations of data, outliers, and extremes in the data. An example is given in the upper part of Fig. 2.6.

G3: Jittered-dot diagram

(*x-axis*: *x* values; *y-axis*: a small interval of random numbers)

The jittered-dot diagram is also a univariate projection of a quantile plot. The values of the sample points are randomly spread out in the *y*-direction, so this diagram gives a clearer view of the local concentration of points [2]. An example is shown in the lower part of Fig. 2.6.

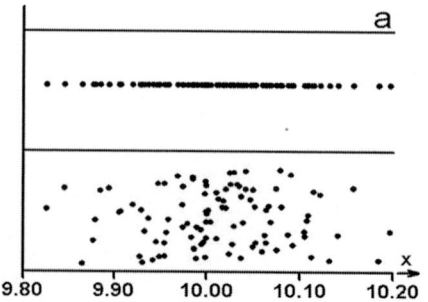
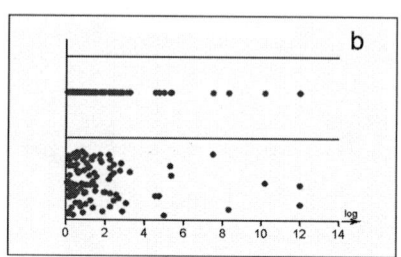

Figure 2.6   Dot diagram G2 above (*x-axis*: *x* values; *y-axis*: selected level, usually *y* = 0) and jittered dot diagram G3 below (*x-axis*: *x* values; *y-axis*: a small interval of random numbers), for samples with (a) *norm,* symmetric (Gaussian, normal), and (b) *log*, asymmetric (log.-normal) distributions, *QC-EXPERT.*

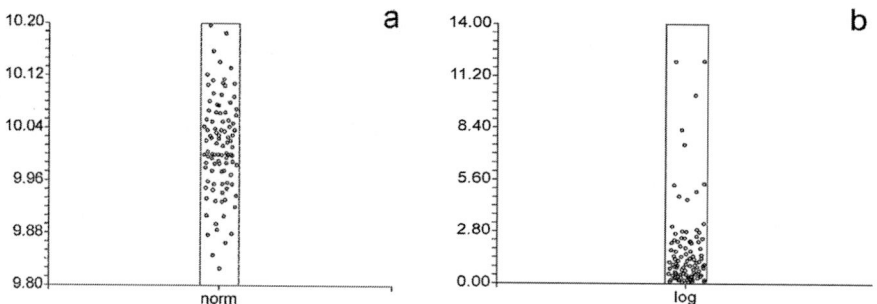

Figure 2.7   Another form of jittered dot diagram G3 (*x-axis*: *x* values; *y-axis*: a small interval of random numbers) for samples with (a) *norm,* symmetric (Gaussian, normal), and (b) *log,* asymmetric (log.-normal) distributions, *NCSS2000.*

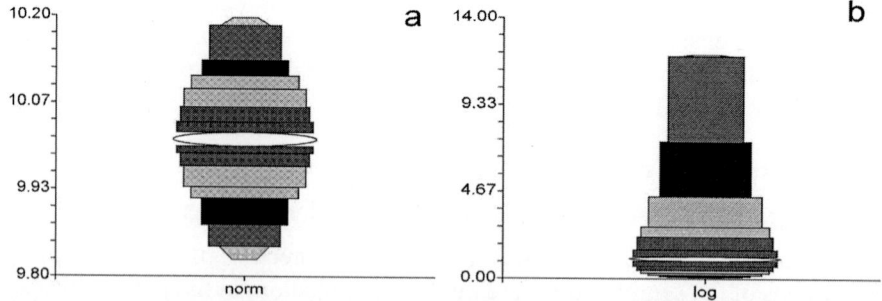

Figure 2.8   A percentiles diagram for samples with (a) *norm,* symmetric (Gaussian, normal), and (b) *log*, asymmetric (log.-normal) distributions, *NCSS2000.*

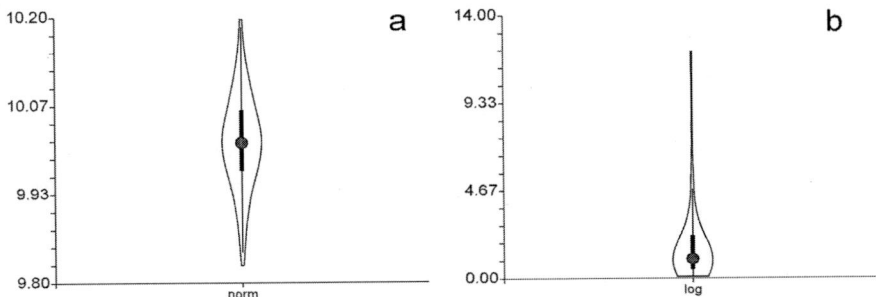

Figure 2.9 A violin diagram for samples with (a) *norm*, symmetric (Gaussian, normal), and (b) *log*, asymmetric (log.-normal) distributions, *NCSS2000.*

G4: Box-and-whisker plot

(x-axis: x values; y-axis: any suitable interval proportional to $\sqrt{n}$)

The box-and-whisker plot shows the *5-number summary overview* of the letter values in the form of the median, two quartiles (hinges) and two extremes. This plot permits the determination of a robust estimate of the median *M*, illustrates the spread and skewness of the sample data, shows the symmetry and length of the tails of the distribution, and aids in the identification of outliers.

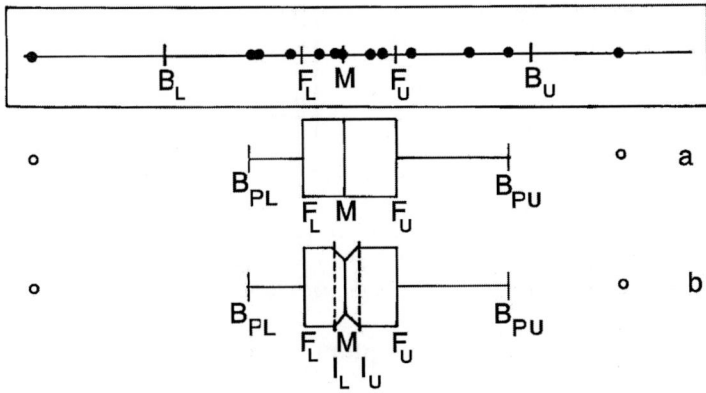

Figure 2.10 Scheme of the box-and-whisker plot (*x-axis*: x values; *y-axis*: any suitable interval).

The letter values are shown graphically in this plot. The skeletal box-and-whisker plot has a length from the lower quartile F_L to the upper F_U quartile equal to $R_F = F_U - F_L = \tilde{x}_{0.75} - \tilde{x}_{0.25}$ and the width is proportional to $\sqrt{n}$. A vertical crossbar inside the box marks the position of the median. The classic box-and-whisker plot is then completed by drawing lines (whiskers) out from each quartile to the corresponding extreme values $x_{(1)}, x_{(n)}$ at the ends of the order statistics. An example is given in Fig. 2.10. This plot is useful in illustrating the skewness of a sample. If the distribution has a long tail to the right (*positive skew*), the right-hand section of the box will be longer than the left, and the upper extreme point will be further from the median than the lower extreme. The converse will be true if the distribution has *negative skew*, with its longer tail to the left. In the modified box-and-whisker plot, the whiskers are terminated by the "adjacent" values B_{PU} and B_{PL}. These values lie just within the inner bounds defined by the cut-offs B_U and B_L, which are given by

$$B_U = F_U + 1.5\, R_F \qquad (2.12a)$$

$$B_L = F_L - 1.5\, R_F \qquad (2.12b)$$

For a sample from a normal distribution, $B_U - B_L \approx 4.2$. The probability that data lie outside this interval is 0.04. Observations outside the inner bounds (smaller than B_L or larger than B_U) are probable outliers, and are marked on the box-and-whisker plot by circles (Fig. 2.10 and Fig. 2.11).

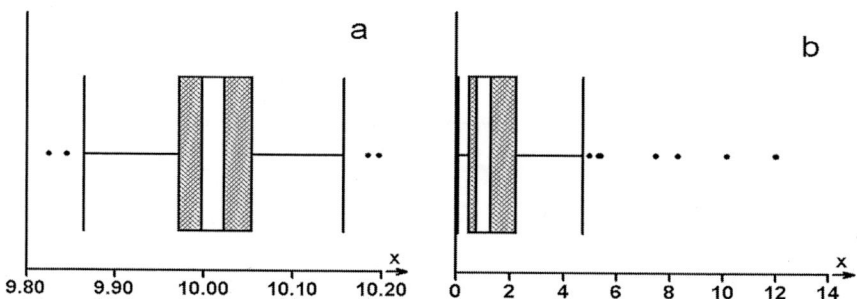

Figure 2.11 Construction of a notched box-and-whisker plot (G5, *x-axis*: x values, *y-axis*: any suitable interval) for samples with (a) *norm*, symmetric (Gaussian, normal), and (b) *log*, asymmetric (log.-normal) distributions. Empty circles indicate outliers, *QC-EXPERT*.

G5: Notched box-and-whisker plot

(*x-axis*: x values, *y-axis*: any suitable interval)

An analogue of the box-and-whisker plot is the notched box-and-whisker plot, which facilitates examination of the variability of the median. The median variability is expressed by notches given by the robust confidence interval $I_L \leq M \leq I_U$, where the lower and upper limits are

$$I_L = M - 1.57\, R_F / \sqrt{n} \qquad (2.13a)$$

$$I_U = M + 1.57\, R_F / \sqrt{n} \qquad (2.13b)$$

The notches I_L and I_U are placed symmetrically around the median. The properties of the notched box-and-whisker plot are similar to those of the G4 plot.

The main statistical features of the sample distribution are examined by comparing the asymmetry and tail lengths with those of a normal (Gaussian) distribution. The skewness and kurtosis can be characterised at various distances from the median through the following statistical characteristics:

the midsum $Z_Q = (L_L + L_U)/2$,

the interquantile range $R_Q = L_U - L_L$

the skewness $S_Q = (M - P_Q)/R_Q$

the pseudosigma $G_Q = R_Q/(-2u_{P_i})$

the length of tails $T_Q = \ln(R_Q/R_F)$.

where u_{P_i} is the quantile of the standardised normal distribution for $P = 2^{-i}$. These characteristics are summarised in Table 2.3.

Table 2.3 Characteristics of a distribution shape

Characteristic	Used for	Valid for L
Midsum Z_Q	Symmetry (at $Z_Q = 0$)	F, E, D
Interquantile range R_Q	Spread	F, E, D,
Skewness S_Q	Symmetry (at $S_Q = 0$)	F, E, D
Pseudosigma G_0	Kurtosis (for Gaussian distribution G_Q = const.)	F, E, D
Tail lengths T_Q	Kurtosis	E, D,

For any symmetric distribution, the theoretical length of the tails, T_E and T_D, can be computed: for a normal distribution, $T_E = 0.534$ and $T_D = 0.822$; for a rectangular distribution $T_E = 0.405$ and $T_D = 0.559$; and for a Laplace distribution, $T_E = 0.693$ and $T_D = 1.098$.

The skewness S_Q has negative values for distributions skewed to higher values, and positive values for distributions skewed to lower values. For distributions with longer tails than a normal distribution, the values of pseudosigma G_Q increase with distance from the median. When the values of pseudosigma G_Q decrease with distance from the median, the sample distribution has shorter tails than a normal distribution.

To examine all the statistical features of the sample, various plots of the characteristics from Table 2.4 are used. For large samples, the letter values are examined, whereas for small samples the quantile $\tilde{x}_{P_i} = x_{(i)}$ usually for $P_i = (i - 1/3)/(n + 1/3)$ is used.

G6: Midsum plot

(x-axis: the order statistic $x_{(i)}$; *y-axis:* the midsum $Z_i = (x_{(n+1-i)} + x_{(i)})/2$)

The midsum plot gives information about the symmetry of a distribution. For a symmetrical distribution, the midsum plot forms a horizontal line $y = M$. This plot is a sensitive indicator of distributional asymmetry. Ideally, the points should lie on a horizontal line. The central horizontal line corresponds to the median, and the dashed lines to its confidence limits. When the data distribution is asymmetric, the plot shows a clear trend (increasing where there is negative skewness and decreasing where there is positive skewness), going far beyond the dashed lines. Pairs of data points (first–last, second–second largest, etc.) are used when constructing the plot, so that when selecting a point on the plot, two data points are marked on the data table.

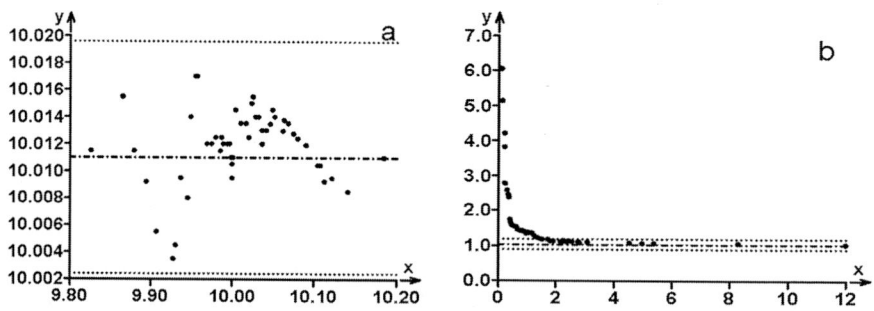

Figuer 2.12   The midsum plot G6 (x-axis: the order statistic $x_{(i)}$; y-axis: the midsum $Z_i = (x_{(n+1-i)} + x_{(i)})/2$) for samples with (a) *norm,* symmetric (Gaussian, normal), and (b) *log,* asymmetric (log.-normal) distributions, QC-EXPERT.

G7: Symmetry plot

(x-axis: the quantile $u_{P_i}^2/2$ for $P_i = i/(n + 1)$; *y-axis:* the midsum $Z_i = (x_{(n+1-i)} + x_{(i)})/2$)

For a symmetrical distribution, the symmetry plot forms the horizontal line $y = M$. When this line has a non-zero slope, the slope gives an estimate of the skewness [3]. It has a similar use to the mid-sum plot described above. The slope of a trend is proportional to the skewness. When the data distribution is asymmetric, the plot shows a clear trend (increasing for a negative skewness and decreasing for a positive skewness), going far beyond the dashed lines. Pairs of data points (first–last, second–second largest, etc.) are used when constructing the plot, so that when selecting a point on the plot, two data points are marked on the data table.

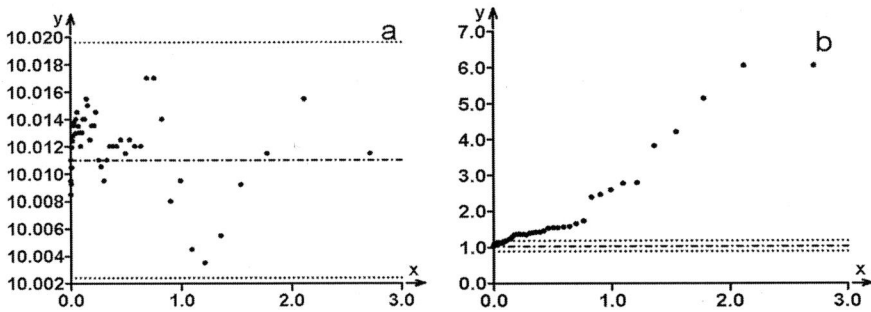

Figure 2.13   The symmetry plot G7 (*x-axis*: the quantile $u_{P_i}^2 / 2$ for $P_i = i/(n + 1)$; *y-axis*: the midsum $Z_i = (x_{(n+1-i)} + x_{(i)})/2$) for samples with (a) *norm*, symmetric (Gaussian, normal), and (b) *log*, asymmetric (log.-normal) distributions, QC-EXPERT.

G8: Kurtosis plot

(*x-axis*: the quantile $u_{P_i}^2 / 2$ for $P_i = i/(n + 1)$; *y-axis*: the quantity $\ln\left[(x_{(n+1-i)} - x_{(i)})/(-2u_{P_i})\right]$

The kurtosis plot indicates the peakedness of a distribution. For a normal distribution, the kurtosis plot gives a horizontal line. When the line has a non-zero slope, the value of the slope gives an estimate of the kurtosis [3]. The meaning is analogous to the previous two plots. The slope of the trend is proportional to the difference (kurtosis-3). When the kurtosis is very different from normal, the plot shows a clear trend. Pairs of data points (first–last, second–second largest, etc.) are used when constructing the plot, so that when selecting a point on the plot, two data points are marked on the data table.

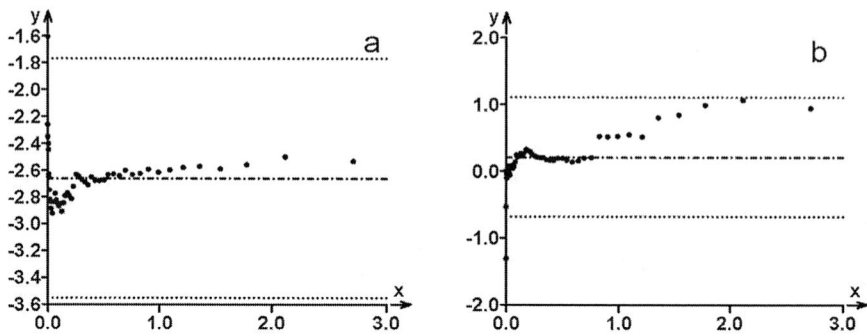

Figure 2.14   The kurtosis plot G8 (x-axis: the quantile $u_{P_i}^2 / 2$ for $P_i = i/(n + 1)$; y-axis: the quantity $\ln[(x_{(n+1-i)} - x_{(i)})/ -2u_{P_i}]$ for samples with (a) *norm*, symmetric (Gaussian, normal), and (b) *log*, asymmetric (log.-normal) distributions, *QC-EXPERT*.

G9: The differential quantile plot

(x-axis: the quantile u_{P_i}; y-axis: the deviation of order statistics $d_{(i)} = x_{(i)} - \tilde{s}u_{P_i}$)

The differential quantile plot compares the sample distribution with a normal one. The statistic $\tilde{s}$ represents a robust estimate of the standard deviation, calculated for example by the use of the interquantile range. A horizontal line indicates a symmetrical distribution with tails similar to the normal distribution.

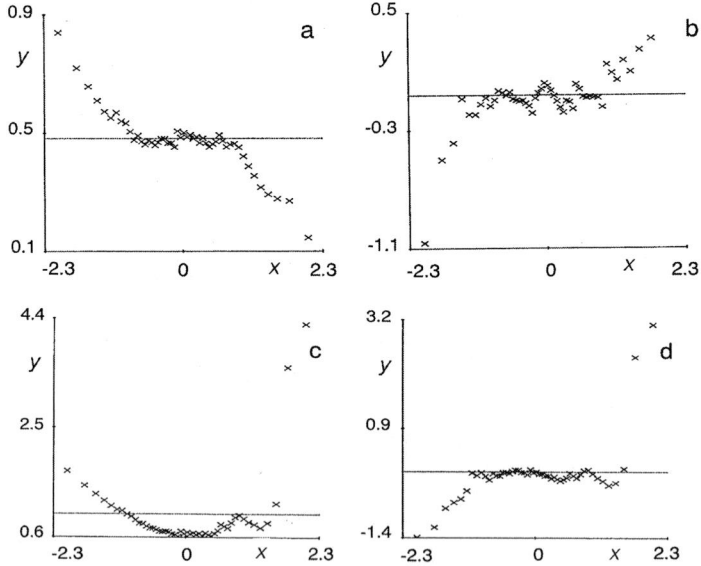

Figure 2.15   The differential quantile plot G9 (x-axis: the quantile u_{P_i}; y-axis: the deviation of order statistics $d_{(i)} = x_{(i)} - \tilde{s}u_{P_i}$) for samples with (a) rectangular, (b) normal, (c) exponential, and (d) Laplace distributions, *QC-EXPERT*.

G10: Quantile-box plot

(*x-axis*: the order probability P_i; *y-axis*: the order statistic $x_{(i)}$)

The quantile-box plot (Fig. 2.16) is a simple and universal tool for examining the statistical features of data. The plot is based on an estimate of sample quantile function formed by connecting points $\{x_{(i)}, P_i\}$ by straight lines. P_i is calculated from $P_i = (i-1/3)/(n+1/3)$. For symmetrical distributions, the sample quantile function has a sigmoid shape, whereas for asymmetrical distributions the quantile function is convex or concave increasing. For ease of interpretation, the following quantile boxes are included on the graph:

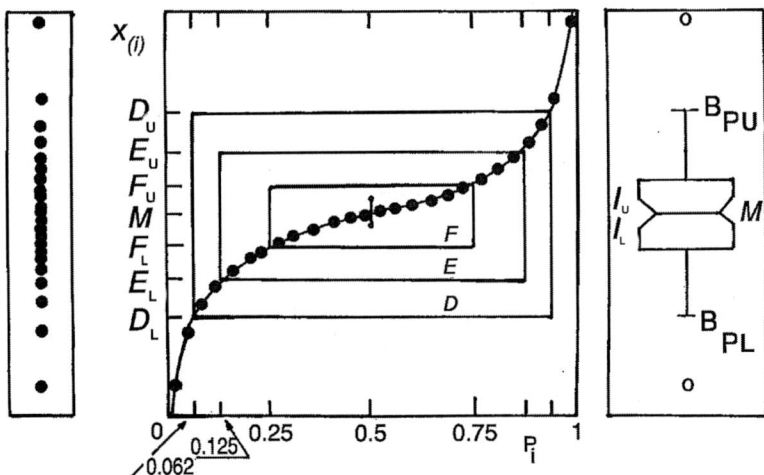

Figure 2.16 Construction of the quantile-box plot G10 (*x-axis*: the order probability P_i, *y-axis*: the order statistic $x_{(i)}$). The dot diagram (left) and the notched box-and-whisker plot (right) are given for comparison of an actual distribution.

(a) *The quartile box F* has two vertices on the y-axis given by the quartiles F_L and F_U with corresponding values on the x-axis equal to the cumulative probability $P_2 = 2^{-2} = 0.25$ and $1 - 2^{-2} = 0.75$.

(b) *The octiles box E* has the octiles E_L and E_U on the y-axis, and the cumulative probabilities $P_3 = 2^{-3} = 0.125$ and $1 - 2^{-3} = 0.875$ on the x-axis.

(c) *The sedeciles box D* has the sedeciles D_L and D_U on the y-axis, and the cumulative probabilities $P_4 = 2^{-4} = 0.0625$ and $1 - 2^{-4} = 0.9375$ on the x-axis.

A horizontal line inside the quartile box marks the position of the *median* M. The robust estimate of the confidence interval of the median $M \pm 1.57$ $R_F / \sqrt{n}$, is drawn as a vertical line at $P = 0.5$. From this plot, and estimates of the midsum Z_Q, the interquantile range R_Q, the relative skewness S_Q and the relative lengths of the tails T_Q, the following may be stated about the sample distribution:

(1) A *symmetric unimodal sample distribution* contains individual boxes arranged symmetrically inside one another, with a value of relative skewness close to zero, $S_Q \approx 0$. When the tail lengths T_Q are approximately equal to their theoretical values for a particular distribution, then the normal distribution, the Laplace (long tails) and the rectangular distribution (short tails) may be distinguished.

(2) An *asymmetric sample distribution* has, in the case of a distribution skewed to higher values, significantly shorter distances between the lower than between the upper parts of the boxes. The skewness S_Q then has a negative value. For a distribution skewed to lower values, the skewness S_Q is positive.

(3) *Outliers* are indicated by a sudden increase of the quantile function outside the F box; the slope may approach infinity.

(4) A *multimodal sample distribution* is indicated by several parts of the quantile function inside box F reaching zero slope.

The quantile-box plot is one of the most useful diagnostics in exploratory data analysis. The sample values are not transformed, and all the original information about the data is available.

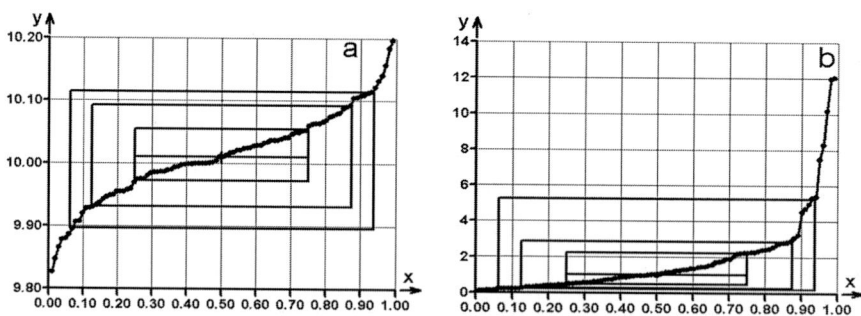

Figure 2.17 The quantile-box plot G10 (*x-axis*: the order probability P_i, *y*-axis: the order statistic $x_{(i)}$) for samples with (a) *norm*, symmetric (Gaussian, normal), and (b) *log*, asymmetric (log.-normal) distributions, *QC-EXPERT*.

2.4 Examining Sample Distribution by EDA

The first step in any data examination is to summarise the information contained in the data. EDA can perform this step in two ways: (a) by use of an appropriate picture or display; or (b) by calculation of characteristics from the data which indicate certain basic features.

Some graphical displays can show overall patterns or trends. They can also reveal surprising, unexpected, or amusing features of the data that might otherwise go unnoticed. When a large number of observations is available, the estimation of the *probability density function* or other function characterising the data distribution can help to elucidate the structure of the sample.

In order to elucidate the structure of a large sample, we can divide the range covered by the sample into a number of classes, usually of equal length, and then count the number of f_i of the sample values falling into each class. In this way the sample is reduced to a grouped sample characterised by frequencies f_i and the mid-point of classes x_i^+, $i = 1,..., k$, $k < n$. Grouping generally leads to a drop in the information content, especially with small and medium sample sizes.

The number of classes taken in forming a grouped sample is to some extent arbitrary. As the class width is reduced, a situation is eventually reached in which there is a frequency f_i of either 1 or 0 in some classes. On the other hand, if the class width is increased, the observations in the sample will fall into fewer and fewer classes and the picture of the structure of the sample presented will become cruder and less informative. As a compromise between these extremes, we use, for example, with a sample of size 100, about 10 classes. Some empirical rules for choosing the optimal number of classes are discussed in section G13.

Usually the class widths are chosen to be the same, but this is not essential. If the tails of the distribution contain only a few members of the sample, it may be convenient to take wider classes in the tails than in the rest of the range.

G11: Stem-and-leaf display

(The stems are written as a column with the smallest at the top, and the leaves are written on the same lines as their stems). The stem-and-leaf display shows (1) the range of values covered by the data; (2) where the values are concentrated; (3) how symmetric the sample is; (4) whether there are gaps where no values were observed; and (5) whether any values stray markedly from the rest. A working stem-and-leaf display is constructed such that the numerical values of the observations in each class of the distribution are divided into two parts: (1) the *stem*, which consists of all the digits common to the members of the class, and (2) the *leaf*, which consists of the remaining digits. The stems are then written as a column with the smallest at the top, and the leaves are written on the same lines as their stems to give an ordered *stem-and-leaf display*. The leaves in each of the rows may be ordered to give an ordered stem-and-leaf display.

Problem 2.3 *Construction of an ordered stem-and-leaf display*

Construct an ordered stem-and-leaf display for a data sample of the weights of 100 aspirin tablets in 8 classes.

Data: The weights of aspirin tablets [10] grouped into 8 classes are to the nearest mg:

328.99 329.75 331.62 333.08 330.61 331.35 328.42 330.63 332.17
330.15 331.28 330.92 329.36

329.62 329.61 329.17 330.39 333.47 330.59 330.52 329.49 329.01
331.63 330.64 330.85 326.06

329.92 330.66 328.57 331.45 331.54 332.20 329.43 327.76 334.06
331.25 328.43 330.57 329.68

330.27 328.81 332.26 332.60 327.32 331.28 330.92 332.66 329.88
329.84 329.92 329.32 333.37

330.28 330.78 333.19 330.84 330.70 329.73 328.87 331.71 329.76
329.82 330.59 328.57 332.20

328.03 330.28 331.02 330.58 333.35 329.86 331.22 329.99 330.34
331.85 332.88 331.99 330.02

328.14 330.03 330.10 330.03 330.47 330.62 331.78 329.33 330.16
329.46 331.89 330.65 329.35

331.84 330.31 331.31 328.06 332.59 327.57 329.10 331.61 331.69
329.47 332.09 330.45 329.41

331.78 330.23 329.89 331.53 331.49 330.52 329.59 334.53 329.04
330.88 330.08 330.11 331.38

331.85 328.56 332.26 330.98 330.91 330.18 325.47 330.99 330.54
329.74 332.55 329.70 328.99

330.63 330.69 331.00 329.29 328.02 330.16 333.56 331.72 325.47
330.72 331.93 329.23 327.87

331.83 330.58 330.94 331.51 330.00 331.21 331.23 330.57 329.59
327.88 328.86

	Class	Class boundary	Class mid-value	Tally marks	Freq.
1	0.324–0.325	0.3235–0.3255	0.3245	111	3
2	0.326–0.327	0.3255–0.3275	0.3265	11111 11	7
3	0.328–0.329	0.3275–0.3295	0.3285	11111 11111 11111 111	18
4	0.330–0.331	0.3295–0.3315	0.3305	11111 11111 11111 1111 111	23
5	0.332–0.333	0.3315–0.3335	0.3325	11111 11111 11111 11111 1111	24
6	0.334–0.335	0.3335–0.3355	0.3345	11111 11111 11111	15
7	0.336–0.337	0.3355–0.3375	0.3365	11111 11	7
8	0.338–0.339	0.3375–0.3395	0.3385	111	3

Solution: The stem-and-leaf display is as follows:

Table 2.4 The ordered stem-and-leaf display for the data sample of the weights of 100 aspirin tablets in 8 classes

Class		
Stem		Leaves
0.32	(4,5)	545
0.32	(6,7)	77777 77
0.32	(8,9)	99998 89988 88998 998
0.33	(0,1)	00100 00100 01010 10111 100
0.33	(2,3)	33322 23332 32233 22222 2222
0.33	(4,5)	54444 55554 45445
0.33	(6,7)	76776 77
0.33	(8,9)	988

Conclusion: The stems are written as a column with the smallest at the top, and the leaves are written on the same lines as their stems to give an ordered *stem-and-leaf display*. The leaves in each of the rows may be ordered to give an ordered stem-and-leaf display.

G12: Kernel estimation of probability density function

(*x-axis*: the variable *x*, *y-axis*: the probability density $f(x)$)

Let x be a continuous random variable. The statistical properties of x may be determined by specifying the *probability density function – pdf of x* (also termed the *frequency function*), $f(x)$ say. A computer may be used to estimate the kernel estimate of the sample probability density function $\hat{f}(x)$ for small and medium samples

$$\hat{f}(x) = \frac{1}{nh}\sum_{i=1}^{n}K\left[\frac{x-x_i}{h}\right] \qquad (2.14)$$

In this equation h is bandwidth, which controls the smoothness of $\hat{f}(x)$, and $K(x)$ is the kernel function, which is symmetric around zero, and also has the properties of a pdf. The actual choice of shape for the kernel function is not important, so here we consider a bi-quadratic kernel estimate

$$K(x) = \begin{cases} 0.9375(1-x^2)^2 & for \ -1 \le x \le 1 \\ 0 & for \ x \ outside \ [-1;1] \end{cases} \qquad (2.15)$$

The quality of the kernel estimate $\hat{f}(x)$ is controlled mainly by the selection of parameter h. If h is too small, the estimate is too rough; if it is too large, the shape of $\hat{f}(x)$ is flattened too much. For samples taken from a normal distribution, the optimal bandwidth h can be calculated from an expression suggested by Scott and Sheater [4]

$$h_{opt} = 2.34 \ \sigma \ n^{-0.2} \qquad (2.16)$$

Lejenne, Dodge and Koelin [5] recommend the following procedure for construction of the kernel estimate of the probability density function:

(1) From Eq. (2.14), calculate an initial guess for the probability density function $\hat{f}(x)^{(0)}$ with a bandwidth $h^{(0)} = 0.75 \times (n/100)^{-0.2} \times [x_{(i+\text{int}(n/2))} - x_{(i)}]$. then calculate the kernel function $K(x)$ from Eq. (2.15).
(2) Find the final estimate of the probability density function with the kernel function (2.15) and non-constant bandwidth from

$$\hat{f}(x)^{(k)} = \frac{1}{n}\sum_{i=1}^{n}K\left[\frac{x-x_i}{h_i}\right] \qquad (2.17)$$

Here the local bandwidth h_i is calculated from

$$h_i = h^{(0)} \times [\hat{f}(x_i)^{(0)} / \max \hat{f}(x_i)^{(0)}]^{-\sigma} \qquad (2.18)$$

Parameter α is defined in the interval $[0, 1]$ and controls the smoothness of $\hat{f}(x)$. Higher values of α lead to a smoother estimate of $\hat{f}(x)$. The parameter

α is usually chosen to be equal to 1/3. For complex sample distributions, it is useful to construct $\hat{f}(x)$ with various values of α and select the one corresponding to maximal visual smoothness.

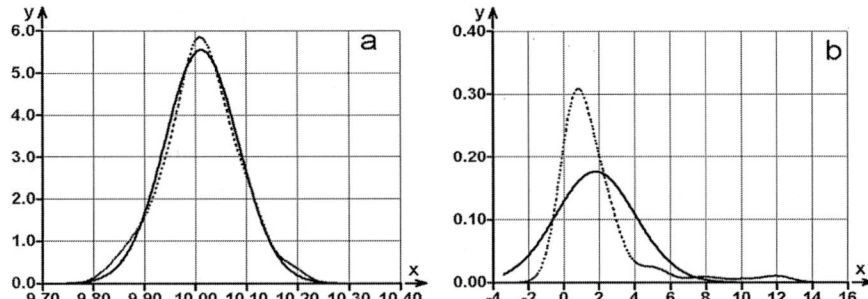

Figure 2.18   A Kernel estimation of probability density G12 (*x-axis*: the variable *x*; *y-axis*: the probability density *f(x)* (...) and the Gaussian (---) function) for samples with (a) *norm*, symmetric (Gaussian, normal), and (b) *log*, asymmetric (log.-normal) distributions, *QC-EXPERT*.

G13: The histogram, frequency polygon, bar chart and rootogram

(*x-axis*: the variable *x; y-axis*: the probability density function $\hat{f}(x)$)

The histogram is one of the oldest classic representations of grouped frequency distributions. The vertical axis represents roughly the class frequency, and the class mid-values x_p $i = 1,...,$ k, are plotted on the horizontal axis. With the class mid-value x_i as the centre of its base, a vertical bar of width equal to the class width and height equal to an empirical relative frequency f_i is drawn for each of the classes.

If the class widths Δx_i are not all equal, a histogram constructed by this method will give a distorted picture of the distribution – it will overemphasise the contributions of the classes with the larger widths. In this situation the correct histogram is constructed with $f_i/\Delta x_p$ along the vertical axis instead of f_i. When the Δx_i are all the same $(= \Delta x$, say) the shape of the histogram will be the same whether f_i or $f_i/\Delta x_i$ is plotted against x_i.

In an *ungrouped* data sample, the class boundaries $x_j^*, j = 1, ..., L+1$, and the number of classes L should be defined. The jth class then has two boundaries, $x_j^* \le x \le x_{j+1}^*$ and their difference represents the class width, $\Delta x_i = x_{j+1}^* - x_j^*$. The quality of the histogram will depend on the width of the classes used. For approximately symmetric distributions, a suitable number of classes L is given by $L = \mathrm{int}(2\sqrt{n})$ where $\mathrm{int}(x)$ is the integer part of a number x. For a large range of sample sizes $L = \mathrm{int}[2.46\,(n-1)^{04}]$ may also be used. For samples from a normal distribution, the optimal class width is $\Delta x_{\mathrm{opt}} = 3.49s\,/\,n^{1/3}$ where s is the standard deviation. A robust estimate of class width for approximately normal data is $\Delta x_{\mathrm{rob}} = 2(F_U - F_L)\,/\,n^{1/3}$ where F_U and F_L are the upper and lower values of the sample quantiles.

For more complicated shapes of sample distribution, the number of classes should be increased, or some special technique of classes with a non-constant length can be used. If the class boundaries for all classes, x*, are known, the histogram is calculated from $\hat{f}(x) = \dfrac{1}{n(x_{j+1}^* - x_j^*)} C(x_j^*, x_{j+1}^*)$ for $x_j^* \le x \le x_{j+1}^*$ where $C(x_j^*, x_{j+1}^*)$ is a function equal to the number of sample observations in the interval $x_j^* \le x \le x_{j+1}^*$.

An alternative method for the graphical representation of a grouped frequency distribution is the *frequency polygon*, whereby the class frequency values are joined by straight lines to form an open polygon which is referred to as the frequency polygon. If the class widths are not all equal, the construction is based on the points $(x_i, f_i\,/\,\Delta x_i)$, as in the case of the histogram.

A *bar chart* is used for the graphical representation of a sample distribution in which all the elements in a given class have the same value. Here the class values are plotted along the x-axis and a vertical line (or bar) of height equal to the class frequency is drawn at the class value.

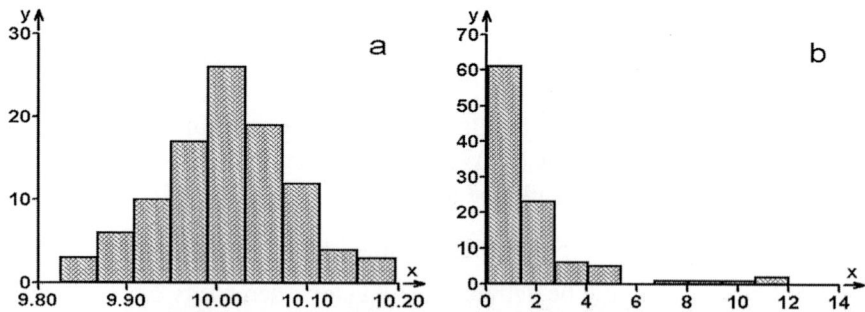

Figure 2.19   Histogram, G13 (*x-axis*: the variable x; *y-axis*: the probability density function $f(x)$) for samples with (a) *norm*, symmetric (Gaussian, normal), and (b) *log*, asymmetric (log.-normal) distributions, QC-EXPERT.

The square-root re-expression of a histogram is the *rootogram*. The class widths have not changed; so we keep the same bar widths as in the histogram, but we now use $\sqrt{f_i} / \Delta x_i$ as the height of the bar for class i. A suitable re-expression can make data more regular and easier to look at.

G14: Quantile-quantile plot (Q-Q plot)

(*x-axis*: the quantile $Q_S(P_i)$; *y-axis*: the order statistic $x_{(i)}$)

Given a random sample, we often need to find out whether the data can be regarded as a sample from a population with a given theoretical distribution. To look at the closeness of the sample distribution to a given theoretical one, the quantile-quantile plot (Q-Q plot) is used [6]. The Q-Q plot allows comparison of the sample distribution described by the empirical $Q_E(P_i)$ quantile function with the given theoretical distribution described by the theoretical $Q_T(P_i)$ quantile function. The empirical Q_E function is approximated by the sample order statistic $x_{(i)}$. If there is close agreement between the sample and theoretical distributions, it must be true that

$$x_{(i)} \simeq Q_T(P_i) \qquad (2.19)$$

Where P_i is the cumulative probability chosen as $P_i = (i - 1/3) / (n + 1/3)$.

To construct this plot, the parameters of the location and spread of the theoretical distribution (or their estimates) must be known. For many theoretical distributions, the standardised variable S may be used

$$S = (x - Q)/R \qquad (2.20)$$

where Q stands for a parameter of location or threshold and R for a parameter of spread. The standardised (theoretical) quantile function $Q_S(P_i)$ then contains only shape parameters (the magnitudes of which may be varied).

When there is agreement between the empirical sample and the standardized theoretical distribution, the Q-Q plot is a straight line

$$x_{(i)} = Q + R\, Q_S(P_i) \qquad (2.21)$$

The x and y co-ordinates of the Q-Q graph for selected theoretical distributions are given in Table 2.5.

Table 2.5 Standardised pdf $f_T(s)$ and distribution $F_T(s)$ functions, and corresponding co-ordinates (x, y) of the Q-Q plot

Distribution	$F_T(s)$	$f_T(s)$	y	X
Rectangular	s	1	$x_{(i)}$	P_i
Exponential	$1 - \exp(-s)$	$\exp(-s)$	$x_{(i)}$	$-\ln(1 - P_i)$
Normal	$\Phi(s)$	$(2\pi)^{-1/2}\exp(0.5s^2)$	$x_{(i)}$	$\Phi^{-1}(P_i)$
Laplace $x < Q$	$0.5\exp(s)$	$0.5\exp(s)$	$x_{(i)}$	$\ln(2P_i)$ for $P_i \le 0.5$
Laplace $x > Q$	$0.5[2 - \exp(-s)]$	$0.5\exp(-s)$	$x_{(i)}$	$-\ln(2(1 - P_i))$ for $P_i > 0.5$
Log-normal	$\Phi[\ln(s)]$	$(2\pi)^{-1/2}\exp(-0.5\ln s^2)$	$x_{(i)}$	$\exp(\Phi^{-1}(P_i))$

In Table 2.5 the normal distribution function $\Phi(s)$ is defined as $\Phi(s) = \dfrac{1}{\sqrt{2\pi}} \int\limits_{-\infty}^{s} \exp(-0.5u^2)du$. To calculate the inverse function $\Phi^{-1}(s)$ the following simple approximate expression may be used

$$\Phi^{-1}(P_i) = -9.4 \times \ln[1 / P_i - 1] / [\mathrm{abs}(\ln(1/P_i - 1))] + 14.$$

G15: Rankit plot

(*x-axis*: the standardised normal quantile u_{P_i} , *y-axis*: the order statistic $x_{(i)}$)

When it is necessary to test whether a given random sample can be regarded as a sample from a normal (Gaussian) distribution, the resulting Q-Q plot is called the rankit or normal probability plot. This plot enables classification of a sample distribution according to its skewness, kurtosis and tail length. A convex or concave shape indicates a skewed sample distribution. A sigmoidal shape indicates that the tail lengths of the sample distribution differ from those of a normal distribution.

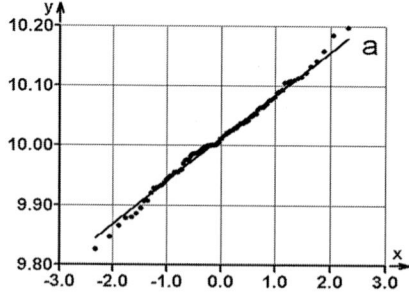
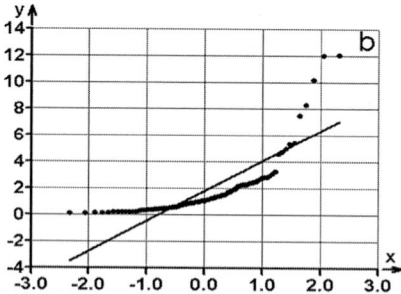

Figure 2.20 The rankit plot G15 (the normal probability plot, *x-axis*: the standardised normal quantile u_{P_i}, *y-axis*: the order statistic $x_{(i)}$) for samples with (a) *norm*, symmetric (Gaussian, normal), and (b) *log*, asymmetric (log.-normal) distributions, QC-EXPERT.

G16: Conditioned rankit plot

(*x-axis*: the function $\Phi^{-1}[U_{(i-1)} + U_{(i+1)}]/2]$; *y-axis*: the order statistic $x_{(i)}$)

Kafander and Spiegelman [7] have recommended the conditioned rankit plot for the examination of the normality of sample distribution. The symbol $\Phi^{-1}(U)$ denotes the quantile function of the standardised normal distribution where, for $U = P_i$, it corresponds to the normal quantile u_{P_i}.

The order statistic $U_{(i)}$ corresponds to the random variable U_i defined by

$$U_i = \Phi[(x_i - \hat{\mu}_R)/\hat{\sigma}_R^2] \qquad (2.22)$$

where the symbol $\Phi(x)$ stands for the distribution function of the standardised normal distribution. The robust estimate of location $\hat{\mu}_R = M$ is equal to the median and the robust estimate of the standard deviation is $\hat{\sigma}_R = 0.75(\tilde{x}_{0.75} - \tilde{x}_{0.25})$. For a complete definition, $U_{(0)} = 0$ and $U_{(n+1)} = 1$ are also required. Approximate linearity in the conditioned rankit plot indicates the normality of the sample distribution.

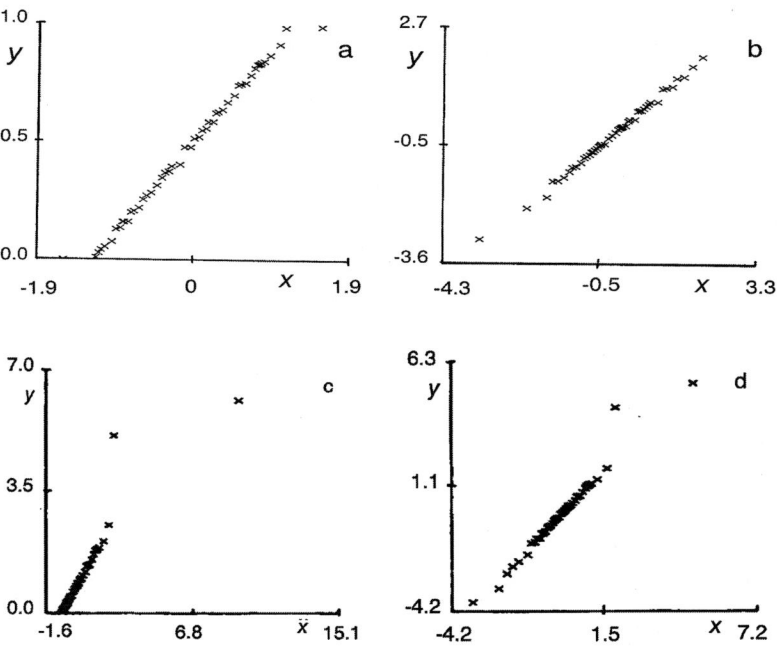

Figure 2.21   The conditioned rankit plot G16 (*x-axis*: the function $\Phi^{-1}[U_{(i-1)} + U_{(i+1)}]/2$; *y-axis*: the order statistic $x_{(i)}$) for samples with (a) rectangular, (b) normal, (c) exponential, and (d) Laplace distributions, *QC-EXPERT*.

G17: P-P plot

(x-axis: P_i, y-axis: $F_T(s_{(i)}$, cf. Table 2.5)

The P-P plot compares data distribution with several theoretical models, using the empirical cumulative distribution function and cumulative distribution functions of normal, Laplace, and uniform distributions. A model which fits the data well should plot approximately as the $y = x$ line. This plot can be used to distinguish among symmetrical distributions according to their kurtosis. Apparent similarity to a uniform distribution suggests that the data were truncated (i.e. both the small and large values have been excluded).

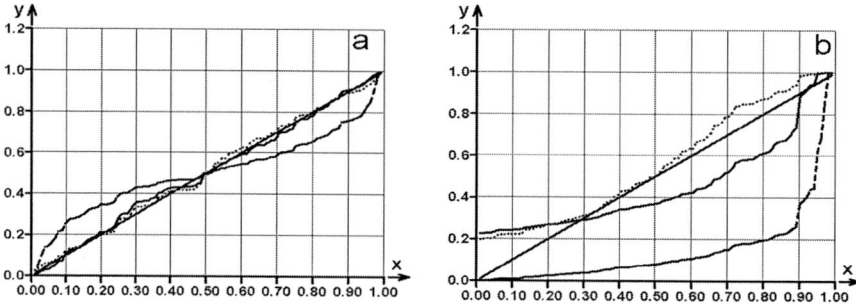

Figure 2.22   The *P-P* plot G17 (x-axis: P_i, y-axis: $F_T(s_{(i)})$) for samples with (a) *norm,* symmetric (Gaussian, normal), and (b) *log,* asymmetric (log.-normal) distributions, *QC-EXPERT.*

G18: Circle plot

(Convex polygon composed from set of vectors with the same length and directions proportional to $\pi \, \Phi \, [x_{(i)}]$)

The circle plot is used for a complex visual assessment of normality, considering skewness and kurtosis simultaneously. One circle is the ideal (for a normal distribution), whilst the other 'convex polygon' is constructed from the data. Both curves should be close to each other for normal data.

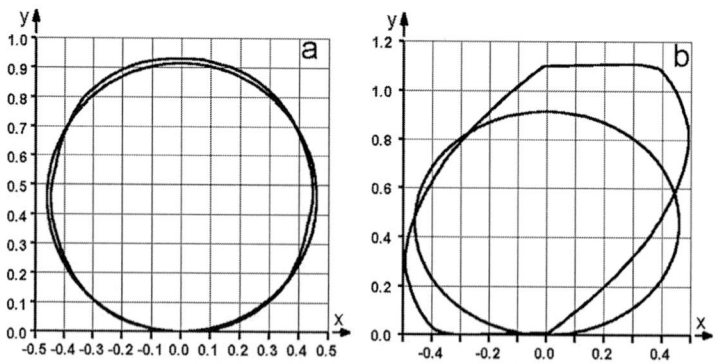

Figure 2.23   The circle plot for samples with (a) *norm,* symmetric (Gaussian, normal), and (b) *log,* asymmetric (log.-normal) distributions, *QC-EXPERT.*

G19: Frequency-ratio plot

(*x-axis*: the level x_i; *y-axis*: the function $x_i p(x_i)/(p(x_i - 1))$

The frequency ratio plot is used to distinguish between various types of discrete distributions, and is based on the expression

$$\frac{x \cdot p(x)}{p(x-1)} = C_0 + C_1 x \qquad (2.23)$$

where the discrete variable levels are $x = 1, 2, ..., k$, and the symbol $p(x)$ stands for the frequency function. Equation (2.23) is valid for many discrete distributions. By comparing the estimated values of the slope C_1 and intercept C_0 of a straight line on the frequency ratio plot with the theoretical values from Table 2.6, the actual type of discrete distribution may be identified.

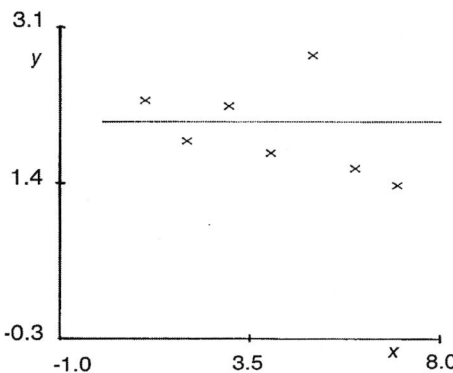

Figure 2.24   The frequency ratio plot G19 (*x-axis*: the variable x_i; *y-axis*: the function $x_i p(x_i)/(p(x_i - 1))$ for a sample with a Poisson distribution.

Table 2.6 The slope C_1 and intercept C_0 of the straight line in the frequency ratio plot

Distribution	Probability function $p(x)$	Slope C_1	Intercept C_0
Poisson	$\exp(-\lambda)\,\lambda^x/x!$	0	λ
Binomial*	$\binom{n}{x} p^x (1-p)^{(n-x)}$	$-p/(1-p)$	$\dfrac{p(n+1)}{(1-p)}$
Negative binomial	$\binom{n+x-1}{x} p^n (1-p)^x$	$1-p$	$(n-1)(1-p)$.
Geometric*	$p(1-p)^{x-1}$	$1-p$	0

where n, p are parameters of the distribution.

G20: Poisson plot

(*x-axis*: the level x_i; *y-axis*: the function $\ln(x_i! \, n_x/n)$)

The Poisson plot is based on the validity of the equation

$$\ln(x! n_x \, / \, n) = -\lambda + x \ln \lambda \qquad (2.24)$$

where the absolute frequency n_x represents the number of sample values reaching the level x, and n is the sample size. If the actual distribution is of a Poisson nature, the Poisson plot is a straight line with slope $\ln \lambda$ and intercept λ. When an estimate of $\hat\lambda$ is known, the "theoretical" straight line $y = -\hat\lambda + x \ln \hat\lambda$ may be drawn.

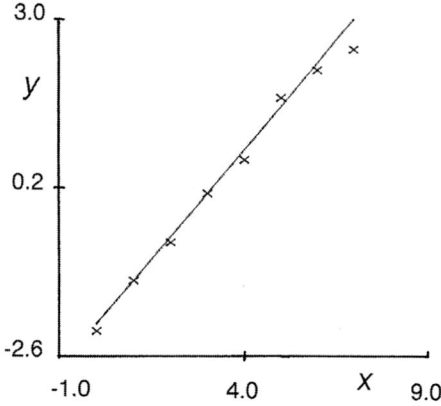

Figure 2.25 The Poisson plot G20 (*x-axis*: the variable x; *y-axis*: the function $\ln(x_i! \, n_x/n)$)

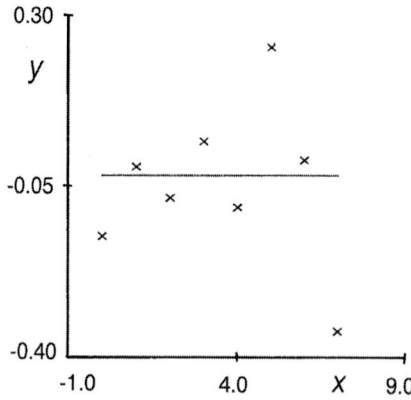

Figuer 2.26 The modified Poisson plot G21, (*x-axis*: the variable x; *y-axis*: the function $\ln(x_i! n_x/n) + (\lambda_0 - x_i \ln \lambda_0)$) for $\lambda_0 = 2$

G21: Modified Poisson plot

(*x-axis*: the variable x_i; *y-axis*: the function $\ln(x_i! n_x/n) + (\lambda_0 - x_i \ln \lambda_0)$)

The modified Poisson plot can be used to examine the suitability of the value selected for parameter λ_0 in the Poisson distribution. When the estimate λ_0 is reasonably suitable, the sample points lie on the horizontal line $y = 0$.

2.5 Data Transformation

When exploratory data analysis proves that the sample distribution strongly differs from a normal one, we are faced with the problem of how to analyse the data. Raw data may require re-expression to produce an informative display, effective summary, or straightforward analysis. We may need to change not only the units in which the data are stated, but also the basic scale of the measurement. To change the shape of a data distribution, we must do more than change the origin and/or unit of measurement: changes of origin and scale mean linear transformations, and they leave shape alone. Non-linear transformations such as the logarithm and square root are necessary to change the shapes.

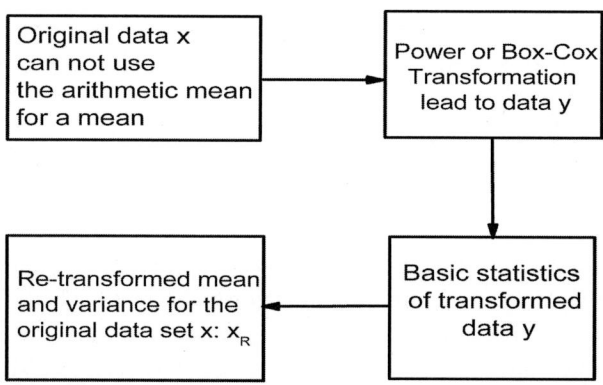

Figure 2.27 Scheme of an application of power and Box-Cox data transformations.

Data must be examined to find the *proper transformation* leading to a symmetric distribution, stabilising the variance, or making the distribution closer to normal. Such a transformation of the original data x to a new variable $y = g(x)$ is based on the assumption that the data represent a non-linear transformation of the normally distributed variable y, according to $x = g^{-1}(y)$.

Transformation for variance stabilisation involves finding a transformation $y = g(x)$ in which the variance $\sigma^2(y)$ is constant. If the variance of the original variable x is a function of type $\sigma^2(x) = f_1(x)$, the variance $\sigma^2(y)$ may be expressed by

$$\sigma^2(y) \approx \left(\frac{d\,g(x)}{dx}\right)^2 f_1(x) = C \tag{2.25}$$

where C is a constant. The chosen transformation $g(x)$ is the solution of the differential equation

$$g(x) \approx C \int \frac{dx}{\sqrt{f_1(x)}} \tag{2.26}$$

In many measuring devices, the constant relative standard deviation $\delta(x)$ of the measured variable is guaranted. This means that the variance $\sigma^2(x)$ is described by the function $\sigma^2(x) = f_1(x) = \delta^2(x)\,x^2 = \text{const } x^2$. The substitution into Eq. (2.26) will be $g(x) = \ln x$, so that suitable form of transformation of the original data is logarithmic. This transformation leads to the use of a geometric mean.

When the dependence $\sigma^2(x) = f_1(x)$ is of a power nature, the optimal transformation will also be a power transformation. Since for a normal distribution the mean is not dependent on the variance, a transformation that stabilises the variance makes the distribution closer to normal.

Transformation for symmetry is carried out by a simple power transformation

x^λ for parameter $\lambda > 0$

$y = g(x) = \ln x$ for parameter $\lambda = 0$ $\tag{2.27}$

$-x^{-\lambda}$ for parameter $\lambda < 0$

which does not retain the scale, is not always continuous, and is suitable only for positive data x. Optimal estimates of parameter $\hat{\lambda}$ are sought by minimising the absolute values of particular characteristics of the asymmetry. In addition to the classic estimate of skewness $\tilde{g}_1(y)$, [Eq. (3.29)] the robust estimate $\tilde{g}_{1,R}(y)$ is used:

$$\tilde{g}_{1,R}(y) = \frac{(\tilde{y}_{0.75} - \tilde{y}_{0.25}) - (\tilde{y}_{0.50} - \tilde{y}_{0.25})}{\tilde{y}_{0.75} - \tilde{y}_{0.25}} \tag{2.28}$$

The relative distance between the arithmetic mean and the median may also be utilised

$$\hat{g}_P(y) = \frac{\bar{y} - \tilde{y}_{0.50}}{\left[\sum_{i=1}^{n}(y_i - \bar{y})^2 / (n-1)\right]^{1/2}} \tag{2.29}$$

because for symmetrical distributions this is equal to zero. An estimate of parameter $\hat{\lambda}$ may also be found from a rankit plot, because for an optimal value of $\hat{\lambda}$ the transformed quantiles $y_{(i)}$ will lie on the straight line.

G22: Hines-Hines selection graph

(*x-axis*: the ratio $\tilde{x}_{0.5} / \tilde{x}_{1-P_i}$, *y-axis*: the ratio $\tilde{x}_{P_i} / \tilde{x}_{0.5}$)

The Hines-Hines selection graph is an excellent diagnostic tool enabling estimation of the parameter λ [8]. This is based on an assumption of the symmetry of the individual quantiles around a median

$$(\tilde{x}_{P_i} / \tilde{x}_{0.5})^{\lambda} + (\tilde{x}_{0.5} / \tilde{x}_{1-P_i})^{-\lambda} = 2 \tag{2.30}$$

where, for the cumulative probability $P_i = 2^{-i}$ the letter values F, E $(i = 2,3)$ are usually chosen.

To compare the empirical dependence of the experimental points with the ideal, patterns for various values of parameter λ are drawn on a selection graph. These patterns λ represent a solution of the equation $y^{\lambda} + x^{-\lambda} = 2$ in the range $0 \le x \le 1$ and $0 \le y \le 1$:

(1) for $\lambda = 0$ the solution is a straight line $y = x$;
(2) for $\lambda \le 0$ the solution takes the form $y \doteq (2 - x^{-\lambda})^{1/\lambda}$
(3) for $\lambda > 0$ the solution takes the form $x = (2 - y^{\lambda})^{-1/\lambda}$.

The estimate $\hat{\lambda}$ is guessed from the selection graph, according to the location of the experimental points near the various theoretical patterns.

In many cases sample distributions can be transformed to approximate normality by use of *Box-Cox family of transformations*, defined as

$$y = g(x) = \begin{cases} (x^{\lambda} - 1) / \lambda & \text{for } \lambda \ne 0 \\ \ln x & \text{for } \lambda = 0 \end{cases} \tag{2.31}$$

where x is a positive variable and λ is real number. A Box-Cox transformation has the following properties:

(1) The curves of transformation $g(x)$ are monotonic and continuous with respect to parameter λ, because

$$\lim_{\lambda \to 0} (x^{\lambda} - 1) / \lambda = \ln x \tag{2.32}$$

(2) All transformation curves share one point $[y = 0, x = 1]$ for all values of λ. The curves nearly coincide at points close to $[0,1]$; that is, they share a common tangent line at that point.

(3) The power transformations with the exponents $-2, -3/2, -1, -1/2, 0,$ $1/2, 1, 3/2$ and 2 have equal spacing between the curves in the Box-Cox transformation graph family.

The Box-Cox transformation defined by Eq. (2.32) can be applied only to positive data. To extend this transformation, the x values are replaced by $(x - x_0)$ values, which are always positive. Here x_0 is the threshold value $x_0 < x_{(1)}$.

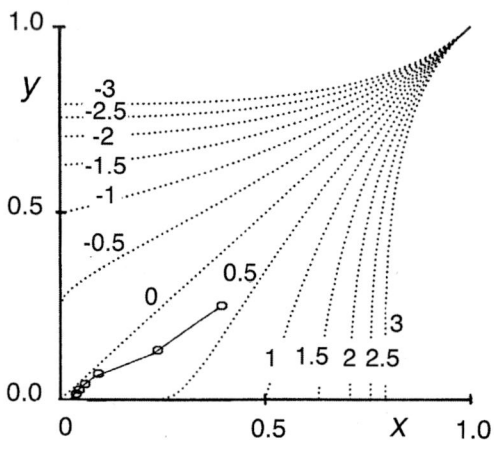

Figure 2.28 Determination of λ from a Hines-Hines selection graph G22 (*x-axis*: the ratio $\tilde{x}_{0.5} / \tilde{x}_{1-P_i}$, *y-axis*: the ratio $\tilde{x}_{P_i} / \tilde{x}_{0.5}$.

G23: Plot of the logarithm of the likelihood function

(*x-axis*: the parameter λ; *y-axis*: the logarithm of the likelihood function $\ln L$)

To estimate the parameter λ in the Box-Cox transformation, Eq. (2.31), the maximum likelihood method may be used, because for $\lambda = \hat{\lambda}$ a distribution of the transformed variable y is considered to be normal, $N[\mu_y, \sigma^2(y)]$. The logarithm of the maximum likelihood function may be written as

$$\ln L(\lambda) = -\frac{n}{2}\ln s^2(y) + (\lambda - 1)\sum_{i=1}^{n}\ln x_i \tag{2.33}$$

where $s^2(y)$ is the sample variance of the transformed data, y. The function $\ln L = f(\lambda)$ is expressed graphically for a suitable interval, for example, $-3 \le \lambda \le 3$. The maximum on this curve represents the maximum likelihood estimate $\hat{\lambda}$.

The asymptotic $100\,(1-\sigma)\%$ confidence interval of parameter λ is expressed by

$$2[\ln L(\hat{\lambda}) - \ln L(\lambda)] \le \chi^2_{1-\alpha}(1) \tag{2.34}$$

where $\chi^2_{1-\alpha}(1)$ is the quantile of the χ^2-distribution with 1 degree of freedom. This interval contains all the λ values for which it is true that:

$$\ln L(\lambda) \ge \ln L(\hat{\lambda}) - 0.5\chi^2_{1-\alpha}(1) \tag{2.35}$$

This Box-Cox transformation is less suitable for wide confidence intervals. When the value $\hat{\lambda} = 1$ is also covered by this confidence interval, the transformation is not efficient.

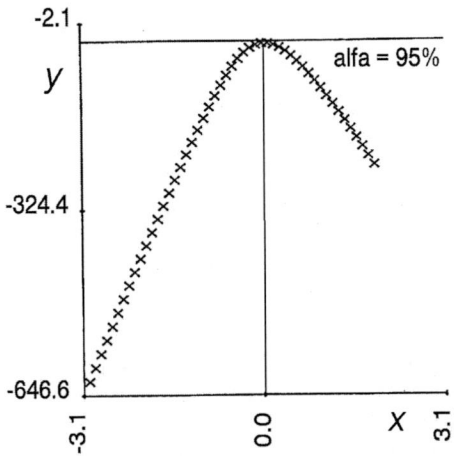

Figure 2.29   Plot of the logarithm of maximum likelihood G23 (*x-axis*: the parameter λ; *y-axis*: the logarithm of the likelihood function ln L).

2.6 The Re-expression of Statistics for Transformed Data

After an appropriate transformation of the original data $\{x\}$ has been found, meaning that the transformed data gives an approximately normal symmetrical distribution with constant variance, the statistical measures of location and spread for the transformed data $\{y\}$ are calculated. These include the sample arithmetic mean $\bar{y}$, the sample variance $s^2(y)$, and the confidence interval of the mean $\bar{y} \pm t_{1-\alpha/2}(n-1)\,s(y)/\sqrt{n}$. These estimates must then be recalculated for the original data $\{x\}$. There are two different approaches to re-expression of the statistics for transformed data.

(1) *Rough re-expressions* represent a single reverse transformation $\bar{x}_R = g^{-1}(y)$. This re-expression for a simple power transformation leads to the general mean

$$\bar{x}_R = \bar{x}_\lambda = \left[\frac{\sum\limits_{i=1}^{n} x_i^\lambda}{n} \right]^{1/\lambda} \tag{2.36}$$

where for $\lambda = 0$, $\ln x$ is used instead of x^λ and e^x instead of $x^{1/\lambda}$. The re-expressed mean $\bar{x}_R = \bar{x}_{-1}$ stands for the *harmonic mean*, $\bar{x}_R = \bar{x}_0$ for the *geometric mean*, $\bar{x}_R = \bar{x}_1$ for the *arithmetic mean* and $\bar{x}_R = \bar{x}_2$ for the *quadratic mean*.

(2) *More correct re-expressions* are based on the Taylor series expansion of the function $y = g(x)$ in the neighbourhood of the value $\bar{y}$. The re-expressed mean $\bar{x}_R$ is then given by

$$\bar{x}_R \approx g^{-1}\left(\bar{y} - \frac{1}{2} \frac{d^2 g(x)}{dx^2} \left(\frac{dg(x)}{dx} \right)^{-2} s^2(y) \right) \tag{2.37}$$

For the variance

$$s^2(\bar{x}_R) = \left(\frac{dg(x)}{dx} \right)^{-2} s^2(y) \tag{2.38}$$

where individual derivatives are calculated at the point $x = \bar{x}_R$. The $100(1 - \alpha)\%$ confidence interval of the re-expressed mean for the original data may be defined as

$$\bar{x}_R - I_L \le \mu \le \bar{x}_R + I_U \tag{2.39}$$

where

$$I_L = g^{-1}[\bar{y} + G - t_{1-\alpha/2}(n-1)\,s(y)/\sqrt{n}] \tag{2.39a}$$

$$I_U = g^{-1}[\bar{y} + G + t_{1-\alpha/2}(n-1)\,s(y)/\sqrt{n}] \tag{2.39b}$$

$$G = -\frac{1}{2} \frac{d^2 g(x)}{dx^2} \left(\frac{dg(x)}{dx} \right)^{-2} s^2(y) \tag{2.39c}$$

On the basis of the (known) actual transformation function $y = g(x)$ and the estimates $\bar{y}$, $s^2(y)$, it is easy to calculate the re-expressed estimates $\bar{x}_R$ and $s^2(\bar{x}_R)$:

(1) For the logarithmic transformation ($\lambda = 0$) and $g(x) = \ln x$, the re-expressed mean and variance will be given by Eq. (2.37), i.e.

$$\bar{x}_R \approx \exp[\bar{y} + 0.5 s^2(y)] \qquad (2.40)$$

and

$$s^2(\bar{x}_R) \approx \bar{x}_R^2 s^2(y) \qquad (2.41)$$

(2) For $\lambda \neq 0$ and the Box-Cox transformation, Eq. (2.31), the re-expressed mean $\bar{x}_R$ will be represented by one of the two roots of the quadratic equation

$$\bar{x}_{R,1,2} = [0.5(1+\lambda\bar{y}) \pm 0.5\left\{1 + 2\lambda(\bar{y} + s^2(y)) + \lambda^2(\bar{y}^2 - 2s^2(y))\right\}^{1/2}]^{1/\lambda} \quad (2.42)$$

which is close to the median $\tilde{x}_{0.5} = g^{-1}(\tilde{y}_{0.5})$. If $\bar{x}_R$ is known, the corresponding variance may be calculated from

$$s^2(x) = \bar{x}^{(-2\lambda+2)} s^2(y) \qquad (2.43)$$

2.7 The Confirmatory Analysis of Assumptions about Data

The statistical treatment of experimental data supposes that the data are independent random variables from the same distribution, which may be normal in nature, and that the sample size is sufficient for precise estimates of location and spread to be obtained [14].

When some of these assumptions about the data are not fulfilled, data analysis becomes rather complicated. These assumptions are examined by *confirmatory data analysis (CDA)* [19].

2.7.1 Examination for Minimum Sample Size

Sample size has an influence on the precision of estimates; e.g., the variance of the parameter estimate is a function of $1/n$. The sample size n controls the size of the confidence intervals, i.e. for larger values of n, the confidence interval is smaller. For very small sample sizes it may happen that the class width and hypothesis tests are affected more by the sample size n than by the variability of data. The procedure for finding a sample size that is sufficient is as follows:

(1) From n_1 starting values, the sample variance $s_0^2(x)$ is calculated. The minimum size n_{min} of a sample taken from a normal distribution is calculated in such a way that for a given probability $(1 - \alpha)$ and value of d, the confidence interval will be $\mu - d \leq \overline{x} \leq \mu + d$ and n_{min} is then given by

$$n_{min} = s_0^2(x)[t_{1-\alpha/2}(n_1 - 1) / d]^2$$

(2.44)

where $t_{1-\alpha/2}(n_1 - 1)$ is the quantile of the Student distribution with $(n_1 - 1)$ degrees of freedom.

(2) The minimum size n_{min} of a sample from a normal distribution may be chosen such that the relative error of the standard deviation $\delta(s)$ has a particular value

$$n_{min} = 1 + [\hat{g}_2(x) - 1] / [4\delta^2(s)]$$

(2.45)

where $\hat{g}_2(x)$ is the estimate of the kurtosis of the sample distribution given by Eq. (3.30). The value of $\delta(s)$ usually chosen is 10%, i.e. 0.1. The minimum size n_{min} is several tens, so the typical sample sizes used in practice n = 5, 10, ... are too small from a statistical point of view.

2.7.2 Examination for Independence of Sample Elements

The basic assumption of good measurement is that the individual measurements (observations) in the sample set are independent. Interdependence of the measurements may be caused by

(1) instability of the measurement equipment, for example, a shift in readings with time;

(2) variable measurement conditions, which might change gradually;

(3) neglect of factor(s) which have a great influence on measurement, for example, sample volume, temperature, purity of chemicals, etc.

(4) a false and non-random (stratified) choice of values in a sample.

When all the experimental conditions change over time, a time dependence in the observations may be indicated if the observations are arranged in order of time. When there is a sudden change in observations, a heterogeneous sample is formed. In both cases, a higher value for the variance will be found than in a homogeneous sample.

Any time dependence or dependence about observations is tested for by examining the significance of the autocorrelation coefficient ρ_α according to

$$t_n = T_1\sqrt{(n+1)} / \sqrt{(1-T_1)} \tag{2.46}$$

where

$$T_1 = (1 - T/2)\sqrt{[(n^2 - 1)/(n^2 - 4)]} \tag{2.46a}$$

and T is the von Neumann ratio defined by

$$T = \frac{\sum\limits_{i=1}^{n-1}(x_{(i+1)} - x_{(i)})^2}{\sum\limits_{i=1}^{n}(x_{(i)} - \overline{x})^2} \tag{2.46b}$$

When the null hypothesis H_0: $\rho_\alpha = 0$ is valid, the test criterion t_n has a Student distribution with $(n + 1)$ degrees of freedom. The alternative hypothesis H_A is $\rho_\alpha \neq 0$. When $|t| > t_{1-\alpha/2}(n_1 + 1)$, the null hypothesis of the independence of sample observations is rejected at the significance level α.

There are other tests for the autocorrelation of higher order and nonparametric tests, which are applied either individually or simultaneously. To find interdependence in the data, the whole measurement and data collection process should be examined.

2.7.3 Testing for the Normality of Sample Distribution

The normality of a sample distribution is a basic assumption in most statistical data treatment, because many statistical tests require normality. When the type of deviation from normality in the sample is known before statistical inference, directional tests are used; when the type of deviation from normality is unknown, omnibus tests are used [11].

In general, statistical tests are less sensitive to deviations from normality than diagnostic graphs. Moreover, deviation from normality can be caused by the presence of outliers. When the normality of a sample distribution has not been proven, the data should be analysed with great care [12,13]. The rankit plot is one of the most useful tools for testing the normality of a sample distribution, but other useful tests are also available.

(1) *Test for combined sample skewness and kurtosis*

The test criterion is defined as

$$C_1 = \frac{\hat{g}_1^2(x)}{D(\hat{g}_1(x))} + \frac{[\hat{g}_2^2(x) - 3]^2}{D(\hat{g}_2(x))} \tag{2.47}$$

where $\hat{g}_1(x)$ is the sample skewness and $D(\hat{g}_1(x))$ its variance, $\hat{g}_2(x)$ is the sample kurtosis and $D(\hat{g}_2(x))$ its variance (calculated from Eqs. 3.19a to 3.20a). For a normal distribution, the test criterion C_1, has approximately the χ^2 distribution, so that when $C_1 > \chi^2_{1-\alpha}(2)$, the null hypothesis about normality of sample distribution is rejected.

(2) *Anderson-Darling test*

This test is based on the empirical distribution function $F_E(x)$. The null hypothesis, $H_0: F_E(x) = F_T(x)$ is tested *vs.* $H_A: F_E(x) \neq F_T(x)$ where $F_T(x)$ is the distribution function of the fully specified distribution. The test criterion is defined as

$$AD = n - \left\{ \sum_{i=1}^{n} (2i - 1)\left[\ln Z_i + \ln(1 - Z_{(n-i+1)}) \right] \right\} / n \tag{2.48}$$

where Z_i is the standardised variable $Z_i = F_T(x_{(i)})$. To test for normality, the null hypothesis is formulated as $H_0: F_E = N(\bar{x}, s^2)$ and the variable $Z_i = \Phi\left[(x_{(i)} - \bar{x} / s\right]$ represents the quantities of the normal distribution. When $AD > D_{1-\alpha}$, the null hypothesis about normality is rejected. The quantile $D_{0.95}$ may, for large samples, be approximated by

$$D_{0.95} = 1.0348(1 - 1.013 / n - 0.93 / n^2) \tag{2.49}$$

2.7.4 Testing for Sample Homogeneity

Sample heterogeneity becomes evident when a sample contains outliers or when the sample can be logically divided into several sub-samples, each of which can be analysed separately. Testing the difference between sub-sample averages can indicate whether the separation into sub-samples can be taken as significant or not. We limit ourselves here to the situation when outliers exist in a data batch. Outliers significantly differ from all other values and can be readily identified by EDA plots. Outliers cause distortion of the estimates $\bar{x}$ and s^2 and may impair the subsequent statistical testing.

There are many different techniques for identifying outliers when a normal distribution of data can be assumed. One of the simplest and most efficient methods seems to be *Hoaglin's modification of inner bounds B* and B** (Fig. 2.7)

$$B_L^* = \tilde{x}_{0.25} - K(\tilde{x}_{0.75} - \tilde{x}_{0.25}) \qquad (2.50a)$$

and

$$B_U^* = \tilde{x}_{0.25} + K(\tilde{x}_{0.75} - \tilde{x}_{0.25}) \qquad (2.50b)$$

where the value of parameter K is selected such that the probability $P(n, K)$ that no observation from a sample of size n will lie outside the modified inner bounds $[B_L^*, B_U^*]$ is sufficiently high, for example, $P(n, K) = 0.95$. For $P(n, K) = 0.95$ and $8 \le n \le 100$, Hoaglin [9] uses the following equation for the calculation of K:

$$K \approx 2.25 - 3.6 / n \qquad (2.51)$$

All elements lying outside the modified inner bounds $[B_L^*, B_U^*]$ are considered to be outliers.

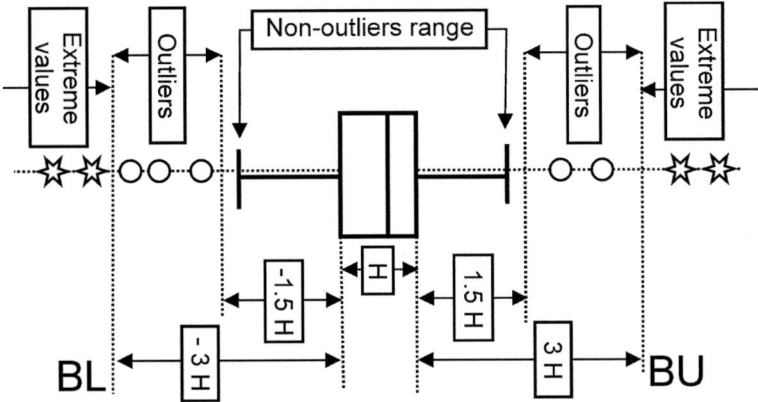

Figure 2.30 Outliers are points beyond the inner bounds BL and BU.

2.8 Summary of the Procedure for the EDA and CDA of Univariate Data

The extent of the exploratory (EDA) and confirmatory (CDA) data analysis of univariate data is best chosen according to experience from previous data analyses. Here, we consider two common situations:

(1) the treatment of routine data, and

(2) the treatment of new data when no preliminary information is available.

(a) *The analysis of routine data*

With routine data, some knowledge of the sample distribution is assumed: it is usually normal, and the data elements are homogeneous and independent. Tests for examining all the assumptions about the data should include (i) a test for minimum sample size; (ii) a test for the independence of the sample elements; (iii) a test for normality; (iv) a test for sample homogeneity. Graphical EDA techniques such as the rankit plot (G15) and quantile-box plot (G10) are often used.

When no preliminary information regarding the data is available, the full range of EDA plots should be applied, followed by determination and construction of the sample distribution. If no suitable distribution is found, power transformation of the data is recommended. The quantile-box plot is *always* used to summarise a batch of experimental data.

(b) *The analysis of new data*

There are several cases in which EDA and CDA will require different strategies.

Case I. No independence of sample elements

When the sample elements are not proved to be independent, the danger arises of systematically biased and over-evaluated estimates for a positive value of ρ_α Eq. (2.46). Therefore, a new logical analysis of the experimental equipment and data measurement procedures is necessary; after improvement of the experimental strategy, the new data should be examined again.

Case II. The sample distribution is not normal

The actual sample distribution is not normal in nature, or outliers are present in the data. When the distribution is not normal, the deviation can be in the lengths of tail or in skewing. When tails differ in length, robust estimates (Section 3.3) may be used, or a power transformation chosen. For skewed distributions, a power transformation should be always used. Once a power transformation is successful and the optimal value λ has been found, estimates of the parameters of location and spread can be calculated and re-expressed in the measure of the original variables. If the power transformation is not successful, exploratory data analysis can be used to find a suitable approximate, theoretical distribution. The estimates of location and scale can then be found as appropriate.

When the actual distribution is strongly skewed, with skewness $\hat{g}_1$, the modified random variable t_c is used,

$$t_c = \left[(\bar{x} - \mu) + \frac{\hat{g}_1}{6\sigma^2 n} + \frac{\hat{g}_1}{3\sigma^4} (\bar{x} - \mu)^2 \right] \frac{\sqrt{n}}{s} \qquad (2.52)$$

where t_c has the Student distribution with $n-1$ degrees of freedom. In practical calculations the variance σ^2 is replaced by its unbiased estimate s^2 and the skewness $\hat{g}_1$ by its unbiased estimate

$$\hat{g}_1 = \frac{n}{(n-1)(n-2)} \sum_{i=1}^{n} (x_i - \bar{x})^3 \qquad (2.53)$$

For construction of the confidence intervals according to Eq. (2.32) $H_L \leq \mu \leq H_U$, the quadratic equation for μ should be solved. The limits H_L and H_U will then be

$$H_L = \bar{x} + \frac{1 - \sqrt{d_1}}{2C_2} \qquad (2.54)$$

$$H_U = \bar{x} + \frac{1 - \sqrt{d_2}}{2C_2} \qquad (2.55)$$

where

$$C_2 = \hat{g}_1 / 3s^4 \qquad (2.56a)$$

$$d_1 = 1 - 4C_2(C_1 - C) \qquad (2.56b)$$

$$d_2 = 1 - 4C_2(C_1 + C) \qquad (2.56c)$$

$$C_1 = \hat{g}_1 / (6s^2 n) \qquad (2.56d)$$

$$C = t_{1-\alpha/2}(n-1) \, s / \sqrt{n} \qquad (2.56e)$$

The confidence interval of the mean, $H_L \leq \mu \leq H_U$, can also be used for statistical inference about this parameter of location.

Case III. The sample is not homogeneous

It should first be considered whether the distribution is skewed or not, because some points which appear to be outliers for a symmetrical (normal) distribution would be accepted in a skewed distribution. When some points may be extremes or outliers there are two alternatives: either (1) to exclude the outliers from the data batch, which for a small sample may lead to the loss of valuable information, or (2) to apply robust methods. In both cases the experimenter should be consulted about the suspect points from the physical point of view, in order to consider the possibility of gross errors.

Case IV. The sample size is insufficient

The best solution is to carry out new experimental measurements. As a general rule, when the variance of the data is small a relatively smaller sample size will be required for any given precision of estimate. When no extra experiments can be carried out, the technique for small sample sizes should be applied (Section 3.3.4). This is convenient for routine data analysis, but for new data, exploratory data analysis should applied firs, so that any statistical peculiarities of the sample can be determined.

2.9 References

[1]   Tukey J. W.: *Exploratory Data Analysis*. Addison Wesley, Reading 1977.

[2]   Chambers J., Cleveland W., Kleiner W., Tukey P.: *Graphical Methods for Data Analysis*, Duxbury Press, Boston 1983.

[3]   Hoaglin D. C., Mosteler F., Tukey J. W.: *Exploring Data Tables, Trends and Shapes*. J. Wiley, New York 1985.

[4]   Scott D. W., Sheater S. J.: *Commun. Statist.* **14**, 1353, 1985.

[5]   Lejenne M., Dodge Y., Koelin E.: *Proc. Conf. COMSTAT 82*, Toulouse, (Vol. III). Str. 173.

[6]   Hoaglin D. C., Mosteler F., Tukey J. W.: *Edits: Understanding Robust and Exploratory Data Analysis*. J. Wiley, New York 1983.

[7]   Kafander K., Spiegelman C. H.: *Comput. Stat. and Data Anal.* **4**, 167, 1986.

[8]   Hines W. G. S., Hines R. J. H.: *Amer. Statist.* **41**, 21, 1987.

[9]   Hoaglin D. C. et al.: *J. Amer. Statist. Assoc.* **81**, 991, 1986.

[10] Royston J. P.: *Appl. Statist.* **31**, 115 (1982)

[11] Gan F. F., Koehler K. J., Thompson J. C.: *Amer. Statist.* **45**, No 1, 14 (1991)

[12] D'Agostino R. B., Belanger A., D'Agostino R. B. J.: *Amer. Statist.* **44**, No 4, 316 (1990)

[13] Potocký R., Kalas J., Komorník J., Lamoš F.: *Zbierka úloh z pravdepodobnosti a matematickej štatistiky.* ALFA-SNTL, Bratislava 1986.

[14] Cyhelský L., Hustopecký J., Závodský P.: *Příklady k základům statistiky.* SNTL, Praha 1988.

[15] Miller J. C., Miller J. N.: *Statistics for Analytical Chemistry.* Ellis Horwood, Chichester 1984.

[16] Anderson R. L.: *Practical Statistics for Analytical Chemists.* van Nostrand Reinhold Comp., New York 1987.

[17] Rice J. A.: *Mathematical Statistics and Data Analysis.* Wadsworth & Brooks, California, 1988.

[18] Dempír J.: *Geolog. průzkum* **26**, 247 (1981).

[19] Meloun M., Militký J.: *Statistické zpracování experimentálních dat*, Plus Praha 1994 (1st ed), EAST PUBLISHING, Praha 1998 (2nd ed), Academia Praha 2004 (3rd ed).

Supplemented material (Review Questions, Exercises, Results of Exercises) to Chapter 2 is on the enclosed CD.

3

Statistical analysis of univariate data

After the exploratory data analysis, the next step is statistical analysis. With small samples the statistical characteristics are estimated directly, but with large samples the data are divided into classes and the statistical characteristics of each class are estimated. For univariate data, a single property or quantitative parameter is examined.

Univariate samples come from a population with an unknown probability distribution. A univariate population (or ensemble) is considered to be a set in which only one property is studied, and one quantity with frequency N is measured. The population is characterized both by measures of the *location,* i.e. the level at which the quantity values vary, by the degree of the *dispersion* (or *spread, scatter, variability*) of the quantity of interest, and by the shape of the distribution.

In practice, the large population of all measured quantities is rarely available. Therefore, statistical analysis examines a *representative random sample* (or *sample*) of n measurements. A representative random sample has the properties:

(1) All sample elements $\{x_i\}$, $i = 1, ..., n$, are random quantities from the *same distribution*; that is the sample is homogeneous.

(2) All sample elements x_i are selected *independently*. The choice of one element does not affect the value of any other element in sample.

The sample is characterized by information about the *mean value* of the sample elements and their *variability* around this mean. In addition, there may be interest in the shape of the sample distribution. Statistical characteristics of location, spread and shape are called the *sample characteristics*. From these sample characteristics, the measures for the population are derived.

In statistical analysis it is assumed that the sample distribution is the same as the population distribution. For continuous random quantities the sample distribution is described by the probability density function $f(x, \theta)$, and for discrete random quantities by the probability function $p(x, \theta)$. The probability density depends on vector θ, which contains the parameters of location, scale and shape. The purpose of an analysis is the estimation of these parameters. Since these estimates are also random quantities, their distribution or at least their characteristics should be estimated.

The main task of statistical analysis is to collect information about a population, so the sample estimates are used to find confidence intervals of parameters. With a given probability, the confidence interval of a population parameter will include the true value of this "unknown" parameter. Statistical testing of hypotheses about "unknown" parameters of the population is also carried out.

A main purpose of laboratory experimentation is to draw inferences about a population from samples of the population. We can identify three different types of inferences, namely:

(1) parameter (point) estimation;
(2) interval estimation,
(3) hypothesis testing.

If we want to make the best estimate of one or more parameters of a probability distribution, the problem is said to be parameter estimation. By parameters we usually mean measures of location, scale and the shape of probability distribution. Estimation of a single value for a parameter is called *point estimation*.

Interval estimation is concerned with estimation of the interval that will include the population parameter with a specified probability. An interval estimate is more informative than a point estimate.

Interval estimation is closely related to *hypothesis testing*. In hypothesis testing, one or more propositions are selected about parameters of population probability distribution. Hypotheses are stated, a criterion of some sort is formulated, and a decision is reached.

"Good" estimates should, if possible, be: (1) unbiased, (2) consistent, (3) efficient, and (4) sufficient.

(1) *Unbiased.* An estimate $\hat{\theta}$ of a parameter θ is said to be unbiased if its expected value, $E(\theta)$, is equal to the population value θ.

(2) *Consistent.* An estimator is said to be consistent if the estimate tends to approach the population value more and more closely as the sample size is increased; that is, if $E[(\hat{\theta} - \theta)^2]$ approaches zero as the sample size n approaches infinity.

(3) *Efficient.* The estimate $\hat{\theta}$ is efficient when its variance around the population value θ is the smallest of all the possible estimates. If two point estimates of a single parameter θ are calculated from the

same sample size n, the one with the smaller variance has the higher efficiency.

(4) *Sufficient*. If $\hat{\theta}$ is a sufficient estimate of population parameter θ, then it contains all sample information. An estimate of θ is denoted as *best unbiased* if it is unbiased, efficient and sufficient simultaneously.

We now turn to methods for estimation of parameters.

3.1 Point Estimates of Location, Spread and Shape

3.1.1 Maximum Likelihood Method

There are many varied methods of point estimation. Regression uses the least-squares method, but for univariate samples the simple method of moments is often used. A well-known and desirable estimation procedure is that of maximum likelihood, which leads asymptotically to estimates that are efficient but not necessarily unbiased. A desirable feature of the maximum likelihood method is that, under certain conditions, the estimated parameters are normally distributed for large samples.

Suppose that $f(x; \theta)$ is a probability density function of known form for the continuous random variable x. The simultaneous probability density function $(x; \theta)$ of sample composed from independent measurements $\{x_i\}$, $i = 1, ..., n$, contains one or more unknown parameters

$$f(x; \theta) = f(x_1; \theta) f(x_2; \theta) ... f(x_n; \theta)$$

One way to estimate the parameters $\theta_1,...,\theta_m$, is to maximize the likelihood function $L(\theta, x)$. The corresponding estimators $\hat{\theta}_1,...,\hat{\theta}_m$ are known as maximum likelihood estimators. The likelihood function based on one observation x_1 is directly equal to the probability density function

$$L(\theta_1, \theta_2,...,\theta_m; x_1) = f(x_1; \theta_1, \theta_2,...,\theta_m) \qquad (3.1)$$

The likelihood function for several independent observations is a product of densities

$$L(\theta_1, \theta_2,...,\theta_m; x_1,..., x_n) = \prod_{i=1}^{n} f(x_i; \theta) \qquad (3.2)$$

The estimates of θ_i that are chosen as the ones that give correspond to the maximum value of L for the given data $(x_1, x_2,..., x_n)$. However, it is more convenient to work with $\ln L$ which has the same position of the extreme not affected by logarithmic transformation.

$$\ln L = \ln f(x_1;\boldsymbol{\theta}) + \ln f(x_2;\boldsymbol{\theta}) + \ldots + \ln f(x_n;\boldsymbol{\theta}) = \sum_{i=1}^{n} \ln f(x_i;\boldsymbol{\theta}) \quad (3.3)$$

The value of $\ln L$ can be maximized with respect to the vector $\boldsymbol{\theta}$ by equating to zero the partial derivatives of $\ln L$ with respect to each of the parameters:

$$\frac{\partial \ln L}{\partial \theta_1} = \frac{\sum_{i=1}^{n} \partial \ln f(x_i;\boldsymbol{\theta})}{\partial \theta_1} = 0$$

$$\frac{\partial \ln L}{\partial \theta_2} = \frac{\sum_{i=1}^{n} \partial \ln f(x_i;\boldsymbol{\theta})}{\partial \theta_2} = 0 \quad (3.4)$$

$$\ldots \qquad \ldots$$

$$\ldots \qquad \ldots$$

(and analogously for $p(x_i;\theta)$ for a discrete random variable). System of eqs. (3.4) yields the maximum likelihood estimates $\hat{\theta}_1, \hat{\theta}_2, \ldots, \hat{\theta}_m$. It can be shown that as sample size n approaches infinity the maximum likelihood estimates have the desired asymptotic properties; that is (1) they are best unbiased; and (2) values of $[\sqrt{n}(\hat{\theta}_i - \theta_i)]$ have the normal distribution $N(0, D(\theta))$. When one parameter θ is estimated, the variance of the maximum likelihood most probable estimate can be expressed by

$$D(\hat{\theta}) = -1 / E\left[\frac{d^2 \ln L}{d\theta^2}\right] \quad (3.5a)$$

where $D()$ is an operator of the variance and $E()$ is an operator of the mean value. For a parameter vector $\boldsymbol{\theta}$ is variability characterized by the covariance matrix $\boldsymbol{C} = \boldsymbol{A}^{-1}$ containing on its diagonal the variances $D(\hat{\theta}_j)$. The elements of matrix $\boldsymbol{A}$ have the form

$$A_{ij} = -E\left[\frac{\partial^2 \ln L}{\partial \theta_j \partial \theta_i}\right] \quad (3.5b)$$

The distribution of the estimates $\hat{\theta}$ is asymptotically normal $N(\theta, A^{-1})$. For real samples of finite size n, the maximum likelihood estimates lose some of their asymptotic properties. They are biased and non-effective.

3.1.2 Sample Characteristics

The maximum likelihood estimates of location $\hat{\theta}_1$ and of dispersion $\hat{\theta}_2$ based on data from a normal distribution are the *sample arithmetic mean* $\bar{x}$, ($\hat{\theta}_1 = \bar{x}$), and the *sample variance* s^2, ($\hat{\theta}_2 = s^2$).

The sample arithmetic mean $\bar{x}$ and sample variance s^2 can be used as location and variance characteristics for data sampled from all other distributions. If the sample comes from a symmetric population distribution with the mean μ, variance σ^2 and kurtosis g_2, it can be proved that

$$E(\bar{x}) = \mu \tag{3.6a}$$

$$D(\bar{x}) = \sigma^2 / n \tag{3.6b}$$

and

$$E(s^2) = \sigma^2 \tag{3.7a}$$

$$D(s^2) = \frac{\sigma^4}{n}\left[g_2 - \frac{n-3}{n-1}\right] \tag{3.7b}$$

In addition to the sample arithmetic mean and the sample variance, other parameters of location and dispersion can be used:

The *sample mode* (or just *mode*) $\hat{x}_M$ is the value corresponding to the maximum concentration of data (in the case of grouped data it is the middle value of interval with maximum absolute frequency) or the most frequently found element value in the sample. The *sample quantiles* are descriptive statistics from exploratory data analysis and are sometimes used to supplement the information obtained from the mean and the variance. The sample values $x_1, ..., x_n$ are first of all arranged in order of ascending magnitude $x_{(1)} < x_{(2)} < ... x_{(n)}$. The quantities $x_{(i)}$ are called the *order statistics*. The pth quantile (or percentile) is defined to be the value of x below which

$p\%$ of the sample values lie. The pth quantile separates the sample data into two parts so that each contains the required percentage of the sample elements, $p\%$ and $(100 - p)\%$.

The *sample median* $\tilde{x}_{0.5}$ is the quantile that separates sample data into two parts: 50% of the elements lie below $\tilde{x}_{0.5}$ and 50% of the elements lie above $\tilde{x}_{0.5}$. The sample median for an odd sample size has the form $\tilde{x}_{0.5} = x_{(k)}$, where $k = (n + 1)/2$. For an even sample size, it is $\tilde{x}_{0.5} = (x_{(k)} + x_{(k+1)})/2$ where $k = n/2$. The mean, mode and median are compared in Fig. 3.1.

The 25th, 50th and 75th percentiles may be called the *first* (or *lower*) *quartile*, *median* (or *second quartile*) and *third* (or *upper*) *quartile* of the sample. The median represents the maximum likelihood estimate of location for the Laplace distribution. For this distribution the variance of the median is expressed by

$$D_L(\tilde{x}_{0.5}) = \sigma^2 / 2n \qquad (3.8)$$

For the normal distribution, however, the sample median is not efficient (Table 3.1). For the rectangular distribution, the efficient estimate of location is the *midsum* $\hat{x}_P$ defined by

$$\tilde{x}_P = (x_{(1)} + x_{(n)})/2 \qquad (3.9)$$

where $x_{(1)}$ is the smallest and $x_{(n)}$ the largest order statistic. The variance of the midsum estimate for the rectangular distribution is defined by

$$D_R(\tilde{x}_P) = \frac{6\sigma^2}{(n-1)(n-2)} \qquad (3.10)$$

Index R denotes the rectangular distribution. The variance of x_P for the normal distribution is much higher.

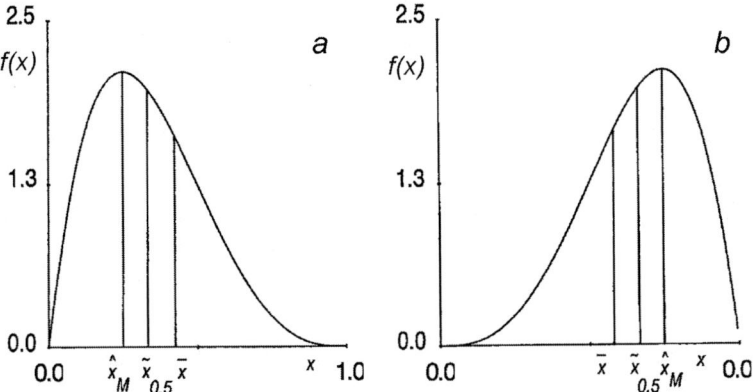

Figure 3.1 Comparison of three measures of location: mean $\overline{x}$, mode $\hat{x}_M$ and median $\tilde{x}_{0.5}$ for (A) negatively and (B) positively skewed distributions.

Often the condition of constant variance of all sample elements is not maintained. If each x_i has a normal distribution with variance σ_i the statistical weights are calculated as $w_i = 1/\sigma_i^2$. Instead of the sample mean $\overline{x}$, the weighted sample mean $\overline{x}_w$ is computed from:

$$\overline{x}_w = \frac{\sum_{i=1}^{n} x_i w_i}{\sum_{i=1}^{n} w_i} = \frac{\sum_{i=1}^{n} x_i / \sigma_i^2}{\sum_{i=1}^{n} 1/\sigma_i^2} \tag{3.11}$$

The variance of the weighted mean is

$$D(\overline{x}_w) = 1/\left[\sum_{i=1}^{n} 1/\sigma_i^2\right] \tag{3.12}$$

If the relative error has a constant value, $\delta = \sigma / x_i = $ const., then the variance $\sigma_i^2 = \overline{x}_i^2 \delta^2$ is nonconstant. Then $w_i = 1/x_i^2$ and the sample mean is calculated from

$$\overline{x}_w = \frac{\sum_{i=1}^{n} 1/x_i}{\sum_{i=1}^{n} 1/x_i^2} \tag{3.13}$$

with variance
$$D(\bar{x}_w) = \delta^2 / \left[\sum_{i=1}^{n} 1/x_i \right]$$
(3.14)

The variance (dispersion) parameters describe the degree of dispersion, (scale, spread, variability or scatter) of the data population elements. The *range* is one of the measures of scale, which represents the difference between the largest and the smallest value of sample. The *interquartile range* can be used as the quantile estimate of population standard deviation a defined by

$$R = 0.7413(\tilde{x}_{0.75} - \tilde{x}_{0.25})$$
(3.15)

where $\tilde{x}_{0.75}$ is the upper and $\tilde{x}_{0.25}$ the lower quartile. Table 3.1 surveys the sample estimates of location and dispersion, with their variances, efficiency and distribution. Sample estimates are for sample size n, and the sample comes from a population with normal distribution $N(\mu, \sigma^2)$.

Table 3.1 Estimates of location and dispersion for sample of size n from a population with normal distribution N (μ, σ^2)

Parameter	Estimate	Variance estimate	Efficiency	Estimate distribution
Mean μ	$\bar{x}$	σ^2 / n	1	$N(\mu, \sigma^2)$
	$\tilde{x}_{0.5}$	$\sigma^2 \pi / (2n)$	0.63	$N(\mu, \sigma^2)$
	$\hat{x}_P$	$\sigma^2 \pi^2 / (24\ln(n))$	$24\ln(n) / (\pi^2 n)$	$N(\mu, \sigma^2)$
Variance σ^2	$\hat{\sigma}^2$	$2\sigma^4 / n$	1*	$N(\sigma^2, D(\sigma^2))$
	s^2	$2\sigma^4 / (n-1)$	1	
Standard deviation σ	$\hat{\sigma}$	$\sigma^2 / (2n)$	$\sim 1^*$	$N(\sigma, D(\sigma))$
	s	$\sigma^2 / [2(n-1)]$	1	
	R	$\sim 1.36\sigma^2 / n$	~ 0.368	
	d	$\sigma^2 / [(\pi - 2)n]$	~ 0.876	

* biased estimate

Another measure of dispersion is the *mean deviation d* defined by

$$d = \sqrt{\frac{\pi}{2}} \left[\frac{1}{n} \sum_{i=1}^{n} |x_i - \mu| \right] \qquad (3.16)$$

where the factor $\sqrt{\pi/2}$ ensures that for normal distribution the value of d approaches that of the standard deviation σ.

The widely used the *coefficient of variation* δ (CV) also known as the *relative standard deviation* s_{rel} or RSD is given by σ/μ and may be estimated by

$$\hat{\delta} = s / \bar{x} \qquad (3.17)$$

The variance of $\hat{\delta}$ is approximately equal to

$$D(\delta) = \hat{\sigma}^2 \left[\frac{n + \hat{\delta}^2 (2n+1)}{2n(n+1)} \right] \qquad (3.18)$$

The error $\hat{\delta}$, units %, is called a *relative error*. Relative errors are frequently used in the comparison of the precision of results with different units or magnitudes, and are again important in calculations of error propagation.

To characterize the shape of a distribution, skewness and kurtosis are used. *Skewness* g_1 is a measure characterizing symmetry, which is equal to zero for a symmetrical distribution. Positive values of g_1 indicate smaller scattering of smalest order statistics values of elements than of the largest order statistics values and negative values of g_1 indicate the opposite case. The *moment estimate of skewness* is defined by

$$\hat{g}_1 = \frac{\sqrt{n} \sum_{i=1}^{n} (x_i - \bar{x})^3}{\left[\sum_{i=1}^{n} (x_i - \bar{x})^2 \right]^{3/2}} \qquad (3.19)$$

Its asymptotic variance is

$$D(\hat{g}_1) \approx \frac{6(n-2)}{(n+1)(n+3)} \qquad (3.19a)$$

The effect of skewness on the shape of the probability density function is shown in Fig. 3.2.

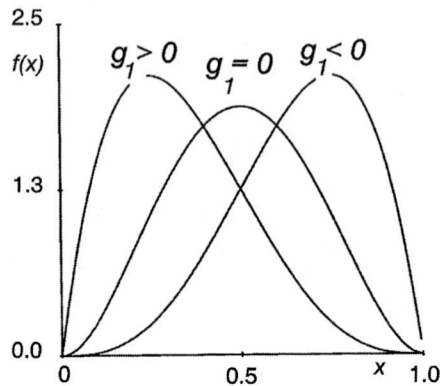

Figure 3.2 The probability density function for various degrees of skewness: $g_1 < 0$, $g_1 = 0$, $g_1 > 0$.

Kurtosis characterizes the shape of the distribution near a modal value, and provides a picture of the shape of the distribution peak. For higher values of kurtosis than 3, the distribution has a sharper peak than the normal distribution, while a flat shape is indicated for values of kurtosis lower than 3 (see Fig. 3.3). The *moment estimate of kurtosis* is defined by

$$\hat{g}_2 = \frac{n\sum_{i=1}^{n}(x_i - \bar{x})^4}{\left[\sum_{i=1}^{n}(x_i - \bar{x})^2\right]^2} \tag{3.20}$$

Its asymptotic variance has the form

$$D(\hat{g}_2) \approx \frac{24n(n-2)(n-3)}{(n+1)^2(n+3)(n+5)} \tag{3.20a}$$

When a point estimate of any parameter is determined, the variance of the parameter must also be calculated. To achieve the same "precision" of estimates when less effective estimates are computed, a greater number of measurements n should be used. To achieve the same parameter precision for data of normal distribution, for example, the calculation of median $\tilde{x}_{0.5}$ needs 1.6 times more measurements that would the arithmetic mean $\bar{x}$.

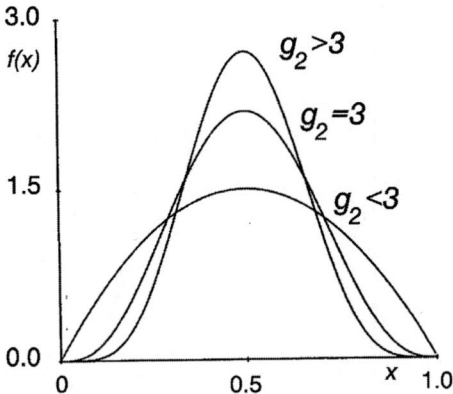

Figure 3.3 The probability density function for various values of the kurtosis: $g_2 > 3$, $g_2 = 3$, $g_2 < 3$.

For samples from a population with a normal distribution, the random variable

$$t = \frac{\bar{x} - \mu}{s} \sqrt{n} \qquad (3.21)$$

has the Student distribution with $(n - 1)$ degrees of freedom (Fig. 3.4). Also, the random variable

$$\chi^2 = \frac{(n-1)s^2}{\sigma^2} \qquad (3.22)$$

has the χ^2 distribution with $(n - 1)$ degrees of freedom. The random variable t and χ^2 are mutually independent. For sufficiently large samples $(n > 40)$ from the normal distribution, for some estimate $\hat{\theta}$ of parameter θ, the random variable

$$U = \frac{\hat{\theta} - \theta}{\sqrt{D(\hat{\theta})}} \qquad (3.23)$$

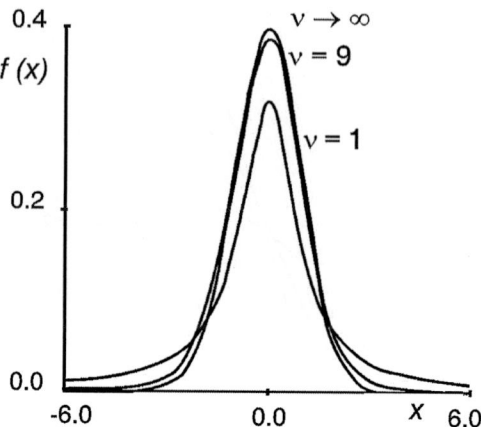

Figure 3.4 The Student distribution for the degrees of freedom $v = 1$, $v = 9$, and $v \to \infty$ (the normal distribution).

has an approximately standard normal distribution $N(0, 1)$. Equation (3.23) is asymptotically valid for any estimate $\hat{\theta}_i$ with variance $D(\hat{\theta}_i)$ determined by the maximum likelihood method, and for any theoretical distribution $f(x, \theta)$.

Instead of maximum likelihood estimates, another statistical characteristic called the *likelihood ratio*

$$L_1 = -2\left[\ln L(\overline{\theta}) - \ln L(\hat{\theta})\right] \tag{3.24}$$

is often used. The *likelihood ratio* L_1 has the $\chi^2(1)$ distribution with 1 degree of freedom. In Eq. (3.24) $\overline{\theta}$ stands for the maximum likelihood estimate $\hat{\theta}$ of parameters θ for which $\hat{\theta}_i = \theta_i$.

It is clear that the distribution of estimators is connected with sample distributions like the Student and χ^2 ones. The Student and χ^2-distributions are both among basic sample distributions which depend only on degrees of freedom, v. For various values of the degree of freedom v, the quantiles of the Student distribution and χ^2-distribution may be found in statistical tables.

3.2 Interval Estimates of Location and Spread

In the previous section, we described ways of obtaining point estimates of location, spread and shape. More informative than these point estimates are *confidence intervals*. The confidence interval is calculated from the sample estimators. It includes the value of the population parameter within the interval limits, termed *confidence limits*, for a specified degree of assurance, called the *confidence coefficient*. Here, the confidence limits are random variables dependent on the sample size [10, 11].

The location is then described not by one value ($\bar{x}$) but by two numerical values L_L and L_U. It is expected that the confidence interval (L_L, L_U) will include the unknown population parameter θ with a preselected probability $(1 - \alpha)$. The degree of trust associated with the confidence statement is called the confidence coefficient; it expresses the degree of certainty or reliability $(1 - \alpha)$ about the unknown population parameter θ.

$$P(L_L < \theta < L_U) = 1 - \alpha \qquad (3.25)$$

where α is called the significance level; the value chosen for α is usually 0.05 or 0.01. It is useful to know that:

(1) the confidence interval is small if the variance of estimate $D(\theta)$ is small,
(2) a large sample size n gives a smaller confidence interval (L_L, L_U),
(3) higher degrees of certainty $(1 - \alpha)$ give broader confidence intervals (L_L, L_U).

Confidence interval (L_L, L_U) is referred to as a two-tailed interval, but one-tailed intervals are also used in specific situations. One-tailed confidence intervals can be:

(1) the left-side or lower-tail interval (L_U, ∞),
(2) the right-side or upper-tail interval ($-\infty$, L_L).

3.2.1 Derivation of the Confidence Interval

Finding the confidence interval $L_{L,U}$ requires knowledge of the distribution of the parameter estimates in question. Let us find the confidence interval of the population mean of the normal distribution $N(\mu, \sigma^2)$. Let $\bar{x}$ be the mean of a sample of n observations on a normally distributed random variable x with unknown mean μ and known variance σ^2. Then the $100(1 - \alpha)\%$ confidence interval for μ may be found from

$$\bar{x} - u_{1-\alpha/2} \frac{\sigma}{\sqrt{n}} \le \mu \le \bar{x} + u_{1-\alpha/2} \frac{\sigma}{\sqrt{n}} \qquad (3.26)$$

where $u_{1-\alpha/2}$ is the $100(1 - \alpha)\%$ quantile of the standardized normal distribution, (e.g., for $u_{0.975} \sim 2$ is $L_{L,U} = \bar{x} \pm 2\sigma / \sqrt{n}$).

In cases where the sample size n is not large enough and the variance σ^2 is not known, the confidence limit for μ may be found from Eq. (3.26), but using quantiles for the Student t-distribution instead of from the normal one. The $100(1 - \alpha)\%$ confidence interval is then given by

$$\bar{x} - t_{1-\alpha/2}(n-1)\frac{s}{\sqrt{n}} \le \mu \le \bar{x} + t_{1-\alpha/2}(n-1)\frac{s}{\sqrt{n}} \qquad (3.27)$$

where $n-1$ is the number of degrees of freedom and $t_{1-\alpha/2}(n-1)$ is the 100(1 $- \alpha/2$)% quantile of the Student distribution. For large sample sizes ($n > 30$) instead of $t_{1-\alpha/2}(n-1)$ the quantile $u_{1-\alpha/2}(n-1)$ can be used.

According to Eq. (3.23), the 100(1 $- \alpha$)% asymptotic confidence interval of any parameter θ may be expressed by

$$\hat{\theta} - u_{1-\alpha/2}(n-1)\sqrt{D(\hat{\theta})} \le \theta \le \hat{\theta} + u_{1-\alpha/2}(n-1)\sqrt{D(\hat{\theta})} \qquad (3.28)$$

The 100(1 $- \alpha$)% two-tailed confidence interval of the variance σ^2 is given by

$$\frac{(n-1)s^2}{\chi^2_{1-\alpha/2}(n-1)} \le \sigma^2 \le \frac{(n-1)s^2}{\chi^2_{\alpha/2}(n-1)} \qquad (3.29)$$

where $\chi^2_{1-\alpha/2}(n-1)$ is the upper and $\chi^2_{\alpha/2}(n-1)$ the lower quantile of the χ^2 distribution, and $n-1$ is the number of degrees of freedom.

Construction of a confidence interval depends on the population distribution from which the sample comes. For example, the variance of the median may be calculated from $D(\tilde{x}_{0.5}) = 1/(4\,nf^2(med))$, where $f(med)$ is the value of the probability density function at the position of the median. For the Laplace distribution, $f(med) = 1/(\sigma\sqrt{2})$ and therefore $D(\tilde{x}_{0.5}) = \sigma^2/2n$, and the confidence interval of the median is given by

$$\tilde{x}_{0.5} - u_{1-\alpha/2}(n-1)\,0.707s/\sqrt{n} \le med \le \tilde{x}_{0.5} + u_{1-\alpha/2}(n-1)\,0.707s/\sqrt{n} \quad (3.30)$$

The Equation (3.30) is valid only if the sample size n is big enough for the median of the Laplace distribution to have approximately normal distribution.

3.3 Point and Interval Estimators for Selected Distributions

Users would like to replace a large volume of experimental data with a few easily grasped numbers. Under favorable circumstances in the EDA, the experimental data are associated with a known function, a probability density function, which corresponds reasonably with the data.

We shall describe some of the most useful probability density functions that the user may meet in the practice [12, 13]. Usual assumption is that data are from normal distribution, but there are some tasks when the random quantity is constrained on one side, or it must be in finite interval. Then the normality assumption is not warranted. In this section the point and interval estimates for one discrete and five continuous distributions are described. These distributions cover all types of data commonly used in practice. Johnson and Kotz may find some details about these distributions in the textbook [3].

3.3.1 The Poisson Distribution

This discrete distribution relates to the number of events that occur in a given interval of time or space when the events occur randomly (in time or space) at a certain average rate. Some examples of random variables for which the Poisson distribution is assumed to apply are: the number of particles emitted from a radioactive source in a given time, the number of typing errors per page of manuscript, the number of calls received at a telephone exchange in a given time period, the number of goals scored by a particular team in a football match. The sample space for the random variable consists of the integers (0, 1, 2, ...).

Suppose a discrete random variable x has a range of possible integer values (levels) 0, 1, 2, ... which has a Poisson distribution with the probability function

$$p(x, \lambda) = \frac{\lambda^x \exp(-\lambda)}{x!} \qquad (3.31)$$

where λ is a positive parameter (Fig. 3.5). For the Poisson distribution it can be shown that $\mu = E(x) = \lambda$ and $\sigma^2 = D(x) = \lambda$. That is, for a Poisson distribution the mean and the variance are equal. For a set of $k + 1$ elements, $x = 0, 1, 2, ...,$ k, the number nx of observations which have a level x is estimated. From Eqs. (3.31) and (3.1), for nx replicated values observations at level x, the likelihood function is

$$L(\lambda) = \prod_{x=1}^{k} \lambda^{n_x x} \frac{\exp(-n_x \lambda)}{n_x x!} \qquad (3.32)$$

After taking logarithms and differentiating

$$\frac{d \ln L(\lambda)}{d\lambda} = \sum_{x=1}^{k} \left(\frac{n_x x}{\lambda} - n_x \right) = 0 \qquad (3.33)$$

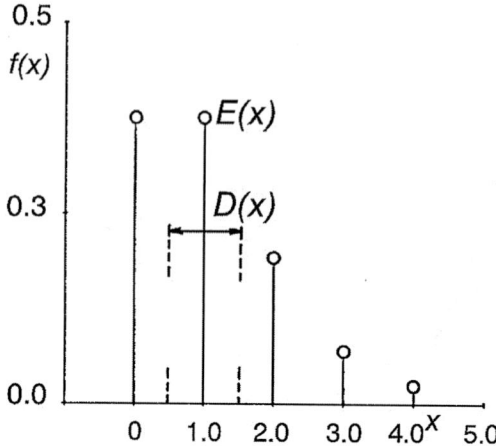

Figure 3.5 The probability density function for the Poisson distribution with $\lambda = 1$.

The estimate $\hat{\lambda}$ can then be calculated

$$\hat{\lambda} = \frac{\sum\limits_{x=1}^{k} n_x x}{\sum\limits_{x=1}^{k} n_x} = \frac{\sum\limits_{x=1}^{k} n_x x}{n} \tag{3.34}$$

where n is the total sample size. The parameter estimate $\hat{\lambda}$ corresponds to the arithmetic mean. To calculate the variance of λ, Eq. (3.33) must be differentiated again $\dfrac{d^2 \ln L(\lambda)}{d\lambda^2} = \dfrac{1}{\lambda^2} \sum\limits_{x=1}^{k} x n_x$ and since $E\left(\sum\limits_{x=1}^{k} x n_x\right) = n\lambda$ use of Eq. (3.4) leads to

$$D(\hat{\lambda}) = \lambda / n \tag{3.35}$$

Construction of the confidence interval of parameter λ for a large sample ($n > 30$) is based on an assumption that the random variable $\sqrt{n}(\hat{\lambda} - \lambda)/\sqrt{\lambda}$ has a standardized normal distribution. Although the square of a random variable with normal distribution has the $\chi^2(1)$ distribution, the confidence limits λ_L and λ_U of $100(1 - \alpha)\%$ confidence interval of parameter λ may be estimated by solving a quadratic equation

$$\hat{\lambda}^2 - \left[2\hat{\lambda} + \frac{\chi_{1-\alpha}^2(1)}{n} \right] \lambda + \hat{\lambda}^2 = 0 \qquad (3.36)$$

The asymptotic confidence interval of parameter λ based on Eq. (3.28) will be

$$\hat{\lambda} - u_{1-\alpha/2}\sqrt{\hat{\lambda}} / \sqrt{n} \le \lambda \le \hat{\lambda} + u_{1-\alpha/2}\sqrt{\hat{\lambda}} / \sqrt{n} \qquad (3.36a)$$

It is more convenient to use the $100(1 - \alpha)\%$ confidence interval of parameter λ, calculated from

$$\frac{\chi_{\alpha/2}^2(2\hat{\lambda}n)}{2n} \le \lambda \le \frac{\chi_{1-\alpha/2}^2(2\hat{\lambda}n+2)}{2n} \qquad (3.36b)$$

For large samples ($n > 100$) and for large values of λ (e.g., $\lambda > 10$) the simple expression (3.36a) is recommended.

Problem 3.1 *Confidence interval of cosmic ray "particles"*

A laboratory counter was set up to measure cosmic ray "particles". For the purpose of this example, the number of particles arriving in 0.1-s intervals was counted. From 200 time intervals the mean of the measurements $\hat{\lambda} = 10.5$ was calculated. With the assumption that the data are described by the Poisson distribution, calculate the 95% confidence interval of the number or particles in 0.1-s intervals.

Data: the numbers of particles k and frequency of detected ray particles nk in sample size $n = 200$ are as follows:

k: 0 1 2 3 4 5 6 7 8 9 10 11 12 13 14 15 16 17 18 19 20 21 22 23
n_k: 0 0 0 2 1 11 12 12 20 22 17 30 20 20 10 18 6 4 3 1 0 0 1 0

Program: ADSTAT or QC-EXPERT: Basic Statistics: Exploratory discrete.

Solution: The 95% confidence interval of parameter λ will be calculated from Eqs. (3.36), (3.36a) and (3.36b):

(1) Equation (3.36): with $n = 200$, $\chi_{0.95}^2(1) = 3.842$ and $\hat{\lambda} = 10.5$ we obtain the quadratic equation $\lambda^2 - 21.0192\lambda + 110.95 = 0$ which has two roots, $\lambda_1 = 10.06$ and $\lambda_2 = 10.96$. The 95% confidence interval of λ will be $10.06 < \lambda < 10.96$.

(2) Equation (3.36a): the 95% confidence interval will be $10.05 \leq \lambda \leq 10.95$.

(3) Equation (3.36b): since the values of $\chi^2_{0.025}$ (4200) and $\chi^2_{0.025}$ (4600) are not available in statistical tables, we use an expression due to Wilson–Wilferty that $\chi^2(v) \approx v(1 - (2/9v + u_{P_i}\sqrt{2/(9v)}))^3$ where u_{P_i} is the $100P\%$ quantile of the standardized normal distribution: For $v = 4200$, $u_{0.025} = -1.96$, we can calculate $\chi^2_{0.025}$ (4200) = 4022.26 and for $v = 4202$, $u_{0.975} = 1.96$, $\chi^2_{0.975}$ (4202) = 4383.57. The 95% confidence interval will be $10.06 < \lambda < 10.96$.

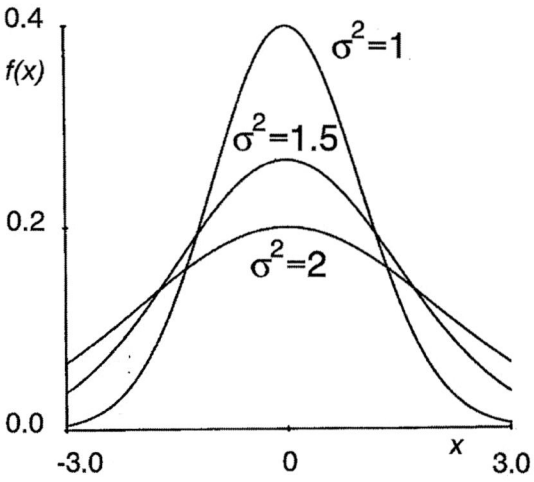

Figure 3.6   The probability density function of the normal distribution for σ^2 = 1, 1.5, 2.

Conclusion: All three equations (3.36), (3.36a) and (3.36b) yield essentially the same confidence interval for parameter λ, $10.06 \leq \lambda \leq 10.96$.

3.3.2 The Normal Distribution

The most important and widely used distribution in practice is the normal distribution. Many continuous random variables encountered in practice follow, at least to a good approximation, this distribution. These include variations in measurement processes; random errors occurring in experiments in physical sciences such as chemistry and physics. In the life sciences, (biology, agriculture, medicine) many directly measured experimental variables do not follow a normal distribution, but transformations can often be made to improve normality (e.g. taking logarithms).

The probability density function of a normally distributed continuous random variable x defined in an infinite interval has a mathematical form

$$f(x) = \frac{1}{\sigma\sqrt{2\pi}} \exp\left[\frac{-(x-\mu)}{2\sigma^2}\right]^2 \qquad (3.37)$$

The mean of variable x is $\mu = E(x)$ and the variance of variable x is $\sigma^2 = D(x)$. The graph of $f(x)$ vs. x forms a bell-shaped curve symmetrical about the mean $x = \mu$ (Fig. 3.6).

Suppose we have a sample $\{x_i\}$, $i = 1, ..., n$, with elements that are independent and come from the same normal distribution. From the logarithm of the likelihood function

$$\ln L = -\frac{n}{2}\ln(2\pi\sigma^2) - \frac{1}{2\sigma^2}\sum_{i=1}^{n}(x_i - \mu)^2 \qquad (3.38)$$

we can calculate the estimate of the *sample mean*

$$\hat{\mu} = \frac{1}{n}\sum_{i=1}^{n}x_i \qquad (3.39)$$

The second derivative of $\ln L$ with respect to parameter μ, together with Eq. (3.4), yield the variance of this sample mean

$$D(\hat{\mu}) = \sigma^2 / n \qquad (3.40)$$

Analogously the estimate of the *sample variance* has the form

$$\hat{\sigma}^2 = \frac{1}{n}\sum_{i=1}^{n}(x_i - \mu)^2 \qquad (3.41)$$

The variance of this estimate is

$$D(\hat{\sigma}^2) = 2\sigma^4 / n \qquad (3.42)$$

In practice, parameter μ is unknown and is replaced by its sample estimate, $\hat{\mu} = \bar{x}$. Then the variance $\hat{\sigma}^2$ defined by Eq. (3.41) is a biased estimate since $E(\sigma^2) = K\,\hat{\sigma}^2$, where $K = (n-1)/n$. For an unbiased estimate of variance, we calculate the sample variance

$$s^2 = \frac{n}{n-1}\hat{\sigma}^2 = \frac{1}{n-1}\sum_{i=1}^{n}(x_i - \bar{x})^2 \qquad (3.43)$$

The confidence interval of the mean is calculated from Eqs. (3.26) and (3.27a) and the confidence interval of variance from Eq. (3.29).

If we know that a random variable x has a normal distribution and the values of μ and σ^2 are given (we write $x \approx N(\mu;\sigma^2)$) then we can calculate a probability $P(a < x < b)$ as the area bounded by the curve $y = f(x)$, the x-axis and the ordinates $x = a$ and $x = b$ for any a and b. To evaluate this area we proceed as follows. The standardized normal random variable u is defined as $u = \dfrac{x - \mu}{\sigma}$. Then $u \approx N(1,0)$, i.e. u is normally distributed with mean zero and variance 1 (Fig. 3.7). Tables of the area under the standard normal distribution (either from $-\infty$ to u or from 0 to u) may be found in standard statistical tables.

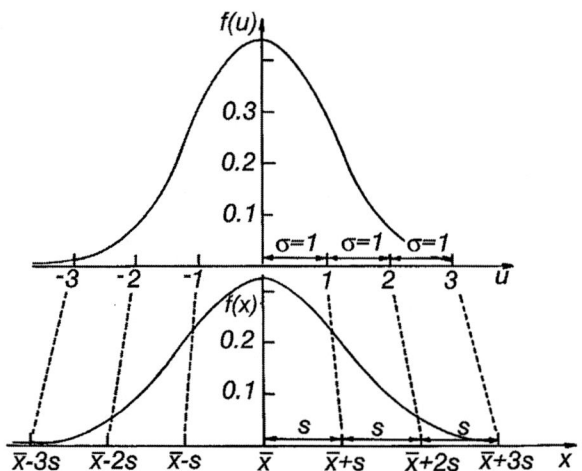

Figure 3.7   The probability density function for the original variable $x \approx N(\mu,\sigma^2)$, and for the standardized normal variable $u \approx N(1,0)$.

Problem 3.2   *Estimation of the mass of aspirin tablets*

A sample of $n = 156$ aspirin tablets were weighed (to the nearest mg). The declared weight of one tablet is $\mu = 330$ mg. The mean calculated from all 156 tablets was $\bar{x} = 330.43$ mg and the variance $s^2 = 2.32$. Subsample A contained $n_A = 32$ tablets, with mean $\bar{x}_A = 330.6$ mg and $s^2 = 2.135$. Subsample B contained $n_B = 10$ tablets and had mean $\bar{x}_B = 330.7$ mg and $s^2 = 2.05$. Estimate the confidence interval of the mean μ and of the variance σ^2, on the assumption that the population has a normal distribution.

Data:

(1) The complete sample of aspirin tablets, $n = 156$.

328.99	329.75	331.62	333.08	330.61	331.35	328.42	330.63	332.17	330.15	331.28	330.92	329.36
329.62	329.61	329.17	330.39	333.47	330.59	330.52	329.49	329.01	331.63	330.64	330.85	326.06
329.92	330.66	328.57	331.45	331.54	332.20	329.43	327.76	334.06	331.25	328.43	330.57	329.68
330.27	328.81	332.26	332.60	327.32	331.28	330.92	332.66	329.88	329.84	329.92	329.32	333.37
330.28	330.78	333.19	330.84	330.70	329.73	328.87	331.71	329.76	329.82	330.59	328.57	332.20
328.03	330.28	331.02	330.58	333.35	329.86	331.22	329.99	330.34	331.85	332.88	331.99	330.02
328.14	330.03	330.10	330.03	330.47	330.62	331.78	329.33	330.16	329.46	331.89	330.65	329.35
331.84	330.31	331.31	328.06	332.59	327.57	329.10	331.61	331.69	329.47	332.09	330.45	329.41
331.78	330.23	329.89	331.53	331.49	330.52	329.59	334.53	329.04	330.88	330.08	330.11	331.38
331.85	328.56	332.26	330.98	330.91	330.18	325.47	330.99	330.54	329.74	332.55	329.70	328.99
330.63	330.69	331.00	329.29	328.02	330.16	333.56	331.72	325.47	330.72	331.93	329.23	327.87
331.83	330.58	330.94	331.51	330.00	331.21	331.23	330.57	329.59	327.88	328.86		

(2) Subsample A, $n_A = 32$.

328.99	329.75	331.62	333.08	330.61	331.35	328.42	330.63	332.17	330.15	331.28	330.92	329.36
329.62	329.61	329.17	330.39	333.47	330.59	330.52	329.49	329.01	331.63	330.64	330.85	326.06
329.92	330.66	328.57	331.45	331.54	332.20							

(3) Subsample B, $n_B = 10$.

328.99	329.75	331.62	333.08	330.61	331.35	328.42	330.63	332.17	330.15

Program: ADSTAT or QC-EXPERT: Basic Statistics: One sample analysis.

Solution: The 95% confidence interval of the mean is calculated from Eqs. (3.26) and (3.27b).

From the original sample, size $n = 156$: $330.19 \leq \mu \leq 330.67$.
From the subsample, size $n_A = 32$: $329.96 \leq \mu \leq 331.09$.
From the subsample, size $n_B = 10$: $329.65 \leq \mu \leq 331.70$.

The 95% confidence interval of the variance σ^2 is calculated from Eq. (3.29).

From the elements of the original sample: $1.879 \leq \sigma^2 \leq 2.937$.
From the subsample, size $n_A = 32$: $1.372 \leq \sigma^2 \leq 3.43$.
From the subsample, size $n_B = 10$: $0.970 \leq \sigma^2 \leq 6.83$.

Conclusion: Small samples from a population with normal distribution may lead to inaccurate results. The hypothesis that $\mu = 330$ and $\sigma^2 = 1$ is not accepted here at level $\alpha = 0.05$, for sample sizes bigger than $n = 100$.

3.3.3 The Laplace Distribution

When random elements are measured under condition of non-constant variance, the Laplace (two-tailed exponential) distribution often occurs. The Laplace probability density function $f(x)$ of random variable x in the interval $(-\infty, \infty)$ is described by

$$f(x) = 0.5\Phi^{-1} \exp\left[-\frac{|x - \theta|}{\Phi} \right] \qquad (3.44)$$

The mean of the Laplace distribution $E(x) = \theta$, the variance $D(x) = 2\Phi^2$, and the skewness and kurtosis are $g_1 = 0$ and $g_2 = 6$. The Laplace distribution has a more peaked shape than the normal distribution, with longer tails. For example, for the Laplace distribution the 1% quantile is equal to $E(x) -2.72 \sqrt{D(x)}$, but for the normal distribution it is $E(x) -2.33 \sqrt{D(x)}$. The Laplace distribution is taken as a natural "robust" alternative for the normal one.

From Eq. (3.2), the logarithm of the maximum likelihood function is

$$\ln L = -n\ln(2\Phi) - \Phi^{-1} \sum_{i=1}^{n} |x_i - \theta| \qquad (3.45)$$

For the known parameter Φ the maximum likelihood estimate of θ, say $\hat{\theta}$, minimizes the sum $\sum_{i=1}^{n} |x_i - \theta|$ and is equal to the sample median $\hat{\theta} = \tilde{x}_{0.5}$. Differentiation of the Eq. (3.45) with respect to Φ and equating to zero yields the maximum likelihood estimate as

$$\hat{\Phi} = \frac{1}{n}\sum_{i=1}^{n}\left|x_i - \hat{\theta}\right| \tag{3.46}$$

Although the median $\tilde{x}_{0.5}$ is unbiased, it does not have minimal variance for small samples. The confidence interval of parameter θ is derived from an asymptotic formula Eq. (3.23) which leads to Eq. (3.30). Since the variance of $\hat{\Phi}$ estimate is

$$D(\hat{\Phi}) = \Phi^2 / n \tag{3.47}$$

the resulting confidence interval of parameter Φ is calculated from Eq. (3.23). When the mean value Φ is known, the $100(1 - \alpha)\%$ confidence interval of Φ is

$$\frac{2n\hat{\Phi}}{\chi^2_{1-\alpha/2}(2n)} \leq \Phi \leq \frac{2n\hat{\Phi}}{\chi^2_{\alpha/2}(2n)} \tag{3.48}$$

Problem 3.3 *Confidence interval of parameters of the Laplace distribution*

From a random sample of size $n = 50$ from the Laplace distribution $L(0, 2)$ the following estimates were calculated: $\hat{\theta} = \tilde{x}_{0.5} = 0.0119$ and $\hat{\Phi} = 1.0596$. Calculate the 95% confidence intervals for these parameters.

Data: taken from Problem 3.9

Program: ADSTAT or QC-EXPERT: Basic Statistics: One sample analysis.

Solution: The variance estimate is $s^2 = 2\hat{\Phi}^2 = 2.246$. Substitution of this value into Eq. (3.30) yields the 95% confidence interval of parameter $\hat{\theta}$ as $-0.282 < \hat{\theta} < 0.306$. With the use of Eq. (3.47), Eq. (3.23) can be rewritten as $\hat{\Phi} - u_{1-\alpha/2} \ \hat{\Phi}/\sqrt{n} \leq \Phi \leq \hat{\Phi} + u_{1-\alpha/2} \ \hat{\Phi}/\sqrt{n}$. The 95% confidence interval of parameter can be calculated as $0.765 \leq \Phi \leq 1.353$. With the assumption that $\hat{\theta} = \tilde{x}_{0.5} = 0$, the estimate $\hat{\Phi}$ according to Eq. (3.46), is 1.0596. Then, from Eq. (3.48), the 95% confidence interval of parameter Φ is $0.818 \leq \Phi \leq 1.428$.

Conclusion: The confidence intervals of parameter Φ calculated from Eqs. (3.23) and (3.48) do not significantly differ because the mean θ is equal to zero.

3.3.4 The Rectangular Distribution

The simplest type of distribution is the rectangular distribution for a random variable constrained on both sides, $a - h \leq x \leq a + h$. When $a = 0$ and $h = 0.5 \ 10^{-k}$, the rectangular distribution describes the distribution of errors that appeared in the rounding-off to k decimal places. The probability density function for a rectangular distribution is

$$f(x) = 1/(2\,h) \tag{3.49}$$

where $a - h \le x \le a + h$. The mean of the rectangular distribution $E(x) = a$, the variance $D(x) = h^2/3$, and the skewness and kurtosis are $g_1 = 0$ and $g_2 = 1.8$. The logarithm of the maximum likelihood function is

$$\ln L = -n \ln (2h) \tag{3.50}$$

for $a - h \le \min (x_1, ..., x_n) \le \max (x_1, ..., x_n) \le a + h$. Equation (3.50) reaches a maximum for a minimum value of h. It is evident that $\min (x_1, ..., x_n) = x_{(1)}$ and $\max (x_1, ..., x_n) = x_{(n)}$ The maximum likelihood estimate of parameter h is given by

$$\hat{h} = 0.5(x_{(n)} - x_{(1)}) \tag{3.51}$$

and the maximum likelihood estimate of parameter a is

$$\hat{a} = 0.5(x_{(n)} + x_{(1)}) \tag{3.52}$$

The estimate $\hat{h}$ is identical to the midsum $\tilde{x}_p$ defined by Eq. (3.9). The estimate $\hat{h}$ is biased. The unbiased estimate $\hat{h}_0$ is calculated by correction of the previous estimate of $\hat{h}$, i.e. $\hat{h}_0 = \hat{h} (n+1)/(n-1)$. The variances of estimates $\hat{h}$ and $\hat{a}$ are calculated from

$$D(\hat{h}) = \frac{2h^2}{(n-1)(n+2)} \tag{3.53}$$

and

$$D(\hat{a}) = \frac{2h^2}{(n+1)(n+2)} \tag{3.54}$$

The variance estimates $D(\hat{h})$ and $D(\hat{a})$ are not correlated but not independent. The confidence intervals for these parameters are calculated for large samples by Eq. (3.25).

Problem 3.4   *Examination of copper content*

For one month the concentration of copper (II) ions ($\mu g/1$) in the cooling water from an electric power station was measured. The sample of size $n = 90$ measurements contained the smallest value $x_{(1)} = 6\ \mu g/1$, and largest $x_{(90)} = 30$ $\mu g/1$. Exploratory data analysis indicated that the sample comes from a rectangular distribution. Calculate the parameters of location and dispersion, and the corresponding 95% confidence intervals.

Data: the concentration of Cu^{2+} in $\mu g/l$:

18.744	22.241	10.107	7.566	26.358	6.000	15.117	24.240	18.578	12.794	26.177	10.118
21.149	21.482	19.605	20.259	9.374	29.860	29.950	11.066	17.696	20.274	24.013	11.533
6.290	17.696	28.167	25.056	14.600	25.751	15.941	8.088	9.528	19.419	7.266	11.207
15.247	24.127	26.467	22.971	12.378	27.074	6.948	15.771	26.146	10.116	13.811	13.072
26.211	21.274	8.838	28.514	29.339	27.463	10.702	11.517	26.881	17.015	15.607	26.432
25.141	21.155	16.466	17.813	9.247	15.693	28.386	28.468	9.946	6.109	25.531	27.227
28.519	22.850	10.568	7.973	19.874	13.189	12.783	23.244	28.047	20.710	30.000	29.166
18.310	14.841	24.431	19.203	21.527	11.599						

Program: ADSTAT or QC-EXPERT: Basic Statistics: One variable analysis.

Solution: From Eq. (3.52), the parameter $\hat{a} = 18$ and for $h = 12$, the variance of $\hat{a}$ from Eq. (3.54) is $D(\hat{a}) = 0.077$. The parameter $\hat{a}$ represents the estimate of the mean. The 95% confidence interval of parameter $\hat{a}$ from Eq. (3.28) is $17.45 \leq a \leq 18.55$ Substitution in Eq. (3.51) of $\hat{h} = 12$ gives the unbiased estimate of h as $\hat{h}_0 = 12.27$. The variance estimate is $\hat{\sigma}^2 = 50.18$. From Eq. (3.53), $D(\hat{h}) = 0.0368$ and the 95% confidence interval of parameter h is $11.62 \leq h \leq 12.38$.

Conclusion: The point estimates of location and dispersion are $\hat{a} = 18$ $\mu g/1$ and $\hat{h} = 12.27 \ \mu g/1$. The 95% confidence intervals of the parameters, a and h, are $17.45 \leq a \leq 18.55 \ [\mu g/l]$ and $11.62 \leq h \leq 12.38 \ [\mu g/l]$.

3.3.5 The Exponential Distribution

The exponential distribution is constrained on one side (upper part) and concerns the time elapsed between consecutive events in a Poisson process. Examples could be listed corresponding to those given for the Poisson distribution. Often, the lifetime of a component in a piece of apparatus is assumed to have such a distribution. Another example of an exponentially distributed random variable could be the distance travelled between successive collisions in a low pressure gas.

3.3.5.1 The One-parameter Exponential Distribution

This distribution describes the behaviour of a continuous random variable for which the sample space is the positive half of the real line, $x \geq 0$. We say that this random variable x has the one-parameter exponential distribution of probability density function $f(x)$ described by

$$f(x) = \theta^{-1} \exp(-x / \theta) \qquad (3.55)$$

The mean of this distribution is defined by $E(x) = \theta$, the variance $D(x) = \theta^2$, the skewness $g_1 = 2$ and kurtosis $g_2 = 9$. The median is $\tilde{x}_{0.5} = \theta \ln 2$. The logarithm of the likelihood function is

$$\ln L = -n \ln \theta - \sum_{i=1}^{n} x_i / \theta \qquad (3.56)$$

From Eq. (3.3), the maximum likelihood estimate is

$$\hat{\theta} = \frac{1}{n} \sum_{i=1}^{n} x_i \qquad (3.57)$$

and from Eq. (3.4) its variance is

$$D(\hat{\theta}) = \theta^2 / n \tag{3.58}$$

Construction of the confidence interval is based on the fact that the random variable $2\hat{\theta}n / \theta$ has the $\chi^2(2n)$ distribution. The 95% confidence interval of parameter θ is

$$\frac{2n\hat{\theta}}{\chi^2_{1-\alpha/2}(2n)} \leq \theta \leq \frac{2n\hat{\theta}}{\chi^2_{\alpha/2}(2n)} \tag{3.59}$$

For large samples the confidence interval defined by Eq. (3.28) can be also used.

Problem 3.5 *Decomposition kinetics of DTBP*

Di-terc-butyl peroxide (DTBP) decomposes at 154.6°C in the gas phase by a first-order process with rate constant $k = 3.46 \times 10^{-4}\text{s}^{-1}$. Calculate the time (called the half-life) when 50% of the molecules have decomposed.

Solution: The number of DTBP molecules remaining at time t is given by $N_t = N_0 \exp(-kt)$, where N_0 is the number of molecules present at $t = 0$. The probability that one of the original molecules will survive for this time $t \leq T \leq t + dt$, is

$$P(t \leq T \leq t + dt) = -\frac{dN_t}{N_0} = k\exp(-kt)dt \tag{3.60}$$

since the probability density function (3.55) of the survival time is $f(t) = k \exp(-kt)$, where in Eq. (3.55) $\theta = 1/k$, the average survival time of the DTBP molecules is $E(x) = 1/k = 10^4\text{s}/3.46 = 2.89 \times 10^3 \text{ s}$.

Conclusion: 50% of the DTBP molecules disappeared in $2.00 \ 10^3$ s, the time that corresponds to the median $\tilde{x}_{0.5}$ (usually called the half-time), the time satisfying $f(t) = 0.5$, that is $t_{0.5} = (\ln 2)/k$. Thus, fewer than half of the molecules survive for that average time.

3.3.5.2 The Two-parameter Exponential Distribution

This distribution describes the statistical behaviour of a constrained random variable which can reach only values $x > n$. The probability density function is defined by

$$f(x) = \theta^{-1} \exp\left[\frac{\mu - x}{\theta}\right]$$ (3.61)

The mean of this distribution is $E(x) = \mu + \theta$. The variance, skewness and kurtosis are the same as for the one-parameter exponential distribution. The logarithm of the likelihood function is

$$\ln L = n \ln \theta - \sum_{i=1}^{n} (x_i - \mu) / \theta$$ (3.62)

and the maximum likelihood estimate $\hat{\mu}$ of Eq. (3.62) is then

$$\hat{\mu} = \min(x_1, x_2, ..., x_n) = x_{(1)}$$ (3.63)

For the maximum likelihood estimate of $\hat{\theta}$ on the basis of Eq. (3.3)

$$\hat{\theta} = \frac{1}{n} \sum_{i=1}^{n} (x_i - \hat{\mu}) \approx (\overline{x} - x_{(1)})$$ (3.64)

The estimate $\hat{\mu}$ has expectation

$$E(\hat{\mu}) = \mu + \frac{\theta}{n}$$ (3.65a)

and variance $D(\hat{\mu}) = \theta^2 / n^2$ (3.65b)

For an estimate $\hat{\theta}$ from Eq. (3.64)

$$E(\hat{\theta}) = \theta(1 - 1/n)$$ (3.66a)

and $D(\hat{\theta}) = \theta^2 (1/n + 1/n^2 + 2/n^3)$ (3.66b)

The maximum likelihood estimates $\hat{\mu}$ and $\hat{\theta}$ are clearly biased. Unbiased estimates $\hat{\mu}_0$ and $\hat{\theta}_0$ take the form

$$\hat{\mu}_0 = \frac{n x_{(1)} - \overline{x}}{n - 1}$$ (3.67a)

with variance

$$D(\hat{\mu}_0) = \frac{\theta^2}{n(n-1)} \qquad (3.67b)$$

$$\text{and } \hat{\theta}_0 = \frac{n(\bar{x} - x_{(1)})}{n-1} \qquad (3.68a)$$

$$\text{with variance } D(\hat{\theta}_0) = \frac{\theta^2}{n-1} \qquad (3.68b)$$

However, the estimates $\hat{\mu}_0$ and $\hat{\theta}_0$ are correlated, with correlation coefficient of $(-1/\sqrt{n})$.

The estimate of a confidence interval is based on the fact that the random variables $2n(x_{(1)} - \mu)/\theta$ and $2(n-1)\hat{\theta}_0/\theta$ are independent and have the χ^2-distribution with 2 and $2(n-1)$ degrees of freedom. Therefore, the $100(1-\alpha)\%$ confidence interval of parameter θ is calculated from

$$\frac{2(n-1)\hat{\theta}_0}{\chi^2_{1-\alpha/2}(2n-2)} \le \theta \le \frac{2(n-1)\hat{\theta}_0}{\chi^2_{\alpha/2}(2n-2)} \qquad (3.69)$$

Since the ratio $n(x_{(1)} - \mu)/\hat{\theta}_0$ has Fisher-Snedecor F distribution with 2 and $2n - 2$ degrees of freedom, the lower limit μ_1 of the $100(1 - \alpha)\%$ confidence interval of parameter μ, is given by

$$\mu_1 = x_{(1)} - \hat{\theta}_0 F_{1-\alpha}(2, 2n-1)/n \qquad (3.70)$$

The upper limit μ_2 is equal to the smallest sample element, $x_{(1)}$. For determination of quantiles of the F-distribution, the following approximate expression may be used

$$F_P(2, 2n-2) = (n-1)[(1-P)^{-1/(n-1)} - 1] \qquad (3.71)$$

Problem 3.6 *Examination of data for biologically cleaned flowing sink-water*

For 4 months, BCO_5 values in g/m³ were recorded for biologically cleaned outflowing sink-water. In the sample of size $n = 125$, the smallest value was $x_{(1)} = 9$ g/m^3 and the arithmetic mean was $\bar{x} = 27$ g/m^3. Exploratory data analysis proved that the distribution was exponential. Estimate values for the parameters of location and dispersion with their 95% confidence intervals.

Data: the value of BCO$_5$ in g/m^3:

20.948	26.818	11.962	10.065	38.752	9.338
31.521	20.717	14.252	37.989	11.970	24.741
22.203	23.229	11.391	90.195	106.359	12.741
23.253	30.910	13.135	9.192	19.543	49.589
16.002	36.321	17.439	10.436	11.509	21.924
12.859	16.678	31.214	39.235	28.378	13.875
9.636	17.249	37.861	11.969	15.213	14.508
24.965	10.986	52.896	65.683	44.458	12.441
41.200	18.693	17.068	39.077	34.205	24.752
19.694	11.294	17.163	51.594	52.413	11.834
35.525	43.054	52.956	28.108	12.332	10.354
14.618	14.241	29.003	48.586	23.978	100.200
20.351	16.250	32.053	21.604	25.430	13.192
43.407	39.788	61.421	10.680	40.342	10.595
21.804	11.014	18.395	46.739	31.923	20.167
13.546	46.334	22.184	24.920	11.151	37.217
9.267	14.745	51.224	16.619	35.590	51.926
17.200	12.859	40.870	48.021	11.344	9.000

Program: ADSTAT or QC-EXPERT: Basic Statistics: One variable analysis.

Solution: From Eq. (3.63), $\hat{\mu} = 9$, and then from Eq. (3.64), $\hat{\theta} = 18$. Equation (3.67) gives the unbiased estimate $\hat{\mu}_0 = 8.854$ and $D(\hat{\mu}_0) = 0.0212$. Analogously, from Eq. (3.68) the unbiased estimate $\hat{\theta}_0 = 18.15$ and $D(\hat{\theta}_0) = 2.655$. The 95% confidence interval of parameter θ is $15.31 \le \theta \le 21.64$. Because for $P = 0.95$, $F_{0.95}(2, 248) = 3.03$, the lower limit of the 95% confidence interval of parameter μ [Eq. (3.70)] is $\hat{\mu}_1 = 8.56$.

Conclusion: The confidence interval of parameter θ is broad compared with that of parameter μ.

3.3.6 The Log-normal Distribution

This distribution is one of the common distributions related to the normal one. For many types of physical measurements and other types of data, the measured values are either positive only (pressure, volume, concentration, weight, absorbance, etc.) or have a defined origin (e.g., absolute zero for temperature). When the measured values are far from an origin, the normal distribution is found to be appropriate, but when measured values are near an origin the approximation by the normal distribution is not convenient, and the log-normal or other distribution should be used. This distribution may be found in the analysis of samples containing low and very low concentrations, i.e. in trace analysis. The log-normal distribution is also applicable to the distribution of powder particles in the atmosphere or the size distribution of powder pigments. A total error that is a product of partial small errors belongs to a log-normal distribution.

The log-normal distribution is derived from the logarithmic transformation of the normal distribution. The random variable x of the log-normal distribution is related to the random variable u of the standardized normal distribution by

$$u = [\ln(x - \theta) - \mu] / \sigma \qquad (3.72)$$

where μ, σ and θ are parameters. The probability density function of the log-normal distribution is defined by

$$f(x) = \frac{1}{(x - \theta)\sigma\sqrt{2\pi}} \exp\left[-\frac{(\ln(x - \theta) - \mu)^2}{2\sigma^2} \right] \qquad (3.73)$$

3.3.6.1 The Two-parameter Log-normal Distribution

This distribution concerns the positive random variable defined in the range $0 \le x \le \infty$. The probability density function is defined by Eq. (3.73) when $\theta = 0$. The random variable x has a two-parameter log-normal distribution if the random variable $\ln x$ has a normal distribution $N(\mu, \sigma^2)$.

The mean and variance of the random variable x are calculated from

$$E(x) = \exp(\mu + 0.5\sigma^2) \qquad (3.74)$$

and

$$D(x) = \exp(2\mu)\omega(\omega - 1) \qquad (3.75)$$

where $\omega = \exp \sigma^2$. The skewness g_1 and kurtosis g_2 depend only on the variable ω according to

$$g_1 = \sqrt{(\omega - 1)(\omega + 2)} \tag{3.76a}$$

and

$$g_2 = \omega^4 + 2\omega^3 + 3\omega^2 - 3 \tag{3.76b}$$

The coefficient of variation δ defined by Eq. (3.17) is a function of parameter ω only for the log-normal distribution,

$$\delta = \sqrt{\omega - 1} \tag{3.77}$$

The mode $\hat{x}_{\mathrm{M}}$ and median $\tilde{x}_{0.5}$ are given by

$$\hat{x}_{\mathrm{M}} = \exp(\mu - \sigma^2) \tag{3.78a}$$

and

$$\tilde{x}_{0.5} = \exp(\mu) \tag{3.78b}$$

The logarithm of the likelihood function of the two-parameter log-normal distribution is

$$\ln L = -\frac{n}{2}\ln(2\pi) - \frac{n}{2}\ln \sigma^2 - \sum_{i=1}^{n} \ln x_i - \frac{1}{2\sigma^2}\sum_{i=1}^{n}(\ln x_i - \mu)^2 \tag{3.79}$$

From Eq. (3.3), the maximum likelihood estimate of parameter μ is

$$\hat{\mu} = \frac{1}{n}\sum_{i=1}^{n} \ln x_i \tag{3.80}$$

and of parameter σ^2

$$\hat{\sigma}^2 = \frac{1}{n}\sum_{i=1}^{n}(\ln x_i - \hat{\mu})^2 \qquad (3.81a)$$

which is biased. The unbiased estimate $\hat{\sigma}_0^2$ is calculated from

$$\hat{\sigma}_0^2 = \hat{\sigma}^2 n / (n-1) \qquad (3.81b)$$

The variance of parameter μ is calculated from Eq. (3.40) and of variance σ^2 from Eq. (3.42). The confidence interval of parameters μ and σ^2 is found in the same as for a normal distribution (Section 3.3.2).

There are some cases when the investigation of data in logarithmic transformation is not convenient, and parameters of location and spread with their confidence intervals should be calculated in the original data scale. The $100(1 - \alpha)\%$ confidence interval of median $\tilde{x}_{0.5}$ is then calculated from

$$\exp(\hat{\mu} - t_{1-\alpha/2}(n-1)\hat{\sigma} / \sqrt{n} \leq \tilde{x}_{0.5} \leq \exp(\hat{\mu} + t_{1-\alpha/2}(n-1)\hat{\sigma} / \sqrt{n} \qquad (3.82)$$

The confidence interval for the coefficient of the variation, the skewness or kurtosis may be calculated analogously as they are functions only of parameter σ^2. The $100(1 - \alpha)\%$ two-tailed confidence interval of the coefficient of variation is estimated by

$$\left[\exp\frac{(n-1)\hat{\sigma}^2}{\chi_{1-\alpha/2}^2(n-1)} - 1\right]^{1/2} \leq \delta \leq \left[\exp\frac{(n-1)\hat{\sigma}^2}{\chi_{\alpha/2}^2(n-1)} - 1\right]^{1/2} \qquad (3.83)$$

The point estimate of the mean $M = E(x)$ and the corresponding estimate of the variance $V = D(x)$ for the original data scale is given by

$$\hat{M} = \exp(\hat{\mu})g(0.5\hat{\sigma}^2) \qquad (3.84)$$

and

$$\hat{V} = \exp(2\hat{\mu})\left[g(2\hat{\sigma}^2) - g\left\{\frac{(n-2)\hat{\sigma}^2}{n-1}\right\}\right] \qquad (3.85)$$

In both expressions the function $g(t)$ is found from an infinite series

$$g(t) = 1 + \frac{n-1}{n}t + \sum_{j=2}^{\infty} \frac{(n-1)^{2j-1}}{n^j(n+1)(n+3)...(n+2j-1)} \times \frac{t^j}{j!} \qquad (3.86)$$

For large samples with $n > 50$, or for sufficiently small values of the variance σ^2, the following approximation [3] may be used

$$g(t) = \exp(t)\left[1 - \frac{t(t+1)}{n} + \frac{t^2(3t^2 + 22t + 21)}{6n^2}\right] \qquad (3.87)$$

The variances of the $\hat{M}$ and $\hat{V}$ estimates are calculated from the expression [3]

$$D(\hat{M}) = \sigma^2 \exp(2\mu)\omega(1 + 0.5\sigma^2)/n \qquad (3.88)$$

and

$$D(\hat{V}) = 2\sigma^2 \exp(\mu)\omega^2 \left[2(\omega-1)^2 + \sigma^2(2\omega-1)^2\right] \qquad (3.89)$$

For large samples, the approximate confidence intervals of $\hat{M}$ and $\hat{V}$ can be calculated from Eqs. (3.23) and (3.28). The confidence interval of the mean M may be calculated with the use of an estimate $\hat{\tau}$ and its variance $D(\hat{\tau})$

$$\hat{\tau} = \hat{\mu} + n\hat{\sigma}^2/(2(n-1)) \qquad (3.90a)$$

and

$$D(\hat{\tau}) = \sigma^2/(n-1) + n^2\sigma^4/(4(n-1)^3) \qquad (3.90b)$$

By using the estimate $\hat{\tau}$ and $D(\hat{\tau})$ the $100(1-\alpha)\%$ *confidence interval of the mean M* may be expressed as

$$\exp\left[\hat{\tau} - \mu_{1-\alpha/2}\sqrt{D(\hat{\tau})}\right] \le M \le \exp\left[\hat{\tau} + \mu_{1-\alpha/2}\sqrt{D(\hat{\tau})}\right] \qquad (3.91)$$

In the case of the log-normal distribution, the data should be analysed in logarithmic transformation or the estimates $\hat{M}$ and $\hat{V}$ should be used, except in cases when $\hat{\sigma}^2 \ll 1$.

Problem 3.7 *Examination of trace concentrations of copper in kaolin*

A set of $n = 32$ samples of kaolin was used to determine the trace concentration of copper in raw kaolin. Exploratory data analysis indicated that the sample came from a two-parameter log-normal distribution. By analysis of logarithms of the measured quantities, the arithmetic mean ($\hat{\mu} = 23$ and the sample variance $\sigma^2 = 0.0004$) were estimated. Calculate point and interval estimates of parameters of location and of dispersion by analysing the original measured quantities.

Data: the trace concentration of copper in raw kaolin in ppm:

9.467	9.785	9.806	9.863	9.867	9.889	9.915
9.933	9.951	9.950	9.969	9.994	10.025	10.059
10.078	10.088	10.093	10.091	10.094	10.097	10.124
10.134	10.185	10.195	10.209	10.222	10.234	10.235
10.317	10.446	10.313	10.502			

Program: ADSTAT or QC-EXPERT: Basic Statistics: One variable analysis.

Solution: From Eq. (3.27b) the 95% confidence interval of parameter μ is calculated to be $2.297 \le \mu \le 2.307$. The confidence limits are the arguments of the exponential in Eq. (3.82). The 95% confidence interval of the median $\tilde{x}_{0.5}$ of the original data is $10.012 \le \tilde{x}_{0.5} \le 10.167$.

The 95% confidence interval of the coefficient of variation δ (Eq. (3.83)) is $0.0160 \le \delta \le 0.0266$. Equation (3.87) gives the values $g(0.5\hat{\sigma}^2) = 1.0002$, $g(2\hat{\sigma}^2) = 1.0008$ and $g((n-2)\hat{\sigma}^2 / (n-1)) = 1.0004$. Then from Eq. (3.84), the mean estimate $\hat{M}$ is calculated to be $\hat{M} = 9.976$ with variance $D(\hat{M}) = 0.0012$. Equation (3.85) gives the variance estimate $\hat{V} = 0.0398$, with its variance $D(\hat{V}) = 1.006 \times 10^{-7}$.

The 95% confidence interval of the mean M Eq. (3.28) is $9.908 \le M \le 10.044$. Equation (3.90) then yields the estimate $\hat{\tau} = 2.30002$ with variance $D(\hat{\tau}) = 1.29 \times 10^{-5}$, and Eq. (3.91) gives the 95% confidence interval of the mean M to be $9.906 \le M \le 10.047$.

Conclusion: For a small value of variance estimates $\hat{\sigma}^2$, the confidence intervals of the mean M are practically identical with the confidence interval of the median.

3.3.6.2 The Three-parameter Log-normal Distribution

This distribution concerns a random variable defined in the interval $0 \leq x \leq \infty$. The probability density function for this distribution is given by Eq. (3.73). The random variable x has the three-parameter log-normal distribution if the random variable $\ln(x - \theta)$ has a normal distribution $N(\mu, \sigma^2)$. The parameters of location are about θ higher than those of the corresponding two-parameter log-normal distribution, whereas the parameters of dispersion and shape are the same.

The logarithm of the likelihood function is

$$\ln L = -\frac{n}{2}\ln(2\pi\sigma^2) - \sum_{i=1}^{n}\ln(x_i - \theta) - \frac{1}{2\sigma^2}\sum_{i=1}^{n}\left[\ln(x_i - \theta) - \mu\right]^2 \quad (3.92)$$

For known values of θ, the maximum likelihood estimates $\hat{\mu}(\theta)$ and $\sigma^2(\theta)$ are calculated from Eqs. (3.80) and (3.81), by using $(x_i - \theta)$ values instead of x_i. It can be proved that for $\theta \to x_{(1)}$, $\ln L$ tends towards infinity.

If the local maximum in the function $\ln L$ is to be found, a restriction must be introduced to avoid this global maximum. An iterative procedure for seeking the maximum likelihood estimates is described by Hill [4].

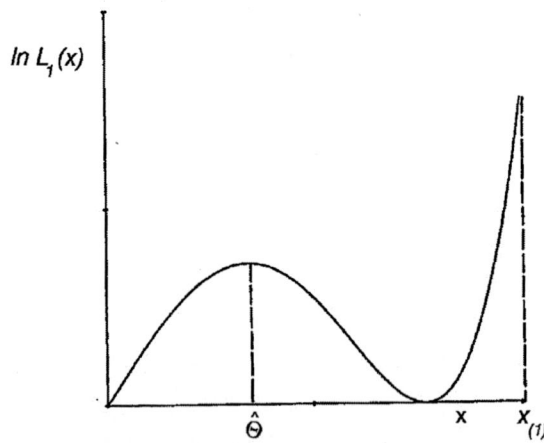

Figure 3.8   A plot of the modified likelihood function $\ln L_1$, vs. θ.

Introduction of the maximum likelihood estimates $\hat{\mu}(\theta)$ and $\sigma^2(\theta)$ into the logarithm of the likelihood function yields a modified maximum likelihood function:

$$\ln L_1 = -n[\hat{\mu}(\theta) + 0.5\ln\hat{\sigma}(\theta)] \qquad (3.92a)$$

This function depends on just a single parameter, θ. A graph of the function $\ln L_1 = f(\theta)$ is shown in Fig. 3.8. The first step is a search over the interval in which the function $\ln L_1$ is unimodal. The second stage is to search iteratively in this interval.

Problem 3.8 *Examination of trace concentration data on antimony in a copper ore*

The set of $n = 40$ samples of copper ore was analysed to determine the trace concentration of antimony (in ppm). Exploratory analysis indicated that the antimony concentration is described by a log-normal distribution. Estimate the point parameters of location and dispersion.

Data: the trace concentration of antimony in ppm:

145.4	151.6	83.7	109.6	124.1	124.4	128.1	139.7
140.3	145.4	156.0	160.3	160.6	163.1	165.6	172.5
182.0	187.7	193.1	199.6	201.9	203.1	204.2	205.0
205.4	206.7	217.0	220.6	240.6	242.4	247.2	253.8
259.6	265.8	266.1	308.8	311.7	401.3	449.3	536.3

Program: ADSTAT or QC-EXPERT: Probability models.

Solution: On maximizing Eq. (3.92a), with the use of Eqs. (3.80) and (3.81) with x_i replaced by $(x_i - \theta)$, the maximum likelihood estimates $\hat{\mu} = 5.065$, $\hat{\sigma} = 0.478$ and $\theta = 36.5$ are calculated. The maximum of the $\ln L$ function reaches the value $\ln L = -258.5$. Figure 3.9 shows a graph of the log-normal distribution with its simple nonparametric estimate.

By taking the minimum value $\theta = 0$, the estimates $\hat{\mu} = 5.29$, $\hat{\sigma}_1 = 0.384$ and $\ln L = -258.8$ are calculated.

Conclusion: The difference between the estimates results from the choice $\theta = 0$; the estimates calculated from original data are not influenced. For the three-parameter model the median is, $\tilde{x}_{0.5} = \hat{\theta} + \exp(\mu) = 194.88$ whereas for the two-parameter model $\tilde{x}_{0.5} = \exp(\hat{\mu}_1) = 198.34$.

3.4 Robust Estimates of Location and Spread

The sample mean $\bar{x}$ and the sample variance s^2 are efficient estimates of the parameters of location and spread only for data from the normal distribution. If the sample is not normally distributed, or if some outliers are present, the efficiency of both $\bar{x}$ and s^2 decreases. We shall introduce many statistical techniques based on normal distribution of the original observations, and these still remain approximately correct for reasonable departures from normality. In this regard they are said to be robust to non-normality. Robustness can relate to the separate effects of deviations from normality, independence, equal variance, and randomness.

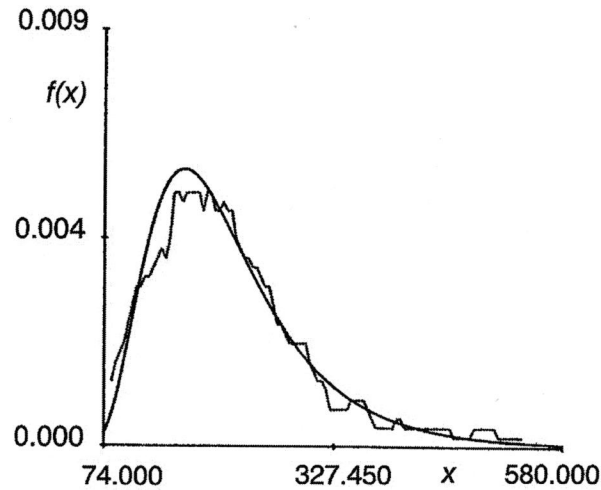

Figure 3.9   The probability density function for the sample log-normal distribution and its nonparametric estimate.

3.4.1 The Median

The median $\tilde{x}_{0.5}$ is the oldest robust estimate of the parameter of location. Despite other robust statistics, the median has a precise interpretation for a symmetrical and also for a non-symmetrical distribution: the median $\tilde{x}_{0.5}$ is the second quartile or 50% percentile, which divides the probability distribution area into two equal areas. Thus, the probability of x being less than $\tilde{x}_{0.5}$ is 1/2 and equal to the probability of x being greater than $\tilde{x}_{0.5}$. When the sample size n is odd, the sample median $\tilde{x}_{0.5}$ is defined to be the middle value in order of size; when n is even it lies between the two middle values, and usually we take the average of these two as its value.

For an unknown distribution and when some outliers are present, the nonparametric estimate of the standard deviation of a median is calculated from

$$s_{\text{M}} = \frac{x_{(n-k+1)} - x_{(k)}}{2u_{\alpha/2}} \tag{3.93}$$

where

$$k = \frac{n+1}{2} - |u_{\alpha/2}|\sqrt{n/4} \tag{3.94}$$

Best results are obtained with $\alpha = 0.05$, for which $|u_{0.025}| = 1.96$.

For small samples, the Marritz-Jarret estimate of the standard deviation of median can be used.

$$s_{\text{M}}^* = \left[\sum_{i=1}^{n} w_i x_{(i)}^2 - \left[\sum_{i=1}^{n} w_i x_{(i)}\right]^2\right]^{1/2} \tag{3.95}$$

where

$$w_i = J\left[\frac{i-0.5}{n}\right]\left[\sum_{i=1}^{n} J\left[\frac{i-0.5}{n}\right]\right]^{-1} \tag{3.95a}$$

and function $J(x)$ is defined by

$$J(x) = \frac{n!}{(m!)^2} x^m (1-x)^m \tag{3.95b}$$

where $m = \text{int}((n-1)/2)$ and $\text{int}(x)$ means the integer part of a number x.

The random variable t_{M} defined by

$$t_{\text{M}} = \frac{\tilde{x}_{0.5} - Me}{s_{\text{M}}} \approx \frac{\tilde{x}_{0.5} - Me}{s_{\text{M}}^*} \tag{3.96}$$

has approximately the Student distribution with $(n-1)$ degrees of freedom. The symbol Me means the median of the population from which the sample comes.

Problem 3.9 *Robust sample estimates of five distributions*

Apply robust analysis to five samples of size $n = 50$, from the normal, rectangular, exponential, Laplace and log-normal distribution, with the use of the median measures.

Data: $n = 50$, Each sample of size 50 was taken from an actual population with known population mean μ and population variance σ^2, denoted $X(\mu, \sigma^2)$:

(a) Sample from the $R(0.5, 1/12)$ distribution, $n = 50$

0.531	0.677	0.171	0.065	0.848	0.021	0.380	0.760	0.524	0.283	0.841	0.631
0.645	0.567	0.594	0.141	0.994	0.998	0.211	0.487	0.595	0.751	0.231	0.012
0.487	0.794	0.358	0.823	0.414	0.087	0.147	0.559	0.053	0.217	0.385	0.755
0.853	0.707	0.266	0.878	0.040	0.407	0.839	0.171	0.325	0.295	0.842	0.636
0.172	0.924										

(b) Sample from the $N(0, 1)$ distribution, $n = 50$

−1.008	−0.500	0.749	1.723	0.076	0.569	−1.389	0.087	1.112	−0.235	0.519	0.279
−0.758	−0.588	−0.594	−0.885	−0.072	−1.980	0.063	0.016	−0.673	−0.993	0.752	0.092
0.236	−2.962	0.109	−1.285	0.634	−0.383	1.134	−0.711	−1.825	2.374	0.500	−1.380
0.046	−0.544	−0.150	−1.129	1.173	1.401	−2.121	0.521	0.280	1.440	−0.415	−0.443
−0.384	0.690										

(c) Sample from the $E(1, 1)$ distribution, $n = 50$

0.757	1.129	0.188	0.067	1.885	0.021	0.478	1.427	0.743	0.333	1.837	0.188
0.998	1.036	0.837	0.902	0.152	5.145	6.170	0.237	0.668	0.903	1.388	0.262
0.012	0.668	2.572	1.580	0.444	1.731	0.535	0.091	0.159	0.819	0.054	0.245
0.487	1.408	1.916	1.228	0.309	2.104	0.040	0.523	1.829	0.188	0.394	0.349
1.846	1.012										

(d) Sample from the $L(0, 2)$ distribution, $n = 50$

0.064	0.436	−1.072	−2.036	1.192	−3.162	−0.275	0.734	0.049	−0.569	1.144	−1.070
0.304	0.343	0.144	0.209	−1.269	4.452	5.477	−0.862	−0.026	0.210	0.695	−0.774
−3.723	−0.026	1.879	0.887	−0.333	1.038	−0.188	−1.749	−1.224	0.126	−2.249	−0.835
−0.261	0.715	1.223	0.535	−0.632	1.411	−2.538	−0.206	1.136	−1.070	−0.429	−0.529
1.153	0.319										

(e) Sample from the $LN(2.718, 47.21)$ distribution, $n = 50$

0.191	2.118	0.380	0.264	3.374	2.490	0.509	0.232	3.482	1.746	2.372	4.657
2.507	2.832	0.150	13.673	0.312	0.810	4.080	0.619	1.691	0.088	1.236	0.726
0.157	1.415	1.002	0.035	0.908	15:880	0.047	1.817	0.078	7.606	1.349	0.267
3.649	0.212	0.397	26.475	0.606	0.440	1.849	27.203	0.545	5.690	48.558	4.732
0.006	2.404										

Program: ADSTAT or QC-EXPERT: Basic Statistics: One sample analysis.

Solution: Results of the robust analysis of the five samples are shown in Table 3.2. These are the median $\tilde{x}_{0.5}$, the square of interquantile deviation R^2 (Eq. (3.15)), the standard deviations of median s_M^* and the limits of the 95% confidence interval L_L and L_U (Eq. (3.96)). The results of Table 3.2 shows that (a) for the normal and rectangular distribution, the median characteristics give nearly the same results as the mean characteristics; (b) for the exponential distribution the confidence interval is nearly the same, but the median is too robust, compared with the mean; (c) for the Laplace distribution the median characteristics are better than the mean ones; (d) for the log-normal distribution the median characteristics are not objective enough and the 95% confidence interval does not contain the mean value.

Table 3.2 The median characteristics of five samples with various population distributions, $n = 50$

Population distribution $X(\mu, \sigma^2)$	$\tilde{x}_{0.5}$	R^2	s_M	s_M^*	L_L	L_U
Normal $N(0; 1)$	0.0309	1.334	0.032	0.423	−0.329	0.391
Rectangular R (0.5; 0.083)	0.506	0.242	0.0061	0.279	0.349	0.663
Exponential $E(1;1)$	0.705	0.923	0.024	0.423	0.395	1.02
Laplace $L(0; 2)$	0.0115	1.964	0.032	0.423	−0.347	0.371
Log-normal $LN(2.71, 47.21)$	1.293	4.109	0.190	0.660	0.415	2.169

Conclusion: The median and other robust characteristics are not suitable for asymmetrical distributions, because robust techniques "cut" good values and make the population look symmetric. Robust techniques cannot be adopted for general use.

3.4.2 The Trimmed Mean

One of the simplest and most efficient robust estimates of location is the *trimmed mean* $\bar{x}(\kappa)$ defined with the use of the order statistics $x_{(i)}$ as

$$\bar{x}(\kappa) = \frac{1}{n-2M} \sum_{i=M+1}^{n-M} x_{(i)} \tag{3.97}$$

where $M = \text{int}(\kappa\, n/100)$. The parameter κ determines the percentage of order statistics $x_{(i)}$ that are to be trimmed off at each (low and high) tail. The usual value of κ is 10%, and this results in the 10% trimmed mean, $\bar{x}(10)$. When there are many outliers, $\bar{x}(25)$ is preferred.

The trimmed mean is used with the *winsorized sum of squared differences*

$$S_w(\kappa) = \sum_{i=M+2}^{n-M-1} (x_{(i)} - \bar{x}_w(\kappa))^2 + (M+1)\left[(x_{(M+1)} - \bar{x}_w(\kappa))^2 + (x_{(n-M)} - \bar{x}_w(\kappa))^2 \right]$$

$$\tag{3.98}$$

where $\bar{x}_w(\kappa)$ is *the winsorized mean* defined by

$$\bar{x}_w(\kappa) = \frac{1}{n}\left[(M+1)(x_{(M+1)} + x_{(n-M)}) + \sum_{i=M+2}^{n-M-1} x_{(i)} \right] \tag{3.99}$$

Tukey and McLaughlin [2] recommend for statistical testing about parameter of location μ the test statistics

$$t_R(\kappa) = \frac{(\bar{x}(\kappa) - \mu)\sqrt{h(h-1)}}{S_w(\kappa)} \tag{3.100}$$

where $h = n - 2M$. The test statistic $t_R(\kappa)$ has approximately the Student distribution with $(n-1)$ degrees of freedom. For small samples from $n = 5$, the recommended value for constant M is 1, i.e. $\kappa = 20\%$. It may be concluded from Eq. (3.100) that the winsorized variance is equal to $s_w^2(\kappa) = S_w(\kappa)/(h-1)$.

For asymmetric and strongly skewed distributions *the asymmetric trimmed mean* $\bar{x}(\kappa_1, \kappa_2)$ is defined by

$$\bar{x}(\kappa_1, \kappa_2) = \frac{\sum_{i=n_1}^{n_2} x_{(i)}}{n_2 - n_1 + 1} \tag{3.101}$$

where $n_1 = \text{int}(\kappa_1\, n/100)$ and $n_2 = n - \text{int}(\kappa_2\, n/100)$. When κ_1 and κ_2 are chosen so that resulting trimmed sample has a symmetrical distribution, the *variance of asymmetrically trimmed mean* may be calculated by

$$s_w^2(\kappa_1,\kappa_2) = \frac{1}{h(h+1)}\Bigg[n_1(x_{(n_1)} - \bar{x}(\kappa_1,\kappa_2))^2$$

$$+ \sum_{i=n_1+1}^{n_2-1}(x_{(i)} - \bar{x}(\kappa_1,\kappa_2))^2 + (n - n_2 + 1)(x_{(n_2)} - \bar{x}(\kappa_1,\kappa_2))^2$$

$$-((n_1-1)(x_{(n_1)} - \bar{x}(\kappa_1,\kappa_2)) + (n - n_2)(x_{(n_2)} - \bar{x}(\kappa_1,\kappa_2))^2\Bigg] \qquad (3.102)$$

where $h = n_2 - n_1 + 1$. If the resulting trimmed sample is already symmetrical, the test criterion

$$t_R(\kappa_1,\kappa_2) = \frac{\bar{x}(\kappa_1,\kappa_1) - \mu}{s_w(\kappa_1,\kappa_1)} \qquad (3.103)$$

has approximately the Student distribution with $v = [(n_2 - n_1 + 1) - 2]$ degrees of freedom.

Various selector criteria are available for choosing the magnitude for trimming off data. These serve as the estimates of the length of tails or of skewness of the sample distribution. The Q_1 criterion for estimation of the *relative tail length of the sample distribution* is defined by

$$Q_1 = \frac{\bar{U}(0.05) - \bar{L}(0.05)}{\bar{U}(0.5) - \bar{L}(0.5)} \qquad (3.104)$$

and also for estimation of the *relative skewness of the sample distribution* by

$$Q_2 = \frac{\bar{U}(0.05) - \bar{M}(0.5)}{\bar{M}(0.05) - \bar{L}(0.05)} \qquad (3.104)$$

where $\bar{U}(\beta)$ is the average of the $n\beta$ largest ordered values, $\bar{L}(\beta)$ is the average of the $n\beta$ smallest ordered values, and $\bar{M}(\beta)$ is the average of ordered values from the middle part of the distribution. When $n\beta$ is not an integer, it is found as a ratio of neighbouring ordered values by linear interpolation.

According to the values of Q_1 and Q_2, the parameter κ of the trimmed mean $\bar{x}(\kappa)$ is chosen as follows.

(a) When $Q_2 \approx 0$ and the sample distribution is symmetric, κ is selected according to sample size (Table 3.3).

(b) When the sample distribution is symmetric and 90% efficiency is desired, parameter κ is chosen such that $\kappa = 15\%$ for $Q_1 < 2.9$, $\kappa = 25\%$ for $2.9 \leq Q_1 \leq 3.5$ and $\kappa = 35\%$ for $Q_1 \geq 3.5$.

Table 3.3 The choice of best trimming parameter, κ in %. Q_1^* is calculated from Eq. (3.104), with difference $[\bar{U}(0.2) - \bar{L}(0.2)]$ in the numerator

n	κ (%)	Q_1^*
<10	6.35	for all Q^*
	12.5	$Q_1^* < 1.84$
10–20	25	$Q_1^* > 1.84$
	9.375	$Q_1^* < 1.81$
20–30	18.75	$1.81 < Q_1^* < 1.87$
	28.125	$Q_1^* > 1.87$

(c) When $Q_2 \neq 0$, the asymmetrically trimmed average may be used, and parameters κ_1 and κ_2 are selected as follows:

for $Q_2 \leq 1.4$ and $Q_1 > 2.68$, $\kappa_1 = 25\%$ and $\kappa_2 = 25\%$,

for $Q_2 > 1.4$ and $Q_1 < 1.98$, $\kappa_1 = 0\%$ and $\kappa_2 = 50\%$,

for $Q_2 > 1.4$ and $Q_1 > 2.68$, $\kappa_1 = 25\%$ and $\kappa_2 = 25\%$.

More detailed information is available [1].

Problem 3.10 *Trimmed mean and winsorized variance of samples from five distributions*

Apply robust analysis to five samples of size $n = 50$ from the normal, rectangular, exponential, Laplace and log-normal distributions, with use of the trimmed mean and winsorized variance.

Data: from Problem 3.9

Program: ADSTAT or QC-EXPERT: Basic Statistics: One sample analysis.

Solution: Table 3.4 lists the trimmed mean $\bar{x}(\kappa)$ and the winsorized variance $s_w^2(\kappa)$ for two values of κ, $\kappa = 0.05$ and $\kappa = 0.10$.

Table 3.4   The trimmed mean $\bar{x}(\kappa)$ and the winsorized variance $s_w^2(\kappa)$ for (a) $\kappa = 0.05$, and (b) $\kappa = 0.10$, and the limits L_L, and L_U of the confidence interval of the mean $\bar{x}_w$.

(a)	$\kappa = 0.05$			
Population distribution $X(\mu;\sigma^2)$	$\bar{x}(\kappa)$	$s_w^{\ 2}(\kappa)$	L_L	L_U
Normal $N(0; 1)$	−0.046	0.825	−0.325	0.232
Rectangular $R(0.5; 0.083)$	0.486	0.088	0.395	0.578
Exponential $E(1; 1)$	0.836	0.519	0.615	1.060
Laplace $L(0;2)$	−0.088	1.271	−0.434	0.259
Log-normal $LN(2.71; 47.21)$	2.551	22.65	1.087	4.020

(b)	$\kappa = 0.10$			
Population distribution $X(\mu;\sigma^2)$	$\bar{x}(\kappa)$	$s_w^{\ 2}(\kappa)$	L_L	L_U
Normal $N(0; 1)$	−0.053	0.787	−0.340	0.234
Rectangular $R(0.5; 0.083)$	0.489	0.092	0.391	0.587
Exponential $E(1; 1)$	0.805	0.512	0.573	1.036
Laplace $L(0;2)$	−0.049	1.119	−0.392	0.293
Log-normal $LN(2.71; 47.21)$	1.796	7.30	0.921	2.671

Table 3.5   The selector statistics Q_1, Q_2^* and Q_2 for five samples from various distributions with $n = 50$

Population distribution $X(\mu;\sigma^2)$	Q_1	Q_1^*	Q_2
Normal $N(0; 1)$	2.856	1.795	0.919
Rectangular $R(0.5; 0.083)$	1.922	1.571	0.054
Exponential $E(1; 1)$	3.576	1.784	6.175
Laplace $L(0;2)$	3.66	1.885	1.358
Log-normal $LN(2.71; 47.21)$	5.02	2.064	25.182

When these results are compared with those of Table 3.2, it is evident that (1) for the symmetric distributions N, R and L, the trimming leads to shortening of the confidence interval, (2) for the asymmetric distribution E and

LN the trimming for $\kappa = 0.10$ causes negative results, and for the log-normal distribution the 95% confidence interval does not contain a true value. A small amount of trimming $\kappa = 0.05$ always leads to confidence intervals which are narrow. Table 3.5 lists some selector statistics Q_1 and Q_2 which indicate the differences in skewness and kurtosis of the samples. Use of selector statistics in the symmetric distributions does not cause significant improvement.

Conclusion: Trimmed means in connection with selector statistics enable determination of robust estimates even for asymmetric distributions.

3.4.3 The Robust *M*-estimates

The robust *M*-estimates represent the maximum likelihood estimates of parameters for some special distributions. Maximization of the likelihood function according to the parameter μ_M leads here to a minimization of the function

$$\sum_{i=1}^{n} \rho \left[\frac{x_i - \mu_M}{\sigma} \right] = \min \qquad (3.106)$$

The shape of the function $\rho(u)$ determines the property of the estimate. Among *M*-estimates are the arithmetic mean and median. The *M-estimate of location parameter* μ_M is generally defined by

$$\hat{\mu}_M = \frac{\sum\limits_{i=1}^{n} w_i x_i}{\sum\limits_{i=1}^{n} w_i} \qquad (3.107)$$

where $w_i = W \left[(x_i - \mu_M)/\sigma \right]$ and $W(u) = \mathrm{d}\rho(u)/\mathrm{d}t$. For a robust estimate the function $W(u)$ must be bounded. The bi-quadratic function $W(u)$ of the following type is recommended

$$W(u) \begin{cases} \left[\left[1 - \left[\frac{u}{4.69} \right]^2 \right]^2 \right] & \text{for } |u| < 4.69 \\ 0 & \text{for } |u| \geq 4.69 \end{cases} \qquad (3.108)$$

where the numerical konstant 4.69 means that for normally distributed data the asymptotic efficiency of estimate $\hat{\mu}_M$ is equal to 0.95. Since the standard deviation σ is usually unknown it is replaced by a suitable robust estimate. Du

Mond and Lenth [5] recommended for the *M-estimate of standard deviation* the expression

$$s_M = -\left[\frac{\sum\limits_{i=1}^{n} V_i(x_i - \hat{\mu}_M)^2}{\sum\limits_{i=1}^{n} V_i}\right]^{1/2} \tag{3.109}$$

where

$$V_i = W\left[\left[\Delta\left[\frac{x_i - \hat{\mu}_M}{s_M}\right]\right]^{1/2}\right] \tag{3.110}$$

The weight function $W()$ is defined by Eq. (3.108) and $\Delta(u)$ is a deviation function, for which:

$$\Delta(u) = \begin{cases} u^2 - \ln(u^2) - 1 & \text{for } u \neq 0 \\ \infty & \text{for } u = 0 \end{cases} \tag{3.111}$$

Du Mond and Lenth [5] procedure for $\hat{\mu}_M$ and $\hat{s}_M$ estimation:

(1) For initial guesses $\hat{\mu}_M^{(0)}$ and $\hat{s}_M^{(0)}$ the median $\tilde{x}_{0.5}$ and the interquantile range $\hat{\sigma}_R = 0.75(\tilde{x}_{0.75} - \tilde{x}_{0.25})$ are computed.

(2) For initial guesses $\hat{\mu}_M^{(0)}$ and $\hat{s}_M^{(0)}$ the the weights w_i and V_i are calculated from Eqs. (3.107) and (3.109), then the refined estimates $\hat{\mu}_M^{(1)}$ and $\hat{s}_M^{(1)}$ are obtained.

(3) In the second iteration new values of weights w_i and V_i are calculated and hence new refined estimates $\hat{\mu}_M^{(2)}$ and $\hat{s}_M^{(2)}$.

(4) Iteration refinement terminates when the estimates from two iterations do not differ significantly.

Because the robust *M*-estimate $\hat{\mu}_M$ represents the weighted arithmetic mean, its variance is expressed by

$$D(\hat{\mu}_M) = s_M^2 / \sum_{i=1}^{n} w_i \tag{3.112}$$

In constructing confidence intervals and statistical testing the random variable

$$t_M = \frac{(\mu_M - \hat{\mu}_M)\left[\sum_{i=1}^{n} w_i\right]^{1/2}}{s_M} \qquad (3.113)$$

which has approximately the Student t-distribution with $\nu = n - 1$ degrees of freedom can be used. Two weight functions are compared in Fig. 3.10.

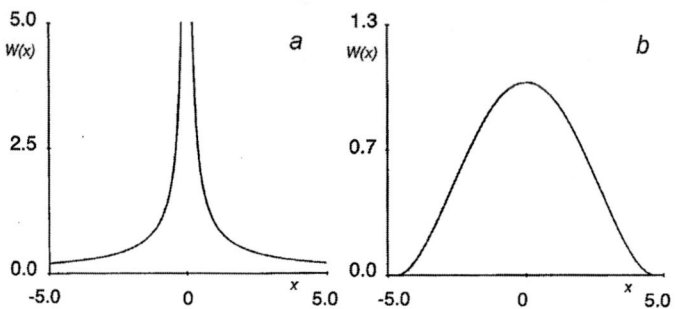

Figure 3.10 Comparison of two weight functions of M-estimates: (a) for the median, and (b) for the bi-quadratic function.

Problem 3.11 *Robust bi-quadratic sample estimates from five distributions*

Apply robust analysis to five samples of size $n = 50$ from normal, rectangular, exponential, Laplace and log-normal distributions with the use of bi-quadratic estimates.

Data: from Problem 3.9

Program: ADSTAT or QC-EXPERT: Basic Statistics: One sample analysis.

Solution: Robust estimates $\hat{\mu}_M$, variances $D(\hat{\mu}_M)$ and the limits of the 95% confidence interval of the mean are listed in Table 3.6. For the symmetric distributions N, R and L the robust analysis gives accurate estimates quite near to the true values, and the confidence interval is narrow. Worse results were achieved with the asymmetric skewed distributions: for the exponential and log-normal distributions the 95% confidence interval does not contain theoretical value μ.

Table 3.6 Robust analysis of samples from five distributions with the use of bi-quadratic estimates

Population distribution $X(\mu;\sigma^2)$	$\hat{\mu}_M$	$D(\hat{\mu}_M)$	L_L	L_U

Normal $N(0; 1)$	−0.0458	1.039	−0.349	0.257
Rectangular $R(0.5; 0.083)$	0.488	0.089	0.399	0.577
Exponential $E(1; 1)$	0.762	0.442	0.561	0.964
Laplace $L(0;2)$	−0.124	1.464	−0.490	0.242
Log-normal $LN(2.71; 47.21)$	1.375	2.378	0.893	1.858

Conclusion: The robust *M*-estimates of this type are not suitable for analysis of skewed distributions.

Problem 3.12 *Comparison of various estimates of location and spread for data with outliers*

The advantage of robust estimates is their lower sensitivity to outliers in data. A sample of size $n = 50$ was generated in which 47 observations came from the normal distribution $N(0, 1)$ with $\mu = 0$ and $\sigma^2 = 1$, and the remaining three from the normal distribution $N(A, 1)$ with the population mean $\mu = A$ and variance $\sigma^2 = 1$. For three different values of A, 100, 50 and 10, three samples were synthesized. Calculate the sample mean $\bar{x}$, variance s^2, median $\tilde{x}_{0.5}$, squared interquantile range R^2, trimmed mean $\bar{x}(10)$, winsorized variance $s_w^2(10)$, bi-quadratic estimate of a mean $\hat{\mu}_M$ with its corresponding variance, and comment on each estimate.

Data:

(1) $n = 50$, 47 elements from $N(0, 1)$ and 3 elements from $N(100, 1)$:

−1.008	−0.500	0.749	1.723	0.076	0.569	−1.389
0.087	1.112	−0.235	0.519	0.279	−0.758	−0.588
−0.594	−0.885	−0.072	1.980	0.063	0.016	−0.673
−0.993	0.752	0.092	0.236	−2.962	−0.383	0.109
−1.285	0.634	0.690	1.134	−0.711	−1.825	2.374
0.500	−1.380	0.046	−0.544	−0.150	−1.129	1.173
1.401	−2.121	0.521	0.280	1.440	99.585	99.557
99.616						

(2) $n = 50$, 47 elements from $N(0, 1)$ and 3 elements from $N(50, 1)$: replace last three points with 49.585, 49.557, 49.616.

(3) $n = 50$, 47 elements from $N(0, 1)$ and 3 elements from $N(10, 1)$:

replace last three points with 9.585, 9.557, 9.616.

Program: ADSTAT or QC-EXPERT: Basic Statistics: One sample analysis.

Solution: The parameters of location and spread are listed in Table 3.7. Whereas $\bar{x}$ and s^2 are quite far from the true values, all the robust methods leads to estimates that are not influenced by the outliers.

Table 3.7   Parameter estimates of location and spread for three samples containing outliers.

Magnitude of outlier	$\bar{x}$	s^2	$\tilde{x}_{0.5}$	R^2	$\bar{x}(10)$	$s_w^2(10)$	$\hat{\mu}_M$	s_M^2
100	5.94	572.2	0.082	1.56	0.092	1.14	−0.020	1.152
50	2.94	142.8	0.082	1.56	0.092	1.14	−0.018	1.136
10	0.54	6.4	0.082	1.56	0.092	1.18	−0.017	1.120

Conclusion: The main advantage of robust estimates is the low sensitivity to outliers. Robust statistics are convenient for statistical data analysis when the data are not homogeneous.

3.4.4 The Analysis of Small Samples

The analysis of small samples is not reliable and results are usually rather uncertain. Small samples are used in cases when experiment repetition is expensive or scarcely possible.

For $n = 2$, statistical analysis is very difficult. If observations are close enough, the arithmetic mean is calculated. If observations do not agree, it is not possible to say which the outlier is. The $100(1 - \alpha)\%$ confidence interval of the mean μ may be calculated by an approximation

$$\frac{x_1 + x_2}{2} - T_\alpha \frac{|x_1 - x_2|}{2} \le \mu \le \frac{x_1 + x_2}{2} + T_\alpha \frac{|x_1 - x_2|}{2}$$

The critical value of T_α depends on the distribution of the data population that the two values come from. For the normal distribution it is $T_\alpha = \cot g(\pi \alpha/2)$, and for $\alpha = 0.05$, T_α is 12.71. For the rectangular distribution $T_\alpha = 1/\alpha - 1$, i.e. $T_{0.05} = 19$ [6].

For $n = 3$ it is also difficult to use statistical analysis. The calculation of the arithmetic mean $\bar{x}$ from two nearer observations is better than the use of the median from all three values. The $100(1 - \alpha)\%$ confidence interval of the mean μ is then calculated by an approximation

$$\bar{x} - s\, T'_\alpha / \sqrt{3} \leq \mu \leq \bar{x} + s\, T'_\alpha / \sqrt{3}$$

For the normal distribution $T'_\alpha \approx 1 / \sqrt{\alpha} - 3\sqrt{\alpha}/4 + \dots$, and when $\alpha = 0.05$, T'_α is 4.30. For the rectangular distribution $T'_{0.05} = 5.74$ [6].

For $4 \leq n \leq 20$ a procedure based on order statistics was introduced by Horn [7]. This is based on the depths which correspond to the sample quartiles (the letter F), cf. Section 2.2. The pivot depth is expressed by $H_L = \text{int}[(n + 1)/2]/2$ or $H_L = \text{int}[(n + 1)/2 + 1]/2$ according to which of the H_L is an integer. The lower pivot is $x_L = x_{(H)}$ and the upper one is $x_U = x_{(n^+1-H)}$. The estimate of the parameter of location is then expressed by the *pivot half sum*

$$P_L = 0.5(x_L + x_U) \tag{3.114}$$

and the estimate of the parameter of spread is expressed by the *pivot range*

$$R_L = x_U - x_L \tag{3.115}$$

The random variable

$$T_L = \frac{P_L}{R_L} = \frac{x_L + x_U}{2(x_U - x_L)} \tag{3.116}$$

has approximately a symmetric distribution and its quantiles are given in Table 3.8.

Table 3.8 The quantile $t_{L,1-\alpha}(n)$ of the T_L-distribution

$1-\alpha$					
n	0.9	0.95	0.975	0.99	0.995
4	0.477	0.555	0.738	1.040	1.331
5	0.869	1.370	2.094	3.715	5.805

Continue

6	0.531	0.759	1.035	1.505	1.968
7	0.451	0.550	0.720	0.978	1.211
8	0.393	0.469	0.564	0.741	0.890
9	0.484	0.688	0.915	1.265	1.575
10	0.400	0.523	0.668	0.878	1.051
11	0.363	0.452	0.545	0.714	0.859
12	0.344	0.423	0.483	0.593	0.697
13	0.389	0.497	0.608	0.792	0.945
14	0.348	0.437	0.525	0.661	0.776
15	0.318	0.399	0.466	0.586	0.685
16	0.299	0.374	0.435	0.507	0.591
17	0.331	0.421	0.502	0.637	0.774
18	0.300	0.380	0.451	0.555	0.650
19	0.288	0.361	0.423	0.502	0.575
20	0.266	0.337	0.397	0.464	0.519

The 95% confidence interval of the mean is expressed by pivot statistics as

$$P_L - R_L \, t_{L,0.975}(n) \le \mu \le P_L + R_L \, t_{L,0.975}(n) \qquad (3.117)$$

and analogously hypothesis testing also may be carried out. For small samples ($4 \le n \le 20$), the pivot statistics lead to more reliable results than the application of Student's F-test or robust t-tests.

Problem 3.13 *Analysis of a small sample (n = 5) taken from five distributions*

Make an analysis of a small sample ($n = 5$) taken from normal, rectangular, exponential, Laplace and log-normal distributions. Try to use Horn's procedure [7].

Data: $n = 5$.

$N(0, 1)$	−1.008	−0.500	0.749	1.723	0.076
$R(0.5, 0.083)$	0.531	0.677	0.171	0.065	0.848
$E(1,1)$	0.757	1.129	0.188	0.067	1.885
$L(0, 2)$	0.064	0.436	−1.072	−2.036	1.192
$LN(2.71,47.21)$	0.191	2.118	0.380	0.264	3.374

Program: ADSTAT or QC-EXPERT: Basic Statistics: One sample analysis.

Solution: The classical and pivot approaches are compared. In the first half of Table 3.9 the classical statistical measures ($\bar{x}$, s^2 and limits L_L, L_U of the 95% confidence interval of the mean) were calculated, and the second half contains the pivot statistics (P_L, R_L and limits L_L, L_U of the 95% confidence interval of the mean).

Table 3.9 Horn's procedure for analysis of a small sample (n = 5), in comparison with the classical statistical approach

Population distribution X(n, 6)²	Classical approach				Pivot approach			
	$\bar{x}$	s^2	L_L	L_U	P_L	R_L	L_L	L_U
Normal N(0; 1)	0.208	1.146	−1.120	1.536	0.125	1.249	−0.084	0.333
Rectangular R(0.5; 0.083)	0.458	0.110	0.046	0.081	0.424	0.506	−1.332	2.180
Exponential E(1, 1)	0.805	0.550	−0.115	1.725	0.658	0.941	−0.806	2.123
Laplace L(0, 2)	−0.283	1.628	−1.866	1.299	−0.318	1.508	−0.760	0.124
Log-normal LN (2.71; 47.21)	1.265	2.029	−0.502	3.033	1.190	1.855	−0.154	2.536

Conclusion: For a small sample size, the pivot approach seems to lead to more reliable parameter estimates.

3.4.5 Computer Intensive Methods

A survey of various computer intensive nonparametric methods for estimating of parametres and their variance was made by Efron [8]. Here only two techniques, the Bootstrap method and Jackknife method, will be demonstrated. Both methods enable a calculation of confidence interval and both are so called distribution free (i.e., the distribution of data are unknown).

The *Bootstrap method* is based on the estimate $\hat{\theta}$ of parameter θ which is a known function of n independent random observations $x_1,...,x_n$, [i.e. $\hat{\theta} = g(x_1,...,x_n)$]. The sample arises from an unknown distribution F_B. The variance of estimate $\hat{\theta}$ depends on the unknown distribution function F_B. The Bootstrap method substitutes the F_B distribution by a discrete distribution with probability $p_i = 1/n$ in points $x_i, i = 1,...,n$.

The Bootstrap procedure:

(1) From the original sample (x_i), $i = 1,...,n$, random sampling with replacement is used to create the B Bootstrap samples of size n. The *i*th element of the original sample may be present in the *i*th Bootstrap sample several times while another element may not be present at all. A selection of Bootstrap sample elements is made on the basis of random index *i*

$$i = \text{int}[\text{rnd}(0) \, n + 1] \qquad (3.118)$$

where rnd(0) is the pseudorandom number from a generator of rectangularly distributed pseudorandom numbers between 0 and 1. For example, when rnd(0) = 0, then $i = 1$ and when rnd(0) = 1, then $i = n$.

(2) Calculation of parameter estimates $\hat{\theta} = g(x_i)$, $i = 1, ..., B$, where x_i stands for the *j*th Bootstrap sample of size *n*.

(3) Calculation of the estimate of variance

$$\hat{\sigma}_B^2 = \frac{\sum\limits_{i=1}^{B} (\hat{\theta}_i - \overline{\theta}_B)^2}{B-1} \qquad (3.119)$$

where $\overline{\theta}_B = \sum\limits_{i=1}^{B} \hat{\theta}_i / B$ represents the estimate of the mean. Usually the number of Bootstrap samples *B* is set between 500 and 1000.

(4) For construction of the confidence interval of the parameter and statistical tests the criterion

$$t_B = \frac{(\overline{\theta} - \mu)\sqrt{B}}{\hat{\sigma}_B} \qquad (3.120)$$

is used; for large *B* values this has a standardized normal distribution.

The Jack-knife procedure:

The Jack-knife method is based on the use of "pseudovalues", y_i defined by

$$y_i = n \, \hat{\theta} - (n-1) \, \hat{\theta}_{(i)} \qquad (3.121)$$

where $\hat{\theta}_{(i)}$ is an estimate computed from all sample elements except the *i*th one. The mean $\overline{\theta}_j$ is calculated from

$$\bar{\theta}_J = \frac{1}{n}\sum_{i=1}^{n} y_i = n\hat{\theta} - \frac{n-1}{n}\sum_{i=1}^{n}\hat{\theta}_{(i)} \qquad (3.122a)$$

With large or medium sample sizes, the pseudovalues y_i are taken to be approximately normally distributed. The random variable

$$t_J = \frac{\bar{\theta}_J - \theta}{\hat{\sigma}_J} \qquad (3.122b)$$

has the standardized normal distribution. The variance of the Jack-knife estimate σ_J^2 is calculated from

$$\hat{\sigma}_J^2 = \frac{1}{(n-1)}\sum_{i=1}^{n}(y_i - \bar{\theta}_J)^2 \qquad (3.122c)$$

Both nonparametric estimates $\hat{\sigma}_B^2$ and $\hat{\sigma}_J^2$ enable determination of a variance of various types of estimators of parameter θ.

Problem 3.14 *Confidence interval of variance by Bootstrap and Jack-knife methods*

Calculate the 95% confidence interval of the sample variance s^2 for a sample of $n = 30$ taken from the Laplace distribution, with the use of the Bootstrap and Jack-knife methods.

Data: $n = 30$

0.064	0.436	−1.072	−2.036	1.192	−3.162	−0.275
0.734	0.049	−0.569	1.144	−1.070	0.304	0.343
0.144	0.209	−1.269	4.452	5.477	−0.862	−0.026
0.210	0.695	−0.774	−3.723	−0.026	1.879	0.887
−0.333	1.038					

Solution: (a) *The Bootstrap method:* 400 simulated Bootstrap samples were selected. The average variance $\bar{s}_B^2 = 3.12$ and Bootstrap variance $\hat{\sigma}_B^2 = 1.53$. From Eq. (3.120) the limits of the 95% confidence interval of the variance are calculated as $L_{L,U} = \bar{s}_B^2 \pm 1.95\hat{\sigma}_B^2 = 3.12 \pm 2.41$. A histogram of the values generated is shown in Fig. 3.11.

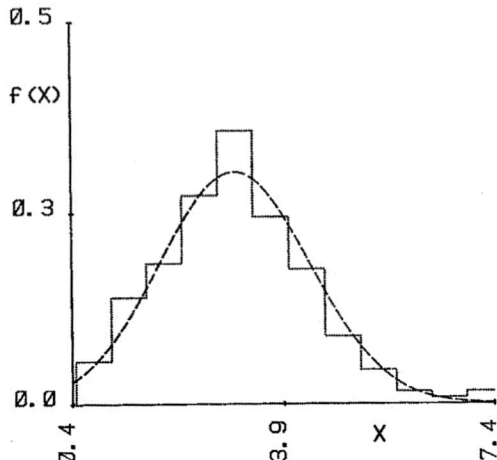

Figure 3.11   The histogram of the 400 Bootstrap sample values simulated from an original sample size of $n = 30$ from the Laplace distribution.

(b) *The Jack-knife method:* the average value of variance $\overline{s}_J^2$ is 3.2 with variance $\hat{\sigma}_J^2 = 1.289$. From Eq. (3.122) the limits of the 95% confidence interval $L_{L,U} = \overline{s}_J^2 \pm 1.95\hat{\sigma}_J^2 = 3.20 \pm 2.51$

Conclusion: The values for the confidence interval of the sample variance estimated by the Bootstrap and Jackknife methods are in reasonable agreement.

3.5 Statistical Hypothesis Testing

In many applications of statistics we are interested in making inferences about *population* (or *ensemble*) characteristics on the basis of observations made on a *random sample* of items from the population. The characteristics of interest may often be expressed in terms of population *parameters*, such as the population mean μ, or variance σ^2, or the proportion p of the population which has a certain characteristic. In other situations we may wish to make inferences about the difference between two (or more) populations, such as the difference between two population means μ_1 and μ_2.

A *statistical hypothesis* is a statement about the population distribution of some random variable or its parameters. Hypothesis testing consists of comparing some statistical measures called *test criteria* (or statistics) deduced from a data sample with the values of these criteria taken on the assumption that a given hypothesis is correct. In hypothesis testing, one examines a *null hypothesis* H_0 against one or more *alternative hypotheses* H_A which are stated explicitly or implicitly. To reach a decision about the hypothesis, we select a value of α, which is termed the *significance level* for the test. Significance level α is usually arbitrarily selected to be fairly small, for example, a might be 0.05 or 0.01. The significance level α is related to the *confidence coefficient* $1 - \alpha$.

Since the alternative hypothesis is the hypothesis which is accepted when a null hypothesis is rejected, the procedure of hypothesis testing seems to be, in fact, a process of rejection of alternative hypotheses.

For hypothesis testing *the test statistic* (or *the test criterion*) is set up. When this statistic falls into the the *range of acceptance*, the null hypothesis is not rejected. When this statistic falls into the *region of rejection* (or the *critical range*) the null hypothesis is rejected. The probability of the test statistic falling in the region of rejection is equal to *the significance level*. It is expressed in %, e.g. 5% or 1%.

A region of rejection can be set up as two-tailed (or two-sided) leading to a two-tailed test, or as a one-tailed test. The two-tailed test is used when test statistic can take values with either positive or negative sign. The significance level α is then split into two equal parts of magnitude $\alpha/2$.

3.5.1 Procedure for Hypothesis Testing

The procedure of statistical hypothesis testing is as follows:

(1) The null hypothesis H_0 and an alternative one H_A are formulated.

(2) The significance level α is selected.

(3) The test statistic is chosen.

(4) The region of rejection of the test statistic, on the basis of its probability distribution and the significance level, is determined.

(5) For the sample, the test statistic is calculated, and the limits of the region of rejection (Fig. 3.12). (a) The null hypothesis is rejected and the alternative one accepted when the value of the test statistic falls into the region of rejection; (b) the null hypothesis is not rejected when the value of the test statistic does not fall into the region of rejection.

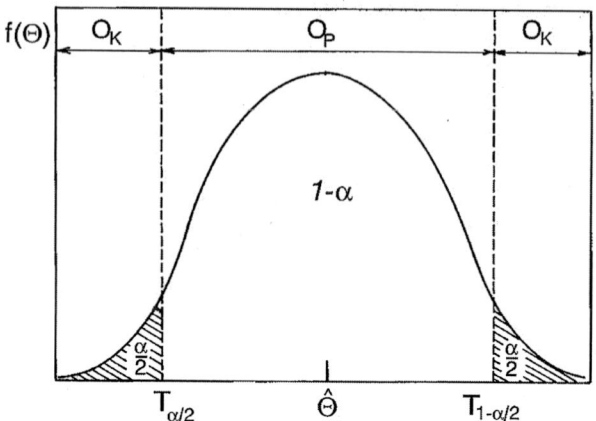

Figure 3.12 Regions of rejection (O_K) and acceptance (O_P) for a symmetrical (two-tailed) hypothesis test.

In judgement, it is necessary to remember that

(i) *rejection* of a null hypothesis H_0 does not necessarily mean that the tested null hypothesis is not valid. The rejection of a null hypothesis H_0 means only that we do not trust its validity because of the statistical test performed. It is understood in the following work that if the H_0 hypothesis is not valid then the alternative one H_A is valid;

(ii) *no rejection* of a null hypothesis H_0 does not imply its acceptance. When we do not reject a null hypothesis H_0 it means only that hypothesis testing did not provide sufficient reason for rejection of the hypothesis. If the H_0 hypothesis is not rejected, it can usually be assumed that either the H_0 (or some other hypothesis close to H_0) is valid.

We illustrate the procedure of hypothesis testing with an example in which we want to test the parameter θ. The sample size is large enough to allow us to use Eq. (3.23). The null hypothesis H_0: $\theta = K$, where K is a known number, is tested against the alternative H_A: $\theta \neq K$. The test statistic $u_s = \dfrac{|\theta\text{-}K|}{\sqrt{D(\theta)}}$ will have a normal distribution if the null hypothesis H_0 is valid. Testing the H_0 hypothesis can lead to the following results:

(1) the test statistic falls into a region of acceptance of the null hypothesis

i.e. $u_{\alpha/2} < u_s < u_{1-\alpha/2}$ and therefore H_0 is not rejected. If H_0 is valid the probability that u_s will fall out of range O_P is equal to the significance level α. The magnitude of α determines the magnitude of the error of the first kind, i.e. wrong rejection of correct hypothesis H_0.

(2) The test statistic falls into a range of rejection O_K i.e. into the interval $u_s < u_{\alpha/2}$ or $u_s > u_{1-\alpha/2}$, respectively.

The null hypothesis H_0 is then rejected in favour of the alternative H_A. The probability that u_s falls into the region of acceptance O_P even if H_0 is wrong represents the magnitude of the error of the second kind, β. This error results when a wrong hypothesis H_A is accepted. The two types of error are illustrated in Fig. 3.13.

Both errors should be minimized. The probability that we will not make an error of the first kind is $(1 - \alpha)$. This is the probability of making the correct decision about the test hypothesis. The second correct decision is made with probability $(1 - \beta)$. This is the probability of not making an error of the second kind. We call the *power of the test* to discriminate $S = 1 - \beta$. It represents the hope of making a correct decision when the hypothesis is actually wrong.

The power of the test is affected by the sample size n: with bigger sample sizes, more information is available, and therefore the wrong hypothesis will be rejected with more confidence in favour of the alternative one. For $n \to \infty$, $S \to 1$.

Attention should be paid to the choice of significance level α in hypothesis testing:

(1) When the null hypothesis H_0 is not rejected at the significance level $\alpha = 0.05$, the difference between the theoretical value K and the estimated parameter $\hat{\theta}$ is not significant.

(2) When the null hypothesis H_0 is rejected also at the significance level $\alpha = 0.01$, the difference between the theoretical value K and the estimated parameter $\hat{\theta}$ is statistically significant.

(3) When the null hypothesis H_0 is rejected at the significance level $\alpha = 0.05$ but not at $\alpha = 0.01$, the sample size did not give sufficient information for a correct decision.

3.5.2 Hypothesis Rests for The Parameters of One Population

Tests on the parameters of a single population enable us to tell if the population mean (or variance) of a say new product or variable (1) is different from, (2) exceeds, or (3) is less than the population mean (or variance) of a standard product or variable.

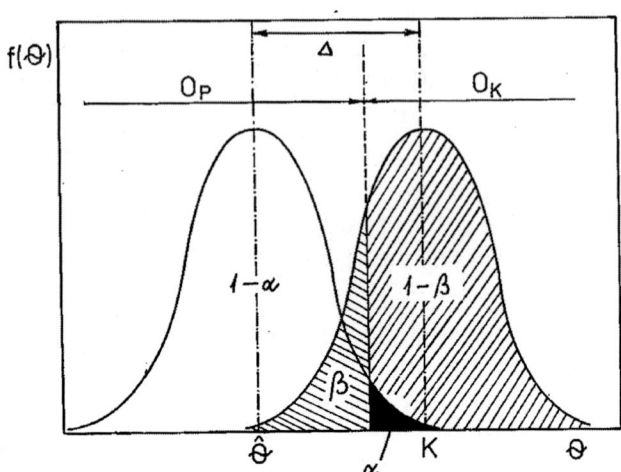

Figure 3.13 Relationship between an error of the first kind and an error of the second kind.

The hypothesis selected assumes that we know the value of the standard population parameters, μ_0, (or σ_0^2, respectively) from past experience or otherwise. As indicated in the third column of Table 3.10, a decision can be reached as follows:

(1) If the inequality proves to be true, i.e. if the calculated difference exceeds the right-hand side of the inequality, the *hypothesis is accepted*.

(2) If the inequality does not prove to be true, i.e. if the calculated difference does not exceed the right-hand side of the inequality, then the *hypothesis is rejected*, and there is little likelihood that the hypothesis is correct.

For a hypothesis test of a normal distribution based on a random sample x_i, $i = 1, ..., n$, with mean $\bar{x}$ and variance s^2, the random variables u (for large samples) and t (for small samples, $n < 30$) are used as test statistics:

$$u = \frac{\bar{x} - \mu_0}{\sigma}\sqrt{n} \approx N(0;1), \text{ and, } t = \frac{\bar{x} - \mu_0}{s}\sqrt{n}$$

where $\bar{x}$ is the sample mean, s is the sample standard deviation and n is the sample size. The null hypothesis $H_0: \mu = \mu_0$ is followed by the alternative hypothesis (1) $H_A: \mu > \mu_0$, (2) $H_A: \mu < \mu_0$, (3) $H_A: \mu \neq \mu_0$, (see Table 3.10).

To test the hypothesis $H_0 : \sigma^2 = \sigma_0^2$ the test statistic

$$\chi^2 = \frac{(n-1)s^2}{\sigma_0^2}$$

is used. The tests assume that the observations are taken randomly from a normal random variable.

(1) *Test the hypothesis* H_0: $\mu = \mu_0$ *vs.* H_A: $\mu > \mu_0$. In this case, large positive values of t (or u) are evidence in favour of H_A. Values of t around zero are evidence in favour of H_0. Large negative values of t (or u), while they are unlikely if H_0 is true, are even more unlikely if H_A is true.

Hypothesis testing may be carried out also by calculating the probability $\alpha = P(t_{1-\alpha} > t)$ as an *one-tailed* (upper tail) *test*:

(a)   If $\alpha > 0.05$ we say that t (or u) *is not significant* and there is *no evidence for rejecting H_0 in favour of H_A*.

(b)   If α lies in the region $0.05 > \alpha > 0.01$ we say that t (or u) *is significant at the 5% level* and there is *some evidence for rejecting H_0 in favour of H_A*.

(c)   If α lies in the region $0.01 > \alpha > 0.001$ we say that t (or u) *is significant at the 1% level* and it is generally interpreted as *strong evidence for rejecting H_0 in favour of H_A*.

(d)   If $0.001 > \alpha$ we say that t (or u) *is significant at the 0.1% level* and there is *almost conclusive evidence for rejecting H_0 in favour of H_A*.

In the formulation of statistical hypothesis testing given above, the results of the tests are presented as evidence at various levels (none, some, strong, almost conclusive) in favour of the alternative hypothesis H_A.

An alternative formulation of statistical hypothesis testing is as follows: before the experiment takes place we decide on a fixed value of α (usually, but not necessarily, one of the values $0.05, 0.01, 0.001$). This will determine a critical region in the tails of the distribution, such that if t falls within the critical region H_A is accepted, otherwise H_0 is accepted. It will be noted that even if H_0 is true there is a probability α that $t_\alpha(\nu)$ will fall into the critical region. Thus α may be interpreted as the probability of accepting H_A when H_0 is true; that is $\alpha = P$ (accepting H_A/H_0). The probability of incorrectly accepting H_A when H_0 is true could obviously be reduced by decreasing α.

(2) *Test of the hypothesis* H_0: $\mu = \mu_0$ *vs.* H_A: $\mu < \mu_0$. The large negative values of t (or u) are evidence in favour of H_A. We calculate $\alpha = P(t_{1-\alpha} < t)$ and the values of a obtained are interpreted as in the previous case. This test is a *one-tailed* (lower tail) *test*.

(3) *Test of the hypothesis* H_0: $\mu = \mu_0$ *vs.* H_A: $\mu \neq \mu_0$. As in (1) and (2) we calculate $t = (\bar{x} - \mu_0)\sqrt{n} / s$ (or $u = (\bar{x} - \mu_0)\sqrt{n} / \sigma$). The both large positive and large negative values of t (or u) are evidence in favour of H_A. We calculate

$$\alpha = P(|t_{1-\alpha}| - |t|) = P(t_{1-\alpha} < |t|) + P(t_{1-\alpha} > |t|) = 2P(t_{1-\alpha} > |t|) \text{ from symmetry.}$$

Again the values of α obtained are interpreted as in (1). This third type of test is referred to as a *two-tailed test.*

The critical regions for the three types of tests are shown in Fig. 3.14.

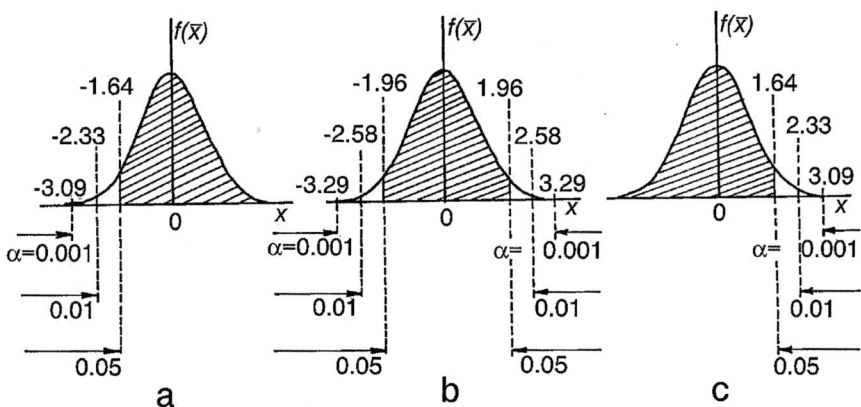

Figure 3.14   Critical regions for (a) a one-tailed test (lower tail), (b) a two-tailed test, (c) a one-tailed test (upper tail).

To investigate whether t (or u) is significant, and if so, at what level, it is not necessary in practice to calculate the value of α. We simply compare the value of t (or u) with the critical values $t_{1-\alpha}$ (or $u_{1-\alpha}$) obtained form statistical tables.

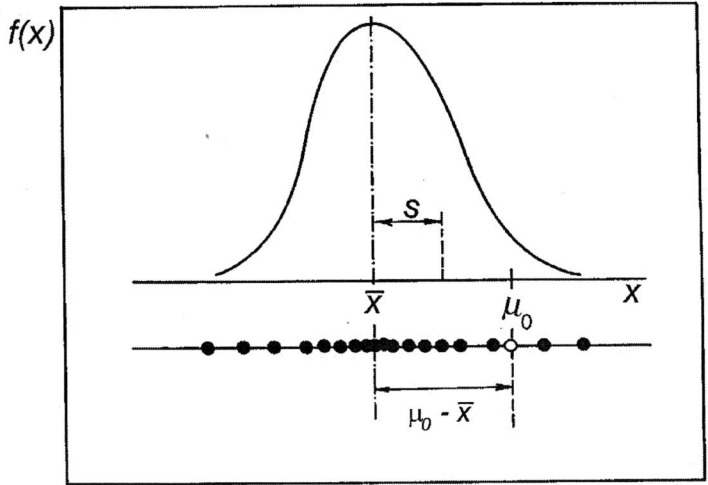

Figure 3.15 The test of accuracy of a result $\mu, H_0 : \mu = \mu_0$ vs. $H_A : \mu \neq \mu_0$.

Table 3.10 Tests of normal data for comparing (a) the mean μ of a new product with a standard μ_0, and (b) these two products with regard to their variability: (a) $H_0 : \mu = \mu_0$ σ unknown, s from sample used $t_{1-\alpha}(n-1)$ or $t_{1-\alpha/2}(n-1)$ is the quantile of the Student distribution. Instead of t, the u quantile may be used (see also Fig. 3.15).

H_A Hypothesis	Region of rejection	t-test
$\mu \neq \mu_0$	$\left\| \dfrac{(\overline{x} - \mu_0)}{s} \sqrt{n} \right\| \geq t_{1-\alpha/2}(n-1)$	Two-tailed
$\mu > \mu_0$	$\dfrac{(\overline{x} - \mu_0)}{s} \sqrt{n} \geq t_{1-\alpha}(n-1)$	One-tailed
$\mu < \mu_0$	$\dfrac{(\overline{x} - \mu_0)}{s} \sqrt{n} < t_{1-\alpha}(n-1)$	One-tailed

(b) $H_0 : \sigma^2 = \sigma_0^2$, s from sample used, $\chi^2_\alpha(n-1)$, $\chi^2_{1-\alpha}(n-1)$, $\chi^2_{\alpha/2}(n-1)$, $\chi^2_{1-\alpha/2}(n-1)$, are the quantiles of the Pearson χ^2 distribution.

H_A Hypothesis	Region of rejection	χ^2-test
$\sigma^2 \neq \sigma_0^2$	$\chi_{\alpha/2}^2 (n-1) \leq \dfrac{s^2(n-1)}{\sigma_0^2} \leq \chi_{1-\alpha/2}^2 (n-1)$	Two-tailed
$\sigma^2 > \sigma_0^2$	$\dfrac{s^2(n-1)}{\sigma_0^2} \geq \chi_{1-\alpha}^2 (n-1)$	One-tailed
$\sigma^2 < \sigma_0^2$	$\dfrac{s^2(n-1)}{\sigma_0^2} < \chi_{\alpha}^2 (n-1)$	One-tailed

In some cases the asymptotic normality of estimators can be accepted for testing.

Problem 3.15   *Test of the sample mean from the log-normal distribution*

A random sample is taken from the log-normally distributed population of copper in kaolin (Problem 3.7). Test whether the sample mean is equal to the expected value $M = 10$, i.e. $H_0: M = 10$ *vs.* $H_A: M \neq 10$.

Data: from Problem 3.7.

Solution: Since $n = 32 > 30$, we use the test statistic for large samples $u = |\hat{M} - 10| / \sqrt{D(\hat{M})}$, which has the standardized normal distribution. Calculation of $u = |9.976 - 10| / \sqrt{0.02} = 1.2$ leads to a lower value than the $u_{1-\alpha/2}$ quantile ($u_{1-0.05/2} = 1.96$ for $\alpha = 0.05$), $1.2 < 1.96$ and hence the hypothesis H_0 is accepted.

Conclusion: The sample mean is not significantly different form the expected value of 10.

3.5.3 Hypothesis Tests for the Parameters of Two Populations

Comparison of two samples $\{x_i\}, i = 1,...,n_1$ and $\{y_i\}, i = 1,...,n_2$ is a frequent problem in the practice, for

(1) comparing results of different methods or testing laboratories,

(2) examining the need to separate heterogeneous samples into homogeneous classes, and

(3) classifying the difference between various materials or various instruments.

Sometimes the problem of two populations may be tested as the problem of one population (paired test). If the elements x_i represent some response before a treatment of a material, and elements y_i the same response after the treatment, the difference between the responses of a pair of values, $d_i = x_i - y_i$ then gives a measure of the effectiveness of the treatment. Let $\bar{d}$ and s_d be the mean and standard deviation of these differences. If the differences d are independently normally distributed (or nearly so) with mean zero and (unknown) variance σ^2 it can be shown that the statistic $t = \bar{d}\sqrt{n}\,/\,s_d$ has Student's t-distribution with $\nu = n - 1$ degrees of freedom. Here n is the number of matched pairs in the experiment (Fig. 3.16). It is assumed here that each set of measurements may be regarded as a sample from a normal population. This statistic may be used to test the hypothesis $H_0: \mu_D = \mu_x - \mu_y = 0$, (no treatment effect) vs. $H_A: \mu_D \neq 0$.

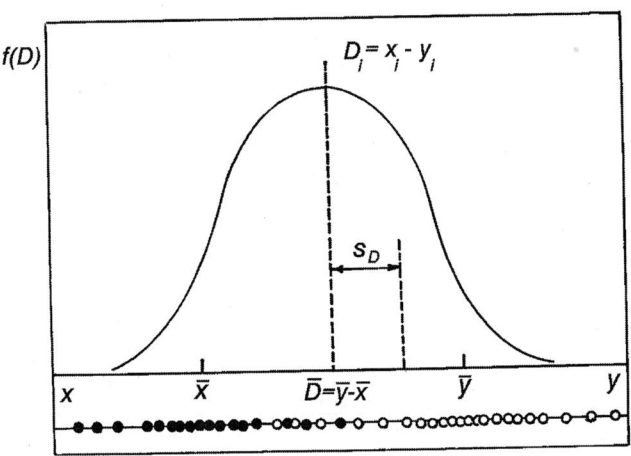

Figure 3.16 The paired test.

Before statistical testing, the methods of exploratory data analysis should be applied to both samples $\{x_i\}$ and $\{y_i\}, i = 1,...,n$ For each sample the box-and-whisker plot and notched box-and-whisker plot are examined. The assumption that both samples have the same distribution is examined by the empirical quantile-quantile plot (Q-Q plot) with the ordered quantities $y_{(i)}$ on the y-axis and $x_{(i)}$ values on the x-axis. When both samples have the same distribution, the points $(x_{(i)}, y_{(i)})$ should lie on a line $y = x$ with slope equal to 1. When the empirical Q-Q plot has the equation $y = kx + q$, mean $\bar{y} = k\bar{x} + q$ and variance $s^2 = k^2 s_x^2$. A nonlinear pattern in the Q-Q plot indicates that the distributions of the samples differ significantly. For different sample sizes, $n_1 > n_2$, the empirical Q-Q plot is drawn for the ordered elements of the smaller sample i.e. for $y_{(i)}$ here. The values of the quantiles of the larger sample $\tilde{x}_j$ are computed according to

$$\tilde{x}_j = (1-z)x_{(k)} + zx_{(k+1)} \qquad (3.123)$$

where $k = \mathrm{int}(v_j)$ is the integer part of real number of $v_j = (j - 0.5)n_1 / n_2 + 0.5$ and $z = v_j - k$. An empirical Q-Q plot is constructed from n_2 pairs of points $(y_{(j)}, \tilde{x}_j)$.

After exploratory data analysis, the classical tests of significance of difference in the parameters of location and spread are applied. Classical tests assume that

(1) the two samples (x_i), $i = 1, \ldots, n_1$, and (y_i), $i = 1, \ldots, n_2$ are independent,

(2) the two populations distributions are normal, $x_i \simeq N(\mu_x, \sigma_x^2)$ and $y_i \simeq N(\mu_y, \sigma_y^2)$.

3.5.3.1 Comparisons of Population Means

Consider a random sample of size n_1, with mean $\bar{x}$ and variance s_x^2, from a population P_1 with unknown mean μ_x and unknown variance σ_x^2, and an independent random sample of size n_2 with mean $\bar{y}$ and variance s_y^2, from population P_2 with unknown mean μ_y and unknown variance σ_y^2. The hypothesis $H_0: \mu_x = \mu_y$ is tested against the alternative $H_A: \mu_x \neq \mu_y$ (Fig. 3.17).

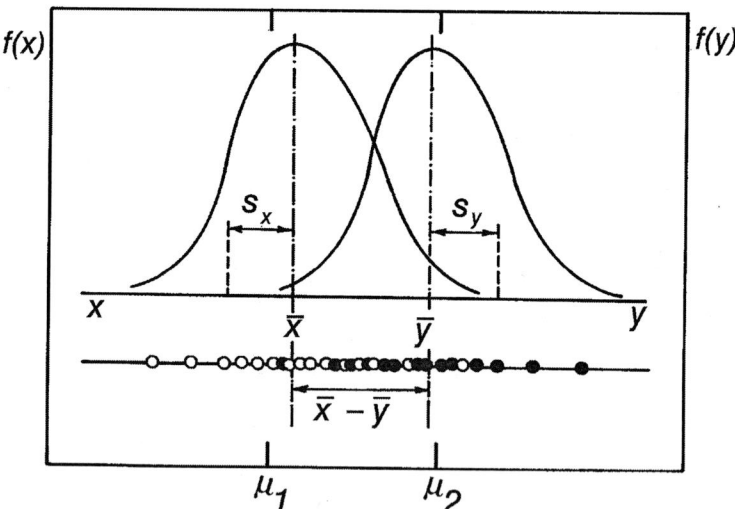

Figure 3.17 The test for comparison of mean values.

To proceed further we distinguish two cases:

(1) If we know that $\sigma_x^2 = \sigma_y^2$, the test statistic

$$t_1 = \frac{|\bar{x} - \bar{y}|}{\sqrt{(n_1 - 1)s_x^2 + (n_2 - 1)s_y^2}} \left[\frac{n_1 n_2 (n_1 + n_2 - 2)}{n_1 + n_2} \right]^{1/2} \tag{3.124}$$

has Student's t-distribution with $v = n_1 + n_2 - 2$ degrees of freedom and significance level α. When $t_1 > t_{1-\alpha/2}(v)$, the null hypothesis H_0 is rejected.

(2) If we know that $\sigma_x^2 \neq \sigma_y^2$, the test statistic

$$t_2 = |\bar{x} - \bar{y}|(s_x^2 / n_1 + s_y^2 / n_2)^{-1/2} \tag{3.125}$$

has Student's t-distribution with v degrees of freedom expressed by

$$v = \frac{\left(s_x^2 / n_1 + s_y^2 / n_2 \right)^2}{\dfrac{s_x^4}{n^2 (n_1 - 1)} + \dfrac{s_y^4}{n^2 (n_2 - 1)}} \tag{3.126}$$

when $t_2 > t_{1-\alpha/2}(v)$, the null hypothesis H_0 is rejected at the significance level α.

In some problems of comparison of means, the variances σ_x^2 and σ_y^2 are unknown. Posten, Yeh and Owen [9] found that for $n_1 - n_2 > 8$ the test statistic t_1 can be applied even for $\sigma_x^2 \neq \sigma_y^2$.

For different sample sizes $n_1 \neq n_2$ and variance ratio $\sigma_x^2 / \sigma_y^2 \approx 1$ the test statistic t_1 can be used. If $n_2 > n_1$ and also n_2 is large enough, the test statistic t_1 can be used provided that

$$0.82 \leq \frac{\dfrac{n_2}{n_1} \dfrac{s_x^2}{s_y^2} + 1}{\dfrac{n_2}{n_1} \dfrac{s_x^2}{s_y^2}} \leq 1.17 \tag{3.127}$$

where $s_x^2 > s_y^2$. Test statistic t_1 is non-robust when the variance of data is not constant. The test criterion t_1 is not robust against heteroscedasticity i.e. against a case of data being measured with different precision. For such case the test criterion t_2 is preferable because it is more robust. However, the number of degrees of freedom v calculated by Eq. (3.116) is less than $n_1 + n_2 - 2$, so that the power of test t_2 is lower than that of t_1 and also the probability of a Type II error β increases.

(3) When both samples are not from a normal distribution, the modified test criterion t_3, is used

$$t_3 = \frac{|\bar{x} - \bar{y}| + C + D(\bar{x} - \bar{y})^2}{\sqrt{s_x^2 / n_1 + s_y^2 / n_2}} \tag{3.128}$$

$$C = \frac{1}{6}\left[\frac{\hat{g}_{1x}}{n_1^2}\frac{s_x^3}{\sqrt{n_1}} - \frac{\hat{g}_{1y}}{n_2^2}\frac{s_y^3}{\sqrt{n_2}}\right]\left[\frac{s_x^2}{n_1} + \frac{s_y^2}{n_2}\right]^{-1} \tag{3.129}$$

$$D = \frac{1}{3}\left[\frac{\hat{g}_{1x}}{n_1^2}\frac{s_x^3}{\sqrt{n_1}} - \frac{\hat{g}_{1y}}{n_2^2}\frac{s_y^3}{\sqrt{n_2}}\right]\left[\frac{s_x^2}{n_1} + \frac{s_y^2}{n_2}\right]^{-2} \tag{3.130}$$

Here $\hat{g}_{1x}$ and $\hat{g}_{2x}$ are sample skewness. In order to use the quantiles of Student's t-distribution for a declared significance level α, another statistic t_3' should be used

$$t_3' = t_2 + B_x - B_y \tag{3.131}$$

$$B_x = \left[\frac{\hat{g}_{1x}s_x^3}{6n_1^2\sqrt{n_1}\left[\frac{s_x^2}{n_1} + \frac{s_y^2}{n_2}\right]} + \frac{\hat{g}_{1x}s_x^3(\bar{x} - \bar{y})^2}{3n_2^2\sqrt{n_2}\left[\frac{s_x^2}{n_1} + \frac{s_y^2}{n_2}\right]}\right]\left[\frac{s_x^2}{n_1} + \frac{s_y^2}{n_2}\right]^{-1/2} \tag{3.132}$$

Quantity B_y is calculated analogously with $\hat{g}_{1y}, \sigma_y^2$ and n_2. The test criterion t_3' for H_0 has the Student t-distribution with $v = n_1 + n_2 - 2$ degrees of freedom. This statistic is robust for distorted samples distributions, also for heteroscedasticity in data and different sample variances $\sigma_x^2 \neq \sigma_y^2$.

The Brown and Forsythe test concerns comparison of multiple population means on the basis of small samples taken from normal populations. The null hypothesis is $H_0: \mu_1 = \mu_2 = ... = \mu_k$. Suppose that there are k samples, each of size n_i from a normal distribution with sample mean $\bar{x}_i$ and variance s_i^2, $i = 1, ..., k$. The test statistic

$$F = \frac{\sum_{i=1}^{k} n_1 (\bar{x}_i - \bar{x})^2}{\sum_{i=1}^{k}\left[1 - \frac{n_i}{n}\right]s_i^2} \tag{3.133}$$

where

$$\bar{x} = \frac{1}{n}\sum_{i=1}^{k} n_i \bar{x}_i \tag{3.134}$$

and

$$\bar{n} = \sum_{i=1}^{k} n_i \qquad (3.135)$$

has in H_0 the Fisher-Snedecor distribution with $(k - 1)$ and v degrees of freedom. When the sample value F is greater than the quantile $F_{1-\alpha}(k-1,v)$ at the 5% level, H_0 is rejected. The number of degrees of freedom is here given by

$$v = 1 / \sum_{i=1}^{n} \frac{o_i}{n_i - 1} \qquad (3.136)$$

where $o_i = \left[(1 - n_i / \bar{n}) s_i^2 \right] / \left[\sum_{i=1}^{n} (1 - n_i / \bar{n}) s_i^2 \right] \qquad (3.137)$

When testing the equivalence of two means, Eqs. (3.123–3.127) with $k = 2, \bar{x}_1 = \bar{x}, \bar{x}_2 = \bar{y}, s_1^2 = s_x^2, s_2^2 = s_y^2$ should be applied.

(4) When there are some outliers present in two samples $\{x_i\}, i = 1,...,n_1$ and $\{y_i\}, i = 1,...,n_2$, the robust test statistic for $\sigma_x^2 = \sigma_y^2$ and for $n_1 = n_2 \geq 7$ is given by

$$t_4 = \frac{(\bar{x}(\kappa) - \bar{y}(\kappa))}{\sqrt{s_{w,x}(\kappa) + s_{w,y}(\kappa)}} \qquad (3.138)$$

where trimmed estimates of mean $\bar{x}(\kappa), \bar{y}(\kappa)$ are calculated from Eq. (3.97) and corresponding $s_{w,x}(\kappa)$ and $s_{w,y}(\kappa)$ are calculated from Eq. (3.98). Then, t_4 has the Student distribution with $2(k - 1)$ degrese of freedom.

(5) The robust test statistic t_4 can be extended to the case of different variance, $\sigma_1^2 \neq \sigma_2^2$, and different sample sizes $n_1 \neq n_2$ by using the form

$$t_5 = \frac{(\bar{x}(\kappa) - \bar{y}(\kappa))}{\sqrt{s_{w,x}^2 / h_1 + s_{w,y}^2 / h_2}} \qquad (3.139)$$

where

$$s_{w,x}^2 = \frac{s_{w,x}^2(\kappa)}{h_1 - 1} \qquad (3.140)$$

$$s_{w,y}^2 = \frac{s_{w,y}^2(\kappa)}{h_2 - 1}$$

(3.141)

$$h_i = n_i - 2 \times \text{int}(\kappa n_i / 100), \, i = 1, 2$$

(3.142)

Then, t_s has the Student distribution with v degrees of freedom defined by

$$\frac{1}{v} = \frac{z^2}{h_1 - 1} + \frac{(1 - z^2)}{h_2 - 1}$$

(3.143)

and

$$z = \frac{s_{w,x}^2 / h_1}{s_{w,x}^2 / h_1 + s_{w,y}^2 / h_2}$$

(3.144)

Both robust test statistics are also suitable for a distribution with long tails and kurtosis greater than 3.

Problem 3.16 *Comparison of the means of two samples*

Samples of size $n = 50$ were taken from two populations of the normal distribution, $N(0, 1)$ and $N(3, 1)$ and from two populations of the Laplace distribution, $L(0, 2)$ and $L(2, 2)$. For the significance level $\alpha = 0.05$, use a suitable test and examine the difference between two sample means.

Data:

(a) The sample from $N(0, 1)$:

−1.008	−0.500	0.749	1.723	0.076	0.569	−1.389
0.087	1.112	−0.235	0.519	0.279	−0.758	−0.588
−0.594	−0.885	−0.072	1.980	0.063	0.016	−0.673
−0.993	0.752	0.092	0.236	−2.962	−0.383	0.109
−1.285	0.634	0.690	1.134	−0.711	−1.825	2.374
0.500	−1.380	0.046	−0.544	−0.150	−1.129	1.173
1.401	−2.121	0.521	0.280	1.440	−0.415	−0.443
−0.384						

(b) The sample from $N(3, 1)$:

1.992	2.500	3.749	4.723	3.076	3.569	1.611
3.087	4.122	2.765	3.519	3.279	2.242	2.412
2.406	2.115	2.928	4.980	3.063	3.016	2.327
2.007	3.752	3.092	3.236	0.038	2.617	3.109
1.715	3.634	3.690	4.134	2.289	1.175	5.374
3.500	1.620	3.046	2.456	2.850	1.871	4.173
4.401	0.809	3.521	3.280	4.440	2.585	2.557
2.616						

(c) The sample from $L(0, 2)$:

0.064	0.436	−1.072	−2.036	1.192	−3.162	−0.275
0.734	0.049	−0.569	1.144	−1.070	0.304	0.343
0.144	0.209	−1.269	4.452	5.477	−0.862	−0.026
0.210	0.695	−0.774	−3.723	−0.026	1.879	0.887
−0.333	1.038	−0.188	−1.749	−1.224	0.126	−2.249
−0.835	−0.261	0.715	1.223	0.535	−0.632	1.411
−2.538	−0.206	−1.136	−1.070	−0.429	−0.529	1.153
0.319						

(d) The sample from $L(2, 2)$:

2.064	2.436	0.928	-0.036	3.192	−1.162	1.725
2.734	2.049	1.431	3.144	0.930	2.304	2.343
2.144	2.209	0.731	6.452	7.477	1.138	1.974
2.210	2.695	1.226	−1.723	1.974	3.879	2.887
1.667	3.038	1.812	0.251	0.776	2.126	−0.249
1.165	1.739	2.715	3.223	2.535	1.368	3.411
−0.538	1.794	3.136	0.930	1.571	1.471	3.153
2.319						

Program: ADSTAT or QC-EXPERT: Basic Statistics: Two sample testing.

Solution: Examining the variance of the two samples by the Fisher-Snedecor test proved that $\sigma_1^2 = \sigma_2^2$ and therefore statistics t_1, t_3 and t_4 can be used. Table 3.11 shows the test statistics with the quantiles of the Student t-test for significance level $\alpha = 0.05$.

Table 3.11   Testing of the sample means of two populations (a) $N(0, 1)$ and $N(0, 3)$, and (b) $L(0, 2)$ and $L(2, 2)$ at significance level $\alpha = 0.05$.

Test statistic	Sample from $N(0, 1)$ and sample from $N(3, 1)$	Sample from $L(0, 2)$ and sample from $L(2, 2)$
t_1 quantile	15.65	4.42
	1.985	1.985
t_3 quantile	14.72	4.7
	1.984	1.985
t_4 quantile	16.11	5.22
	1.988	1.988

Conclusion: All the test statistics indicate a significant difference between the means.

3.5.3.2 Comparisons of the Variances of Two Populations

The comparison of variances is particularly important for analysis of responses obtained from experimental design. A random sample of size n_1 is taken from a normal population with unknown mean μ_x and variance σ_x^2 and an independent random sample of size n_2 from a second normal population with unknown mean μ_y and variance σ_y^2. We take $\hat{\sigma}_x^2 = s_x^2$ and $\hat{\sigma}_y^2 = s_y^2$ where s_x^2 and s_y^2 are the sample variances.

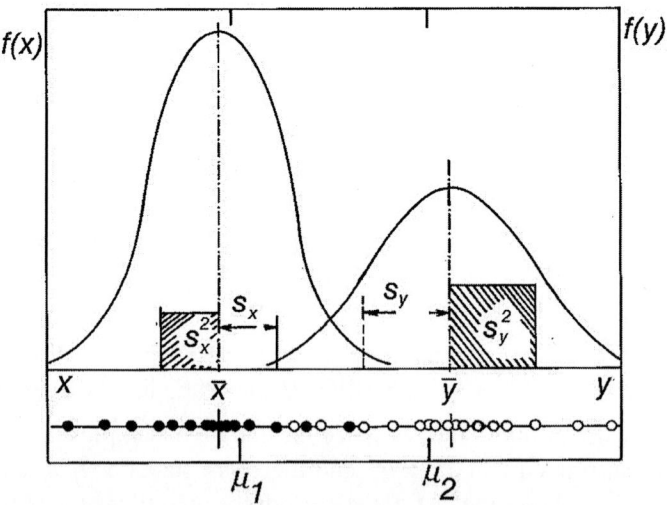

Figure 3.18   The Fisher-Snedecor test for identity of two variances.

For normal and independent populations, the hypothesis $H_0: \sigma_x^2 = \sigma_y^2$ against $H_A: \sigma_x^2 \neq \sigma_y^2$ can be tested by using the statistics

$$F = \max\left(\frac{s_x^2}{s_y^2}, \frac{s_y^2}{s_x^2}\right) \tag{3.145}$$

When $s_x^2 > s_y^2$, the F statistic has the Fisher-Snedecor F-distribution with $v_1 = n_x - 1$ and $v_2 = n_y - 1$ degrees of freedom. In the case when $s_y^2 > s_x^2$ the order of degrees of freedom must be changed. The test is illustrated in Fig. 3.18.

Fisher-Snedecor test is sensitive to the assumption of normality: if the kurtosis of the samples differs from the normal distribution, for the quantile $F_{1-\alpha/2}(v_1, v_2)$ the numbers of degrees of freedom v_1 and v_2 should be calculated from

$$v_1 = (n_1 - 1) / (1 + \hat{g}_{2c} / 2) \tag{3.146}$$

$$v_2 = (n_2 - 1) / (1 + \hat{g}_{2c} / 2) \tag{3.147}$$

where

$$\hat{g}_{2c} = \frac{2(n_1 + n_2)\left[\sum\limits_{i=1}^{n_1}(x_i - \bar{x})^4 + \sum\limits_{i=1}^{n_2}(y_i - \bar{y})^4\right]}{\left[\sum\limits_{i=1}^{n_1}(x_i - \bar{x})^2 + \sum\limits_{i=1}^{n_2}(y_i - \bar{y})^2\right]^2} - 3 \tag{3.148}$$

When more than two sample variances are to be tested, $H_0: \sigma_1^2 = \sigma_2^2 = ... = \sigma_k^2$, the Jack-knife test is recommended. The Jack-knife statistic is calculated by

$$F_j = \frac{\left[n_1(\bar{z}_1 - \bar{z})^2 + n_2(\bar{z}_2 - \bar{z})^2\right](n_1 + n_2 - 2)}{\sum\limits_{i=1}^{n_1}(z_{1i} - \bar{z}_1)^2 + \sum\limits_{i=1}^{n_2}(z_{2i} - \bar{z}_2)^2} \tag{3.149}$$

where

$$\bar{z} = \frac{n_1\bar{z}_1 + n_2\bar{z}_2}{n_1 + n_2} \tag{3.150}$$

$$z_j = \frac{\sum\limits_{i=1}^{n_j} z_{ji}}{2}, j = 1, 2 \tag{3.151}$$

$$z_{1i} = n_1 \ln s_x^2 - (n_1 - 1) \ln s_{1(i)} \tag{3.152}$$

$$s_{1(i)}^2 = \frac{1}{n_1 - 2} \sum_{j \neq i}^{n_1} (x_j - \bar{x}_{(i)})^2 \tag{3.153}$$

In Eq. (3.153) the sample mean $\bar{x}_{(i)}$ is calculated from a sample with the ith element omitted

$$\bar{x}_{(i)} = \frac{1}{n_1 - 1} \sum_{j=i}^{n_1} x_j \tag{3.154}$$

Analogously, for the sample y_i, $i = 1,..., n_2$ the z_{2i}, s_y^2,... etc. are calculated. The Jack-knife statistic has the Fisher-Snedecor distribution with $v_1 = 2$ and $v_2 = n_1 + n_2 - 2$ degrees of freedom. When the sample value F is greater than the quantile $F_{1-\alpha/2}(v_1, v_2)$, the H_0 is rejected at the significance level α.

Problem 3.17 The variances of two samples

For the two pairs of samples from Problem 3.16, test whether both samples in the pair have the same variance. Do the test at significance level $\alpha = 0.05$.

Data: As for Problem 3.16

Program: ADSTAT or QC-EXPERT: Basic Statistics: Two sample testing.

Solution: The variances of two samples are tested with the use of two statistical tests, the Fisher-Snedecor test (F) and Jack-knife test (F_j). Table 3.12 compare the results of the tests.

Table 3.12   Test statistics and quantiles for the 5% level, for data from Problem 3.16

Test statistics	Sample of $N(0, 1)$ with sample of $N(3, 1)$	Sample of $L(0, 2)$ with samples of $L(2, 2)$
F quantile	1.006	1.003
	1.762	1.762
F_j quantile	0.0003	2.21×10^{-5}
	3.831	3.831

Conclusion: Both tests prove the validity of the H_0 hypothesis, the variances are the same. Generally, F-tests are more sensitive to the classical assumptions of an actual distribution than the t-tests.

3.6 Summary of Univariate Data Analysis

Univariate data analysis involves the following steps [20]:

(**1**) *Confirmatory data analysis, examining assumptions about data*

    1.1 Examining the independe nce of sample elements.

    1.2 Examining the normality of the sample distribution.

    1.3 Examining the outliers by the use of modified external hinges $\left[V_L^*, V_U^* \right]$.

    1.4 Examining the minimum sample size n_{min}.

    In addition to these four tasks, other diagnostics of exploratory data analysis may be used to verify the actual sample distribution.

(**2**) *Determination of point and interval estimates of parameters.*

The choice of the type of statistical characteristics depends on the distribution of the population the sample comes from. With an assumption of normality, independence and homogeneity of sample, the moment characteristics are calculated. If outliers are present in the data, robust characteristics are calculated.

    (2.1) The moment characteristics of location and spread include the arithmetic mean $\bar{x}$, Eq. (3.39), with its variance $D(\bar{x})$, Eq. (3.40), and the confidence interval, Eq. (3.27), for the true value μ; the estimate of sample variance s^2, Eq. (3.43), and corresponding confidence interval, Eq. (3.29), the mean absolute deviation $\bar{d}$, Eq. (3.16), the variation coefficient δ, Eq. (3.17), with corresponding variance $D(\delta)$, Eq. (3.18); the weighted arithmetic mean $\bar{x}_w$, Eq. (3.11), with its variance $D(\bar{x}_w)$, Eq. (3.12).

    (2.2) The characteristics of shape give information about the distribution shape; they include the skewness $\hat{g}_1$, Eq. (3.19), with corresponding variance $D(\hat{g}_1)$, Eq. (3.19); the kurtosis $\hat{g}_2$, Eq. (3.20), with corresponding variance $D(\hat{g}_2)$, Eq. (3.20a); and also the selector characteristics Q_1, Eq. (3.104), and Q_2, Eq. (3.105).

    (2.3) The quantile and robust characteristics of location and spread are less sensitive to outliers than the moment characteristics. These characteristics include the median $\tilde{x}_{0.5}$ (Section 3.4.1) with its nonparametric estimate of variance s_M^2, Eq. (3.93), or s_M^*, Eq. (3.95), and the confidence interval of the median; the mode $\hat{x}_M$ (Section 3.1); the upper $\tilde{x}_{0.75}$ and the lower $\tilde{x}_{0.25}$ quartile, which are useful for the calculation of the interquantile range R, Eq. (3.15). The simplest and the most effective robust estimate of location is the trimmed mean $\bar{x}(\kappa)$, Eq. (3.97), with winsorized variance $s_w(\kappa)$, Eq. (3.98), and also the asymmetric trimmed mean

$\bar{x}(\kappa_1, \kappa_2)$, Eq. (3.101), with its variance $s^2(\kappa_1, \kappa_2)$, Eq. (3.103), which is suitable for asymmetric skewed distributions. To select the extent of trimming, some selector criteria are used: the relative tail length Q_1, Eq. (3.104), or the estimate of relative skewness Q_2, Eq. (3.105). When a sample contains outliers the robust M-estimates with bi-quadratic function are used. This leads to the value of location $\hat{\mu}_M$ (3.107) with its variance $D(\hat{\mu}_M)$, Eq. (3.112), and the confidence interval calculated with use of the random variable t_M, Eq. (3.113). For small samples, the pivot halfsum P_L, Eq. (3.114), and the pivot range R_L, Eq. (3.115), are used. The random variable T_L, Eq. (3.116), is used for calculation of the confidence interval of the mean μ.

(2.4) The computer intensive distribution free estimates permit calculation of an estimate of the variance of any parameter of distribution θ and also construction of the corresponding confidence interval. The estimate of parameter $\hat{\theta}_B$ and variance $\hat{\sigma}_B^2$, Eq. (3.119), by the Bootstrap method or $\hat{\theta}_J$, Eq. (3.122a) with its variance $\hat{\sigma}_J^2$, Eq. (3.122c), by the Jack-knife method, are used.

(2.5) The maximum likelihood estimates for distributions other than normal are calculated as follows:

(1) For the Poisson distribution $\hat{\lambda}$, Eq. (3.34), $D(\hat{\lambda})$, Eq. (3.35), and the confidence interval of parameter λ, Eq. (3.36a,b);

(2) For the Laplace distribution $\hat{\Phi}$, Eq. (3.46), $D(\hat{\Phi})$, Eq. (3.47), and the confidence interval of parameter Φ, Eq. (3.48).

(3) For the rectangular distribution $\ln L$, Eq. (3.50), $\hat{h}$, Eq. (3.51), $\hat{a}$, Eq. (3.52), with their variances $D(\hat{h})$, Eq. (3.53), $D(\hat{a})$, Eq. (3.54), and the confidence interval, Eq. (3.28).

(4) For the exponential distribution (one-parameter) $\hat{\theta}$, Eq. (3.57), $D(\hat{\theta})$ Eq. (3.58) and the confidence interval for θ, Eq. (3.59).

(5) For the exponential distribution (two-parameters) $\hat{\theta}$, Eqs. (3.64, 3.65a, 3.66a), $D(\hat{\theta})$, Eqs. (3.65b, 3.66b) and unbiased estimates, Eqs. (3.67, 3.68), the confidence interval of parameter θ, Eq. (3.69) and Eq. (3.70).

(6) For log-normal distribution (two-parameters) $E(x)$, Eq. (3.74), $D(x)$, Eq. (3.75), $\hat{g}_1$, Eq. (3.76a), $\hat{g}_2$, Eq. (3.76b), δ, Eq. (3.77), $\hat{x}_M$, Eq. (3.78a), $\tilde{x}_M$, Eq. (3.78b), the confidence interval $\tilde{x}_M$, Eq. (3.82) and δ, Eq. (3.83). The estimate of the mean value is $\hat{M}$, Eq. (3.84), with variance $\hat{V}$, Eq. (3.85), with variances $D(\hat{M})$, Eq. (3.88) and $D(\hat{V})$, Eq. (3.89). The confidence interval of parameter M is calculated from Eq. (3.91).

(7) For the log-normal distribution (three-parameters) $\hat{\mu}(\theta)$, Eq. (3.80), and $\sigma^2(\theta)$, Eq. (3.81).

(3) *Statistical hypothesis testing*

The simple test of the parameters of population on the basis of one sample uses the $100(1 - \alpha)\%$ confidence interval of parameter θ. If the given value θ_0 lies in this interval, the null hypothesis H_0: $\theta = \theta_0$ is accepted, otherwise the alternative one, H_A: $\theta \neq \theta_0$.

For testing hypotheses about two populations on the basis of two samples, the first step is the test of homogeneity of variances of the two samples by the Fisher-Snedecor F-test, Eq. (3.145). However, this test is rather sensitive to any deviation of the distribution from normality; the Jack-knife test F_J, Eq. (3.149) or some robust test of location t_4, Eq. (3.138) or t_5, Eq. (3.139) are more suitable.

The classical Student t-test t_1 Eq. (3.124), or t_2, Eq. (3.125), is robust enough when the actual distribution deviates from normality, but the two sample sizes are the same. When both samples deviate in skewness from the normal distribution, the test characteristic t_3, Eq. (3.128), is more convenient.

3.7 References

[1]   Hensgaard. D.: *Commun. Statist.* **B8**, 359 (1979).

[2]   Tukey J. W., McLaughlin: *Sankya* **125**, 331 (1963).

[3]   Johnson N. L., Kotz S.: *Continuous Univariate Distributions*. Mifflin 1970.

[4]   Hogg R. V.: *J. Amer. Statist. Assoc.* **69**, 909 (1964).

[5]   Du Mond Ch., Lenth R. V.: *Technometrics* **29**, 211 (1987).

[6]   Blackman N. M., Machol R. E.: *IEEE Trans. on Inform. Theory* **IT-33**, 373 (1987).

[7]   Horn J.: *J. Amer. Statist. Assoc.* **78**, 930 (1983).

[8]   Efron B.: *Canad. J. Statist.* **9**, 139 (1981).

[9]   Posten H. O., Yeh H. C., Owen D. B.: *Commun. Statist.* **A11**, 109 (1982).

[10] Cressie N. A. C., Whitford H. J.: *Biom. J.* **28**, 131 (1986).

[11] Yuen K., Dixon W. J.: *Biometrika* **60**, 369 (1973).

[12] Owen D. B.: *Handbook of Statistical Tables*. Addison Wesley Publ., Reading 1963.

[13] Green J. R., Margerison D.: *Statistical Treatment of Experimental Data*, Elsevier, Amsterdam 1978.

[14] Miller J. C., Miller J. N.: *Statistics for Analytical Chemistry*. Ellis Horwood, Chichester, 1984.

[15] Himmelblau D. M.: *Process Analysis by Statistical Methods*. Wiley, New York 1969.

[16] Liteanu C., Rica I.: *Statistical Theory and Methodology of Trace Analysis*. Ellis Horwood, Chichester 1980.

[17] Anderson R. L.: *Practical Statistics for Analytical Chemists*. van Nostrand Reinhold Comp., New York 1987.

[18] Eason G. a kol.: *Mathematics and Statistics for the Bio-Sciences*. Ellis Horwood, Chichester 1980.

[19] Stoodly K.: *Applied and Computation Statistics*. Ellis Horwood, Chichester 1984.

[20] Meloun M., Militký J.: *Statistické zpracování experimentálních dat,* Plus Praha 1994 (1st edition), EAST PUBLISHING, Praha 1998 (2nd edition), Academia Praha 2004 (3rd edition).

Supplemented material (Review Questions, Exercises, Results of Exercises) to Chapter 3 is on the enclosed CD.

4

Statistical analysis of multivariate data

The multivariate statistical methods are defined as a collection of procedures for analyzing associations between two or more sets of measurements that have been made on one or more multivariate samples. The so-called linear multivariate methods for the cases of linear associations are discussed [1,2].

4.1 Objectives of Multivariate Data Matrix

It is a fact that "nature is multivariate", i.e. that *any one* phenomenon is usually characterized by *several* factors or variables. The amount of information in data will depend on how well a problem is defined, and whether the observations, the measurements or the experiments have been performed accordingly. The data analyst has a very clear responsibility to provide *meaningful* data. Some examples of multivariate data sets [3–24] are

- (a) various properties of raw materials (wheat, crude oil, fibers, ore), semi-products or products (food, petrol, clothing, alloys, etc.) expressed with the use of test methods;
- (b) evaluation of spectra using the wavelength and magnitude of absorption bands serving to identify chemical compounds;
- (c) the results of applying a battery of cognitive tests to a sample of patients with a disease;
- (d) the taxonomic characteristics of bacteria, flowers, animals, etc.;
- (e) the relative proportions of several constituents of different types of natural or artificial materials (rocks,wood, printing paste etc.);
- (f) the parameters of a technological processes (time, temperature, pressure, mixing intensity).

4.1.1 Description of Data

Multivariate data matrix – The data with which we are primarily concerned consist of a series of *variables* (sometimes called characteristics or properties)

$\xi_1,..\xi_m$ in m columns (i.e. measurements or observations, power of engine, weight of car, petrol consumption, length of car, etc.) made on a number of *objects* or *cases* in n rows (patients, individuals, molecules or other entities of interest). A typical non-structured *multivariate data matrix*, $\mathbf{X}$, has the form

$$\mathbf{X} = \begin{pmatrix} x_{11} & x_{12} & \cdots & x_{1m} \\ x_{21} & x_{22} & \cdots & x_{2m} \\ \cdots & \cdots & \cdots & \cdots \\ x_{n1} & x_{n2} & \cdots & x_{nm} \end{pmatrix}$$

where the typical element, x_{ij}, is the value of the jth *variable* ξ_j for the ith *object*. If there are several distinct groups of individuals, one of the columns (usually the first) might be a categorical variable with values of 1, 2, etc. to distinguish these groups. The number of objects under investigation is n and the number of variables taken on each of these n objects is m. In the case of structured data is the matix $\mathbf{X}$ divided into submatrix $\mathbf{Y}$ $(n \times p)$ containing dependent variables (goal variables) and submatrix $\mathbf{Z}$ $(n \times m\text{-}p)$ of explanatory variables. From a statistical point of view are $\xi_1,...,\xi_m$ random variables [28]

Classification of variables – Broadly speaking, there are three main measurement scales of data encountered in practice: *nominal*, *ordinal* and *cardinal*. The most common way of distinguishing between these types of variables is the following [25]:

(a) *Nominal scale variables* are unordered categorical variables. Each observation belongs to one of set of mutually exclusive and collectively exhaustive categories. These categories have no natural or necessary order relative to each other. For these variables, the operators of equality ($=$) and non equality ($\neq$) are defined. Nominal variables are often binary (1 – presence, 0 – absence). Examples include treatment allocation, the sex of the respondent, hair colour, presence or absence of depression, and so on. In computerized data analysis, numbers are often used as symbols.

(b) *Ordinal scale variables* are ranked data where there is an ordering of categories. For these variables the operators less than ($<$) or higher than ($>$) are defined as well. Each observation belongs to one of set of mutually exclusive and collectively exhaustive categories but these categories are naturally ordered (usually the higher category is better, the first category being the worse and last category the best). Categories are ordered, but the differences between them are no quantified. Categories are characterized by symbols with natural ordering as letters (e.g. A, B, C..) or numbers. Examples include social class and self- perception of health (each coded from 1 to 5, say), hand feeling (not acceptable = 1, bad = 2, medium = 3, good = 4, fully acceptable = 5) and educational level (no schooling = 1, primary = 2, secondary = 3 or tertiary = 4 education).

(c) *Cardinal scale variables* are data containing quantitative information where the distance (or norm) is defined. The basic mathematical operations such as summation, substraction, multiplication and division can be used. From a statistical point of view, the cardinal scale data are discrete (categorized) or continuous. In fact all measured data are discrete and the number of categories is dependent on the rounding. The cardinal scale is divided into two subcategories

Interval scale variables – where there are equal differences between successive points on the scale, but the position of zero is arbitrary. The classic example is the measurement of temperature using the Celsius or Fahrenheit scales. In some cases a variable such as a measure of depression, anxiety or intelligence, for example, might be treated as if it were interval-scaled when this, in fact, might be difficult to justify.

Ratio scale variables are interval variables with a natural point representing the origin of measurement, i.e. a natural zero point. It also represents the highest level of measurement, where one can investigate the *relative magnitude* of scores as well as the differences between them. The position of zero is fixed. The classic example is the absolute measure of temperature (in kelvin, for example) but other common ones include age (or any other time from a fixed event), weight and length.

Higher scale type includes naturally the properties of lower scales and can be transformed (with loss of information) into lower scale.

Equality, difference and similarity – concepts of *equality, difference and similarity* seem trivial; but each data set requires careful consideration about the meanings of these three words, because the result of data analysis greatly depends on these [26]. The concept of *similarity* depends on the aim of data analysis. MDA searches for relationships among objects, among variables, and among objects and variables. Objects can be equal, similar, dissimilar and proportional mixtures. Variables can be equal, similar, dissimilar, proportional and linear combinations. This is *clustering* analysis, the search for clusters of similar objects (clustering of objects) or of similar variables (clustering of variables). When from cluster analysis we know that some clusters (categories) of objects are present, we use *classification* analysis to assign objects to one of these categories.

Dissimilarity between entities forms the starting point of various techniques, so it is worth gathering together the basic ideas in one section. Since dissimilarity is so closely linked to the idea of distance, one natural way of measuring it is by the use of a familiar metric such as Euclidean distance. If s is the *similarity* between two entities (usually in the range $0 \le s \le 1$), then the *dissimilarity d* is the direct opposite of s and hence may be obtained by using any monotonically *decreasing* transformation of s. The most common such transformation is $d = 1 - s$. Secondly, all dissimilarity measure formulae naturally assume that a set of data is available on which to apply them.

Dissimilarity d_{kl} between objects k and l: The numerical value x_{kj} is observed for the *j*th variable on the *k*th objects in the sample.

(a) *Euclidean distance (metric)*: $d_{\mathrm{E}}(x_k, x_l) = \sqrt{\sum_{j=1}^{m}(x_{kj} - x_{lj})^2}$.

(b) *Hamming or manhattan distance (metric)*: $d_{\mathrm{H}}(x_k, x_l) = \sum_{j=1}^{m}\left|x_{kj} - x_{lj}\right|$.

(c) *General Minkowski distance (metric)*: $d_{\mathrm{M}}(x_k, x_l) = \sqrt{\sum_{j=1}^{m}\left|x_{kj} - x_{lj}\right|^p}$

where for $p = 1$ it is Hamming distance and for $p = 2$ it is Euclidean metric. The consequence of increasing p is increasingly to exaggerate the more dissimilar units relative to the similar ones.

(d) *Mahalanobis distance (metric)* can be thought of as an appropriate statistical distance for use in sample space similarly as the Euclidean distance but where there exist different variances and covariance between variables expressed by covariance matrix C, $d_{\mathrm{MA}}(x_k, x_l) = \sqrt{(\mathbf{x}_k - \mathbf{x}_l)^{\mathrm{T}} \mathbf{C}^{-1}(\mathbf{x}_k - \mathbf{x}_l)}$.

Measures of similarity: x_{ij} can take just two values, which can be arbitrarily coded 0 and 1, say. To compute a similarity between two entities, therefore, the relevant data can be reduced to the 2 × 2 table:

		Entity 2	
		0	1
Entity 1	0	a	b
	1	c	d

This table shows that, out of all the possible pairwise comparisons between the two entities, a show 0-0 in both positions, d show 1–1 in both positions, b show the disagreement 1–0 , while c show the reverse disagreement 0–1.

Similarities between two objects: In this case we compare the p variable values for the two objects, so $a + b + c + d = p$. The four most common measures of similarity are as follows:

(a) *Sokal-Michener coefficient of association:* $S_{SM} = \dfrac{a+d}{p}$,

(b) *Russel-Rao coefficient of association:* $S_{RR} = \dfrac{d}{p}$,

(c) *Haman coefficient of association:* $S_{H} = \dfrac{a+b+c+d}{p}$,

(d) *Correlation coefficient:* $r_{B} = \dfrac{ad-bc}{\sqrt{(a+b)(c+d)(a+c)(b+d)}}$.

More details about practical applications of these measures are given in the chap. 4.9 and 4.10.

4.1.2 Data Preprocessing

Preprocessing is a very important step in many data treatment techniques; in some cases it must be done before a method is used. In other cases, a self-adjusting procedure forms part of the method, so that raw data can apparently be used [27].

Scaling is one step in the preliminary treatment of multivariate data, a treatment that can be quite wide in scope. The general name, *scaling*, is used because the treatment concerns both the measurement unit of the values and the origin of the scale. In addition, scaling can be applied to variables or objects or both. A scaling of variables is a transform for objects. After the translation of the origin and column centring, the distance between the two objects is unchanged, but the two variables, well separated in object space before scaling, are identical after scaling. For this reason, some scaling techniques may turn out to be true transforms of data. The common scaling techniques (y_{ij} refers to scaled variables) are as follows:

(1) *Column centring*: the new origin of the scale of the variable in the jth column is zero. The previous centre of the variable has been the mean of the variable $\bar{x}_{j}$ before centring. Column centring data y_{ij} are formed according to $y_{ij} = x_{ij} - \bar{x}_{j}$ where $\bar{x}_{j}$ is the mean of elements in the jth column calculated with $\bar{x}_{j} = \sum\limits_{i=1}^{n} \dfrac{x_{ij}}{n}$.

(2) *Column standardization*: the elements of original data in the jth column/variable are divided by its corresponding standard deviation

$y_{ij} = x_{ij} / s_j$, where s_j is the standard deviation calculated from

elements of the jth column with $s_j = \sqrt{\dfrac{\sum\limits_{i=1}^{n} (x_{ij} - \bar{x}_j)^2}{n-1}}$. The standard

deviation of the new standardized variable is 1.

(3) *Autoscaling* is the name currently used to signify the combination of column centring and standardization, i.e. the use of the *t*-transform (studentized variables) $y_{ij} = (x_{ij} - \bar{x}_j)/s_j$, analogous to the Z-transform when both statistics μ_j and σ_j^2 are supposed to be known $y_{ij} = (x_{ij} - \mu_j)/\sigma_j$. Autoscaling uses the estimates both of the mean and of the standard deviation.

(4) *Column range scaling*: variables are scaled to make the minimum of each variable equal to 0 and the maximum 1 using $y_{ij} = \dfrac{x_{ij} - \min x_{ij}}{\max x_{ij} - \min x_{ij}}$.

(5) *Row centring* using $y_{ij} = x_{ij} - \bar{x}_i$.

(6) *Row standardization* using $y_{ij} = x_{ij} / s_i$.

(7) *Global centring* using $y_{ij} = x_{ij} - \bar{x}$ where $\bar{x}$ is the overall mean computed over the $n \times m$ data in the matrix.

(8) *Global standardization* using $y_{ij} = x_{ij} / s$ where s is the standard deviation of the $n \times m$ data about x.

(9) *Double centring* means that data elements are scaled with the use of column centring and then with row centring.

(10) *Row profiles* using $y_{ij} = x_{ij} / (\bar{x}_i \, m)$. This is a rather particular case of scaling, but one frequently used in chemistry when a variable is given as a percentage, so that the sum of a row is 1 or 100.

(11) *Column profiles* using $y_{ij} = x_{ij} / (\bar{x}_j \, n)$.

4.2 Descriptive statistics

Multivariate data (also called multidimensional or *n*-dimensional) consists of some number of cases (objects), *n*, each of which is defined by an *m*-vector of variables $\xi_1, .. \xi_m$. Such data can be viewed as a $n \times m$ *source matrix*, where each row represents a case and each column represents a variable (also called a observation). In order to summarize a multivariate data set we need to produce summaries for each of the variables separately and also summarize the relationship between them.

Mean: For m variables, the population mean vector is given as $\mu = \begin{bmatrix} i_1, i_2, ..., i_m \end{bmatrix}^T$ where $\mu_j = E(\xi_j)$ is population mean of the j th variable.

An *estimate* $\hat{\boldsymbol{\mu}}$, based on sample of *n cases*, is $\hat{\boldsymbol{\mu}} = \overline{\mathbf{x}} = \left[\overline{x}_1, \overline{x}_2, ..., \overline{x}_m\right]^T$, where

$\overline{x}_j$ is the sample mean of the *j th* variable given with $\overline{x}_j = \dfrac{1}{n}\sum\limits_{i=1}^{n} x_{ij}$.

Variance: The vector of population variances can be represented by $\boldsymbol{\sigma}^2 = \left[\sigma_1^2, \sigma_2^2, ..., \sigma_m^2\right]^T$ where $\sigma_j^2 = D(\xi_j)$ is population variance of the *j th*

variable. An estimate of $\boldsymbol{\sigma}^2$ is $\mathbf{s}^2 = \left[s_1^2, s_2^2, ..., s_m^2\right]^T$ where s_j^2 is the sample variance of of the *j th* variable.

Pairwise Linear Association: Association means that the *j*th variable ξ_j bear some clear relationship to the *k*th variable ξ_k. A fundamental measure of linear association between the *j*th and the *k*th variable is defined by $\sigma_{jk} = Cov(\xi_j, \xi_k) = E\left[(\xi_j - \mu_j)(\xi_k - \mu_k)\right]$.

Covariance has the following properties:

(a) Its sign show the direction of stochastic relation between ξ_j and ξ_k . . .If large values of variable ξ_j occur together with large values of variable ξ_k, the covariance sign is positive. Conversely, if large values of variable ξ_j occur together with small values of variable ξ_k, and vice versa (small values of variable ξ_j together with small values of variable ξ_k), the covariance sign is negative.

(b) It is in absolute value above-limited with a product $\sigma_j \sigma_k$, i.e. $\left|\sigma_{jk}\right| \le \sigma_j \sigma_k$.

(c) It is a symmetric function of its arguments i.e. $\sigma_{jk} = \sigma_{kj}$.

(d) It does not change with a shift of the origin of ξ_j and ξ_k . The change of its scale will make a change of its magnitude, $cov(a_1 \xi_i + b_1, a_2 \xi_j + b_2) = a_1 a_2 \, cov(\xi_i, \xi_j)$.

(e) For non-correlated random values it is valid $cov(\xi_i, \xi_j) = 0$ and two cases may exist:

(1) $E(\xi_i \xi_j) = 0$ and also $E(\xi_i) = E(\xi_j) = 0$ is the case of the *centred orthogonal random variables*.

(2) $E(\xi_i \xi_j) = E(\xi_i)E(\xi_j)$ is the case of independent random variables.

This disadvantage of covariance is a fact that its values depend on a scale in which variables ξ_i and ξ_j are expressed.-The sample *covariance* has the form:

$$S_{jk} = \frac{1}{n-1}\sum_{i=1}^{n}(x_{ij} - \overline{x}_j)(x_{ik} - \overline{x}_k)$$

The magnitude of covariance is often difficult to interpret because it depends on the units in which the two variables are measured; consequently, it is often standardized by dividing by the product of the standard deviations of the two variables to give a quantity called the *correlation coefficient,* ρ_{jk}

$$\text{where } \rho_{jk} = \frac{\sigma_{jk}}{\sqrt{\sigma_{jj}\sigma_{kk}}} = \frac{\sum_{i=1}^{n}(x_{ij}-\overline{x}_j)(x_{ik}-\overline{x}_k)}{\sqrt{\sum_{i=1}^{n}(x_{ij}-\overline{x}_j)^2 \sum_{i=1}^{n}(x_{ik}-\overline{x}_k)^2}}.$$

The correlation coefficient lies between -1 and $+1$ and gives a measure of the *linear* relationship of the variables ξ_j and ξ_k. It is positive if high values of ξ_j are associated with high values of ξ_k and negative if high values of ξ_j are associated with low values of ξ_k. The correlation coefficient has the following properties:

(a) Equality $|\rho_{ij}| = 1$ shows that between ξ_i and ξ_j the exact linear relation exists.

(b) If both random variables ξ_i and ξ_j are mutually non-correlated, it is $\rho_{ij} = 0$.

(c) In case that ξ_i and ξ_j come from multivariate normal distribution and $\rho_{ij} = 0$, it means that they are mutually independent variables.

(d) It is valid that for nonlinear random variables it may be $\rho_{ij} = 0$.

(e) The correlation coefficient ρ_{ii} of the random variable ξ_i mutually with itself is equal to one.

(f) The correlation coefficient ρ_{ij} is invariant with linear transformation of random variables ξ_i, ξ_j. For parameters a_1, a_2, b_1, b_2 it is valid $\rho(a_1\xi_i + b_1, a_2\xi_j + b_2) = \text{sign}(a_1 a_2)\rho(\xi_i\xi_j)$ where sign(x) is the sign function for which it is valid $\text{sign}(x) = \begin{cases} -1 & \text{for} \quad x < 0 \\ 0 & \text{for} \quad x = 0 \\ 1 & \text{for} \quad x > 0 \end{cases}$

Correlation is thus simply a *unitless, scaled* covariance measure. Sample *correlation coefficient r* is defined with $r_{ij} = \dfrac{\text{cov}(x_i, x_j)}{s_i s_j}$. The correlation coefficient r_{ij} is sometimes referred to as the *Pearson's product-moment correlation coefficient.* If the values x_{ij} are replaced by their rank ordering, and if the values x_{ik} are similarly replaced by their rank ordering, then r_{jk} applied to the two sets of ranks results in *Spearman's rank correlation coefficient* $r_{s,ij}$. This measure is most useful when the raw data are themselves in the form of ranks.

With m variables, $\xi_1, \xi_2, ..., \xi_m$, there are m variances and $m(m-1)/2$ covariances (note that $\sigma_{ij} = \sigma_{ji}$).

In general these quantities are arranged in a $m \times m$ symmetric matrix, Σ, where

$$\Sigma = \begin{pmatrix} \sigma_{11} & \sigma_{12} & \cdots & \sigma_{1m} \\ \sigma_{21} & \sigma_{22} & \cdots & \sigma_{2m} \\ \cdots & \cdots & \cdots & \cdots \\ \sigma_{m1} & \sigma_{m1} & \cdots & \sigma_{mm} \end{pmatrix}$$

This matrix is generally known as the *variance-covariance matrix* or simply the *covariance matrix*. The matrix Σ is estimated by the *sample covariance matrix* **S**, given by

$$\mathbf{S} = \sum_{i=1}^{n} (\mathbf{x}_i - \bar{\mathbf{x}})(\mathbf{x}_i - \bar{\mathbf{x}})^{\mathrm{T}} / (n-1)$$

where $\mathbf{x}_i^{\mathrm{T}} = [x_{i1}, x_{i1}, ..., x_{im}]^{\mathrm{T}}$ is the vector of observations for the ith object.

With m variables there are $m(m-1)/2$ distinct correlations which may be arranged in a $m \times m$ matrix, **R**, whose diagonal elements are unity. This matrix may be written in terms of the covariance matrix, Σ, as follows: $\mathbf{R} = \mathbf{D}^{-1/2} \Sigma$ $\mathbf{D}^{-1/2}$, where $\mathbf{D}^{-1/2} = \mathrm{diag}(1/\sqrt{\sigma_{ii}})$. In most situations we will be dealing with covariance and correlation matrices of full rank, m, so that both matrices will be non-singular (i.e. invertible) [136].

Linear combinations of variables: Many of the methods of analysis to be described in this text involve *linear combinations* of the original variables, $\xi_1, \xi_2, ... \xi_m$, that is, a variable constructed thus:

$y = a_1 \xi_1 + a_2 \xi_2 + ... + a_m \xi_m$ where $a_1, a_2, ..., a_m$, are a set of scalars. This can be written more simply as $y = \mathbf{a}^{\mathrm{T}} \xi$ where $\mathbf{a}^{\mathrm{T}} = [a_1, a_2, ..., a_m]$. The variable y has a mean given by $E(y) = \mathbf{a}^{\mathrm{T}} E(\xi) = \mathbf{a}^{\mathrm{T}} \mathbf{\mu}$ and variance $D(y) = E\left[\mathbf{a}^{\mathrm{T}} (\xi - \mathbf{\mu})^2\right]$. A little algebra shows that this can be written as $D(y) = \mathbf{a}^{\mathrm{T}} \Sigma \mathbf{a}$.

4.3 Exploratory Analysis of Multivariate Data Structure

Information visualization and visual data analysis can help to deal with the flood of information. The advantage of visual data exploration is that the user is directly involved in the data analysis process. According to Chambers et al. [7] "there is no statistical tool that is as powerful as a well-chosen graph". The basic idea of visual data mining is to present the data in some visual form, allowing the user to gain insight into the data, draw conclusions, and directly interact with the data. Visual data analysis techniques have proven to be of high value in exploratory data analysis [30].

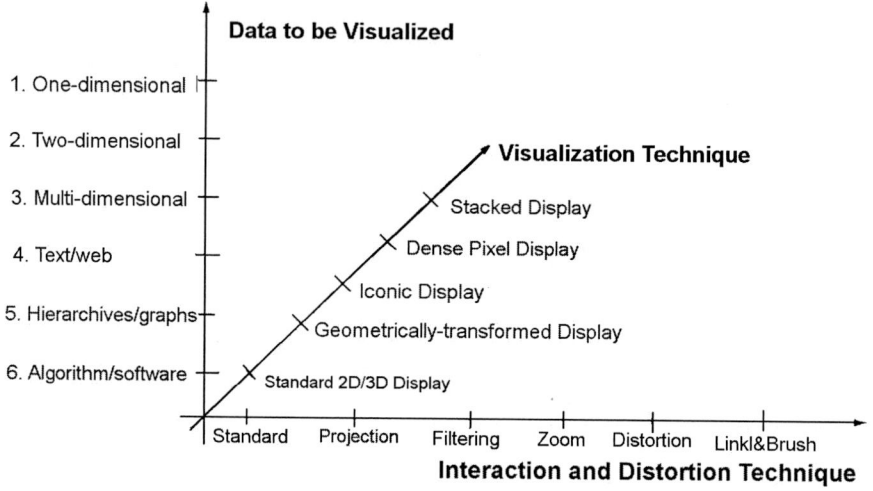

Figure 4.1   Classification of information visualization techniques [29, 31].

Information visualization (Fig. 4.1) focuses on data sets lacking inherent 2D or 3D semantics and therefore also lacking a standard mapping of the abstract data onto the physical screen space [32].

The techniques can be classified based on three criteria [33]. The data type to be visualized, the visualization technique used, and the interaction technique used are commented:

(1) The *data type used* [34] may be:

    1.1 One-dimensional data, such as temporal (time-series) data;

    1.2 Two-dimensional data, such as scatter graphs;

    1.3 Multi-dimensional data, such as relational tables.

(2) The *visualization technique used* may be classified as:

    2.1 *Geometric techniques* where the methods described are based on the Cartesian coordinates, leading to standard 2D/3D displays, such as bar charts and *x-y* scatterplots.

    2.2 *Icon* or *glyphs techniques* where the main tool and basic unit of the visualization technique is an icon and other icon-based displays, such as needle icons and star icons.

2.3 *Pixel oriented techniques* where the pixels are used as the basic visualization unit such as recursive patterns and circle segments.

2.4 *Hierarchical techniques* where the data items and the query results are presented in hierarchical displays;

2.5 *Stacked displays*, such as treemaps and dimensional stacking.

2.6 *Distortion techniques* where the three-dimensional space is distorted to allow more information to be visualized; and

2.7 *Graph based techniques* where the information is presented using nodes and edges. Geometrically-transformed displays, such as landscapes and parallel coordinates.

(3) The *interaction technique used* allows users to directly navigate and modify the visualizations:

3.1 *Dynamic projection* allows smooth navigations through the data space.

3.2 *Interactive filtering* enables users to isolate subsets of data for focused analysis.

3.3 *Zooming* enables to enlarge data for detailed analysis.

3.4 *Distortion* may increase the screen space allocated to areas of interest while preserving the context of the entire data set.

3.5 *Linking and brushing*, to enable users to select data of interest in one view and see it highlighted in other views.

4.3.1 Visualization Techniques

There are a large number of visualization techniques that can be used for visualizing data. In addition to standard 2D/3D techniques such as x-y (x-y-z) scatterplots, scatterplots matrix, bar charts, line graphs, and maps, there are a number of more sophisticated classes of visualization techniques [35].

Visual data exploration usually allows a faster data exploration and often provides more interesting results, especially in cases where automatic algorithms fail. While visualization is a quite powerful tool, there are two basic things connected to it that limits the possibilities: the human ability to distinguish image details and memorize them, and the available computer power. Three distinctive visualization goals can be mentioned:

(1) *Explorative data analysis*, where the visualization of data or data objects provides hypotheses about the data.

(2) *Confirmative data analysis*, where the visualization of data provides confirmation for already existing hypotheses.

(3) *Presentation of data*, where a priori fixed facts are being visualized [32].

Looking at the taxonomies the following stand out as high-dimensional visualizations: *2D* and *3D scatterplots, Satterplots matrix, Heat maps, Height maps, Table lens, Survey plots, Iconographic displays, Dimensional stacking* (or *eral logic diagrams*), *Parallel coordinates, Line graph, multiple line graph, Pixel techniques, circle segments, Multi-dimensional scaling and Sammon plots, Polar charts, RadViz, PolyViz, Projection pursuit, Kohonen self-organizing maps.* Several of these are quite similar and related. We give a brief description and visualization for each, along with key references (see [36–38]). We use the Fisher Iris flower data set [39] or the car data set from UC Irvine Machine Learning Repository, whenever possible. The Iris flower data set contains 50 specimens from each of the three species of Iris flowers: Iris setosa, I. Versicolor, and I. Virginica. The dimensions of the data set are sepal length, sepal width, petal length and petal width, measured in millimeters.

4.3.1.1 Geometrically Transformed Displays

Geometrically transformed display techniques [40–42] aim at finding "interesting" transformations of multi-dimensional data sets. The class of geometric display methods includes techniques from exploratory statistics such as:

- *Scatterplots*: A scatterplot (Fig. 4.2) is a point projection (usually affine) of the data into a 2D or 3D dimensional space represented on the screen in classic (X, Y) or (X, Y, Z) format. This is the most commonly utilized data visualization method which links at least two variables, encouraging and even imploring the viewer to access the possible causal relationship between the plotted variables.

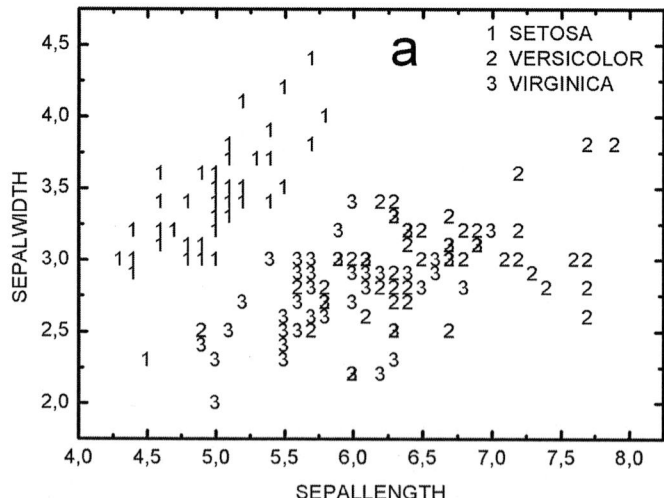

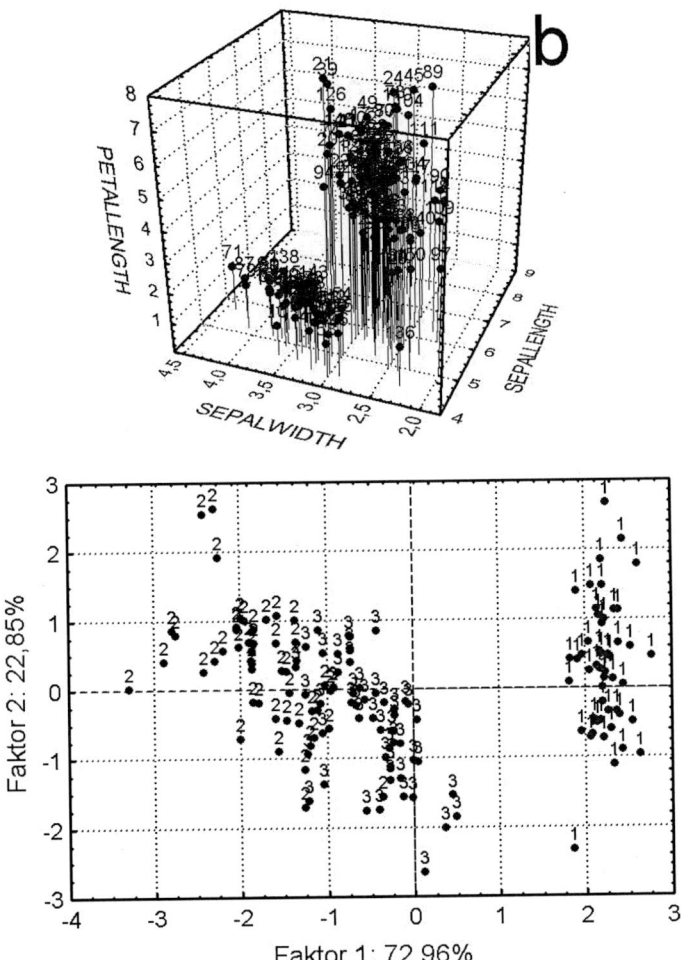

Figure 4.2   EDA of Iris data set: (a) 2D-scatterplot, (b) 3D-scatterplot, (c) 2D-factor plot.

It confronts causal theories that x causes y with empirical evidence as to the actual relationship between x and y. Numerous mappings or transformations can be applied to it. The displayed points can have numerous attributes such as color, size, shape, texture, motion and even sound (when interacted with). To interpret the 3D projection interaction, it is necessary to resolve ambiguities, although other techniques have been used (animation). In its most general form this method is related to iconographic and pixel displays.

- *Scatterplot matrices* [43], [44] and techniques that can be sumed under the term "projection pursuit" (Fig. 4.2). In a set of multivariate data with more than two variables, viewing the scatterplots of each pair of variables is often a useful way to begin to examine the data. A matrix of scatterplots is an array of scatterplots displaying all possible pairwise combinations of dimensions or coordinates. For n-dimensional data this yields $n(n-1)/2$ scatterplots with shared scales, although most often n^2 scatterplots are displayed. The number of scatterplots, however, quickly becomes daunting: for 10 variables, for example, there are 45 plots to consider. Arranging the pairwise scatterplots in the form of a square grid, usually known as a *draughtsman's plot* or *scatterplot matrix*, can help in assessing all scatterplots at the same time. Because the scatterplot matrix is symmetric about its diagonal, despite the apparent redundancy, it enables a row and column to be visually scanned to see one variable against all others, with the scale for the one variable lined up along the horizontal or the vertical. One can visually link features of one scatterplot with features on another, which greatly increases its power. Several variations on the theme of a matrix of scatterplots have since been developed:

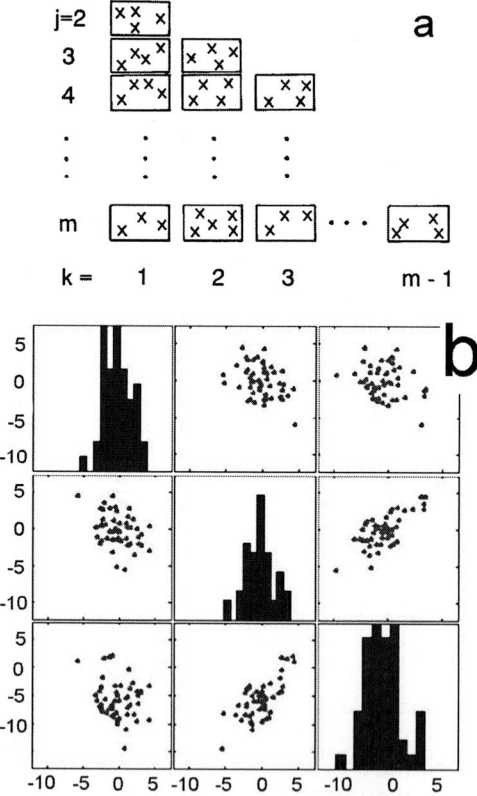

Figure 4.3 (a) Lower half of scatterplot matrix, (b) Full scatterplot matrix [45].

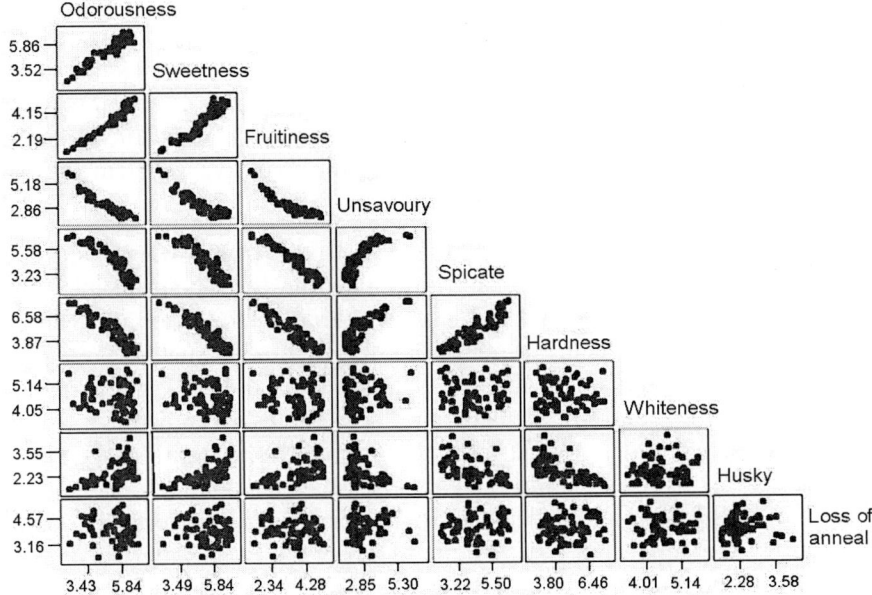

Figure 4.4   Scatterplot of correlation matrix [46].

- *Locate projections* a class of techniques that satisfy some computable quality of interestingness.

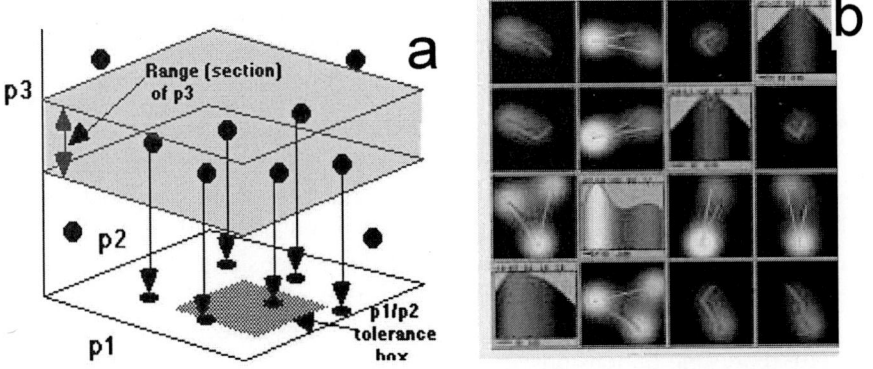

Figure 4.5   (a) Prosection views, (b) hyperslice, both taken from [30].

- *Prosection Views*, where only user-selected slices of the data are projected [47]. Prosection is a method more suitable for data mining, since it does not project all points onto the scatterplot matrix, but projects only points within a certain range of each dimension, similar to brushing and dynamic queries [1]. The projection of a graph section of two variables where a range has been imposed on a third variable so as to cut a section of the distribution is presented in a space fillings plot (Fig. 4.5a). A matrix is formed by all the projection plots and every variable is represented in the matrix. A colour filling technique is then applied to smoothen out the granularity of the graphs. The method has been developed to externalize mathematical models and the authors claim to have transformed a very difficult cognitive problem into a much easier perceptual task.

- *Hyperslice* [48], [49] (Fig. 4.5b), is a matrix of panels where "slices" of multivariate function are shown at a certain focal point of interest. Hyperslice could be considered as an evolution of the scatterplot matrix. The basic concept is the same. In the diagonal of the matrix, there are scalar functions of single variables represented and the other cells show the scalar relationships of many variables. The user can define a focus point $c = (c_1, c_2, ..., c_n)$ and a set of scalar widths, setting a range slice such as in the cell (2,3) in Fig. 4.5b. Only the data within the range are viewed. By steering the focus point, the user can fast explore other data near the set range and visualize the trends in the cells where the linked data can be found.

- *N-vision* [50], where the matrix panel accommodates for interactive exploration of a multivariate function.

- *Hyperbox* [51] uses the same pairwise projections of the data, but projects onto panels of an *n*-dimensional box. Each of the panels has a different orientation and the dimensions can be cut in order to show histograms on the panels, according to ranges of the dimensions being cut.

- *Parallel Coordinates* or *Parallel Graphs* the well-known visualization technique [52] which maps the *k*-dimensional space onto the two display dimensions by using *k* axes that are parallel to each other (either horizontally or vertically oriented) and are evenly spaced across the display. The axes correspond to the dimensions and are linearly scaled from the minimum to the maximum value of the corresponding dimension. Parallel coordinates use parallel axes instead of perpendicular to represent dimensions of a multidimensional data set [53], [54], [55–57]. A vertical line is used for the projection of each dimension or attribute, with the maximum and minimum values of each dimension usually scaled to the upper and lower bounds on those vertical lines. A polyline made up of $n - 1$ lines at the appropriate dimensional values connects the axes to represent an *n*-dimensional point. Each data item is presented as a chain of connected line segments, intersecting each of the axes at the location corresponding to the value of the dimension considered. Parallel graphs are useful in that they allow for an arbitrary number of dimensions and datasets to be shown. However, it is apparent, even with as few as thirty datasets, that parallel graphs quickly become cumbersome even though useful data can still be easily extracted visually. For instance clusters, minima, and maxima in a single dimension are still quite apparent, but following a single dataset through the graph can be difficult. The first problem that is apparent is that with so many data points, it can be hard to see which values go where and what it all means.

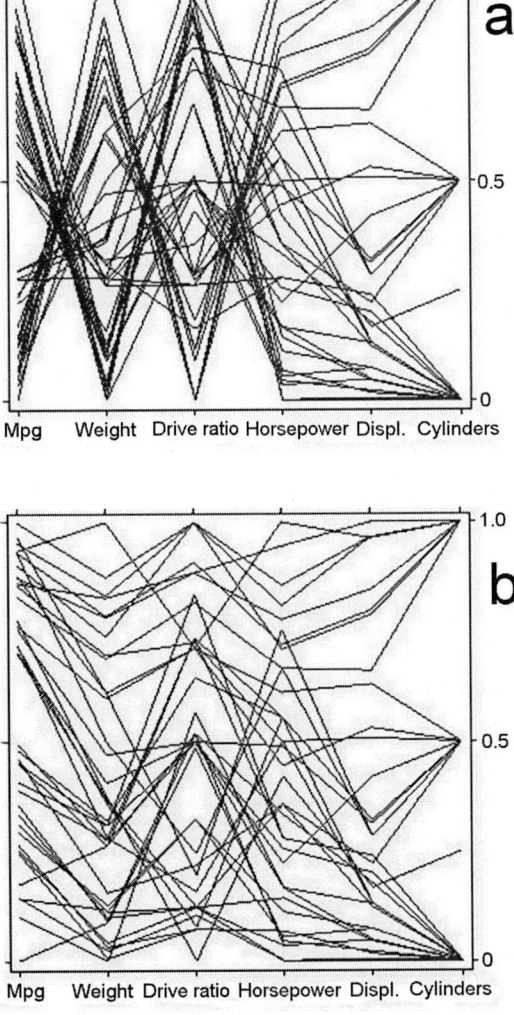

Figure 4.6 Parallel coordinates (a) original, (b) correlated, of Iris data set [58, 45].

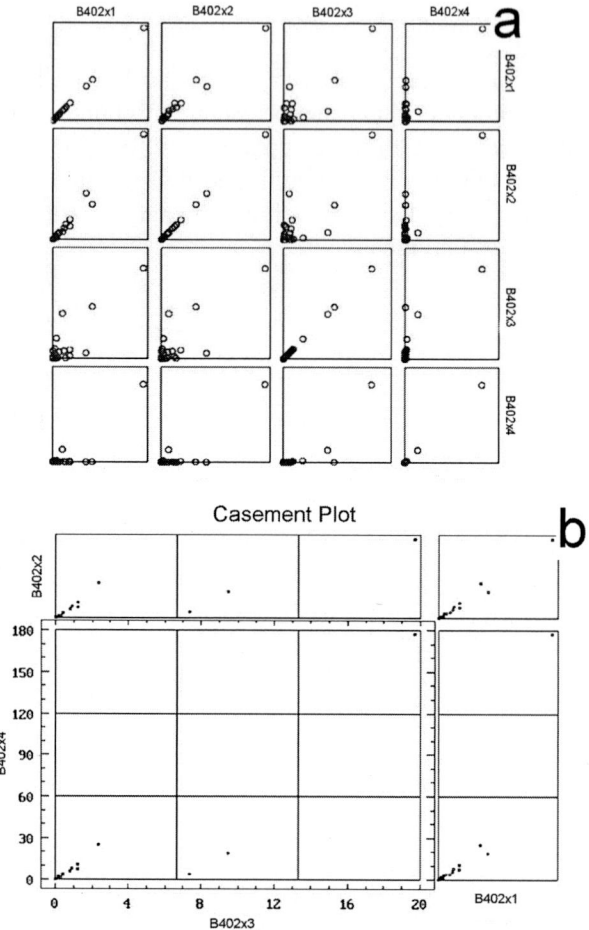

Figure 4.7   (a) The scatterplot matrix, (b) The locate projections in the casement plot [59].

4.3.1.2 Iconic Displays or Glyphs

Glyphs (also referred to as *icons*) are graphical entities that convey one or more data values via attributes such as shape, size, color, and position.

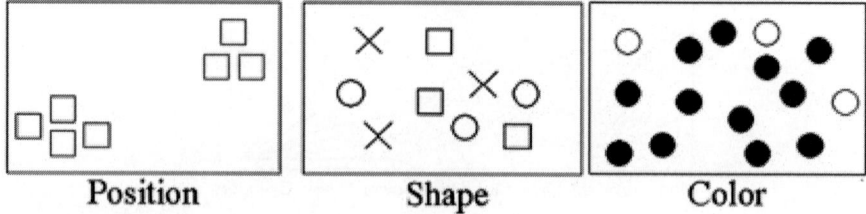

Figure 4.8   Position, shape and color in visualizations tools.

They have been widely used in the visualization of data and information, and are especially well suited for displaying complex, multivariate data sets. A glyph consists of a graphical entity with p components, each of which may have r geometric attributes and s appearance attributes. Typical geometric attributes include shape, size, orientation, position, and direction/magnitude of motion, while appearance attributes include color, texture, and transparency. Attributes can be discrete or continuous, scalar or vector, and may or may not have a distance metric, ordering relation, or absolute zero. The process of creating a glyph thus becomes one of mapping one or more data dimensions for a data point to one or more geometric and/or appearance attributes of one or more components of a graphical entity. A glyph may also contain components or attributes that are independent of the data being used in its formation. From a point-of-view of the glyph placement issues for creation of glyphs it is valid that

(1) the first consideration when selecting a placement strategy is whether the placement will be data-driven (e.g., based on two or more data dimensions) or structure-driven (e.g., based on an explicit or implicit order or other relationship between data points).

(2) A second consideration is whether overlaps between glyphs will be allowed. This can have a significant impact on the size of the data set that can be displayed, the size of the glyphs used, and the interpretability of the resulting images.

(3) A third consideration is the trade-off between optimized screen utilization (e.g., space-filling algorithms) versus the use of white space to reinforce distances between data points.

(4) A fourth consideration is whether the glyph positions can be adjusted after initial placement to improve visibility at the cost of distorting the computed position (how accurately must the position reflect a value?).

The list below describes various glyphs from the literature in terms of the graphical entities and attributes controlled by the multivariate data point:

- *Profiles* [61]: height and color of bars.

- *Stars* [62]: A *star* (or *sun*) is composed of equally spaced radii, as many as the number of attributes in the data table, stemming from the centre. The length of the rightmost spike is proportional to the value of the first attribute for a given row; the remaining attributes are assigned to their spikes counter clockwise in this manner. The clarity of a star display will suffer as the number of attributes increases, and grouping correlated attributes to provide smooth transitions between spikes might be beneficial. The similarity or dissimilarity of a pair of stars can be appreciated visually; however, gaining a proper overview of a large data table can become a tedious task. This sort of processing is best left to the computer, so that proximity between rows of a data table can be represented in a direct spatial form. *Star plots* are a useful way to display multivariate object or observations with an arbitrary number of variables (Fig. 4.9). Each object is represented as a star-shaped figure with one ray for each variable. For a given object, the length of each ray is made proportional to the size of that variable. Each ray length of the star (or of sun) x_{ij}^* is transformed into an interval $[a,1]$ where a is the

lower limit usually equal to $a = 0$, $x_{ij}^* = \dfrac{(1-a)(x_{ij} - \min x_{ij})}{\max x_{ij} - \min x_{ij}} + a$, where

$\min x_{ij}$ is the minimal and $\max x_{ij}$ value of the jth variable of object $\mathbf{x}^{\mathrm{T}}$

over all objects $x_i^{\mathrm{T}}, i = 1,...,n$. The angle of individual rays is defined

with $\alpha_j = \dfrac{2\pi(j-1)}{m}$, $j = 1,...,m$. When R is the maximal length of rays,

the co-ordinate of polygon is $p_{ij} = (x_{ij}^* R \cos \alpha_j; x_{ij}^* R \sin \alpha_j)$.

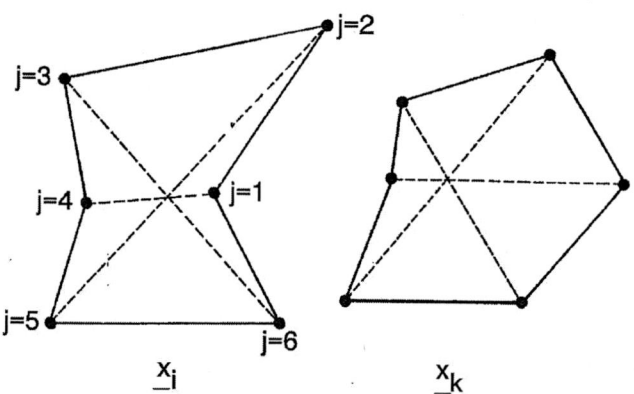

Figure 4.9 Description of variables in stars (or suns).

Star plots differ from glyph plots in that all variables are used to construct the plotted star figure; there is no separation into foreground and background variables. Instead, the star-shaped figures are usually arranged in a rectangular array on the page. It is somewhat easier to see patterns in the data if the observations are arranged in some non-arbitrary order, and if the variables are assigned to the rays of the star in some meaningful order.

- *Suns* [63]: length of evenly spaced rays emanating from center (Fig. 4.9). Suns are a form of glyphs, or shapes. Like parallel stars, suns take each dimension of data and plot it on a separate axis, however instead of plotting these axes parallel to each other, they are radially spread out from a central point.. Also each dataset is a separate sun, although all datasets on a single sun is possible.

- *Anderson/metroglyphs* [64–65]: length of rays.
- *Stick figure icons* [66]: length, angle, color of limbs.
- *Trees* [67]: length, thickness, angles of branches; branch structure derived from analyzing relati-ons between dimensions.
- *Autoglyph* [68]: color of boxes.
- *Boxes* [69]: height, width, depth of first box; height of successive boxes.
- *Hedgehogs* [70]: spikes on a vector field, with variation in orientation, thickness, and taper.
- *Piles* [71] are a method of stacking icon/glyph based visualization techniques. Columns make it easier to view relationships between datasets on a single dimension. However they introduce other problems of overlapping and ordering of the data and the dimensions to allow for optimal viewing of the column.
- *Chernoff faces* [72] take advantage of the natural familiarity and recognition of human faces.

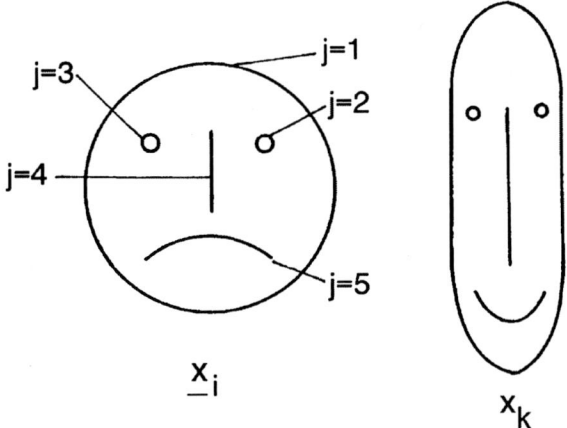

Figure 4.10   Description of variables in Chernoff faces.

Each facial feature represents one variable: area of the face, shape of the face, length of the nose, location of the mouth, curve of the smile, width of the mouth, location, separation, angle, shape, and width of the eyes, location of the pupil, location, angle, and width of the eyebrows. In total, 15 attributes can be represented, and additional variables could be encoded by making faces asymmetric. The trouble is that the appearance of a face will vary with the order of assignment of variables to facial expressions, and perceived similarity of faces will be affected.

- *Arrows* [73]: length, width, taper, and color of base and head.

- *Polygons* [74]: conveying local deformation in a vector field via orientation and shape changes.
- *Dashtubes* [75]: texture and opacity to convey vector field data.
- *Weathervanes* [76]: level in bulb, length of flags.
- *Circular profiles* [77]: distance from center to vertices at equal angles.
- *Color icons* [78]: colored lines across a box.
- *Bugs* [79]: wing shapes controlled by time series; length of head spikes (antennae); size and color of tail; size of body markings.
- *Wheels* [80]: time wheels create ring of time series plots, value controls distance from base ring; 3D wheel maps time to height, variable value to radius.
- *Boids* [81] : shape and orientation of primitives moving through a time-varying field.
- *Procedural shapes* [82–83]: blobby objects controlled by up to 14 dimensions:
- *Glyphmaker* [84]: user-controlled mappings.
- *Icon Modeling Language* [85]: attributes of a 2D contour and the parameters that extrude it to 3D and further transform/deform it.

The glyph placement strategies [86] are shown in Fig. 4.11

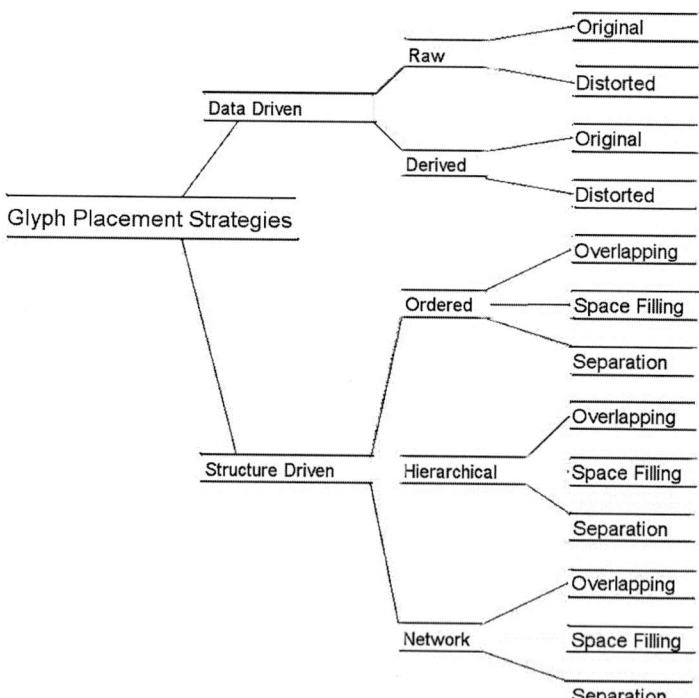

Figure 4.11   Glyph placement strategies [86].

Examples of various exploratory techniques are shown in the Figs. 4.12–4.17.

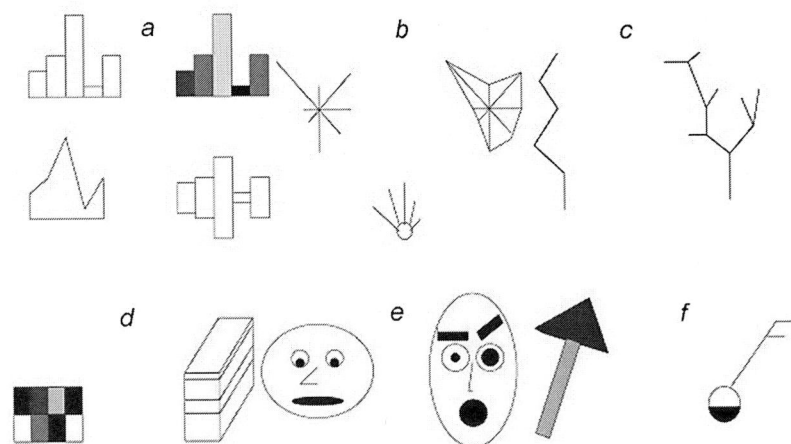

Figure 4.12 Examples of various glyphs: (a) Variations on profiles; (b) Stars/Metroglyphs; (c) Stick figure icons and Trees; (d) Autoglyphs and Boxes; (e) Chernoff faces; (f) Arrows and Weathervanes.

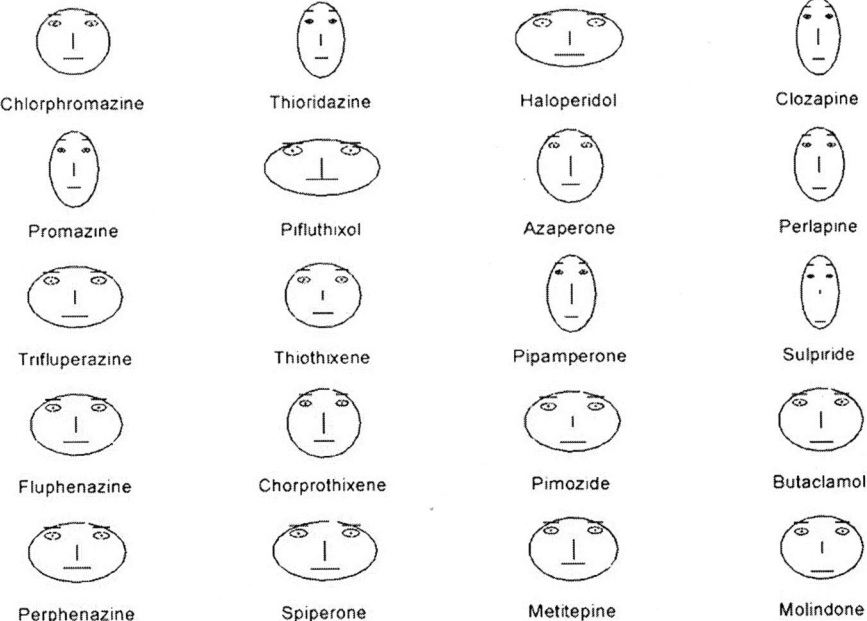

Figure 4.13 Chernoff faces of the original data [59].

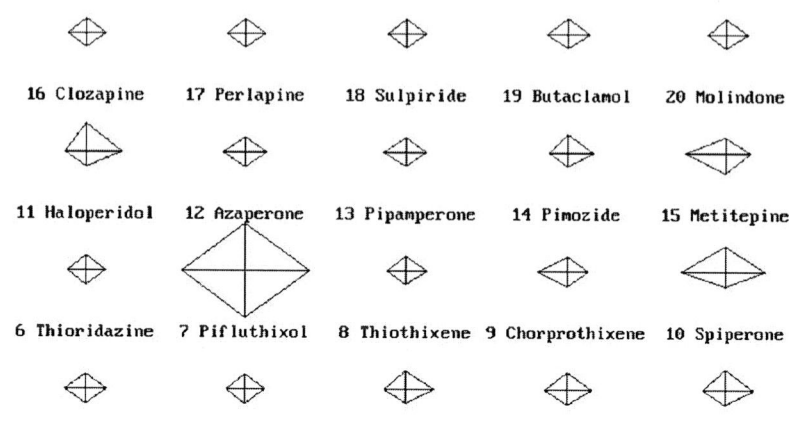

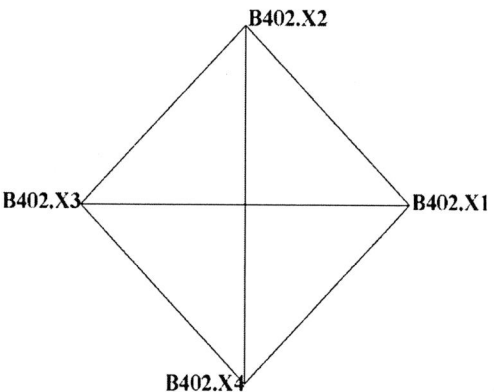

Figure 4.14 (Left): Star plot of neuroleptics. Each star represents one object (here neuroleptics), each ray in the star is proportional to one variable. (Right): Variable assignment key for star plot [59].

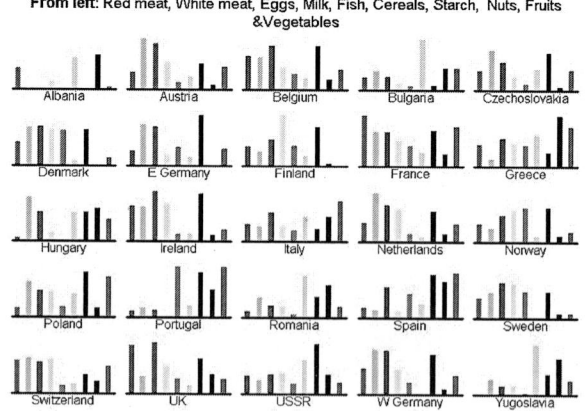

From left: Red meat, White meat, Eggs, Milk, Fish, Cereals, Starch, Nuts, Fruits &Vegetables

From left: Red meat, White meat, Eggs, Milk, Fish, Cereals, Starch, Nuts, Fruits &Vegetables

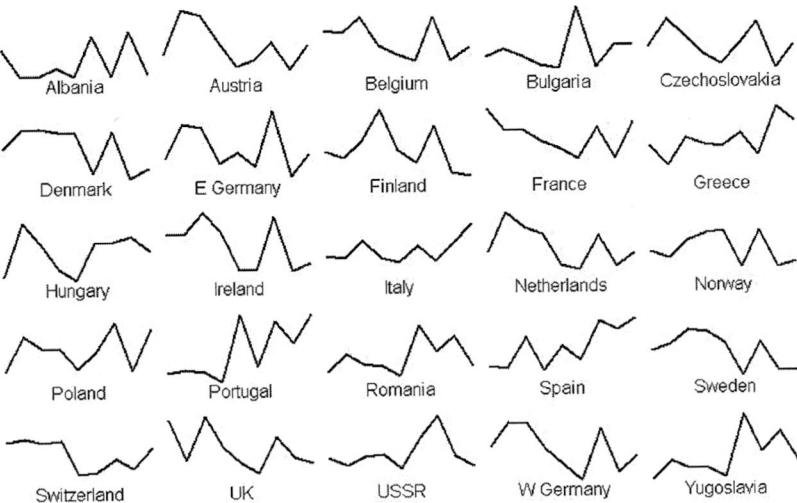

Figure 4.15 Examples of (a) profile columns, (b) profile curves, [59].

From left: face/w=Red meat, ear/lev=White meat, halfface/h=Eggs, upface/ecc=Milk, loface/ecc=Fish, nose/l=Cereals, mouth/cent= Starch, mouth/curv= Nuts, mouth /l=Fruits&Vegetables

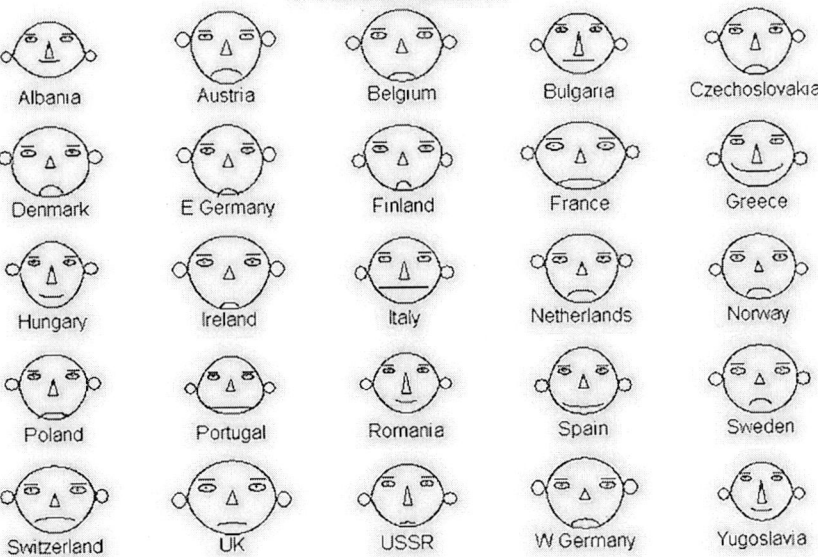

From left: Red meat, White meat, Eggs, Milk, Fish, Cereals, Starch, Nuts, Fruits &Vegetables

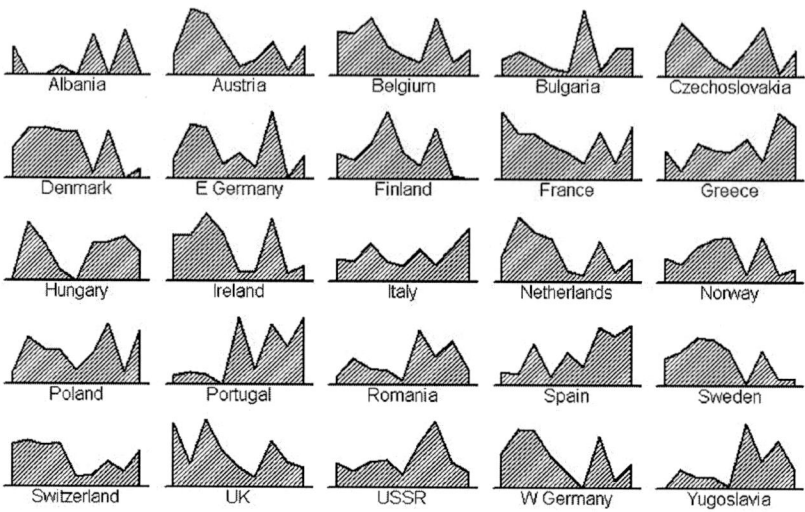

Figure 4.16   Examples of (a) Chernoff faces, (b) Profiles, [59].

From left: Red meat, White meat, Eggs, Milk, Fish, Cereals, Starch, Nuts, Fruits &Vegetables

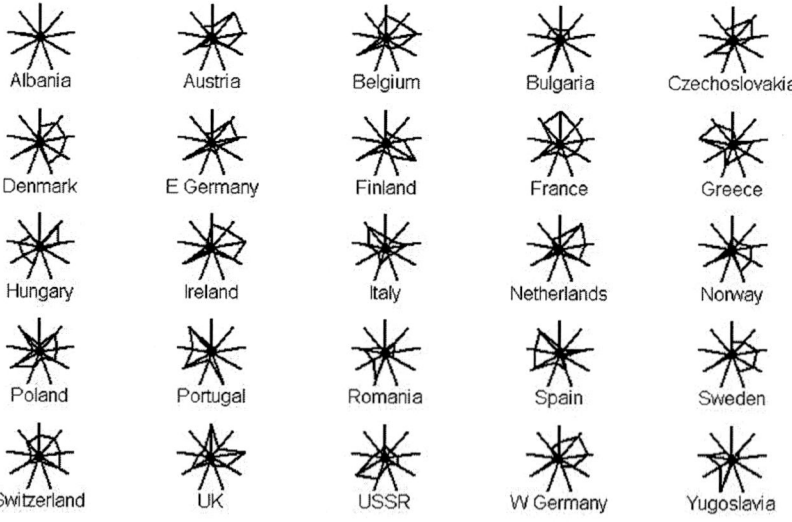

From left: Red meat, White meat, Eggs, Milk, Fish, Cereals, Starch, Nuts, Fruits &Vegetables

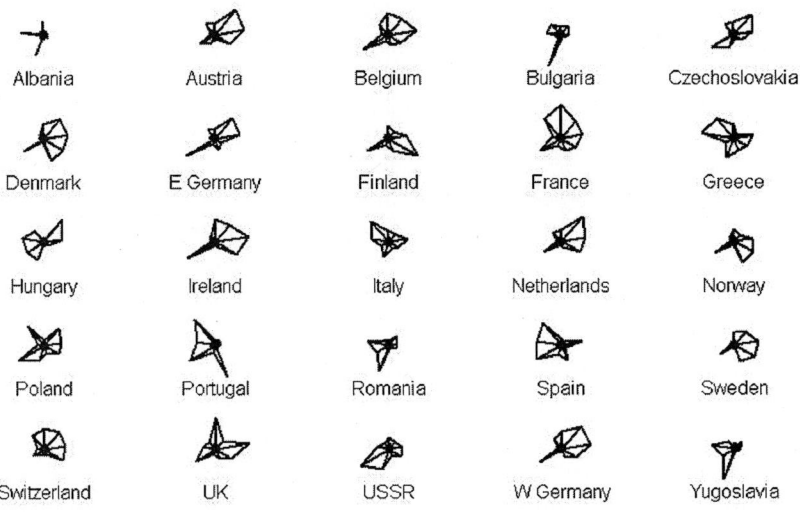

Figure 4.17 Examples of (a) Sun polygons, (b) Stars polygons, [59].

4.3.1.3 Dense Pixel Displays

The basic idea of dense pixel techniques is to map each dimension value to a colored pixel and group the pixels belonging to each dimension into adjacent areas [87]. Since in general dense pixel displays use one pixel per data value, the techniques allow the visualization of the largest amount of data possible on current displays (up to about 1,000,000 data values). If each data value is represented by one pixel, the main question is how the pixels are arranged on the screen. Dense pixel displays (Fig. 4.18) use different arrangements to provide detailed information on local correlations, dependencies, and hot spots. Well-known examples are the recursive pattern technique [88] and the circle segments technique [89]. The recursive pattern technique is based on a generic recursive back-and-forth arrangement of the pixels and is particularly aimed at representing data sets with a natural order according to one attribute (e.g. time-series data). The user may specify parameters for each recursion level and thereby control the arrangement of the pixels to form semantically meaningful substructures. The basic element on each recursion level is a pattern of height h_i and width w_i as specified by the user. First, the elements correspond to single pixels that are arranged within a rectangle of height h_i and width w_i from left to right, then below backwards from right to left, then again forward from left to right, and so on. The same basic arrangement is done on all recursion levels with the only difference that the basic elements that are arranged on level i are the patterns resulting from level $(i - 1)$.

Figure 4.18a   Example of pixel display of an eight-dimensional data set 1000 objects using 2D arrangement [45].

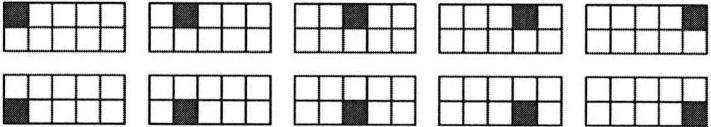

Figure 4.18b   Example of pixel display of the 10-dimensional unit vector data [45].

4.3.1.4 Graphs

Graphs are inherently two-dimensional (Fig. 4.19). Some ingenuity is therefore required to display the relationships of three or more variables on a flat piece of paper. All multivariate graphics require changing or expanding the familiar visual metaphors we use for two variables, and a wide variety of methods have been developed.

The primary goals of the book are to survey the kinds of graphic displays that are most useful for different questions. A graph consists of a set of objects,

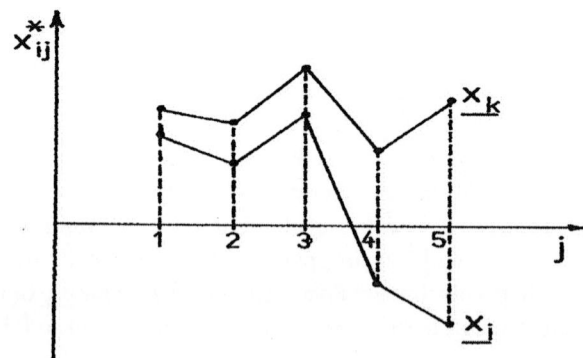

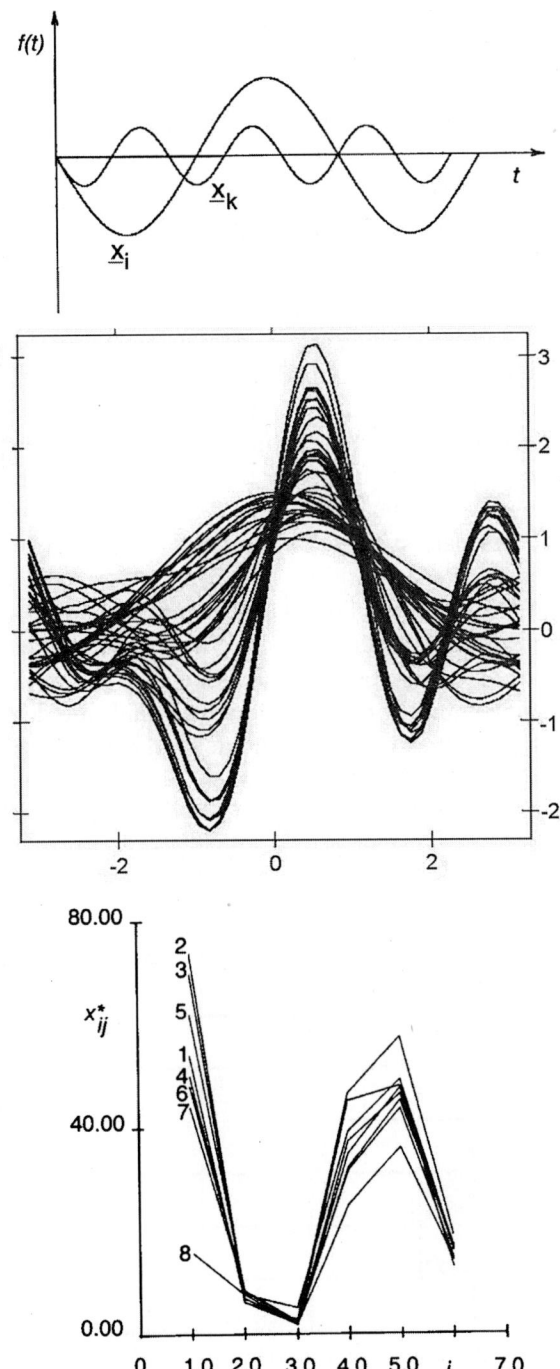

Figure 4.19 Various types of curves

called nodes, and connections between these objects, called edges or links. An overview of hierarchical information visualization techniques can be found in [90], an overview of Web visualization techniques is presented in [91], and an overview book on all aspects related to graph drawing is [92].

4.3.1.5 Stacked Displays

Stacked display techniques are tailored to present data partitioned in a hierarchical fashion. In the case of multi-dimensional data, the data dimensions to be used for partitioning the data and building the hierarchy have to be selected appropriately. An example of a stacked display technique is *Dimensional Stacking* [93]. The basic idea is to embed one coordinate system inside another coordinate system, i.e. two attributes form the outer coordinate system, two other attributes are embedded into the outer coordinate system, and so on. The display is generated by dividing the outermost level coordinate system into rectangular cells. Within the cells, the next two attributes are used to span the second level coordinate system. This process may be repeated multiple times. The usefulness of the resulting visualization largely depends on the data distribution of the outer coordinates and therefore the dimensions that are used for defining the outer coordinate system have to be selected carefully. A rule of thumb is to choose the most important dimensions first. A Dimensional Stacking visualization of mining data with longitude and latitude mapped to the outer x and y axes, as well as ore grade and depth mapped to the inner x and y axes is shown in Figure 4.20. Other examples of stacked display techniques include Worlds-within-Worlds [94], Treemap [95–96], and Cone Trees [97].

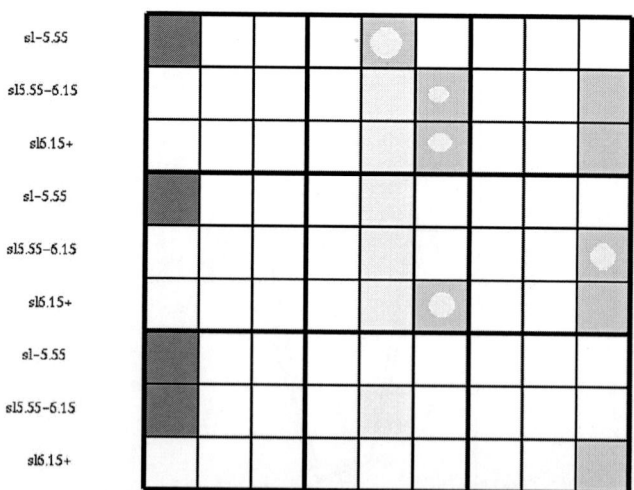

Figure 4.20 Example of the Dimensional Stacking of the Iris data set [45].

4.3.2 Interaction Techniques

In addition to the visualization technique, for an effective data exploration it is necessary to use one or more interaction techniques. *Interaction techniques* allow the data analyst to directly interact with the visualizations and dynamically change the visualizations according to the exploration objectives. In addition, they also make it possible to relate and combine multiple independent visualizations.

Interaction techniques can be categorized based on the effects they have on the display. *Navigation techniques* focus on modifying the projection of the data onto the screen, using either manual or automated methods. *View enhancement methods* allow users to adjust the level of detail on part or all of the visualization, or modify the mapping to emphasize some subset of the data. *Selection techniques* provide users with the ability to isolate a subset of the displayed data for operations such as highlighting, filtering, and quantitative analysis. Selection can be done directly on the visualization (*direct manipulation*) or via dialog boxes and other query mechanisms (*indirect manipulation*). Some examples of interaction techniques are described below.

4.3.2.1 Dynamic Projection

Dynamic projection is an automated navigation operation. The basic idea is to dynamically change the projections in order to explore a multi-dimensional data set. A well-known example is the GrandTour system [98] which tries to show all interesting – i.e., those exhibiting desireable properties such as well-separated clusters – two-dimensional projections of a multi-dimensional data set as a series of scatterplots. Note that the number of possible projections is exponential in the number of dimensions, i.e., it is intractable for large dimensionality. The sequence of projections shown can be random, manual, precomputed, or data driven. Systems supporting dynamic projection techniques include XGobi [99–100], XLispStat [101], and ExplorN [102].

4.3.2.2 Interactive Filtering

Interactive filtering is a combination of selection and view enhancement. In exploring large data sets, it is important to interactively partition the data set into segments and focus on interesting subsets. This can be done by a direct selection of the desired subset *(browsing)* or by a specification of properties of the desired subset *(querying)*. Browsing is difficult for very large data sets and querying often does not produce the desired results. Therefore, a number of interactive selection techniques have been developed to improve interactive filtering in data exploration. An example of a tool that can be used for interactive filtering is the Magic Lens [103–104]. The basic idea of Magic Lenses is to use a tool similar to a magnifying glass to filter the data directly in the visualization. The data under the magnifying glass is processed by the filter and displayed in a different way than the remaining data set. Magic Lenses show a modified view of the selected region, while the rest of the visualization remains unaffected. Note that several lenses with different filters may be used; if the filter overlap, all filters are combined. Other examples of interactive filtering techniques and tools are InfoCrystal [105], Dynamic Queries [106–108], and Polaris [109].

4.3.2.3 Zooming

Zooming is a well known view modification technique that is widely used in a number of applications. In dealing with large amounts of data, it is important to present the data in a highly compressed form to provide an overview of the data but at the same time allow a variable display of the data at different resolutions. Zooming does not only mean displaying the data objects larger, but also that the data representation may automatically change to present more details on higher zoom levels. The objects may, for example, be represented as single pixels at a low zoom level, as icons at an intermediate zoom level, and as labeled objects at a high resolution. An interesting example applying the zooming idea to large tabular data sets is the TableLens approach [110]. Getting an overview of such data sets is difficult if the data are displayed in textual form. The basic idea of TableLens is to represent each numerical value by a small bar. All bars have a one-pixel height and the lengths are determined by the attribute values. This means that the number of rows on the display can be nearly as large as the vertical resolution and the number of columns depends on the maximum width of the bars for each attribute. The initial view allows the user to detect patterns, correlations, and outliers in the data set. In order to explore a region of interest the user can zoom in, with the result that the affected rows (or columns) are displayed in more detail, possibly even in textual form. Other examples of techniques and systems that use interactive zooming include PAD++ [111–113], IVEE/Spotfire [114], and DataSpace [115]. A comparison of fisheye and zooming techniques can be found in [116].

4.3.2.4 Distortion

Distortion is a view modification technique that supports the data exploration process by preserving an overview of the data during drill-down operations. The basic idea is to show portions of the data with a high level of detail while others are shown with a lower level of detail. Popular distortion techniques are hyperbolic and spherical distortions; these are often used on hierarchies or graphs but may also be applied to any other visualization technique. An overview of distortion techniques is provided in [117] and [118]. Examples of distortion techniques include Bifocal Displays [119], Perspective Wall [120], Graphical Fisheye Views [121–122], Hyperbolic Visualization [123–124], and Hyperbox [125]. Figures usually show the effect of distorting part of a scatterplot matrix to display more detail from one of the plots while preserving context from the rest of the display.

4.3.2.5 Brushing and Linking

Brushing is an interactive selection process that is often, but not always, combined with *linking,* a process for communicating the selected data to other views of the data set. There are many possibilities to visualize multi-dimensional data, each with their own strengths and weaknesses. The idea of linking and brushing is to combine different visualization methods to overcome the shortcomings of individual techniques. Scatterplots of different projections, for example, may be combined by coloring and linking subsets of points in all projections. In a similar fashion, linking and brushing can be applied to visualizations generated by all visualization techniques described above. As a result, the brushed points are highlighted in all visualizations, making it possible to detect dependencies and correlations. Interactive changes made in one visualization are automatically reflected in the other visualizations. Note that connecting multiple visualizations through interactive linking and brushing provides more information than considering the component visualizations independently [126–130].

4.4 Principal Components Analysis (PCA)

PCA is among the oldest and most widely used multivariate technique, originated by Pearson (1901) and independently by Hotelling (1933). The objective of PCA is the representation of the objects in the new principal component PC-coordinate space. The PCA performs a *dual objective:* a transformation into a more relevant co-ordinate system which lies directly in the center of the data swarm of objects, and a dimensionality reduction using only the few principal components which reflect the *structure* in the data.

The m-D $\rightarrow$ 2-D reduction is only a particularly useful *conceptual image* of the dimensionality reduction potential of PCA. In fact, a large number of variables can often be compressed into relatively small number, e.g., 2, 3, 4, 5 PCs or so, which allows us to actually *see* the data structure regadless of the original dimensionality of the data matrix with but a few plots. Principal components are thought to reflect underlying processes that have created the correlations among variables. In PCA the variables are treated equally, i.e., they are not divided into dependent and independent variables, as in regression analysis. The specific goals of PCA are to summarizee patterns of correlations among observed variables, to reduce a large number of observed variables to a smaller number of latent variables, to provide an operational definition (a regression equation) for an underlying process by using observed variables, or to test a theory about the nature of underlying processes. Some or all these goals may be the focus of a particular research project.

Latent variables: PCA and FA have considerable utility in reducing numerous variables down to a few latent variables, i.e., principal components (PCA) or components (FA). Mathematically, PCA produce several linear combinations of observed variables Because the number of principal components is usually far fewer than the number of observed variables, there is considerable parsimony in using the principal component analysis.

Steps in PCA: steps in PCA include selecting and measuring a set of variables, preparing the correlation matrix to perform PCA, extracting a set of principal components or components from the correlation matrix, determining the number of principal components, probably rotation principal components or components to increase interpretability, and, finally, interpreting the results. Although there is relevant statistical consideration to most of these steps, an important test of the analysis is its interpretability.

The difference between FA and PCA: theoretically, the difference between FA and PCA lies in the reason that variables are associated with a principal component or component. Components are thought to "cause" variables -the underlying construct the component is what produces scores on the variables. Principal components are simply aggregates of correlated variables. In that sense, the variables "cause"-or produce- the principal components. There is no underlying theory about which variables should be associated with which components; they are simply empirically associated. It is understood that any labels applied to derived components are merely convenient descriptions of the combination of variables associated with them, and do not necessarily reflect some underlying process [131].

4.4.1 Basics of Principal Components

The PCA method can be summarized as a technique of transforming the original variables $x_1, x_2, ..., x_m$ into new, uncorrelated latent variables $y_1, y_2, ..., y_m$ called the *principal components*. Each principal component y_i is a linear combination of the original variables $x_1, x_2, ..., x_m$ [25]. One measure of the amount of information conveyed by each principal component is its variance. For this reason the principal components are arranged in order of decreasing variance. Thus the most informative principal component is the first, y_1, and the least informative is the last, y_m.

To reduce the dimensionality of the actual problem means to reduce the number of variables without losing much of the information. This can be achieved by chosing to analyze only the first few principal components. The principal components not analyzed convey only a small amount of information since their variances are small. As the principal components are not intercorrelated, thus instead of analyzing a large number of original variables $x_i, i = 1,..., m$, with complex interrelatonships, one can analyze a small numer of uncorrelated principal components. PCA is considered to be an exploratory technique that may be useful in gaining a better understanding of the interrelationships among the variables.

The first principal component y_1 is expressed with $y_1 = a_{11}x_1 + a_{12}x_2 + ... + a_{1m}x_m = \mathbf{a}_1^T\mathbf{x}$. To find the coefficients defining the first principal component we need to choose the elements of the vector $\mathbf{a}_1$ so as to maximize the variance of y_1, i.e. $D(y_1)$ subject to the constraint $\mathbf{a}_1^T\mathbf{a}_1 = 1$. The variance of y_1 is equal to $D(y_1) = D(\mathbf{a}_1^T\mathbf{x}) = \mathbf{a}_1^T\mathbf{S}\mathbf{a}_1$.

To maximize a function $D(y_1)$ of several variables subject to one or more constraints, the method of Lagrange multipliers is used. This leads to the solution that $\mathbf{a}_1$ is the eigenvector of the covariance matrix $\mathbf{S}$ corresponding to the largest eigenvalue [25]. The second principal component y_2 is defined as linear combination of the original variables $y_2 = a_{21}x_1 + a_{22}x_2 + ... + a_{2m}x_m = \mathbf{a}_2^T\mathbf{x}$ that accounts for a maximal proportion of the remaining variance subject to being uncorrelated with the first principal component. The y_2 has the greatest variance subject to the following conditions, $\mathbf{a}_2^T\mathbf{a}_2 = 1$ and $\mathbf{a}_2^T\mathbf{a}_1 = 0$. The second condition ensures here that y_1 and y_2 are uncorrelated.

Similarly, the jth principal component is that linear combination $y_j = \mathbf{a}_j^T\mathbf{x}$ which has greatest variance subject to the conditions $\mathbf{a}_j^T\mathbf{a}_j = 1$ and $\mathbf{a}_j^T\mathbf{a}_i = 0$ for $i < j$. The other components are derived in similar fashion, with $\mathbf{a}_j$ being the eigenvector of $\mathbf{S}$ associated with the jth largest eigenvalue. If the eigenvalues of the covariance matrix $\mathbf{S}$ are $\lambda_1, \lambda_2,..., \lambda_m$, then since $\mathbf{a}_i^T\mathbf{a}_i = 1$, the variance of the ith principal component is given by λ_i.

The total variance of m principal components will equal the total variance of the original variables that $\sum_{i=1}^{m} \lambda_i = \text{trace}(\mathbf{S})$. Consequently, the jth principal component accounts for a proportion P_j of the total variation on the original data, where $P_j = \dfrac{\lambda_j}{\text{trace}(\mathbf{S})}$. The first p principal components, where $p < m$

account for P^* of the total variation in the original data, where $P^* = \dfrac{\sum\limits_{i=1}^{p} \lambda_i}{\text{trace}(\mathbf{S})}$.

In practice, however, it is often usual to extract the components from the correlation matrix, $\mathbf{R}$.

If the vectors $\mathbf{a}_1, \mathbf{a}_2, ..., \mathbf{a}_m$, which define the principal components, are used to form a $m \times m$ matrix, $\mathbf{A} = [\mathbf{a}_1, \mathbf{a}_2, ..., \mathbf{a}_m]$ and the eigenvalues $\lambda_1, \lambda_2, ..., \lambda_m$ are arranged in a diagonal matrix, $\ddot{\mathbf{e}}$, then it is easy to express the covariance matrix of the original variables in compact form $\mathbf{S} = \mathbf{A}\lambda\mathbf{A}^T$.

By rescaling the vectors $\mathbf{a}_1, \mathbf{a}_2, ..., \mathbf{a}_m$ so that the sum of squares of their elements is equal to the corresponding eigenvalue, λ_i, rather than unity, that is calculating $\mathbf{a}_i^* = \lambda_i^{1/2}\mathbf{a}_i$, then the covariance matrix $\mathbf{S}$ may be written more simply as $\mathbf{S} = \mathbf{A}^*\mathbf{A}^{*T}$, where $\mathbf{A}^* = [\mathbf{a}_1^*, \mathbf{a}_2^*, ..., \mathbf{a}_m^*]$. The elements of $\mathbf{A}^*$ are such that the coefficients of the more important components are scaled up compared to those of the less important components, a scaling which is intuitively reasonable. The rescaled vectors have many other advantages, since their elements are analogous to *component loadings*. The rescaled coefficients give correlations between the components and the original variables. These rescaled coefficients are often presented as the own results of a PCA.

4.4.2 Geometrical Interpretation of Graphical Tools of PCA

Most researchers seldom use a single criterion only in determining how many components to extract. Instead, they initially use a criterion such as the latent roots as a guideline for the first attempt at interpretation. After the components have been interpreted, the practicality of the components is assessed. Components identified by other criteria are also interpreted. Thus, several components solutions with differing numbers of components are examined before the structure is well defined [60].

4.4.2.1 Criteria for the Number of Components to Extract

When a large set of variables is analyzed, the method first extracts the combinations of variables explaining the greatest amount of variance and then proceeds to combinations that account for smaller and smaller amounts of variance. In deciding how many components to extract, one generally begins with some predetermined criterion, such as the percentage of variance or latent root criterion, to arrive at a specific number of components to extract. Even an exact quantitative basis for deciding the number of components to extract has not been developed, the following stopping criteria for the number of components to extract are currently being utilized.

Latent Root Criterion (Kaiser's Rule): This technique is simple to apply to any components analysis. The rationale for this criterion is that any individual component should account for the variance of at least a single variable if it is to be retained for interpretation. Each variable contributes a value of 1 to the total eigenvalue. Thus, only the components having latent roots or eigenvalues greater than 1 are considered significant, and all components with latent roots less than 1 are considered insignificant and are disregarded. Using the eigenvalue for establishing a cutoff is most reliable when the number of variables is between 20 and 50.

A Priori Criterion: When applying this criterion, one already knows how many components to extract before undertaking PCA. This approach is useful when testing a theory or hypothesis about the number of components to be extracted.

Percentage of Variance Criterion: This criterion is based on achieving cumulative percentage of total variance extracted by successive components. The purpose is to ensure practical significance for derived components by ensuring that they explain at least a specified amount of variance. In natural sciences the procedure usually should be stopped until the extracted components account for at least 95 percent of the variance. In contrast, in the social science and medicine where information is often less precise, it is uncommon to consider a solution that accounts for 60 percent of the total variance as satiscomponenty.

Scree Test Criterion: This test identifies the optimum number of components that can be extracted before the amount of unique variance begins to dominate the common variance structure. The scree test is derived by plotting the latent roots against the number of components in their order of extraction, and the shape of resulting curve is used to evaluate the cutoff point (Fig. 4.21). Starting with the first component, the plot of eigenvalues slopes steeply downward initially and then slowly becomes an approximately horizontal line. The point of eigenvalues at which the curve first begins to straighten out is considered to indicate the maximum number of components to extract. It is the index value i corresponding to an "elbow" in the curve, this point being considered to be where "large" eigenvalues cease and "small" eigenvalues begin. The word *scree*, first used by Cattell (1966), is usually defined as the rubble at the bottom of a cliff. When using the scree plot, we must determine which eigenvalues form the "cliff" and which form the "rubble". We keep the components that make up the cliff plus the first component of the rubble. As a general rule, the scree test results in at least one and sometimes two or three *more* components being considered for inclusion than does the latent root criterion [60].

4.4.2.2 Plot of the Components Loadings – Map of Variables

Component extraction yields a solution in which observed variables are vectors that run from the origin to the point indicated by the coordinate system (Fig. 4.21). The components serve as axes for the system. The coordinates of each point are the entries from the loading matrix for the variable. If there are three components, then the space has three axes and three dimensions, and each observed variable is positioned by three coordinates. The components are orthogonal, the component axes are all at right angles to one another and the coordinates of the variable points are correlation between the common components and the observed variables. Loadings vectors, or simply loadings, can be viewed as the bridge between the variable space and the principal component space. As with score vectors, loadings vectors can be plotted against each other. These plots are equally valuable for interpretation in their own right, but specially when compared with their corresponding score plots. The loading plot provides a projection view of the *inter-variable relationships* (variable similarities). The loading plot shows how much each variable contributes to each PC. Recall tha PCs can be represented as linear combinations of the original unit vectors. The loading themselves are the coefficients in these linear combinations. Each variable can contribute to more than one PC. Correlations (components loadings) are read directly from these graphs by projecting perpendicular lines from each variable point to each of the component axes.

One of the primary goals of PCA and FA, and the motivation behind extraction, is to discover the minimum number of component axes needed to reliably position variables. A second major goal, and motivation behing rotation, is to discover the meaning of the components that underline response to observed variables.

The variables form a "swarm" in which variables that are correlated with one another form a cluster of points. The goal is to shoot an axis to the swarm of points. With luck, the swarms are about 90° away from one another so that an orthogonal solution is indicated. And with lots of luck, the variables cluster in just a few swarms with empty spaces between them so that the components axes are novely defined.

Components loadings a_i is the means of interpreting the role each variable plays in defining each component y_i. Components loadings a_i are the correlation of each variable and the component. Loadings indicate the degree of correspondence between variable and the component, with higher loadings making the variable representative of the component. The ones assesses the need to respecify the PCA model owing to (1) the deletion of a variable(s) from the analysis, (2) the desire to employ a different rational metod for interpretation, (3) the need to extract a different numer of components, or (4) the desire to change from one extraction metod to another. Respecification of a component model is accomplished by returning to the extraction stage, extracting components, and interpreting them again.

To summarize the criteria for the significance of components loadings, the following guidelines can be stated: (1) the larger the sample size, the smaller the loading to be considered significant; (2) the larger the number of variables been analyzed, the smaller the loading to be considered significant; (3) the larger the number of components, the larger size of the loading on later components to be considered significant for interpretation.

Interpreting the complex interrelationships represented in a component matrix is no simple matter. The columns of numbers are the components loadings for each variable on each component. The interpretation should start with the first variable on the first component looking for the highest loading for that variable on any component. When the highest loading is identified, attention then focuses on the second variable and, again looking for the highest loading for that variable on any component. When each variable has only one loading on one component that is considered significant, the interpretation of the meaning of each component is simplified considerebly. In practice, however, many variables may have several moderate-size loadings, all of which are significant, and the job of interpreting the components is much more difficult. The difficulty arises because a variable with several significant loadings must be considered in interpreting all the components on which it has a significant loading. Most component solutions do not result in a simple structure solution i.e. a single high loading for each variable on only one component. Ultimately, the objective is to minimize the number of significant loadings on each row of the components matrix. A variable with several high loading is a candidate for deletion.

Interpretation of plot of component loadings may be written in following items:

(1) Variables with a high degree of systematic variation typically have large absolute variances, and consequently large loadings. In a two-vector loading plot they lie far away from the origin.

(2) Variables of little importance lie near origin. This is a general statement, which is always scaling dependant however.

(3) When assessing importance, it is mandatory also to consider the proportions of the total explained variance along each component.

If e.g. PC1 explains 75% and PC2 only 5%, then variables with large loadings in PC1 are *much more important* than those with large loadings in PC2 - in fact 15 times as important.

(4) Variables close to each other, situated out towards the periphery of the loading plots, covary strongly, proportionally to the degree distanced from the PC-origin.

(5) If the variables lie on the same side of the origin, they covary in a positive sense, i.e. they have a positive correlation. If they lie on opposite sides of the origin, *more less* (same latitude here) along a straight line through the PC-origin, they are negatively correlated.

(6) Correlation is not automatically reflecting a causal relation, interpretation is always necessary.

(7) Loadings which are at 90 degrees to each other through the origin, are independent. Loadings close to a PC axis are significant only to that PC. Variables with a large loading on two PCs are significant on both PCs.

4.4.2.3 Scatterplot of Components Scores – Map of Objects

Scatterplot of components scores is simply any two pair of components scores vectors plotted against each other (Fig. 4.22). It is important to remember that the score vectors are only the "footprints" of the objects projected down onto the Principal Components. Plotting the components scores corresponds to plotting the objects in a pertinent PC sub-space. Components score plots are typically referred to by their score designation, for example, for the PC1-PC2 score sub-space. Components score plots can be viewed as particularly useful 2-D "windows" into PC-space, where one observes how the objects are related to one another. The PC-space may certainly not always be fully visualized in just one 2-D plot, in which case two, or more components score plots is all what we need. We are of course necessarily restricted to 2- or 3-dimensional representation when plotting on paper or working computer screen.

The most commonly used plot in multivariate data analysis is the components score vector for PC1 versus the score for PC2. This is easy to understand, since these are two dimensions along which the data swarm exhibits the largest and the second largest variances. Components scores for PC1 are along the "x-axis" abscissa and the components scores for PC2 are along the "y-axis" ordinate. Notice that *objects* are plotted in the score plot in their relative dispositions with respect to this score plane, and that we have here used the very powerful option of having one, two (or more) of the object name characters serving as plotting symbol. This option will greatly facilitate *interpretation* of the meaning of the inter-objects dispositions.

Components scores are, in fact, composite measures of each component computed for each object. Conceptually the components score represents the degree to which each individual scores high on the group of items that have high loadings on a component. Thus, higher values on the variables with high loadings on a component will result in a higher component score. The one key characteristic that differentiates a component score from a summated scale is that the component score is computed based on the component loadings of all variables on the component, whereas the summated scale is calculated by combining only selected variables. Therefore, although the one is able to characterize a component by the variables with the highest loadings, consideration must also be given to the loadings of other variables, albeit lower, and their influence on the components score.

Rules of thumb concerning scatterplots of component scores may be written in several items. These rules of thumb are very general and there are many exceptions. An advice is to start all data analysis following these simple rules, but always look for posible deviations. After an initial analysis, one may for example find that higher-order score plots are necessary for interpretation after all. There are also many interesting cases in which the problem-specific information is to be found "swamped" in many "other" components. Sometimes, both the first, largest components, and the truly insignificant higher-order ones have to be discarded: the particular subject matter can be revealed in one, of the intermediate components.

(1) *Location of objects:* Objects close to the cordinate origin are the most typical, the most "average". The ones far away from the origin may be extremes, or even outliers, but they may also be legitimate end-members. It is researcher problem-specific responsibility to decide in these issues.

(2) *Similarity of objects:* Objects close to each other are similar, those far away from each other are dissimilar.

(3) *Objects in a cluster:* Objects located in a clear groups are similar to each other and dissimilar to other groups. Well-separated groups may indicate that a model for each separate group will be appropriate.

(4) *Isolated, secluded objects:* Isolated objects may be outliers – objects that do not fit in with the rest.

(5) *Outlying objects:* In the ideal case, objects typically should be "well-spread" over the whole plot. If they are not, researcher problem/specific, domain knowledge must be brought in.

(6) *Labeling objects:* By using well-reflected object descriptive names or labels that are related to the most important external properties of

the different objects, one may better understand the meaning of the principal components as directly related to the problem context.

(7) *Interpretation of an object location:* The layout of the overall object structure in score plots must be interpreted by studying the corresponding loading plots in so called a *biplot* (Fig. 4.23) which is a reflection that the technique allows the display of both information about the variables as indicated by their variances and covariance *and* the relationships between objects as indicated by particular measures of interobjects distance. The length of the arrows reflect the variances of the corresponding variable, and the angles between them indicate the size of their correlations, small angles corresponding to high correlations. The relative positions of the points corresponding to objects indicate similarities and differences.

(8) Always the same principal component should be used as abscissa on the x-axis in all the score plots. In this way all the other PC-phenomena against the same yardstick will be "measured". This will greatly help getting the desired *overview* of the compound data structure.

(9) The pr incipal component that has the largest "problem relevant" variance as this basis (x-axis) plotting component should be used. For many applications this will turn out to be PC1, but it is entirely possible in other cases that PC1 lies along a direction that for some problem-specific reason is not interesting. In general PC1 describes the largest structural variation in any data set, and in many situations this – *per se* – is often an "interesting" feature, but this does not necessarily mean that this variation always is the most important for our particular interpretation purpose. Correlation is neither *per se* equivalent to causality.

4.4.2.4 Biplot

The biplot is graph that represents both variables and objects together in two dimensions (Fig. 4.23). Information about the variables is provided by the variable projections or axes. The orientation of the axes displays the relationship between the variables and the components. For instance, a variable that lines up with a component axis will load heavily on that component and a variable axis that is almost perpendicular to a component axis will not load heavily on that component. In addition variable axes in close proximity indicate a high correlation among the variables. The common biplot is based upon the raw or unstandardized components. This graph is good at representing distances between the data points while the biplot based on standardized components is better at representing the relationships among the variables.

Useful potential pifalls of PCA application: Hopefully we have not been put off completely by this list of possible errors, some of which cannot even be detected when they arise. *Experience – and still more experience* is the only thing that will help us through many of these pitfalls.

(1) *Data set does not contain the information expected/needed.* It can happen that in a direct PCA the lack of reflection of the entire problem domain can easily lead to misinterpretations. It all boils down to knowing the maximum about the problem and data set otherwise the interpretations could not make sense.

(2) *Too few PCs in the model are used.* It means that the potential information is not fully exploiting in data. The total potential information is lost out. It is not the worst mistake what can be made, but should of course be avoided through careful analysis.

(3) *Too many PCs in the model are used.* It can brings a serious mistake as noise in the suggested model may be included. The noise contribution must lead to erroneous interpretations, the analysis will always be wrong, at least partly.

(4) *Outliers which are truly due to erroneous data were not removed.* Obviously this also gives an *invalid* model. Errors instead of the interesting variations were modeled to some significant extent.

(5) *Outliers that contain important information were removed.* To put it bluntly,we can miss something important. The model will be inferior, as it will not describe all the phenomena hidden in the data set in this case.

(6) *The score plots were not explored sufficiently.* If the score plots were not studied carefully, we may miss important clues. Errors 4 and 5 above are connected to this mistake.

(7) *Lodings are interpreted with the wrong number of PCs.* This may give rise to serious misinterpretations. We may even remove variables that are important because they seen to show up as outliers. We should remember that the loadings constitute the bridge between variable space and PC space. If we have chosen the "wrong" PC space, the "bridge" will not take us to the right place. The bridge will be the wrong one, then the loadings cannot be trusted in any way.

(8) *The standard diagnostics in computer output are relied too much without thinking.* This is a very common mistake – and the most serious of them all. The diagnostics may be adequate and helpful most of the time, but one must always use one's own problem understanding and check the consequences. We should remember that the computer program has no knowledge of our *specific problem*

– it runs along standard procedures that may not apply to our specific data set.

(9) *The "wrong" data pretreatment was used.* This is rather tricky point. Pre-processing/pre-treatment of the data is essential for relevant and valid modeling results. The correct type of pre-treatment is generally given by the type of problem, but this is certainly up to us to decide. The software unfortunately cannot be made clairvoyant. The wrong pre-processing may well nearly always give rise to misinterpretations.

Problem 4.1 *Protein consumption in 25 countries of Europe*

These data measure protein consumption in twenty-five European countries for nine food groups as stated in the **Exercise B4.18**. It is possible to use multivariate methods to determine whether there are groupings of countries and whether meat consumption is related to that of other foods. PCA of all nine variables shows that the diets of the countries can be divided into several groups according to their geographic locations.

Data: Data measure protein consumption in twenty-five European countries for nine food groups. The variables in this data set represent food groups: *Country* – country in Europe, *RedMeat* – % of protein from red meat, *WhiteMeat* – % of protein from white meat, *Eggs* – % of protein from eggs, *Milk* – % of protein from milk, *Fish* – % of protein from fish, *Cereals* – % of protein from cereals, *Starch* – % of protein from starches (e.g. potatoes), *Nuts* – % of protein from nuts, *Fr&Veg* – % of protein from fruits and vegetables.

Country	ReadMeat	WhiteMeat	Eggs	Milk	Fish	Cereals	Starch	Nuts	Fr&Veg
Albania	10.1	1.4	0.5	8.9	0.2	42.3	0.6	5.5	1.7
Austria	8.9	14.0	4.3	19.9	2.1	28.0	3.6	1.3	4.3
Belgium	13.5	9.3	4.1	17.5	4.5	26.6	5.7	2.1	4.0
Bulgaria	7.8	6.0	1.6	8.3	1.2	56.7	1.1	3.7	4.2
Czechoslovakia	9.7	11.4	2.8	12.5	2.0	34.3	5.0	1.1	4.0
Denmark	10.6	10.8	3.7	25.0	9.9	21.9	4.8	0.7	2.4
Germany	8.4	11.6	3.7	11.1	5.4	24.6	6.5	0.8	3.6
Finland	9.5	4.9	2.7	33.7	5.8	26.3	5.1	1.0	1.4
France	18.0	9.9	3.3	19.5	5.7	28.1	4.8	2.4	6.5
Greece	10.2	3.0	2.8	17.6	5.9	41.7	2.2	7.8	6.5
Hungary	5.3	12.4	2.9	9.7	0.3	40.1	4.0	5.4	4.2
Ireland	13.9	10.0	4.7	25.8	2.2	24.0	6.2	1.6	2.9
Italy	9.0	5.1	2.9	13.7	3.4	36.8	2.1	4.3	6.7

Netherlands	9.5	13.6	3.6	23.4	2.5	22.4	4.2	1.8	3.7
Norway	9.4	4.7	2.7	23.3	9.7	23.0	4.6	1.6	2.7
Poland	6.9	10.2	2.7	19.3	3.0	36.1	5.9	2.0	6.6
Portugal	6.2	3.7	1.1	4.9	14.2	27.0	5.9	4.7	7.9
Romania	6.2	6.3	1.5	11.1	1.0	49.6	3.1	5.3	2.8
Spain	7.1	3.4	3.1	8.6	7.0	29.2	5.7	5.9	7.2
Sweden	9.9	7.8	3.5	4.7	7.5	19.5	3.7	1.4	2.0
Switzerland	13.1	10.1	3.1	23.8	2.3	25.6	2.8	2.4	4.9
UK	17.4	5.7	4.7	20.6	4.3	24.3	4.7	3.4	3.3
USSR	9.3	4.6	2.1	16.6	3.0	43.6	6.4	3.4	2.9
WGermany	11.4	12.5	4.1	18.8	3.4	18.6	5.2	1.5	3.8
Yugoslavia	4.4	5.0	1.2	9.5	0.6	55.9	3.0	5.7	3.2

The data set *Protein*:

Software: NCSS2000, STATISTICA, SCAN and UNSCRAMBLER.

Solution:

Descriptive statistics section: This report lets us compare the relative sizes of the standard deviations. In this data set, they are all about the same size, so we could analyze either the correlation or the covariance matrix. We will analyze the correlation matrix. Count, mean, and standard deviation are the familiar statistics of each variable. They are displayed to allow us to make sure that we have specified the correct variables. Using missing value imputation or robust estimation will change these values. The communality shows how this variable is predicted by the retained components. It is the R^2 that would be obtained if this variable were regressed on the components that were kept. Here all components were kept, so the R^2 is one.

Variables	Count	Mean	Standard Deviation	Communality
ReadMeat	25	9,83	3,35	1,00
WhiteMeat	25	7,90	3,69	1,00
Eggs	25	2,94	1,12	1,00
Milk	25	17,11	7,11	1,00
Fish	25	4,28	3,40	1,00
Cereals	25	32,25	10,97	1,00
Starch	25	4,28	1,63	1,00
Nuts	25	3,07	1,99	1,00
Fr&Veg	25	4,14	1,80	1,00

Correlation section: The report gives the correlations for a test of the overall correlation structure in the data. In this example, we notice several high correlation values.

Correlations: The simple correlations between each pair of variables. Note that using the missing value imputation or robust estimation options will affect the correlations in this report. When the above options are not used, the correlations are constructed from those observations having no missing values in any of the specified variables.

Φ value: Φ is the Gleason-Staelin redundancy measure of how interrelated the variables are. A zero value of Φ means that there is no correlation among the variables, while a value of 1 indicates perfect correlation among the variables. This coefficient may have a value less than 0.5 even when there is obvious structure in the data, so care should to be taken when using it. This statistic is especially useful for comparing two or more sets of data. The formula for

computing is $\Phi = \sqrt{\dfrac{\sum\limits_{i=1}^{m}\sum\limits_{j=1}^{m} r_{ij}^{2} - m}{m(m-1)}}$. The Gleason-Staelin redundancy measure,

Φ is 0.4137, which is sufficiently high to prove correlation in data. There is apparently some correlation structure in this data set that can be modeled. If all the correlations were small, there would be no need for a PCA.

Value ln(Det$|R|$): This is the logarithm (base e) of the determinat of the correlation matrix. If you used the covariance matrix, this is the logarithm (base e) of the determinat of the covariance matrix.

Bartlett test, DF, Prob: This is Bartlett's sphericity test for testing the null hypothesis that the correlation matrix is an identity matrix, i.e. all correlations are zero. If a calculated significance level p (Prob) value is greater than 0.05, a PCA should not be performed on the data. The test is valid for large samples, $n > 150$. It uses χ^2 distribution with $m(m-1)$ degrees of freedom. Note that this test is only available when a correlation matrix is analyzed. The formula for computing this test is $\chi^2 = \dfrac{(11+2m-6n)}{6}\ln|R|$.

Variables	Read Meat	White Meat	Eggs	Milk	Fish	Cereals	Starch	Nuts	Fr&Veg
ReadMeat	1,000	0,153	0,586	0,503	0,061	−0,500	0,135	−0,349	−0,074
WhiteMeat	0,153	1,000	0,620	0,282	−0,234	−0,414	0,314	−0,635	−0,061
Eggs	0,586	0,620	1,000	0,576	0,066	−0,712	0,452	−0,560	−0,046

Milk	0,503	0,282	0,576	1,000	0,138	−0,593	0,222	−0,621	−0,408
Fish	0,061	−0,234	0,066	0,138	1,000	−0,524	0,404	−0,147	0,266
Cereals	−0,500	−0,414	−0,712	−0,593	−0,524	1,000	−0,533	0,651	0,047
Starch	0,135	0,314	0,452	0,222	0,404	−0,533	1,000	−0,474	0,084
Nuts	−0,349	−0,635	−0,560	−0,621	−0,147	0,651	−0,474	1,000	0,375
Fr&Veg	−0,074	−0,061	−0,046	−0,408	0,266	0,047	0,084	0,375	1,000

$\Phi = 0,4138$; $\mathrm{Ln(Det|R|)} = -5,705$; Bartlett Test $= 115,05$; DF $= 36$; Prob $= 0,000$.

Bar chart of absolute correlation section: This chart graphically displays the absolute values of the correlations. It lets us quickly find high and low correlations.

Variables	Read Meat	White Meat	Eggs	Milk	Fish	Cereals	Starch	Nuts	Fr&Veg
ReadMeat		‖‖	‖‖‖‖‖	‖‖‖‖‖	‖	‖‖‖‖	‖	‖‖‖	‖
WhiteMeat	‖‖		‖‖‖‖‖	‖‖	‖‖	‖‖‖‖	‖‖‖	‖‖‖‖‖	‖
Eggs	‖‖‖‖‖	‖‖‖‖‖		‖‖‖‖‖	‖	‖‖‖‖‖‖	‖‖‖‖	‖‖‖‖‖	‖
Milk	‖‖‖‖	‖‖	‖‖‖‖‖		‖	‖‖‖‖	‖‖	‖‖‖‖‖	‖‖‖
Fish	‖	‖‖	‖	‖		‖‖‖‖	‖‖‖	‖	‖‖
Cereals	‖‖‖‖	‖‖‖	‖‖‖‖‖‖	‖‖‖‖‖	‖‖‖‖		‖‖‖‖	‖‖‖‖‖	‖
Starch	‖	‖‖‖	‖‖‖‖	‖‖	‖‖‖‖	‖‖‖‖		‖‖‖‖	‖
Nuts	‖‖‖	‖‖‖‖‖	‖‖‖‖‖	‖‖‖‖‖	‖	‖‖‖‖‖	‖‖‖‖		‖‖‖
Fr&Veg	‖	‖	‖	‖‖‖‖	‖‖	‖	‖	‖‖‖	

Eigenvalues: Often, these are used to determine how many components to retain. When the PCA is run on the correlations, one rule-of-thumb is to retain those components whose eigenvalues are greater than 1. The sum of the eigenvalues is equal to the number of variables. When the PCA is run on the covariances, the sum of the eigenvalues is equal to the sum of the variances of the variables.

Individual and cumulative percents: The first column gives the percentage of the total variation in the variables accounted for by this component. The second column is the cumulative total of the percentage. Some authors suggest that the user pick a cumulative percentage, such as 80% or 90%, and keep enough components to attain this percentage.

Scree plot: This is a rough bar plot of the eigenvalues. It enables us to quickly note the relative size of each eigenvalue. Many authors recommend it as a method of determining how many components to retain.

Individual No.	Cumulative Eigenvalue	Individual Percent	Cumulative Percent	Scree Plot
1	4,006	44,5	44,5	|||||||||
2	1,635	18,2	62,7	||||
3	1,128	12,5	75,2	|||
4	0,955	10,6	85,8	|||
5	0,464	5,2	91,0	||
6	0,325	3,6	94,6		
7	0,272	3,0	97,6		
8	0,116	1,3	98,9		
9	0,099	1,1	100,0		

Eigenvectors: The eigenvectors $a_{ij}, j = 1,...,m$, are the weights that relate the scaled original variables, $x_i = (X_i - Mean_i) / Sigma_i$, to the components. For example, the first component, $Comp_i$, is the weighted average of the scaled variables, the weight of each variable given by the corresponding element of the first eigenvector. Mathematically, the relationship is given by $Comp_1 = a_{11}x_{11} + a_{12}x_{12} + ... + a_{1m}x_{1m}$. These coefficients $a_{ij}, j = 1,...,m$, may be used to determine the relative importance of each variable in forming the component. Often, the eigenfactors are scaled so that the variances of the component are equal to one. These scaled eigenvectors are given in the *Score Coefficients Section* described later.

Variables	PC1	PC2	PC3	PC4	PC5	PC6	PC7	PC8	PC9
ReadMeat	−0,303	−0,056	−0,298	0,646	0,322	0,460	0,150	0,020	0,246
WhiteMeat	−0,311	−0,237	0,624	−0,037	−0,300	0,121	−0,020	0,028	0,592
Eggs	−0,427	−0,035	0,182	0,313	0,079	−0,361	−0,443	0,491	−0,333
Milk	−0,378	−0,185	−0,386	−0,003	−0,200	−0,618	0,462	−0,081	0,178
Fish	−0,136	0,647	−0,321	−0,216	−0,290	0,137	−0,106	0,449	0,313
Cereals	0,438	−0,233	0,096	−0,006	0,238	−0,081	0,405	0,703	0,152
Starch	−0,297	0,353	0,243	−0,337	0,736	−0,148	0,153	−0,115	0,122
Nuts	0,420	0,143	−0,054	0,330	0,151	−0,447	−0,407	−0,184	0,518
Fr&Veg	0,110	0,536	0,408	0,462	−0,234	−0,119	0,450	−0,092	−0,203

Bar chart of absolute eigenvectors: This chart graphically displays the absolute values of the eigenvectors. It lets us quickly interpret the eigenvector structure. By looking at which variables correlate highly with a component, we can determine what underlying structure it might represent.

Variables	PC1	PC2	PC3	PC4	PC5	PC6	PC7	PC8	PC9
ReadMeat	|||||||	||	|||||	||||||||||||	||||||	||||||||||	||||			|||||
WhiteMeat	|||||||	|||||	||||||||||||			||||||	|||					|||||||||||
Eggs	|||||||||			||||	||||||	||	||||||||	|||||||||	||||||||||	||||||
Milk	||||||||	||||	|||||||			|||||	||||||||||||	|||||||||	||	||||
Fish	|||	||||||||||||	|||||||	|||||	||||||	|||	|||	||||||||	||||||
Cereals	|||||||||	|||||	||			||||	||	|||||||||	|||||||||||||	|||
Starch	||||||	||||||||	|||||	|||||||	||||||||||||||	|||	||||	|||	|||
Nuts	|||||||||	|||	||	||||||||	||||	||||||||	||||||||	||||	|||||||||||
Fr&Veg	|||	||||||||||	||||||||	||||||||||	|||||	|||	||||||||	||	|||||

Component loadings section: These are the correlations between the variables and principal components which are available in the numerical form and in the bar chart form.

Variables	PC1	PC2	PC3	PC4	PC5	PC6	PC7	PC8	PC9
ReadMeat	−0,606	−0,072	−0,316	0,632	0,219	0,262	0,078	0,007	0,077
WhiteMeat	−0,622	−0,303	0,663	−0,036	−0,204	0,069	−0,010	0,010	0,186
Eggs	−0,854	−0,045	0,193	0,306	0,054	−0,206	−0,231	0,168	−0,105
Milk	−0,756	−0,236	−0,410	−0,003	−0,136	−0,353	0,241	−0,028	0,056
Fish	−0,272	0,827	−0,341	−0,211	−0,198	0,078	−0,055	0,153	0,098
Cereals	0,876	−0,299	0,102	−0,006	0,162	−0,046	0,211	0,240	0,048
Starch	−0,595	0,451	0,258	−0,329	0,501	−0,084	0,080	−0,039	0,038
Nuts	0,841	0,183	−0,058	0,323	0,103	−0,255	−0,212	−0,063	0,163
Fr&Veg	0,221	0,686	0,433	0,451	−0,159	−0,068	0,235	−0,031	−0,064

Bar chart of absolute component loadings: This chart graphically display the absolute values of the principal components loadings. It lets us quickly interpret the correlation structure. By looking at which variables correlate highly with a principal component, we can determine what underlying structure it might represent.

Variables	PC1	PC2	PC3	PC4	PC5	PC6	PC7	PC8	PC9
ReadMeat	||||||||||||	||	||||||	||||||||||||	|||||	||||||	||			||
WhiteMeat	||||||||||||	||||||	||||||||||||			|||||	||					||||
Eggs	||||||||||||||||			||||	||||||	||	||||	|||||	||||	|||
Milk	||||||||||||||	|||||	|||||||||			|||	|||||||	|||||			||
Fish	|||||	|||||||||||||	||||||	|||||	||||	||	||	||||	||

(Continued)

Cereals																																														
Starch																																																									
Nuts																																																							
Fr&Veg																																																								

Components structure summary section: This report is provided to summarize the principal component structure. Variables with an absolute loading greater than the amount set in the *Minimum Loading* option are listed under each principal component. Using this report, we can quickly see which variables are related to each principal component. Notice that it is possible for a variable to have high loadings on several components.

	PC1	PC2	PC3	PC4	PC5	PC6	PC1	PC7	PC8	PC9
Cereals	Fish	WhiteMeat	ReadMeat	Starch						
Eggs	Fr&Veg	Fr&Veg	Fr&Veg							
Nuts	Starch	Milk								
Milk										
WhiteMeat										
ReadMeat										
Starch										

Score coefficients section: These are the coefficients that are used to form the components scores. The component scores are the values of the components for a particular row of data. These score coefficients are similar to the eigenvectors. They have been scaled so that the scores produced have a variance of one rather than a variance equal to the eigenvalue. This causes each of the factors to have the same variance.

Variables	PC1	PC2	PC3	PC4	PC5	PC6	PC7	PC8	PC9
ReadMeat	−0,151	−0,044	−0,280	0,662	0,473	0,807	0,288	0,058	0,781
WhiteMeat	−0,155	−0,185	0,587	−0,038	−0,441	0,212	−0,038	0,082	1,882
Eggs	−0,213	−0,028	0,171	0,321	0,116	−0,634	−0,851	1,440	−1,059
Milk	−0,189	−0,144	−0,363	−0,003	−0,294	−1,085	0,887	−0,239	0,566
Fish	−0,068	0,506	−0,303	−0,221	−0,426	0,240	−0,204	1,316	0,994
Cereals	0,219	−0,183	0,090	−0,006	0,350	−0,142	0,777	2,061	0,484
Starch	−0,149	0,276	0,229	−0,345	1,081	−0,259	0,293	−0,336	0,387
Nuts	0,210	0,112	−0,051	0,338	0,221	−0,784	−0,781	−0,539	1,646
Fr_Veg	0,055	0,419	0,384	0,473	−0,343	−0,208	0,863	−0,270	−0,645

Residual section: This report is useful for detecting outliers i.e. observations that are very different from the bulk of the data. To do this, two quantities are displayed: T^2 and Q_k. We will now define these two quantities: T^2 measures the combined variability of all the variables in a single observation:

$T^2 = [\mathbf{x} - \overline{\mathbf{x}}]^T \, \mathbf{S}^{-1} \, [\mathbf{x} - \overline{\mathbf{x}}]$ where $\mathbf{x}$ represents a m-variable observation, $\overline{\mathbf{x}}$ represents the m-variable mean vector, and $\mathbf{S}^{-1}$ represents the inverse of the covariance matrix. T^2 is not affected by a change in scale. It is the same whether the analysis is performed on the covariance or the correlation matrix. T^2 gives a scaled Mahallanobis distance measure of an individual observation from the overall mean. The closer an observation is to its mean, the smaller will be the value of T^2. If the variables follow a multivariate normal distribution, then the probability distribution of T^2 may be related to

the common F distribution using the formula: $T^2_{m,n,\alpha} = \dfrac{m(n-1)}{m-n} F_{m,n-m,\alpha}$. Using

this relationship, we can perform a statistical test at a given level of significance to determine if the observation is significantly different from the vector of means at α significance level. The other quantity shown on this report is Q_k which represents the sum of squared residuals when an observation is predicted using the first k principal components. Mathematically, the formula for Q_k is

$Q_k = [\mathbf{x} - \hat{\mathbf{x}}]^T \, [\mathbf{x} - \hat{\mathbf{x}}] = \sum\limits_{i=1}^{m} (x_i - {}_k\hat{x}_i)^2 = \sum\limits_{i=k+1}^{m} \lambda_i \, (mc_i)^2$ where ${}_k x_i$ refers to the

value of variable i predicted from the first k principal components, λ_i refers to the ith eigenvalue, and pc_i is the score of the ith principal component for this particular observation. An upper limit for Q_k is given by the formula:

$Q_\alpha = a \left[\dfrac{z_\alpha \sqrt{2bh^2}}{a} + \dfrac{bh(h-1)}{a^2} + 1 \right]^{1/h}$ where $a = \sum\limits_{i=k+1}^{m} \lambda_i$, $b = \sum\limits_{i=k+1}^{m} \lambda_i^2$,

$c = \sum\limits_{i=k+1}^{m} \lambda_i^3$, $h = 1 - \dfrac{2ac}{3b^2}$ and z_α is the upper normal deviate of area α if h

is positive or the lower normal deviate of area α if h is negative. The limit is valid for any value of k, whether too many or too few principal components are kept. Note that these formulas are for the case when the correlation matrix is being used. When the analysis is being run on the covariance matrix, the pc_i's must be adjusted. Notice that significant (starred) values of Q_k indicate observations that are not duplicated well by the first k principal components. These should be checked to see if they are valid. T^2 and Q_k provide an initial data screening tool.

Row	T_2	$T_2 Prob$	Q_0	Q_1	Q_2	Q_3	Q_8
1	16,67	0,3412	19,84	7,70	5,04	1,94	0,01
2	5,95	0,8930	5,99	3,96	2,88	1,09	0,05
3	2,99	0,9865	3,71	1,08	1,05	1,01	0,00
4	12,87	0,5094	13,26	3,44	1,75	1,72	0,09
5	4,57	0,9483	3,01	2,87	2,51	1,08	0,02
6	10,07	0,6644	8,44	2,85	2,76	2,20	0,23
7	7,21	0,8293	6,21	4,19	3,99	2,29	0,07
8	11,35	0,5909	10,30	7,85	7,50	3,30	0,02
9	13,85	0,4610	8,73	6,51	5,89	5,89*	0,30
10	12,71	0,5175	11,76	6,74	5,74	4,96	0,15
11	10,40	0,6454	7,69	5,57	4,90	1,24	0,29
12	5,33	0,9200	9,14	2,04	1,46	1,46	0,01
13	9,35	0,7067	5,28	2,93	2,77	2,75	0,54*
14	5,76	0,9016	5,08	2,39	1,55	0,97	0,07
15	5,29	0,9218	6,03	5,08	4,41	1,50	0,02
16	10,05	0,6655	4,70	4,69	4,41	2,23	0,05
17	15,70	0,3790	22,85	19,94*	1,55	1,55	0,07
18	4,56	0,9487	9,49	1,89	0,63	0,63	0,11
19	10,82	0,6209	9,99	8,27	1,75	1,48	0,23
20	5,41	0,9168	5,87	3,21	3,16	1,52	0,01
21	6,94	0,8435	4,04	3,20	2,64	2,62	0,00
22	9,87	0,6759	9,04	6,03	6,02	4,69	0,02
23	8,62	0,7494	4,78	4,17	4,16	4,02	0,01
24	3,74	0,9720	5,51	1,13	1,04	0,39	0,00
25	5,93	0,8937	15,25	2,13	1,05	1,01	0,02

Component score section: This report presents the individual principal component scores scaled so each column has a mean zero and a standard deviation of one. These are the values that are plotted in the plots to follow. Remember, that there is one row of score values for each object, observation and one column for each principal component that was kept.

Row	PC1	PC2	PC3	PC4	PC5	PC6	PC7	PC8	PC9
1	1,74	−1,28	−1,66	0,24	0,03	1,81	−0,91	−2,23	−0,33
2	−0,71	−0,81	1,26	0,17	−1,37	−0,38	−0,35	0,74	−0,69
3	−0,81	0,12	0,20	0,53	1,11	0,51	−0,38	0,60	−0,11

4	1,57	−1,02	0,14	0,22	−0,71	1,22	0,89	2,37	−0,95
5	−0,19	−0,47	1,13	−0,47	0,38	1,44	0,60	−0,04	−0,47
6	−1,18	0,22	−0,71	−0,99	−1,10	0,30	−0,43	1,82	1,53
7	−0,71	0,35	1,23	−1,16	0,62	1,14	−1,06	0,48	−0,83
8	−0,78	−0,47	−1,93	−1,45	0,05	−1,46	1,39	−0,66	−0,42
9	−0,74	0,61	0,00	2,00	0,37	1,58	1,82	0,07	1,73
10	1,12	0,78	−0,83	1,84	−0,59	−2,01	−0,28	0,90	1,23
11	0,73	−0,64	1,80	−0,22	−0,06	−0,95	−1,47	−0,43	1,71
12	−1,33	−0,60	−0,02	0,44	1,49	−0,85	−0,06	−0,07	−0,25
13	0,77	0,31	0,12	1,25	−1,18	−0,38	0,29	0,24	−2,33
14	−0,82	−0,71	0,72	−0,13	−1,12	−0,52	−0,12	−1,35	0,83
15	−0,49	0,64	−1,60	−1,16	−0,61	0,10	−0,08	0,31	−0,47
16	−0,06	0,42	1,39	−0,47	−0,03	−1,03	2,42	−0,56	−0,70
17	0,85	3,35	0,04	−0,91	−0,57	1,22	0,09	−0,60	0,84
18	1,38	−0,88	0,07	−0,63	0,47	−0,23	−0,26	0,08	1,07
19	0,66	2,00	0,49	0,37	0,76	−1,17	−1,15	−0,69	−1,51
20	−0,82	−0,16	−1,21	−0,75	−1,20	−0,08	−1,04	0,21	−0,34
21	−0,46	−0,59	−0,15	1,20	−1,22	0,16	0,98	−1,55	0,21
22	−0,87	−0,07	−1,09	1,77	1,59	0,17	−1,25	0,70	−0,42
23	0,39	−0,09	−0,35	−0,95	2,45	−0,33	1,10	0,15	0,29
24	−1,05	−0,23	0,76	0,11	−0,10	0,35	−0,88	−1,05	−0,08
25	1,81	−0,81	0,19	−0,84	0,55	−0,62	0,12	0,57	0,47

Plot of the components loadings: This set of plots shows each of the principal component loading columns plotted against each other and often makes the interpretation of principal components much easier. Cattell suggests to find the place where the smooth decrease of eigenvalues appears to level off to the right of the plot. To the right of this point, presumably, one finds only "componetal scree". Scree is the geological term referring to the debris that collects on the lower part of a rocky slope. Thus, no more than the number of components (here perhaps 2 or 3) to the left of this point should be extracted. Note that by default this graph will also show a *unit circle*. Because the current analysis is based on correlations, the largest component coordinate (variable-component correlation) that can occur is equal to 1.0; also, the sum of all squared component coordinates for a variable (i.e., squared correlations between the variable and all components) cannot exceed 1.0. Hence, all component coordinates must fall within the unit circle indicated in the graph, and this circle can provide a visual indication (scale) of how well each variable is represented by the current set of components (the closer a

variable in this plot is located to the unit circle, the better is its representation
by the current coordinate system). Figure 4.21b shows that, for example, the
variable *Milk* and *Cereals* do not correlate at all. Strong correlation exhibit
variables *WhileMeat, Milk, Eggs* and *RedMeat*.

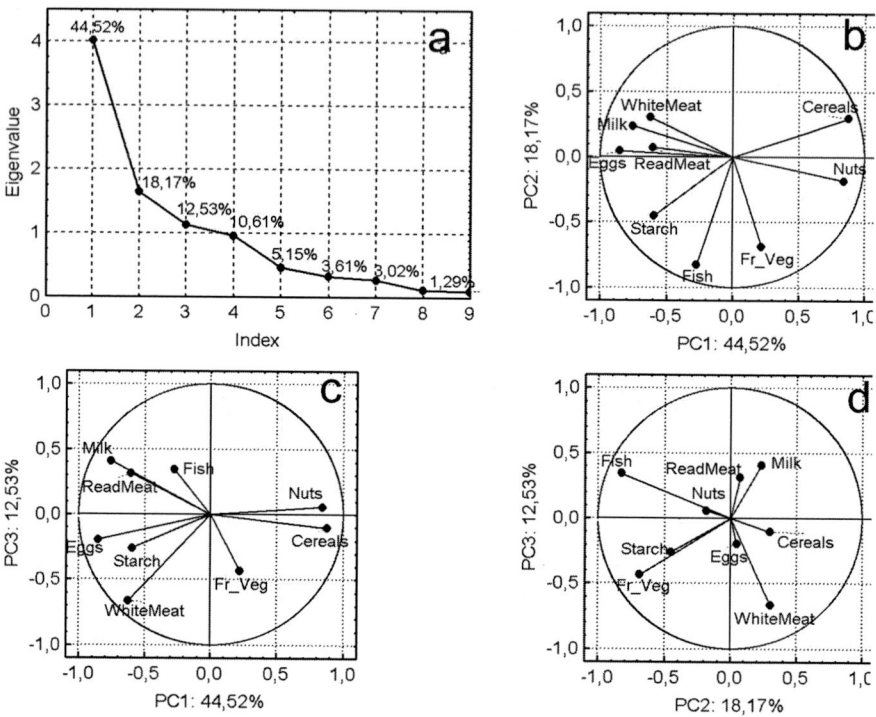

Figure 4.21 The scree plot and three plots of the components loadings: (a) The scree
plot, (b) PC1 vs. PC2 plot, (c) PC1 vs. PC3 plot, (d) PC2 vs. PC3 plot,
(STATISTICA).

Scatterplot of principal component scores: This plot shows the principal
component coordinates for all objects (or all observations). One interesting
result that is apparent in this plot pertains to the clustering of objects. This set
of plot shows each principal component plotted against every other component.
The first k components (where k is the number of large eigenvalues) usually
shows the major structure that will be found in the data. The rest of the
components show outliers and linear dependencies. This scatterplot separates
countries according to proteins consumption into several clusters. One obvious
cluster is formed with countries Bulgaria, Romania, Albania and Yugoslavia,
which are located at Balkan. Positive correlation exhibit countries Poland,
Greece, USSR, Czechoslovakia, Italy and Hungary. Spain correlates with
Portugal.

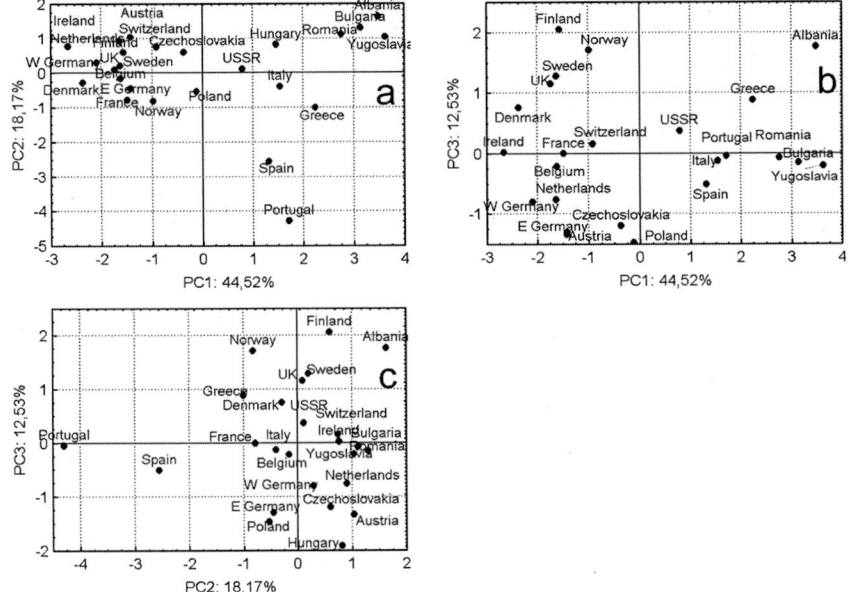

Figure 4.22   The set of three scatterplots of components score: (a) PC1 vs. PC2 plot,
(b) PC1 vs. PC3 plot, (c) PC2 vs. PC3 plot, (STATISTICA).

Biplot: In the biplot here is shown that some variables are highly correlated
with the index of an actual object. Information about the cases can also be
obtained. Distances between points, clusters of points, outliers, etc. can be
quickly visualized with this graph as they would be with a regular scatterplot.

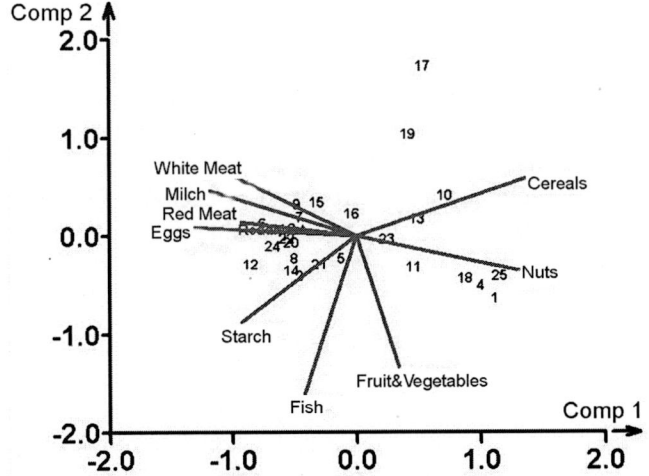

Figure 4.23   Biplot

Analysis of residuals: The plot on Fig. 4.24 indicate five influential points, *Italy, France, Bulgaria, Albania* and *Portugal*. All these countries are outliers. The rest of countries fit suggested PCA model quite well.

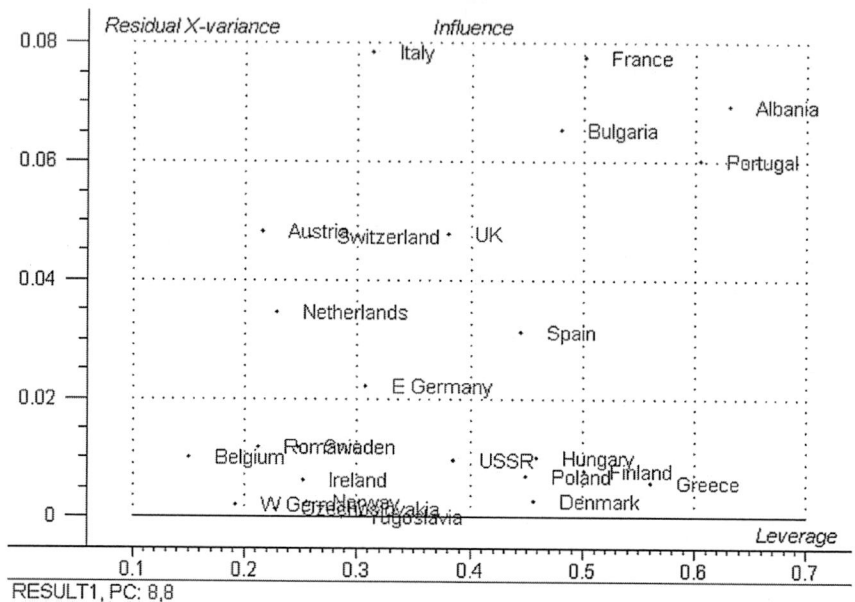

Figure 4.24   Plot of the statistical analysis of residuals indicate the influential point.

4.5 Factor Analysis (FA)

Factor analysis is a generic name given to a class of multivariate statistical methods whose primary purpose is to define the underlying structure in a data matrix. It addresses the problem of analyzing structure of the interrelationships (i.e., correlations) among a large number of variables by defining a set of common underlying dimensions, known as *factors*. With factor analysis, one can first identify the separate dimensions of the structure and then determine the extent to which each variable is explained by each dimension. Once these dimensions and the explanation of each variable are determined, the two primary uses for factor analysis – summarization and data reduction – can be achieved [131, 132].

Factor analysis can be presented within the framework of the multivariate linear models for data analysis. The multivariate linear factor model has several variants as found by Gorsuch [133]: one class of variants is the *full component model*, which is based on perfect calculation of the variables from the components. The other class of variants, the *common factor model,* included sources of variance not attributable to the common factors. Factor analysis as a term is occasionally used with only the common factor model, but then there is no term to use with the broader model of which both components and common factors are a part. Therefore, factor analysis can be approached as one method of analyzing data within the broader multivariate linear model.

The full component model: The variables, $x_i, i = 1,...,m$, can be directly calculated from the factors, $F_1, F_2,..., F_p$ by applying the weights, $a_{i1}, a_{i2},..., a_{ip}$ and the resulting mathematical model

$$x_i = a_{i1}F_1 + a_{i2}F_2 + ... + a_{ip}F_p + e_i,$$ is referred to as the *full component* model. When one factors for all components, the existence of a set of factor scores that produce the original variables *exactly* is assumed. Any observed error is a reflection of the inaccuracy of the model in that particular sample. *Truncated components* occur whenever some, but not all, of the components are used to estimate the original variables. In truncated components the smaller components are dismissed as due to inaccuracy in the model's fit for a particular sample. Truncated components are the usual form of a principal component analysis.

In the multivariate linear model, the variables are defined as linear functions of the factors. Knowing the theoretical relationships will enable us to select appropriate starting points for a factor analysis (i.e., to find weights and factors when only scores on the variables are known). If the correlations between variables are available and suitable restrictions are placed on the weights and correlations among factors, the correlations can be analyzed to find the weights. Analyzing correlations is the basis for most factor-analytic calculation procedures presently in use. A variable's correlations with a set of factors is called its *factor structure*. Note that the factor structure becomes identical to the factor pattern (i.e., the weights given the factors to compute the variable, when the variance of the variable is 1.0 and the correlations among factors are zero).

The common factor model: In the common factor model, the factors are divided into two groups. The first group, the *common factors* $F_1, F_2,..., F_p$ themselves, consists of those factors which contribute to two or more of the variables i.e., several variables have these factors in common. However, the variables will seldom be calculable from the common factors alone, for each variable will be influenced by sources independent of the other variables. These sources may be legitimate causal influence that are not affecting the other variables, systematic distortion in the scores (usually called *bias*), or random error from inaccuracies of measurement. For most factor-analytic purposes, all of these sources simply detract from the accuracy of computing the variable scores from the common factors. The noncommon factor variance for each variable is summarized in a *unique factor* $e_i, i = 1,...,m$, that contains the scores necessary to complete the prediction of that variable. The unique factor e_i includes the random errors of measurement, influences that affect only this variable and all other sources of error and bias that prevent the common factors from completely defining that particular variable. The unique contribution to the variable can be subdivided into one part arising from unreliability of measurement and another part arising from systematic sources in addition to the common factors. The latter component is a composite of all of that variable's reliable variance that does not overlap with a factor in that analysis, and is called a *specific factor* because it is specific to one variable. Both of these sources are assumed to be uncorrelated with the common factors and with each other.

4.5.1 Principles of Factor Analysis [25]

We assume the data matrix $\mathbf{X}_R$ of the standardized data $x_{Rij} = (x_{ij} - \bar{x}_j)/s_j$ of size $n \times m$ where x_{ij} are the elements of the original data matrix $\mathbf{X}$. The matrix $\mathbf{X}$ contains m columns of elements $\mathbf{X}_1, \mathbf{X}_2, ..., \mathbf{X}_m$ and n rows of objects. Supposing that $\mathbf{x}^T = [x_1, x_2, ..., x_m]$ is assumed to be linked to a smaller number of unobserved latent variables, factors, $F_1, F_2, ..., F_m$, where $n < m$, by a regression model of the form

$$x_1 = a_{11}F_1 + a_{12}F_2 + ... + a_{1p}F_p + e_1,$$
$$x_2 = a_{21}F_1 + a_{22}F_2 + ... + a_{2p}F_p + e_2,$$
.....
$$x_m = a_{m1}F_1 + a_{m2}F_2 + ... + a_{mp}F_p + e_m$$

We assume that the 'residual' terms $e_1, e_2, ..., e_m$ are uncorrelated with each other and with factors $F_1, F_2, ..., F_p$. This implies that, given the values of the factors, the observed variables are independent, that is, the correlations of the observed variables arise from their relationships with the factors. In factor analysis the regression coefficients in $\mathbf{a}$ are more usually known as *factor loadings*.

Since the factors are unobserved we can fix their location and scale arbitrarily. We shall assume they occur in standardized form with mean zero and standard deviation one. We shall also assume, initially at least, that the factors are uncorrelated with one another. With these assumptions the factor model given above implies that the variance of variable x_i, i.e., σ_i^2 is given by relation $\sigma_i^2 = \sum_{j=1}^{p} a_{ij}^2 + \psi_i$ where ψ_i is the variance of e_i. So the factor analysis model implies that the variance of each observed variable can be split into two parts.

(1) The first h_i^2, given by $h_i^2 = \sum_{j=1}^{p} a_{ij}^2$ is known as the *communality* of the variable and represents the part of the variance shared with the other variables that is due to the common factors.

(2) The second part, ψ_i is called the *specificity* or the *specific* or *unique* variance (or often also variable's *uniqueness*) relates to the part of the variability in x_i not shared with other variables but due to the unique factor e_i. It is proportion of the variance excluding the variance attributed to the common factors or the extracted components in a truncated component analysis, $\psi_i = 1 - h^2$. In addition, the factor model leads to the following expression for the covariance of variables

$$x_i \text{ and } x_j : \sigma_{ij} = \sum_{l=1}^{p} a_{il} a_{jl} .$$

The numerical aspects of factor analysis are concerned with finding estimates of the *factor loadings* $\left(a_{ij}\right)$ and the *communalities* $\left(h_i^2\right)$. There are many ways available to numerically solve for these quantities. The solution process is called *initial factor extraction*. Once a set of initial factors is obtained, the next major step in the analysis is to obtain new factors, called the *rotated factors*, in order to improve interpretation.

Summarizing, the factor analysis model implies that the covariance of two observed variables depends solely on their relationship with the common factors as measured by the sizes of the relevant factor loadings. When the factors are assumed to be uncorrelated or *orthogonal*, the factor loadings are simply the correlations between factors and observed variables. The residual term e_i, often known in factor analysis parlance as *specific variates* or *unique factors*, is each relating to one of the original variables and play no part in determining the covariances of the observed variables. The matrix of factor loadings is sometimes called the *pattern matrix*. When the factor loadings are correlations between the x_i's and F_j's, it is also called the *factor structure matrix*. It can be shown mathematically that the communality of x_i is $h_i^2 = a_{i1}^2 + a_{i2}^2 + ... + a_{im}^2$.

So the factor analysis model implies the following expression for the covariance matrix, Σ, of the observed variables: $\Sigma = \mathbf{aa}^{\mathrm{T}} + \psi$ where $\psi = \mathrm{diag}(\psi_i)$. The converse also holds: if Σ can be decomposed into the form given above, then the p-factor model holds for $\mathbf{x}$. In practice, Σ will be estimated by the sample covariance matrix $\mathbf{S}$ (alternatively, the model will be applied to the correlation matrix $\mathbf{R}$), and we will need to obtain estimates of $\mathbf{a}$ and Ψ so that the observed covariance matrix takes the form required by the model.

4.5.2 Geometric Representation of Factor Models

Gorsuch[133] suggested a brief overview of the factor model from the geometric viewpoint which can also add greatly to one's intuitive understanding.

Presenting variables and factors geometrically – To represent factors and variables geometrically, one identifies the *factors* as the axes and the *variables* as vectors (lines), drawn in what is called a Cartesian coordinate space. When this *factor loadings plot* shows the variable to be physically close to a common factor, they are highly related. The direction and length for the vector representing the variable are determined by the factor pattern of the variable. The *factor pattern* gives the coordinates of each vector's point (Figure 4.25). To keep matters simple, we will assume, that the factors are uncorrelated with each other. Because uncorrelated factors form 90° angles with each other, they are *orthogonal* to each other. The number of factors determines the number of axes, or geometric dimensions, required to plot the variables.

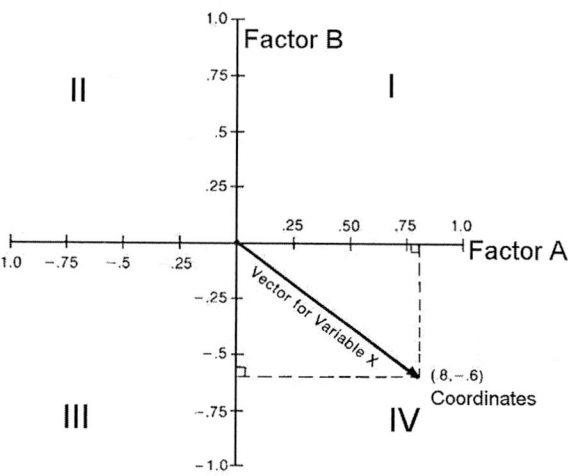

Figure 4.25   Factor loadings plot of one variable $X \varepsilon$ (0.8, −0.6) in the two orthogonal factor space of Factor A and Factor B [26].

Plotting a variable – Once the orthogonal common factors have been drawn arbitrarily, each variable is drawn as a vector (or line) from the *point of origin* (i.e., from the point where the factors intersect). The location of the end point for a variable is determined by its factor weights or coordinates, Fig. 4.25. The factor weights form a row of P (here $X \in$ (0.8, −0.6)), wherein each of the elements is one of the coordinates necessary to define the end point (for Factor A it is 0.8 and for Factor B it is −0.6). The line itself is referred to as the *vector* for the variable.

Length of the vectors – The length of the vector can be computed from the Pythagorean theorem: the square of the length of the hypotenuse of a right triangle is equal to the sum of the squares of the other two sides. The length of each vector is also its *communality* (symbolized by h). Because the two factors are the sole basis for this variable, the variable is drawn to *unit length*. If the length of the variable were less than unity but the variable scores had an observed standard deviation of 1.0, then these two factors would not perfectly reproduce the *variable scores* [133].

Presenting more than two factors – Three factors would require three-dimensional (denoted 3D-) representation, which could be made by projecting the new factor off the paper or by building a 3D-model. Regardless of the limitations of space, geometric principles can be utilized for more than three dimensions. Indeed, an infinite number of dimensions can be conceptualized and the m-multidimensional space is then referred to a *hyperspace*.

The correlation between a vector and factor – When variables are represented in hyperspace as vectors, the correlation between that vector and one of the axes (i.e., factors), or with another vector (i.e., variable), can be readily determined as a function of the angle between them. The formula is $r_{12} = h_1 h_2 \cos \alpha_{12}$, that is, the *correlation* between any two vectors in space (including a factor as a vector) is equal to the length of the first times the length of the second times the cosine of the angle (α) between the two vectors. Because each factor 1 represented as a coordinate and all its variance is included in the dimension that it defines, its length is always 1.0. Hence, when the correlation of a variable with a factor is being determined, the resulting formula is $r_{1k} = h_1 \cos \alpha_{1k}$ where r_{1k} is the correlation of variable 1 with factor k. Further simplification is possible by substituting the definition of cosine of the angle. $r_{1k} = h_1 \dfrac{\text{length of the adjacent side}}{\text{length of the hypotenuse}}$. However, the

hypotenuse is the same as h_1 in this case, so $r_{1k} = $ length of the adjacent side . The length of the adjacent side is none then the factor weight. Therefore, the correlation of the variable with its factor is its factor pattern element. This applies only to rectangular Cartesian coordinates (i.e., uncorrelated factors). The length of the variable, h, may be any positive value. The variable's length squared, h^2, is also called the variables's *communality*. The cosine of the angle between two vectors is determined by, first, drawing a line from the end point of one vector to the other vector in such a manner that the line is perpendicular to the second vector. The result is a right triangle (Fig. 4.25). The cosine is then the length of the adjacent side of the right triangle divided by the length of the hypotenuse. If the angle between the two variables were greater than 90°, one vector would be extended through the point of origin. The usual right triangle would then be drawn with the extended vector. The cosine would be negative. Equation $r_{12} = h_1 h_2 \cos \alpha_{12}$ can be also used to determine the correlation between any two variables being represented as vectors. In the simplest case where each variable is completely represented in the hyperspace, the length of each vector is then 1.0 and the formula becomes simply $r_{12} = \cos \alpha_{12}$. *The correlation coefficient therefore equals the cosine of the angle between the two variables.* A positive correlation is represented as an angle ranging from 0 to 90° and a negative correlation by angles from 90° to 180°. Because no other information is given, the first vector can be drawn in any direction. The second vector must be drawn to be consistent with the first.

Developing a geometric presentation from only the variables – The fact that a plot of variable vectors can be developed independently of the factors and weights leads to factor analysis itself. The analysis begins by considering the correlations as cosines and plotting the resulting vectors. After the plot is made, coordinates (factors) are added according to a rationale determined by the researcher. The factor weights can be read off the plot. When the correlation are such that all of the angles cannot be plotted in two-dimensional space, then one of two courses of action can be followed: (1) More dimensions may be added until all the coefficients are represented accurately. Adding more dimensions means that the number of factors are increased by moving to a space with three or more coordinates but that the variables would be completely represented. Complete representation is the *full component model*. (2) Not all of the vector need be represented. Instead, only that part compatible with representing the other variables as vectors may be included. Because the variable's original communality was assumed to be 1.0, it will now be less than 1.0. Therefore, the length of the variable will be less than unity in the hyperspace. The common factor model generally represents only that portion of the variable which is associated with the common factors.

The uncorrelated (orthogonal) component model – In the model the entire variable is represented in space (i.e., the variable's communality is 1.0 and is represented in hyperspace by a unit-length vector). Therefore, the correlations in the correlation matrix should give the cosines between the vectors directly. When the variables have been plotted, the factors can be added as deemed best so long as they are all orthogonal. Usually, a factor is placed near the center of a cluster of variables so that it can be identified by the conceptual central thrust of the variables. Once several variables have been added, however, they need not remain in the same position. Instead, they can be shifted or *rotated* to any position for which the researcher can present an appropriate rationale, Fig. 4.26ab.

The correlated (oblique) component model – In representing correlated components, the general Cartesian coordinate system is used instead of the rectangular Cartesian coordinate system. The former allows the axes (or factors) to form an oblique angle and thus are correlated. The latter allows only orthogonal factors that are therefore uncorrelated. Because both factors are defined as unit length, this equation becomes $r_{12} = \cos \alpha_{12}$ (i.e., the cosine of the angle is equal to the correlation between two factors). Because the angle between the factors is oblique, they are referred to as *oblique* factors. Each correlation coefficient is converted to a cosine and, if two dimensions are adequate, the variable vectors are plotted in a plane. The use of correlated coordinates does not alter the configuration of the variables.

4.5.3 Interpreting Factors

The interpretation of a factor is based upon the variables that are and are not related to that factor. When the factors are uncorrelated, one matrix summarizes factor-variable relationship. A problem in interpreting the factors is deciding when a borderline loading should be considered significant or salient [133].

Interpretation of factor matrices – The relationship between variables and factors can be described in several ways [133]:

(1) The weight matrix to calculate variable standard scores from factor standard scores: P.

(2) The correlations of the variables with the factors: S.

(3) The correlations between the variables with the reference vectors, V, which are the correlations between the variables and the factors when the variance attributable to all the other factors has been removed.

The basic matrix for interpreting the factors is the *factor structure*. By examining which variable correlate high with the factor and correlate low, it is possible to draw some conclusions as to the nature of the factor. Factor structure has several advantages [133]:

First, investigators are practiced in interpreting correlation coefficients. The coefficients have a limited range and raters have a feel for what practical importance each part of the range has.

Second, matrix **S** shows the relationship of the variable to the full factor. The elements of matrices **P** and **V** systematically exclude overlap among the factors and represent only their unique contributions even the overlap is theoretically important.

Third, matrix **S** gives the relationship of the variables to the factors even if the factors are taken out of this specific context. Regardless of what other factors occur in the next study, the variables should correlate at the same level with a particular factor. The factor pattern and reference vector structure shift with the context of factors in which they are included and therefore can be interpreted only in that context.

Through the use of matrices **V** and **P** , the unique contribution of each factor to each variable can be evaluated. That information is often highly useful, particularly when the other factors in the solution are well known. Indeed, proper interpretation of a set of factors can probably only occur if at least **S** and **P** are both examined. Ideally, all three matrices **S, V** and **P** would be presented so that the reader would have the greatest amount of information to guide conclusions.

Salient loadings: "Salient" has been used previously on an intuitive basis to identify high loadings. More technically, a *salient loadings* is one that is sufficiently high to assume that a relationship exist between variable and the factor. In addition, it usually means that the relationship is high enough so that the variable can aid in interpreting the factor and vice versa. What is a salient level for **S** may not, of course, be a salient level for **P.** Unfortunately, there is no exact way to determine salient loadings in any of the three matrices used for interpreting factors. Only rough guidelines can be given.

Cluster analysis in grouping – The purpose of a *cluster analysis* of variables is to group together those variables that are most alike. Such grouping can aid in the interpretation of factors by bringing together variables of a similar composition and nature. It can also aid in interpreting the variables. In a simple procedure for clustering, one first identifies those variables that are salient on one and only one factor. These variables then form a cluster that is identified with the name of the factor. If there is a clear simple structure in the data, then it would be expected that there would be as many of these clusters as there are factors. A cluster analysis generally gives more clusters than there are factors. The only real advantage of cluster analysis is in the interpretation. By grouping variables together, the solution is simplified and thus made more comprehensible to the finite human mind.

4.5.4 Factor Rotations

Recall that the main purpose of FA is to derive from the data easily interpretable common factors. The initial factors, however, are often difficult to interpret. The factor interpretations usually are not clear/cut and such situation is often the case in practice, regardless of the method used to extract the initial factors [26].

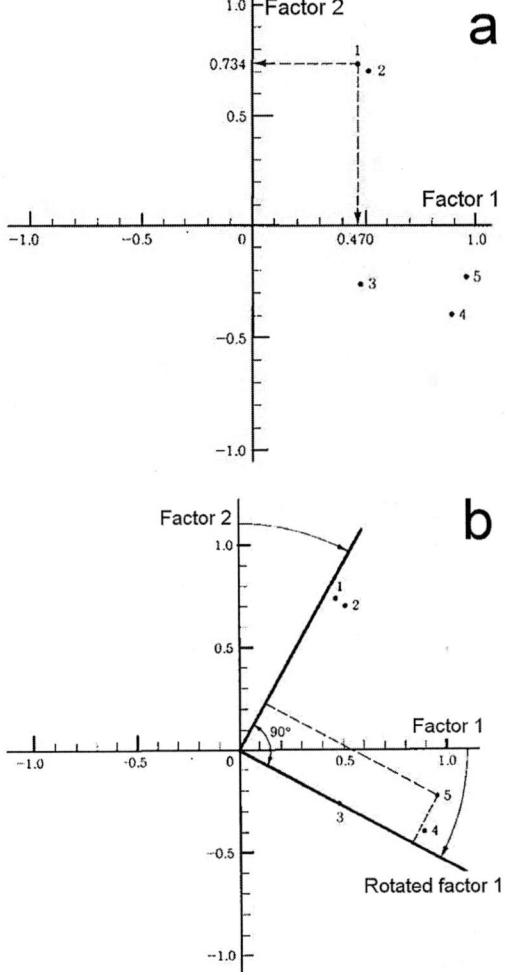

Figure 4.26 (a) Factor loadings plot after an iterated factor extraction, (b) Varimax orthogonal rotation for factor plot of Fig. 4.26a when rotated axes go through clusters of variables and are perpendicular to each other [26].

Fortunately, it is possible to find new factors whose loadings are easier to interpret. These new factors, called the *rotated factors* are selected so that some of the loadings are very large (near ±1) and the remaining loadings are very small (near zero). Conversely, we would ideally wish, for any given variable, that it have a high loading on only one factor. If this is the case, it is easy to give each factor interpretation arising from the variables with which it is highly correlated (high loadings). If we are interested in detailed description of factor rotations methods, some of the books listed in the reference should be consulted [26]. The most commonly used technique is the *varimax rotation* [134]. The other technique is called *oblique rotation* or the *direct quartimin rotation*. Interpretation of factors is more straightforward if each variable is highly loaded on at most one factor, and if all factor loadings are either large and positive or near zero, with few intermediate values. The variables are thus split into disjoint sets, each of which is associated with a single factor [25]. This aim is essentially what Thurstone [135] referred to as *simple structure*. In more detail such structure has the following properties [25]:

(a) Each row or the factor loading matrix should contain at least one zero.

(b) Each column of the loading matrix should contain at least k zeros.

(c) Every pair of columns of the loading matrix should contain several variables whose loadings vanish in one column but not in the other.

(d) If the number of factors is four or more, every pair of columns should contain a large number of variables with zero loadings in both columns.

(e) Conversely, for every pair of columns of the loading matrix only a small number of variables should have non-zero loadings in both columns.

When simple structure is achieved the observed variables will fall into mutually exclusive groups whose loadings are high on single factors, perhaps moderate to low on a few factors, and of negligible size on the remaining factors.

4.5.5 Interpreting Factor Scores

Once the initial extraction of factors and the factor rotations are performed, it may be of interest to obtain the score an object has for each factor. The original data matrix of observed variables is reproduced from two other matrices: the factor pattern and the factor scores. Given the data and factor pattern, it is possible to calculate or estimate the factor scores, that is, the scores for each of the n objects on the p factors and m variables. Two situations often arise in which factor scores are desired. On the one hand, it may be desirable to have factor scores to relate the factors to other variables, principally nominal variables, that cannot be related to the factors by estension analysis or by any other convenient method. On the other hand, a procedure of scoring for the factors may be desired so that a formal analysis is not necessary in future research. The latter situation arises when a limited subset of the variables is to be carried in additional studies to measure the factor so that the entire set of variables is not needed.

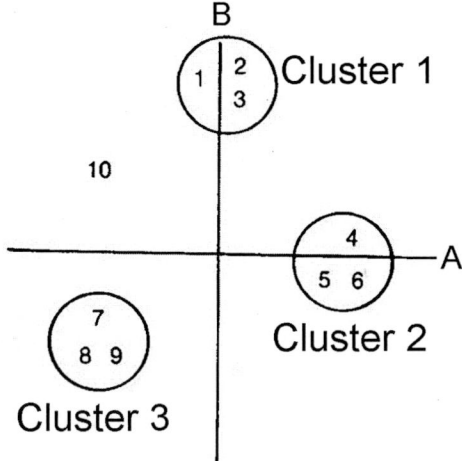

Figure 4.27   Scatterplot of two factor scores indicates ten objects separated into three clusters and one outlier when each cluster contains 3 objects [133].

Clusters of objects: It is often useful to identify types of objects (or individuals, subjects), i.e., to establish categories where the object is either completely in or completely out of each category. Types can be formed in two ways. First, one can place together those objects whose scores fall within the same general area in hyperspace. Empirical analyses for clusters generally start by calculating the distance between each of the individuals for whom factor scores are available. Factoring before clustering allows one to clarify the basis on which the objects are clustered. These procedures provide empirical methods of producing typologies. A second approach to clustering objects builds from the fact that a typology is basically a nominal scoring system. In a nominal scoring system, each type is defined by a vector where all those within the type receive one score (usually a one) and all those who are not of that type receive a different score (usually a zero). The goal of a typological analysis in a factor-analytic context would be to find a set of factor scores that resemble the form of typological standard scores.

Evaluating the factor scores: Regardless of whether the scores are calculated, estimated, or approximated, the quality of the resulting factor scores needs to be checked. In the component model, the correlation of the factor scores with the factor itself should be unity and any departures from unity indicate an error in the procedure. In the estimation and approximation approaches, the check is crucial because some factors may be estimated quite well and others estimated or approximated poorly. The information also suggests which factors can be interpreted because a factor with poorly defined estimates is not as likely to replicate.

4.5.6 Factor Analysis (FA) and Principal Components Analysis (PCA) Compared

Factor analysis FA, like principal components analysis PCA, is an attempt to explain a set of multivariate data using a smaller number of dimensions than one begins with, but the procedures used to achieve this goal are essentially quite different in two approaches.

First, PCA is merely a transformation of the data. No assumptions are made about the form of the covariance matrix from which the data comes. On the other hand, FA supposes that the data comes from the well-defined model, where the underlying factors satisfy the assumptions. If these assumptions are not met, then FA may give spurious results.

Second, in PCA the emphasis is on a transformation from the observed variables to the principal components whereas in FA the emphasis is on ab transformation *from* the underlying factors *to* the observed variables.

Everitt and Dunn [25] commented some principal differences between the two methods which are as follows:

(1) FA postulates a model for the data – PCA does not.

(2) FA tries to explain the covariances or correlations of the observed variables by means of a few common factors. PCA is primarily concerned with explaining the variance of the observed variables.

(3) If the number of retained components is increased, say, from m to $m + 1$, the first m components are unchanged. This is not the case in FA, where there can be substantial changes in *all* factors if the number of factors is changed.

(4) The calculation of principal components score is straightforward while the calculation of factor scores is more complex, and a variety of methods have been suggested.

(5) There is usually no relationship between the principal components of the sample correlation matrix and the sample covariance matrix. For maximum likelihood factor analysis, however, the results of analysing either matrix are essentially equivalent.

Despite these differences, the results from both types of analysis are frequently very similar. Certainly if the specific variances are small we would expect both forms of analysis to give similar results. However, if the specific variances are large they will be absorbed into all the principal components, both retained and rejected, whereas FA makes special provision for them.

Lastly, it should be remembered that both PCA and FA are similar in one important respect – they are both pointless if the observed variables are almost uncorrelated. In this case FA has nothing to explain and PCA will simply lead to components which are similar to the original variables.

Problem 4.2 *A sociologic study of life satisfaction*

This example is based on a fictitious sociologic data set describing a study of life satisfaction. Suppose that a questionnaire is administered to a random sample of 100 adults. The questionnaire contains 10 items that are designed to measure satisfaction at work, satisfaction with hobbies, satisfaction at home, and general satisfaction in other areas of life. Responses to all questions are recorded via computer and scaled so that the mean for all items is approximately 100. The goal is to learn more about the relationships between satisfaction in the different domains. Specifically, it was desired to learn about the number of factors "behind" these different domains of satisfaction, and their meaning.

Data: The results for all respondents are entered into the *Factor.sta* data file. The items were designed to measure satisfaction at work *Work1, Work 2, Work 3*, satisfaction with hobbies *Hobby1, Hobby2*, satisfaction at home *Home1, Home2, Home3*, and general satisfaction in other areas of life *Miscel1, Miscel2*.

Object	Work1	Work2	Work3	Hobby1	Hobby2	Home1	Home2	Home3	Miscel1	Miscel2
1	105.13	101.66	115.06	101.00	95.18	100.28	101.67	85.55	104.03	110.28
..	...	...	...	...	...	...	...	...	...	...
..	...	...	...	...	...	...	...	...	...	...
100	106.05	120.71	119.81	101.85	94.96	75.86	93.17	93.38	109.36	83.79

Software: NCSS2000, STATISTICA.

Solution:

- **Descriptive statistics section:** Here the descriptive statistics are behind the spreadsheet also reviewed graphically in the box-and-whisker in Fig. 4.28. Count, mean and standard deviation allow to make sure that the correct variables have been specified. Using missing value imputation or robust estimation will change these values. The communality shows how well this variable is predicted by the retained factors. It is similar to the R^2 that would be obtained if this variable were regressed on factors that were kept. However, we should remember that this is not based directly on the correlation matrix. Instead, calculations are based on an adjusted correlation matrix.

Variables	Count	Mean	Standard Deviation	Communality
Work1	100	97,03	15,52	0,565
Work2	100	98,17	11,27	0,784
Work3	100	98,94	12,49	0,705
Hobby1	100	98,03	15,94	0,884
Hobby2	100	100,11	19,94	0,828
Home1	100	99,51	11,98	0,614
Home2	100	101,60	11,06	0,880
Home3	100	101,37	12,73	0,624
Miscel1	100	99,16	16,97	0,917
Miscel2	100	98,22	19,11	0,797

The diagram of the box-and-whisker plot in Fig. 4.28 shows that variables *Hobby1, Hobby2, Misc1,* and *Misc2* reach the highest variability in data and therefore they are quite important here.

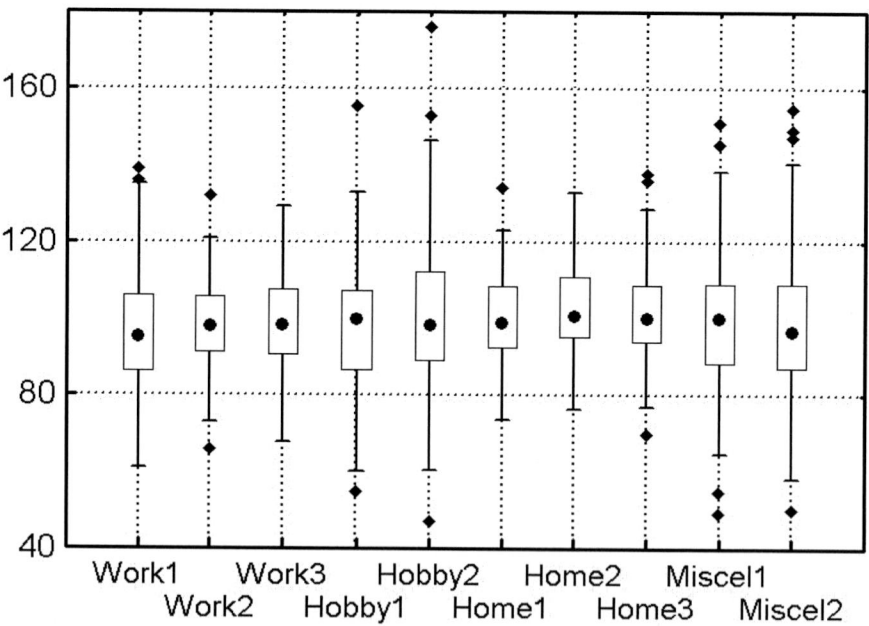

Figure 4.28 Box-and-whisker plot of all variables represents the graphical measure of a variability of individual variables.

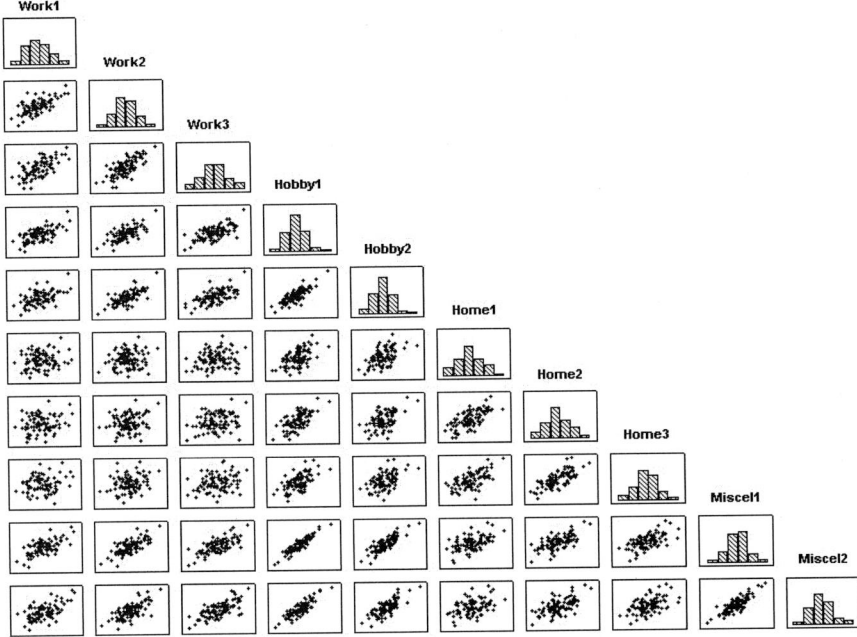

Figure 4.29 Scatter plot of correlation matrix indicates which variables correlate and which do not.

- **Correlation section:** This report gives the correlations alone for a test of the overall correlation structure in the data (Fig. 4.29). If all correlations are small, say less then 0.3, there would be no need for a factor analysis. All correlations in the the spreadsheet are positive; some correlations are of substantial magnitude. For example, variables *Hobby1, Miscel1* and *Miscel2* are correlated at the level of 0.90. Some correlations (for example the ones between work satisfaction and home satisfaction) seem comparatively small. So, it looks like there is some clear structure in this correlation matrix plot on Fig. 4.29. Note that using the missing value imputation or robust estimation options will affect the correlations in this report. When the above options are not used, the correlations are constructed from those observations having no missing values in any of the specified variables.

Variables	Work1	Work2	Work3	Hobby1	Hobby2	Home1	Home2	Home3	Miscel1	Miscel2
Work1	1,000									
Work2	0,647	1,000								
Work3	0,653	0,732	1,000							
Hobby1	0,598	0,689	0,637	1,000						
Hobby2	0,521	0,698	0,630	0,805	1,000					

Home1	0,143	0,143	0,164	0,536	0,506	1,000				
Home2	0,145	0,182	0,238	0,634	0,496	0,658	1,000			
Home3	0,138	0,236	0,255	0,583	0,482	0,590	0,731	1,000		
Miscel1	0,611	0,709	0,698	0,904	0,811	0,498	0,644	0,586	1,000	
Miscel2	0,549	0,685	0,671	0,843	0,756	0,425	0,593	0,518	0,841	1,000

$\varphi = 0.5902$, $Ln(Det|R|) = -9.462$, Bartlett Test $= 897.35$, DF $= 45$, Prob $= 0.000$

The value of φ is the Gleason-Staelin redundancy measure of how interrelated the variables are. A zero value of φ means that there is no correlation among the variables, while a value of one indicates perfect correlation among the variables. The value of $\ln(Det|R|)$ is the logarithm (base e) of the determinant of the correlation matrix. If the covariance matrix was used, this is the ln of the determinant of the covariance matrix. Bartlett's sphericity test for testing the null hypothesis that the correlation matrix in an identify matrix in which all correlations are zero. If we get a calculated probability level of statistical significance (Prob) value greater than $\alpha = 0.05$, we should not perform a factor analysis on the data. This is only available when we analyze a correlation matrix. The formula for computing this χ^2-test is $\chi^2 = \dfrac{(11 + 2m - 6n)}{6} \ln|R|$, where m is the number of variables and $m(m-1)/2$ is the numer of degrees of freedom.

- **Bar chart of absolute correlation section:** This chart graphically displays the absolute values of the correlations. It lets us quickly find high and low correlations.

Variables	*Work1*	*Work2*	*Work3*	*Hobby1*	*Hobby2*	*Home1*	*Home2*	*Home3*	*Miscel1*	*Miscel2*
Work1										
Work2	‖‖‖‖‖‖									
Work3	‖‖‖‖‖‖	‖‖‖‖‖‖‖								
Hobby1	‖‖‖‖‖	‖‖‖‖‖‖	‖‖‖‖‖‖							
Hobby2	‖‖‖‖‖	‖‖‖‖‖‖	‖‖‖‖‖	‖‖‖‖‖‖‖						
Home1	‖‖	‖‖	‖‖‖	‖‖‖‖‖	‖‖‖‖‖					
Home2	‖‖	‖‖‖	‖‖‖	‖‖‖‖‖‖	‖‖‖‖‖	‖‖‖‖‖‖				
Home3	‖‖	‖‖‖	‖‖‖	‖‖‖‖‖	‖‖‖‖‖	‖‖‖‖‖	‖‖‖‖‖‖			
Miscel1	‖‖‖‖‖‖	‖‖‖‖‖‖‖	‖‖‖‖‖‖	‖‖‖‖‖‖‖‖ ‖‖‖‖‖‖‖ ‖‖‖‖		‖‖‖‖‖	‖‖‖‖‖		
Miscel2	‖‖‖‖‖	‖‖‖‖‖‖	‖‖‖‖‖‖	‖‖‖‖‖‖‖ ‖‖‖‖‖‖‖ ‖‖‖‖		‖‖‖‖‖	‖‖‖‖‖	‖‖‖‖‖‖	

- **Eigenvalues section:** One rule-of-thumb is to retain those factors whose eigenvalues are greater than one. The sum of the eigenvalues is equal to the number of variables. Individual percent column gives the percentage of the total variation in the variables accounted for by this factor. Cumulative percent column is the cumulative total of the percentage. Some authors suggest that the user pick a cumulative percentage, such as 80% or 90% and keep enough factors to attain this percentage. The meaning of eigenvalues and how they help you decide how many factors to retain (interpret) is used for a calculation of the percent of total variance, cumulative eigenvalues, and cumulative percent. As we can see, the eigenvalue for the first factor is equal to 5,910217, which means that the proportion of variance accounted for by the first factor is approximately 77.82%. These values happen to be easily comparable here because there are 10 variables in the analysis, and thus the sum of all eigenvalues is equal to 10. The second factor accounts for about 19.86% of the variance. The remaining eigenvalues each account for less than 3% of the total variance.

No.	Eigenvalue	Individual Percent	Cumulative Percent	Scree Plot
1	5,910	77,8	77,8	|||||||||||||
2	1,508	19,9	97,7	||||
3	0,179	2,4	100,0		
4	0,083	1,1	101,1		
5	0,040	0,5	101,7		
6	0,021	0,3	101,9		
7	0,003	0,0	102,0		
8	−0,021	−0,3	101,7		
9	−0,049	−0,6	101,1		
10	−0,080	−1,1	100,0		

- **Number of factors section.** This table briefly describes how these eigenvalues can be used to decide how many factors to retain, that is, to interpret. According to the Kaiser criterion, we would retain factors with an eigenvalue greater than 1. Based on the eigenvalues in this table the spreadsheet shown above, that criterion would suggest we choose 2 factors.
- **Eigenvectors and bar chart of absolute eigenvectors:** This chart graphically displays the absolute values of the eigenvectors. It lets us quickly interpret the eigenvector structure. By looking at which variables correlate highly with a factor, we can determine what underlying structure it might represent.

Variables	Factor1	Factor2	Factor3	Factor1	Factor2	Factor3
Work1	−0,254	−0,349	−0,086	‖‖‖‖	‖‖‖‖	‖
Work2	−0,306	−0,391	0,087	‖‖‖‖‖	‖‖‖‖‖‖	‖
Work3	−0,297	−0,341	−0,220	‖‖‖‖	‖‖‖‖	‖‖‖
Hobby1	−0,387	0,016	0,025	‖‖‖‖‖‖	‖	‖
Hobby2	−0,356	−0,048	0,646	‖‖‖‖‖‖	‖	‖‖‖‖‖‖‖‖‖‖
Home1	−0,226	0,421	0,496	‖‖‖	‖‖‖‖‖‖	‖‖‖‖‖‖
Home2	−0,276	0,514	−0,425	‖‖‖	‖‖‖‖‖‖‖‖	‖‖‖‖‖‖
Home3	−0,252	0,404	−0,104	‖‖‖	‖‖‖‖‖‖‖	‖‖
Miscel1	−0,393	−0,012	−0,159	‖‖‖‖‖‖	‖	‖‖‖
Miscel2	−0,364	−0,039	−0,235	‖‖‖‖‖‖	‖	‖‖‖‖

- **Scree test section.** Now, to produce a line graph of the eigenvalues in order to perform Cattell's scree test the *Scree plot* is displayed Fig. 4.30. The *Plot of Eigenvalues* graph shown below has been "enhanced" to clarify the test. In our example, that point could be at factor 2 or factor 3. Therefore, we should try both solutions and see which one will yield the most interpretable factor pattern.

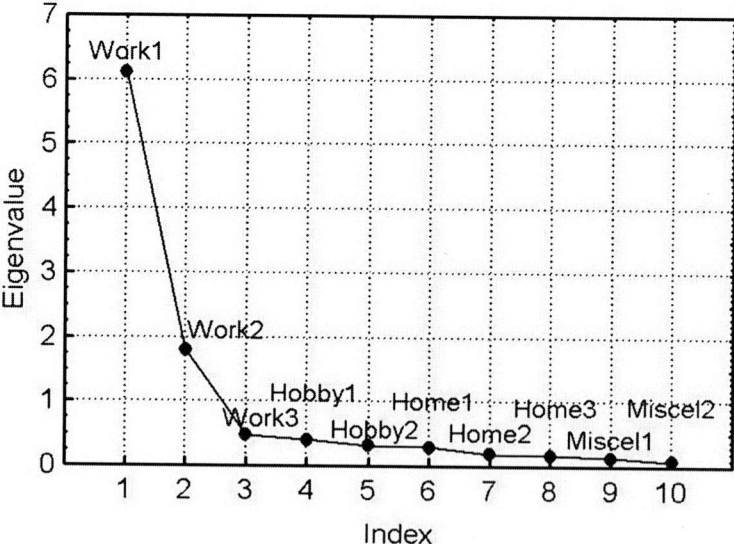

Figure 4.30 Cattell's scree plot of an eigenvalue against the index being equal to the number of factors used here indicates 2 or 3 significant factors.

- **Factor loadings section.** Factor loadings can be interpreted as the correlations between the factors and the variables. Thus, they represent the most important information on which the interpretation of factors is based. First look at the (unrotated) factor loadings for all 10 factors at Fig. 4.31a. Remember that factors are extracted so that successive factors account for less and less variance. Therefore, it is not surprising to see that the first factor shows most of the highest loadings. Also note that the sign of the factor loadings only counts insofar as variables with opposite loadings on the same factor relate to that factor in opposite ways. However, we could multiply all loadings in a column by -1 (i.e., reverse all signs), and the results would not be affected in any way.
- **Factor loadings and bar chart of absolute factor loadings:** This chart graphically displays the absolute values of the factor loadings. It lets us quickly show the correlation structure.

Variables	Factor1	Factor2	Factor3	Factor1	Factor2	Factor3
Work1	−0,617	−0,428	−0,036	||||||||||	|||||||		
Work2	−0,743	−0,481	0,037	||||||||||||	|||||||||		
Work3	−0,722	−0,419	−0,093	||||||||||||	|||||||	||
Hobby1	−0,940	0,020	0,011	|||||||||||||||||				
Hobby2	−0,866	−0,059	0,274	||||||||||||||	||	|||||
Home1	−0,549	0,517	0,210	||||||||||	||||||||||	|||||
Home2	−0,670	0,631	−0,180	|||||||||||	|||||||||||	||||
Home3	−0,613	0,496	−0,044	||||||||||	|||||||||		
Miscel1	−0,955	−0,015	−0,067	|||||||||||||||||			||
Miscel2	−0,886	−0,047	−0,099	||||||||||||||			||

- **Rotating the factor solution section.** The actual orientation of the factors in the factorial space is arbitrary, and all rotations of factors will reproduce the correlations equally well. This being the case, it seems natural to rotate the factor solution to yield a factor structure that is simplest to interpret; in fact, the formal term *simple structure* was coined and defined by Thurstone to basically describe the condition when factors are marked by high loadings for some variables, low loadings for others, and when there are few high cross-loadings, that is, few variables with substantial loadings on more than one factor. The most standard computational method of rotation to bring about simple structure is the *varimax* rotation [134]; others that have been proposed are *quartimax*, *biquartimax*, and *equamax* and are implemented. In specifying a rotation, first, the number of factors to be used to rotate should be considered, that is, to retain and to interpret. It was previously decided that two is most likely the appropriate number of factors; however, based on the results of the scree plot, it was also decided to look at the three factor solution. We will start with three factors. In the factor loadings spreadsheet above, substantial loadings on the first factor appear for all but the home-related items. *Factor2* shows fairly substantial factor loadings for all but the work-related satisfaction items. *Factor3* only has one substantial loading for variable *Home1*. The fact that only one variable shows a high loading on the third factor makes one wonder whether one cannot do just as well without it (the third factor). Two-factor rotated solution leads to conclusion that the *Factor1* shows the highest loadings for the items pertaining to work-related satisfaction. The smallest loadings on that factor are for home-related satisfaction items. The other loadings fall in-between. *Factor2* shows the highest loadings for the home-related satisfaction items, lowest loadings for work-related satisfaction items, and loadings in-between for the other items. It looks like the two factors are best identified as the work satisfaction factor (*Factor1*) and the home satisfaction factor (*Factor2*). Satisfaction with hobbies and miscellaneous other aspects of life seem to be related to both factors. This pattern makes some sense in that satisfaction at work and at home may be independent from each other in this sample, but both contribute to leisure time (hobby) satisfaction and satisfaction with other aspects of life. The graph below simply shows the two loadings for each variable. Note that this scatterplot nicely illustrates the two independent factors and the 4 variables (*Hobby1*, *Hobby2*, *Miscel1*, *Miscel2*) with the cross-loadings.

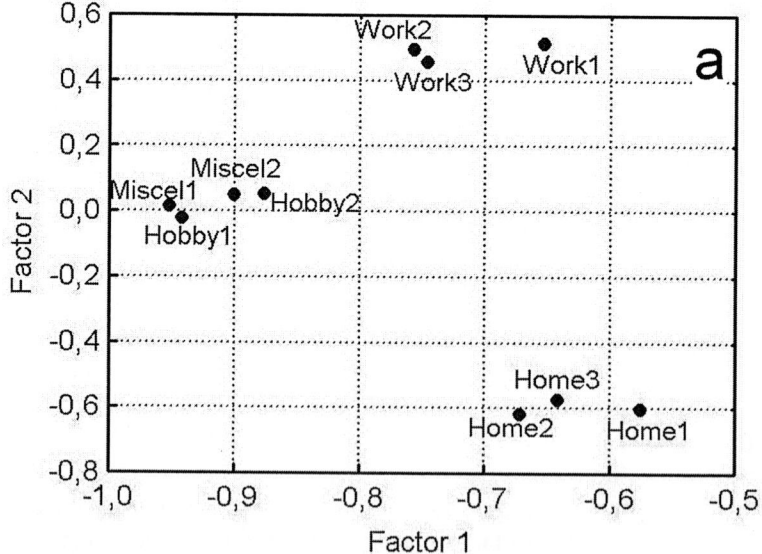

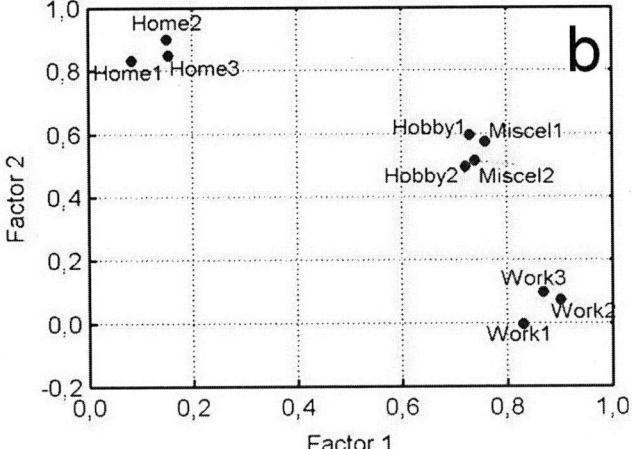

Figure 4.31   The plot of factor loadings (a) unrotated and (b) after normalized varimax rotation.

- **The "Secret" to the perfect example.** The example we have reviewed does indeed provide a nearly perfect two-factor solution. It accounts for most of the variance, allows for ready interpretation, and reproduces the correlation matrix with only minor disturbances (remaining residual correlations). Of course, nature rarely affords one such simplicity, and, indeed, this fictitious data set was generated via the normal random number generator accessible in the spreadsheet formulas. Specifically, two orthogonal (independent) factors were "planted" into the data, from which the correlations between variables were generated. The factor analysis example retrieved those two factors as intended (i.e., the work satisfaction factor and the home satisfaction factor); thus, had nature planted the two factors, you would have learned something about the underlying or latent structure of nature.
- **Communalities section.** To view the communalities for the current solution, that is, current numbers of factors we should remember that the communality of a variable is the portion that can be reproduced from the respective number of factors; the rotation of the factor space has no bearing on the communalities. Very low communalities for one or two variables (out of many in the analysis) may indicate that those variables are not well accounted for by the respective factor model. It is similar to the R^2 value that would be achieved if this variable were regressed on the retained factors. This table value gives the amount added to the communality by each factor. Bar chart graphically displays the values of the communalities.

Variables	Factor1	Factor2	Factor3	Communality	Factor1	Factor2	Factor3	Communality
Work1	0,380	0,184	0,001	0,565	|||||||	||||			|||||||||
Work2	0,552	0,231	0,001	0,784	|||||||||||	|||||			|||||||||||||
Work3	0,521	0,175	0,009	0,705	||||||||||	||||			||||||||||||
Hobby1	0,884	0,000	0,000	0,884	|||||||||||||||||					|||||||||||||||
Hobby2	0,750	0,003	0,075	0,828	|||||||||||||			||	||||||||||||||
Home1	0,302	0,268	0,044	0,614	||||||	|||||			|||||||||||
Home2	0,449	0,398	0,032	0,880	|||||||	|||||||			|||||||||||||||
Home3	0,376	0,246	0,002	0,624	|||||||	|||||			||||||||||||
Miscel1	0,912	0,000	0,005	0,917	||||||||||||||||||					||||||||||||||||||
Miscel2	0,785	0,002	0,010	0,797	|||||||||||||					||||||||||||||

- **Factor score coefficients section.** The factor score coefficients can be used to compute factor scores. These coefficients represent the weights that are used when computing factor scores from the variables. The coefficient matrix itself is usually of little interest; however, factor scores are useful if one wants to perform further analyses on the factors. These score coefficients are similar to the eigenvectors. They have been scaled so that the scores produced have a variance of one rather a variance equal to the eigenvalue. This causes each of the factors to have the same variance.

Variables	Factor1	Factor2	Factor3
Work1	−0,104	−0,284	−0,203
Work2	−0,126	−0,319	0,205
Work3	−0,122	−0,278	−0,520
Hobby1	−0,159	0,013	0,060
Hobby2	−0,147	−0,039	1,527
Home1	−0,093	0,343	1,172
Home2	−0,113	0,419	−1,005
Home3	−0,104	0,329	−0,246
Miscel1	−0,162	−0,010	−0,375
Miscel2	−0,150	−0,031	−0,555

- **Factor scores section.** Factor scores (values) can be thought of as the actual values for each respondent on the underlying factors in a particular row of data that we discovered. These scores can be saved and used later in other graphical data analyses.

Row	Factor1	Factor2	Factor3
1	−0,26	−1,00	−1,16
2	1,86	0,94	−0,69
3	1,15	0,82	−0,62
4	0,29	−0,81	3,23
5	0,94	−1,48	3,36
6	0,90	1,29	0,53
7	0,40	−0,20	0,04
8	−0,08	−1,57	−0,93
9	0,10	−0,19	−0,28
10	0,84	0,10	2,30
11	1,22	−1,36	−2,26
.....	.....	.....	.....
.....	.....	.....	.....
98	−2,19	−0,90	−2,26
99	−0,35	0,39	−1,98
100	−0,17	−2,44	−2,16

- **Factor structure summary section:** This report is provided to summarize the factor structure. Variables with an absolute loading greater than the amount set in the minimum loading option are listed under each factor. Using this report, we can quickly see which variables are related to each factor. Note that it is possible for a variable to have high loadings on several factors, although varimax rotation makes this very unlikely.

Factor1	Factor2	Factor3
Miscel1	*Home2*	
Hobby1	*Home1*	
Miscel2	*Home3*	
Hobby2	*Work2*	
Work2	*Work1*	
Work3	*Work3*	
Home2		
Work1		
Home3		
Home1		

- **Plots section:**

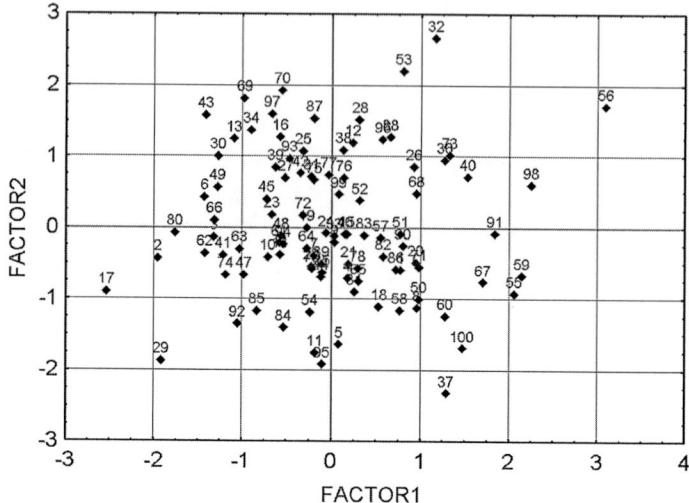

Figure 4.32 Scatterplot of the 1st and 2nd factor score.

- *Conclusion:* Factor analysis is a not a simple procedure. Anyone who is routinely using factor analysis with many (e.g., 50 or more) variables has seen a wide variety of "pathological behaviors" such as negative eigenvalues, un-interpretable solutions, ill-conditioned matrices, and such adverse conditions. We are usually interested in using factor analysis in order to detect structure or meaningful factors in large numbers of variables. Also, because many crucial decisions in factor analysis are by nature subjective (number of factors, rotational method, interpreting loadings), be prepared for the fact that experience is required before we feel comfortable making those judgments. The factor analysis is specifically designed to make it easy to switch interactively between different numbers of factors, rotations, etc., so that different solutions can be tried and compared.

4.6 Canonical Correlation Analysis (CCA)

Canonical correlation analysis CCA is a statistical technique developed by H. Hotelling [137] for measuring linear relationship between two sets of variables. It may be useful to think of one set of variables called *the left set variables* and the other set called *the right set variables*. Like Principal Component Analysis (PCA) and Linear Discriminant Analysis (LDA), CCA also reduces the dimensionality of the original variables, since only a few factor pairs are normally needed to represent the relevant information. However, they serve different purposes: whilst PCA aims to minimize the reconstruction error, and LDA derives a discriminant function that maximizes between-class scatter and minimize within-class scatter, CCA seeks directions for two variables to maximize their correlations, so it is better suited for regression tasks. Compared to other linear regression methods such as Partial Least Squares and Multivariate Linear Regression, CCA has some attractive properties. For example, CCA is invariant to affine transformations of the input variables [138].

4.6.1 The Objectives of CCA

The technique of canonical correlation analysis is best understood by considering it as an extension of multiple regression and correlation analysis. In multiple regression analysis we find the best linear combination of p variables, $x_1, x_2, ..., x_p$, to predict one variable y only. The multiple correlation coefficient R represents here the *simple correlation* between y and its predicted value $\hat{y}_{pred}$. In CCA we examine the linear relationships between a set of x variables $x_1, x_2, ..., x_p$ on the right side, i.e., $V_1 = b_1 x_1 + b_2 x_2 + ... + b_p x_p$, and a set of more than one y variable $y_1, y_2, ..., y_q$ on the left hand side, i.e., $U_1 = a_1 y_1 + a_2 y_2 + ... + a_q y_q$. The technique consists of finding several linear combinations of the x variables and the same number of linear combinations of the y variables in such a way that these linear combinations best express the correlations between these two sets. Those linear composites V_1 and U_1 are known as *canonical variates,* and the correlations between corresponding pairs of canonical variates are called *canonical correlation R_1*. CCA applies to situations in which regression techniques are appropriate and where there exists more than one left side variable. A good deal of the difficulty with CCA is due to jargon: first, there are original variables, then there are canonical variates, and, finally, there are pairs of canonical variates. *Variables* refers to the original variables measured in research. *Canonical variates* are linear composites of original variables, one combination on the left side and a second combination on the right side. These two composites form *a pair of canonical variates*. However, there may be more than one reliable pair of canonical variates only [60].

In canonical correlation, sets of variables on each side are combined to produce, for each side, a predicted value that has the highest correlation with the predicted value on the other side. The combination of original variables on each side can be thought of as a dimension that relates the original variables on one side to the original variables on the other. There is a complication, however. In multiple regression, there is only one combination of variables because there is only a single variable to predict on the other side of the equation. In canonical correlation, there are several original variables on both sides and there may be several ways to recombine the variables on both sides to relate them to each other. There are potentially as many ways to recombine the original variables as there are variables in the smaller set, usually only the first two or three combinations are reliable and need to be interpreted. CCA can address a wide range of objectives. These objectives may be any or all of the following:

(1) Determining whether two sets of variables (measurements made on the same objects) are independent of one another or, conversely, determining the magnitude of the relationships that may exist between the two sets. It is to ask how strongly the variate on one side of the equation relates to the variate on the other side of the equation; that is, how strong is the correlation between both variates in a pair?

(2) Deriving a set of weights for each set of left set and right set variables so that the linear combinations of each set are maximally correlated. Additional linear functions that maximize the remaining correlation are independent of the preceding set(s) of linear combinations.

(3) Explaining the nature of whatever relationships exist between the left sets and right set variables, generally by measuring the relative contribution of each variable to the canonical functions that are extracted.

If canonical variates U_i and V_i are interpretable, scores on them might be useful as left side variables or right side variables in other analyses. CCA has several important theoretical limitations that help explain its scarcity in the literature. Perhaps the most critical limitation is interpretability: procedures that maximize correlation do not necessarily maximize interpretation of pairs of canonical variates. Therefore, canonical solutions are often mathematically elegant but uninterpretable.

The algorithm also computes pairs of canonical variates that are independent of all other pairs. In canonical analysis, only the orthogonal solution is routinely available. An important concern is the sensitivity of the solution in one set of variables to the variables included *in the other set*. In canonical analysis, the solution depends both on correlations among variables in each set and on correlations among variables between sets. Changing the variables in one set may markedly alter the composition of canonical variates in the other set.

4.6.2 Deriving the Canonical Functions

Suppose we wish to study the relationship between a right set of variable $x_1, x_2, ..., x_p$ and left set $y_1, y_2, ..., y_q$. We assume that in any given sample the mean of each variable has been subtracted from the original data so that the sample means of all **x** and **y** variables are zero.

The canonical correlation analysis addresses two primary objectives: (1) the identification of dimensions among the dependent and independent variables that (2) maximize the relationship between the dimensions. Figure 4.33 shows, in general, the relationships among original variables, canonical variates, and the first pair of canonical variates.

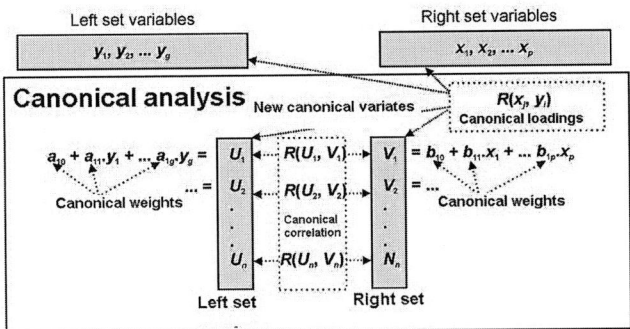

Figure 4.33 Scheme of canonical correlation analysis CCA: left and right sets of original variables, canonical variates, canonical weights, canonical loadings.

For any particular choice of the coefficients (a's and b's) we can compute values of U_1 and V_1 for each individual in the sample. From the n individuals in the sample we can then compute the simple correlation between the n pairs of U_1 and V_1 values in the usual manner. The resulting correlation depends on the choice of the a's and b's. In CCA we select values of a and b coefficients so as to *maximize* the correlation between U_1 and V_1. With this particular choice the resulting linear combination U_1 is called the *first canonical variate* of the **y**'s and V_1 is called the first canonical variate of the **x**'s. The resulting correlation between U_1 and V_1 is called the *first canonical correlation*. The square of the first canonical correlation R_1^2 is often called the first *eigenvalue*. The first canonical correlation R_1^2 is thus the highest possible correlation between a linear combination of the **x**'s and a linear combination of the **y**'s. In this sense it is the maximum linear correlation between the set of **x** variables and the set of **y** variables. The first canonical correlation R_1^2 is analogous to the multiple correlation coefficient between a single y variable and the set of **x** variables. The difference is that in canonical correlation analysis we have several **y** variables and we must find a linear combination of them also.

The first step of canonical correlation analysis is to derive one or more canonical functions. Each function consists of a pair of variates, one representing the right set variables and the other representing the left set variables. The maximum number of canonical variates (or canonical functions) that can be extracted from the sets of variables equals the number of variables in the smallest data set, right or left set. The CCA is based on the correlation between two sets of variables which are denoted **y** and **x**. The correlation matrix **R** of all the variables is divided into four parts:

(1) $\mathbf{R}_{xx}$ concerns the correlations among the x variables.

(2) $\mathbf{R}_{yy}$ concerns the correlations among the y variables.

(3) $\mathbf{R}_{xy}$ concerns the correlations between the x and y variables.

(4) $\mathbf{R}_{yx}$ concerns the correlations between the y and x variables.

The general equations for performing a canonical correlation are relatively simple. First, a correlation matrix **R** is formed as the product of the inverse of the correlation matrix $\mathbf{R}_{yy}$, a correlation matrix $\mathbf{R}_{yx}$, the inverse of correlation matrix $\mathbf{R}_{xx}$, and the other correlation matrix $\mathbf{R}_{xy}$, i.e., leading to $\mathbf{R} = \mathbf{R}_{yy}^{-1}\mathbf{R}_{yx}\mathbf{R}_{xx}^{-1}\mathbf{R}_{xy}$. Canonical analysis proceeds by solving the above equation for eigenvalues and eigenvectors of the matrix **R**. Eigenvalues consolidate the variance of the matrix, redistributing the original variance into a few composite variates. Eigenvectors, transformed into coefficients, are used to combine the original variables into these composites.

The diagonal matrix Λ_m of the singular values of **R** is made up of the eigenvalues of **R**. The ith eigenvalue λ_i of the matrix **R** is equal to the square of the ith canonical correlation which is called $\lambda_i = R_{ci..}^2$. Hence, the ith canonical correlation is the square root of the ith eigenvalue of **R**, i.e., $\sqrt{\lambda_i}$. That is, each eigenvalue equals the squared canonical correlation R_i^2 for each pair of canonical variates. The significance of one or more canonical correlations is tested as a χ^2-square criterion using the formula:

$$\chi^2 = -\left[n-1-\left(\frac{k_x + k_y + 1}{2}\right)\right]\ln\Lambda_m, \quad \text{where} \quad \Lambda_m = \prod_{i=1}^{m}(1-\lambda) \quad \text{and} \quad n \text{ is the}$$

number of cases, k_x is the number of variables in **x** set, k_y is the number of variables in **y** set, df is equal to $k_x k_y$, m is the number of canonical correlations. Significant results indicate that the overlap in variability between variables in the two sets is significant; this is evidence of significance in the first canonical correlation. This process (of finding a canonical correlation and testing for significance) is then repeated with the first pair of variates removed to see if any of the other pairs are significant. Only pairs that test significant are interpreted. Canonical coefficients **a** and **b** (also referred to as canonical weights) are analogous to the beta values in regression. One set of canonical coefficients is required for each set of variables for each canonical correlation. To facilitate comparisons, these values are usually reported for standardized variables (i.e., z transformed variables). The coefficients reflect differences in the contribution of the variables to the canonical correlation. Two sets of canonical coefficients (like regression coefficients) are used for each canonical correlation: one for the **x** variables and another for the **y** variables. These coefficients are defined as follows: $\mathbf{a} = \mathbf{R}_{yy}^{-1/2}\hat{\mathbf{B}}_y$ and $\mathbf{b} = \Lambda_m \mathbf{R}_{xx}^{-1}\mathbf{R}_{xy}\mathbf{B}_y$ where $\hat{\mathbf{B}}_y$ is the normalized matrix of eigenvectors, **a** is the matrix formed by the coefficients for **y** set each divided by their corresponding canonical correlation and **R** is the matrix of correlations. The average squared loadings are given as the sum of the squared correlations divided by the number of variables in the set, $pv_{xc} = \sum_{i=1}^{k_x} \dfrac{a_{ixc}^2}{k_x}$,

and $pv_{yc} = \sum_{i=1}^{k_y} \dfrac{a_{iyc}^2}{k_y}$ where a is the loading correlations, k is the number of variables in the set. How strongly does a variate relate to the variables on the other side of the equation? *Redundancy* is the amount of variance that the canonical variates from one set of variables extract from the other set. It is the product of the percentage of variance extracted from one set of variables and the squared canonical correlation for the pair, $rd = (pv)r_c^2$.

4.6.3 Designing CCA and Basic Issues

CCA is subject to several limitations. It is mathematically elegant but difficult to interpret because solutions are not unique. Procedures that maximize correlation between canonical variate pairs do not necessarily lead to solutions that make logical sense. It is the canonical variates that are actually being interpreted and they are interpreted in pairs. A variate is interpreted by considering the pattern of variables that are highly correlated (loaded) with it. Variables in one set of the solution can be very sensitive to the identity of the variables in the other set; solutions are based upon correlation within and between sets, so a change in a variable in one set will likely alter the composition of the other set. There is no implication of causation in solutions. The pairings of canonical variates must be independent of all other pairs. Only linear relationships are appropriate. Some of the basic issues that must be dealt with during a canonical correlation analysis are:

(1) Determining the number of canonical variate pairs to use. The number of pairs possible is equal to the smaller of the number of variables in each set.

(2) The canonical variates themselves often need to be interpreted. As in factor analysis, we are dealing with mathematically constructed variates which are usually difficult to interpret. However, in this case, we must relate two constructed variates to each other.

(3) The importance of each variate must be evaluated from two points of view. We have to determine the strength of the relationship between the variate and the variables from which it was created. We also need

to study the strength of the relationship between the corresponding V and U variates.

(4) Do we have a large enough sample size? In social science work we will often need a minimum of ten cases per variable. In fields with more reliable data, we can get by with a little less.

4.6.4 Assumptions in CCA

The generality of CCA also extends to its underlying statistical assumptions.

Normality, Linearity, and Homoscedasticity: Although there is no requirement that the variables be normally distributed when canonical correlation is used descriptively, the analysis is enhanced if they are. However, inference regarding number of significant canonical variate pairs proceeds on the assumption of multivariate normality. Multivariate normality is the assumption that all variables and all linear combinations of variables are normally distributed. Linearity is important to canonical analysis in at least two ways. The first is that the analysis is performed on correlation or variance-covariance matrices that reflect only linear relationships. If the relationship between two variables is nonlinear, it is not "captured" by these statistics. The second is that canonical correlation maximizes the linear relationship between a variate from one set of variables and a variate from the other set. Canonical analysis misses potential nonlinear components of relationships between canonical variate pairs. Finally, canonical analysis is best when relationships among pairs of variables are homoscedastic, that is, when the variance of one variable is about the same at all levels of the other variable. Alternatively, distributions of canonical variate scores produced by a preliminary canonical analysis are examined for normality, linearity, and homoscedasticity and, if found, screening of the original variables is not necessary. In the event of persistent heteroscedasticity, we might consider weighting cases based on variables producing unequal variance or adding a variable that accounts for unequal variance.

Multicollinearity and Singularity: Multicollinearity occurs when one variable is almost a weighted average of the others. Singularity occurs when this relationship is exact. Since inverse matrices are needed during the analysis, we must check for this. Try running a principal components analysis on each set of variables, separately. If we have eigenvalues at or near zero, we have multicollinearity problems. We must omit the offending variables.

4.6.5 Interpretation of Canonical Variates

As in PCA, two kinds of graphical representations can be displayed to visualize and interpret the results of CCA: scatter plots for the variates U and V and scatter plots for the experimental units. Graphical representations can be drawn for every pair of the variates U and V of axes. Like in PCA, it is advocated to choose a small value for *number of significant pairs of variates* d from an interval $(1, p)$. In practice, this value is very often 2, 3 or 4. Note that small canonical correlations are not relevant: they do not express linear relationships between columns of U and V and can be neglected. For great values of p, we suggest an empirical approach for choosing the dimension d based on the joint examination of two graphical representations: the scree graph of canonical correlations and the scatter plots of variables. The scree graph is the plot of canonical correlations versus the dimension; a clear gap between two successive values suggest to select for d the rank of the greatest one. The original variables plot is of interest because it allows to discern the structure of correlation between the two sets of variables **x** and **y**.

A useful option available in some programs is a plot of the canonical variable scores U_i versus V_i. For multivariate normal data the graph would approximate an ellipse of concentration. Such a plot can be useful in highlighting unusual cases in the sample as possible outliers or blunders. The plot of U_1 versus V_1 does not result in an apparently nonlinear scatter diagram, nor does it look like a bivariate normal distribution (elliptical in shape). Canonical correlation creates linear combinations of variables, canonical variates, that represent mathematically viable combinations of variables. However, although mathematically viable, they are not necessarily interpretable. A major task for the researcher is to discern, if possible, the meaning of pairs of canonical variates. Interpretation of reliable pairs of canonical variates is based on the loading matrices, **a** and **b**. Each pair of canonical variates is interpreted as a pair, with a variate from one set of variables interpreted vis-a-vis the variate from the other set. A variate is interpreted by considering the pattern of variables highly correlated (loaded) with it. Because the loading matrices contain correlations, and because squared correlations measure overlapping variance, variables with correlations of 0.30 (9% of variance) and above are usually interpreted as part of the variate, and variables with loadings below 0.30 are not. Deciding on a cutoff for interpreting loadings is, however, somewhat a matter of taste.

Another useful optional output is the set of correlations between the canonical variates and the original variables used in deriving them. This output provides a way of interpreting the canonical variables when some of the variables within either the set of independent or the set of dependent variables are highly intercorrelated with each other. These correlations are sometimes called *canonical variable loadings*. Other terms are *canonical loadings* and *canonical structural coefficients*. Since the canonical variable loadings can be interpreted as simple correlations between each variable and the canonical variate, they are useful in understanding the relationship between the original variables and the canonical variates. When the set of variables used in one canonical variate are uncorrelated, the canonical variable loadings are equal to the standardized canonical variable coefficients. When some of the original variables are highly intercorrelated, the loadings and the coefficients can be quite different. It is in these cases that some statisticians find it simpler to try to interpret the canonical variable loadings rather than the canonical variate coefficients. For example, suppose that there are two x variables that are highly positively correlated, and that each is positively correlated with the canonical variate. Then it is possible that one canonical variate coefficient will be positive and one negative, while the canonical variable loadings are both positive, the result one expects.

Canonical functions which should be interpreted: The most common practice is to analyze functions whose canonical correlation coefficients are statistically significant beyond some level, typically 0.05 or above. Three criteria are recommended to be used in conjunction with one another to decide which canonical functions should be interpreted. The three criteria are (1) level of statistical significance of the function, (2) magnitude of the canonical correlation, and (3) redundancy measure for the percentage of variance accounted for from the two data sets.

(1) *Level of significance:* The level of significance of a canonical correlation generally considered to be the minimum acceptable for interpretation is the 0.05 level, which (along with the 0.01 level) has become the generally accepted level for considering a correlation coefficient statistically significant. The most widely used test, and the one normally provided by computer packages, is the F statistic, based on Rao's approximation [3].

(2) *Magnitude of the canonical relationships:* The practical significance of the canonical functions, represented by the size of the canonical correlations, also should be considered when deciding which functions to interpret. No generally accepted guidelines have been established regarding suitable sizes for canonical correlations. It seems logical that the guidelines suggested for significant factor loadings in factor analysis might be useful with canonical correlations, particularly when one considers that canonical correlations refer to the variance explained in the canonical variates (linear composites), not the original variables.

(3) *Redundancy analysis:* The average of the squared canonical variable loadings for the first canonical variate, V_1 gives the proportion of the variance in the X variables explained by the first canonical variate. The same is true for U_1 and y. Similar results hold for each of the other canonical variates. Sometimes the proportion of variance explained is quite low, even though there is a large canonical correlation. This may be due to only one or two variables having a major influence on the canonical variate. The above computations provide one aspect of what is known as *redundancy analysis*. To overcome the inherent

bias and uncertainty in using canonical roots (squared canonical correlations) as a measure of shared variance, a *redundancy index* has been proposed [139]. It is the equivalent of computing the squared multiple correlation coefficient between the total independent variable set and each variable in the dependent variable set, and then averaging these squared coefficients to arrive at an average R^2. This index provides a summary measure of the ability of a set of independent variables (taken as a set) to explain variation in the dependent variables (taken one at a time). As such, the redundancy measure is perfectly analogous to multiple regression's R^2 statistic, and its value as an index is similar.

The *Stewart-Love index of redundancy* calculates the amount of variance in one set of variables that can be explained by the variance in the other set. This index serves as a measure of accounted-for variance, similar to the R^2 calculation used in multiple regression. The calculation of the redundancy index is a three-step process. The first step involves calculating the amount of shared variance from the set of dependent variables included in the dependent canonical variate. The second step involves calculating the amount of variance in the dependent canonical variate that can be explained by the independent canonical variate. The final step is to calculate the redundancy index, found by multiplying these two components.

4.6.6 Validation and Diagnosis of CCA

Validation methods to ensure that the results are not specific only to the sample data and can be generalized to the population. The most direct procedure is to create two subsamples of the data (if sample size allows) and perform the analysis on each subsample separately. Then the results can be compared for similarity of canonical functions, variate loadings, and the like. Another approach is to assess the sensitivity of the results to the removal of a left and/ or right variable. Because the canonical correlation procedure maximizes the correlation and does not optimize the interpretability, the canonical weights and loadings may vary substantially if one variable is removed from either variate. The first problem we encountered was that the correlation matrix for the original variables was singular. This is a common problem when the number of variables is large – it simply means that some of the variables are redundant. It is difficult, however, to determine by inspection alone which variables are redundant. We used several methods to attack this problem, but the best overall solution was to use cluster analysis prior to the CCA. Cluster analysis is a way of analyzing a correlation matrix which is complementary to CCA. Whereas CCA emphasizes global patterns, cluster analysis works "bottom up" by aggregating the most highly inter-correlated sets of variables first, and then working up to larger clusters which are less tightly inter-related.

As a result, the first clusters identified the most likely sources of redundancy. As a side-benefit, the largest clusters allowed us to check the robustness of the CCA results (since cluster analysis and CCA are quite different mathematically). We tried to find graphical methods which would help us understand and explain the multidimensional patterns found by CCA. These patterns are important because they help the analyst define, in a data-driven way, the most important conditions and their corresponding effects. One of the most helpful suggestions we found was due to Cliff, who suggests interpreting structure correlations rather than weights. Structure correlations are the correlations of the V canonical variate with each of the original independent variables, and of the U canonical variate with each of the original dependent variables. In this way, the somewhat mysterious canonical variates can be interpreted in terms of their correlations with the original variables. We then used two graphical methods to depict the pattern of structure correlations and to highlight deviations from the pattern and outliers.

The user should be aware of the following points: The sample should be representative of the population to which the investigator wishes to make inferences. A simple random sample has this property. If this is not attainable, the investigator should at least make sure that the cases are selected in such a way that the full range of observations occurring in the population can occur in the sample. If the range is artificially restricted, the estimates of the correlations will be affected.

(1) Poor reliability of the measurements can result in lower estimates of the correlations among the x's and among the y's.

(2) A search for outliers should be made by obtaining histograms and scatter diagrams of pairs of variables.

(3) Stepwise procedures are not available in the programs described in this chapter. Variables that contribute little and are not needed for theoretical models should be candidates for removal. It may be necessary to run the programs several times to arrive at a reasonable choice of variables.

(4) The investigator should check that the canonical correlation is large enough to make examination of the coefficients worthwhile. In particular, it is important that the correlation not be due to just one dependent variable and the independent variable. The proportion of variance should be examined, and if it is small then it may be sensible to reduce the number of variables in the model.

(5) If the sample size is large enough, it is advisable to split it, run a canonical analysis on both halves, and compare the results to see if they are similar.

(6) If the canonical coefficients and the canonical variable loadings differ considerably, (i.e. if they have different signs), then both should be

examined carefully to aid in interpreting the results. Problems of interpretation are often more difficult in the second or third canonical variates than in the first. The condition that subsequent linear combinations of the variables be independent of those already obtained places restrictions on the results that may be difficult to understand.

(7) Tests of hypotheses regarding canonical correlations assume that the joint distribution of the x's and y's is multivariate normal. This assumption should be checked if such tests are to be reported.

(8) Since canonical correlation uses both a set of U variables and a set of V variables, the total number of variables included in the analysis may be quite large. This can increase the problem of many cases being not used because of missing values. Either careful choice of variables or imputation techniques may be required.

Problem 4.3 *Comparison and relationship of three new psychological tests with three standard IQ tests*

Suppose we have given a group of 15 students two tests of ten questions each and wish to determine the overall correlation between these two tests. Canonical correlation finds a weighted average of the questions from the first test and correlates this with a weighted average of the questions from the second test. The weights are constructed to maximize the correlation between these two averages. Try to evaluate and compare first three psychological tests with three standard IQ tests and to find a relationship (*Test1, Test2, Test3*) = (*Test4, Test5, IQ*), [131].

Data: Each of used tests contained 10 questions and each question was answered with grades of the range from 0 to 100. Resulting matrix *Test1* to *Test5, IQ* was of size 15 × 10.

Object	Test1	Test2	Test3	Test4	Test5	IQ
1	83	34	65	63	64	106
2	73	19	73	48	82	92
...	...	...	...	...	...	...
15	50	75	72	64	45	103

Software: NCSS2000, STATISTICA

Solution:

- **Exploratory data analysis:** In order to visualize the distribution of the variables, the Box-and-whisker plots (Fig. 4.34) and the histogram in the correlation diagram (Fig. 4.35) are available. The distribution of most variables follows the symmetrical distribution

and shows that the data are homoscedastic. The matrix plot of
correlations should be examined for outliers, which may greatly
bias the computation of the correlation coefficients, and thus the
canonical analysis.

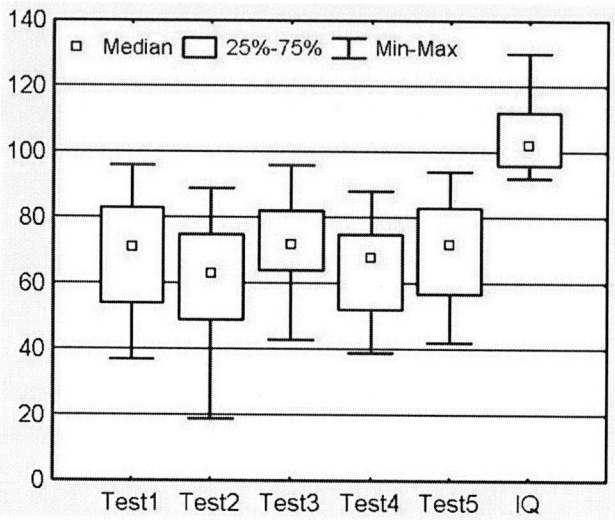

Figure 4.34 Box-and-whisker plot for 6 original variables and 15 respondents
(STATISTICA).

- **Descriptive statistics section:** This report displays the descriptive
  statistics for each variable. We should check that the mean is reasonable
  and that the number of nonmissing rows is accurate.

Type	Variable	Mean	Deviation	Rows
U	Test1	67.93	17.39	15
U	Test2	61.40	19.39	15
U	Test3	72.33	14.73	15
V	Test4	65.53	13.95	15
V	Test5	69.93	16.15	15
V	IQ	104.33	11.02	15

- **Correlation section:** This report presents the simple correlations among all
  variables specified. Matrix plot (Fig. 4.35) and the spreadsheet of correlation
  coefficients indicates that only two Person's correlation coefficients are
  statistically significant (bold letters) and therefore an application of CCA
  will be more difficult. Only two couples of original variables *Test1-Test4* and
  Test2-Test4 indicate significant correlation.

	Test1	Test2	Test3	Test4	Test5	IQ
Test1	1.000					
Test2	0.100	1.000				
Test3	−0.261	0.057	1.000			
Test4	**0.754**	**0.719**	−0.141	1.000		
Test5	0.014	−0.281	0.347	−0.173	1.000	
IQ	0.225	0.241	0.074	0.371	−0.058	1.000

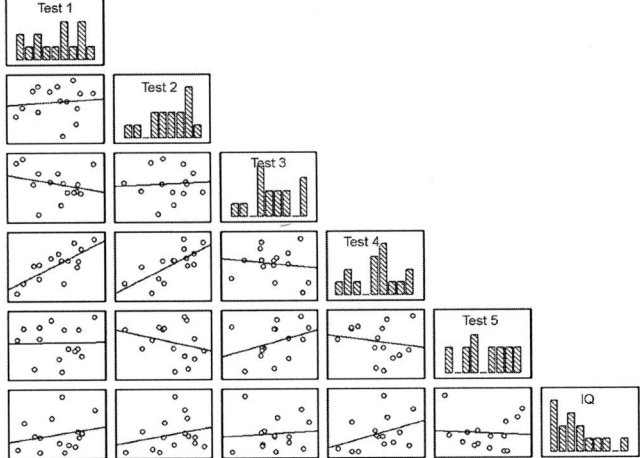

Figure 4.35 Matrix plot of correlations also contains the histogram indicating an actual sample distribution (STATISTICA).

- **Deriving canonical variates:** This report presents the canonical correlations plus supporting material to aid in their interpretation. If the canonical relationship is statistically significant and the magnitudes of the canonical variate and the redundancy index are acceptable, the researcher still needs to make substantive interpretations of the results. Making these interpretations involves examining the canonical functions to determine the relative importance of each of the original variables in the canonical relationships.

Variate Number	Canonical Correlation	R^2	F-Value	Num DF	Den DF	Prob Level	Wilks' Lambda
1	0.996	0.991	16.58	9	22	0.000	0.007
2	0.467	0.218	0.67	4	20	0.617	0.776
3	0.080	0.006	0.07	1	11	0.795	0.993

F-value tests whether this canonical correlation and those following are zero. **Variate Number:** This is the sequence number of the canonical correlation. The first correlation will always be the largest, the second will be the next to largest, and so on. **Canonical Correlation:** The value of the canonical correlation coefficient R_i. This coefficient has the same properties as any other correlation: it ranges between minus one and one, a value near zero indicates low correlation. and an absolute value near one indicates near perfect correlation. **R^2:** The square of the canonical correlation coefficient. This gives the **R^2** value of fitting the *U* canonical variate to the corresponding *V* canonical variate. ***F*-Value:** The value of the *F* approximation for testing the significance of the Wilks' lambda λ corresponding to this row and those below it. In this example. the first *F*-value tests the significance of the first. Second, and third canonical correlations while the second *F*-value tests the significance of only the second and third. **Num DF:** The numerator degrees of freedom of the above *F*-ratio. **Den DF:** The denominator degrees of freedom of the above *F*-ratio. **Prob Level:** This is the probability value for the above *F* statistic. A value near zero indicates a significant canonical correlation. A cutoff value of 0.05 or 0.01 is often used to determine significance. **Wilks' Lambda:** The Wilks' lambda λ value for the canonical correlation on this report row. Wilks' lambda λ is the multivariate generalization of **R^2**. The Wilks' lambda λ statistic is interpreted just the opposite of **R^2**: a value near zero indicates high correlation while a value near one indicates low correlation.

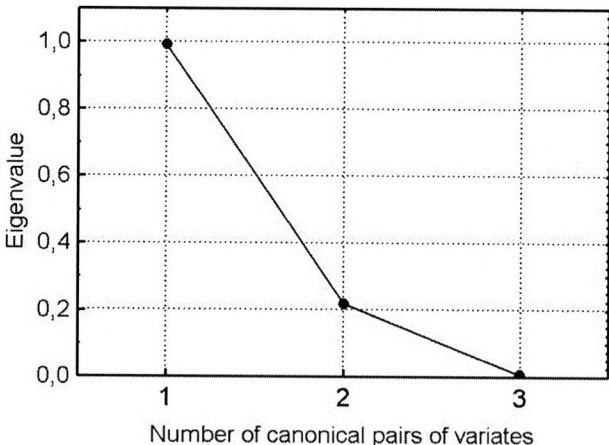

Figure 4.36   Cattel's scree plot of eigenvalue versus the number of canonical pairs of variates (STATISTICA).

- **Variation explained section:** This report displays the percent of the variation in each set of variables explained by other sets of variables.

Canonical Variate Number	Variation in these Variables	Explained by these Variates	Individual Percent Explained	Cumulative Percent Explained	Canonical Correlation Squared
1	U	U	37.4	37.4	0.9912
2	U	U	24.8	62.2	0.2185
3	U	U	37.8	100.0	0.0064
.	.	.	.	.	.
1	U	V	37.1	37.1	0.9912
2	U	V	5.4	42.5	0.2185
3	U	V	0.2	42.8	0.0064
.	.	.	.	.	.
1	V	U	37.2	37.2	0.9912
2	V	U	7.0	44.3	0.2185
3	V	U	0.2	44.5	0.0064
.	.	.	.	.	.
1	V	V	37.6	37.6	0.9912
2	V	V	32.1	69.7	0.2185
3	V	V	30.3	100.0	0.0064

Canonical Variate Number: This is the sequence number of the canonical variable being reported on. The maximum number of variates is the minimum of the number of variables in each set. **Variation in these Variables:** Each row of the report presents the results of how well a set of variables is explained by a particular canonical variate. This column designates which set of variables is being reported on. **Explained by these Variates:** Each row of the report presents the results of how well a set of variables is explained by a particular canonical variate. This column designates which set of canonical variates is being reported on. **Individual Percent Explained:** This column indicates the percentage of the variation in the designated set of variables that is explained by this canonical variate. **Cumulative Percent Explained:** This column indicates the cumulative percentage of the variation in the designated set of variables that is explained by this canonical variate and those listed above it. **Canonical Correlation Squared:** The square of the canonical correlation coefficient. This is repeated from an earlier report.

- **Standardized canonical weights:** interpreting canonical functions involves examining the sign and the magnitude of the canonical weight assigned to each variable in its canonical variate (or canonical root). Original variables with relatively larger weights contribute more to the variates, and vice versa. Similarly, original variables whose weights have opposite signs exhibit an inverse relationship with each other, and original variables with weights of the same sign exhibit a direct relationship. However, interpreting the relative importance or contribution of a original variable by its canonical weight is subject to the same criticisms associated with the interpretation of beta weights in regression techniques. A small weight may mean either that its corresponding variable is irrelevant in determining a relationship or that it has been partialed out of the relationship because of a high degree of multicollinearity. Another problem with the use of canonical weights is that these weights are subject to considerable instability (variability) from one sample to another. This instability occurs because the computational procedure for canonical analysis yields weights that maximize the canonical correlations for a particular sample of observed left set and right set variables [140]. These problems suggest considerable caution in using canonical weights to interpret the results of a canonical analysis. These coefficients are used to estimate the standardized scores for the V and U variates. They aid the interpretation of the variates by showing the weight given each variable in the construction of the variate. They are analogous to standardized beta coefficients in multiple regression.

.	V_1	V_2	V_3
Test4	−1,021	−0,105	−0,371
Test5	0,006	−0,990	−0,224
IQ	0,065	−0,230	1,050
.	U_1	U_2	U_3
Test1	−0,691	−0,592	−0,510
Test2	−0,656	0,428	0,636
Test3	0,009	−0,920	0,485

- **Canonical loadings:** Canonical loadings have been increasingly used as a basis for interpretation because of the deficiencies inherent in canonical weights. *Canonical loadings,* also called canonical structure correlations, measure the simple linear correlation between an original observed variable in the left set or right set variables and the set's canonical variate (Figure 4.37). The canonical loading reflects the variance that the observed variable shares with the canonical variate and can be interpreted like a factor loading in assessing the relative contribution of each variable to each canonical function. The larger the coefficient, the more important it is in deriving the canonical variate. The criteria for determining the significance of canonical structure correlations are the same as with factor loadings in factor analysis. Canonical loadings, like weights, may be subject to considerable variability from one sample to another. This variability suggests that loadings, and hence the relationships ascribed to them, may be sample-specific, resulting from chance or extraneous factors [140]. Although canonical loadings are considered re80latively more valid than weights as a means of interpreting the nature of canonical relationships, the researcher still must be cautious when using loadings for interpreting canonical relationships, particularly with regard to the external validity of the findings. The canonical loadings approach is somewhat more representative than the use of weights, just as was seen with factor analysis and discriminant analysis. Therefore, whenever possible the loadings approach is recommended as the best alternative to the canonical cross-loadings method. This spreadsheet shows the correlations between the variables and the variates. By determining which variables are highly correlated with a particular variate. it is hoped that we can determine its interpretation. For example, we can see that variate U_1 is highly correlated with *Test4*. Hence. we assume that U_1 has the same interpretation as *Test4*.

	U_1	U_2	U_3	V_1	V_2	V_3
Test1	−0,759	−0,310	−0,573	−0,755	−0,145	−0,046
Test2	−0,724	0,316	0,613	−0,721	0,148	0,049
Test3	0,152	−0,741	0,655	0,151	−0,346	0,052
Test4	−0,994	−0,009	0,005	−0,998	−0,019	0,058
Test5	0,178	−0,448	−0,018	0,179	−0,959	−0,221
IQ	−0,313	−0,099	0,074	−0,314	−0,211	0,926

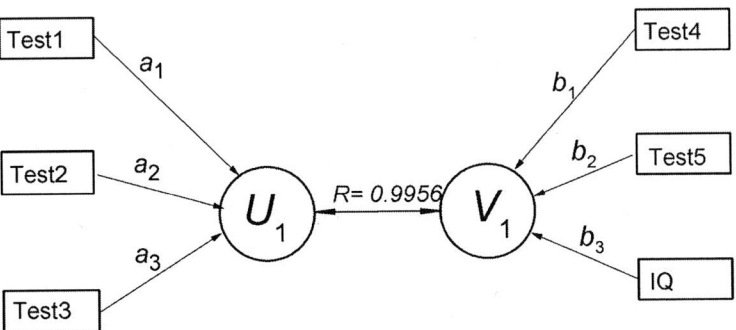

Figure 4.37 The first pair of canonical correlation coefficients U_1 and V_1 sufficiently describe the association of three original variables *Test1, Test2, Test3* on other three original variables *Test4, Test5, IQ* of data matrix (STATISTICA, ORIGIN).

- **Scores section:** This spreadsheet provides the canonical scores of each set of variates for each row of non-missing data. These are the values that are plotted in the score plots shown next.

Row	U_1	U_2	U_3	V_1	V_2	V_3
1	0,323	−0,660	−1,582	0,193	0,348	0,308
2	1,232	−1,150	−1,517	1,215	−0,351	−0,877
3	−0,103	0,304	1,370	0,026	−0,135	−0,251
4	−1,461	−1,887	0,139	−1,537	−1,992	0,658
5	−0,354	−0,712	−0,758	−0,190	−0,710	−0,455
6	−1,081	0,201	−0,490	−0,987	0,678	−0,115
7	−0,346	0,259	−0,491	−0,299	0,491	−0,709
8	0,955	2,032	−0,964	0,923	−0,503	−1,011
9	1,862	−0,580	0,952	1,882	0,288	−0,308
10	1,294	−0,757	1,298	1,334	−0,829	1,016
11	−0,188	1,200	−0,707	−0,112	1,151	2,742
12	−0,229	0,342	0,613	−0,329	−1,555	0,579
13	−0,699	−0,207	0,930	−0,736	1,038	−0,634
14	−1,457	0,684	0,247	−1,477	0,514	−1,202
15	0,252	0,932	0,961	0,095	1,568	0,259

- **Plots:** These reports show the relationship between each pair of canonical variates. The correlation coefficient of the data in the first plot (U_1 versus V_1) is the first canonical correlation coefficient (Figure 4.38).

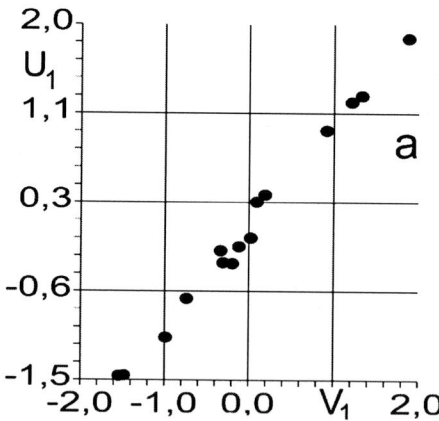

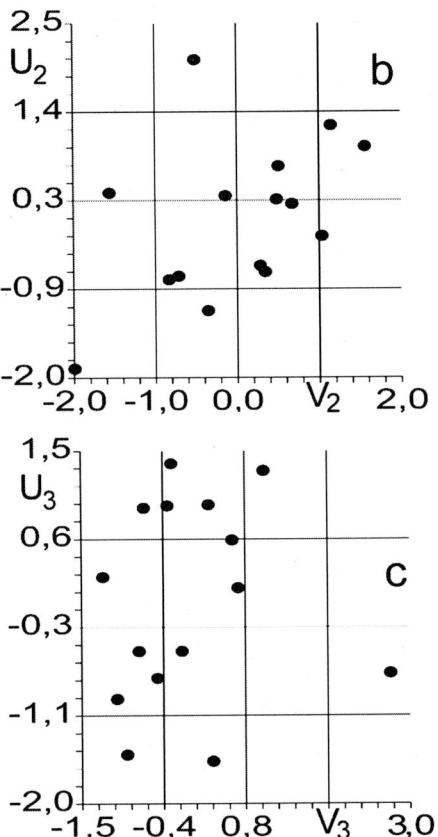

Figure 4.38   The scatterplot of first three pairs of canonical variates: (a) for the first pair U_1 and V_1, (b) for the second pair U_2 and V_2, (c) for the third pair U_3 and V_3, (STATISTICA).

The canonical correlation analysis addresses two primary objectives [60]: (1) the identification of dimensions among the left set and right set variables that (2) maximize the relationship between the dimensions. From a managerial perspective, this provides the researcher with some insight into the structure of the different variable sets as they relate to a dependence relationship. First, the results indicate only a single relationship exists, supported by the low practical significance of the second canonical function. In examining this relationship, we first see that the three left set variables are quite closely related and create a well-defined dimension for representing the right set variables. Second, this outcome dimension is fairly well predicted by the set of right set variables when acting as a set. It may be concluded that a relationship can be reduced to the form (*Test1, Test2*) = f(*Test4*).

4.7 Discriminant Analysis (DA)

Discriminant analysis techniques are used to classify individuals into one of two or more alternative groups on the basis of a set of measurements. The groups are known to be distinct, and each individual belongs to one of them. These techniques can also be used to identify which variables $x_1,...,x_m$ contribute to making the classification. Thus, as in regression analysis, we have two uses, prediction and description. Multiple regression predicts and explains metric variables but is not suitable for non-metric (categoric) variables. The investigator has one set of multivariate observations, the *training sample,* for which group membership is *known with certainty a priori*, and a second set, the *test sample*, consisting of observations for which group is *unknown* and which have to be assigned to one of the known groups as accurately as possible. Discriminant analysis provides classification of cases into groups where group membership is known, at least for the sample from whom the classification equations are derived. Cluster analysis is a similar procedure except that group membership is not known. Instead, the analysis develops groups on the basis of similarities among cases [131].

4.7.1 The Purpose and Objectives of DA

The basic purpose of discriminant analysis is to estimate the relationship between a single *nonmetric (categorical) dependent variable y* and a set of *metric independent variables* $x_1,...,x_m$, in this general form:

$$y = x_1 + x_2 + x_3 + ... + x_m$$

In many cases, the dependent variable consists of two groups or classifications, for example, male versus female or high versus low. In other instances, more than two groups are involved, such as low, medium, and high. Discriminant analysis is capable of handling either two groups or multiple (three or more) groups. When two groups are involved, the technique is referred to as two-group discriminant analysis. When three or more groups are identified, the technique is referred to as *multiple discriminant analysis (MDA)*, Fig. 4.39.

A review of the objectives for applying discriminant analysis should further clarify its nature. Discriminant analysis can address any of the following research objectives [60]:

(1) Determining whether statistically significant differences exist between the average score profiles on a set of variables for two (or more) a priori defined groups.

(2) Determining which of the independent variables account the most for the differences in the average score profiles of the two or more groups.

(3) Establishing procedures for classifying objects (individuals, firms, products, and so on) into groups on the basis of their scores on a set of independent variables.

(4) Establishing the number and composition of the dimensions of discrimination between groups formed from the set of independent variables.

Discriminant analysis, therefore, can be considered either a type of profile analysis or an analytical predictive technique.

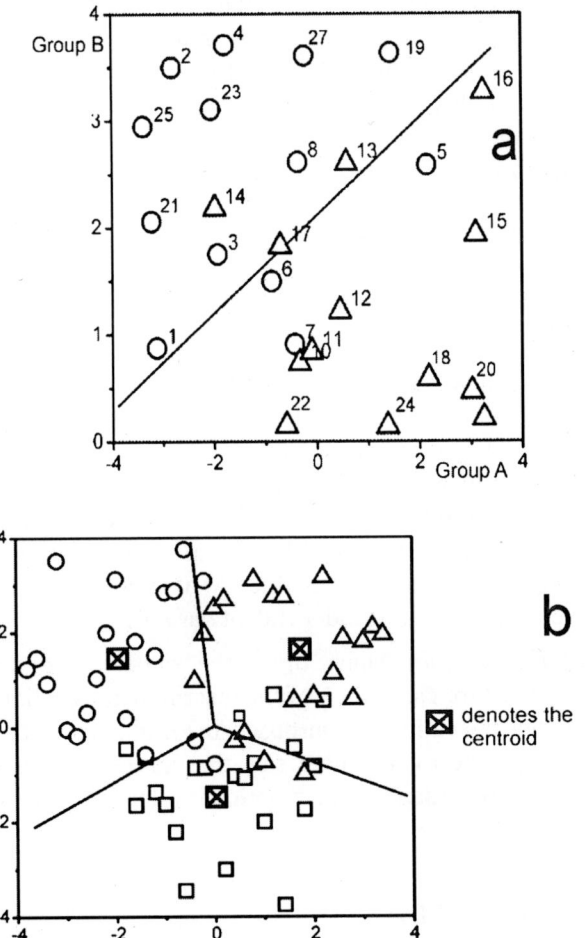

Figure 4.39 Teritorial map of the multiple discriminant analysis (MDA) (a) with two groups, (b) with three groups.

4.7.2 Research Design for DA

The primary goals of DA are to find the dimension or dimensions along which groups differ, and to find classification functions to predict group membership. The degree to which these goals are met depends, of course, on choice of predictors. The successful application of discriminant analysis requires consideration of several issues. These issues include the selection of both the dependent and the independent variables, the sample size needed for estimation of the discriminant functions, and the division of the sample for validation purposes [60].

Selection of variables: When using discriminant analysis, a question of interest is whether some subset of the original variables could provide a classification rule equal to the rule based on using all available variables. A variety of procedures have been suggested for identifying such subsets of variables. The general principle behind the majority of these methods is to choose a measure of 'separability' of the groups and then either sequentially accumulate the variables that maximize the measure, or beginning with all the available variables, sequentially eliminate those whose removal leads to the least reduction in separation. For Fisher's linear discriminant approach the criterion generally used to assess variables for entry (or removal) is their contribution to Hotelling's T^2-statistic, judged by an appropriate F-statistic. As before, the user may specify a value for the F-to-enter and F-to-remove values. F-to-enter is usually equal to a P of 0.15 and for the F-to-remove value a P of 0.30.

Sample size: Unequal sample sizes are acceptable. As a "rule of thumb", the smallest sample size should be at least 20 for a few (4 or 5) predictors. The maximum number of independent variables is $n - 2$, where n is the sample size. Discriminant analysis is quite sensitive to the ratio of sample size to the number of predictor variables. Many studies suggest a ratio of 20 observations for each predictor variable. The minimum size recommended is five observations per independent variable. At a minimum, the smallest group size must exceed the number of independent variables.

Division of the sample: Many times the sample is divided into two subsamples, one used for estimation of the discriminant function, the *training sample,* and another for validation purposes, the *testing sample.* It is essential that each subsample be of adequate size to support conclusions from the results. This method of validating the function is referred to as the *split-sample* or *cross-validation* approach.

4.7.3 Assumptions of DA

It is desirable to meet certain conditions for proper application of discriminant analysis. The key assumptions for deriving the discriminant function are multivariate normality of the independent variables and equal dispersion and covariance matrices for the groups as defined by the dependent variable [60].

Normal distribution: It is assumed that the data for the variables represent a sample from a multivariate normal distribution. However, violations of the normality assumption are not "fatal" and the resultant significance test are still reliable as long as non-normality is caused by skewness and not outliers [131].

Homogeneity of variances/covariances: DA is very sensitive to heterogeneity of variance-covariance matrices. Before accepting final conclusions for an important study, it is a good idea to review the within-groups variances and correlation matrices.

Outliers: DA is highly sensitive to the inclusion of outliers. We shall run a test for univariate and multivariate outliers for each group, and transform data or eliminate outliers. If one group in the study contains extreme outliers that impact the mean, they will also increase variability. Overall significance tests are based on pooled variances, that is, the average variance across all groups. Thus, the significance tests of the relatively larger means with the large variances would be based on the relatively smaller pooled variances, resulting erroneously in statistical significance.

Linearity: The DA model assumes linear relationships among all pairs of predictors within each group. The assumption is less serious (from some points of view) than others, however, in that violation leads to reduced power rather than increased Type I error.

Absence of multicollinearity and singularity: Multicollinearity or singularity may occur with highly redundant predictors, making matrix inversion unreliable. Fortunately, most computer programmes for DA protect against this possibility by testing tolerance. Predictors with insufficient tolerance are excluded. If one of the independent variables is very highly correlated with another, or one is a function (e.g., the sum) of other independents, then the tolerance value for that variable will approach 0 and the matrix will not have a unique discriminant solution. There must also be low multicollinearity of the independents. To the extent that independents are correlated, the standardized discriminant function coefficients will not reliably assess the relative importance of the predictor variables.

4.7.4 Fundamental Equations of DA

If predictors discriminate among groups, it is important to report just how the groups differ on those variables. The best estimate of central tendency in a population is the sample mean. The discrimination problem may be stated formally as follows: Suppose we have data for g groups, with N_j observations per group. Let n represent the total number of observations $y_i, i = 1,...,n$. Each observation consists of the measurements of m variables, $x_1,...,x_m$. The ith observation y_i is represented by x_{ji}. Let $\bar{Z}$ represent the vector of means of these variables across all groups, $\bar{Z}_j$ the vector of means of observations in the jth group and S_j the corresponding covariance matrix. We define three sums of squares and cross products matrices S_T, S_W and S_A as follows

$$S_T = \sum_{j=1}^{g}\sum_{i=1}^{N_j} \left(x_{ji} - \bar{Z}\right)\left(x_{ji} - \bar{Z}\right)^{\mathrm{T}}, \; S_W = \frac{\sum_{j=1}^{g}(N_j - 1)S_j}{(n-g)}, \; S_A = S_T - S_W, \text{ and two}$$

degrees of freedom values, $df1 = g - 1$ and $df2 = n - g$. The discrimination problem is how best to allocate an individual or observations $y_i, i = 1,...,n$, to one of these groups on the basis of a set of variables $x_1,...,x_m$. If associated with each group Z_j there is a probability density function of the measurements of the form $f_j(\mathbf{x})$, where $\mathbf{x}^{\mathrm{T}} = [x_1,...,x_m]^{\mathrm{T}}$ then a sensible rule for the allocation process would be to allocate the individual with vector of scores $\mathbf{x}$ to Z_j if $f_j(\mathbf{x}) = \max_{i \in 1,2,...,g} f_i(\mathbf{x})$. Suppose we have a single binary variable, x, and two groups G_1 and G_2. In the first assume that in G_1 it is $\Pr(x = 0) = \Pr(x = 1) = \dfrac{1}{2}$ while in $\Pr(x = 0) = \dfrac{1}{4}$, $\Pr(x = 1) = \dfrac{3}{4}$. The rule described will now allocate an individual with $x = 0$ to G_1 and an individual with $x = 1$ to G_2 [25, 26].

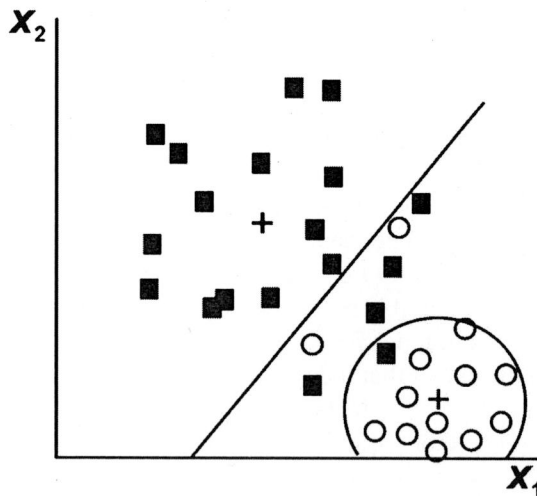

Figure 4.40 Teritorial map of the Fisher linear discriminant function (straight line) and the quadratic discriminant function (curve around circles) classify observations into two groups. Crosses denote the centroids [60].

The linear combination for a discriminant analysis, also known as the *Fisher discriminant Z-score function* Z_{ij}, is derived from an equation that takes the following form:

$$Z_{ij} = w_0 + w_1 x_{1i} + w_2 x_{2i} + ... + w_m x_{mi}, \quad i = 1,...,n, \text{ and } j = 1,...,g,$$

where Z_{ij} is the discriminant Z score of the discriminant function j for observation i, w_0 is the intercept, w_k is the discriminant weight for the kth independent variable, and x_{ki} is the kth independent variable for the ith observation, $i = 1,...,n$. The result is a single composite *discriminant Z score* for each individual or observation in the analysis. By averaging the discriminant scores for all the individuals within a particular group, we arrive at the group mean which is referred to as a *centroid* $\overline{Z}_j$. When the analysis involves two groups, there are two centroids, $\overline{Z}_1$ and $\overline{Z}_2$; with three groups, there are three centroids $\overline{Z}_1, \overline{Z}_2$ and $\overline{Z}_3$; and so forth. The centroids indicate the most typical location of any individual observation from a particular group, and a comparison of the group centroids shows how far apart the groups are along the dimension being tested. The Fisher discriminant Z-score function Z_{ij} is a weighted average of the values of the independent variables. The weights are selected so that the resulting weighted average separates the observations into the groups. High values of the average come from one group, low values of the average come from another group. The problem reduces to one of finding the weights which, when applied to the data, best discriminate among groups according to some criterion. The solution reduces to finding the eigenvectors, V, of $S_W^{-1} S_A$. The canonical coefficients are the elements of these eigenvectors [142].

A goodness-of-fit parameter, *Wilks' lambda*, is defined as follows:

$$\lambda = \frac{|S_W|}{|S_T|} = \prod_{j=1}^{m} \frac{1}{1 + \lambda_j},$$

where λ_j is the jth eigenvalue corresponding to the eigenvector described above and m is the minimum of K-1 and p.

The *canonical correlation* between the jth discriminant function and the independent variables is related to these eigenvalues as follows:

$$r_{cj} = \sqrt{\frac{\lambda_j}{1 + \lambda_j}}.$$

Various other matrices are often considered during a discriminant analysis.

The *overall covariance matrix, T,* is given by:

$$T = \left(\frac{1}{n-1}\right)S_T$$

The *within-group covariance matrix, W,* is given by:

$$W = \left(\frac{1}{n-g}\right)S_W$$

The *among-group (or between-group) covariance matrix, A,* is given by:

$$A = \left(\frac{1}{g-1}\right)S_A$$

In deriving his linear discriminant function Z_{ij}, Fisher [39] did not have to make any distributional assumptions for the variables used in classification. We denote the two mean values of Z by $\bar{Z}_I$ and $\bar{Z}_{II}$ and the pooled sample variance of Z by s_Z^2 (what is the statistic similar to the pooled variance used in the standard two-sample t test). To measure how 'far apart' the two groups are in terms of values of Z, we compute $D^2 = \frac{\left(\bar{Z}_I - \bar{Z}_{II}\right)^2}{s_Z^2}$. Fisher selected the coefficients $w_0, w_1, ..., w_m$ so that D^2 has the maximum possible value. The term D^2 can be interpreted as the squared distance between the means of the standardized value of Z. A larger value of D^2 indicates that it is easier to discriminate between the two groups. The quantity D^2 is called the *Mahalanobis distance.* Both w_i, and D^2 are functions of the group means and the pooled variances and covariances of the variables.

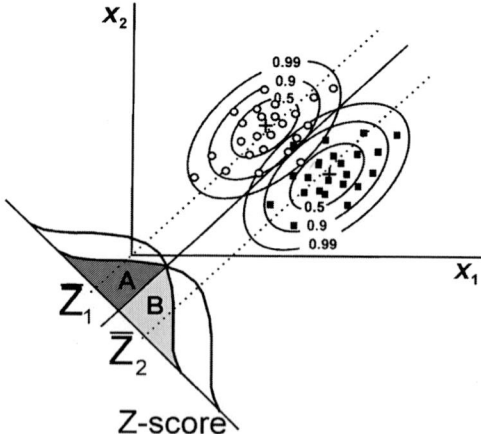

Figure 4.41 Graphical illustration of two-variable profiles of two-group discriminantion analysis [60].

A graphic illustration of two-group analysis will help to further explain the nature of discriminant analysis [60]. Fig. 4.42 demonstrates what happens when a two-group discriminant function is computed. Assume we have two groups, A and B, and two measurements, x_1 and x_2, on each member of the two groups. We can plot in a scatter diagram the association of variable x_1 with variable x_2 for each member of the two groups. In Figure 4.42 the full dots represent the variable measurements for the members of group B and the empty circles those for group A. The ellipses drawn around the full dots and empty circles would enclose some prespecified proportion of the points, usually 95 percent or more in each group. If we draw a straight line through the two points at which the ellipses intersect and then project the line to a new Z axis, we can say that the overlap between the univariate distributions (represented by the shaded area) is smaller than would be obtained by any other line drawn through the ellipses formed by the scatterplots. The important thing to note about Fig. 4.42 is that the Z axis expresses the two-variable profiles of groups A and B as single numbers (discriminant scores). By finding a linear combination of the original variables x_1 and x_2, we can the results as a discriminant function. For example, if the empty circle and full dots are projected onto the new Z axis as discriminant Z scores, the result condenses the information about group differences (shown in the x_1 and x_2 plot) into a set of points (*Z scores*) on a single axis. To summarize, for a given discriminant analysis problem, a linear combination of the independent variables is derived, resulting in a series of discriminant scores for each object in each group. The discriminant scores are computed according to the statistical rule of maximizing the variance between the groups and minimizing the variance within them. If the variance between the groups is large relative to the variance within the groups, we say that the discriminant function separates the groups well.

4.7.5 Adjusting the Value of the Dividing Point

The *dividing point C* or also called *cutting score* is the score criterion against which each object's discriminant score is compared to determine into which group the object should be classified. In constructing classification matrices, the researcher will want to determine the *optimum cutting score* (it is called a *critical Z value*). The optimal cutting scores will differ depending on whether the sizes of the groups are equal or unequal [26].

The linear combination for a discriminant analysis, also known as the *Fisher discriminant Z-score function* Z_{ij}, is derived from an equation that takes the following form: $Z_{ij} = w_0 + w_1 x_{1i} + w_2 x_{2i} + ... + w_m x_{mi}$, $i = 1,...,n$, and $j = 1,...,g$, where Z_{ij} is the discriminant Z score of the discriminant function *j* for object *i*, w_0 is the intercept, w_k is the discriminant weight for the *k*th independent variable, and x_{ki} is the *k*th independent variable for the *i*th observation, $i = 1, ..., n$. The result is a single composite *discriminant Z score* for each individual or observation in the analysis. By averaging the discriminant scores for all the individuals within a particular group, we arrive at the group mean which is referred to as a *centroid* $\bar{Z}_j$. When the analysis involves two groups, there are two centroids, $\bar{Z}_1$ and $\bar{Z}_2$; with three groups, there are three centroids $\bar{Z}_1, \bar{Z}_2$ and $\bar{Z}_3$; and so forth.

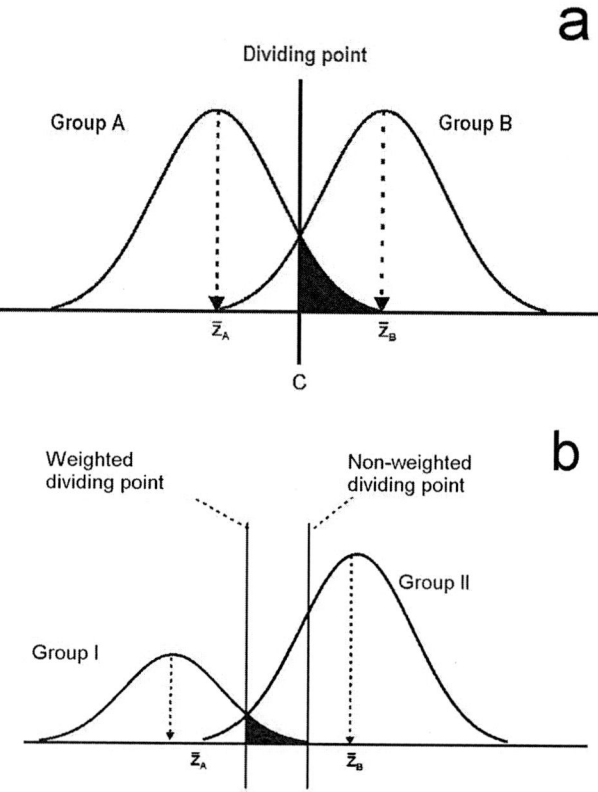

Figure 4.42 Dividing point *C* and distribution of variables *x* in group I and II for (a) equal variances, and (b) unequal variances.

If the groups are of equal size, the optimal cutting score will be halfway between the two group centroids, defined as $C = \dfrac{\bar{Z}_I + \bar{Z}_{II}}{2}$ where *C* is the critical cutting score value for equal group sizes, $\bar{Z}_I$ the centroid for group I, and $\bar{Z}_{II}$ is the centroid for group II (Figure 4.42). The dividing point (or cutting score) *C* has been used as the point producing an equal percentage of errors of both types, i.e. the probability of misclassifying an individual from group I into group II, or *vice versa*. To explain how this choice is made, we must introduce the concept of *prior probability*. Since the two groups constitute an overall group, it is of interest to examine their relative size. The prior probability of group I is the probability that an individual selected at random actually comes from group I. In other words, it is the proportion of individuals in the overall group who fall in group I. This proportion is denoted by q_I.

The theoretical choice of the dividing point C is made so that the total probability of misclassification is minimized. This total probability is defined as q_1 times the probability of misclassifying an individual from group I into group II plus q_{II} times the probability of misclassifying an individual from group II into group I: q_I Prob(II given I) + q_{II} Prob(I given II). The optimal choice of the dividing point C is $C = \dfrac{\overline{Z}_I + \overline{Z}_{II}}{2} + \ln \dfrac{q_{II}}{q_I}$. If $q_I = q_{II} = \dfrac{1}{2}$, then

$q_{II} / q_I = 1$ and $\ln q_{II} / q_I = 0$. In this case C is $C = \dfrac{\overline{Z}_I + \overline{Z}_{II}}{2}$ if $q_I = q_{II}$.

4.7.6 Discriminant Model and Assessing Overall Fit

To derive the discriminant function, the researcher must decide on the method of estimation and then determine the number of functions to be retained. With the functions estimated, overall model fit can be assessed in several ways. First, *discriminant Z scores* can be calculated for each object. Comparison of the group means $\overline{Z}_I$ and $\overline{Z}_{II}$ on the Z scores provides one measure of discrimination between groups. Predictive accuracy is measured as the number of observations classified into the correct groups. A number of criteria are available to assess whether the classification process achieves practical and/or statistical significance.

Computational methods: Two computational methods can be utilized in deriving a discriminant function: the simultaneous method and the stepwise method.

Simultaneous estimation involves direct computing the discriminant function so that all of the independent variables are considered concurrently. Thus the discriminant function is computed based upon the entire set of independent variables, regardless of the discriminating power of each independent variable. The simultaneous method is appropriate when, for theoretical reasons, the researcher wants to include all the independent variables in the analysis and is not interested in seeing intermediate results based only on the most discriminating variables.

Stepwise estimation is an alternative to the simultaneous approach. It involves entering the independent variables into the discriminant function one at a time on the basis of their discriminating power. The stepwise approach begins by choosing the single best discriminating variable. The initial variable is then paired with each of the other independent variables one at a time, and the variable that is best able to improve the discriminating power of the function in combination with the first variable is chosen. The third and any subsequent variables are selected in a similar manner. As additional variables are included, some previously selected variables may be removed if the information they contain about group differences is available in some combination of the other variables included at later stages. Related to criteria for statistical inference is the choice among methods to direct the progression of entry of predictors in stepwise discriminant function analysis. Different methods of progression maximize group differences along different statistical criteria. Selection of stepping method depends on the availability of programmes and choice of statistical criterion. If, for example, the statistical criterion is Wilks' lambda, it is beneficial to choose the stepping method that minimizes lambda. Or, if the statistical criterion is "change in Rao's V," the obvious choice of stepping method is Rao's V.

Statistical significance: Criteria for evaluating overall statistical reliability in DA are Wilks' lambda, Roy's gcr, Hotelling's trace, and Pillai's criterion. Two additional statistical criteria, Mahalanobis' D^2 and Rao's V, are especially relevant to stepwise DA. Mahalanobis' D^2 is based on distance between pairs of group centroids which is then generalizable to distances over multiple pairs of groups. Rao's V *is* another generalized distance measure that attains its largest value when there is greatest overall separation among groups. These two criteria are available both to direct the progression of stepwise discriminant function analysis and to evaluate the reliability of a set of predictors to predict group membership. After the discriminant function has been computed, the researcher must assess its calculated level of significance. The conventional significance criterion of 0.05 or beyond is often used. Many researchers believe that if the function is not significant at or beyond the 0.05 level, there is little justification for going further. The rule for continuing to a higher significance level (e.g., 0.10 or more) is the cost versus the value of the information. If the higher levels of risk for including nonsignificant results (e.g., significance levels > 0.05) are acceptable, discriminant functions may be retained that are significant at the 0.2 or even the 0.3 level. If the number of groups is three or more, then the researcher must decide not only if the discrimination between groups overall is statistically significant but also if each of the estimated discriminant functions is statistically significant [25].

Number of discriminant functions: In discriminant function analysis with more than two groups, a number of discriminant functions are extracted. The maximum number of functions is the lesser of either degrees of freedom for groups or, as in canonical correlation, principal components analysis and factor analysis, equal to the number of predictors. As in these other analyses, some functions often carry no worthwhile information. It is frequently the case that the first one or two discriminant functions account for the lion's share of discriminating power, with no additional information forthcoming from the remaining functions.

Assessing overall fit: Once the significant discriminant functions have been identified, attention shifts to ascertaining the overall fit of the retained discriminant function(s). This assessment involves three tasks: calculating discriminant Z scores for each observation, evaluating group differences on the discriminant Z scores, and assessing group membership prediction accuracy. With the retained discriminant functions defined, the basis for calculating the discriminant Z scores has been established. This score, a metric variable, provides a direct means of comparing observations on each function. Observations with similar Z scores are assumed more alike on the variables constituting this function than those with disparate scores. We should note that the discriminant function differs from the *classification function,* also known as *Fisher's linear discriminant function.* The classification functions, one for each group, can be used in classifying observations. In this method of classification, an observation's values for the independent variables are inserted in the classification functions and a classification score for each group is calculated for that observation. The observation is then classified into the group with the highest classification score. We use the discriminant function as the means of classification because it provides a concise and simple representation of each discriminant function, simplifying the interpretation process and the assessment of the contribution of independent variables.

4.7.7 Interpretation of the Discriminant Function

If a primary goal of analysis is to discover and interpret the combinations of predictors (in form of the discriminant functions) that separate groups in various ways. If the discriminant function is statistically significant and the classification accuracy is acceptable, the researcher should focus on making substantive interpretations of the findings. This process involves examining the discriminant functions to determine the relative importance of each independent variable in discriminating between the groups. Three methods of determining the relative importance have been proposed: (1) standardized discriminant weights, (2) discriminant loadings (structure correlations), and (3) partial F values [131].

Discriminant function plots: Groups are spaced along the various discriminant functions according to their centroids. Discriminant functions form axes and the centroids of the groups are plotted along the axes. If there is a big difference between the centroid of one group and the centroid of another along a discriminant function axis, the discriminant function separates the two groups. If there is not a big distance, the discriminant function does not separate the two groups. Many groups can be plotted along a single axis (Fig. 4.42).

Discriminant weights: The traditional approach to interpreting discriminant functions examines the sign and magnitude of the standardized *discriminant weight* (sometimes referred to as a *discriminant coefficient*) assigned to each variable in computing the discriminant functions. When the sign is ignored, each weight represents the relative contribution of its associated variable to that function. Independent variables with relatively larger weights contribute more to the discriminating power of the function than do variables with smaller weights. The sign denotes only that the variable makes either a positive or a negative contribution [141]. The interpretation of discriminant weights is analogous to the interpretation of beta weights in regression analysis and is therefore subject to the same criticisms. For example, a small weight may indicate either that its corresponding variable is irrelevant in determining a relationship or that it has been partialed out of the relationship because of a high degree of multicollinearity. Another problem with the use of discriminant weights is that they are subject to considerable instability. These problems suggest caution in using weights to interpret the results of discriminant analysis.

Discriminant loadings: In recent years, loadings have increasingly been used as a basis for interpretation because of the deficiencies in utilizing weights. Discriminant loadings, referred to sometimes as *structure correlations,* measure the simple linear correlation between each independent variable and the discriminant function. The discriminant loadings reflect the variance that the independent variables share with the discriminant function and can be interpreted like factor loadings in assessing the relative contribution of each independent variable to the discriminant function. Discriminant loadings (like weights) may be subject to instability. Loadings are considered relatively more valid than weights as a means of interpreting the discriminating power of independent variables because of their correlational nature. The researcher still must be cautious when using loadings to interpret discriminant functions. Plots of centroids tell us how groups are separated by a discriminant function, but they do not reveal the meaning of the discriminant function. The meaning of the function is inferred by a researcher from the pattern of correlations between the function and the predictors. Correlations between predictors and functions are called loadings in both discriminant function analysis and factor analysis. If predictors x_1, x_2 and x_3 load (correlate) highly with the function but predictors x_4 and x_5 do not, the researcher attempts to understand what x_1, x_2 and x_3 have in common with each other that is different from x_4 and x_5; the meaning of the function is determined by this understanding.

Partial F values: Two computational approaches–simultaneous and stepwise–can be utilized in deriving discriminant functions. When the stepwise method is selected, an additional means of interpreting the relative discriminating power of the independent variables is available through the use of partial F values. This is accomplished by examining the absolute sizes of the significant F values and ranking them. Large F values indicate greater discriminatory power. In practice, rankings using the F-values approach are the same as the ranking derived from using discriminant weights, but the F values indicate the associated level of significance for each variable.

4.7.8 Validation of the Results

The final stage of a discriminant analysis involves validating the discriminant results to provide assurances that the results have external as well as internal validity. With the propensity of discriminant analysis to inflate the hit ratio if evaluated only on the analysis sample, cross-validation is an essential step. Most often the cross-validation is done with the original *training sample*, but it is possible to employ an additional sample as the *testing sample*. In addition to cross-validation, the researcher should use group profiling to ensure that the group means are valid indicators of the conceptual model used in selecting the independent variables.

Split-sample or *cross-validation procedures:* Recall that the most frequently utilized procedure in validating the discriminant function is to divide the groups randomly into training and testing samples. This involves developing a discriminant function with the training sample and then applying it to the testing sample. The justification for dividing the total sample into two groups is that an upward bias will occur in the prediction accuracy of the discriminant function if the observations used in developing the classification matrix are the same as those used in computing the function; that is, the classification accuracy will be higher than is valid for the discriminant function, if it was used to classify a separate testing sample. The implications of this upward bias are particularly important when the researcher is concerned with the external validity of the findings [131].

Profiling group differences: Another validation technique is to profile the groups on the independent variables to ensure their correspondence with the conceptual bases used in the original model formulation. When the researcher has identified the independent variables that make the greatest contribution in discriminating between the groups, the next step is to profile the characteristics of the groups based on the group means. This profile enables the researcher to understand the character of each group according to the predictor variables. Another approach is to profile the groups on a separate set of variables that should mirror the observed group differences. This separate profile provides an assessment of external validity in that the groups vary on both the independent variable(s) and the set of associated variables.

Problem 4.4 *Taxonometric classification of iris plants with discriminant analysis (Fisher, [39])*

Discriminant analysis finds a set of prediction equations, based on sepal and petal measurements, that classify additional irises into one of these three varieties. Here *iris* is the dependent variable, while *Sepallength, Sepalwidth,*

Petallength, and Petalwidth are the independent variables. It contains the lengths and widths of sepals and petals of three types of irises (*Setosa, Versicol,* and *Virginic*). The purpose of the analysis is to learn how one can discriminate between the three types of plants, based on the four measures of width and length of petals and sepals and build a "classification model" of how one can best predict to which group a case belongs. In the following discussion we will use the term "in the model" in order to refer to variables that are included in the prediction of group membership, and we will refer to variables as being "not in the model" if they are not included.

Data: Data are measurements in centimeters of sepal length, sepal width, petal length, and petal width of fifty plants for each of three varieties of iris: (1) Iris setosa, (2) Iris versicolour, and (3) Iris virginica. Iris versicolour is a polyplid hybrid of the two other species. Iris setosa is a diploid species with 38 chromosomes, Iris virginica is a tetraploid, and Iris versicolour is a hexaploid with 108 chromosomes. The first two variables in this file (*Sepallength, Sepalwidth*) pertain to the length and width of sepals; the next two variables (*Petallength, Petalwidth*) pertain to the length and width of petals. The last variable in this file is a grouping or coding variable that identifies to which type of iris each plant belongs (*Setosa, Versicol,* and *Virginic*). In all, there are 150 plants in this sample, 50 of each type. If missing values are found in any of the independent variables being used, the row is omitted. If they occur only in the dependent (categorical) variable, the row is not used during the calculation of the prediction equations, but a predicted group (and scores) is calculated. This allows you to classify new observations.

Sepallength	Sepallwidth	Petallength	Petalwidth	Types of iris
5	3,3	1,4	0,2	Setosa
6,4	2,8	5,6	2,2	Virginic
.....	.....	.....	.....	.....
.....	.....	.....	.....	.....
5	2,3	3,3	1	Versicol

(the spreadsheet continues for all 150 iris plants)

Softwar   ution:

- **Group means report:** This spreadsheet shows the means of each of the independent variables across each of the iris groups. The last row shows the count (number of iris plants) in the group. The column headings come from the use of value labels for the iris group variable.

Variable	Setosa	Versicol	Virginic	Overall
Sepallength	5,006	5,936	6,588	5,843
Sepalwidth	3,428	2,77	2,974	3,057
Petallength	1,462	4,26	5,552	3,758
Petalwidth	0,246	1,326	2,026	1,199
Count	50	50	50	150

- **Group standard deviations report:** It is assumed that the variance/covariance matrices of variables are homogeneous across groups. Minor deviations are not that important; however, before accepting final conclusions for an important study it is probably a good idea to review the within-groups variances and correlation matrices. This spreadsheet shows the standard deviations of each of the independent variables across each of the iris groups. The last row shows the count or number of iris plants in the group. Discriminant analysis makes the assumption that the covariance matrices are identical for each of the iris groups. This spreadsheet lets us glance at the standard deviations to check if they are about equal.

Variable	Setosa	Versicol	Virginic	Overall
Sepallength	0,352	0,516	0,636	0,828
Sepalwidth	0,379	0,313	0,322	0,436
Petallength	0,173	0,470	0,552	1,765
Petalwidth	0,105	0,198	0,275	0,762
Count	50	50	50	150

- **Total correlation\covariance report:** This spreadsheet shows the correlation and covariance matrices that are formed when the grouping variable is ignored. The correlations are on the lower left and the covariances are on the upper right. The variances are on the diagonal.

Variable	Sepallength	Sepalwidth	Petallength	Petalwidth
Sepallength	0,686	−0,042	1,274	0,516
Sepalwidth	−0,117	0,190	−0,329	−0,122
Petallength	0,872	−0,428	3,116	1,296
Petalwidth	0,818	−0,366	0,963	0,581

- **Between-group correlation\covariance report:** This spreadsheet displays the correlations and covariances formed using the group means as the individual iris plants. The correlations are shown in the lower-left half of the matrix. The within-group covariances are shown on the diagonal and in the upper-right half of the matrix. If there are only two groups, all correlations will be equal to one since they are formed from only two rows (the two group means).

Variable	Sepallength	Sepalwidth	Petallength	Petalwidth
Sepallength	31,61	−9,98	82,62	35,64
Sepalwidth	−0,75	5,67	−28,62	−11,47
Petallength	0,99	−0,81	218,55	93,39
Petalwidth	1,00	−0,76	1,00	40,21

- **Within-group correlation\covariance report:** This spreadsheet shows the correlations and covariances that would be obtained from data in which the group means had been subtracted. The correlations are shown in the lower-left half of the matrix. The within-group covariances are shown on the diagonal and in the upper-right half of the matrix.

Variable	Sepallength	Sepalwidth	Petallength	Petalwidth
Sepallength	0,265	0,093	0,168	0,038
Sepalwidth	0,530	0,115	0,055	0,033
Petallength	0,756	0,378	0,185	0,043
Petalwidth	0,365	0,471	0,484	0,042

- **Box-and-whisker plot:** The box-and-whisker plot is useful to indicate the potential outliers. We can also view a symmetry of the distribution of the variables within each level of the grouping variable (Fig. 4.43).

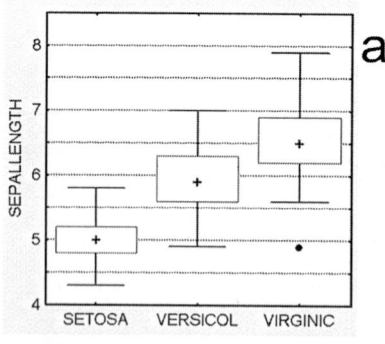
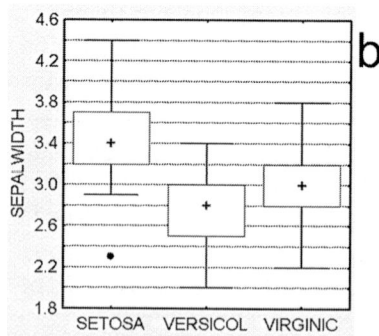

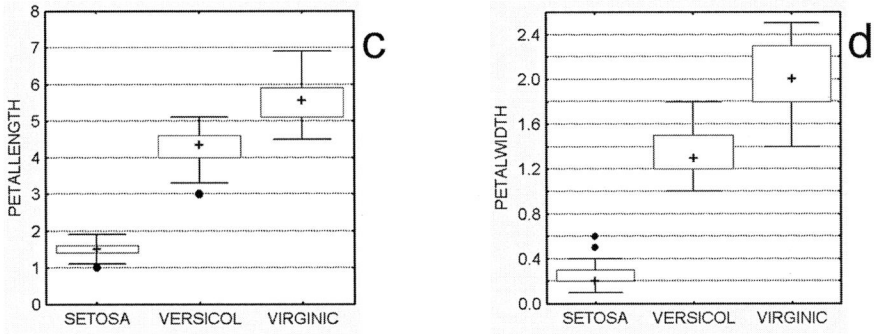

Figure 4.43   The box-and-whisker plot by group for all four variables: (a) Sepallength, (b) Sepalwidth, (c) Petallength, and (d) Petalwidth, (STATISTICA).

- **Histogram:** A histogram of the probability density function of an actual sample distribution for each variable helps to examine the sample distribution (Fig. 4.44).

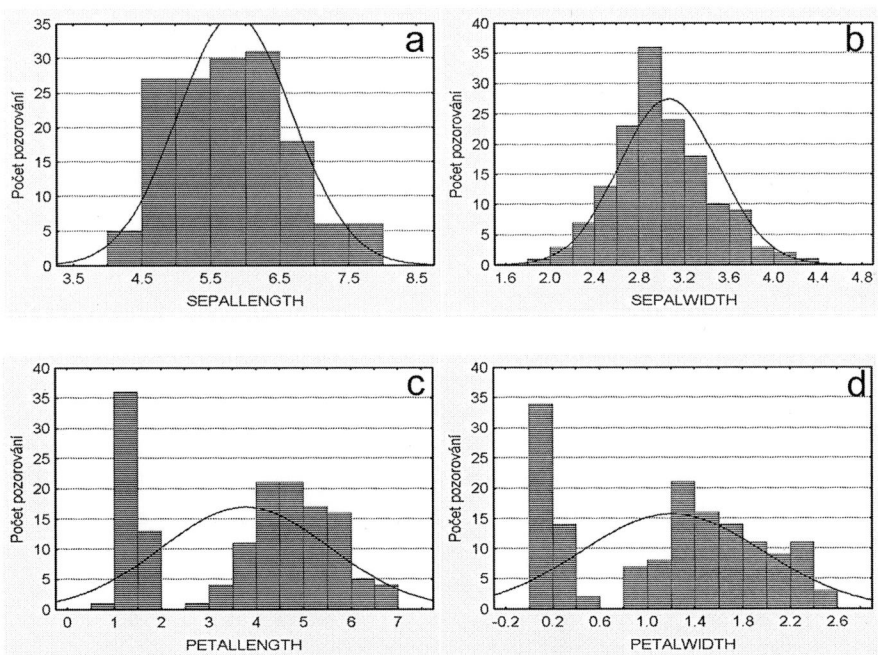

Figure 4.44   Histogram and the graph of the kernel estimate of the probability density function for all four variables: (a) Sepallength, (b) Sepalwidth, (c) Petallength, and (d) Petalwidth, (STATISTICA).

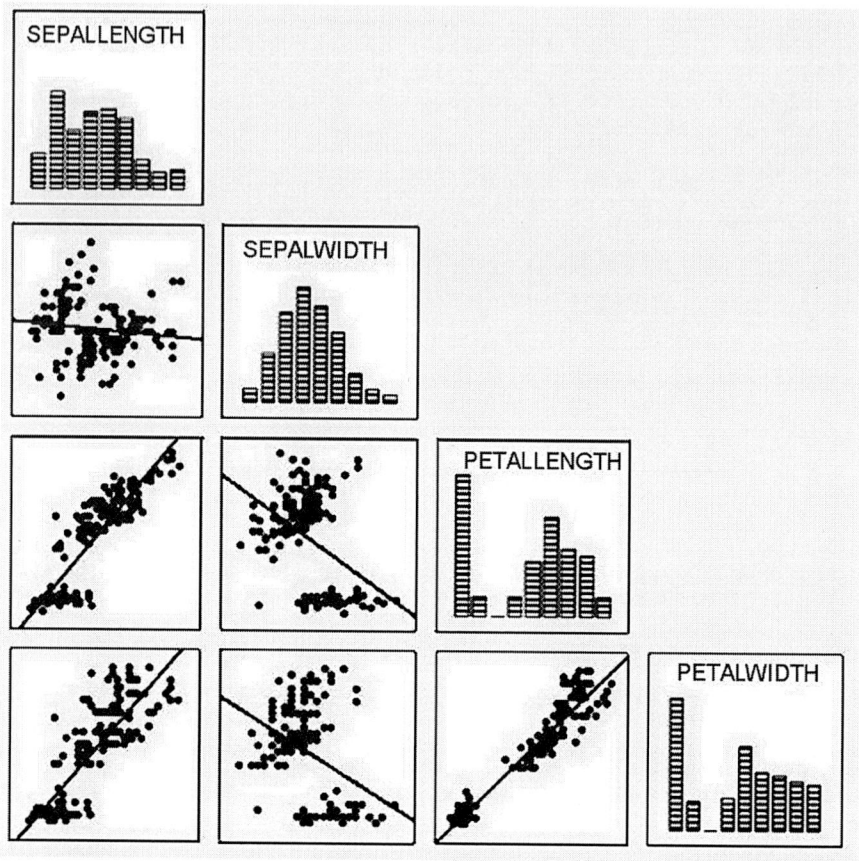

Figure 4.45 Total correlation matrix plot contains scatterplots and histograms of all four variables, Sepallength, Sepalwidth, Petallength, and Petalwidth, (STATISTICA).

- **Correlation matrix scatterplot.** Another type of graph of interest would be the scatterplot of correlations between variables included in the analysis. We graphically view the correlations between variables together in a matrix scatterplot (Fig. 4.45). It appears that there are two "clouds" of points in this plot. Perhaps the points in the lower-left corner of this plot all belong to one iris type. If so, then there is good "hope" for this discriminant analysis. However, if not, then the possibility that the underlying distribution for these two variables is not bivariate normal, but rather multimodal with more than one "peak," would have to be considered. This scatterplot shows the correlation between variables *Sepallength* and *Petallength* within groups. Thus, it can be concluded that the assumption of a bivariate normal distribution within each group is probably not violated for this particular pair of variables.

- **Specifying discriminant function analysis.** We perform a stepwise analysis in order to see what happens at each step of the discriminant analysis. The programme will keep "stepping" until one of four things happen: 1. All variables have been entered or removed, or 2. The maximum number of steps has been reached, as specified, or 3. No other variable that is not in the model has an F value greater than the F *to enter* that is specified in this dialog and when no other variable in the model has an F value that is smaller than the F *to remove* specified in this dialog, or 4. Any variable after the next step would have a tolerance value that is smaller than that specified. When stepping forward, programme will select the variable for inclusion that makes the most significant unique (additional) contribution to the discrimination between groups; that is, programme will choose the variable with the largest F value (greater than the respective user-specified F *to enter* value). When stepping backward, programme will select the variable for exclusion that is least significant, that is, the variable with the smallest F value (less than the respective user-specified F *to remove* value). Therefore, if we want to enter all variables in a forward stepwise analysis, we set the F *to enter* value as small as possible (and the F *to remove* to 0). If we want to remove all variables from a model, one by one, we set F *to enter* to a very large value (e.g., 9999), and also set F *to remove* to a very large value that is only marginally smaller than the F *to enter* value (e.g., 9998). We should remember that the F *to enter* value must always be set to a larger value than the F *to remove* value.
- **Tolerance:** At each step the programme will compute the multiple correlation R^2 for each variable with all other variables that are currently included in the model. The *tolerance* value of a variable is then computed as $1 - R^2$. Thus, the tolerance value is a measure of the redundancy of a variable. If a variable that is about to enter into the model has a tolerance value of 0.01, then this variable can be considered to be 99% redundant with the variables already included. At one point, when one or more variables become too redundant, the variance-/covariance matrix of variables included in the model can no longer be inverted, and the discriminant function analysis cannot be performed. It is generally recommended that we leave the tolerance setting at its default value of 0.01. If a variable is included in the model that is more than 99% redundant with other variables, then its practical contribution to the improvement of the discriminatory power is dubious. More importantly, if we set the tolerance to a much smaller value, round-off errors may result, leading to unstable estimates of parameters. Since no other variables have been chosen yet, all R^2 are equal to 1.0. Overall, the discrimination between types of irises is

highly significant (*Wilks' Lambda* $= 0.037$; $F = 307.1$, $p < 0.0001$). Now we look at the independent contributions to the prediction for each variable in the model. As we can see, both variables that are not yet in the model have *F to enter* values that are larger than 1; thus, we know that the stepping will continue and that the next variable that will enter into the model is the variable *Petalwidth*.

- **Wilks' lambda** is the standard statistic that is used to denote the statistical significance of the discriminatory power of the current model. Its value will range from 1.0 (no discriminatory power) to 0.0 (perfect discriminatory power). Each value in the first column of the spreadsheet shown above denotes the Wilks' lambda after the respective variable is entered into the model.

- **Partial Wilks' lambda** is the Wilks' lambda for the unique contribution of the respective variable to the discrimination between groups. In a sense, one can look at this value as the equivalent to the partial correlation coefficients reported in multiple regression. Because a *lambda* of 0.0 denotes perfect discriminatory power, the lower the value in this column, the greater is the unique discriminatory power of the respective variable. Because no variable has been entered into the model yet, the Partial Wilks' lambda at step 0 is equal to the Wilks' lambda after the variable is entered, that is, the values reported in the first column of the spreadsheet.

- **F to enter and p-level**. Wilks' lambda can be converted to a standard *F* value, and we can compute the corresponding *p-levels* for each *F*. One is always capitalizing on chance when including several variables in an analysis without having any a priori hypotheses about them, and choosing to interpret only those that happen to be "significant" is not appropriate. In short, there is a big difference between predicting *a priori* a significant effect for a particular variable and then finding that variable to be significant, as compared to choosing from among 100 variables in the analysis the one that happens to be significant. Without going into details, in purely practical terms, in the latter case, it is not very likely that we would find the same variable to be significant if we were to replicate the study. When reporting the results of a discriminant function analysis, we should be careful not to leave the impression as if only the significant variables were chosen in the first place, when, in fact, they were chosen because they happened to "work." Looking at the spreadsheet above, you can see that the largest *F to enter* is shown for variable *Petallen*. Thus, that variable will be entered into the model at the next (first) step.

- **Automatic variable selection:** A common task in discriminant analysis is variable selection. Often we have a large pool of possible independent variables from which we want to select a smaller set (up to about eight variables) which will do almost as well at discriminating as the complete set. The automatic variable selection is run so that it will first find the best discriminator and then the second best. After it has found two, it checks whether the discrimination would be almost as good if one were removed. This stepping process of adding the best remaining variable and then checking if one of the active variables could be removed continues until no new variable can be found whose F-value has a probability smaller than the Probability Enter value p.

Iteration	Action This Step	Independent Variable	Pct Chg In Lambda	F-Value	Prob Level	Wilks' Lambda
0	None					1.000
1	Entered	Petal Length	94.14	1180.16	0.000	0.058
2	Entered	Sepal Width	37.09	43.04	0.000	0.037
3	Entered	Petal Width	32.23	34.57	0.000	0.025
4	Entered	Sepal Length	6.15	4.72	0.010	0.023

This spreadsheet shows what action was taken at each step. **Iteration:** This gives the number of this step. **Action This Step:** This tells what action (if any) was taken during this step. "Entered" means that the variable was entered into the set of active variables. "Removed" means that the variable was removed from the set of active variables. **Pct Chg In Lambda:** This is the percentage decrease in lambda that resulted from this step. Wilks' lambda is analogous to $1 - R^2$ in multiple regression. Hence, we want to *decrease* Wilks' lambda to improve our model. For example, going from iteration 2 to iteration 3 results in lambda decreasing from 0.036884 to 0.024976. This is a 32.29% decrease in lambda. **F-Value:** This is the F-ratio for testing the significance of this variable. If the variable was "Entered," this tests the hypothesis that the variable should be added. If the variable was "Removed," this tests whether the variable should be removed. **Prob Level:** The significance level of the above F-Value. **Wilks' Lambda:** The multivariate extension of R^2. Wilks' lambda reduces to $1 - R^2$ in the two-group case. It is interpreted just backwards from R^2. It varies from one to zero. Values near one imply low predictability, while values close to zero imply high predictability. Note that this Wilks' lambda value corresponds to the currently active variables.

- **Variable influence report:** This spreadsheet analyzes the influence of each of the independent variables on the discriminant analysis. **Variable:** The name of the independent variable (or discriminator). **Removed Lambda:**

This is the value of a Wilks' lambda computed to test the impact of removing this variable. **Removed *F*-Value:** This is the *F*-ratio that is used to test the significance of the above Wilks' lambda. **Removed *F*-Prob:** This is the probability (significance level) of the above *F*-ratio. It is the probability to the right of the *F*-ratio. The test is significant (the variable is important) if this value is less than the value of α that we are using, such as $\alpha = 0.05$. **Alone Lambda:** This is the value of a Wilks' lambda that would be obtained if this were the only independent variable used. **Alone *F*-Value:** This is an *F*-ratio that is used to test the significance of the above Wilks' lambda. **Alone *F*-Prob:** This is the probability (significance level) of the above *F*-ratio. It is the probability to the right of the *F*-ratio. The test is significant (the variable is important) if this value is less than the value of α that we are using, such as $\alpha = 0.05$. **R^2 Other *X*'s:** This is the R^2 value that would be obtained if this variable were regressed on all other independent variables. When this R^2 value is larger than 0.99, severe multicollinearity problems exist. We should remove variables (one at a time) with large R^2 and rerun our analysis.

Variable	Removed Lambda	Removed *F*-Value	Removed *F*-Prob	Alone Lambda	Alone *F*-Value	Alone *F*-Prob	Alone R-Squared Other *X*'s
Sepallength	0,9384	4,72	0,010	0,381	119,26	0,000	0,859
Sepalwidth	0,7665	21,94	0,000	0,599	49,16	0,000	0,524
Petallength	0,6692	35,59	0,000	0,058	1180,16	0,000	0,968
Petalwidth	0,7430	24,90	0,000	0,071	960,01	0,000	0,938

- **Fisher linear discriminant functions report:** This spreadsheet presents the estimates of the Fisher linear discriminant function coefficients. These are often called the discriminant coefficients. They are also known as the "plug-in" estimators, since the true variance-covariance matrices are required but their estimates are plugged-in. This technique assumes that the independent variables in each group follow a multivariate-normal distribution with equal variance-covariance matrices across groups. Studies have shown that this technique is fairly robust to departures from either assumption. The spreadsheet represents three classification functions, one for each of the three iris groups. Each function is represented vertically. When a weighted average of the independent variables is formed using these coefficients as the weights (and adding the constant), the discriminant scores result. To determine which group an individual belongs to, we should select the iris group with the highest score.

Variable	Setosa	Versicol	Virginic
Constant	−85,21	−71,75	−103,27
Sepallength	23,54	15,69	12,44
Sepalwidth	23,59	7,07	3,68
Petallength	−16,43	5,21	12,77
Petalwidth	−17,40	6,43	21,08

- **Regression coefficients estimates report:** This spreadsheet presents the estimates of the regression coefficients when using a multiple linear regression model. These coefficients are determined as follows: 1. We create three indicator variables, one for each of the three varieties of iris. Each indicator variable is set to one when the row belongs to that iris group and zero otherwise. 2. We fit a multiple regression of the independent variables on each of the three indicator variables. 3. The regression coefficients obtained are those shown in this spreadsheet. Hence, predicted values generated by these coefficients will be between zero and one. To determine which iris group an individual plant belongs to, we select the iris group with the highest score.

Variable	Setosa	Versicol	Virginic
Constant	0,118	1,577	−0,695
Sepallength	0,066	−0,020	−0,046
Sepalwidth	0,242	−0,445	0,203
Petallength	−0,224	0,221	0,003
Petalwidth	−0,057	−0,494	0,552

- **Classification count table report:** This spreadsheet presents a matrix that indicates how accurately the estimated discriminant functions classify the iris plants. If perfect classification has been achieved, there will be zeros on the off-diagonals. The rows of the table represent the actual, given iris groups, while the columns represent the predicted, found iris group. **Percent Reduction:** The percent reduction is the classification accuracy achieved by the estimated current discriminant functions over what is expected if the iris plants were randomly classified.

Predicted, found Actual, given	Setosa	Versicol	Virginic	Total
Setosa	50	0	0	50
Versicol	0	48	2	50
Virginic	0	1	49	50
Total	50	49	51	150

Reduction in classification error due to X's $= 97,0\%$.

- **Misclassified rows report:** This spreadsheet shows the actual or given group and the predicted or found iris group of each iris plant that was misclassified. It also shows 100 times the estimated probability, $P(i)$, that the row is in each group. For easier viewing, we have multiplied the probabilities by 100 to make this a percent probability (between 0 and 100) rather than a regular probability (between 0% and 1%). A value near 100% gives a strong indication that the observation belongs in that group. **$P(i)$:** If the linear discriminant classification technique was used, these are the estimated probabilities that this row belongs to the ith group. The algorithm used is briefly outlined here. Let f_i $(i = 1, 2, ..., K)$ be the linear discriminant function value. Let $\max(f_k)$ be the maximum score of all groups. Let $P(G_i)$ be the overall probability of classifying an individual into group i. The values of $P(i)$ are generated using the following equation: $P(i) = \dfrac{\exp\left[f_i - \max\left(f_k\right)\right]P(G_i)}{\displaystyle\sum_{j=1}^{K}\exp\left[f_j - \max\left(f_k\right)\right]P(G_j)}$. If the regression classification technique was used, this is the predicted value of the regression equation. The implicit y value in the regression equation is one or zero, depending on whether this observation is in the ith group or not. Hence, a predicted value near zero indicates that the observation is not in the ith group, while a value near one indicates a strong possibility that this observation is in the ith group. There is nothing to prevent these predicted values from being greater than one or less than zero. They are not estimated probabilities.

| | | | Percent chance of each group | | |
Row	Actual	Predicted	P(1) %	P(2) %	P(3) %
5	*Virginic*	*Versicol*	0,0	72,9	27,1
9	*Versicol*	*Virginic*	0,0	25,3	74,7
12	*Versicol*	*Virginic*	0,0	14,3	85,7

- **Predicted classification report:** This spreadsheet shows the actual (or given) group, the predicted (or found) group, and the percentage probabilities of each row. Once we have computed the classification scores for a case, it is easy to decide how to classify the case: in general we classify the case as belonging to the group for which it has the highest classification score unless the *a priori* classification probabilities are widely disparate.

Row	Actual, given	Predicted, found	Percent chance of each group		
			P(1) %	P(2) %	P(3) %
1	Setosa	Setosa	100,0	0,0	0,0
2	Virginic	Virginic	0,0	0,0	100,0
3	Versicol	Versicol	0,0	99,6	0,4
4	Virginic	Virginic	0,0	0,0	100,0
5	Virginic	Versicol	0,0	72,9	27,1
6	Setosa	Setosa	100,0	0,0	0,0
7	Virginic	Virginic	0,0	0,0	100,0
8	Versicol	Versicol	0,0	96,0	4,0
9	Versicol	Virginic	0,0	25,3	74,7
10	Setosa	Setosa	100,0	0,0	0,0
...	...	...	...	...	...
	...	...	...	...	...
150	Versicol	Versicol	0,0	100,0	0,0

(the spreadsheet continues for all 150 iris plants)

- **Canonical variate analysis report:** The F-value tests whether this function and those below it are significant. This spreadsheet provides a canonical correlation analysis of the discriminant problem. Recall that canonical correlation analysis is used when we want to study the correlation between two sets of variables. In this case, the two sets of variables are defined in the following way. The independent variables comprise the first set. The group variable defines another set, which is generated by creating an indicator variable for each group except the last one. **Inv(W)B Eigenvalue:** The eigenvalues of the matrix $W^{-1}A$. These values indicate how much of the total variation explained is accounted for by the various discriminant functions. Hence, the first discriminant function corresponds to the first eigenvalue, and so on. Note that the number of eigenvalues is the minimum of the number of variables and g − 1, where g is the number of groups. **Ind'l Prcnt:** The percent that this eigenvalue is of the total. **Total Prcnt:** The cumulative percent of this and all previous eigenvalues. **Canon Corr:** The canonical correlation coefficient. **Canon Corr2:** The square of the canonical correlation. This is similar to R^2 in multiple regression. **F-Value:** The value of the approximate F-ratio for testing the significance of the Wilks' lambda corresponding to this row and those below it. Hence, in this example, the first F-value tests the significance of both the first and second canonical correlations, while the second F-value tests the significance of the second correlation only. **Num DF:** The numerator degrees of freedom for this F-test. **Denom DF:** The denominator degrees of freedom for this F-test. **Prob Level:** The significance level of the F-test. This is the area under the F-distribution to the right of the F-value. Usually, a value less than 0.05 is considered significant. **Wilks' Lambda:** The value of Wilks' lambda for this row. This Wilks' lambda is used to test the significance of the discriminant function corresponding to this row and those below it. Recall that Wilks' lambda is a multivariate generalization of R^2. The above F-value is an approximate test of this Wilks' lambda.

	Inv(W)B	Ind'l	Total	Canon	Canon		Numer	Denom	Prob	Wilks'
									Level	
Fn	Eigenvalue	Pcnt	Pcnt	Corr	Corr2	F-Value	DF	DF	Level	Lambda
1	32,19	99,1	99,1	0,9848	0,9699	199,1	8,0	288,0	0,00	0,023439
2	0,29	0,9	100,0	0,4712	0,2220	13,8	3,0	145,0	0,00	0,777973

The F-value tests whether this function and those below it are significant.

- **Canonical coefficients report:** This spreadsheet gives the estimates of coefficients used to create the canonical scores. The canonical scores are weighted averages of the observations, and these coefficients are the weights (with the constant term added).

Variable	Variate1	Variate2
Constant	−2,105	6,661
Sepallength	−0,829	−0,024
Sepalwidth	−1,534	−2,164
Petallength	2,201	0,932
Petalwidth	2,810	−2,839

- **Canonical variates at group means report:** This spreadsheet gives the results of applying the canonical coefficients to the means of each of the groups and the graphical presentation is on Fig. 4.46.

Group	Function1	Function2
Setosa	−7,607	−0,215
Versicol	1,825	0,7279
Virginic	5,782	−0,513

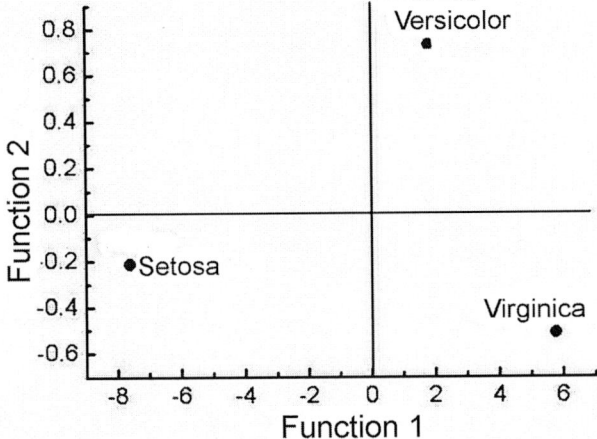

Figure 4.46 Plot of two discriminant functions shows centroids of three types of iris plants.

- **Std. canonical coefficients report:** This spreadsheet gives the standardized canonical coefficients.

Variable	Variate1	Variate2
Sepallength	−0,427	−0,012
Sepalwidth	−0,521	−0,735
Petallength	0,947	0,401
Petalwidth	0,575	−0,581

- **Variable-variate correlations report:** This spreadsheet gives the loadings (correlations) of the variables on the canonical variates. That is, each entry is the correlation between the canonical variate and the independent variable. This spreadsheet can help you interpret a particular canonical variate.

Variable	Variate1	Variate2
Sepallength	0,222	−0,311
Sepalwidth	−0,119	−0,863
Petallength	0,706	−0,168
Petalwidth	0,633	−0,737

- **Linear discriminant scores report:** This spreadsheet gives the individual values of the linear discriminant scores which are plotted on Fig. 4.47.

Row	Group	Score1	Score2	Score3
1	Setosa	83,87	38,66	−6,79
2	Virginic	1,23	91,86	104,57
3	Versicol	32,19	83,71	78,29
4	Virginic	11,89	99,98	113,62
5	Virginic	19,27	83,18	82,19
6	Setosa	75,07	33,73	−9,29
7	Virginic	26,55	99,87	107,62
8	Versicol	12,62	74,24	71,07
9	Versicol	19,00	80,09	81,18
150	Versicol	15,14	46,64	30,64

(the spreadsheet continues for all 150 iris plants)

- **Regression scores report:** This spreadsheet gives the individual values of the predicted scores based on the regression coefficients. Even though these values are predicting indicator variables, it is possible for a value to be less than zero or greater than one which are plotted on Fig. 4.47.

(the spreadsheet continues for all 150 iris plants)

- **Canonical scores report:** This spreadsheet gives the scores of the canonical variates for each row which are plotted on Fig. 4.47.

Row	Group	Score1	Score2	Score3
1	Setosa	0,924	0,216	−0,140
2	Virginic	−0,164	0,349	0,815
3	Versicol	0,108	0,472	0,420
4	Virginic	−0,083	0,110	0,973
5	Virginic	−0,018	0,586	0,431
6	Setosa	0,916	0,130	−0,046
7	Virginic	0,049	0,045	0,906
8	Versicol	−0,035	0,723	0,312
9	Versicol	0,103	0,202	0,695
10	Setosa	1,060	0,002	−0,062
...	...	...	...	...
...	...	...	...	...
150	Versicol	0,208	0,685	0,107

Row	Group	*Score1*	*Score2*
1	*Setosa*	−7,672	0,135
2	*Virginic*	6,800	−0,581
3	*Versicol*	2,549	0,472
4	*Virginic*	6,653	−1,805
5	*Virginic*	3,815	0,943
6	*Setosa*	−7,213	−0,356
7	*Virginic*	5,106	−1,992
8	*Versicol*	3,498	1,685
9	*Versicol*	3,716	−1,045
10	*Setosa*	−8,681	−0,878
...	...	...	...
...	...	...	...
150	*Versicol*	0,293	1,799

(the spreadsheet continues for all 150 iris plants)

- **A priori versus post hoc classification:** When classifying iris plants from which the discriminant functions were computed, we usually obtain a fairly good discrimination (although usually not as good as in this example). However, we should only look at those classifications as a diagnostic tool for identifying areas of strengths and weaknesses in the current classification functions, because these classifications are not *a priori* predictions but rather *post hoc* classifications. Only if we classify different (new) plants can we interpret this table in terms of predictive discriminatory power. Thus, it would be unjustified to claim that we can successfully predict the type of iris in 98 percent of all plants, based on only four measurements. Because we capitalized on chance, we could expect much less accuracy if we were to classify new plants.

(a) *Classification of plants using Mahalanobis distances:* The Mahalanobis distance is a measure of the distance that can be used in the multivariate space defined by the variables in the model. We can compute the distance between each iris plant and the centre of each group (i.e., the group centroid, defined by the respective group means for each variable). The closer the plant is to a group centroid, the more confidence we can have that it belongs to that group.

	Actual, given	P(1), $\alpha = 0{,}3333$	P(2), $\alpha = 0{,}3333$	P(3), $\alpha = 0{,}3333$
1	Setosa	0,24	90,66	181,56
2	Virginic	208,57	27,32	1,89
3	Versicol	105,27	2,23	13,07
4	Virginic	207,92	31,75	4,45
5*	Virginic	133,07	5,25	7,24
6	Setosa	1,33	84,01	170,06
7	Virginic	173,18	26,56	11,05
8	Versicol	131,66	8,43	14,76
9*	Versicol	130,86	8,67	6,51
10	Setosa	2,29	113,65	210,02
.....	.....	.....	.....	.....
.....	.....	.....	.....	.....
150	Versicol	68,47	5,48	37,47

*a misclassification

(b) *Classification of plants using posterior probabilities:* The probability that an iris plant belongs to a particular group is basically proportional to the Mahalanobis distance from that group centroid. Because we compute the location of each iris from our prior knowledge of the values for that iris on the variables in the model, these probabilities are called *posterior* probabilities. The posterior probability is the probability, based on our knowledge of the values of other variables, that the respective iris belongs to a particular group. This is a conditional probability, that is, it is contingent on our knowledge of the values for the variables in the model.

	Actual, given	P(1), $\alpha = 0{,}3333$	P(2), $\alpha = 0{,}3333$	P(3), $\alpha = 0{,}3333$
1	Setosa	1,000	0,000	0,000
2	Virginic	0,000	0,000	1,000
3	Versicol	0,000	0,996	0,004
4	Virginic	0,000	0,000	1,000

* 5	Virginic	0,000	0,729	0,271
6	Setosa	1,000	0,000	0,000
7	Virginic	0,000	0,000	1,000
8	Versicol	0,000	0,960	0,040
* 9	Versicol	0,000	0,253	0,747
10	Setosa	1,000	0,000	0,000
.....	.....	.....	.....	.....
.....	.....	.....	.....	.....
150	Versicol	0,000	1,000	0,000

*a misclassification

- **Classification of iris plants – Conclusion:** The classifications are ordered into a first, second, and third choice. The column under the header P(1) contains the first classification choice, that is, the group for which the respective case had the highest posterior probability. The rows marked by the asterisk (*) are cases that are misclassified. Again, in this example, the classification accuracy is very high, even considering the fact that these are all *post hoc* classifications. Such accuracy is rarely attained in research in the social sciences. This spreadsheet illustrates the basic ideas of discriminant function analysis. In general, in many cases where there are naturally occurring groups that you would like to be able to discriminate, this technique is appropriate. However, as stated at various points in the preceding discussion, if correct predictive classification is the goal of the research, then at least two studies must be conducted: one in order to build the classification functions and another to validate them.

	Actual, given	P(1), $\alpha = 0{,}3333$	P(2), $\alpha = 0{,}3333$	P(3), $\alpha = 0{,}3333$
1	Setosa	Setosa	Versicol	Virginic
2	Virginic	Virginic	Versicol	Setosa
3	Versicol	Versicol	Virginic	Setosa
4	Virginic	Virginic	Versicol	Setosa
5*	Virginic	Versicol	Virginic	Setosa
6	Setosa	Setosa	Versicol	Virginic
7	Virginic	Virginic	Versicol	Setosa

(Continued)

8	Versicol	Versicol	Virginic	Setosa
9*	Versicol	Virginic	Versicol	Setosa
10	Setosa	Setosa	Versicol	Virginic
…..	…..	…..	…..	…..
…..	…..	…..	…..	…..
150	Versicol	Versicol	Virginic	Setosa

*a misclassification

- **Scores plot(s):** We may select plots of the linear discriminant scores, regression scores, or canonical scores to aid in your interpretation. These plots are usually used to give a visual impression of how well the discriminant functions are classifying the data. This chart plots the values of the first and second canonical scores. By looking at this plot we can see what the classification rule would be. Also, it is obvious from this plot that only the first canonical function is necessary in discriminating among the varieties of iris since the groups can easily be separated along the vertical axis. This plot confirms the interpretation so far. Clearly, the plants of type *Setosa* are plotted much further to the right in the scatterplot. Thus, the first discriminant function mostly discriminates between that type of iris and the two others. The second function seems to provide some discrimination between the plants of type *Versicol* (which mostly show negative values for the second canonical function) and the others (which have mostly positive values). However, the discrimination is not nearly as clear as that provided by the first canonical function (root).
- ***Conclusion:*** In general *Discriminant Analysis* is a very useful tool (1) for detecting the variables that allow the researcher to discriminate between different (naturally occurring) groups, and (2) for classifying cases (here iris plants) into different groups with a better than chance accuracy. To summarize the findings so far, it appears that the most significant and clear discrimination is possible for plants of the type *Setosa* by the first discriminant function. This function is marked by the negative coefficients for the width and length of petals and positive weights for the width and length of sepals. Thus, the longer and wider the petals, and the shorter and smaller the sepals, the less likely it is that the plant is of iris type *Setosa* (remember that in the scatterplot of the canonical functions, the plants of the type *Setosa* were plotted to the right, that is, they were distinguished by high values on this function).

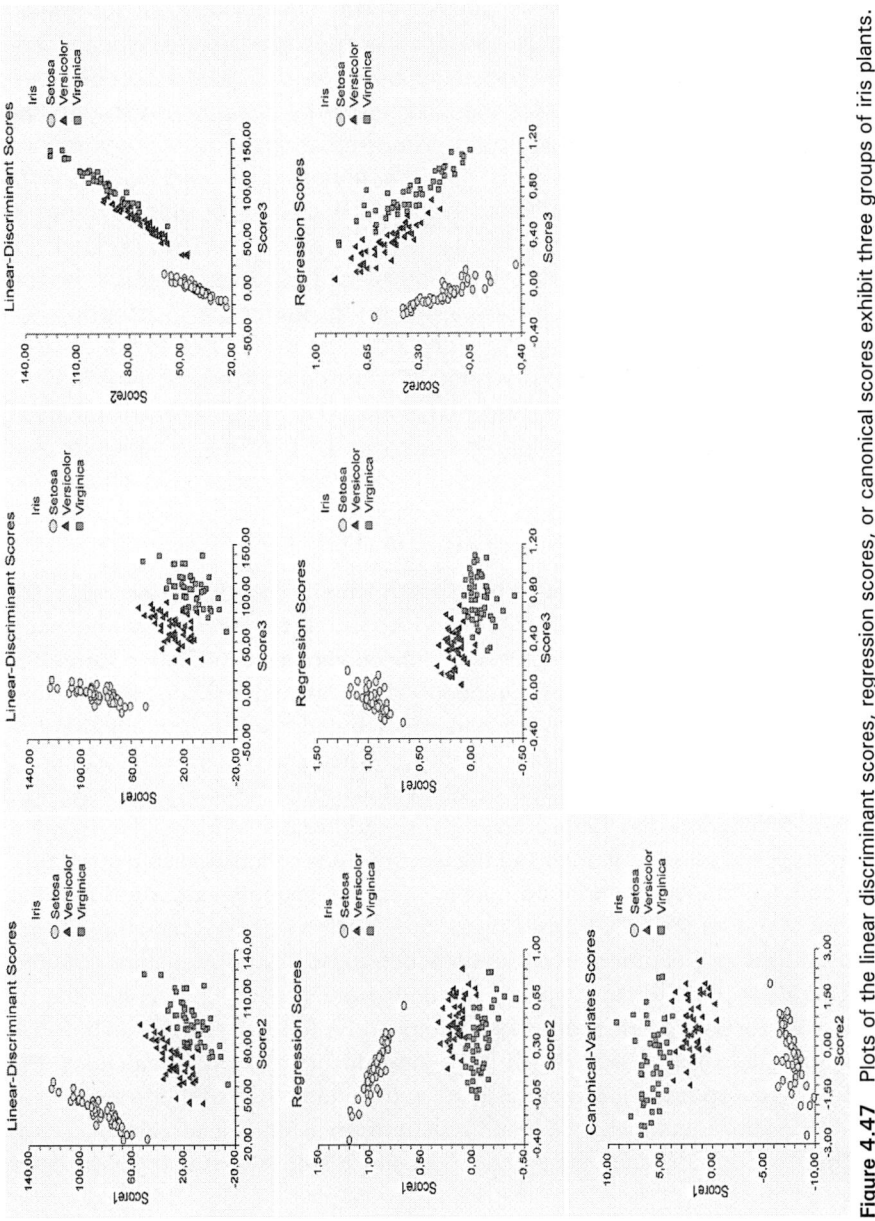

Figure 4.47 Plots of the linear discriminant scores, regression scores, or canonical scores exhibit three groups of iris plants.

4.8 Logistic Regression (LR)

Logistic regression analysis (LR) studies the association between a categorical dependent variable and a set of independent (explanatory) variables. Explanatory variables may be continuous, discrete, dichotomous, or a mix. The name logistic regression (LR) is often used when the dependent variable has only two values. The name *multiple-group logistic regression* (MGLR) is usually reserved for the case when the dependent variable has three or more unique values. Multiple-group logistic regression is sometimes called *multinomial, polytomous, polychotomous,* or *nominal logistic regression.* Although the data structure is different from that of multiple regression, the practical use of the procedure is similar. Logistic regression is more flexible than the other techniques. Unlike discriminant function analysis, logistic regression has no assumptions about the distributions of the predictor variables. In logistic regression, the predictors do not have to be normally distributed, linearly related, or of equal variance within each group. Unlike multiway frequency analysis, the predictors do not need to be discrete; the predictors can be any mix of continuous, discrete and dichotomous variables. Unlike multiple-regression analysis, which also has distributional requirements for predictors, logistic regression cannot produce negative predicted probabilities [131].

Logistic regression competes with discriminant analysis as a method for analyzing discrete response variables. In fact, the current feeling among many statisticians is that logistic regression is more versatile and better suited for most situations than is discriminant analysis because it does not assume that the independent variables are normally distributed, as discriminant analysis does. Discriminant analysis is appropriate when the dependent variable is nonmetric. However, when the dependent variable has only two groups, logistic regression may be preferred for several reasons. First, discriminant analysis relies on strictly meeting the assumptions of multivariate normality and equal variance-covariance matrices across groups—assumptions that are not met in many situations. Logistic regression does not face these strict assumptions and is much more robust when these assumptions are not met, making its application appropriate in many more situations. Second, even if the assumptions are met, many researchers prefer logistic regression because it is similar to regression. Both have straightforward statistical tests, the ability to incorporate nonlinear effects, and a wide range of diagnostics. For these and more technical reasons, logistic regression is equivalent to two-group discriminant analysis and may be more suitable in many situations [60, 131].

4.8.1 The Binary Response Variable and Logistic Regression Model

Consider the situation where the response variable in a regression problem takes on only two possible values, 0 and 1. These could be arbitrary assignments resulting from observing a qualitative response. Suppose that the model has the form $y_i = \mathbf{x}_i^T \beta + \varepsilon_i$ where $\mathbf{x}_i^T = [1, x_{i1}, x_{i2}, ..., x_{im}]$, $\hat{\mathbf{a}}^T = [\beta_0, \beta_1, ..., \beta_m]$ and the response variable y_i takes on the value either 0 or 1. We will assume that the response variable y_i is a *Bernoulli random variable* with probability distribution as follows [143]:

y_i	Probability Pr
1	$\Pr(y_i = 1) = p$
0	$\Pr(y_i = 0) = 1 - p$

Now since $E(\varepsilon_i) = 0$, the expected value of the response variable is $E(y_i) = 1(p_i) + 0(1 - p_i) = p_i$. This implies that $E(y_i) = \mathbf{x}_i^T \beta = p_i$. This means that the expected response given by the response function $E(y_i) = \mathbf{x}_i^T \beta$ is just the probability that the response variable takes on the value 1.

While in multiple regression, a mathematical model of a set of explanatory variables is used to predict the mean of the dependent variable, in logistic regression, a mathematical model of a set of explanatory variables is used to predict a transformation of the dependent variable. This is the *logit* transformation: Suppose the numerical values of 0 and 1 are assigned to the two categories of a binary variable y. Often, the 0 represents a negative response and the 1 represents a positive response. The mean of this variable will be the proportion of positive responses. Because of this, we might try to model the relationship between the probability (proportion) of a positive response and the explanatory variables. If p is the proportion of observations with a response of 1, then $1 - p$ is the probability of a response of 0. The ratio $p/(1 - p)$ is call the *odds* and the *logit* is the logarithm of the odds, or just *log odds*. Mathematically, the logit transformation is written $L_{(1)} = \text{logit}(p) = \ln\left(\dfrac{p}{1 - p}\right)$. The following table shows the logit for various values of probability p:

p	Logit(p)	p	Logit(p)
0.001	−6.907	0.999	6.907
0.01	−4.595	0.99	4.595
0.05	−2.944	0.95	2.944

(Continued)

0.10	−2.197	0.90	2.197
0.20	−1.386	0.80	1.386
0.30	−0.847	0.70	0.847
0.40	−0.405	0.60	0.405
0.50	0.000		

While p ranges between zero and one, the logit ranges between minus and plus infinity and the zero logit occurs when p is 0.50. The *logistic* transformation is the inverse of the logit transformation and may be written as $p = \dfrac{\exp(L_{(1)})}{1 + \exp(L_{(1)})}$.

Because the model produced by logistic regression is nonlinear, the equations used to describe the outcomes are slightly more complex than those for multiple regression. The outcome variable, y, is the probability of having one outcome or another based on a nonlinear function of the best linear combination of predictors: with two outcomes: $\hat{p}_i = \dfrac{\exp(L_{(1)})}{1 + \exp(L_{(1)})}$, where $\hat{p}_i$ is the estimated probability that the ith case ($i = 1, ..., n$) is in one of the categories and $L_{(1)}$ is the usual linear regression equation: $L_{(1)} = \beta_0 + \beta_1 x_1 + \beta_2 x_2 + ... + \beta_m x_m$ with constant β_0, coefficients $\beta_1, \beta_2, ..., \beta_m$, and predictors x_j for m predictors ($j = 1, 2, ..., m$). This linear regression equation creates the *logit* or log of

odds: $\ln\left(\dfrac{\hat{p}}{1 - \hat{p}}\right) = \beta_0 + \sum_{j=1}^{m} \beta_j x_{ij}$. That is, the linear regression equation the

the natural logarithms of the probability of being in one group divided by the probability of being in the other group. The procedure for estimating coefficients is maximum likelihood, and the goal is to find the best linear combination of predictors to maximize the likelihood of obtaining the observed outcome frequencies. Maximum likelihood estimation is an iterative procedure that starts with arbitrary values of coefficients and determines the direction and size of change in the coefficients that will maximize the likelihood of obtaining the observed frequencies. Then residuals are tested and another determination of direction and size of change in coefficients is made, and so on, until the coefficients change very little, i.e. convergence is reached. Logistic regression, like multiway frequency analysis, can be used to fit and also compare models. The simplest (and worst-fitting) model includes only the constant and none of the predictors. The most complex (and *best'-fitting*) model includes the constant, all predictors and, perhaps, also interactions among predictors. Often, however, not all predictors (and interactions) are related to the outcome. The researcher uses goodness-of-fit tests to choose the model that does the best job of prediction with the fewest predictors.

Generally, when the response variable y is binary, there is considerable empirical evidence indicating that the shape of the response function should be nonlinear. A monotonically increasing (or decreasing) S-shaped (or reverse S-shaped) function is usually employed. The difference between two log odds can be used to compare two proportions, such as that of males versus females. Mathematically, this difference is written

$$L_{(1)} - L_{(0)} = \text{logit}(p_1) - \text{logit}(p_0)$$

$$= \ln\left(\frac{p_1}{1-p_1}\right) - \ln\left(\frac{p_0}{1-p_0}\right) = \ln\left(\frac{p_1(1-p_0)}{p_0(1-p_1)}\right) = \ln\left(OR_{1,0}\right)$$

This difference is often referred to as the *log odds ratio.* The odds ratio is often used to compare proportions across groups. The logistic transformation is closely related to the odds ratio which may be written as the reverse relationship as $OR_{1,2} = \exp(L_{(1)} - L_{(0)})$.

4.8.2 Kinds of Research Questions

The goal of logistic regression analysis (LR) is to correctly predict the category of the outcome for individual cases. The first step is to establish that there is a relationship between the outcome and the set of predictors. Once a reduced set of predictors is found, the equation can be used to predict outcomes for new cases on a probabilistic basis [131].

Prediction of group membership or outcome: A reliable difference between the models indicates a relationship between the predictors and the outcome. An alternative is to test a LR model with only some predictors against the model with all predictors (called a full model). The goal of LR is to find a nonsignificant *yr,* indicating no reliable difference between the LR model with only some predictors and the full model.

Importance of predictors: Which variables predict the outcome? How do variables affect the outcome? Does a particular variable increase or decrease the probability of an outcome, or does it have no effect on outcome?

Interactions among predictors: As in multiway frequency (logit) analysis, a LR model can also include interactions among the predictor variables: two-way interactions and, if there are many predictor variables, higher-order interactions. Like individual predictors, interactions may complicate a model without reliably improving the prediction. Decisions about including interactions are made in the same way as decisions about including individual predictors.

Parameter estimates: The parameter estimates in logistic regression are the coefficients of the predictors included in a model.

Classification of cases: How good is a reliable model at classifying cases for whom the outcome is known?

Significance of prediction with covariates: The researcher may consider some of the predictors covariates and others independent variables.

Strength of association: The logic of assessing strength of association is different in routine statistical hypothesis testing from situations where LR models are being evaluated. In routine statistical hypothesis testing, one does not report strength of association for a nonsignificant effect.

4.8.3 Estimating the Multiple-Group Logistic Regression Models

In multiple-group logistic regression, a discrete dependent variable y having g unique values $(g > 2)$ is regressed on a set of m independent variables $x_1, x_2, ..., x_m$. Here y represents a way of partitioning the population of interest. Since the names of these partitions are arbitrary, we refer to them by consecutive numbers. That is, in the discussion below, y will take on the values 1, 2, ..., g.

Let $\mathbf{x} = (x_1, x_2, ..., x_m)$ and $\boldsymbol{\beta}_g = \begin{pmatrix} \beta_{g1} \\ ... \\ \beta_{gm} \end{pmatrix}$. The *multiple logistic regression model* is

given by the g equations $\ln\left(\dfrac{L_{(1)}}{L_{(0)}}\right) = \ln\left(\dfrac{P_g}{P_1}\right) + b_1 x_1 + b_2 x_2 + ... + b_m x_m$. Here, $L_{(1)}$

is the probability that an individual with values $x_1, x_2, ..., x_m$ is in group g. That is, $L_{(1)} = \Pr(y = g | \mathbf{x})$. Usually $x_1 = 1$ (that is, an intercept is included), but this is not necessary. The quantities $p_1, p_2, ..., p_g$ represent the prior probabilities of group membership. If these prior probabilities are assumed equal, then the term $\ln(P_g / P_1)$ becomes zero and drops out. If the priors are not assumed equal, they change the values of the intercepts in the logistic regression equation. Group one is called the *reference group*. The regression coefficients $\beta_{11}, \beta_{12}, ..., \beta_{1m}$ for the reference group are set to zero. The choice of the reference group is arbitrary. Usually, it is the largest group or a control group to which the other groups are to be compared. This leaves $g - 1$ logistic regression equations in the multinomial logistic model. The β's are population regression coefficients that are to be estimated from the data. Their estimates are denoted by b's. The β's represents the unknown parameters, while the b's are their estimates. These equations are linear in the logits of p. However, in terms of the probabilities, they are nonlinear. The corresponding nonlinear equations are

$$p_g = \Pr(y = g \mid \mathbf{x}) = \frac{\exp(\mathbf{x}\boldsymbol{\beta}_g)}{1 + \exp(\mathbf{x}\boldsymbol{\beta}_2) + \exp(\mathbf{x}\boldsymbol{\beta}_3) + ... + \exp(\mathbf{x}\boldsymbol{\beta}_g)}$$

since $\exp(\mathbf{x}\boldsymbol{\beta}_1) = 1$ because all of its regression coefficients are zero. Often, all of these models are referred to as *logistic regression models*. However, when the independent variables are coded as ANOVA type models, they are sometimes called *logit models*.

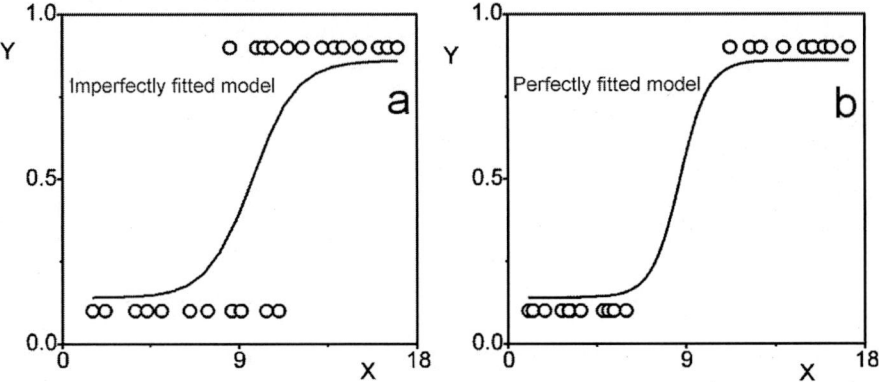

Figure 4.48 Logistic curve for the logistic regression model of (a) poorly fitted, and (b) well-defined and sufficiently fitted.

The nonlinear nature of the logistic transformation requires that another procedure, the maximum likelihood procedure, be used in an iterative manner to find the "most likely" estimates for the coefficients. This results in the use of the *likelihood value* instead of the sum of squares when calculating measure of overall model fit. The process of estimating the coefficients, however, is still quite similar in many regards to that of linear regression. The logit model has the specific form of the *logistic curve*. To estimate a logistic regression model, this curve is fitted to the actual data [131]. Figure 4.48 portrays two hypothetical examples of fitting a logistic relationship to sample data. The actual data, an event either happening or not (0 or 1), are represented as observations at either the top or bottom of the graph. These are the events that occur at each value of the independent variable (the x-axis). In part a, the logistic curve cannot fit the data well because there are a number of values of the independent variable that have both events and nonevents (i.e., high overlap of the distributions). However, in part b, there is a much more well-defined relationship, and the logistic curve fits the data quite well. This simple example, similar to a scatterplot of dependent and independent variables in regression with a line representing the "best fit" of the correlation, can be extended to include multiple independent variables just as in regression.

4.8.4 Analysis and Interpretation of Regression Coefficients

One of the advantages of logistic regression is that we need to know only whether an event occurred to then use a dichotomous value as our dependent variable. From this dichotomous value, the procedure predicts its estimate of the probability that the event will or will not occur. If the predicted probability is greater than 0.50, then the prediction is yes, otherwise no. Logistic regression derives its name from the *logistic transformation* used with the dependent variable. When this transformation is used, however, the logistic regression and its coefficients take on a somewhat different meaning from those found in regression with a metric dependent variable. The interpretation of the estimated regression coefficients is not as easy as in multiple regression. In multinomial logistic regression, not only is the relationship between x and y nonlinear, but also, if the dependent variable has more than two unique values, there are several regression equations. Consider the simple case of a binary dependent variable, y, and a single independent variable, x. Assume that y is coded so it takes on the values 0 and 1. In this case, the logistic regression equation is

$\ln\left(\dfrac{p}{1-p}\right) = \beta_0 + \beta_1 x$. Now consider impact of a unit increase in x. The logistic

regression equation becomes

$$\ln\left(\frac{p'}{1-p'}\right) = \beta_0 + \beta_1(x+1) = \beta_0 + \beta_1 x + \beta_1.$$

We can isolate the slope by taking the difference between these two equations. We have

$$\beta_1 = \beta_0 + \beta_1(x+1) - \beta_0 + \beta_1 x = \ln\left(\frac{p'}{1-p'}\right) - \ln\left(\frac{p}{1-p}\right) = \ln\left(\frac{odds(x+1)}{odds(x)}\right).$$

That is, β_1 is the log of the ratio of the odds at $x + 1$ and x. Removing the logarithm by exponentiating both sides gives $\exp(\beta_1) = \dfrac{odds(x+1)}{odds(x)}$. The regression coefficient β_1 is interpreted as the log of the odds ratio comparing the odds after a one unit increase in x to the original odds. Unlike multiple regression, the interpretation of β_1 depends on the particular value of x

since the probability values, the p's, will vary for different x. The procedure that calculates the logistic coefficient compares the probability of an event occurring with the probability of its not occurring. This odds ratio can be expressed as

$$\frac{\text{Pr(event)}}{\text{Pr(no-event)}} = \exp(b_0 + b_1 x_1 + b_2 x_2 + \ldots + b_m x_m).$$

The estimated coefficients $b_0, b_1, b_2, \ldots, b_m$ are actually measures of the changes in the ratio of the probabilities, termed the *odds ratio*. Moreover, they are expressed in logarithms, so they need to be transformed back (the antilog of the value has to be taken) so that their relative effect on the probabilities is assessed more easily. Algorithms perform this procedure automatically and give both the actual coefficient and the transformed coefficient. Use of this procedure does not change in any manner the way we interpret the sign of the coefficient. A positive coefficient increases the probability, whereas a negative value decreases the predicted probability.

Binary X: When x can take on only two values, say 0 and 1, the above interpretation becomes even simpler. Since there are only two possible values of x, there is a unique interpretation for β_1 given by the log of the odds ratio. In mathematical terms, the meaning of β_1 is then $\beta_1 = \ln\left(\dfrac{odds(x=1)}{odds(x=0)}\right)$. If we take antilogs, we obtain the *odds ratio* as $\exp(\beta_1) = \dfrac{odds(x=1)}{odds(x=0)}$. The odds is itself the ratio of two probabilities, p and $1-p$. The odds ratio can be interpreted as the estimated increase in the probability of success associated with a one-unit change in the value of the predictor variable. In general, the estimated increase in the odds ratio associated with a change of d units in the predictor variable is $\exp(\beta_1)$. We consider the following table of odds values for various values of p. Here 9:1 is read '9 to 1'.

Value of p	Odds of p
0.9	9:1
0.8	4:1
0.6	1.5:1
0.5	1:1
0.4	0.67:1
0.2	0.25:1
0.1	0.11:1

To completely understand β_1, we must take the logarithm of the odds ratio. It is difficult to think in terms of logarithms. However, we can remember that the log of one is zero. So a positive value of β_1 indicates that the odds of the numerator are large while a negative value indicates that the odds of the denominator are larger. It is probability easiest to think in terms of $\exp(\beta_1)$ rather than β_1, because $\exp(\beta_1)$ is the odds ratio while β_1 is the log of the odds ratio.

Multiple independent variables: When there are multiple independent variables, the interpretation of each regression coefficient becomes more difficult, especially if interaction terms are included in the model. In general, however, the regression coefficient is interpreted the same as above, except that the caveat 'holding all other independent variables constant' must be added. That is, can the value of this independent variable be increased by one without changing any of the other variables. If it can, then the interpretation is as before. If not, then some type of conditional statement must be added that accounts for the values of the other variables.

Multinomial dependent variable: When the dependent variable has more than two values, there will be more than one regression equation. In fact, the number of regression equations is equal to one less than the number of values. This makes interpretation more difficult because there are several regression coefficients associated with each independent variable. In this case, care must be taken to understand what each regression equation is predicting. Once this is understood, interpretation of each of the $m - 1$ regression coefficients for each variable can proceed as above.

4.8.5 Statistical Tests and Confidence Intervals

Logistic regression has two types of inferential tests: tests of models and tests of individual predictors. Statistical inference in logistic regression is based on certain properties of maximum-likelihood estimators and on likelihood ratio tests. These are large-sample or *asymptotic results*. Inferences about individual regression coefficients, groups of regression coefficients, goodness-of-fit, mean responses, and predictions of group membership of new observations are all of interest. These inference procedures can be treated by considering hypothesis tests and/or confidence intervals. The inference procedures in logistic regression rely on large sample sizes for accuracy. Two procedures are available for testing the significance of one or more independent variables in a logistic regression: likelihood ratio tests and Wald tests. Simulation studies usually show that the likelihood ratio test performs better than the Wald test. However, the Wald test is still used to test the significance of individual regression coefficients because of its ease of calculation.

There are numerous models in logistic regression: an intercept-only model that includes no predictors, an incomplete model that includes the intercept plus some predictors, a full model that includes the intercept plus all predictors (including, possibly, interactions and variables raised to a power), and a perfect (hypothetical) model that would provide an exact fit of expected frequencies to observed frequencies if only the right set of predictors were measured. As a consequence, there are several comparisons possible: between the intercept-only model and the full model, between the intercept-only model and an incomplete model, between an incomplete model and the full model, between two incomplete models, between a chosen model and the perfect model, etc. Not only are there numerous possible comparisons among models but also there are several tests to evaluate goodness-of-fit. No single test is universally preferred, so the computer programmes use different tests for the models.

Likelihood ratio and deviance: The overall measure of how well the model fits, similar to the residual or error sums of squares value for multiple regression, is given by the *likelihood ratio* test statistic *LR*. (It is actually −2 times the log of the likelihood value and is referred to as −2LL or −2 log likelihood.) A well-fitting model will have a small value for −2LL. The minimum value for −2LL is 0. (A perfect fit has a likelihood of 1, and −2LL is then 0.) The likelihood value can be compared between equations as well, with the difference representing the change in predictive fit from one equation to another. Statistical programmes have automatic tests for the significance of these differences. The likelihood ratio *LR* is −2 times the difference between the log likelihoods of two models, one of which is a subset of the other. The distribution of the *LR* statistic is closely approximated by the chi-square distribution for large sample sizes. The degrees of freedom (*DF*) of the approximating chi-square distribution is equal to the difference in the number of regression coefficients in the two models. The test is named as a ratio rather than a difference since the difference between two log likelihoods is equal to the log of the ratio of the two likelihoods. That is, if L_{full} is the log likelihood of the full model and L_{subset} is the log likelihood of a subset of the full model, the

likelihood ratio is defined as $LR = -2\left[L_{subset} - L_{full}\right] = -2\left[\ln\left(\dfrac{L_{subset}}{L_{full}}\right)\right]$. The

−2 adjusts *LR* so the chi-square distribution can be used to approximate its distribution. The likelihood ratio test is the test of choice in logistic regression. Various simulation studies have shown that it is more accurate than the Wald test in situations with small to moderate sample sizes. In large samples, it performs about the same. Unfortunately, the likelihood ratio test requires more calculations than the Wald test, since it requires that two maximum-likelihood models must be fit.

Deviance: When the full model in the likelihood ratio test statistic is the saturated model, LR is referred to as the *deviance*. A saturated model is one which includes all possible terms (including interactions) so that the predicted values from the model equal the original data. The formula for the deviance is $D = -2\left[L_{\text{Reduced}} - L_{\text{Saturated}}\right]$. The deviance may be calculated directly using the formula for the deviance residuals. This expression may be used to calculate the log likelihood of the saturated model without actually fitting a saturated model. The formula is $L_{\text{Saturated}} = L_{\text{Reduced}} + \dfrac{D}{2}$. The deviance in logistic regression is analogous to the residual sum of squares in multiple regression. In fact, when the deviance is calculated in multiple regression, it is equal to the sum of the squared residuals. Deviance residuals, to be discussed later, may be squared and summed as an alternative way to calculate the deviance, D. The change in deviance, ΔD, due to excluding (or including) one or more variables is used in logistic regression just as the partial F test is used in multiple regression. The formula for ΔD for testing the significance of the regression coefficient(s) associated with the independent variable x_1 is

$$\Delta D_{x_1} = D_{\text{without } x_1} - D_{\text{with } x_1}$$
$$= -2\left[L_{\text{without } x_1} - L_{\text{saturated}}\right] + 2\left[L_{\text{with } x_1} - L_{\text{saturated}}\right]$$
$$= -2\left[L_{\text{without } x_1} - L_{\text{with } x_1}\right]$$

This formula looks identical to the likelihood ratio statistic. Because of the similarity between the change in deviance test and the likelihood ratio test, their names are often used interchangeably.

Testing for significance of the coefficients: Logistic regression can also test the hypothesis that a coefficient is different from zero (zero means that the odds ratio does not change and the probability is not affected), as is done in multiple regression. In multiple regression, the t value is used to assess the significance of each coefficient. Logistic regression uses a different statistic, the *Wald statistic*. It provides the statistical significance for each estimated coefficient so that hypothesis testing can occur just as it does in multiple regression. The Wald test will be familiar to those who use multiple regression. In multiple regression, the common t-test for testing the significance of a particular regression coefficient is a Wald test. In logistic regression, the Wald test is calculated in the same manner. The formula for the Wald statistic is $z_j = \dfrac{b_j}{s_{b_j}}$

where s_{b_j} is an estimate of the standard deviation of b_j provided by the square root of the corresponding diagonal element of the covariance matrix, $V(\hat{\beta})$. With large sample sizes, the distribution of z_j is closely approximated by the normal distribution. With small and moderate sample sizes, the normal approximation is described as 'adequate.'

Confidence intervals: Confidence intervals *CI* for the regression coefficients are based on the Wald statistics. The formula for the limits of a $100(1 - \alpha)\%$ two-sided confidence interval *CI* is $b_j \pm |z_{\alpha/2}| s_{b_j}$. When the confidence interval *CI* includes zero, so at the 5% significance level, we would not reject the null hypothesis $H_0 : \beta_j = 0$ that this model coefficient is zero. The regression coefficient β_j is also the logarithm of the odds ratio. Because we know how to find a confidence interval *CI* for β_j, it is easy to find a CI for the odds ratio. The point estimate of the odds ratio is $O_R = \exp(b_j)$ and the $100(1 - \alpha)$ percent

CI for the odds ratio is $\exp\left[b_j - z_{\alpha/2} \times s_{b_j}\right] \le O_R \le \exp\left[b_j + z_{\alpha/2} \times s_{b_j}\right]$. The *CI* for the odds ratio is generally not symmetric around the point estimate. Furthermore, the point estimate $\hat{O}_R = \exp(b_j)$ actually estimates the median of the sampling distribution of $\hat{O}_R$.

R^2: In multiple regression, R_M^2 represents the proportion of variation in the dependent variable accounted for by the independent variables. (The subscript "M" emphasizes here that this statistic is for multiple regression.) It is the ratio of the regression sum of squares to the total sum of squares. When the residuals from the multiple regression can be assumed to be normally distributed, R_M^2 can be calculated as $R_M^2 = \dfrac{L_p - L_0}{L_0}$ where L_0 is the

log likelihood of the intercept-only model and L_p is the log likelihood of the model that includes the independent variables. Here L_p varies from L_0 to 0 and R_M^2 varies between zero and one. This quantity has been proposed for use in logistic regression. Unfortunately, when R_L^2 (the R^2 for logistic regression) is calculated using the above formula, it does not necessarily range between zero and one. This is because the maximum value of L_p is not always 0 as it is in multiple regression. Instead, the maximum value of L_p is L_s, the log likelihood of the saturated model. To allow R_L^2 to vary from zero to one, it is calculated

as follows $R_L^2 = \dfrac{L_p - L_0}{L_0 - L_s}$. The introduction of L_s into this formula causes a

degree of ambiguity with R_L^2 that does not exist with R_M^2. This ambiguity is due to the fact that the value of L_s depends on the configuration of independent variables. The following example will point out the problem. Consider a logistic regression problem consisting of a binary dependent variable and a pool of four independent variables. The data for this example are given in the following table.

y	x_1	x_2	x_3	x_4
0	1	1	2.3	5.9
0	1	1	3.6	4.8
1	1	1	4.1	5.6
0	1	2	5.3	4.1
0	1	2	2.8	3.1
1	1	2	1.9	3.7
1	1	2	2.5	5.4
1	2	1	2.3	2.6
1	2	1	3.9	4.6
0	2	1	5.6	4.9
0	2	2	4.2	5.9
0	2	2	3.8	5.7
0	2	2	3.1	4.5
1	2	2	3.2	5.5
1	2	2	4.5	5.2

If only x_1 and x_2 are included in the model, the dataset may be collapsed because of the number of repeats. In this case, the value of L_s will be less than zero. However, if x_3 or x_4 are used there are no repeats and the value of L_s will be zero. Hence, the denominator of R_L^2 depends on which of the independent variables is used. This is not the case for R_M^2. This ambiguity comes into play especially during subset selection. It means that as we enter and remove independent variables, the target value L_s can change. Hosmer and Lemeshow [144] recommend against the use R_L^2 as a goodness-of-fit measure. However, we have included it in our output because it does provide a comparative measure of the proportion of the log likelihood that is accounted for by the model. Just we should remember than an R_L^2 value of 1.0 indicates that the logistic regression model achieves the same log likelihood as the saturated model. However, this does not mean that it fits the data perfectly. Instead, it means that it fits the data as well as could be hoped for.

4.8.6 Residual Diagnostics Checking in Logistic Regressions

Residuals are the discrepancies between the data values and the their predicted values from the fitted model. Residuals can also be used for diagnostic checking and investigating model adequacy in logistic regression. A residual analysis detects outliers, identifies influential observations, and diagnoses the appropriateness of the logistic model, cf. Chapter 6. An analysis of the residuals should be conducted before a regression model is used. Unfortunately, the residuals are more difficult to define in logistic regression than in regular multiple regression because of the nonlinearity of the logistic model and because more than one regression equation is used. The discussion that follows provides an introduction to the residuals that are produced by the logistic regression procedure. Pregibon [145] presented this material for the case of the two-group logistic regression. Extensions of Pregibon's results to the multiple-group case are provided in an article by Lesaffre and Albert [146] and in the book by Hosmer and Lemeshow [144].

Data Configuration: When dealing with residuals, it is important to understand the data configuration. Often, residual formulations are presented for the case when each observation has a different combination of values of the independent variables. When some observations have identical independent variables or when we have specified a frequency variable, these observations are combined to form a single row of data. The n original observations are combined to form j unique rows. The response indicator variables y_{gj} for the original observations are replaced by two variables: w_{gj} and n_j. The variable n_j is the total number of observations with this independent variable configuration. The variable w_{gj} is the number of the n_j observations that are in group g. The residuals are reported in the original observation order. Thus, if two identical observations have been combined, the residual is shown for each. If corrective action needs to be taken because a residual is too large, both observations must be deleted. Also, if we want to calculate the deviance or Pearson chi-square from the corresponding residuals, care must be taken that we use only the j collapsed rows, not the n original observations.

Simple Residuals: Each of the g logistic regression equations can be used to estimate the probabilities that an observation of independent variable values given by x_j belongs to the corresponding group. The actual values of these probabilities were defined earlier as $p_{gj} = \Pr(y = g | x_j)$. The estimated values of these probabilities are called $\hat{p}_{gj}$ and hat symbol is used to represent an estimated parameter. These estimated probabilities $\hat{p}_{gj}$ can be compared to the actual probabilities p_{gj} occurring in the database by subtracting the two quantities, forming a residual. The actual values were defined as the indicator variables y_{gj}. Thus, simple residuals may be defined as $r_{gj} = y_{gj} - p_{gj}$. Unlike multiple regression, there are g residuals for each observation instead of just one. This makes residual analysis much more difficult. If the logistic regression model fits an observation closely, all of its residuals will be small. Hence, when y_{gj} is one, p_{gj} will be close to one and when y_{gj} is zero, p_{gj} will be close to zero. Unfortunately, the simple residuals have unequal variance equal to $n_j p_{gj}(1 - p_{gj})$, where n_j is the number of observations with the same values of the independent variables as observation j. This unequal variance makes comparisons among the simple residuals difficult and alternative types of residuals are necessary.

Pearson Residuals: One popular alternative to the simple residuals are the *Pearson residuals* which are so named because they give the contribution of each observation to the Pearson chi-square goodness of fit statistic. When the values of the independent variables of each observation are unique, the

formula this residual is $\chi'_j = \pm \sqrt{\sum_{g=1}^{G} \dfrac{(y_{gj} - p_{gj})^2}{p_{gj}}}, j = 1, 2, ..., n$. The negative

sign is used when $y_{gj} = 0$ and the positive sign is used when $y_{gj} = 1$. When some of the observations are duplicates and the database has been collapsed

the formula is $\chi_j = \pm \sqrt{\sum_{g=1}^{G} \dfrac{(w_{gj} - n_j p_{gj})^2}{n_j p_{gj}}}, j = 1, 2, ..., J$, where the plus (minus)

is used if w_{gj} / n_j is greater (less) than p_{gj}. By definition, the sum of the squared Pearson residuals is the Pearson chi-square goodness of fit statistics.

That is, $\chi^2 = \displaystyle\sum_{j=1}^{J} \chi_j^2$.

Deviance Residuals: Remember that the deviance is -2 times the difference between log likelihoods of a reduced model and the saturated model. The deviance is calculated using

$$D = -2\left[L_{\text{Reduced}} - L_{\text{Saturated}}\right] = -2\left[\sum_{j=1}^{n}\sum_{g=1}^{G} y_{gj} \ln(p_{gj}) - \sum_{j=1}^{n}\sum_{g=1}^{G} y_{gj} \ln(y_{gj})\right]$$

$$= -2\left[\sum_{j=1}^{n}\sum_{g=1}^{G} y_{gj} \ln(p_{gj})\right]$$

This formula uses the fact that the saturated model reproduces the original data exactly and that, in these sums, the value of $0 \ln(0)$ is defined as 0 and that the $\ln(1)$ is also 0. The deviance residuals are the square roots of the contribution of each observation to the overall deviance. Thus, the formula for the deviance residual is $d'_j = \pm\sqrt{2\sum_{g=1}^{G} y_{gj} \ln(\frac{1}{p_{gj}})}, j = 1, 2, ..., n$. The negative

sign is used when $y_{gj} = 0$ and the positive sign is used when $y_{gj} = 1$. When some of the observations are duplicates and the database has been collapsed the

formula is $d_j = \pm\sqrt{2\sum_{g=1}^{G} w_{gj} \ln(\frac{w_{gj}}{n_j p_{gj}})}, j = 1, 2, ..., J$ where the plus (minus) is

used if $w_{Ref(g),j} / n_j$ is greater (less) than $p_{Ref(g),j}$. By definition, the sum of the

squared deviance residuals is the deviance. That is, $D = \sum_{j=1}^{J} d_j^2$.

Hat Matrix Diagonal: The diagonal elements of the hat matrix can be used to detect points that are extreme in the independent variable space. These are often called *leverage* design points. The larger the value of this statistic, the more the observation influences that estimates of the regression coefficients. An observation that is discrepant, but has low leverage, should not cause much concern. However, an observation with a large leverage and a large residual should be checked very carefully. The use of these hat diagonals is discussed further in the multiple regression chapter 6. The formula for the hat diagonal associated with the *j*th

observation and *g*th group is $h_{gj} = n_j p_{gj}(1 - p_{gj})\sum_{i=1}^{p}\sum_{k=1}^{p} x_{ij} x_{kj} \hat{V}_{gik}, j = 1, 2, ..., J$
,
where $\hat{V}_{gik}$ is the portion of the covariance matrix of the regression coefficients associated with the *g*th regression equation. The interpretation of this diagnostic is not as clear in logistic regression as in multiple regression because it involves the predicted values which in turn involve the dependent variable. In multiple regression, the hat diagonals only involve the independent variables. This formula matches Pregibon [145] in the two-group case. In the multiple-group case, the two-group formula is applied to each group.

DFBETA: One way to study the impact of an observation on each regression coefficient is to determine how much that coefficient changes when the observation is deleted. The DFBETA statistic is the standardized difference between a regression coefficient before and after the removal of the *j*th observation. The formula for DFBETA is approximated by

$DFBETA_{gij} = \left(\dfrac{w_{gj} - n_j p_{gj}}{(1 - h_{gj})\sqrt{\hat{V}_{gii}}}\right)\sum_{k=1}^{p} x_{kj}\hat{V}_{gik}, j = 1, 2, ..., J$, where $\hat{V}_{gik}$ is the portion

of the covariance matrix associated with the *g*th regression equation. This formula matches Pregibon [145] in the two-group case, but is different from Lesaffre [146] in the multi-group case.

Cooks Distance C and Cbar: C and *Cbar* are extensions of Cooks distance for logistic regression. Quoting from Pregibon [145], page 719: "Cbar measures the overall change in fitted logits due to deleting the lth observation for all points excluding the one deleted. Conversely, C includes the deleted point. Although C will usually be the preferred diagnostic to measure overall coefficients changes, in the examples examined to date, the one-step approximations were more accurate for *Cbar* than C". The formulas for C and *Cbar* are $C_{gj} = \dfrac{\chi_j^2 h_{gj}}{\left(1 - h_{gj}\right)^2}, 1 = 1, 2, ..., J$

and $\overline{C}_{gj} = \dfrac{\chi_j^2 h_{gj}}{\left(1 - h_{gj}\right)}, 1 = 1, 2, ..., J$ [145].

DFDEV and DFCHI2 are statistics that measure the change in deviance and in Pearson's chi-square, respectively, that occurs when an observation is deleted from the dataset. Large values of these statistics indicate observations that have not been fitted well. The formulas for these statistics are

$$DFDEV_{gj} = d_j^2 + \overline{C}_{gj}, j = 1, 2, ..., J \text{ and } DFCHI2_{gj} = \frac{\overline{C}_{gj}}{h_{gj}}, j = 1, 2, ..., J \text{ [145]}.$$

4.8.7 Predicted Probabilities

This section describes how to calculate the predicted probabilities of group membership and associated confidence intervals [142]. Recall that the regression equation is linear when expressed in logit form. That is,

$$\ln\left(\frac{P_g}{P_1}\right) = \beta_{g0} + \beta_{g1}x_1 + \beta_{g2}x_2 + ... + \beta_m x_m$$

The adjustment for the prior probabilities changes the value of the intercepts, so this expression may be simplified to $\ln\left(\dfrac{P_g}{P_1}\right) = \beta_{g1}x_1 + \beta_{g2}x_2 + ... + \beta_m x_m$ if we assume that the intercepts have been appropriately adjusted. Assuming that the estimated matrix of regression coefficients is distributed asymptotically as a multivariate normal, the point estimates of this quantity for a specific set of x values is given by $l_j = \ln\left(\dfrac{P_g}{P_1}\Big| x_j\right) = x_j \hat{\mathbf{b}}_g$ and the corresponding confidence interval is given by $l_j \pm z_{\alpha/2}\left(\mathbf{x}_j^T \mathbf{V}_g \mathbf{x}_j\right)$ where $\mathbf{V}_g$ is that portion of the covariance matrix $\mathbf{V}\left(\hat{\mathbf{b}}\right)$ that deals with the gth regression equation.

4.8.8 Subset Selection

Subset selection refers to the task of finding a small subset of the available independent variables that does a good job of predicting the dependent variable. Because logistic regression must be solved iteratively, the task of finding the best subset can be very time consuming. Hence, techniques that search all possible combinations of the independent variables are not feasible. Instead, algorithms, for example NCSS2007, that add or remove a variable at each step must be used. Two such searching algorithms are available in this module: forward selection and forward selection with switching. Before discussing the details of these two algorithms, it is important to comment on a couple of issues that can come up. First of all, since there is more than one regression equation when there are more than two categories in the dependent variable, it is possible that a variable is important in one of the equations and not in the others. The algorithms presented here are based on the overall likelihood. This means that if an independent variable is important in at least one of the regression equations, it will be kept. A second issue is what to do with the individual-degree of freedom variables that are generated for a categorical independent variable. If such a variable has six categories, five binary variables are generated. We can see that with two or three categorical variables, a large number of binary variables may result, which greatly increases the total number of variables that must be searched. To avoid this problem, the algorithms search on model terms rather than on the individual binary variables. Thus, the whole set of binary variables associated with a given term are considered together for inclusion in, or deletion from, the model. It is all or none. Because of the time consuming nature of the algorithm, this is the only feasible way to deal with categorical variables. If we want the subset algorithm to deal with them individually, we also can generate the set of binary variables manually.

Hierarchical Models: A third issue is what to do with interactions. Usually, an interaction is not entered in the model unless the individual terms that make up that interaction are also in the model. For example, the interaction term A*B*C is not included unless the terms A, B, C, A*B, A*C, and B*C are already in the model. Such models are said to be *hierarchical*. We have the option during the search to force the algorithm to consider only hierarchical models during its search. Thus, if C is not in the model, interactions involving C are not even considered. Even though the option for non-hierarchical models is available, we recommend that we only consider hierarchical models.

Forward Selection: The method of forward selection proceeds as follows:

(1) Begin with no terms in the model.

(2) Find the term that, when added to the model, achieves the largest value of the log likelihood. Enter this term into the model.

(3) Continue adding terms until a target value for the log-likelihood is achieved or until a preset limit on the maximum number of terms in the model is reached. These terms can be limited to those keeping the model hierarchical.

This method is comparatively fast, but it does not guarantee that the best model is found except for the first step when it finds the best single term. We might use it when we have a large number of observations and terms so that other, more time consuming, methods are not feasible.

Forward Selection with Switching: This method is similar to the previous method of Forward Selection. However, at each step when a term is added, all terms in the model are switched one at a time with all candidate terms not in the model to determine if they increase the value of the log likelihood. If a switch can be found, it is made and the pool of terms is again searched to determine if another switch can be made. This switching can be limited to those keeping the model hierarchical. When the search for possible switches does not yield a candidate, the subset size is increased by one and a new search is begun. The algorithm is terminated when a target subset size is reached or all terms are included in the model.

Discussion: These algorithms usually require two runs. In the first run, we set the maximum subset size to a large value such as 10. By studying the Subset Selection reports from this run, we can quickly determine the optimum number of terms. We reset the maximum subset size to this number and make the second run. This two-step procedure works better than relying on some F-to-enter and F-to-remove tests whose properties are not well understood to begin with.

4.8.9 Receiver Operating Characteristic Plot (ROC)

Sometimes the investigator would want to know how much to trust such predictions. In other words, can the equation predict correctly a high proportion of the time? This question is different from asking about the statistical significance, as it is possible to obtain statistically significant results that do not predict very well. If the investigator wishes to classify cases, a cutoff point on the probability of being depressed, for example, must be found. This cutoff point is denoted by P_c. The user can quickly try a set of cutoff points and zero in on a good one.

It is possible to create an *ROC* (*receiver operating characteristic*) curve. *ROC* was originally proposed for signal-detection work when the signals were not always correctly received. This is often called the *true positive fraction*, or *sensitivity* in medical research. On the horizontal axis is the proportion of nondepressed persons classified as depressed (called *false positive fraction* or one minus *specificity* in medical studies). Three *ROC* curves are drawn in Fig. 4.49. The top one represents a hypothetical curve which would be obtained from a logistic regression equation that resulted in excellent prediction.

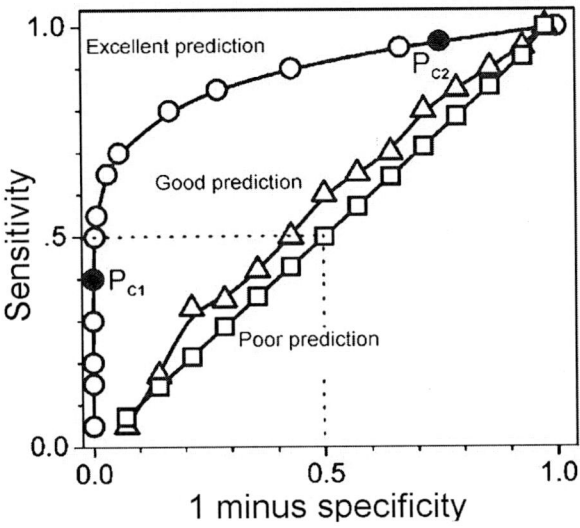

Figure 4.49 ROC Curve from Logistic Regression with two positions of the cutoff point: P_{C1} is a strict threshold cutoff point, P_{C2} is a lax threshold cutoff point.

The lower hypothetical curve (straight line) represents the chance-alone assignment (i.e. flipping a coin). The closeness of the middle curve to the lower curve shows that we perhaps need other predictor variables in order to obtain a better logistic regression model. This is the case even though our model was significant at the $P = 0.009$ level. This is called a *strict threshold*. The downside of this approach is that many depressed persons would be missed. This type of cutpoint is called a *lax threshold*. The curve must pass through the points (0,0) and (1,1). The maximum area under the curve is one. The numerical value of this area would be close to one if the prediction is excellent and close to one-half if it is poor. The ROC curve can be useful in deciding which of two logistic regression models to use. All else being equal, the one with the greater area would be chosen. Alternatively, we might wish to choose the one with the greatest height to the ROC curve at the desired cutoff point.

Problem 4.5   *The relationship between selected variables CELL, LI, and TEMP on the leukimia remission occurred variable REMISS*

In the first step of an analysis, a logistic regression would be run to determine the relationship between all six variables on the binary dependent variable

Remiss while in the second step between three selected variables *Cell, Li*, and *Temp* on the *Remiss*.

Data: Data measurements file *Leukemia* given below are the first few rows of a set of data about leukemia patients published in Lee [147]. The response variable is whether leukemia remission occurred (*Remiss*). The independent variables are cellularity of the marrow clot section (*Cell*), smear differential percentage of blasts (*Smear*), percentage of absolute marrow leukemia cell infiltrate (*Infil*), percentage labelling index of the bone marrow leukemia cells (*Li*), absolute number of blasts in the peripheral blood (*Blast*), and the highest temperature prior to start of treatment (*Temp*). This dataset is stored in the *leukemia* database in the *Data* directory. If missing values are found in any of the independent variables being used, the row is omitted. If only the dependent variable is missing, the row will not be used in the formation of the coefficient estimates, but a predicted value will be generated for that row.

LEUKEMIA dataset:

Remiss	Cell	Smear	Infil	Li	Blast	Temp
1	80	83	66	190	11.6	996 I
1	90	36	32	140	4.5	992 I
...	...	...	...	...	...	...
...	...	...	...	...	...	...
...	...	...	...	...	...	...
...	...	...	...	...	...	...
1	95	97	92	100	16.0	992 I
0	95	87	83	190	21.6	1020	

*Software:*NCSS2007, STATISTICA

Solution:

First step: Subset selection – This section presents an example of how to conduct a subset selection. This analysis will search for the best model from among a pool of the six numeric independent variables. We may follow along here by making the appropriate entries.

Parameter	Value	Parameter	Value
Dependent Variable	REMISS	Rows Processed	29
Reference Value	1	Rows Used	27
Number of Values	2	Rows for Validation	0
Frequency Variable	None	Rows X's Missing	2

Numeric Ind. Variables	6	Rows Freq Miss. or 0	0
Categorical Ind. Vriables	0	Rows Prediction Only	0
Final Log Likelihood	−10.87752	Unique Row Patterns	27
Model R^2	0.36707	Sum of Frequencies	27
Actual Convergence	2.081623E-06	Likelihood Iterations	9
Target Convergence	0.000001	Maximum Iterations	20
Model D.F.	6	Max Like Message	Quasi-Separation
Model	Cell\Smear\Infil\Li\Blast\Temp		

The first thing we notice is the warning message about quasi-separation. If quasi-separation occurs, the maximum likelihood estimates do not exist and all results are suspect. We note that 9 likelihood iterations occurred and the Actual Convergence is near the Target Convergence. We decide to rerun the analysis after resetting the Max Terms in Subset box from 6 to 5. This error message often occurs when a small set of data is fit with a model with too many terms.

- **Run summary section:** The warning message has disappeared and the algorithm finished normally.

Parameter	Value	Parameter	Value
Dependent Variable	REMISS	Rows Processed	29
Reference Value	1	Rows Used	27
Number of Values	2	Rows for Validation	0
Frequency Variable	None	Rows X's Missing	2
Numeric Ind. Variables	6	Rows Freq Miss. or 0	0
Categorical Ind. Variables	0	Rows Prediction Only	0
Final Log Likelihood	−10.92900	Unique Row Patterns	27
Model R^2	0.36407	Sum of Frequencies	27
Actual Convergence	7.136538E-07	Likelihood Iterations	7
Target Convergence	0.000001	Maximum Iterations	20
Model D.F.	5	Max Like Message	Normal Completion
Model		Cell\Smear\Infil\Li\Blast\Temp	

- **Subset selection summary section:** This report shows the best log-likelihood value for each subset size. In this example, it appears that four terms (the intercept and three independent variables) provides the best model. Adding the fifth independent variable does not increase the R^2 value very much. **No. Terms:** The number of terms. This includes the intercept. **No. X's:** The number of x's that were included in the model. In this case, the number of terms matches the number of x's. This would not be the case if some of the terms were categorical variables. **Log Likelihood:** This is the value of the log likelihood function evaluated at the maximum likelihood estimates. Our goal is to find a subset size above which little is gained by adding more variables. **R^2 Value:** This is the value of R^2 calculated using the formula $R_L^2 = \dfrac{L_p - L_0}{L_0 - L_S}$. We are looking for the subset size at which this value does not increase by a meaningful amount. **R^2 Change:** This is the increase in R^2 that occurs when each new subset size is reached. Search for the subset size below which the R^2 value does not increase by more than 0.02 for small samples or 0.01 for large samples. In this example, the optimum subset size appears to be four terms.

No. Terms	No. X's	Log Likelihood	R^2 Value	R^2 Change
1	1	−17.186	0.000	0.000
2	2	−13.036	0.241	0.241
3	3	−12.170	0.292	0.050
4	4	−10.977	0.361	0.069
5	5	−10.929	0.364	0.003

- **Subset selection detail section:** This report shows the highest log likelihood for each subset size. In this example, it appears that four terms (the intercept and three independent variables) provide the best model. Adding the fifth independent variable does not increase the R^2 value very much. **Action:** This item identifies the action that was taken at this step. A term was added, removed, or two were switched. **No. Terms:** The number of terms. This includes the intercept. **No. X's:** The number of x's that were included in the model. In this case, the number of terms matches the number of x's would not be the case if some of the terms were categorical variables. **Log Likelihood:** This is the value of the log likelihood function after the completion of this step. Our goal is to find a subset size above which little is gained by adding more variables. **Terms Entered and Removed:** These columns identify the terms added, removed, or switched.

Step	Action	No. Terms	No. X's	Log Likelihood	Term Entered or Removed
1	Add	1	1	-17.18	Intercept
2	Add	2	2	-13.03	*Li*
3	Add	3	3	-12.17	*Cell*
4	Add	4	4	-10.97	*Temp*
5	Add	5	5	-10.93	*Smear*

Conclusion of the 1st step: After considering these reports, it was decided to include CELL, LI, and TEMP in the final logistic regression model. Another run should now take place using only these independent variables. A complete residual analysis would be necessary before the equation is finally adopted.

Second Step: Subset selection – It should be studied to make sure that the data were read in properly and that the logistic regression procedure terminated normally. We will only discuss those parameters that need special explanation. **Reference Group:** The reference group is that category of the dependent variable that is defined implicitly in terms of the other categories. This is the category that is skipped on much of the output. **Number of Groups:** This is the number of unique categories that were found for the dependent variable. **Final Log Likelihood:** This is the log likelihood of the model that is reported on here. **Model R^2:** This is the R^2 that was achieved by our regression. Read the discussion of R^2 that was given earlier to better understand how to interpret R^2 in the case of logistic regression. **Actual and Target Convergence:** The Target Convergence is the amount that is used to stop the iterative fitting of the maximum likelihood algorithm. If the Actual Convergence amount is larger than the Target amount, the algorithm ended before converging and care must be taken in using any of the results. If this happens, the usual remedy is to increase the maximum number of iterations. If this does not solve the problem, we will have to change the variables in the model. **Rows Processed, Used, etc.:** These values record how many of each type of observation were encountered when the database was read. We should make sure that these amounts are what we expect. **Unique Row Patterns:** This gives the number of unique patterns found in the variables. Both the dependent and independent variables are considered in forming this count. **Likelihood and Maximum Iterations:** The Likelihood Iterations are the number of iterations necessary to solve the likelihood equations. Usually, fewer than ten iterations are necessary. If the number of Likelihood

Iterations is equal to the Maximum Iterations, the maximum likelihood algorithm did not converge and we should take some remedial action such as increasing the Maximum Iterations or changing the regression model. **Max Like Message:** This is the message that was returned when the maximum likelihood algorithm ended. Unless the message "Normal Completion" is received, we should take appropriate corrective action. **Model D.F.:** This is the number of degrees of freedom in the $G - 1$ logistic regression models. **Model:** This is an abbreviated representation of the regression model that was fit to the data.

Run Summary Section:

Parameter	Value	Parameter	Value
Dependent Variable	REMISS	Rows Processed	27
Reference Value	1	Rows Used	27
Number of Values	2	Rows for Validation	0
Frequency Variable	None	Rows X's Missing	0
Numeric Ind. Variables	3	Rows Freq Miss. or 0	0
Categorical Ind. Variables	0	Rows Prediction Only	0
Final Log Likelihood	−10.97669	Unique Row Patterns	26
Model R^2	0.36130	Sum of Frequencies	27
Actual Convergence	2.94261E-07	Likelihood Iterations	7
Target Convergence	0.000001	Maximum Iterations	20
Model D.F.	4	Max Like Message	Normal Completion
Model	$Cell|Li|Temp$		

- **Response analysis section:** This report describes the dependent variable y. We use it to understand the dependent variable y and how well the regression model approximates it. **Categories:** These are the unique values found for the dependent variable y. We check to make sure that no unexpected categories were found. **Count:** This is the sum of the frequencies (counts) for each category of the dependent variable y. **Unique Rows:** This is the number of unique rows in each category as determined by the values of the independent variables x. **Prior:** This is the prior probability of each category as given by the user in the Prior Probabilities option box. **Act vs Pred R^2:** This is the R^2 that is achieved when the indicator variable for this category is regressed on the predicted probability of being in this category. **% Correctly Classified:** This is the percent of the observations from this category that were correctly classified as such by the multinomial logistic regression model.

Response Analysis Section:

REMISS Categories	Count	Unique Rows	Prior	Act vs Pred R^2	% Correctly Classified
0	18	17	0.667	0.389	83.3
1	9	9	0.333	0.389	66.6
Total	27	26			77.7

- **Parameter significance tests section:** This report gives the estimated logistic regression equation and associated significance tests. The reference group of the dependent variable y is shown in the title. If the dependent variable has more than two categories, the appropriate information is displayed for each of the G-1 equations. **Parameter:** This is the variable from the model that is displayed on this line. If the independent variable is continuous, it is displayed directly. If the independent variable is discrete, the definition of the binary variable that was generated is given. **Regression Coefficient (B or Beta):** This is the estimated value of the corresponding regression coefficient, sometimes referred to as b or $\hat{\beta}$. The interpretation of the regression coefficients is often difficult. **Standard Error:** This is s_{b_j}, the large-sample estimate of the standard deviation of the regression coefficient. This is an estimate of the precision of the regression coefficient. It is used as the denominator of the Wald test. **Wald Z-Value (Beta=0):** This is the z value of the Wald test used for testing the hypothesis that $\beta_{gj} = 0$ against the alternative $\beta_{gj} \neq 0$. The Wald test is calculated using the formula $z_{gj} = \dfrac{b_{gj}}{s_{b_j}}$. The distribution of the Wald statistic is closely approximated by the normal distribution in large samples. However, in small samples, the normal approximation may be poor. For small samples, the deviance tests should be used instead to test significance since they perform better. One problem that occurs in multiple-group logistic regression is that the test may be significant for the regression coefficient associated with one category, but not for the same coefficient associated with another category. In this case, we recommend that the independent variable be kept in the model if it is significant in at least one of the G-1 regression equations. **Wald Prob Level:** This is the calculated significance level of the Wald test. If this value is less than some predefined α level, say $\alpha = 0.05$, the independent variable is said to be statistically significant.

Otherwise, the independent variable is not significant. **Odds Ratio Exp(B):** This is the estimated odds ratio associated with this regression coefficient. It is only useful for binary independent variables in which the two values are zero and one. These are the values that are generated for categorical independent variables. The formula used is $OR = \exp(b)$.

Parameter Significance Tests Section (Reference Value: *Remiss* = 1):

Parameter	Regression Coefficient (B or Beta)	Standard Error	Wald Z-Value (Beta=0)	Wald Prob Level	Odds Ratio Exp(B)
B0: Intercept	−67.63	56.88	−1.189	0.234	0.000
B1: *Cell*	−9.65	7.75	−1.245	0.213	0.000
B2: *Li*	−3.87	1.77	−2.175	0.030	0.021
B3: *Temp*	82.07	61.71	1.330	0.183	100

- **Parameter confidence limits section:** This report gives the estimated logistic regression equation and associated confidence limits. The reference group of the dependent variable is shown in the title. If the dependent variable has more than two categories, the information is displayed for each of the G-1 equations. **Parameter:** This is the independent variable that is displayed on this line. If the independent variable is continuous, it is displayed directly. If the independent variable is discrete, the definition of the binary variable that was generated is given. **Regression Coefficient (B or Beta):** This is the estimated value of the regression coefficient, sometimes referred to as b or β. The interpretation of the regression coefficients is difficult. **Standard Error:** This is s_{b_j}, the large-sample estimate of the standard deviation of the regression coefficient. This is an estimate of the precision of the regression coefficient. It is used as the denominator of the Wald test. **Confidence Limits:** These are the lower and upper confidences limits for β_{gj} based on the Wald statistic. These confidence limits are use the formula $b_{gj} \pm z_{1-\alpha/2} \times s_{b_{gj}}$. Since they are based on the Wald test, they are only valid for large samples. **Odds Ratio Exp(B):** This is the estimated odds ratio associated with this regression coefficient. It is only useful for binary independent variables in which the two values are zero and one. These are the values that are generated for categorical independent variables. The formula used is $OR = \exp(b)$. Because of formatting limitations, the value is not displayed if it is larger than 10000.

Parameter Confidence Limits Section (Reference Value: *Remiss* = 1):

Parameter	Regression Coefficient (B or Beta)	Standard Error	Lower 95% Confidence Limit	Upper 95% Confidence Limit	Odds Ratio Exp(B)
B0: Intercept	−67.63	56.88	−179.13	43.86	0.000
B1: *Cell*	−9.65	7.75	−24.84	5.54	0.000
B2: *Li*	−3.87	1.78	−7.35	−0.382	0.021
B3: *Temp*	82.07	61.71	−38.88	203.03	1000

- **Odds ratio estimation section:** This report presents estimates of the odds ratios *OR* and associated confidence limits associated with each independent variable in the model. **Parameter:** This is the independent variable that is displayed on this line. If the independent variable is continuous, it is displayed directly. If the independent variable is discrete, the definition of the binary variable that was generated is given. The interpretation of the regression coefficients is often difficult. **Odds Ratio Exp(B):** This is the estimated odds ratio *OR* associated with this regression coefficient. It is only useful for binary independent variables in which the two values are zero and one. These are the values that are generated for categorical independent variables. The formula used is $OR = \exp(b)$. Because of formatting limitations, the value is not displayed if it is larger than 10000. **Confidence Limits:** The lower and upper confidence limits yield an interval estimate of the odds ratio. The confidence coefficient is one minus alpha. Thus, when alpha is $\alpha = 0.05$, the confidence coefficient is 0.95 or 95%. The formula used is $b_{gj} \pm z_{1-\alpha/2} \times s_{b_{gj}}$. Since these confidence limits are based on Wald statistics, they are only valid for large samples.

Odds Ratios Section (Reference Value: *Remiss* = 1):

Parameter	Regression Coefficient (B or Beta)	Odds Ratio Exp(B)	Lower 95% Confidence Limit	Upper 95% Confidence Limit
B0: Intercept	−67.63	0.000	0.000	10000
B1: *Cell*	−9.65	0.000	0.000	254.59
B2: *Li*	−3.87	0.021	0.000	0.683
B3: *Temp*	82.07	1000	0.000	10000

- **Estimated logistic regression model(s):** This report gives the logistic regression model in a regular text format that can be used as a transformation formula. A separate model is displayed for each of the G-1 categories of the dependent variable.

Estimated Logistic Regression Model(s): Model For *Remiss*=0:

$-67.63 -9.65*Cell -3.87*Li + 82.07*Temp$

Each model estimates B for a specific group, where Logit(y) = $\mathbf{x}\beta$.

To calculate a probability, transform the logit using Prob($y <>$ group) = $1/(1 + \exp(-\mathbf{x}\beta))$

or Prob($y =$ group) = $\exp(-\mathbf{x}\beta)/(1 + \exp(-\mathbf{x}\beta))$.

- **Analysis of deviance section:** This report is the logistic regression analog of the analysis of variance table. It displays the results of a chi-square test used to test whether each of the individual terms in the regression are statistically significant after adjusting for all other terms in the model. This report is not produced during a subset selection run. **Term Omitted:** This is the model term that is being tested. The test is formed by comparing the deviance statistic when the term is removed with the deviance of the complete model. Thus, the deviance when the term is left out of the model is shown. The "All" line refers to the intercept-only model. This line tests the significance of the full model. The "None(Model)" refers to the complete model with no terms removed. It is usually not advisable to include an interaction term in a model when one of the associated main effects is missing – which is what happens here. However, in this case, we believe this to be a useful test. The name may become very long, especially for interaction terms. These long names may misalign the report. **DF:** This is the degrees of freedom of the chi-square test displayed on this line. *DF* is equal to (G-1) *DFt* where *DFt* is the degrees of freedom of the term. **Deviance:** The deviance is equal to minus two times the log likelihood achieved by the model being described on this line of the report. A useful way to interpret the deviance is as the analog of the residual sum of squares in multiple regression. This value is used to create the difference in deviance that is used in the chi-square test. **Increase From Model Deviance (Chi-Square):** This is the difference between the deviance for the model described on this line and the deviance of the complete model. This value follows the chi-square distribution in medium to large samples. This value can be thought of as the analog of the residual sum of squares in multiple regression. Thus, we can think of this value as the increase in the residual sum of squares that occurs when this term is removed from the model. Another way to interpret this test is as a redundancy test because it tests whether this term is redundant after considering all of the other terms in the model. The first line gives a test for the whole model. **Prob Level:** This is the significance level of the chi-square test. This is the probability that a chi-square value with degrees of freedom *DF* is equal to this value or greater. If this value is less than 0.05 (or other appropriate value), the term is said to be statistically significant.

Analysis of Deviance Section:

Term Omitted	DF	Deviance	Increase From Model Deviance (Chi-Square)	Prob Level
All	3	34.37	12.42	0.006
Cell	1	24.65	2.69	0.100
Li	1	30.83	8.87	0.003
Temp	1	24.34	2.38	0.122
None(Model)	3	21.95		

The Prob Level is for testing the significance of that term after considering all other terms.

- **Log likelihood and R^2 section:** This report provides the log likelihoods and R^2 values of various models. This report is not produced during a subset selection run. This report requires that a separate logistic regression be run for each line. Thus, if the running time is too long, we might consider omitting this report. **Term Omitted:** This is the term that is omitted from the model. The "All" line refers to the intercept-only model. The "None(Model)" refers to the complete model with no terms removed. The "None(Saturated)" line gives the results for the saturated model. The name may become very long, especially for interaction terms. These long names may misalign the report. **DF:** This is the degrees of freedom of the term displayed on this line. *DF* is equal to $(G - 1)DFt$ where *DFt* is the degrees of freedom of the term. **Log Likelihood:** This is the log likelihood of the model displayed on this line. This is the log likelihood of the logistic regression without the term listed. **R^2 of Remaining Term(s):** This is the R^2 of the model displayed on this line, R_L^2. The model does not include the term listed at the beginning of the line. This R^2 is analogous to the R^2 in multiple regression, but it is not the same. We repeat the summary of the interpretation of R^2 in logistic regression. Hosmer and Lemeshow [144] recommend against the use R_L^2 as a goodness-of-fit measure. However, we have included it in our output because it does provide a comparative measure of the proportion of the log likelihood that is accounted for by the model. We just remember than an R_L^2 value of 1.0 indicates that the logistic regression model achieves the same log likelihood as the saturated model. However, this does not mean that it fits the data perfectly. Instead, it means that it fits the data as well as could be hoped for. **Reduction From Model R^2**: This is amount that R^2 is reduced when the term is omitted from the regression model. This reduction is calculated from the R^2 achieved by the full model. This quantity is used to determine if removing a term causes a large reduction in R^2. If it does not, then the term can be safely removed from the model. **Reduction From Saturated R^2**: This is the amount that R^2 is reduced when the term is omitted from the regression model. This reduction is calculated from the R^2 achieved by the saturated model. This item is included because it shows how removal of this term impacts the best R^2 that is possible.

Log Likelihood and R^2 Section:

Term(s) Omitted	DF	Log Likelihood	R^2 Of Remaining Term(s)	Reduction From Model R^2	Reduction From Saturated R^2
All	1	−17.19	0.00		
Cell	1	−12.32	0.28	0.078	0.717
Li	1	−15.41	0.10	0.258	0.897
Temp	1	−12.17	0.29	0.069	0.708
None(Model)	3	−10.98	0.36	0.000	0.639
None(Saturated)	26	0.00	1.00		0.000

- **Classification table:** This table displays the results of classifying the data based on the logistic regression equations. The table presents the counts for each category. The Percent Correctly Classified is also presented. This is the percent of the total count that fall on the diagonal of the table.

Classification table:

	Estimated		
Actual	0	1	Total
0	15	3	18
1	3	6	9
Total	18	9	27

Percent Correctly classified = 77.8%

- **ROC section:** One ROC report is generated for each category 0 and 1 but only the report for category 0 is displayed here. ROC curves can be used to determine appropriate cutoff values for classification by letting us compare the sensitivity and specificity of various cutoff values. When classifying, we usually classify a row into that category that has the highest membership probability. However, this is not always the optimum strategy. This table shows us what happens when various cutoff values are selected. Classifying an observation can have any one of four possible results. An observation from the group can be correctly classified as being from that group (state A) or incorrectly classified as being from another group (state C). An observation from another group can be incorrectly classified as being from the group (state B) or correctly classified as being from another group (state D). The number of observations in each state is computed for each cutoff value between zero and one. A number of measures can be calculated from these values. The measures used in ROC analysis are called *sensitivity* and *specificity*. Sensitivity is the proportion of those from this group that are correctly identified as such. In terms of the four states, *sensitivity* $= A/(A + C)$. Specificity is the proportion of those from other groups that are correctly identified as such. In terms of four states, *specificity* $= D/(B + D)$. Thus, the optimum cutoff value is that one for which the sum of sensitivity and specificity is the maximum. This may be found be investigating the report. An ROC plot is also generated for each report that gives a graphical display of this report. An ROC analysis is most useful in the two-group case. In the multiple-group case, it is of only marginal usefulness, since a cutoff value is not specified. Rather, each observation is classified into that group which has the highest membership probability. **Prob Cutoff:** This is the probability cutoff for classification into this group. If an observation's predicted probability for membership in this group is greater than this amount, the observation is classified in this group. Otherwise, it is classified as being in some other group. **A B C D:** The counts for each of the four states. These counts are represented using the notation $N(i|j)$ where i is the classified group and j is the actual group. **Sensitivity:** Sensitivity is the proportion of those from this group that are correctly identified as such. In terms of the four states, *sensitivity* $= A/(A + C)$. **Specificity:** Specificity is the proportion of those from other groups that are correctly identified as such. In terms of four states, *specificity* $= D/(B + D)$. **Sensitivity + Specificity:** A common rule for selecting an appropriate cutoff value is to choose the cutoff with the largest total of sensitivity and specificity. This column allows us to do this very quickly. **Proportion Correct:** Another rule for selecting an appropriate cutoff value is to choose that cutoff which maximizes the number of observations that are correctly classified. This column of the report allows us to quickly find the optimum cutoff value. Unfortunately, when one group has many more rows than the others, this rule may not be useful since it will lead us to classify everyone into the most prevalent group. **Area Under ROC Curve:** The area under the ROC curve is a popular measure associated with ROC curves. When applied to classification in logistic regression, its maximum value of one occurs when all rows are correctly classified. Its minimum value of zero occurs when all rows are incorrectly classified. Thus, the nearer this value is to one, the better the classification.

ROC Section for Value 0:

Prob Cutoff	N(1|1) A	N(1|0) B	N(0|1) C	N(0|0) D	Sensitivity A/(A+C)	Specificity D/(B+D)	Sensitivity +Specificity	Proportion Correct
0,05	18	9	0	0	1,000	0,000	1,000	0,667
0,10	18	8	0	1	1,000	0,111	1,111	0,704
0,15	17	8	1	1	0,944	0,111	1,056	0,667
0,20	17	8	1	1	0,944	0,111	1,056	0,667
0,25	17	8	1	1	0,944	0,111	1,056	0,667
0,30	17	5	1	4	0,944	0,444	1,389	0,778
0,35	17	5	1	4	0,944	0,444	1,389	0,778
0,40	16	4	2	5	0,889	0,556	1,444	0,778
0,45	15	3	3	6	0,833	0,667	1,500	0,778
0,50	15	3	3	6	0,833	0,667	1,500	0,778
0,55	15	2	3	7	0,833	0,778	1,611	0,815
0,60	15	2	3	7	0,833	0,778	1,611	0,815
0,65	15	2	3	7	0,833	0,778	1,611	0,815
0,70	15	1	3	8	0,833	0,889	1,722	0,852
0,75	12	0	6	9	0,667	1,000	1,667	0,778
0,80	11	0	7	9	0,611	1,000	1,611	0,741
0,85	9	0	9	9	0,500	1,000	1,500	0,667
0,90	8	0	10	9	0,444	1,000	1,444	0,630
0,95	7	0	11	9	0,389	1,000	1,389	0,593

Area Under ROC Curve = 0.892

ROC Section for Value 1:

Prob Cutoff	N(1|1) A	N(1|0) B	N(0|1) C	N(0|0) D	Sensitivity A/(A+C)	Specificity D/(B+D)	Sensitivity +Specificity	Proportion Correct
0,05	9	11	0	7	1,000	0,389	1,389	0,593

0,10	9	10	0	8	1,000	0,444	1,444	0,630
0,15	9	9	0	9	1,000	0,500	1,500	0,667
0,20	9	7	0	11	1,000	0,611	1,611	0,741
0,25	9	6	0	12	1,000	0,667	1,667	0,778
0,30	8	3	1	15	0,889	0,833	1,722	0,852
0,35	7	3	2	15	0,778	0,833	1,611	0,815
0,40	7	3	2	15	0,778	0,833	1,611	0,815
0,45	7	3	2	15	0,778	0,833	1,611	0,815
0,50	6	3	3	15	0,667	0,833	1,500	0,778
0,55	6	3	3	15	0,667	0,833	1,500	0,778
0,60	5	2	4	16	0,556	0,889	1,444	0,778
0,65	4	1	5	17	0,444	0,944	1,389	0,778
0,70	4	1	5	17	0,444	0,944	1,389	0,778
0,75	1	1	8	17	0,111	0,944	1,056	0,667
0,80	1	1	8	17	0,111	0,944	1,056	0,667
0,85	1	1	8	17	0,111	0,944	1,056	0,667
0,90	1	0	8	18	0,111	1,000	1,111	0,704
0,95	0	0	9	18	0,000	1,000	1,000	0,667

Area Under ROC Curve $= 0.892$

- **Row classification section:** This report displays the actual and predicted group and membership probability for each row of the report. It also provides confidence limits for the predicted group-membership probability. **Row:** This is the row from the database. Rows that are starred are misclassified. **Actual Group:** This is the group to which this row belongs (if known). **Estimated Group:** This is the group with the largest membership probability. **Estimated Probability:** This is the estimated probability that the row belongs to the group listed in the Estimated Group column. These values allow us to determine how certain the classification is. When the value is near one (above 0.7), the logistic regression is convinced that the observation belongs in the designated group. When the value is near 0.5 or less, the classification was not as clear. **Lower and Upper Confidence Limits:** These values provide a confidence interval for the estimated membership probability. This confidence interval is only approximate in the multiple-group case. Formulas and technical details are given above in the section entitled Predicted Probabilities.

Row Classification Section:

Row	Actual Remiss	Estimated Remiss	Estimated Remiss Probability	Lower 95% Confidence Limit	Upper 95% Confidence Limit
1	1	1	0.723	0.169	0.971
2	1	1	0.578	0.268	0.837
3	0	0	0.895	0.366	0.992
4	0	0	0.717	0.343	0.925
5	1	1	0.714	0.252	0.949
...	...	...	...	...	...
...	...	...	...	...	...
26*	1	0	0.539	0.215	0.834
27	0	0	0.717	0.343	0.925

- **Row classification probabilities**: This report displays the actual group and the membership probabilities for each group and each row. This allows us investigate how certain each classification is. **Row:** This is the row from the database. Rows that are starred are misclassified. **Actual Group:** This is the group to which this row belongs (if known). **Estimated Prob. in Group:** This is the estimated probability that the row belongs in each group. These values allow us to determine how certain the classification is.

Row Classification Probabilities:

Row	Actual Remiss	Estimated Prob. in 0	Estimated Prob. in 1
1	1	0.277	0.723
2	1	0.421	0.579
3	0	0.895	0.105
4	0	0.717	0.282
5	1	0.286	0.714
...	...	...	...
...	...	...	...
26*	1	0.539	0.460
27	0	0.717	0.283

- **Simple Residual Report:** This report displays the simple residuals for each group. Each of the g logistic regression equations can be used to estimate the probabilities that each observation belongs to the corresponding group. **Row:** This is the row from the database. Rows that are starred are misclassified. **Actual Group:** This is the group to which this row belongs (if known). **Residual for Group:** These residuals are defined as $r_{gj} = y_{gj} - p_{gj}$ where p_{gj} is the estimated membership probability and y_{gj} is an indicator variable that is one if the actual group is g and zero otherwise. Unlike multiple regression, there are g residuals for each observation instead of just one. This makes residual analysis much more difficult. If the logistic regression model fits an observation closely, all of its residuals will be small, but never zero. Unfortunately, the simple residuals have unequal variance equal to $n_j p_{gj}(1 - p_{gj})$, where n_j is the number of observations with the same values of the independent variables as observation j. This unequal variance makes comparisons among the simple residuals difficult and alternative types of residuals are necessary.

Simple Residual Report:

Row	Actual *Remiss*	Residual for Group 0	Residual for Group 1
1	1	−0.277	0.277
2	1	−0.421	0.421
3	0	0.105	−0.105
4	0	0.283	−0.283
5	1	−0.286	0.286
...	...	...	...
...	...	...	...
26*	1	−0.539	0.539
27	0	0.283	−0.283

- **Residual Report:** This report displays the Pearson residuals, the deviance residuals, and the hat diagonal for each row. **Row:** This is the row from the database. Rows that are starred are misclassified. **Actual Group:** This is the group to which this row belongs (if known). **Pearson Residual:** The *Pearson residuals* give the contribution of each row to the Pearson chi-square goodness of fit statistic. When the values of the independent variables of each observation are unique, the formula for this residual is

$$\chi_j = \pm\sqrt{\sum_{g=1}^{G}\frac{\left(w_{gj}-n_j p_{gj}\right)^2}{n_j p_{gj}}}, j=1,2,...,J,$$

where the plus (minus) is used if w_{gj}/n_j is greater (less) than p_{gj}. By definition, the sum of the squared Pearson residuals is the Pearson chi-square goodness of fit statistics. **Deviance Residuals:** We should remember that the deviance is -2 times the difference between log likelihoods of a reduced model and the saturated model. The formula for a deviance residual is $d_j = \pm\sqrt{2\sum_{g=1}^{G}w_{gj}\ln\left(\frac{w_{gj}}{n_j p_{gj}}\right)}, j=1,2,...,J,$

where the plus (minus) is used if $w_{REF(g),j}/n_j$ is greater (less) than $p_{REF(g),j}$. By definition, the sum of the squared deviance residuals is the deviance. **Maximum Hat Diagonal:** The diagonal elements of the hat matrix can be used to detect points that are extreme in the independent variable space. These are often called *leverage* design points. The larger the value of the hat diagonal, the more the observation influences estimates of the regression coefficients. There is a separate hat diagonal defined for each category. The value reported here is the maximum of all G of the hat diagonals for each row. An observation that has a large residual, but has low leverage, does not cause much concern. However, an observation with a large leverage and a large residual should be checked very carefully. The formula for the hat diagonal associated with the jth observation and gth group is

$$h_{gj} = n_j p_{gj}\left(1-p_{gj}\right)\sum_{i=1}^{P}\sum_{k=1}^{P}x_{ij}x_{kj}\hat{V}_{gik}, j=1,2,...,J$$

where $\hat{V}_{gik}$ is the portion of the covariance matrix of the regression coefficients associated with the gth regression equation. The interpretation of this diagnostic is not as clear in logistic regression as in multiple regression because it involves the predicted values which in turn involve the dependent variable. In multiple regression, the hat diagonals only involve the independent variables. This formula matches Pregibon [145] in the two-group case. In the multiple-group case, the two-group formula is applied to each group.

Residual Report:

Row	Actual Remiss	Pearson Residual		Deviance Residual		Maximum Hat Diagonal	
1	1	−0.619	|||.........	−0.806	|||||.........	0.339	|||||||||....

2	1	−0.853	||||..........	−1.046	||||||........	0.111	|||...........
3	0	0.342	|..............	0.470	|||............	0.177	|||||.........
4	0	0.887	|||||.........	1.153	|||||||.......	0.264	||||||||......
5	1	−0.633	|||...........	−0.820	|||||.........	0.213	||||||.......
...	...	...	...	...	...	...	...
...	...	...	...	...	...	...	...
26*	1	−1.082	||||||........	−1.245	||||||||......	0.137	||||...........
27	0	0.887	|||||..........	1.153	|||||||.......	0.264	||||||||......

- **DFBetas:** One way to study the impact of an observation on each regression coefficient is to determine how much that coefficient changes when the observation is deleted. The *DFBETA* statistic is the standardized difference between a regression coefficient before and after the removal of the *j*th observation. **Row:** This is the row from the database. Rows that are starred are misclassified. **Actual Group:** This is the group to which this row belongs (if known). **DFbeta:** The *DFBeta* statistic is the standardized difference between a regression coefficient before and after the removal of the *j*th observation. The formula for *DFBeta* is approximated by

$$DFBeta_{gij} = \left(\frac{w_{gj} - n_j p_{gj}}{(1 - h_{gj})\sqrt{\hat{V}_{gii}}} \right) \sum_{k=1}^{p} x_{kj} \hat{V}_{gik}, j = 1, 2, ..., J,$$

where $\hat{V}_{gik}$ is the portion of the covariance matrix associated with the *g*th regression equation. This formula matches Pregibon [145] in the two group case, but is different from Lesaffre [146] in the multi-group case.

DFBetas Report For REMISS = 0:

Row	Actual Remiss	Dfbeta Intercept		Dfbeta Cell		Dfbeta Li	
1	1	−0.113	|..............	0.239	||||||........	−0.257	|||...........
2	1	−0.105	|..............	0.060	|..............	−0.134	|..............
3	0	−0.016	|..............	−0.146	|||...........	−0.096	|..............
4	0	0.054	|..............	0.149	|||...........	−0.244	|||...........
5	1	−0.306	|||||..........	−0.047	|..............	−0.223	|||...........
...	...	...	...	...	...	...	...
...	...	...	...	...	...	...	...
26*	1	−0.186	|||...........	−0.233	||||||........	−0.024	|..............
27	0	0.054	|..............	0.149	|||...........	−0.244	|||...........

- **Residual Diagnostics Report:** This report gives statistics that help detect observations that have not been fitted well by the model. **Row:** This is the row from the database. Rows that are starred are misclassified. **Actual Group:** This is the group to which this row belongs (if known). **Hat Diagonal:** The diagonal elements of the hat matrix can be used to detect points that are extreme in the independent variable space. **Deviance Change (DFDev)** and **Chi-Square Change (DFChi2):** *DFDEV* and *DFCHI2* are statistics that measure the change in deviance and in Pearson's chi-square, respectively, that occurs when an observation is deleted from the dataset. Large values of these statistics indicate observations that have not been fitted well. The formulas for these statistics are $DFDEV_{gj} = d_j^2 + \overline{C}_{gj}, j = 1, 2, ..., J$ and $DFCHI2_{gj} = \dfrac{\overline{C}_{gj}}{h_{gj}}, j = 1, 2, ..., J$. This formula matches Pregibon [145] in the two-group case. In the multiple-group case, the two-group formula is applied to each group.

Residual Diagnostics Report For *Remiss* = 0:

Row	Actual *Remiss*	Hat Diagonal		Deviance Change (*DFDev*)		Chi-Square Change (*DFChi2*)	
1	1	0.339	‖‖‖‖‖‖‖‖....	0.847	‖............	0.581	‖.............
2	1	0.111	‖‖............	1.185	‖‖............	0.819	‖.............
3	0	0.177	‖‖‖‖‖..........	0.246	‖............	0.142	‖.............
4	0	0.264	‖‖‖‖‖‖‖.......	1.610	‖‖‖‖............	1.070	‖‖.............
5	1	0.213	‖‖‖‖‖‖‖........	0.782	‖............	0.509	‖.............
...	...	...	...	...	...	...	...
...	...	...	...	...	...	...	...
26*	1	0.137	‖‖‖‖............	1.737	‖‖‖‖............	1.357	‖‖............
27	0	0.264	‖‖‖‖‖‖‖‖.......	1.611	‖‖‖‖............	1.070	‖‖............

- **Influence Diagnostics Report:** This report gives two distance measures similar to Cook's distance in multiple regression. **Row:** This is the row from the database. Rows that are starred are misclassified. **Actual Group:** This is the group to which this row belongs (if known). **Hat Diagonal:** The diagonal elements of the hat matrix can be used to detect points that are extreme in the independent variable space. They are discussed in more detail in the Residual Report. **Cook's Distance (C) and (CBar):** C and *Cbar* are extensions of Cooks distance for logistic regression. Quoting from Pregibon, page 719 [145]: "Cbar measures the overall change in fitted logits due to deleting the *l*th observation for all points excluding the one deleted. Conversely, C includes the deleted point. Although C will usually be the preferred diagnostic to measure overall coefficients' changes, in the examples examined to date, the one-step approximations were more accurate for *Cbar* than C."

The formulas for C and *Cbar* are $C_{gj} = \dfrac{\chi_j^2 h_{gj}}{\left(1 - h_{gj}\right)^2}, j = 1, 2, ..., J$

and $\overline{C}_{gj} = \dfrac{\chi_j^2 h_{gj}}{\left(1 - h_{gj}\right)^2}, j = 1, 2, ..., J$. This formula matches Pregibon

[145] in the two-group case. In the multiple-group case, the two-group formula is applied to each group.

Influence Diagnostics Report For *Remiss* **= 0:**

Row	Actual Remiss	Hat Diagonal		Cook's Distance (C)		Cook's Distance (CBar)	
1	1	0.339	|||||||||....	0.299	||............	0.197	||............
2	1	0.111	|||...........	0.102	|............	0.091	|............
3	0	0.177	|||||.........	0.031	|............	0.025	|............
4	0	0.264	|||||||.......	0.383	|||...........	0.282	|||...........
5	1	0.213	||||||........	0.138	|............	0.108	|............
...	...	...	...	...	...	...	...
...	...	...	...	...	...	...	...
26*	1	0.137	||||...........	0.215	|............	0.186	||............
27	0	0.264	|||||||.......	0.383	|||...........	0.282	|||...........

- **Y versus X Plots:** This section shows scatter plots with the dependent variable on the vertical axis and each of the independent variables on the horizontal axis. The plot is useful for finding typos, outliers, and other anomalies in that data. **Vertical Axis:** The categories of the dependent variable are shown on the vertical axis. Each category is assigned a whole number, beginning with the number one. The numbers are assigned in sorted order. **Horizontal Axis:** The independent variables are shown on the horizontal axis. When the independent variable is categorical, binary variables are generated for each of the categories and a separate scatter plot is generated for each binary variable.

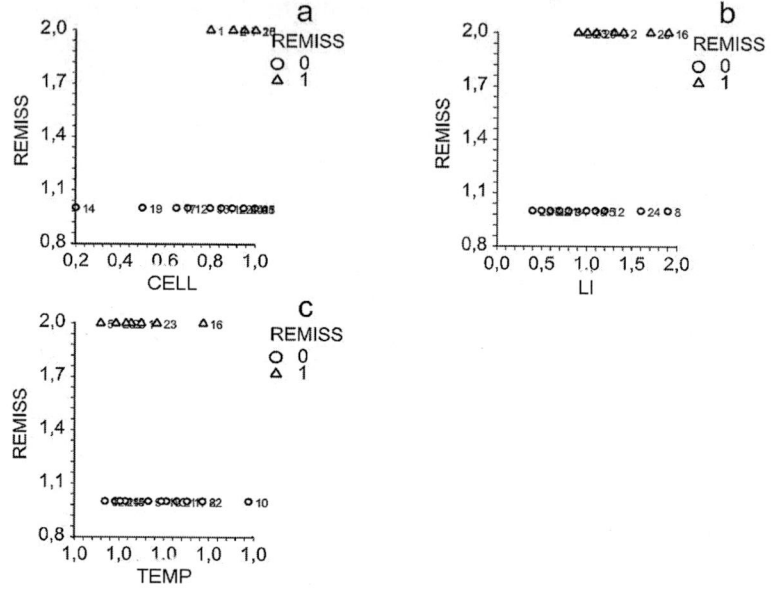

Figure 4.50 Scatter plots with the dependent variable *Remiss* on the vertical axis and each of the independent variables on the horizontal axis (a) *Remiss vs. Cell*, (b) *Remiss vs. Li*, (c) *Remiss vs. Temp*.

- **Simple Residuals versus X Plots:** This section shows scatter plots with the simple residuals on the vertical axis and each of the independent variables on the horizontal axis. The plots are useful for finding outliers and other anomalies in the data. **Vertical Axis:** The residuals are displayed on the vertical axis. The G residuals for each row corresponding to the simple residuals are displayed. Thus, if we have n rows, we will have Gn points displayed on the plot. **Horizontal Axis:** The independent variables are shown on the horizontal axis. When the independent variable is categorical, binary variables are generated for each of the categories and a separate scatter plot is generated for each binary variable.

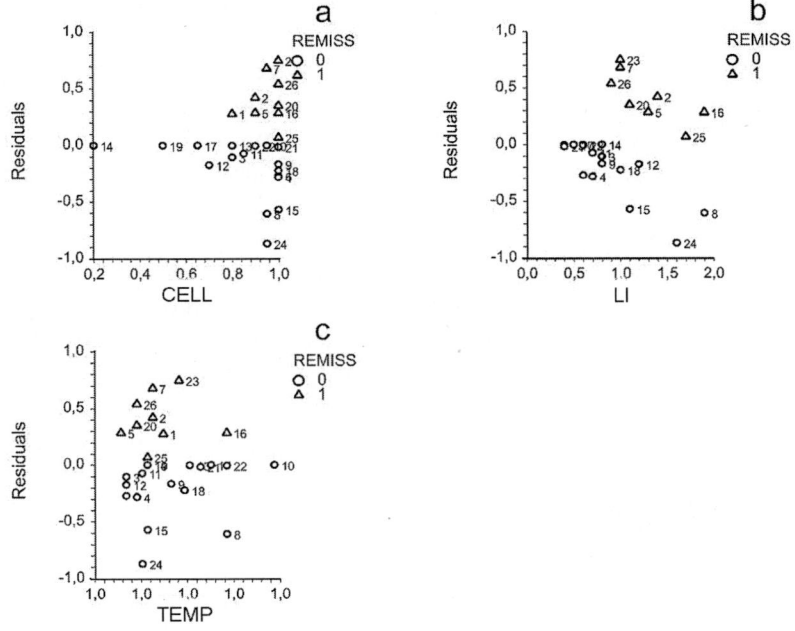

Figure 4.51 Scatter plots with the simple residuals on the vertical axis and each of the independent variables on the horizontal axis. (a) residuals *vs. Cell*, (b) residuals *vs. Li*, (c) residuals *vs. Temp*.

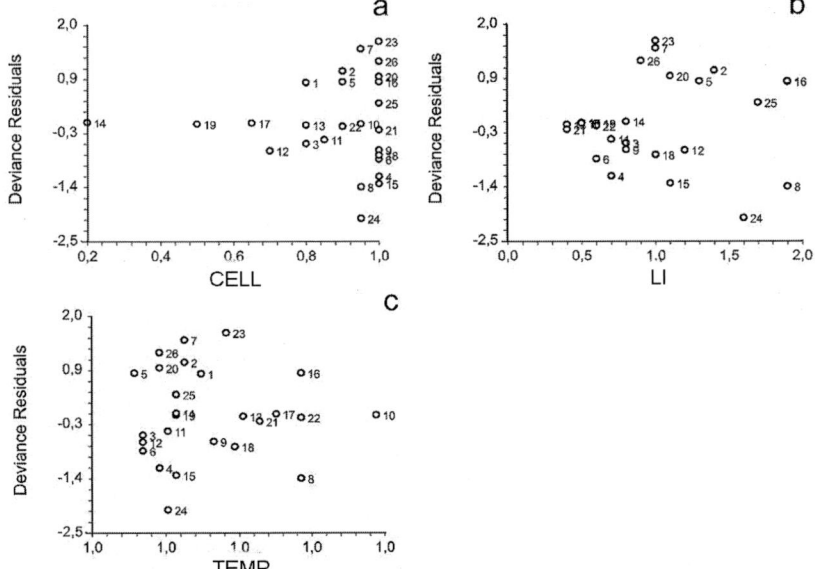

Figure 4.52 Scatter plots with the deviance residuals on the vertical axis and each of the independent variables on the horizontal axis. (a) Deviance residuals *vs Cell*, (b) Deviance residuals *vs Li*, (c) Deviance residuals *vs Temp*.

- **Deviance Residuals versus X Plots:** This section shows scatter plots with the deviance residuals on the vertical axis and each of the independent variables on the horizontal axis. The plots are useful for finding outliers and other anomalies in the data. **Vertical Axis:** The deviance residuals are displayed on the vertical axis. **Horizontal Axis:** The independent variables are shown on the horizontal axis. When the independent variable is categorical, binary variables are generated for each of the categories and a separate scatter plot is generated for each binary variable.
- **Pearson Residuals versus X Plot:** This section shows scatter plots with the Pearson residuals on the vertical axis and each of the independent variables on the horizontal axis. The plots are useful for finding outliers and other anomalies in the data. **Vertical Axis:** The Pearson residuals are displayed on the vertical axis. **Horizontal Axis:** The independent variables are shown on the horizontal axis. When the independent variable is categorical, binary variables are generated for each of the categories and a separate scatter plot is generated for each binary variable.

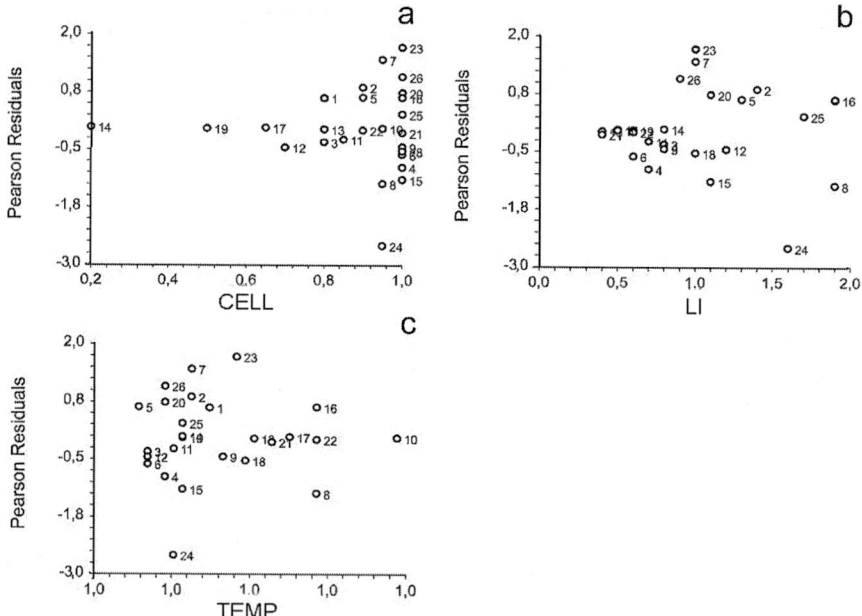

Figure 4.53 Scatter plots with the Pearson residuals on the vertical axis and each of the independent variables on the horizontal axis. (a) Pearson residuals *vs Cell*, (b) Pearson residuals *vs Li*, (c) Pearson residuals *vs Temp*.

- **ROC Curves—Combined and Separate:** This section displays the ROC curves that can be used to help we find the best cutoff points to use for classification. The cutoff point nearest the top-left corner of the plot is the optimum cutoff. We will have to refer to the ROC Report to determine the exact value of the cutoff. **Vertical Axis:** The sensitivity is displayed on the vertical axis. **Horizontal Axis:** One minus the specificity is displayed on the horizontal axis.

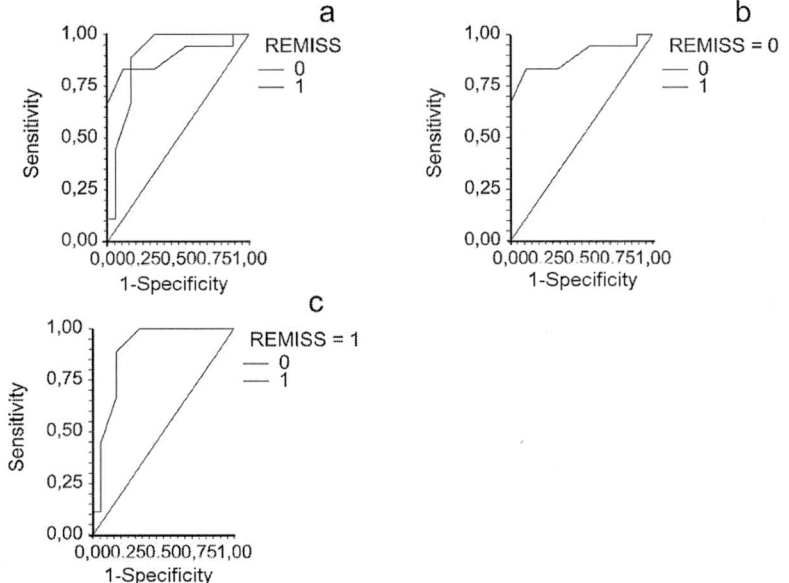

Figure 4.54   ROC curves (a) for both values 0 and 1 of *Remiss*, (b) for value 0 of *Remiss*, (c) for value 1 of *Remiss*.

- **Prob Correct versus Cutoff Plot:** This section displays a plot that shows the proportion correct versus the cutoff. It is useful to help determine the cutoff point used in classification. This plot may be difficult to use with three or more categories because of the ambiguity in the plot. **Vertical Axis:** The proportion correctly classified for various cutoff values are displayed on the vertical axis. **Horizontal Axis:** The cutoff values are displayed on the horizontal axis. These cutoff values are in terms of the estimated group-membership probabilities. Thus a cutoff of 0.4 means that any rows with a group-membership probability of 0.4 or more are classified into this group.

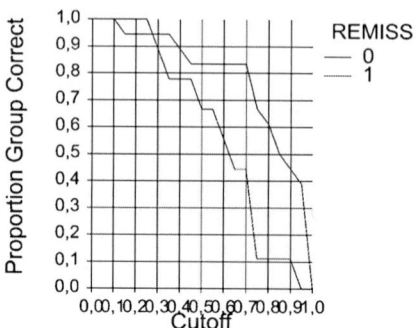

Figure 4.55 The plot of the proportion correct versus the cutoff for both values 0 and
1 of *Remiss.*

4.9 Cluster Analysis (CLU)

Cluster analysis is the name for a group of multivariate techniques whose
primary purpose is to group objects based on the characteristics they possess.
Cluster analysis classifies *objects* so that each is very similar to the others
in the cluster with respect to some predetermined selection criterion. The
resulting clusters of objects should then exhibit high internal (within-cluster)
homogeneity and high external (between-cluster) heterogeneity. Thus, if the
classification is successful, the objects within clusters will be close together
when plotted geometrically, and different clusters will be far apart. Cluster
analysis grouping individuals or objects into *unknown groups* differs from
other methods of classification, such as discriminative analysis, in that in
cluster analysis the number and characteristics of the groups are to be derived
from the data and are not usually known prior to the analysis. It should ne
noted that cluster analysis is highly empirical. Different methods can lead
to different groupings, both in number and in content. Furthermore, since
the groups are not known *a priori,* it is usually difficult to judge whether
the results make sense in the context of the problem being studied. It is also
possible to cluster the variables rather than the objects. *Clustering of variables*
is sometimes used in analyzing the items in a scale to determine which items
tend to be close together in terms of the object's response to them. Clustering
variables can be considered an alternative to factor analysis, although the
output is quite different. The concept of the variate is again a central issue,
but in a quite different way from other multivariate techniques. The *cluster
variate* is the set of variables representing the characteristics used to compare
objects in the cluster analysis. Because the cluster variate includes only the
variables used to compare objects, it determines the "character" of the objects.
The focus of cluster analysis is on the comparison of objects based on the
variate, not on the estimation of the variate itself. This makes the researcher's
definition of the variate a critical step in cluster analysis.

4.9.1 Objectives of Cluster Analysis

The primary goal of cluster analysis is to partition a set of objects into two or more groups based on the similarity of the objects for a set of specified characteristics (cluster variate).

Clustering for understanding: Classes, or conceptually meaningful groups of objects that share common characteristics, play an important role in how people analyze and describe the world. In the context of understanding data, clusters are potential classes and cluster analysis is the study of techniques for automatically finding such classes. In forming homogeneous groups, the researcher can achieve any of three objectives [60]:

(1) *Taxonomy description.* The most traditional use of cluster analysis has been for exploratory purposes and the formation of a *taxonomy*—an empirically based classification of objects. If a proposed structure can be defined for a set of objects, cluster analysis can be applied, and a proposed *typology* (i.e. theoretically based classification) can be compared to that derived from the cluster analysis.

(2) *Data simplification.* In the course of deriving a taxonomy, cluster analysis also achieves a simplified perspective on the objects. With a defined structure, the objects can be grouped for further analysis.

(3) *Relationship identification.* With the clusters defined and the underlying structure of the data represented in the clusters, the researcher has a means of revealing relationships among the objects that was perhaps not possible with the individual objects.

Clustering for utility: Cluster analysis provides an abstraction from individual data objects to the clusters in which those data objects reside. Additionally, some clustering techniques characterize each cluster in terms of a cluster prototype, i.e. a data object that is representative of the other objects in the cluster. Therefore, in the context of utility, cluster analysis is the study of techniques for finding the most representative cluster prototypes.

(1) *Summarization:* Many data analysis techniques, such as regression or PCA, have a time or space complexity and thus are not practical for large data sets.

(2) *Compression:* Cluster prototypes can also be used for data compression. In particular, a table is created that consists of the prototypes for each cluster, i.e. each prototype is assigned an integer value that is its position (index) in the table. Each object is represented by the index of the prototype associated with its cluster.

(3) *Efficiently finding the nearest neighbours:* Finding nearest neighbours can require computing the pairwise distance between all points. Often

clusters and their cluster prototypes can be found much more efficiently. If objects are relatively close to the prototype of their cluster, then we can use the prototypes to reduce the number of distance computations that are necessary to find the nearest neighbours of an object.

Selection of clustering variables: Selecting the variables to be included in the cluster variate must be done with regard to theoretical and conceptual as well as practical considerations. Any application of cluster analysis must have some rationale upon which variables are selected. Whether the rationale is based on an explicit theory, past research, or supposition, the researcher must realize the importance of including only those variables that (1) characterize the objects being clustered, and (2) relate specifically to the objectives of the cluster analysis. Cluster analysis technique has no means of differentiating relevant from irrelevant variables. It only derives the most consistent, yet distinct, groups of objects across *all* variables. The inclusion of an irrelevant variable increases the chance that outliers will be created on these variables, which can have a substantive effect on the results. Thus, one should never include variables indiscriminately but instead carefully choose the variables with the research objective as the criterion for selection.

4.9.2 Basic Concepts and Research Design

Cluster analysis groups data objects based only on information found in the data that describes the objects and their relationships. The goal is that the objects within a group be similar (or related) to one another and different from (or unrelated to) the objects in other groups. Classification in the sense of DA or LR is *supervised classification,* i.e. new, unlabeled objects are assigned a class label using a model developed from objects with known class labels. For this reason, cluster analysis is sometimes referred to as *unsupervised classification.* When the term classification is used without any qualification within data mining, it typically refers to supervised classification.

Also, while the terms *segmentation* and *partitioning* are sometimes used as synonyms for clustering, these terms are frequently used for approaches outside the traditional bounds of cluster analysis. The term *partitioning* is often used in connection with techniques that divide graphs into subgraphs and that are not strongly connected to clustering. *Segmentation* often refers to the division of data into groups using simple techniques; e.g., an image can be split into segments based only on pixel intensity and colour, or people can be divided into groups based on their income. But to accomplish this task, we must address three basic questions [60]:

First, how do we measure similarity? We require a method of simultaneously comparing objects on the two clustering variables. Several methods are possible, including the correlation between objects, a measure of association used in other multivariate techniques, or perhaps measuring their proximity in two-dimensional space such that the distance between objects indicates similarity.

Second, how do we form clusters? No matter how similarity is measured, the procedure must group those objects that are most similar into a cluster. This procedure must determine the group membership of each object.

Third, how many groups do we form? Any number of "rules" might be used, but the fundamental task is to assess the "average" similarity across clusters such that as the average increases, the clusters become less similar. Yet as the number of clusters decreases, the homogeneity within the clusters necessarily decreases. Thus, a balance must be made between defining the most basic structure (fewer clusters) that still achieves the necessary level of similarity within the clusters.

With the objectives defined and variables selected, the researcher must address three questions before starting the clustering process: (1) Can outliers be detected and, if so, should they be deleted? (2) How should object similarity be measured? (3) Should the data be standardized?

The importance of these issues and the decisions made in later stages becomes apparent when we realize that although cluster analysis is seeking structure in the data, it must actually impose a structure through a selected methodology. Cluster analysis cannot evaluate all the possible partitions because, even for the relatively small problem of partitioning 25 objects into 5 nonoverlapping clusters, there are 2.4×10^{15} possible partitions [148]. Instead, based on the decisions of the researcher, the technique identifies one of the possible solutions as "correct". From this viewpoint, the research design issues and the choice of methodologies made by the researcher have greater impact than perhaps with any other multivariate technique.

Detecting outliers: In its search for structure, cluster analysis is very sensitive to the inclusion of irrelevant variables (Fig. 4.56). But cluster analysis is also sensitive to outliers i.e., objects that are very different from all others. Outliers can represent either (1) truly "aberrant" objects that are not representative of the general population, or (2) an undersampling of an actual group or groups in the population that causes an underrepresentation of the group(s) in the sample. In both cases, the outliers distort the true structure and make the derived clusters unrepresentative of the true population structure. For this reason, a preliminary screening for outliers is always necessary. Probably the easiest way to conduct this screening is to prepare graphic scatter diagrams, a profile diagram and other diagrams of exploratory data analysis. Obviously, such a procedure becomes cumbersome with a large number of objects (observations) or variables.

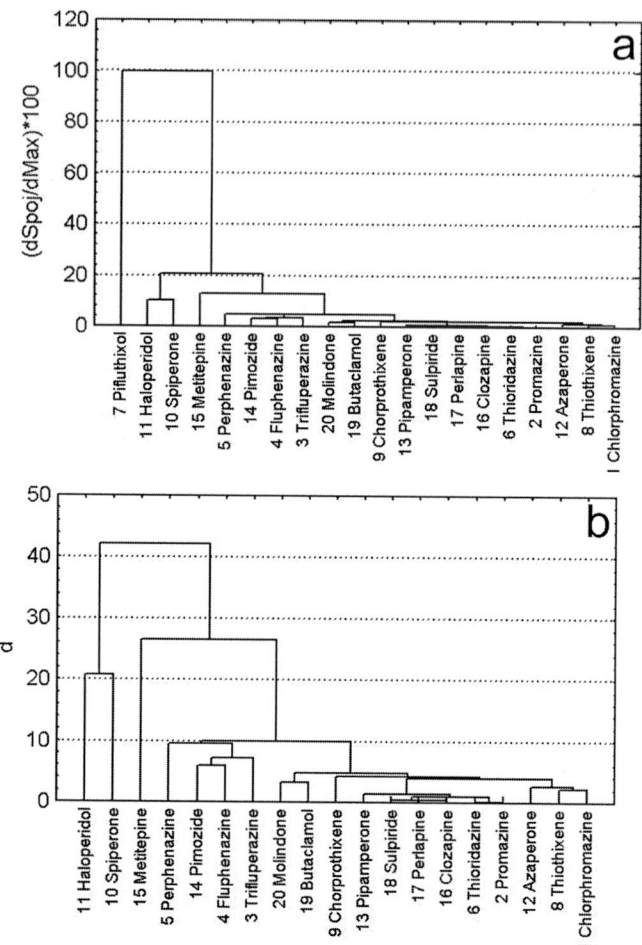

Figure 4.56 Influence of one outlier on the dendrogram formed for 20 neuroleptics: (a) Dendrogram contains one outlier detected, i.e., *7-Pifluthixol,* (b) Dendrogram with one outlier (*7-Pifluthixol*) removed.

Scatter diagrams: Prior to using any of the analytical clustering procedures, most investigators begin with simple graphical exploratory displays of their data. In the case of two variables a scatter diagram can be very helpful in displaying some of the main characteristics of the underlying clusters. If the number of variables is small, it is possible to examine scatter diagrams of each pair of variables and search for possible clusters. But this technique may become unwieldy if the number of variables exceeds four, particularly if the number of points is large.

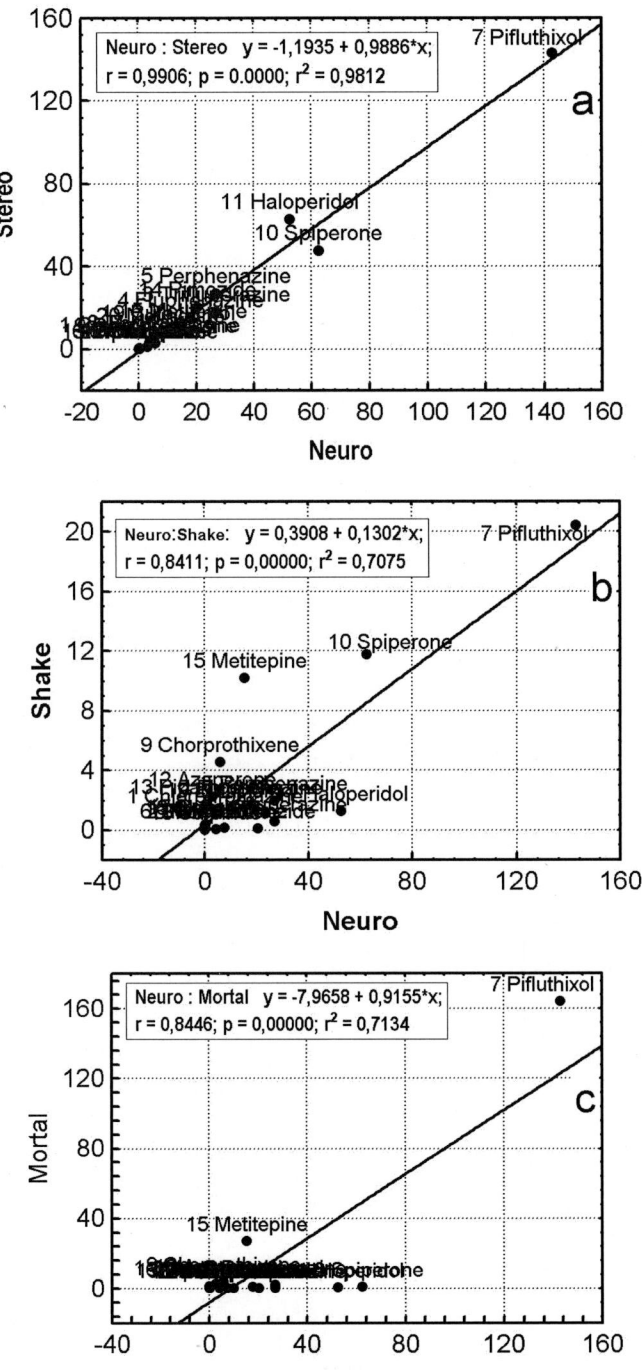

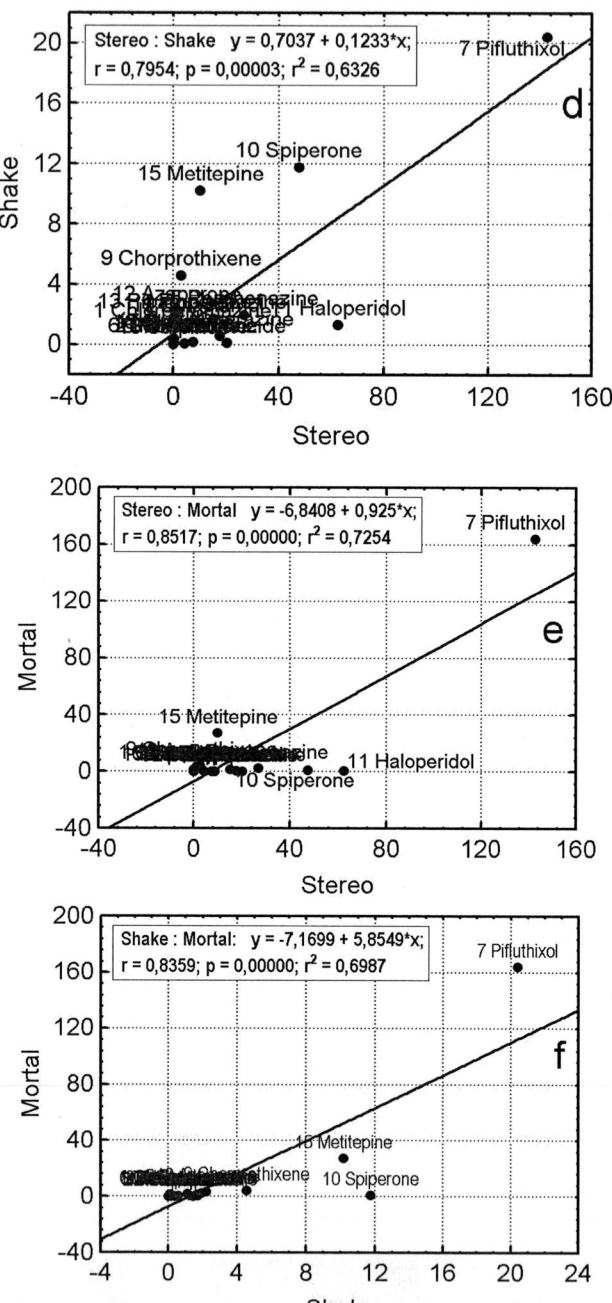

Figure 4.57 Six scatter diagrams for 4 variables (*Neuro, Stereo, Shake, Mortal*) of data set *Neuroleptics* present a plot for each pair of variables: (a) *Neuro – Stereo*, (b) *Neuro – Shake*, (c) *Neuro – Mortal*, (d) *Stereo – Shake*, (e) *Stereo – Mortal*, (f) *Shake – Mortal*.

Profile diagram: A helpful technique for a moderate number of variables is a *profile diagram.* To plot a profile of an individual object in the sample, the investigator customarily first standardizes the data by subtracting the mean and dividing by the standard deviation for each variable. However, this step is omitted by some researchers, especially if the units of measurement of the variables are comparable. The profile diagram lists the variables along the horizontal axis and the variable values along the vertical axis. Each point on the graph represents the value of the corresponding variable, and the points are connected to facilitate visual interpretation. Profiles for all objects are then plotted on the graph, a line or curve for each object. Outliers are those objects with very different profiles, most often characterized by extreme values on one or more variables. A quite different profile is exhibited by *7-Pifluthixol* which is detected as the outlier.

From left: Neuro, Stereo, Shake, Mortal   a

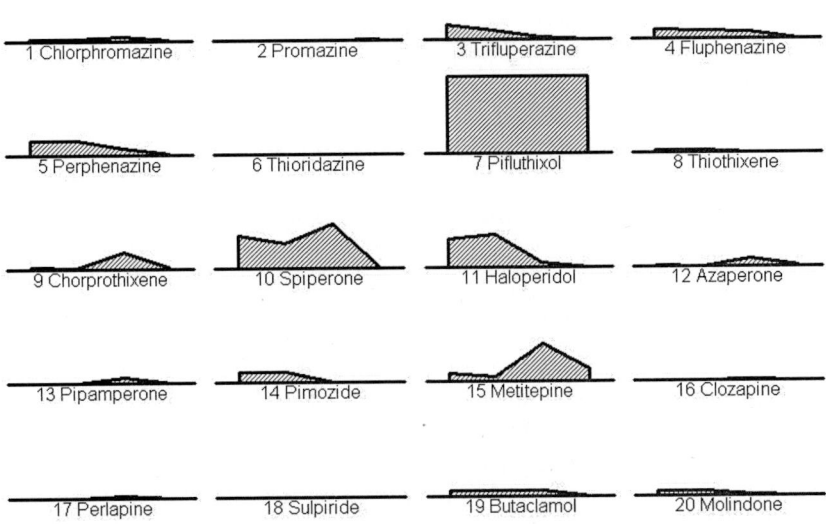

Similarity measures: The concept of similarity is fundamental to cluster analysis. *Interobject similarity* is a measure of correspondence, or resemblance, between objects to be clustered. Here, the characteristics defining similarity are first specified. Then, the characteristics are combined into a similarity measure calculated for all pairs of objects, just as we used correlations between variables in factor analysis. In this way, any object can be compared to any other object through the similarity measure. The cluster analysis procedure then proceeds to group similar objects together into clusters. Interobject similarity can be measured in a variety of ways, but three methods dominate the applications of cluster analysis: correlation measures, distance measures, and association measures. Each of the methods represents a particular perspective on similarity, dependent on both its objectives and type of data.

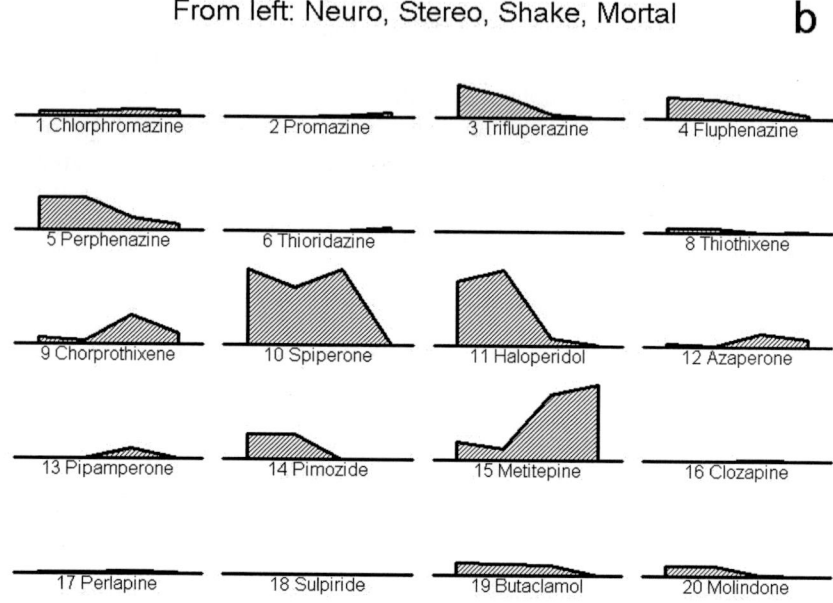

From left: Neuro, Stereo, Shake, Mortal   b

Figure 4.58   The profile diagram lists for variables along the horizontal axis (from left) *Neuro, Stereo, Shake, Mortal* and the value of each variable along the vertical axis. Points are connected to facilitate visual profile for each neuroleptics: (a) Detection of one outlier, i.e., *7-Pifluthixol,* (b) One outlier (*7-Pifluthixol)* is removed.

(1) *Correlation measures:* The inter-object measure of similarity that probably comes to mind first is the correlation coefficient between a pair of objects measured on several variables. In effect, instead of correlating two sets of variables, we invert the objects' x variables matrix so that the columns represent the objects and the rows represent the variables. Thus, the correlation coefficient between the two columns of numbers is the correlation (or similarity) between the profiles of the two objects. High correlations indicate similarity and low correlations denote a lack of it. A correlation measure of similarity does not look at the magnitude but instead at the patterns of the values. Thus, correlations represent patterns across the variables much more than the magnitudes. Correlation measures, however, are rarely used because emphasis in most applications of cluster analysis is on the magnitudes of the objects, not the patterns of values.

(2) *Distance measures of similarity*, which represent similarity as the proximity of objects to one another across the variables in the cluster variate, are the similarity measure most often used. Distance measures are actually a measure of *dissimilarity*, with larger values denoting lesser similarity. Distance is converted into a similarity measure by using an inverse relationship. Before defining distance measures, though, we would warn the investigator that many of the analytical techniques are particularly sensitive to outliers. Some preliminary checking for outliers and blunders is therefore advisable. This check may be facilitated by the graphical methods just described here. The most commonly used is the *Euclidian distance*. Unfortunately, the Euclidian distance is not invariant to changes in scale and the results can change appreciably by simply changing the units of measurement. Since the square root operation does not change the order of how close the points are to each other, some procedures use the sum of the *squared differences* instead of the Euclidian distance (i.e. they don't take the square root).

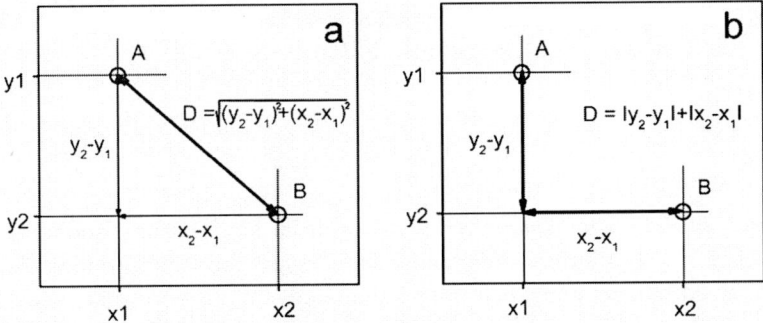

Figure 4.59 Distance measures: (a) Euclidian distance, (b) City blocks distance.

Another option is to replace the squared differences by another power of the absolute differences. For example, if the power of 1 is chosen, the distance is the sum of the absolute differences of the coordinates. The distance is the so-called *city-block distance* or *Manhattan distance* (Fig. 4.59). In effect the *Mahalanobis distance* is a generalization of the idea of standardization. The squared Euclidian distance based on standardized variables is the sum of the squared differences, each divided by the appropriate variance. When the variables are correlated, a distance can be defined to take this correlation into account. For two variables x_1 and x_2, suppose that the sample variances are s_1^2 and s_2^2, respectively, and that the correlation is r. The squared Euclidian distance based on the original values is (*Euclidian distance*)$^2 = [(x_{11} - x_{12})^2 + (x_{21} - x_{22})^2]$. The same quantity based on standardized variables is

$$(standardized\ Euclidian\ distance)^2 = \frac{(x_{11} - x_{12})^2}{s_1^2} + \frac{(x_{21} - x_{22})^2}{s_2^2} \quad \text{If } r = 0, \text{ then}$$

the last quantity is also the Mahalanobis distance.

If $r \neq 0$, the Mahalanobis distance is

$$D^2 = \frac{1}{1-r^2}\left[\frac{(x_{11} - x_{12})^2}{s_1^2} + \frac{(x_{21} - x_{22})^2}{s_2^2} - \frac{2r(x_{11} - x_{12})(x_{21} - x_{22})}{s_1 s_2}\right]. \text{ For more}$$

than two variables the Mahalanobis distance is easily defined in terms of vectors and matrices [152].

(3) *Association measures:* Association measures of similarity are used to compare objects whose characteristics are measured only in nonmetric terms (nominal or ordinal measurement). As an example, respondents could answer yes or no on a number of statements. An association measure could assess the degree of agreement or matching between each pair of respondents. The simplest form of association measure would be the percentage of times there was agreement (both respondents said yes or both said no to a question) across the set of questions. Extensions of this simple matching coefficient have been developed to accommodate multicategory nominal variables and even ordinal measures. Reviews of the various types of association measures can be found in several sources [150, 151].

Standardizing the data: With the similarity measure selected, the researcher must address only one more question: Should the data be standardized before similarities are calculated? In answering this question, the researcher must consider several issues. First, most distance measures are quite sensitive to differing scales or magnitude among the variables. In general, variables with larger dispersion (i.e., larger standard deviations) have more impact on the final similarity value. The most common form of standardization is the conversion of each variable to standard scores (also known as Z scores) by subtracting the mean and dividing by the standard deviation for each variable. This is the general form of a *normalized distance function,* which utilizes a Euclidean distance measure amenable to a normalizing transformation of the raw data. This process converts each raw data score into a standardized value with a mean of 0 and a standard deviation of 1. This transformation, in turn, eliminates the bias introduced by the differences in the scales of the several attributes or variables used in the analysis. First, it is much easier to compare between variables as they are on the same scale (a mean of 0 and standard deviation of 1). Where positive values are above the mean, and negative values are below; the magnitude represents the number of standard deviations the original value is from the mean. Second, there is no difference in the standardized values when only the scale changes. Thus, using standardized variables truly eliminates the effects due to scale differences not only across variables, but for the same variable as well.

Assumptions in cluster analysis: Cluster analysis is not a statistical inference technique in which parameters from a sample are assessed as possibly being representative of a population. Instead, cluster analysis is an objective methodology for quantifying the structural characteristics of a set of objects. As such, it has strong mathematical properties but not statistical foundations. The requirements of normality, linearity, and homoscedasticity that were so important in other techniques really have little bearing on cluster analysis. The researcher must focus, however, on two other critical issues: representativeness of the sample and multicollinearity.

4.9.3 Clustering Techniques and Assessing Overall Fit

The first major question to answer in the partitioning phase is, what procedure should be used to place similar objects into groups or clusters? That is, what clustering algorithm or set of rules is the most appropriate? This is not a simple question because hundreds of procedures using different algorithms are available, and more are always being developed. The essential criterion of all the algorithms, however, is that they attempt to maximize the differences between clusters relative to the variation within the clusters, as shown in Fig. 4.60. The ratio of the between-cluster variation to the average within-cluster variation is then comparable to (but not identical to) the F ratio in analysis of variance.

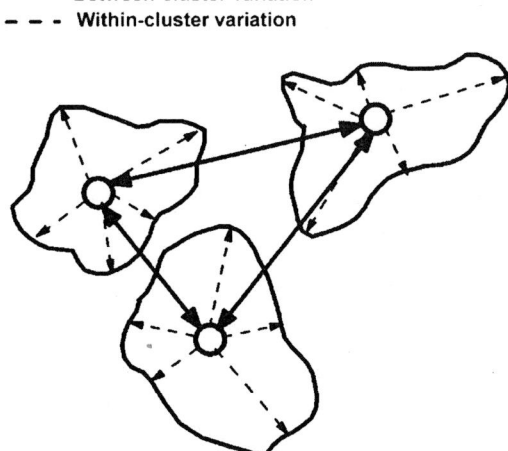

Figure 4.60 Cluster diagram showing between- and within-cluster variation

An entire collection of clusters is commonly referred to as a *clustering,* and in this section, we distinguish various types of cluster: hierarchical (nested) *versus* partitioned (unnested), exclusive *versus* overlapping *versus* fuzzy, and complete *versus* partial.

(1) *Hierarchical versus partitioned:* The most commonly discussed distinction among different types of cluster is whether the set of clusters is nested or unnested, or in more traditional terminology, hierarchical or partitioned. A *partitioned clustering* is simply a division of the set of data objects into non-overlapping subsets (clusters) such that each data object is in exactly one subset. If we permit clusters to have subclusters, then we obtain *hierarchical clustering,* which is a set of nested clusters that are organized as a tree or dendrogram. Each node (cluster) in the tree (except for the leaf nodes) is the union of its children (subclusters), and the root of the tree is the cluster containing all the objects.

(2) *Exclusive versus overlapping versus fuzzy:* The clusters are *exclusive* when they assign each object to a single cluster. There are many situations in which a point could reasonably be placed in more than one cluster, and these situations are better addressed by non-exclusive clustering. In the most general sense, an *overlapping* or *non-exclusive clustering* is used to reflect the fact that an object can *simultaneously* belong to more than one group (class).

(3) *In a fuzzy clustering,* every object belongs to every cluster with a membership weight that is between 0 (absolutely doesn't belong) and 1 (absolutely belongs). In other words, clusters are treated as fuzzy sets. Mathematically, a fuzzy set is one in which an object belongs to any set with a weight that is between 0 and 1. In fuzzy clustering, we often impose the additional constraint that the sum of the weights for each object must equal 1.

(4) *Complete versus partial:* A complete clustering assigns every object to a cluster, whereas *a partial clustering* does not. The motivation for a partial clustering is that some objects in a data set may not belong to well-defined groups.

Hierarchical clustering: Hierarchical methods can be either agglomerative or divisive. In the *agglomerative methods* we begin with n clusters, i.e. each object constitutes its own cluster. In successive steps (Fig. 4.61) we combine the two closest clusters, thus reducing the number of clusters by one in each step. In the final step all objects are grouped into one cluster. In the *agglomerative methods,* each object starts out as its own cluster. In subsequent steps, the two closest clusters (or objects) are combined into a new aggregate cluster, thus reducing the number of clusters by one in each step. In some cases, a third object joins the first two in a cluster. In others, two groups of objects formed at an earlier stage may join together in a new cluster. Eventually, all objects are grouped into one large cluster; for this reason, agglomerative procedures are sometimes referred to as buildup methods.

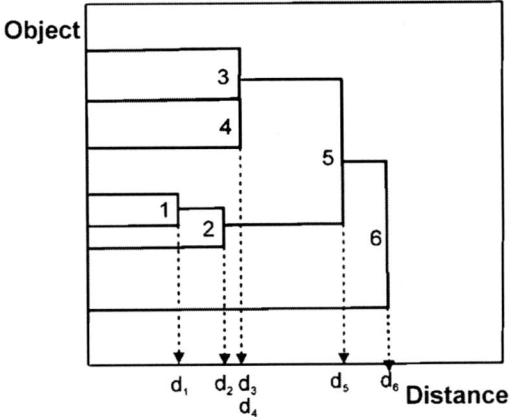

Figure 4.61 A stepwise building dendrogram.

The results at an earlier stage are always nested within the results at a later stage, creating a similarity to a tree. Because clusters are formed only by joining existing clusters, any member of a cluster can trace its membership in an unbroken path to its beginning as a single object. This process is shown in Fig. 4.61; the representation is referred to as a *dendrogram* or *tree graph*. Dendrograms may be constructed vertically (Fig. 4.62a) or horizontally (Fig. 4.62b). Another popular graphical method is the *vertical icicle diagram* (Fig. 4.62c).

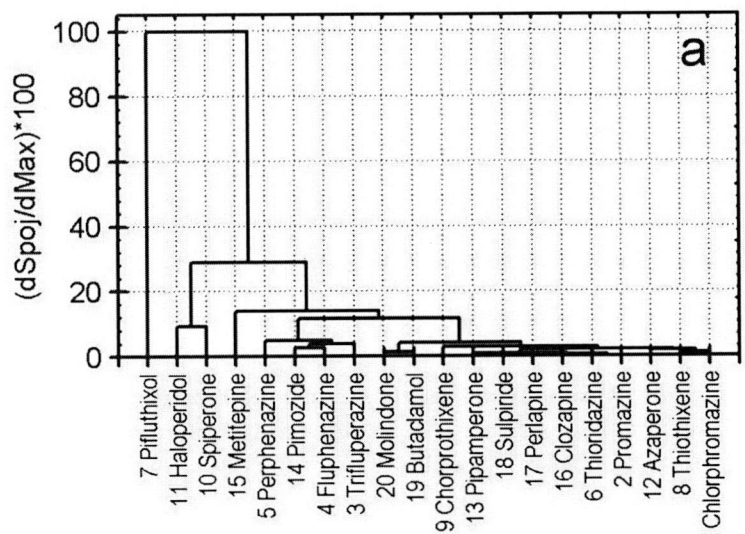

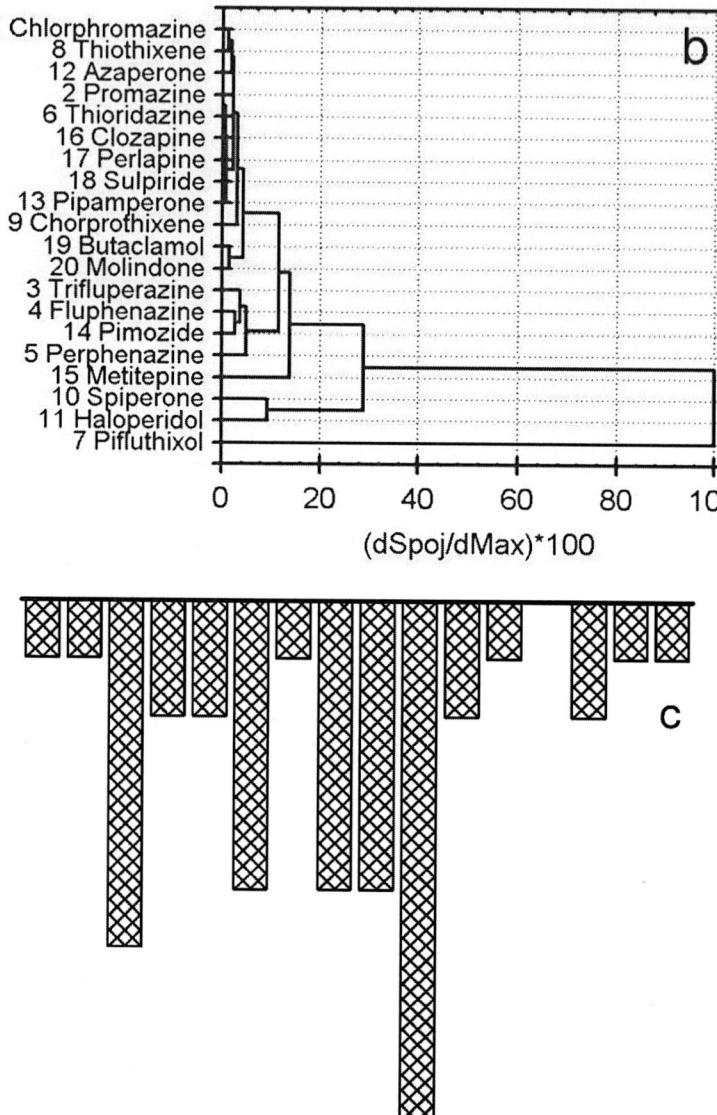

Figure 4.62 Three shapes of dendrogram: (a) Vertical tree, (b) Horizontal tree, (c) Icicle tree.

In the *divisive methods* we begin with one cluster containing all of the objects. In successive steps we split off the cases that are most dissimilar to the remaining ones. *Linkage methods* uses the linkage distance, which is the distance between two clusters defined according to one of these methods. If the number of variables is more than two, such a graph is not feasible. A clever device called the *dendrogram* or *tree graph* has therefore been incorporated into packaged computer programmes to summarize the clustering at successive steps. When the clustering process proceeds in the direction opposite to agglomerative methods, it is referred to as a *divisive method*. In divisive methods, we begin with one large cluster containing all the objects. In succeeding steps, the objects that are most dissimilar are split off and made into smaller clusters. This process continues until each object is a cluster in itself. Because most commonly used computer packages use agglomerative methods, and divisive methods act almost as agglomerative methods in reverse, we focus here on the agglomerative methods.

The hierarchical algorithms used by all eight of the clustering methods is outlined as follows. Let the distance between clusters i and j be represented as d_{ij} and let cluster i contain n_i objects. Let $\boldsymbol{D}$ represent the set of all remaining d_{ij}. Suppose there are N objects to cluster.

(1) Find the smallest element d_{ij} remaining in $\boldsymbol{D}$.

(2) Merge clusters i and j into a single new cluster, k.

(3) Calculate a new set of distances d_{km} using the following distance formula

$$d_{km} = \alpha_i d_{im} + \alpha_j d_{jm} + \beta d_{ij} + \gamma |d_{im} - d_{jm}|$$ where m represents any cluster other

than k. These new distances replace d_{im} and d_{jm} in $\boldsymbol{D}$. Also let $n_k = n_i + n_j$. Note that the eight algorithms available represent eight choices for α_i, α_j, β, and γ.

Repeat steps 1–3 until $\boldsymbol{D}$ contains a single group made up off all objects. This will require N-1 iterations.

Single linkage (Nearest neighbor)

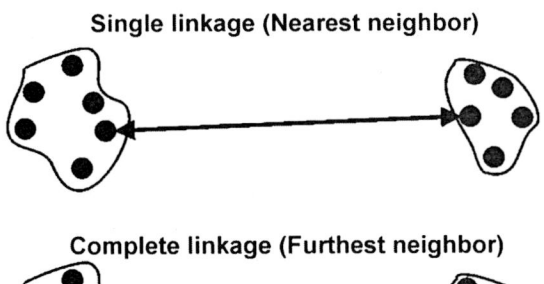

Complete linkage (Furthest neighbor)

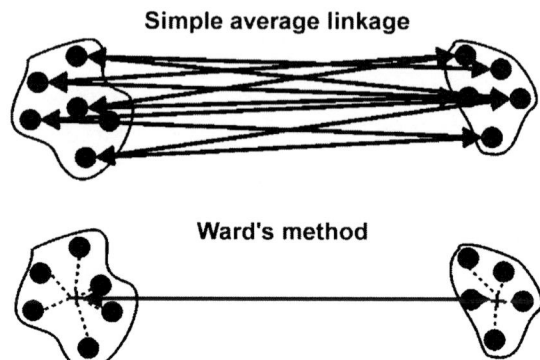

Figure 4.63   Four various technique of distance between two clusters calculus:

We will now give brief comments about each of the eight techniques (Fig. 4.63).

(1) *Single linkage* (*Nearest Neighbour*): The single-linkage procedure is based on minimum distance. It finds the two objects separated by the shortest distance and places them in the first cluster. Then the next-shortest distance is found, and either a third object joins the first two to form a cluster, or a new two-member cluster is formed. The process continues until all objects are in one cluster. This procedure has also been called the *nearest-neighbour approach*. The distance between any two clusters is defined as the shortest distance from any point in one cluster to any point in the other. The coefficients of the distance equation are $\alpha_i = \alpha_j = 0.5, \beta = 0$, and $\gamma = -0.5$.

(2) *Complete linkage* (*Furthest Neighbour*): The complete linkage procedure is similar to single linkage except that the cluster criterion is based on maximum distance. For this reason, it is sometimes referred to as the *farthest-neighbour approach* or as a *diameter method*. The maximum distance between objects in each cluster represents the smallest (minimum-diameter) sphere that can enclose all objects in both clusters. This method is called complete linkage because all objects in a cluster are linked to each other at some maximum distance or by minimum similarity. This technique eliminates the snaking problem identified with single linkage. The coefficients of the distance equation are $\alpha_i = \alpha_j = 0.5, \beta = 0$, and $\gamma = 0.5$.

(3) *Simple average linkage* (*Weighted Pair-Group*): The simple average linkage method starts out the same as that of single linkage or complete linkage, but the cluster criterion is the average distance from all objects in one cluster to all objects in another. Such techniques do not depend on extreme values, as do single linkage or complete linkage, and partitioning is based on all members of the clusters rather than on a single pair of extreme members. Simple average linkage approaches tend to combine clusters with small within-cluster variation. They also tend to be biased toward the production of clusters with approximately the same variance. The coefficients of the distance equation are $\alpha_i = \alpha_j = 0.5, \beta = 0$, and $\gamma = 0$.

(4) *Ward's method:* In Ward's method, the distance between two clusters is the sum of squares between the two clusters summed over all variables. At each stage in the clustering procedure, the within-cluster sum of squares is minimized over all partitions (the complete set of disjoint or separate clusters) obtainable by combining two clusters from the previous stage. This procedure tends to combine clusters with a small number of objects. It is also biased toward the production of clusters with approximately the same number of objects. With this method, groups are formed so that the pooled within-group sum of squares is minimized. That is, at each step, the two clusters are fused which result in the least increase in the pooled within-group sum of squares. The coefficients of the distance equation are

$$\alpha_i = \frac{n_i + n_m}{n_k + n_m}, \alpha_j = \frac{n_j + n_m}{n_k + n_m}, \beta = \frac{-n_m}{n_k + n_m}, \gamma = 0$$.

(5) *Centroid method (Unweighted Pair-Group Centroid):* In the centroid method the distance between two clusters is the distance (typically squared Euclidean or simple Euclidean) between their centroids. Cluster centroids are the mean values of the objects on the variables in the cluster variate. In this method, every time objects are grouped, a new centroid is computed. Cluster centroids migrate as cluster mergers take place. In other words, there is a change in a cluster centroid every time a new object or group of objects is added to an existing cluster. The advantage of this method is that it is less affected by outliers than are other hierarchical methods. Also referred to as the unweighted pair-group centroid method, this method defines the distance between two groups as the distance between their centroids (centre of gravity or vector average). The method should only be used with Euclidean distances. The coefficients of the distance equation are

$$\alpha_i = \frac{n_i}{n_k}, \alpha_j = \frac{n_j}{n_k}, \beta = -\alpha_i\alpha_j, \gamma = 0$$.

(6) *Median method (Weighted Pair-Group Centroid):* Also called the weighted pair-group centroid method, this defines the distance between two groups as the weighted distance between their centroids, the weight being proportional to the number of objects in each group. The method should only be used with Euclidean distances. The coefficients of the distance equation are $\alpha_i = \alpha_j = 0.5, \beta = -0.25$ and $\gamma = 0$.

(7) *Group average method (Unweighted Pair-Group):* Also called the unweighted pair-group method, this is perhaps the most widely used of all the hierarchical cluster techniques. The distance between two groups is defined as the average distance between each of their members. The coefficients of the distance equation are

$$\alpha_i = \frac{n_i}{n_k}, \alpha_j = \frac{n_j}{n_k}, \beta = 0, \gamma = 0$$
.

(8) *Flexible strategy method:* Lance and Williams [153] suggested that a continuum could be made between single and complete linkage. The programme lets you try various settings of these parameters which do not conform to the constraints suggested by Lance and Williams. The coefficients of the distance equation are $\alpha_i = 1 - \beta - \alpha_j, \alpha_j = 1 - \beta - \alpha_i, -1 \leq \beta \leq 1, \gamma = 0$
.

One interesting exercise is to vary these values, trying to find the set that maximizes the cophenetic correlation coefficient. When in doubt, we suggest trying the Group Average method. It seems to be the most popular and most recommended in the cluster literature.

Different types of clusters: Clustering aims to find useful groups of objects (clusters), where usefulness is defined by the goals of the data analysis [149].

(1) *Well-separated:* A cluster is a set of objects in which each object is closer (or more similar) to every other object in the cluster than to any object not in the cluster. Sometimes a threshold is used to specify that all the objects in a cluster must be sufficiently close (or similar) to one another.

(2) *Prototype-based:* A cluster is a set of objects in which each object is closer (more similar) to the prototype that defines the cluster than to the prototype of any other cluster. For data with continuous attributes, the prototype of a cluster is often a centroid, i.e., the average (mean) of all the points in the cluster.

(3) *Graph-based:* If the data is represented as a graph, where the nodes are objects and the links represent connections among objects, then a cluster can be defined as a *connected component;* i.e., a group of objects that are connected to one another, but that have no connection to objects outside the group. An important example of graph-based clusters are *contiguity-based clusters*, where two objects are connected only if they are within a specified distance of each other.

(4) *Density-based:* A cluster is a dense region of objects that is surrounded by a region of low density.

(5) *Shared-property (Conceptual clusters):* More generally, we can define a cluster as a set of objects that share some property. This definition encompasses all the previous definitions of a cluster; e.g., objects in a centre-based cluster share the property that they are all closest to the same centroid or medoid.

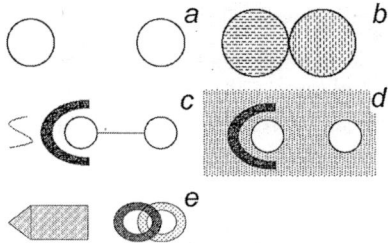

Figure 4.64 Different types of clusters as illustrated by sets of two-dimensional points: (a) Well-separated clusters. Each point is closer to all of the points in its cluster than to any point in another cluster. (b) Centre-based clusters. Each point is closer to the centre of its cluster than to the centre of any other cluster. (c) Contiguity-based clusters. Each point is closer to at least one point in its cluster than to any point in another cluster. (d) Density-based clusters. Clusters are regions of high density separated by regions of low density. (e) Conceptual clusters. Points in a cluster share some general property that derives from the entire set of points. Here, points in the intersection of the circles belong to both. [149]

Goodness-of-fit: Given the number of eight different techniques, it is often difficult to decide which is best. One criterion that has become popular is to use the result that has largest value of the *cophenetic correlation coefficient*. This stands for the correlation between the original distances and those that result from the cluster configuration. Values above 0.75 are felt to be good. The Group Average method usually appears to produce high values of this statistic. This may be one reason that it is so popular. A second measure of goodness of fit called *delta* is described in Mather [154]. These statistics measure degree of distortion rather than degree of resemblance as with the cophenetic correlation. The two delta coefficients are given by where A is either 0.5 or 1 and d^*_{ij} is the distance obtained from the cluster configuration

$$\Delta_A = \left[\frac{\sum\limits_{j \prec k}^{N} \left| d_{jk} - d^*_{jk} \right|^{1/A}}{\sum\limits_{j \prec k}^{N} (d^*_{jk})^{1/A}} \right]^A$$

where d_{ij} is the distance in an original matrix of distances and d_{ij}^* is the distance calculated from the dendrogram. Values close to zero are desirable. Mather [154] suggests that the Group Average method is the safest to use as an exploratory method, although he goes on to suggest that several methods should be tried and the one with the largest cophenetic correlation be selected for further investigation.

Number of clusters: These techniques do not let us explicitly set the number of clusters. Instead, we pick a distance value that will yield an appropriate number of clusters.

Limitations and criticisms: We have attempted problems with up to 1,000 objects. Running times will vary with computer speed, with larger problems running several hours. Problems with 100 objects or less should run in a few seconds. Hierarchical clustering methods are popular because they are relatively simple to understand and implement. However, this simplicity yields one of their strongest criticisms. Once two objects are joined, they can never be separated.

Problem 4.6   *Classification of neuroleptics using the cluster analysis*

Neuroleptics are used in psychiatry for the control of psychotic states. The introduction of neuroleptic treatment has revolutionized the management of severe psychotic disorders, such as schizophrenia and mania. It has done away with shackles and straitjackets. Nowadays, patients remain hospitalized for only a fraction of the time that was required before the discovery of neuroleptics. Neuroleptics all exert their influence on in the central nervous system, where they attach to very specialized proteins that mediate in the transmission of brain signals. Normally, receptors are activated in a delicately balanced way by so-called neurotransmitter substances, primarily dopamine, norepinephrine (adrenaline) and serotonin. An excess of dopamine is known to cause mania, delusions and other characteristics of psychosis. Abnormal stimulation by norepinephrine and related compounds are the cause of anxiety and agitation. Serotonin seems to play a harmonizing and regulating function and influences sleep patterns. All known neuroleptics attach to the dopamine receptor, and thus block its interaction with natural dopamine. Hence, neuroleptics protect the brain from exposure to an excess of dopamine and thus prevent delusions and manic states. Neuroleptics also attach to various extents to other receptors, among which are those sensitive to norepinephrine and serotonin. Each neuroleptic is known to possess a typical spectrum of activity. Some are predominantly dopamine-blockers, others have additional serotonin- or norepinephrine-blocking properties. Some interfere with all three types of receptor at the same time. In the laboratory, one can mimic excess stimulation of the receptors in animals by administration of a fixed dose of apomorphine (a dopamine agonist), tryptamine (a serotonin agonist) and norepinephrine itself. These compounds are called agonists because they stimulate the receptors. In rats, apomorphine causes stereotyped behaviour and agitation (apo-agitation and apo-stereotypy), tryptamine causes seizures and tremors (try-seizures), and a high dose of norepinephrine is lethal (nep-mortality). These effects are easily reproducible in rats, unless they have been pretreated by a protective dose of a neuroleptic drug. The goal is to classify 20 neuroleptics with the use of a cluster analysis.

Data: Neuroleptic profiles – *in vivo* pharmacology in dataset *Neuroleptics*.

| | Apo-agitation | Apo-stereotypy | Try-seizure | Nep-mortality | |
Drug	*Neuro*	*Stereo*	*Shake*	*Mortal*	Total
1 Chorphromazine	3.846	3.333	1.111	1.923	10.219
2 Promazine	0.323	0.213	0.108	1.429	2.073
3 Trifluperazine	27.027	17.857	0.562	0.140	45.586
4 Fluphenazine	17.857	15.385	1.695	1.075	36.012
5 Perphenazine	27.027	27.027	1.961	2.083	58.098
6 Thioridazine	0.244	0.185	0.093	1.333	1.855
7 Pifluthixol	142.857	142.857	20.408	163.934	470.056
8 Thiothixene	4.348	4.348	0.047	0.345	9.088
9 Chorprothixene	5.882	2.941	4.545	4.167	17.535
10 Spiperone	62.500	47.619	11.765	0.847	122.731
11 Haloperidol	52.632	62.500	1.282	0.568	116.982
12 Azaperone	2.941	1.282	2.222	3.030	9.475
13 Pipamperone	0.327	0.187	1.724	0.397	2.635
14 Pimozide	20.408	20.408	0.107	0.025	40.948
15 Metitepine	15.385	10.204	10.204	27.027	62.820
16 Clozapine	0.161	0.093	0.327	0.323	0.904
17 Perlapine	0.323	0.323	0.370	0.067	1.083
18 Sulpiride	0.047	0.047	0.003	0.001	0.098
19 Butaclamol	10.204	9.091	1.471	0.025	20.791
20 Molindone	7.692	7.692	0.140	0.006	15.530
Total	402.031	373.592	60.145	208.745	1044.513

Software: NCSS2007, STATISTICA

Solution:

- **Cluster Detail Section:** This report displays the cluster number associated with each row. The report is sorted by row number within cluster number. The cluster number of rows that cannot be classified is left blank.

Row	Cluster	Neuroleptics
1	1	1 Chlorphromazine
9	1	9 Chorprothixene
12	1	12 Azaperone
15	1	15 Metitepine
2	2	2 Promazine
6	2	6 Thioridazine
13	2	13 Pipamperone
16	2	16 Clozapine
17	2	17 Perlapine
3	3	3 Trifluperazine
8	3	8 Thiothixene
14	3	14 Pimozide
19	3	19 Butaclamol
20	3	20 Molindone
4	4	4 Fluphenazine
5	4	5 Perphenazine
7	4	7 Pifluthixol
10	4	10 Spiperone
11	4	11 Haloperidol
18		18 Sulpiride

- **Linkage Section:** This report displays the subgroup that is formed at each fusion that took place during the cluster analysis. The links are displayed in reverse order so that we can quickly determine an appropriate number of clusters to use. It displays the distance level at which the fusion took place. It will let us precisely determine the best value of the Cluster Cutoff value. The cophenetic correlation and the two delta goodness of fit statistics are reported at the bottom of this report. As discussed earlier, these values let to compare the fit of various cluster configurations. **Link:** This is the sequence number of the fusion. **Number Clusters:** This is the number of clusters that would result if the Cluster Cutoff value were set to the corresponding Distance Value or higher. This number includes outliers. **Distance Value:** This is distance value between the two joining clusters that is used by the algorithm. Normally, this value is monotonically increasing. When backward linking occurs, this value will no longer exhibit a strictly increasing behaviour. These values are used to determine an appropriate number of clusters. **Distance Bar:** This is a bar graph of the Distance Values. Choose the number of clusters by finding a jump in the decreasing pattern shown in this bar chart. **Rows Linked:** These are the rows that were joined at this step. Remember that the links are presented in reverse order. **Cophenetic Correlation:** This is the Pearson correlation between the actual distances and the predicted distances based on this particular hierarchical configuration. A value of 0.75 or above needs to be achieved in order for the clustering to be considered useful. **Delta (0.5, 1):** These are the values of the goodness of fit deltas. When comparing to clustering configurations, the configuration with the smallest delta value fits the data better.

Linkage Section:

Link	Number Clusters	Distance Value	Distance Bar	Rows Linked
19	1	1,57	|IIIIIIIIIIIIIIIIIIIIIIIIIIIIIIII	1,9,12,15,4,5,11,10,7,2,6,13,16 ,17,18,3,14,19, 20,8
18	2	1,31	|IIIIIIIIIIIIIIIIIIIIIIIIIII	2,6,13,16,17,18,3,14,19,20,8
17	3	1,17	|IIIIIIIIIIIIIIIIIIIIIII	1,9,12,15,4,5,11,10,7
16	4	1,01	|IIIIIIIIIIIIIIIIIII	2,6,13,16,17,18
15	5	0,95	|IIIIIIIIIIIIIIIIII	4,5,11,10,7
14	6	0,85	|IIIIIIIIIIIIIIII	3,14,19,20,8
13	7	0,82	|IIIIIIIIIIIIIIII	2,6,13,16,17
12	8	0,76	|IIIIIIIIIIIIII	1,9,12,15
11	9	0,67	|IIIIIIIIIIII	13,16,17
10	10	0,67	|IIIIIIIIIIII	4,5,11,10
9	11	0,66	|IIIIIIIIIIII	3,14,19,20
8	12	0,62	|IIIIIIIIIII	1,9,12
7	13	0,61	|IIIIIIIIIII	4,5,11
6	14	0,53	|IIIIIIIII	19,20
5	15	0,50	|IIIIIIIII	4,5
4	16	0,47	|IIIIIIII	3,14

Continue

3	17	0,43	|IIIIIIII	16,17
2	18	0,32	|IIIIII	9,12
1	19	0,27	|IIIII	2,6

Cophenetic Correlation	0,639
Delta(0.5)	0,230
Delta(1.0)	0,301

- **Dendrogram Section:** This report displays the dendrogram which visually displays a particular cluster configuration. Rows that are close together (have small dissimilarity) will be linked near the right side of the plot. Rows that link up near the left side are very different. The number of clusters the will be formed at a particular Cluster Cutoff value may be quickly determined from this plot by drawing a vertical line at that value and counting the number of lines that the vertical line intersects. We strongly recommend comparing the dendrograms from several different methods and on several different datasets with known cluster patterns in order to get the feel of the technique.

Distance Section:

First Row	Second Row	Actual Distance	Dendrogram Distance	Actual Difference	Percent Difference
1	2	0,62	1,57	−0,96	−155,53
1	3	1,20	1,57	−0,37	−30,82
1	4	0,79	1,17	−0,38	−48,15
1	5	1,10	1,17	−0,08	−6,94
1	6	0,88	1,57	−0,69	−78,42
1	7	1,68	1,17	0,51	30,27
1	8	0,92	1,57	−0,65	−70,58
1	9	0,68	0,62	0,05	8,02
1	10	1,47	1,17	0,29	20,10
1	11	1,27	1,17	0,10	7,91
1	12	0,57	0,62	−0,05	-9,55
1	13	0,78	1,57	−0,80	−102,24
1	14	1,40	1,57	−0,17	−12,18
1	15	0,96	0,76	0,20	20,77
1	16	1,06	1,57	−0,51	−48,30

1	17	0,97	1,57	−0,60	−62,11
1	18	1,63	1,57	0,06	3,39
1	19	1,08	1,57	−0,49	−45,78
1	20	1,15	1,57	−0,42	−36,37
2	3	1,54	1,31	0,23	14,70
2	4	1,33	1,57	−0,24	−18,16
2	5	1,68	1,57	0,11	6,46
2	6	0,27	0,27	0,00	0,00
2	7	2,28	1,57	0,71	31,01
2	8	0,81	1,31	−0,50	−61,90
2	9	1,20	1,57	−0,38	−31,37
2	10	2,04	1,57	0,46	22,77
2	11	1,74	1,57	0,17	9,71
2	12	1,00	1,57	−0,57	−57,55
2	13	0,89	0,82	0,06	7,05
2	14	1,59	1,31	0,27	17,06
2	15	1,55	1,57	−0,02	−1,24
2	16	0,71	0,82	−0,11	−16,02
2	17	0,83	0,82	0,01	1,12
2	18	1,19	1,01	0,18	15,23
2	19	1,38	1,31	0,06	4,54
2	20	1,23	1,31	−0,09	−7,00
		....			
18	19	1,622	1,315	0,307	18,93
18	20	1,267	1,315	−0,047	−3,74
19	20	0,530	0,530	0,000	0,00

- **Dendrogram:** *7-Pifluthixol* is still seen to be the most potent compound. The more potent compounds appear separated as an outlier on the lower side of the display, while the least potent ones are found in the upper corner. This is partly due to the important size component in the data. A compound that scores heavily in one test can be expected to score heavily in most of the others. In dendrogram two compounds (Spiperone, Haloperidol) are displayed closely together when their results in the four tests are numerically close. In most dendrograms, for example, the numerical profiles of Thioridazine, Clozapine, Perlapine,

Sulpiride and Pipamperone are very much in agreement, as well as those of Chlorphromazine, Thiothixene and Azaperone. Evidently, the clustering of compounds that can be formed in this way follows from their absolute or quantitative properties. As eight different clustering techniques were applied, it is difficult to decide which is best (four techniques are on Fig. 4.65). The largest value of the cophenetic correlation coefficient expressing the correlation between the original distances and those that result from the cluster configuration leads to conclusion that the Simple Average method appears to bring the best clustering. A second measure of goodness of fit called *delta* measures degree of distortion rather than degree of resemblance as with the cophenetic correlation. The two delta coefficients are given by where A is either 0.5 or 1 and d_{ij}^* is the distance obtained from the cluster configuration. Values close to zero are desirable.

(1) Single Linkage, *Cophenetic correlation CC:* 0.988, *Delta(0.5):* 0.474, *Delta(1.0):* 0.392.

(2) Complete Linkage: *Cophenetic correlation CC:* 0.983, *Delta(0.5):* 0.178, *Delta(1.0):* 0.183;

(3) Simple Average, *Cophenetic correlation CC:* 0.989, *Delta(0.5):* 0.178, *Delta(1.0):* 0.189;

(4) Group Average, *Cophenetic correlation CC:* 0.987, *Delta(0.5):* 0.137, *Delta(1.0):* 0.125;

(5) Centroid, *Cophenetic correlation CC:* 0.985, *Delta(0.5):* 0.175, *Delta(1.0):* 0.166;

(6) Median, *Cophenetic correlation CC:* 0.984, *Delta(0.5):* 0.452, *Delta(1.0):* 0.428;

(7) Ward's method, *Cophenetic correlation CC:* 0.979, *Delta(0.5):* 0.549, *Delta(1.0):* 0.493.

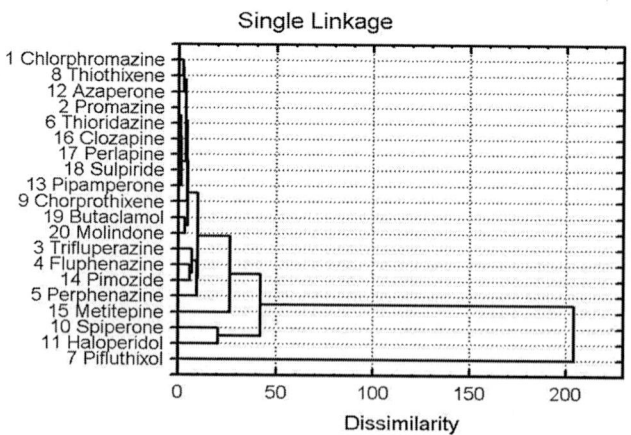

Single Linkage

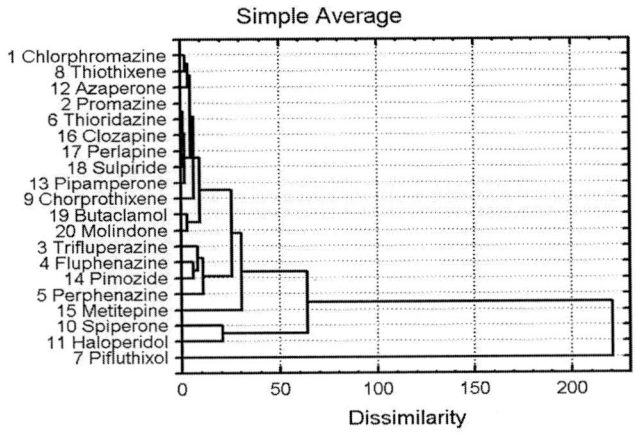

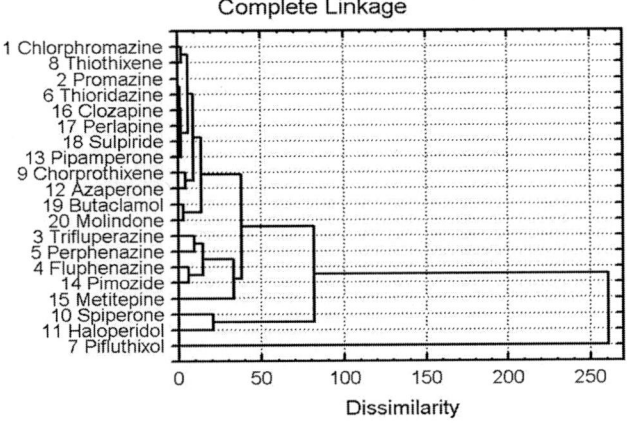

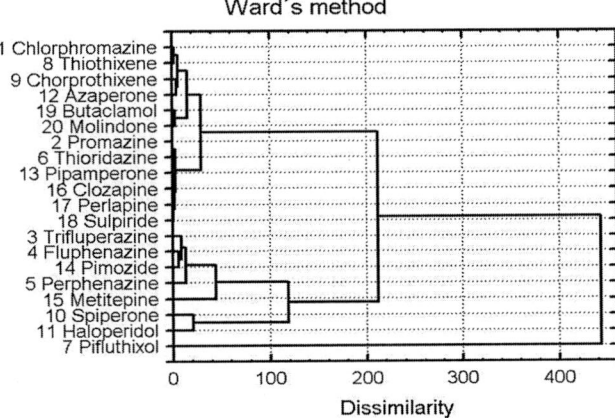

Figure 4.65 Four techniques of dendrogram building shows clusters when analyzing the data set *Neuroleptics*.

4.10 Multidimensional Scaling (MDS)

Multidimensional scaling (MDS), also known as *perceptual mapping*, is a procedure that allows a researcher to determine the *perceived relative image* of a set of objects (firms, objects, ideas, or other items associated with commonly held perceptions). MDS creates a map displaying the relative positions of objects, given only a table of the distances between them. The map may consist of one, two, three, or even more dimensions. The metric or the non-metric solution are used. The table of distances is known as the *proximity* matrix.

The purpose of MDS is to transform consumer judgments of similarity or preference into distances represented in multidimensional space. Assume that objects E and F in a set A, B, C, D, E, F are judged by respondents to be the most similar compared with all other possible pairs of objects. MDS techniques will position objects E and F so that the distance between them in multidimensional space is smaller than the distance between any other two pairs of objects. The resulting *perceptual map,* also known as a *spatial map,* shows the relative positioning of all objects, as shown in Fig. 4.66, [60].

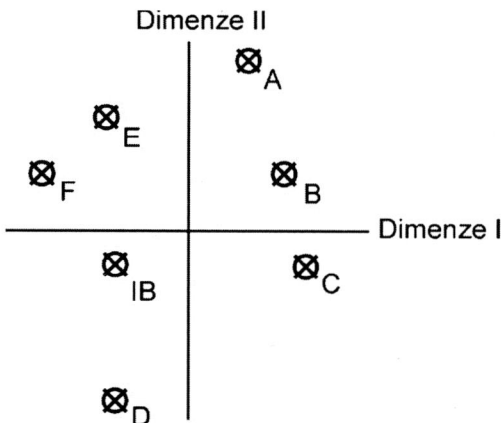

Figure 4.66   The *perceptual map,* also known as a *spatial map,* shows the relative positioning of all six objects and one respondent's *ideal point IB.*

Multidimensional scaling is based on the comparison of *objects.* Any object (e.g., substance, image, aroma) can be thought of as having both *perceived* and *objective dimensions. Perceived dimensions* are also known as *subjective dimensions.* Two objects may have the same physical characteristics (objective dimensions) but be viewed differently because the different brands are perceived to differ in quality (a perceived dimension) by many customers. Thus, the following two differences between objective and perceptual dimensions are very important:

(1) *DIndividual differences:* The dimensions perceived by customers may not coincide with the objective dimensions assumed by the researcher. We expect that each individual may have different perceived dimensions, but the researcher must also accept that the objective dimensions may also vary substantially.

(2) *DInterdependence:* The evaluations of the dimensions may not be independent and may not agree. Both perceived and objective dimensions may interact with one another to create unexpected evaluations.

MDS can be compared to the other interdependence techniques such as factor and cluster analysis based on its approach to defining structure. Factor analysis groups variables into variates that define underlying dimensions in the original set of variables. Variables that highly correlate are grouped together. Cluster analysis groups observations according to their profile on a set of variables (the cluster variate) in which observations in close proximity to each other are grouped together. MDS differs from cluster analysis in two key aspects: (1) a solution can be obtained for each individual, and (2) it does not use a variate.

In MDS, each respondent provides evaluations of all objects being considered, so that a solution can be obtained for each individual that is not possible in cluster analysis or factor analysis. Once the perceptual dimensions are defined, the relative comparisons among objects can also be made.

MDS, unlike the other multivariate techniques, does not use a variate. Instead, the "variables" that would make up the variate (i.e., the perceptual dimensions of comparison) are inferred from global measures of similarity among the objects.

4.10.1 Proximity Matrix

The *proximity matrix,* arising either directly from experiments in which subjects are asked to assess the similarity of pairs of stimuli, or indirectly, as a measure of the correlation or covariance of the pair of stimuli derived from their raw profile data is another frequently encountered type of data. The investigator who collects such data is generally interested in uncovering any structure or pattern that may be present in the observed proximity matrix, in particular to identify the dimensions on which subjects make their similarity judgements. To understand how the proximity matrix may be observed directly, consider the following marketing research example:

Suppose ten subjects rate the similarities of six automobiles, A, B, C, D, E and F. That is, each subject rates the similarity of each of the fifteen possible pairs, AB, AC, AD, AE, AF, BC, BD, BE, BF, CD, CE, CF, DE, DF and EF.

The ratings are on a scale from 1 to 10, with "1" i.e., AB = 1 meaning that the cars A and B are very identical in every way or very similar and "10" i.e., AB = 10 meaning that the cars A and B are as different as possible or very dissimilar. The ratings are averaged across subjects, forming a similarity matrix. MDS provides the marketing researcher with a perceptual map (scatter plot) of the six cars that summarizes the results visually. This map shows the perceived differences between six cars.

The unifying feature of the MDS is that they seek to represent the proximity matrix by a simple geometrical model or map. A geometrical or *spatial model* for the observed proximity matrix consists of a set of points,

$\mathbf{x}_1, \mathbf{x}_2, ..., \mathbf{x}_n$ in d dimensions (each point representing one of the stimuli of interest) and a measure of distance between pairs of points. The objective of a multidimensional scaling approach is to determine both the dimensionality of the model (i.e., the value of d) that provides a satisfactory 'fit', and the positions of the points in the resulting d-dimensional space. Fit will be judged by some numerical index of the correspondence between the observed proximities and the inter-point distances. In general, this simply means that the larger the perceived dissimilarity between two stimuli (or the smaller their similarity), the further apart should be the points representing them in the final geometrical model.

Proximity measures: Proximity measures quantify how "close" two objects are. Three forms of proximity values are usually used: dissimilarities, similarities, and correlations [142].

Dissimilarities represent the distance between two objects. They may be measured directly, as in the distance between two towns, or approximated, as in "Bill is five points different from Joe on a ten-point scale." A dissimilarity matrix is symmetrical.

Similarities represent how close (in some sense) two objects are. Similarities must obey the rule: $similarity_{ij} \leq similarity_{ii}$ and $similarity_{jj}$ for all i and j. Similarity matrices are symmetrical. Similarities are converted to dissimilarities using the formula: $d_{ij} = \sqrt{s_{ii} + s_{jj} - 2s_{ij}}$, where d_{ij} represents a dissimilarity and s_{ij} represents a similarity.

When our data consists of standard measures rather than dissimilarities or similarities, we can create a dissimilarity matrix by first creating the correlation matrix and then using the above formula to convert the correlations to dissimilarities.

A number of inter-point distance measures might be used, but by far the most common is *Euclidean distance* (see sec. 4.1).

Directly observed proximities arise in experiments where people are asked to judge the 'psychological distance' (or 'closeness') of the stimuli of interest.

Such experiments are usually designed to uncover rather than impose the dimensions on which human subjects make judgements, so that the attributes on which the stimuli are to be judged are not generally specified. The following illustrative example will help explain what MDS does. Consider the following set of data.

Label	X	Y
A	1	5
B	1	4
C	1	1
D	3	3

A scatter plot of the multidimensional spatial map of these data appears as follows:

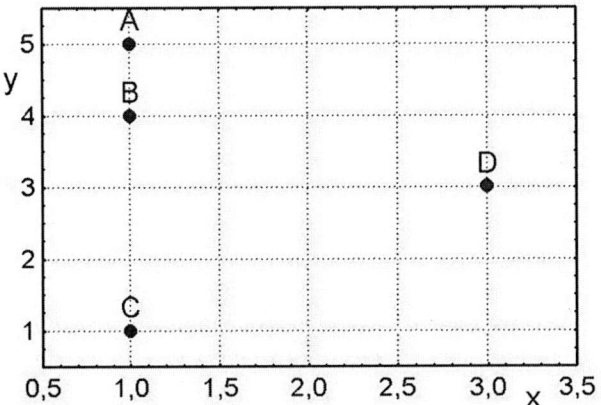

Figure 4.67   Illustration of a multidimensional spatial map of perception (scatter plot) or perceptual map of four objects A, B, C, D.

This scatter plot lets us visually assess the distance between each pair of points (Fig. 4.67). We can see that A is near B, but far from C and D. We can also see that C and D each seem to be by themselves. The actual distance between two points i and j may be computed numerically using the Euclidean distance formula. For example, the distance from A to D is calculated as follows: $d_{AD} = \sqrt{(1-3)^2 + (5-3)^2} = 2.83$
. These distances are arranged in matrix format as follows:

	A	B	C	D
A	0.00	1.00	4.00	2.83
B	1.00	0.00	3.00	2.24
C	4.00	3.00	0.00	2.83
D	2.83	2.24	2.83	0.00

Since the distance from A to D is the same as the distance from D to A, the distance matrix is symmetric. We only need to consider half of the matrix. In the calculation, we only require the upper half. The final distance matrix will be:

	A	B	C	D
A	0.00	1.00	4.00	2.83
B		0.00	3.00	2.24
C			0.00	2.83
D				0.00

The task attempted by MDS is given only a distance matrix, to find the original data so that a multidimensional perceptual map (scatter plot) of the data may be drawn. Some of the difficulties facing MDS may be seen even in this simple example. First, as the number of objects increases, the possible number of dimensions increases as well. If we have three objects, these will at most define a two-dimensional plane. With four objects, we will usually find a three-dimensional space, and so on, with each new object adding one more possible dimension. Also, if the data are shifted in such a way that their positions relative to each other are maintained (rotated, translated, or transposed), the computed distance matrix will be the same. Hence, the distance matrix could have come from numerous sets of data. A third challenge comes when the distances themselves are not actually known. We might only be given knowledge of their relative size. MDS techniques have proved useful because circumstances often occur where the actual coordinates of the objects are not known, but some type of distance matrix is available. This is especially the case in psychology where people cannot draw an overall picture of a group of objects, but they can express how different individual pairs of objects are. From these pair-wise differences MDS often can provide a useful picture.

4.10.2 Objectives of MDS

Perceptual mapping, and MDS in particular, is most appropriate for achieving two objectives [60]:

(1) *D*As an exploratory technique to identify unrecognized dimensions affecting behaviour.

(2) *D*As a means of obtaining comparative evaluations of objects when the specific bases of comparison are unknown or undefined.

In MDS, it is not necessary to specify the attributes of comparison for the respondent. All that is required is to specify the objects and make sure that the objects share a common basis of comparison.

Key decisions in setting objectives: A common characteristic of each objective is the lack of specificity in defining the standards of evaluation of the objects. The strength of perceptual mapping is its ability to "infer" dimensions without the need for defined attributes. To ensure this success, the researcher must define an MDS analysis through three key decisions: selecting the objects that will be evaluated, deciding whether similarities or preferences are to be analyzed, and choosing whether the analysis will be performed at the group or individual level.

Identification of all relevant objects to be evaluated: The most basic, but important, issue in perceptual mapping is the defining of the objects to be evaluated. The researcher must ensure that all "relevant" firms, substances, chemicals, services, properties or other objects are included, because perceptual mapping is a technique of relative positioning.

Similarities versus preference data: Having selected the objects for study, the researcher must next select the basis of evaluation: *similarity* versus *preference. Similarity-based perceptual maps* represent attribute similarities and perceptual dimensions of comparison but do not reflect any direct insight into the determinants of choice. *Preference-based perceptual maps* do reflect preferred choices but may not correspond in any way to the similarity-based positions, because respondents may base their choices on entirely different dimensions or criteria from those on which they base comparisons.

Aggregate versus disaggregate analysis: In considering similarities or preference data, we are taking respondents' perceptions of stimuli and creating outputs of representations of stimulus proximity in q-dimensional space (where the number of dimensions q is less than the number of stimuli). The researcher can generate this output on a subject-by-subject basis (producing as many maps as subjects), a method known as a *disaggregate analysis.* MDS techniques can also combine respondents and create fewer perceptual maps by some process of *aggregate analysis.* The choice of aggregate or disaggregate analysis is based on the study objectives. If the focus is on an understanding of the overall evaluations of objects and the dimensions employed in those evaluations, an aggregate analysis is the most suitable. But if the objective is to understand variation among individuals, then a disaggregate approach is the most helpful.

4.10.3 Basic Concept and Research Design of MDS

Although MDS looks quite simple computationally, the results, as with other multivariate techniques, are heavily influenced by a number of key issues that must be resolved before the research can proceed. We cover four of the major issues, ranging from discussions of research design (selecting the approach and objects or stimuli for study) to specific methodological concerns (*metric* versus *nonmetric methods*) and data collection methods [60].

Selection of either a decompositional or compositional approach: Perceptual mapping techniques can be classified by the nature of the responses obtained from the individual concerning the object. One approach, the *decompositional method,* measures only the overall impression or evaluation of an object and then attempts to derive spatial positions in multidimensional space that reflect these perceptions. This technique is typically associated with MDS. The *compositional method* is an alternative approach, which employs several of the multivariate techniques already discussed that are used in forming an impression or evaluation based on a combination of specific attributes.

(1) *Decompositional or attribute-free approach:* Commonly associated with the techniques of MDS, decompositional methods rely on global or overall measures of similarity, from which the perceptual maps and relative positioning of objects are formed. They have two distinct advantages. First, they require only that respondents give their overall perceptions of objects; respondents do not detail the attributes used in this evaluation. Second, because each respondent gives a full assessment of similarities among all objects, perceptual maps can be developed for individual respondents or aggregated to form a composite map.

(2) *Compositional or attribute-based approach:* Compositional methods include some of the more traditional multivariate techniques (e.g., discriminative analysis or factor analysis), as well as methods specifically designed for perceptual mapping, such as correspondence analysis. A principle common to all of these methods, however, is the assessment of similarity in which a defined set of attributes is considered in developing the similarity between objects.

Comparison of metric and non-metric MDS: The original MDS procedures were truly nonmetric, meaning that they required only nonmetric input but they also provided only nonmetric (rank-order) output. The nonmetric output, however, limited the interpretability of the perceptual map. Therefore, all MDS programmes used today produce metric output. The metric multidimensional positions can be rotated about the origin, the origin can be changed by adding a constant, the axes can be flipped (reflection), or the entire solution can be uniformly stretched or compressed, all without changing the relative positions of the objects.

Because nonmetric methods contain less information for creating the perceptual map, they are more likely to result in degenerate or suboptimal solutions. Metric methods assume that input as well as output is metric. This assumption allows us to strengthen the relationship between the final output dimensionality and the input data.

Although the computations are simpler for the metric method than for the non-metric method, both seem to yield similar results when applied to well-known examples. When we have true distance data, the classical method yields a solution that can be used directly. When we only have dissimilarities, the non-metric approach is somewhat more appealing.

(1) *Metric MDS:* Classical MDS procedures stem back to Torgenson [155], who was one of the pioneers of the technique. His algorithm is explained: Suppose a distance matrix D approximates the inter-point distances of a configuration of points x in a space of low dimensionality p (usually $p = 1, 2,$ or 3). That is, the elements of D, denoted d_{ij}, may be calculated from x using the Euclidean distances. The steps in the classical MDS algorithm are as follows:

(a) From D calculate $A = \left\{ -\dfrac{1}{2} d_{ij}^2 \right\}$.

(b) From A calculate $B = \left\{ a_{ij} - a_{i.} - a_{.j} + a_{..} \right\}$ where $a_{i.}$ is the average of all a_{ij} across j.

(c) Find the p largest eigenvalues $\lambda_1 > \lambda_2 > ... > \lambda_p$ of B and corresponding eigenvectors $L = \left(L_{(1)}, L_{(2)}, ..., L_{(p)} \right)$ which are normalized so that $L_{(i)}^T L_{(i)} = \lambda_i$ (We are assuming that p is selected so that the eigenvalues are all relatively large and positive.)

(d) The coordinates of the objects are the *rows* of L.

The classical solution is optimal in the least-squares sense. That is, when a direct solution is possible (i.e., when D is truly a Euclidean distance matrix), the solution, L, minimizes the sum of squared differences between the actual d_{ij}'s (elements of $D)$ and the $\hat{d}_{ij}$'s based on L. Another way of saying this is that it minimizes the value of *stress*.

The stress measure is simply a measure of how well (or poorly) the ranked distances on the map agree with the ranks given by the respondents.

(2) *Non-metric MDS:* Implicit in the above is the assumption that there is a true configuration in p dimensions, i.e., that D is a distance matrix. Often, however, it is more realistic to assume a less stringent relationship between the observed distances (or dissimilarities) d_{ij} and the true distances, denoted δ_{ij}. That is, suppose we assume that $d_{ij} = f(\delta_{ij} + e_{ij})$ where e_{ij} represents errors of measurements, distortions, etc. Also, we assume that $f(x)$ is an unknown, monotonically increasing function. For this model, the only information we can use is the rank order of

the d_{ij}. Usually, this approach is used when D is simply a dissimilarity matrix rather than a true distance matrix. This assumption is often more plausible in practical situations. An algorithm to produce a solution based only on the rank order information was provided by Kruskal [156]. Kruskal's algorithm minimizes stress when using steepest descent to find a local minimum from a given starting configuration. The choice of the starting configuration is important to finding the global rather than a local minimum. Many authors recommend using the solution of the metric MDS as the starting configuration. This is the default starting configuration in most programmes.

4.10.4 Assumptions of MDS Analysis

Multidimensional scaling, while having no restraining assumptions on the methodology, type of data, or form of the relationships among the variables, does require that the researcher accept several tenets about perception, including the following:

(1) *Variation in dimensionality*—each respondent will not perceive a stimulus to have the same dimensionality (although it is thought that most people judge in terms of a limited number of characteristics or dimensions). For example, some might evaluate a car in terms of its horsepower and appearance, whereas others do not consider these factors at all but instead assess it in terms of cost and interior comfort.

(2) *Variation in importance*—respondents need not attach the same level of importance to a dimension, even if all respondents perceive this dimension. For example, two respondents perceive a cola drink in terms of its level of carbonation, but one may consider this dimension unimportant whereas the other may consider it very important.

(3) *Variation over time*—judgments of a stimulus in terms of either dimensions or levels of importance need not remain stable over time. In other words, one may not expect respondents to maintain the same perceptions for long periods of time.

4.10.5 Deriving the MDS Solution and Assessing Overall Fit

The variety of computer programmes for MDS is rapidly expanding. The basic MDS programmes are available in all of the major statistical packages.

Determining an object's position in the perceptual map: The first task of this stage involves the positioning of objects to best reflect the similarity evaluations provided by the respondents. MDS programmes follow a common process for determining the optimal positions. This process can be described in four steps:

Step 1: Select an initial configuration of stimuli (S_k) at a desired *initial dimensionality* (q). A number of options for obtaining the initial configuration are available.

Step 2: Compute the distances between the stimuli points and compare the relationships (observed versus derived) with a measure of fit. Once a configuration is found, the interpoint distances between stimuli (d_{ij}) in the starting configurations are compared with distance measures ($\hat{d}_{ij}$) derived from the similarity judgments (s_{ij}) The two distance measures are then compared by a measure of fit, typically a measure of *stress*.

Step 3: If the measure of fit *stress* does not meet a selected predefined stopping value, find a new configuration for which the measure of fit is further minimized. The programme determines the directions in which the best improvement in fit can be obtained and then moves the points in the configuration in those directions in small increments.

Step 4: Once satisfactory stress has been achieved, the dimensionality is reduced by one, and the process is repeated until the lowest dimensionality with an acceptable measure of fit has been reached.

The primary criterion in all instances for finding the best representation of the data is preservation of the ordered relationship between the original rank data and the derived distances between points.

Selecting the dimensionality of the perceptual map: The determination of how many dimensions are actually represented in the data is generally reached through one of two approaches: subjective evaluation and scree plots of the stress measures. The spatial map is a good starting point for the evaluation. The number of maps necessary for interpretation depends on the number of dimensions. A map is produced for each combination of dimensions. One objective of the researcher should be to obtain the best fit with the smallest possible number of dimensions. The researcher typically makes a *subjective evaluation* of the perceptual maps and determines whether the configuration looks reasonable. This evaluation is important because at a later stage the dimensions will need to be interpreted and explained. A second approach is to use a *stress* measure, which indicates the proportion of the variance of the *disparities* not accounted for by the MDS model. Kruskal's *stress* [156] is the most commonly used measure for determining a model's goodness of fit defined as

$$stress = \sqrt{\frac{\sum \left(d_{ij} - \hat{d}_{ij} \right)^2}{\sum d_{ij}^2}}$$

, where $\hat{d}_{ij}$ is predicted distance based on

the MDS model. This predicted value depends on the number of dimensions kept and the algorithm that we used, metric versus non-metric. As we can see from this equation, MDS fits with *stress* values near zero are the best. In his original paper on MDS, Kruskal [156] gave the following advice about *stress* values based on his experience:

stress	Goodness-of-fit
0.200	poor
0.100	Fair
0.050	good
0.025	excellent
0.000	perfect

More recent articles caution against using a table like this since acceptable values of stress depends on the quality of the distance matrix and the number of objects in that matrix.

Ideal points: We can assume that if we locate on the derived perceptual map the point that represents the most preferred combination of perceived attributes, we have identified the position of an ideal object or ideal point. Equally, we can assume that the position of this ideal point (relative to the other objects on the derived perceptual map) defines relative preferences so that objects farther from the ideal should be less preferred. An ideal point is positioned so that the distance from the ideal conveys changes in preference. There are several procedures for implicitly positioning ideal points.

Number of dimensions: One of the main tasks the analyst has is determining the number of dimensions in the MDS model. Each dimension represents a different underlying factor. One of the goals of the MDS analysis is to keep the number of dimensions as small as possible. Usually, the analyst will anticipate select two or, at most, three dimensions. If more are required, we may decide that MDS is not appropriate for our data. The usual technique is to solve the MDS problem for a number of dimension values and adopt the smallest number of dimensions that achieves a reasonably small value of stress. Some researchers also consider the relative size of the eigenvalues that are generated during the solution process. These eigenvalues are then used to determine the number of dimensions just as they are used in factor analysis to determine the number of factors.

4.10.6 Interpreting the MDS Results

Once the perceptual map is obtained, the two approaches—compositional and decompositional—again diverge in their interpretation of the results.

For compositional methods, the perceptual map must be validated against other measures of perception, because the positions are totally defined by the attributes specified by the researcher. For example, discriminative analysis results might be applied to a new set of objects or respondents, assessing the ability to differentiate with these new observations.

For decompositional methods, the most important issue is the description of the perceptual dimensions and their correspondence to attributes. A number of descriptive techniques to "label" the dimensions, as well as to integrate preferences (for objects and attributes) with the similarity judgments, are discussed later. Again, in line with their objectives, the decompositional methods provide an initial look into perceptions from which more formalized perspectives may emerge.

Multidimensional scaling techniques have no built-in procedure for labelling the dimensions. The researcher, having developed the maps with a selected dimensionality, can adopt several procedures, either subjective or objective:

(1) *Subjective procedures:* A quite simple, yet effective, method is labelling (by visual inspection) the dimensions of the perceptual map by the respondent. Although there is no attempt to quantitatively link the dimensions to attributes, this approach may be the best available if the dimensions are believed to be highly intangible, or affective or emotional, in content. The researcher may describe the dimensions in terms of known (objective) characteristics.

(2) *Objective procedures:* As a complement to the subjective procedures, a number of more formalized methods have been developed.

Selecting between subjective and objective procedures: A more common procedure is to collect data on several attributes, associate them either subjectively or empirically with the dimensions where applicable, and determine labels for each dimension using multiple attributes, in a similar way to factor analysis. The researcher must select the type of procedure that best suits both the objectives of the research and the available information. Thus, the researcher must plan for the derivation of the dimensional labels as well as the estimation of the perceptual map.

4.10.7 Validating the MDS Results

Validation in MDS is as important as in any other multivariate technique. Owing to the highly inferential nature of MDS, efforts should be directed generalising of the results both across objects and to the population. But validation efforts are problematic. Thus, although the positions can be compared, the underlying dimensions have no basis for comparison. The most direct approach is a split- or multisample comparison, in which either the original sample is divided or a new sample is collected. In either instance, the researcher must then find a means of comparing the results.

Problem 4.7   *Comparison of six ball sports*

A classification and mutual comparison of six ball sports using a multidimensional perceptual map with both methods, CMDS and NNMDS, is to carry out.

Data: An upper-triangular distance matrix of the symmetrical *Sports* database is available. The ratings are on a scale from 1 to 6, with "1" meaning that two sports are very similar in every way and "6" meaning that two sports are very dissimilar. The ratings are averaged across all subjects, so forming a similarity matrix.

Sport	Hockey	Football	Basketball	Tennis	Golf	Croquet
Hockey	0	2	3	4	5	5
Football		0	3	5	6	5
Basketball			0	5	4	6
Tennis				0	4	3
Golf					0	2
Croquet						0

Software: NCSS2007, STATISTICA

Solution:

Metric Multidimensional Scaling: This report is produced by classical MDS denoted as CMDS.

- **Eigenvalue Section:** This report is produced by classical MDS denoted as CMDS. In this particular example, the first two dimensions account for 77% of the variation while the first three dimensions account for 88%. We would probably use two or perhaps three dimensions. **Eigenvalues:** These are the eigenvalues found during CMDS. The eigenvalues are helpful in determining the number of dimensions that are necessary to represent the dissimilarity matrix accurately. As in factor analysis, the task is to select enough dimensions to approximate the data, but few enough to keep the interpretation simple. The eigenvalue report allows us quickly to determine the impact of each new dimension. In MDS, some of the eigenvalues can be negative. We do not keep these dimensions. The basic rule is to find the number of relatively large, positive eigenvalues. This report provides a bar graph and percentages to help us determine the number of dimensions. **Individual and Cumulative Percents:** The first column gives the percentage of the total of the absolute value of the eigenvalues accounted for by this dimension. The second column is the cumulative total of the percentage. **Bar Chart:** This is a rough bar plot of the eigenvalues. It enables us to quickly note the relative size of each eigenvalue. Many authors recommend it as a method of determining how many dimensions to retain.

Dim No.	Eigenvalue	Individual Percent	Cumulative Percent	Bar Chart
1	30,73	54,28	54,28	|IIIIIIIIIIIIIIIIIIIIIIIIIIII
2 (Used)	12,85	22,69	76,97	|IIIIIIIIIII
3	6,38	11,27	88,24	|IIIIII
4	1,68	2,97	91,21	|I
5	0,00	0,00	91,21	|
6	-4,98	8,79	100,00	|IIII
Total	56,62			

- **Fit Summary Section (Metric solution):** This report provides information useful in determining the number of dimensions that are necessary and assessing the goodness-of-fit of the CMDS model. **No. Dim's:** The number of dimensions used in calculating this row of statistics. **Squared Differences:** The sum of the squared differences between the actual dissimilarity values and those predicted by the solution. **Stress:** This is the value of the stress goodness-of-fit statistic. It is equal to the square root of the Squared Differences divided by the square root of the Sum of the Squared Dissimilarities. It is one of the most popular measures of accuracy of the fit. A value below 0.05 is acceptable. A value below 0.01 is considered good. **Pseudo R-Squared:** This is an index, similar to the R-squared value in regression analysis, which indicates what percentage of the sum of squared dissimilarities (corrected for the mean) is accounted for by this number of dimensions. A value above 80% is hoped for. **Number of Dissimilarities:** This is the number of dissimilarity values. **Mean of Dissimilarities:** This is the mean of the dissimilarity values. **Sum of Squared Dissimilarities:** This is the sum of the squared dissimilarities. It is the denominator of the stress statistic. **Mean Corrected Sum of Squared Dissimilarities:** This is the sum of the squared dissimilarities about their mean. It is the denominator of the Pseudo R-Squared statistic.

No. Dim's	Squared Differences	Stress	Pseudo R-Squared
1	37,10	0,36	0,00
2	6,95	0,16	70,73
3	2,41	0,09	89,83
4	2,47	0,09	89,60

Number of Dissimilarities	15
Mean of Dissimilarities	4,13
Sum of Squared Dissimilarities	280,00
Mean Corrected Sum of Squared Dissimilarities	23,73

- **Solution Section:** This report presents the solution of the MDS procedure. These are the data that are plotted in the MDS map. They have been scaled so that the sum of squares for each column is equal to the eigenvalue for that dimension. Note that these data were constructed so that the distance between two rows is close to the original dissimilarity value. Although some interpretation of these numbers may be made directly, usually the data are displayed on scatter plots.

Variables	Dim1	Dim2	Dim3	Dim4
Hockey	1,93	−0,68	0,38	1,04
Football	2,62	−1,13	−1,13	−0,47
Basketball	2,11	2,09	0,42	−0,40
Tennis	−1,48	−1,36	1,81	−0,39
Golf	−2,38	2,01	−0,27	0,23
Croquet	−2,80	−0,93	−1,20	−0,01

- **Dissimilarity Section:** We might think of this as a residual analysis report since it highlights the differences between the actual and the predicted dissimilarities. It will let us focus on those dissimilarities that are not fitted well by the model. **Row:** The variable associated with this row of the dissimilarity matrix. **Column:** The variable associated with this column of the dissimilarity matrix. **Actual Dissimilarity:** The value from the input (or calculated) dissimilarity matrix for this row and column. **Predicted Dissimilarity:** The predicted dissimilarity value based on the number of dimensions that we have selected. **Actual Difference:** The Actual Dissimilarity minus the Predicted Dissimilarity. This value shows the size of the error in predicting this element of the dissimilarity matrix. **Percent Difference:** The percentage the Actual Difference is of the Actual Dissimilarity. This value highlights the outliers— those dissimilarities that are not fitted well by the MDS model. **Dimensions:** The number of dimensions used in calculating the statistics. **Sum of Squared Dissimilarities:** This is the sum of the squared dissimilarities. It is the denominator of the stress statistic. **Sum of Squared Differences:** This is the sum of the squared differences. It is the numerator of the stress statistic. **Stress:** This is the value of the stress goodness-of-fit statistic. It is equal to the Squared Differences divided by the Sum of the Squared Dissimilarities. It is one of the most popular measures of accuracy of the fit. A value below 0.05 is acceptable. A value below 0.01 is considered good. **Pseudo R-Squared:** This is an index, similar to the R-squared value in regression analysis, which indicates what percentage of the sum of squared dissimilarities (corrected for the mean) is accounted for by this number of dimensions. A value above 80% is hoped for.

Row	Column	Actual Dissimilarity	Predicted Dissimilarity	Actual Difference	Percent Difference
1 Hockey	*2 Football*	2,00	0,82	1,18	58,83
5 Golf	6 Croquet	2,00	2,97	−0,97	−48,39
1 Hockey	3 Basketball	3,00	2,77	0,23	7,57
2 Football	3 Basketball	3,00	3,26	−0,26	−8,63
4 Tennis	6 Croquet	3,00	1,39	1,61	53,78
1 Hockey	4 Tennis	4,00	3,48	0,52	13,08
4 Tennis	5 Golf	4,00	3,49	0,51	12,84
3 Basketball	5 Golf	4,00	4,50	−0,50	−12,41
3 Basketball	4 Tennis	5,00	4,98	0,02	0,38
1 Hockey	5 Golf	5,00	5,08	−0,08	−1,58
1 Hockey	6 Croquet	5,00	4,73	0,27	5,31
2 Football	6 Croquet	5,00	5,42	−0,42	−8,38
2 Football	4 Tennis	5,00	4,10	0,90	17,94
3 Basketball	6 Croquet	6,00	5,77	0,23	3,90
2 Football	5 Golf	6,00	5,90	0,10	1,63

Dimensions	2
Sum of Squared Dissimilarities	280,00
Sum of Squared Differences	6,95

Stress	0,158
Pseudo R-Squared	70,73

- **MDS Map:** This plot is the chief objective of an MDS analysis. It is often referred to as the MDS *map*. It allows us to interpret the dissimilarity matrix on a two-dimensional scatter plot. There is no real orientation to this map. We could legitimately rotate the values around the plot's centre. The main characteristics of interest are the relative positions of the points and any clusters that are apparent. In this example, we see that the respondents considered hockey and football to be similar. They also considered croquet and tennis to be quite similar. Football appears quite different from golf. And so on. We should notice how easy it is to draw conclusions about the similarities among the sports. A second task of the MDS analyst is to find the underlying factors that respondents used when they created these dissimilarities. For example, a vertical line down the centre of the plot would divide team sports on the right from individual sports on the left. We would hypothesize this as one interpretation of the Dim1 (horizontal) axis.

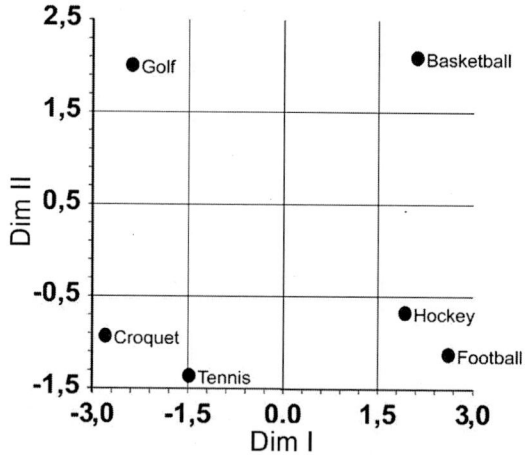

Figure 4.68 Interpretation of the dissimilarity matrix of six ball sports using the two-dimensional spatial map of perception CMDS (scatter plot).

(b) Non-Metric Multidimensional Scaling: This section presents an example of how to run an analysis of the data contained in the *Sports* database using NMMDS.

- **Eigenvalue Section:** This report is produced by CMDS which was used as the starting configuration. Its definitions were given above and will not be repeated here.

Dim No.	Eigenvalue	Individual Percent	Cumulative Percent	Bar Chart
1	30,73	54,28	54,28	|IIIIIIIIIIIIIIIIIIIIIIIII
2 (Used)	12,85	22,69	76,97	|IIIIIIIIIII
3	6,38	11,27	88,24	|IIIIII
4	1,68	2,97	91,21	|I
5	0,00	0,00	91,21	|
6	−4,98	8,79	100,00	|IIII
Total	56,62			

- **Non-Metric Iteration Summary Section:** This report provides information about the number of dimensions that are necessary and the goodness-of-fit of the solution. **No. Dim's:** The number of dimensions used in calculating this row of statistics. **Percent Rank Maintained:** The non-metric solution tries to maintain the rank ordering of the dissimilarities. This is the percentage of the dissimilarities whose rank order was maintained. The higher this value is, the better the quality of the solution. **Stress:** Defined earlier, this is one of the most popular measures of accuracy of the fit. A value below 0.05 is acceptable. A value below 0.01 is considered good. **Why Terminated:** This field explains which stopping rule caused the iterative procedure to stop. This is important to watch since the solution is not optimal if the maximum iterations were reached before the algorithm converged. When this happens, we should change some of the iteration control parameters, especially the number of iterations. **Bar Chart of Stress:** This column graphically portrays the stress values. We want to choose the fewest number of dimensions that give us a small stress value.

No. Dim's	Percent Rank Maintained	Stress	Why Terminated	Bar Chart of Stress
1	57.14	0.213	Stress Change	|IIIIIIIIIIIIIIIIIIIIIIII IIIIIIIIIIIIIIIIIIIIII
2	64.29	0.052	Stress Change	IIIIIIIIIIIIIIIIIIIIIIIIII
3	78.57	0.000	Max Iterations	|
4	64.29	0.000	Min Stress	|

- **Solution Section:** This report presents the final configuration of the NMMDS procedure. These are the data that are plotted in the

MDS map. Note that these data were not constructed so that the distance between two rows is close to the original dissimilarity value. Instead, the non-metric solution attempts to maintain the same rank ordering of the calculated distances as occur in the original dissimilarity matrix. Although some interpretation of these numbers may be made directly, usually the data are displayed on scatter plots.

Variables	Dim1	Dim2
Hockey	0,325	−0,162
Football	0,422	−0,038
Basketball	0,278	0,273
Tennis	−0,267	−0,344
Golf	−0,339	0,236
Croquet	−0,419	0,035

- **Dissimilarity Section:** This plot is the chief objective of an MDS analysis. It is often referred to as the MDS *map*. It allows us to interpret the dissimilarity matrix on a two-dimensional scatter plot. There is no real orientation to this map. We could legitimately rotate the values around the plot's centre. The main characteristics of interest are the relative positions of the points and any clusters that are apparent. In this example, we see that the respondents considered hockey and football to be similar. They also considered croquet and golf to be similar. Football appears quite different from croquet. And so on. We should notice how easy it is to draw conclusions about the similarities among the sports. A second task of the MDS analyst is to find the underlying factors that respondents used when they created these dissimilarities. For example, a vertical line down the centre of the plot would divide team sports on the right from individual sports on the left. We might hypothesize this as one interpretation of the Dim1 (horizontal) axis. It is interesting to compare this map with the map produced by the metric solution. The main difference appears to be that golf and croquet are now much closer together (as they were rated in the original data). Again, football and Basketball appear to be closer together in this plot as we might expect from the original data. In this case, the NMMDS map appears to be more accurate than the CMDS map. This is as we might expect since, the NMMDS procedure refined the CMDS map.

Row	Column	Actual Dissimilarity	Predicted Dissimilarity
1 Hockey	2 Football	2,0	0,16
5 Golf	6 Croquet	2,0	0,22
1 Hockey	3 Basketball	3,0	0,44
2 Football	3 Basketball	3,0	0,34
4 Tennis	6 Croquet	3,0	0,41
1 Hockey	4 Tennis	4,0	0,62
4 Tennis	5 Golf	4,0	0,58
3 Basketball	5 Golf	4,0	0,62
3 Basketball	4 Tennis	5,0	0,82
1 Hockey	5 Golf	5,0	0,77
1 Hockey	6 Croquet	5,0	0,77
2 Football	6 Croquet	5,0	0,84
2 Football	4 Tennis	5,0	0,75
3 Basketball	6 Croquet	6,0	0,74
2 Football	5 Golf	6,0	0,81

NNMDS Plot:

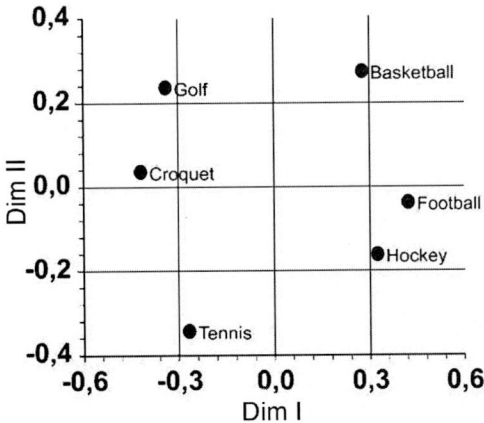

Figure 4.69 Interpretation of the dissimilarity matrix of six ball sports using the two-dimensional spatial NNMDS map of perception (scatter plot).

- **Dissimilarity Fit Plot:** This graph plots the dissimilarity values d_{ij} on the vertical axis against the predicted dissimilarity values $\hat{d}_{ij}$ on the horizontal axis (Fig. 4.70). The calibre of the solution depends upon this plot showing an upward-sloping trend.

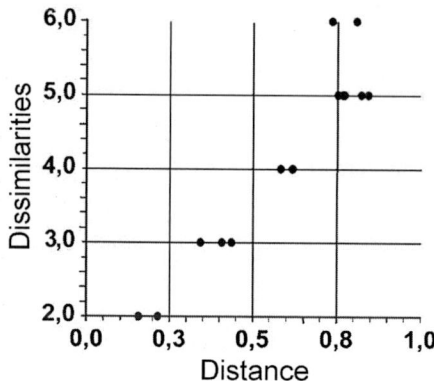

Figure 4.70 Interpretation of the dissimilarity d_{ij} versus predicted distance $\hat{d}_{ij}$ plot NNMDS of six ball sports.

Conclusion:

If the solution was perfect, then as we moved across the plot from left to right, we would never go down from one point to the next. We notice in this case that the solution confuses the large distances. This may be due to the large number of ties in this area.

4.11 Correspondence Analysis (CA)

Correspondence analysis (CA) is a technique for displaying the associations among a set of categorical variables in a type of scatterplot or map, thus allowing a visual examination of any pattern or structure in the data. It is an interdependence technique that has become increasingly popular for dimensional reduction and perceptual mapping [157, 158, 159, 160] and also a compositional technique because the perceptual map is based on the association between objects and a set of descriptive characteristics or attributes specified by the researcher. Its most direct application is portraying the "correspondence" of categories of variables, particularly those measured in nominal measurement scales. This correspondence is then the basis for developing perceptual maps. The *benefits of CA* lie in its unique abilities for representing rows and columns, for example, brands *and* attributes, in joint space:

(1) The simple cross-tabulation of multiple categorical variables, such as product attributes versus brands, can be represented in a perceptual space. This approach allows the researcher either to analyze existing responses or to gather responses at the least restrictive measurement type, the nominal or categorical level.

(2) CA portrays not only the relationships between the rows and columns, but also the relationships between the categories of either the rows or the columns.

(3) The most important advantage of CA is that it can provide a joint display of row and column categories in the same dimensionality.

However, there are also some *disadvantages* or *limitations* of CA:

(1) The technique is descriptive and not at all appropriate for hypothesis testing. If the quantitative relationship of categories is desired, methods such as log-linear models are suggested. CA is best suited for exploratory data analysis.

(2) CA has no method for conclusively determining the appropriate number of dimensions. As with similar methods, the researcher must balance interpretability versus parsimony of the data representation.

(3) The technique is quite sensitive on outliers, in terms of either rows or columns (e.g., attributes or brands). Also, for purposes of generalizability, the problem of omitted objects or attributes is critical.

Quintessentially, however, correspondence analysis is a technique for displaying multivariate categorical data graphically, by deriving coordinates to represent the categories of the variables involved, which may then be plotted to provide a 'picture' of the data.

4.11.1 Objectives of CA

Researchers are constantly faced with the need to "quantify the qualitative data" found in nominal variables. CA differs from other interdependence techniques in its ability to accommodate both nonmetric data and nonlinear relationships. It performs dimensional reduction similar to multidimensional scaling or factor analysis. CA can address either of two basic objectives:

(1) *DAssociation among row or column categories.* CA can be used to examine the association among the categories of just a row or column. A typical use is the examination of the categories of a scale, such as the Likert scale (five categories from "strongly agree" to "strongly disagree") or other qualitative scales (e.g., excellent, very good, good poor, bad). The categories can be compared to see if two can be combined (i.e., they are in close proximity on the map) or if they do provide discrimination (i.e., they are located separately in the perceptual space).

(2) *DAssociation between row and column categories.* In this application, interest lies is portraying the association between categories of the rows and columns.

4.11.2 Computational Details

Correspondence analysis requires only a rectangular data matrix (cross-tabulation) of nonnegative entries. The rows and columns do not have predefined meanings (i.e., attributes do not always have to be rows, and so on) but instead represent the responses to one or more categorical variables.

Consider a contingency table with n rows and m columns. A correspondence analysis of the table produces two sets of coordinate values—a set of n coordinates corresponding to the rows and a set of m coordinates corresponding to the columns. A correspondence analysis is interpreted by examining the positions of the row categories and the column categories as reflected by their respective coordinate values. The values of the coordinates reflect associations between the categories of the row variable and those of the column variable. Column points that are close together indicate columns with similar profiles (conditional distributions) down the rows. Finally, row points that are close to column points represent combinations that occur more frequently than would be expected under an independence model, that is, one in which the categories of the row variable are unrelated to the categories of the column variable.

Computation procedure: We will now present an outline of the computational methods used to perform the analysis. We will use standard matrix terminology to present the steps.

(1) Read in the n (rows) by m (columns) data matrix, $\mathbf{K}$. Note that the elements of $\mathbf{K}$ must be non-negative and that none of the row or column totals is zero.

(2) Compute the proportion matrix, $\mathbf{P}$, by dividing the elements of $\mathbf{K}$ by the total of all numbers in $\mathbf{K}$. Mathematically, $\mathbf{P} = \{p_{ij}\} = \{k_{ij} / k_{..}\}$ is the matrix of relative frequencies, i.e., each element of $\mathbf{P}$ is computed as the respective frequency from the input table, divided by the grand total of all values.

(3) Compute the totals of the rows of matrix $\mathbf{P}$ and the columns of matrix $\mathbf{P}$, putting the results in the vectors r and c. Using standard matrix notation, we write $\mathbf{r} = \mathbf{P1}$ and $\mathbf{c} = \mathbf{P}^T \mathbf{1}$ where 1 is an appropriately dimensioned vector of ones.

(4) Change the square roots of the vectors r and c into diagonal matrices and take the inverse of the resulting square matrices, $\mathbf{D}_r = \left[diag\left(\mathbf{r}\right) \right]^{-1/2}$ and $\mathbf{D}_c = \left[diag\left(\mathbf{c}\right) \right]^{-1/2}$.

(5) Compute the scaled matrix, $\mathbf{A}$ being defined $\mathbf{A} = \mathbf{D}_r \mathbf{P} \mathbf{D}_c$. The computation of the row and column coordinates is based on the generalized singular value decomposition of $\mathbf{P}$, as: $\mathbf{P} = \mathbf{A} \mathbf{D}_u \mathbf{B}'$ so that $\mathbf{A}$ $inverse(\mathbf{D}_r)\mathbf{A} = \mathbf{B}'$ $inverse(\mathbf{D}_c)\mathbf{B} = \mathbf{I}$, where $\mathbf{A}$ is the matrix of the left-side generalized singular vectors, B is the matrix of the right-side generalized singular vectors, $\mathbf{D}_u$ is a diagonal matrix with the diagonal elements equal to the generalized singular values, and $\mathbf{I}$ stands for the identity matrix (a diagonal matrix with 1's in the diagonal).

(6) Compute the Singular Value Decomposition (SVD) of A, as follows: $\langle \mathbf{B}, \mathbf{W}, \mathbf{C} \rangle = SVD(\mathbf{A})$.

(7) Compute the coordinate matrices, $\mathbf{F}$ and $\mathbf{G}$, as follows: $\mathbf{F} = \mathbf{D}_r \mathbf{BW}$ and $\mathbf{G} = \mathbf{D}_c \mathbf{BW}^{\mathrm{T}}$.

(8) Compute the eigenvalues, $\mathbf{V}$ as follows: $\mathbf{V} = \mathbf{WW}^{\mathrm{T}}$.

(9) Compute the row distances, d_i, and the column distances, d_j, as follows:

$$d_i = \sum_j \left(\frac{1}{p_{.j}}\right)\left(\frac{p_{ij}}{p_{i.}} - p_{.j}\right)^2 \quad \text{and} \quad d_j = \sum_i \left(\frac{1}{p_{i.}}\right)\left(\frac{p_{ij}}{p_{.j}} - p_{i.}\right)^2.$$

(10) The weights, w_i and w_j, come from the vectors r and c that were formed in step 3, as follows:

$$w_i = \{r_i\} \quad \text{and} \quad w_j = \{c_j\}.$$

(11) The reported statistics are computed as follows:

Mass	w_i
Inertia	$\dfrac{w_i d_i^2}{\sum_k w_k^2 d_k^2}$
Distance	d_i^2
Row Factor	f_{ij}
Column Factor	g_{ij}
Angle	$\operatorname{Arc\,cos}\left(\sqrt{\mathrm{COR}_{ij}}\right)$

Row & column profiles. The row coordinates are computed based on the row profile matrix $\mathbf{R} = inverse(\mathbf{D}_r)\mathbf{P}$, and the column coordinates are computed based on the column profile matrix computed analogously. Specifically, the row coordinates are computed as $\mathbf{F} = inverse(\mathbf{D}_r)\mathbf{AD}_u$, and the column coordinates as $\mathbf{G} = inverse(\mathbf{D}_c)\mathbf{BD}_u$.

Row profiles (interpret row distances): The row coordinates are computed based on the row profile matrix $\mathbf{R} = inverse(\mathbf{D}_r)\mathbf{P}$. Specifically, the (principal) row coordinates are computed as $\mathbf{F} = inverse(\mathbf{D}_r)\mathbf{AD}_u$, and the standard column coordinates as $\mathbf{G} = inverse(\mathbf{D}_c)\mathbf{B}$. This option is appropriate when we are interested in interpreting the distances between row points; the column coordinates should not be interpreted.

Column profiles (interpret col. distances): When reviewing the results for column points in multiple correspondence analysis, the column coordinates are computed based on the column profile matrix. Specifically, the (principal) column coordinates are computed as $\mathbf{F} = inverse(\mathbf{D}_c)\mathbf{BD}_u$, and the standard row coordinates as $\mathbf{G} = inverse(\mathbf{D}_r)\mathbf{A}$. This option is appropriate when we are interested in interpreting the distances between column points; the row coordinates should not be interpreted.

"Model" equation. When using the method of standardization, the following "model" on $\mathbf{P}$ in k dimensions shows how the relative frequencies are approximated: $\mathbf{P} \gg \mathbf{rc'} + \mathbf{D}_r\mathbf{F}\ inverse(\mathbf{D}_u)\mathbf{G'D}_c$. In this formula $\mathbf{F}$ and $\mathbf{G}$ stand for the row and column coordinates, respectively.

4.11.3 Assessing Overall Fit

With a cross-tabulation table, the frequencies for any row–column combination of categories are related to other combinations based on the marginal frequencies. This procedure yields a conditional expectation (a chi-square value). Once obtained, these chi-square values are standardized and converted to a distance metric, and then a process much like multidimensional scaling defines lower-dimensional solutions. These "factors" simultaneously relate the rows and columns in a single joint plot. The result is a representation of categories of rows and/or columns in the same plot.

To assess overall fit, the researcher must first identify the appropriate number of dimensions and their importance. The maximum number of dimensions that can be estimated is one less than the smaller of the number of rows or columns. For example, with six columns and eight rows, the maximum number of dimensions would be five, which is six (the number of columns) minus one. Eigenvalues are derived for each dimension and indicate the relative contribution of each dimension in explaining the variance in the categories. Some programs introduce a measure termed *inertia*, which also measures explained variation and is directly related to the eigenvalue. The researcher selects the number of dimensions based on the overall level of explained variance desired and the incremental explanation gained by adding another dimension. A rule of thumb is that dimensions with inertia (eigenvalues) greater than 0.2 should be included in the analysis. As discussed with regard to perceptual mapping, using a three-dimensional or lower representation facilitates interpretation.

4.11.4 Interpretation of the Results

Once the dimensionality has been established, the researcher can identify a category's association with other categories by proximity after the appropriate normalization. The researcher must select the type of normalization, and determine whether comparisons are to be made between row categories, column categories, or row and column categories. In most instances, the researcher wishes to compare between row and column categories. There may be instances, however, in which the focus is on only rows or columns, such as when examining the categories of a scale to see if they can be combined. If only a row or column normalization is available, alternative procedures are proposed to make all categories comparable, but there is still disagreement as to their success. In the cases for which direct comparisons are not possible, the general correspondence still holds and specific patterns can be distinguished.

If the researcher is interested in defining the character of one or more dimensions in terms of the row or column categories, there are descriptive measures that indicate the association of each category with a specific dimension. From the collective measures, an assessment of the fit for each category can also be made.

Problem 4.7 *Correspondence analysis of the smoking habits of different employees in a company*

Correspondence analysis (CA) is a technique for graphically displaying a two-way table by calculating coordinates representing its rows and columns. The data shows the results of a survey relating the smoking habits of the employees of a fictitious company to their position withing the company.

Data: The following data set Smoking.sta (STATISTICA, StatSoft) contains the entries which represent the counts of the number of employees falling into each cell.

Start Group	Smoking Category				
	(1) None	(2) Light	(3) Medium	(4) Heavy	Row Totals
(1) Senior Managers (SM)	4	2	3	2	11

Continue

(2) Junior Managers (JM)	4	3	7	4	18

(Continued)

(3) Senior Employees (SE)	25	10	12	4	51
(4) Junior Employees (JE)	18	24	33	13	88
(5) Secretaries (SC)	10	6	7	2	25
Column Totals	61	45	62	25	193

Solution: We may think of the 4 column values in each row of the table as coordinates in a 4-dimensional space, and one could compute the (Euclidean) distances between the 5 row points in the 4-dimensional space. The distances between the points in the 4-dimensional space summarize all information about the similarities between the rows in the table above. Now suppose one could find a lower-dimensional space, in which to position the row points in a manner that retains all, or almost all, of the information about the differences between the rows. We could then present all information about the similarities between the rows (types of employees in this case) in a simple 1, 2, or 3-dimensional graph. While this may not appear to be particularly useful for small tables like the one shown above, one can easily imagine how the presentation and interpretation of very large tables (e.g., differential preference for 10 consumer items among 100 groups of respondents in a consumer survey) could greatly benefit from the simplification that can be achieved via correspondence analysis (e.g., represent the 10 consumer items in a two-dimensional space).

- **Analyzing rows and columns.** This simple example began with a discussion of the row-points in the table shown above. However, one may rather be interested in the column totals, in which case one could plot the column points in a small-dimensional space, which satisfactorily reproduces the similarity (and distances) between the relative frequencies for the columns, across the rows, in the table shown above. In fact it is customary to simultaneously plot the column points and the row points in a single graph, to summarize the information contained in a two-way table.
- **Row Profiles Section:**

Staff	None	Light	Medium	Heavy	Total
SM	36.36	18.18	27.27	18.18	100.00
JM	22.22	16.67	38.89	22.22	100.00
SE	49.02	19.61	23.53	7.84	100.00

JE	20.45	27.27	37.50	14.77	100.00
SC	40.00	24.00	28.00	8.00	100.00
Total	31.61	23.32	32.12	12.95	100.00

This report shows the row profiles (percentages). These are the values the will be plotted on the row oriented plot. Note that since there are five rows, these data would require five dimensions to be plotted in the standard fashion. CA investigates the differences between each individual row profile and the average row profile (the row labeled "Total").

- **Column Profiles Section:**

Staff	None	Light	Medium	Heavy	Total
SM	6.56	4.44	4.84	8.00	5.70
JM	6.56	6.67	11.29	16.00	9.33
SE	40.98	22.22	19.35	16.00	26.42
JE	29.51	53.33	53.23	52.00	45.60
SC	16.39	13.33	11.29	8.00	12.95
Total	100.00	100.00	100.00	100.00	100.00

This report shows the column profiles (percentages). These are the values the will be plotted in a column oriented CA plot. Note that since there are four columns, these data would require four dimensions to be plotted in the standard fashion. CA investigates the differences between each individual column profile and the average column profile (the column labeled "Total").

- **Eigenvalue Section:**

Let us now look at some of the results for the table shown above. First, shown below are the so-called *Singular Values, Eigenvalues, Percentages of Inertia Explained, Cumulative Percentages*, and the contribution to the overall *Chi-squares*.

Factor No.	Eigenvalue	Individual Percent	Cumulative Percent	Bar Chart
1	0.074759	87.76	87.76	|IIIIIIIIIIIIIIIIIIIIIIIIIIIIIIIIIIIIIIIIIIIIIIIIIIIIIIIIIIIIIIIII
2	0.010017	11.76	99.51	|IIIIIIII
3	0.000414	0.49	100.00	|
Total	0.085190			

Since CA projects the row (or column) profiles onto a two-dimensional subspace, a critical issue is how well this projection works. The eigenvalues gives us important information regarding this. The Cumulative Percent column tells us how much of the total information is reproduced by each number of dimensions. In this example, the CA plot using the first two factors accounts for 99.5% of the variation in data. In other words, the dimension reduction is only costing us a 0.5% loss in informatik of data. We can be confident that the patterns we see in the CA plot represent the patterns that we would see if we could peer into n-dimensional space. **Factor No.:** This is the number of the factor (coordinate or dimension) that is reported about on this row of the report. **Eigenvalue:** This is the eigenvalue associated with this dimension. It gives a relative size (importance) of this dimension. **Individual and Cumulative Percents:** The first column gives the percentage of the total of the eigenvalues accounted for by this dimension. The second column is the cumulative total of the percentage. In ideal situations, the first two dimensions will account for over 90% of the variation. If the cumulative percentage is less than 50%, CA is not appropriate. **Bar Chart:** This is a rough bar plot of the eigenvalues. It enables us to quickly note the relative size of each eigenvalue. The dimensions are "extracted" so as to maximize the distances between the row or column points, and successive dimensions (which are independent of or orthogonal to each other) will "explain" less and less of the overall *Chi*-square value. Thus, the extraction of the dimensions is similar to the extraction of principal components in FA: first, it appears that, with a single dimension, 87.76% of the inertia can be "explained," that is, the relative frequency values that can be reconstructed from a single dimension can reproduce 87.76% of the total *Chi*-square value (and, thus, of the inertia) for this two-way table; two dimensions allow us to explain 99.51%.

- **Maximum number of dimensions.** Since the sums of the frequencies across the columns must be equal to the row totals, and the sums across the rows equal to the column totals, there are in a sense only (nnumber of columns minus 1) independent entries in each row, and (number of rows minus 1) independent entries in each column of the table (once we know what these entries are, we can fill in the rest based on our knowledge of the column and row marginal totals). Thus, the maximum number of eigenvalues that can be extracted from a two-way table is equal to the minimum of the number of columns minus 1, and the number of rows minus 1. If we choose to extract (i.e., interpret) the maximum number of dimensions that can be extracted, then we can reproduce exactly all information contained in the table.

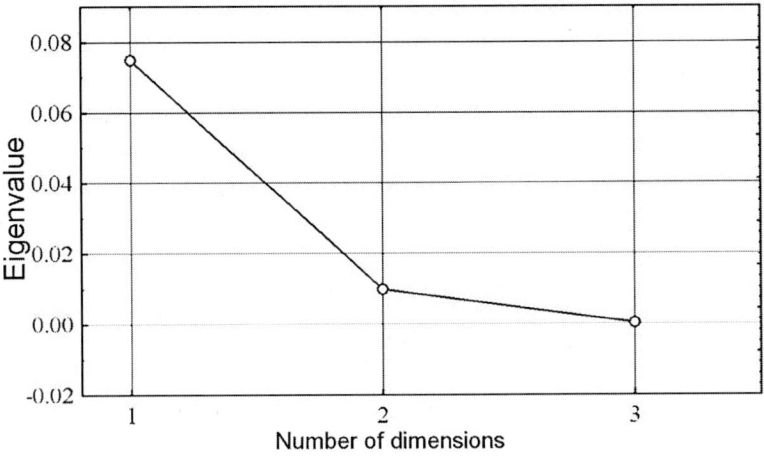

Figure 4.71 Cattel's free plot shows the optimal number of dimensions

- **Row and column coordinates.** Next look at the coordinates for the two-dimensional solution:

Row Name	*Dim. 1*	*Dim. 2*
(1) Senior Managers SM	−0.065768	0.193737
(2) Junior Managers JM	0.258958	0.243305
(3) Senior Employees SE	−0.380595	0.010660
(4) Junior Employees JE	0.232952	−0.057744
(5) Secretaries SC	−0.201089	−0.078911

We can plot these coordinates in a two-dimensional scatterplot. Remember that the purpose of correspondence analysis is to reproduce the distances between the row and/or column points in a two-way table in a lower-dimensional display; note that, as in factor analysis, the actual rotational orientation of the axes is arbitrarily chosen so that successive dimensions "explain" less and less of the overall *Chi*-square value (or inertia). We could, for example, reverse the signs in each column in the table shown above, thereby effectively rotating the respective axis in the plot by 180°. What is important are the distances of the points in the two-dimensional display, which are informative in that row points that are close to each other are similar with regard to the pattern of relative frequencies across the columns. If we have produced this plot we will see that, along the most important first axis in the plot, the *Senior employees* and *Secretaries* are relatively close together on the left side of the origin (scale position 0). If we looked at the table of relative row frequencies (i.e., frequencies standardized, so that their sum in each row is equal to 100%), we will see that these two groups of employees indeed show very similar patterns of relative frequencies across the categories of smoking intensity.

Percentages of Row Totals

	Smoking Category				
Start Group	(1) None	(2) Light	(3) Medium	(4) Heavy	Row Totals
(1) Senior Managers	36.36	18.18	27.27	18.18	100.00
(2) Junior Managers	22.22	16.67	38.89	22.22	100.00
(3) Senior Employees	49.02	19.61	23.53	7.84	100.00
(4) Junior Employees	20.45	27.27	37.50	14.77	100.00
(5) Secretaries	40.00	24.00	28.00	8.00	100.00

Obviously the final goal of correspondence analysis is to find theoretical interpretations (i.e., meaning) for the extracted dimensions. One method that may aid in interpreting extracted dimensions is to plot the column points. Shown below are the column coordinates for the first and second dimension.

Smoking category	*Dim. 1*	*Dim. 2*
None	−0.393308	0.030492
Light	0.099456	−0.141064
Medium	0.196321	−0.007359
Heavy	0.293776	0.197766

It appears that the first dimension distinguishes mostly between the different degrees of smoking, and in particular between category *None* and the others. Thus one can interpret the greater similarity of *Senior Managers* with *Secretaries*, with regard to their position on the first axis, as mostly deriving from the relatively large numbers of *None* smokers in these two groups of employees. For more complex tables, with many levels, some of the point labels may overlap in the scatterplots. We can use the brushing facilities to turn off the points that are of less interest, and only display those points that clearly "mark" the respective axes.

- **Compatibility of row and column coordinates.** It is customary to summarize the row and column coordinates in a single plot. However, it is important to remember that in such plots, one can only interpret the distances between row points, and the distances between column points, but not the distances between row points and column points. To continue with this example, it would not be appropriate to say that the category *None* is similar to *Senior Employees* (the two points are very close in the simultaneous plot of row and column coordinates). However, as was indicated earlier, it is appropriate to make general statements about the nature of the dimensions, based on which side of the origin particular points fall. For example, because category *None* is the only column point on the left side of the origin for the first axis, and since employee group *Senior Employees* also falls onto that side of the first axis, one may conclude that the first axis separates *None* smokers from the other categories of smokers, and that *Senior, Employees* are different from, for example, *Junior Employees*, in that there are relatively more non-smoking *Senior Employees.*
- **Scaling of the coordinates (standardization options).** Another important decision that the analyst must make concerns the scaling of the coordinates. The computations following from the choice of the different available options. The nature of the choice pertains to whether or not we want to analyze the relative row percentages, column percentages, or both. In the context of the example described above, the row percentages were shown to illustrate how the patterns of those percentages across the columns are similar for points which appear more closely together in the graphical display of the row coordinates. Put another way, the coordinates are based on the analysis of the row profile matrix, where the sum of the table entries in a row, across all columns, is equal to 1.0 (each entry r_{ij} in the row profile matrix can be interpreted as the conditional probability that a case belongs to column j, given its membership in row i). Thus, the coordinates are computed so as to maximize the differences between the points with respect to the row profiles (row percentages). Conversely, if we are interested in the similarities and differences between the columns of the table, we should select the *Column profiles (interpret col. dist.)* option in the *Standardization of coordinates* group; the resulting column coordinates are then derived from the analysis of the column profile matrix (the matrix of column proportions, where the sum of the table entries in each column is equal to 1.0). This standardization will maximize the distances between the column points in the final coordinate system.
- **Metric of coordinate system.** In several places in this introduction, the term distance was (loosely) used to refer to the differences

between the pattern of relative frequencies for the rows across the columns, and columns across the rows, which are to be reproduced in a lower-dimensional solution as a result of the correspondence analysis. Actually, these distances represented by the coordinates in the respective space are not simple Euclidean distances computed from the relative row or column frequencies, but rather, they are weighted distances. Specifically, the weighting that is applied is such that the metric in the lower-dimensional space is a *Chi-square* metric, provided that 1) we are comparing row points, and chose either row-profile standardization or both row- and column-profile standardization, or 2) we are comparing column points, and chose either column-profile standardization or both row- and column-profile standardization. In that case (but not if we chose the canonical standardization), the squared Euclidean distance between, for example, two row points i and i' in the respective coordinate system of a given number of dimensions actually approximates a weighted (i.e., *Chi*-square) distance between the relative frequencies:

$$d_{ii'}^2 = S_j \left(1/c_j \left[\frac{p_{ij}}{r_i} - \frac{p_{i'j}^2}{r_{i'}} \right] \right)$$

In this formula, $d_{ii'}^2$ stands for the squared distance between the two points, c_j stands for the column total for the j'th column of the standardized frequency table (where the sum of all entries or mass is equal to *1.0*), p_{ij} stands for the individual cell entries in the standardized frequency table (row i, column j), r_i stands for the row total for the i'th column of the relative frequency table, and the summation (S) is over the columns of the table. To reiterate, only the distances between row points, and correspondingly, between column points are interpretable in this manner; the distances between row points and column points cannot be interpreted.

- **Judging the quality of a solution.** A number of auxiliary statistics are reported, to aid in the evaluation of the quality of the respective chosen numbers of dimensions. The general concern here is that all (or at least most) points are properly represented by the respective solution, that is, that their distances to other points can be approximated to a satisfactory degree. Shown below are all statistics reported for the row coordinates for the example table discussed so far, based on a one-dimensional solution only (i.e., only one dimension is used to reconstruct the patterns of relative frequencies across the columns).

Row Coordinates and Contributions to Inertia

Staff Group	Coordin. Dim.1	Mass	Quality	Relative Inertia	Inertia Dim.1	Cosine² Dim.1
(1) Senior Managers	−0.065768	0.056995	0.092232	0.031376	0.003298	0.092232
(2) Junior Managers	0.258958	0.093264	0.526400	0.139467	0.083659	0.526400
(3) Senior Employees	−0.380595	0.264249	0.999033	0.449750	0.512060	0.999033
(4) Junior Employees	0.232952	0.455959	0.941934	0.308354	0.330974	0.941934
(5) Secretaries	−0.201089	0.129534	0.865346	0.071053	0.070064	0.865346

The first numeric column shown in the table above contains the coordinates, as discussed in the previous paragraphs. To reiterate, the specific interpretation of these coordinates depends on the standardization chosen for the solution. The number of dimensions is chosen by the user (in this case we chose only one dimension), and coordinate values will be shown for each dimension (i.e., there will be one column with coordinate values for each dimension). **Mass.** The *Mass* column contains the row totals (since these are the row coordinates) for the table of relative frequencies (i.e., for the table where each entry is the respective mass. The coordinates are computed based on the matrix of conditional probabilities shown in the *Mass* column. **Quality.** The *Quality* column contains information concerning the quality of representation of the respective row point in the coordinate system defined by the respective numbers of dimensions, as chosen by the user. In the table shown above, only one dimension was chosen, and the numbers in the *Quality* column pertain to the quality of representation in the one-dimensional space. To reiterate, computationally, the goal of the correspondence analysis is to reproduce the distances between points in a low-dimensional space. If we extracted (i.e., interpreted) the maximum number of dimensions (which is equal to the minimum of the number of rows and the number of columns, minus 1), we could reconstruct all distances exactly. The quality of a point is defined as the ratio of the squared distance of the point from the origin in the chosen number of dimensions, over the squared distance from the origin in the space defined by the maximum number of dimensions (remember that the metric here is *Chi*-square). By analogy to Factor Analysis, the quality of a point is similar in its interpretation to the communality for a variable in factor analysis. Note that the *Quality* measure is independent of the chosen method of standardization, and always pertains to the standardization (i.e., the distance metric is *Chi*-square, and the quality measure can be interpreted as the "proportion of *Chi*-square accounted for" for the respective row, given the respective number of dimensions). A low *Quality* means that the current number of dimensions does not well represent the respective row (or column). In the table shown above, the quality for the first row (*Senior Managers*) is less than *0.1*, indicating that this row point is not well represented by the one-dimensional representation of the points. **Relative inertia.** The *Quality* of a point represents the proportion of the contribution of that point to the overall inertia (*Chi*-square) that can be accounted for by the chosen number of dimensions. However, it does not indicate whether or not, and to what extent, the respective point does in fact contribute to the overall inertia (*Chi*-square value). The relative inertia represents the proportion of the total inertia accounted for by the respective point, and it is independent of the number of dimensions chosen by the user. Note that a particular solution may represent a point very well (high *Quality*), but the same point may not contribute much to the overall inertia (e.g., a row point with a pattern of relative frequencies across the columns that is similar to the average pattern across all rows). **Relative inertia for each dimension.** This column contains the relative contribution of the respective (row) point to the inertia "accounted for" by the respective dimension. Thus, this value will be reported for each (row or column) point, for each dimension. **Cosine² (quality or squared correlations with each dimension).** This column contains the *quality* for each point, by dimension. The sum of the values in these columns across the dimensions is equal to the total *Quality* value discussed above (since in the example table above, only one dimension was chose, the values in this column are identical to the values in the overall *Quality* column). This value may also be interpreted as the "correlation" of the respective point with the respective dimension. The term *Cosine²* refers to the fact that this value is also the squared cosine value of the angle the point makes with the respective dimension.

- **Statistical significance.** It should be noted at this point that correspondence analysis is an exploratory technique. Actually, the method was developed based on a philosophical orientation that emphasizes the development of models that fit the data, rather than the rejection of hypotheses based on the lack of fit Therefore, there are no statistical significance tests that are customarily applied to the results of a correspondence analysis; the primary purpose of the technique is to produce a simplified (low-dimensional) representation of the information in a large frequency table (or tables with similar measures of correspondence).

- **Multiple Correspondence Analysis (MCA)-Plots:** This plot is the main objective of a CA. The plot on the left shows the column profiles and the plot on the right shows the row profiles. It is important to remember that each point represents a profile projected onto the plane defined by the two axes.

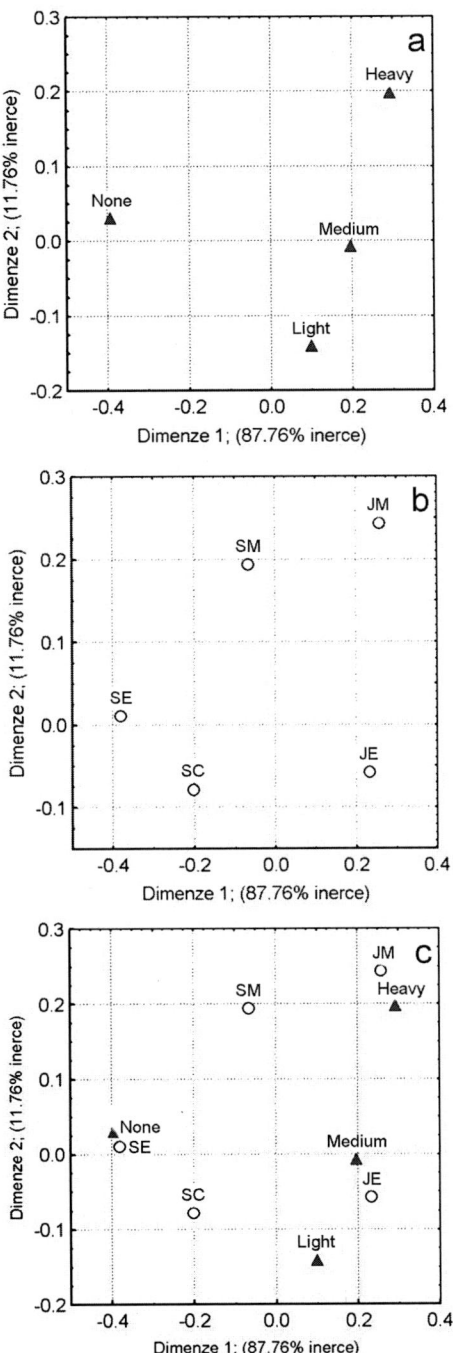

Figure 4.72a,b Correspondence Analysis Plots: the plot on the left shows the column profiles and the plot on the right shows the row profiles.

Lets begin by discussing the left plot, the one presenting the four column profiles. These profiles represented the proportions belonging to each staff category. We can see from this plot that the first dimension seems to separate those who smoke from those who do not. The second dimension seems to separate the three types of smokers: light, medium, and heavy.

The right plot presents the five row profiles. The first dimension appears to separate junior people from senior people. The second dimension seems to separate managers (near the bottom) from non-managers (near the top). Note that the distances between points on these plots are Chi-square distances between the profiles those of two points. Hence, the closer two points appear, the closer their profile patterns are to each other.

Finally, we come to the most popular CA plot in which we overlay the two plots shown above onto one plot. Extreme caution must be used when interpreting this plot. The critical point to remember is that this is a combination of two independent plots. Distances between the row profile points and column profile points are not defined. Hence, the distance between the categories SE and None (although this appear near each other on the plot) is not defined.

Conclusion: The point SE is a projection of the SE profile from the four dimensional space to the two dimensional subspace defined by our axes. The point None is a projection of the None profile from the five dimensional space to the two-dimensional subspace defined by the two axes. The original spaces are different. They represent different things. In the case of the row profiles, each of the four axes represent a smoking pattern (none, light, medium, and heavy). In the case of column profiles, each of the five axes represent a staff category. The point is that the meaning of the original spaces was completely different. Their axes have completely different definitions. This is the classical apples and oranges situation.

4.12 References

[1] M. Siotani, T. Hayakawa, Y. Fujikoshi: *Modern Multivariate Statistical Analysis,* A Graduate Course and Handbook, American Science Press, Columbia 1985.

[2] M. G. Kendall, A. Stuart: *The Advanced Theory of Statistics,* Vol. III., New York 1966.

[3] W. James, C. Stein: *Estimation with Quadratic Loss,* Proceed. 4th Berkeley Symp. on Math. Statist., p. 361, 1961.

[4] R. Guanadeskian, J. R. Kettenring: *Biometrics* **28**, 80 (1972).

[5] N. A. Campbell: *Appl. Statist.,* 29, 231 (1980).

[6] J. Hu, P. Skrabal, H. Zollinger: *Dyes and Pigments,* **8**, 189 (1987).

[7] J. M. Chambers, W. S. Cleveland, B. Kleiner, P. A. Tukey: *Graphical Methods for Data Analysis,* Duxburg Press, Belmont, California 1983.

[8] V. Barnett (ed.): *Interpreting Multivariate Data, Wiley,* Chichester 1981, chapter 6.

[9] I. T. Jolliffe: *Principal Component Analysis,* Springer Verlag, New York 1986.

[10] V. B. S. Barnett: *Graphical Techniques for Multivariate Data,* London 1978.

[12] D. F. Andrews: *Biometrics,* **28**, 125 (1972).

[13] S. R. Kulkarni, S. R. Paranjape: *Commun. Statist.,* **13**, 2511 (1984).

[14] R. Guanadeskian: *Methods for Statistical Data Analysis of Multivariate Observations,* Wiley, New York 1977.

[15] B. Kleiner, J. A. Hartigan: *J. Amer. Statist. Assoc.,* **76**, 260 (1981).

[16] H. Kres: *Statistical Tables for Multivariate Analysis,* Springer, New York 1983.

[17] G. A. F. Seber: *Multivariate Observations,* Wiley, New York 1984.

[18] E. Stryjewska, S. Rubel, A. Henrion, G. Henrion: *Z. Anal. Chem.,* **327**, 679 (1987).

[19] G. S. Mudholkar, M. S. Trivedi, T. C. Lin: *Technometrics,* **24**, 139 (1982).

[20] R. A. Johnson, D. W. Wichern: *Applied Multivariate Statistical Analysis,* Prentice Hall, 1998.

[21] S. Ajvjazin, Z. Bežajeva, O. Staroverov: *Metody vícerozměrné analýzy,* SNTL Praha 1981.

[22] M. Meloun, J. Militký, M. Forina: *Chemometrics for Analytical Chemistry, Volume 1, PC-Aided Statistical Data Analysis,* Ellis Horwood, Chichester 1992.

[23] R. G. Brereton: *Multivariate Pattern Recognition in Chemometrics,* Illustrated by Case Studies, Elsevier 1992.

[24] W. J. Krzanowski: *Principles of Multivariate Analysis, A User's Perspective,* Oxford Science Publications, 1988.

[25] B. S. Everitt, G. Dunn: *Applied Multivariate Data Analysis,* Arnold, London 2001.

[26] A. A. Afifi, V. Clark: *Computer-Aided Multivariate Analysis*, Chapman & Hall, CRC New York 1997.

[27] M. Meloun, J. Militký, M. Forina: *Chemometrics for Analytical Chemistry, Volume 2, PC-Aided Regression and Related Methods,* Ellis Horwood, Chichester 1994.

[28] K. Esbensen, S. Schonkopf, T. Midtgaard: *Multivariate Analysis in Practice,* CAMO Computer-Aided Moddeling AS, N-7011 Trondheim, Norway.

[29] D. A. Keim: Information visualization and visual data mining, *Transactions on Visualization and Computer Graphics,* **8**(1), 1–8, 2002.

[30] A. Sachinopoulou: *Multidimensional Visualization,* Research Notes 2114, VTT Technical Research Centre of Finland, Espoo 2001.

[31] D. A. Keim, Visual exploration of large databases, *Communications of the ACM,* **44**(8):38–44 (2001).

[32] D. A. Keim: Visual Techniques for Exploring Databases, Invited Tutorial, Int. Conference on Knowledge Discovery in Databases (KDD'97), Newport Beach, CA, 1997.

[33] D. A. Keim: Information visualization and visual data mining, *Transactions on Visualization and Computer Graphics,* **8**(1), 1–8 (2002).

[34] B. Shneiderman: The eye have it: A task by data type taxonomy for information visualizations, In *Proc. Visual Languages,* 1996, W. M. Siebert, *Circuits, Signals, and Systems,* MIT Press, 1986.

[35] B. Shneiderman: "The Eyes Have It: A Task by Data Type Taxonomy of Information Visualization," presented at IEEE Symposium on Visual Languages '96' Boulder, CO, 1996.

[36] P. E. Hoffman, G. Grinstein: "Multidimensional Information Visualizations for Data Mining with Applications for Machine Learning Classifiers," in *Information Visualization in Data Mining and Knowledge Discovery, The Morgan Kaufmann Series in Data Managament Systems,* U. Fayyad, G. Grinstein, and A. Wierse, Eds., 1st ed: Morgan-Kaufmann Publishers, 2001.

[37] G. Grinstein, P. E. Hoffman, S. Laskowski, R. Pickett: "Benchmark Development for the Evaluation of Visualization for Data Mining," in *Information Visualization in Data Mining and Knowledge Discovery, The Morgan Kaufmann Series in Data Managament Systems,* U. Fayyad, G. Grinstein, and A. Wierse, Eds., 1st ed: Morgan-Kaufmann Publishers, 2001.

[38] U. Fayyad, G. Grinstein, A. Wierse: *Information Visualization in Data Mining and Knowledge Discovery*, 1st ed: Morgan-Kaufmann Publishers, 2001.

[39] R. A. Fisher: "The Use of Multiple Measurements in Taxonomic Problems," *Annals of Eugenics*, vol. 7. pp. 179 - 188 (1936).

[40] D. F. Andrews: Plots of high-dimensional data, *Biometrics,* **29**:125–136 (1972).

[41] W. S. Cleveland: *Visualizing Data,* AT&T Bell Laboratories, Murray Hill, NJ, Hobart Press, Summit NJ, 1993.

[42] P. J. Huber: The annals of statistics, *Projection Pursuit,* **13**(2):435–474 (1985).

[43] D. F. Andrews: Plots of high-dimensional data, *Biometrics,* **29**:125–136 (1972).

[44] W. S. Cleveland: *Visualizing Data,* AT&T Bell Laboratories, Murray Hill, NJ.Hobart Press, Summit NJ, 1993.

[45] G. Grinstein, M. Trutschl, U. Cvek: *High-Dimensional Visualizations*, in Proceedings of the Visual Data Mining Workshop, KDD'2001, http://genome.uml.edu/pub_cs.htm

[46] L. Tweedie, R. Spence, H. Dawkes & H. Su: Externalising Abstract mathematical Models, Proceedings of the Conference on Human Components in Computing Systems (SIGCHI '96), 406–412, (1996).

[47] G. W. Furnas and A. Buja: Prosections views: Dimensional inference throughsections and projections, *Journal of Computational and Graphical Statistics* 3(4):323–353 (1994).

[48] P. C. Wong & R. D. Bergeron: 30 Years of Multidimensional Multivariate Visualization, In: G. Nielson. Scientific visualization: overviews, methodologies, and techniques, IEEE Computer Society, Washington, 1997.

[49] R. Becker, J. M. Chambers, and A. R. Wilks: *The New S Language,* Wadsworth & Brooks/Cole Advanced Books and Software, Pacific Grove, CA, 1988.

[50] J. R. A. Allwright and D. B. Carpenter: A distributed implementation of simulated annealing for the traveling salesman problem. *Parallel Computing.* 10:335–338 (1989).

[51] H. D. I. Abarbanel: *Analysis of Observed Chaotic Data.* Springer. 1995.

[52] A. Inselberg and B. Dimsdale: Parallel coordinates: A tool for visualizing multidimensional geometry. In *Proc. Visualization 90. San Francisco. CA.* 361–370. 1990.

[53] A. Inselberg: "The Plane with Parallel Coordinates." *Special Issue on Computational Geometry: The Visual Computer.* vol. 1. 69–91. 1985.

[54] A. Inselberg, B. Dimsdale: "Parallel Coordinates for Visualizing Multidimensional Geometry." presented at Computer Graphics International '87. Tokyo. 1987.

[55] Y. H. Fua, M. O. Ward: Hierarchical Parallel Coordinates for Exploration of Large Datasets, E. A. Rundensteiner, Proc. of Visualization '99, Oct. 1999.

[56] A. Goel, C. Baker, C. A. Shaffer, B. Grossman, R. F. Haftka, W. H. Mason, L. T. Watson, VizCraft: A Multidimensional Visualization Tool for Aircraft Configuration Design. Proceedings of IEEE, Visualization '99, San Francisco, CA, October 1999.

[57] P. Hoffman, G. Grinstein, K. Marx, I. Gorse, E. Stanley: DNA Visual And Analytic Data Mining, IEEE Visualization 1997, October 1997, Phoenix, Arizona, USA.

[58] A. Inselberg: The plane with parallel coordinates, The Visual Computer, 1:69–91 (1985).

[59] M. Meloun, J. Militký: *Statistické zpracování experimentálních dat,* Plus Praha 1994 (1st edit.), EAST PUBLISHING, Praha 1998 (2nd edit.), Academia Praha 2004 (3rd edit).

[60] J. F. Hair, R. E. Anderson, R. L. Tatham, W. C. Black: *Multivariate Data Analysis,* Prentice Hall, Upper Saddle River, New Jersey 07458, USA, 1998.

[61] S. du Toit, A. Steyn, R. Stumpf: *Graphical Exploratory Data Analysis,* Springer-Verlag: Berlin; 1986

[62] J. Siegel, E. Farrell, R. Goldwyn, H. Friedman: The surgical implication of physiologic patterns in myocardial infarction shock, *Surgery* 1972;72: 126–141.

[63] M. Berthold, D. J. Hand: *Intelligent Data Analysis,* Springer-Verlag Berlin 1998.

[64] E. Anderson: A semigraphical method for the analysis of complex problems, *Proceedings of the National Academy of Science* 13: 923–927 (1957).

[65] R. Gnanadesikan: *Methods of Statistical Data Analysis of Multivariate Observations,* John Wiley and Sons, New York; 1977.

[66] R. Pickett, G. Grinstein: *Iconographic displays for visualizing multidimensional data*, Proc. IEEE Conference on Systems, Man. and Cybernetics (1988); 164–170.

[67] B. Kleiner, J. Hartigan: Representing points in many dimension by trees and castles, *Journal of the American Statistical Association*; **76**, 260–269 (1981).

[68] J. Beddow: *Shape coding of multidimensional data on a microcomputer display*, Proc. Visualization '90 (San Francisco, CA), IEEE Computer Society Press: Los Alamitos, CA; 238–246 (1960).

[69] J. Hartigan: Printer graphics for clustering, *Journal of Statistical Computing and Simulation*; 4: 187–213 (1975).

[70] R. Klassen, S. Harrington: *Shadowed hedgehogs: a technique for visualizing 2D slices of 3D vector fields*, Proc. Visualization '91 (San Diego, CA, 1991), IEEE Computer Society Press: Los Alamitos, CA; 148–153 (1991).

[71] D. B. Heise: *Multidimensional Visualization technice Viewer (MVTView),* http://filer.case.edu/~dbh10 /eecs466/report.html.

[72] H. Chernoff: The use of faces to represent points in k-dimensional space graphically, *Journal of the American Statistical Association* **68**: 361–368 (1973).

[73] C. Wittenbrink, A. Pang, S. Lodha: Glyphs for visualizing uncertainty in vector fields, *IEEE Transactions on Visualization and Computer Graphics* **2**: 266–279 (1996).

[74] W. Schroeder, C. Volpe, W. Lorensen: *The Stream Polygon: a technique for 3D vector field visualization*, Proc. Visualization '91 (San Diego, CA,), IEEE Computer Society Press: Los Alamitos, CA; 126–132 (1991).

[75] A. Fuhrmann, E. Groller: *Real-time techniques for 3D flow visualization*, Proc. Visualization '98 (Research Triangle Park, NC), IEEE Computer Society Press: Los Alamitos, CA; 305–312, 1998.

[76] J. Friedman, E. Farrell, R. Goldwyn, M. Miller, J. Sigel: *A graphic way of describing changing multivariate patterns*, Proc. Sixth Interface Symposium on Computer Science and Statistics 56–59 (1972).

[77] J. Mezzich, D. Worthington: A comparison of graphical representations of multidimensional psychiatric diagnostic data, In: P. Wang (Ed.) *Graphical Representation of Multivariate Data*, Academic Press: New York; 1978.

[78] H. Levkowitz: *Color icons: merging color and texture perception for integrated visualization of multiple parameters*, Proc. Visualization '91 (San Diego, CA), IEEE Computer Society Press: Los Alamitos, CA; 164–170 (1991).

[79] M. Chuah, S. Eick: Information rich glyphs for software management data, *IEEE Computer Graphics and Applications* **18**: 24–29 (1998).

[80] M. O. Ward: *Multivariate Data Glyphs: Principles and Practice* in *Handbook of Data Visualization,* Springer-Verlag Berlin 2008.

[81] G. Kerlick: *Moving iconic objects in scientific visualization*, Proc. Visualization '90 (San Francisco, CA), IEEE Computer Society Press: Los Alamitos, CA; 124–130 (1990).

[82] R. Rohrer, D. Ebert, J. Sibert: *The shape of Shakespeare: visualizing text using implicit surfaces*, Proc. Visualization '98 (Research Triangle Park, NC), IEEE Computer Society Press: Los Alamitos, CA; 121–129 (1998).

[83] D. Ebert, R. Rohrer, C. Shaw, P. Panda, J. Kukla, D. Roberts: *Procedural shape generation for multi-dimensional data visualization*, Proc. Data Visualization '99, Springer-Verlag: Berlin; 3–12 (1999).

[84] W. Ribarsky, E. Ayers, J. Eble, S. Mukherjea, Glyphmaker: creating customized visualizations of complex data, *Computer* **27**: 57–64 (1994).

[85] F. Post, T. Walsum, F. Post, D. Silver: *Iconic techniques for feature visualization*, Proc. Visualization '95 (Atlanta, GA), IEEE Computer Society Press: Los Alamitos, CA; 288–295, 1995.

(http://www.elsevier.nl/locate/ida).

[86] M. Ward: *A Taxonomy of Glyph Placement Strategies for Multidimensional Data Visualization*, NSF Grants Iis-9732897.

[87] D. A. Keim: Designing pixel-oriented visualization techniques: Theory and applications, *Transactions on Visualization and Computer Graphics,* **6**(1):59–78, Jan–Mar 2000.

[88] D. A. Keim, H. - P. Kriegel, and M. Ankerst: Recursive Pattern: A technique for visualizing very large amounts of data, In *Proc. Visualization 95, Atlanta, GA*, pages 279–286, 1995.

[89] M. Ankerst, D. A. Keim, and H. - P. Kriegel: Circle Segments: A Technique for Visually Exploring Large Multidimensional Data Sets, In *Proc. Visualization 96, Hot Topic Session, San Francisco, CA,* 1996.

[90] C. Chen: *Information Visualisation and Virtual Environments,* Springer-Verlag, London, 1999.

[91] M. Dodge, R. Kitchin: *Atlas of Cyberspace,* Addison Wesley, Aug 2001.

[92] G. D. Battista, P. Eades, R. Tamassia, and I. G. Tollis: *Graph Drawing,* Prentice Hall, 1999.

[93] J. LeBlanc, M. O. Ward, N. Wittels: Exploring *n*-dimensional databases, In *Proc. Visualization '90. San Francisco, CA,* pages 230–239, 1990.

[94] S. Feiner, C. Beshers: Visualizing *n*-dimensional virtual worlds with *n*-vision,*Computer Graphics,* **24**(2):37–38, 1990.

[95] B. Johnson and B. Shneiderman: Treemaps: A space-filling approach to the visualization of hierarchical information, In *Proc. Visualization '91 Conf., San Diego,CA,* pages 284–291, 1991.

[96] B. Shneiderman: Tree visualization with treemaps: A 2D space-filling approach, *ACM Transactions on Graphics,* **ll**(l): 92–99, 1992.

[97] G. G. Robertson, J. D. Mackinlay and S. K. Card: Cone trees: Animated 3Dvisualizations of hierarchical information, In *Proc. Human Components in ComputingSystems CHI '91 Conf., New Orleans, LA,* 189–194, 1991.

[98] D. Asimov: The grand tour: A tool for viewing multidimensional data, *SIAM Journal of Science & Stat. Comp.,* **6**:128–143, 1985.

[99] A. Buja, D. F. Swayne and D. Cook: Interactive high-dimensional data visualization, *Journal of Computational and Graphical Statistics,* **5**(l):78–99, 1996.

[100] D. F. Swayne, D. Cook, and A. Buja: *User's Manual for XGobi: A Dynamic Graphics Program for Data Analysis,* Bellcore Technical Memorandum, 1992.

[101] L. Tierney: *LispStat: An Object-Oriented Environment for Statistical Computingand Dynamic Graphics,* Wiley, New York, NY, 1991.

[102] D. B. Carr, E. J. Wegman, and Q. Luo: Explorn: Design considerations pastand present, In *Technical Report, No. 129, Center for Computational Statistics,George Mason University,* 1996.

[103] E. A. Bier, M. C. Stone, K. Pier, W. Buxton, and T. DeRose: Toolglass and magic lenses: The see-through interface, In *Proc. SIGGRAPH '93, Anaheim, CA,* 73–80, 1993.

[104] K. Fishkin and M. C. Stone: Enhanced dynamic queries via movable filters, In *Proc. Human Components in Computing Systems CHI '95 Conf., Denver, CO*, 415–420, 1995.

[105] A. Spoerri, InfoCrystal: A visual tool for information retrieval, In *Proc. Visualization '93, San Jose, CA*, pages 150–157, 1993.

[106] C. Ahlberg and B. Shneiderman: Visual information seeking: Tight coupling of dynamic query filters with starfield displays, In *Proc. Human Components in Computing Systems CHI '94 Conf., Boston, MA*, 313–317, 1994.

[107] S. G. Eick: Data visualization sliders, In *Proc. ACM VIST*, 119–120, 1994.

[108] J. Goldstein and S. F. Roth: Using aggregation and dynamic queries for exploringlarge data sets, In *Proc. Human Components in Computing Systems CHI '94 Conf., Boston, MA*, 23–29, 1994.

[109] C. Stolte, D. Tang, and P. Hanrahan: Polaris: A system for query, analysis and visualization of multidimensional relational databases, *Transactions on Visualization and Computer Graphics*, **8**(1), 2002.

[110] R. Rao and S. K. Card: The table lens: Merging graphical and symbolic representation in an interactive focus + context visualization for tabular information, In *Proc. Human Components in Computing Systems CHI '94 Conf., Boston, MA*, 318–322, 1994.

[111] B. Bederson: PAD++: Advances in multiscale interfaces, In *Proc. Human Components in Computing Systems CHI '94 Conf, Boston, MA*, 315, 1994.

[112] B. B. Bederson and J. D. Hollan: PAD++: A zooming graphical interface for exploring alternate interface physics, In *Proc. VIST*, 17–26, 1994.

[113] K. Perlin and D. Fox, PAD: An alternative approach to the computer interface,In *Proc. SIGGRAPH, Anaheim, CA*, 57–64, 1993.

[114] C. Ahlberg and E. Wistrand: IVEE: An information visualization and exploration environment, In *Proc. Int. Symp. on Information Visualization, Atlanta, GA*, 66–73, 1995.

[115] V. Anupam, S. Dar, T. Leibfried, and E. Petajan: Dataspace: 3D visualization of large databases, In *Proc. Int. Symp. on Information Visualization, Atlanta, GA*, 82–88, 1995.

[116] Schaffer, Doug, Zuo, Zhengping, Bartram, Lyn, Dill, John, Dubs, Shelli, Greenberg, Saul, and Roseman, Comparing fisheye and full-zoom techniques for navigation of hierarchically clustered networks,

In *Proc. Graphics Interface (GI '93),Toronto, Ontario, 1993, in: Canadian Information Processing Soc. Toronto, Ontario, Graphics Press, Cheshire, CT,* 87–96, 1993.

[117] Y. Leung and M. Apperley: A review and taxonomy of distortion-oriented presentation techniques, In *Proc. Human Components in Computing Systems CHI '94 Conf., Boston, MA,* 126–160, 1994.

[118] M. S. T. Carpendale, D. J. Cowperthwaite and F. D. Fracchia: IEEE computergraphics and applications, special issue on information visualization, *IEEE Journal Press,* **17**(4):42–51, 1997.

[119] R. Spence and M. Apperley: Data base navigation: An office environment for theprofessional, *Behaviour and Information Technology,* **1**(1):43–54, 1982.

[120] J. D. Mackinlay, G. G. Robertson, and S. K. Card: The perspective wall: Detailand context smoothly integrated, In *Proc. Human Components in Computing SystemsCHI '91 Confi, New Orleans, LA,* 173–179, 1991.

[121] G. Furnas: Generalized fisheye views, In *Proc. Human Components in ComputingSystems CHI '86 Conf., Boston, MA,* 18–23, 1986.

[122] M. Sarkar and M. Brown: Graphical fisheye views, *Communications of the ACM,* **37**(12): 73–84, 1994.

[123] J. Lamping, R. Rao, and P, Pirolli: A focus + context technique based on hyperbolic geometry for visualizing large hierarchies, In *Proc. Human Components inComputing Systems CHI '95 Conf. Denver, CO,* 401–408, 1995.

[124] T. Munzner and P. Burchard: Visualizing the structure of the world wide webin 3D hyperbolic space, In *Proc. VRML '95 Symp, San Diego, CA,* 33–38, 1995.

[125] B. Alpern and L. Carter: Hyperbox, In *Proc. Visualization '91, San Diego, CA,* 133–139, 1991.

[126] R. Becker, J. M. Chambers, and A. R. Wilks: *The New S Language,* Wadsworth & Brooks/Cole Advanced Books and Software, Pacific Grove, CA, 1988.

[127] R. A. Becker, W. S. Cleveland, and M.-J. Shyu: The visual design and control of trellis display, *Journal of Computational and Graphical Statistics,* 5(2):123–155, 1996.

[128] D. F. Swayne, D. Cook, and A. Buja: *User's Manual for XGobi: A Dynamic Graphics Program for Data Analysis,* Bellcore Technical Memorandum, 1992.

[129] M. O. Ward, XmdvTool: Integrating multiple methods for visualizing multivariate data, In *Proc. Visualization 94, Washington, DC,* 326–336, 1994.

[130] A. Wilhelm, A. Unwin, and M. Theus: Software for interactive statistical graphics- a review, In *Proc. Softstat '95 Conf., Heidelberg, Germany,* 1995.

[131] B. G. Tabachnick, L. S. Fidell: *Using Multivariate Statistics,* Allyn and Bacon, A Pearson Education Company, 160 Gould Street, Needham Heights, MA 02494, USA (2001).

[132] E. R. Malinowski: *Factor Analysis in Chemistry,* Wiley (1980) New York.

[133] R. L. Gorsuch: *Factor analysis,* Lawrence Erlbaum Associates, Publishers, Hillsdale, New Jersey, 1983.

[134] H. F. Kaiser: The varimax criterion for analytic rotation in factor analysis, *Psychometrika* **23**, 187–200 (1958).

[135] L. L. Thurstone: Multiple factor analysis, *Psychology Review,* **39**, 406–427 (1931).

[136] K. V. Mardia, J. T. Kent, J. M.Bibby: *Multivariate Analysis,*Academic Press, London, 2003.

[137] H. Hotelling: Relations between two sets of variates, *Biometrika* **8**, 321–377 (1936).

[138] R. Donner, M. Reiter, G. Langs, P. Peloscheck, and H. Bischof: Fast active appearance model search using canonical correlation analysis, *IEEE Transactions on Pattern Analysis and Machine Intelligence,* **28**, 1690–1694 (2006).

[139] S. Douglas, W. Love: A General Canonical Correlation Index, *Psychological Bulletin* **70**, 160–163 (1968).

[140] Z. Lambert, R. Durand: Some Precautions in Using Canonical Analysis, *Mathematical Tools for Applied Multivariate Analysis,* Academic Press, New York (1978).

[141] W. R. Dillon, M. Goldstein: *Multivariate Analysis: Methods and Applications,* Wiley, New York 1984.

[142] J. L. Hintze: *NCSS2000, Statistical System for Windows*, Kaysville, Utah USA, 1999.

[143] D. C. Montgomery, E. A. Peck, G. G. Vining: *Introduction to Linear Regression Analysis*, John Wiley & Sons, New York 2006.

[144] D. Hosmer, S. Lemeshow: *Applied Logistic Regression*, John Wiley & Sons, New York 1989.

[145] D. Pregibon: *Logistic Regression Diagnostics, Annals of Statistics,* Volume 9, 705–725 (1981).

[146] E. Lesaffre, A. Albert: *Multigroup Logistic Regression Diagnostics, Applied Statistics,* Volume 38, 425–440 (1989).

[147] E. T. Lee: *Statistical Methods for Survival data Analysis,* Lifetime Learning Publications, Belmont, California, 1980.

[148] M. Anderberg: *Cluster Analysis for Applications,* Academic Press, New York 1973.

[149] P. N. Tan, M. Steinbach, V. Kumar: *Introduction to Data Mining,* Addison Wesley, 2006.

[150] B. Everitt: *Cluster Analysis,* 2d ed. Halsted Press, New York 1980.

[151] P. H. A. Sneath, R. R. Sokal: *Numerical Taxonomy,* Freeman Press, San Francisco 1973.

[152] A. A. Afifi, S. P. Azen: *Statistical Analysis: A Computer Oriented Approach,* 2nd edn., Academic Press, New York 1979.

[153] G. N. Lance, W. T. Williams: *A general theory of classificatory sorting strategies. I. Hierarchical systems,* Comput. J., 9, 373–380 (1967).

[154] P. Mather: *Computational Methods of Multivariate Analysis in Physical Geography*, Wiley, 1976.

[155] W. S. Torgenson: *Multidimensional scaling. I. Theory and method*, Psychometrika 17, 401–419 (1952).

[156] J. Kruskal: *Multidimensional scaling by optimizing goodness of fit to a nonmetric hypothesis,* Psychometrika 29, 1–27, 115–129 (1964).

[157] J. Carroll, J. Douglas, P. E. Green, V. M. Schaffer, *Interpoint Distance Comparisons in Crrespondence Analysis*, J. of Marketing Research 23, 271–80 (1986).

[158] M. J. Greenacre, *Theory and Applications of Correspondence Analyses.*, Academic Press, London, 1984.

[159] D. L. Hoffman, G. R. Franke, *Correspondence Analysis: Graphical Representation of Categorical Data in Marketing Research*, J. of Marketing Research 23, 213–227 (1986).

[160] L. Lebart, A. Morineau, K.M.Warwick, *Multivariate Descriptive Statistical Analysis: Correspondence Analysis and Related Techniques for Large Matrices,* Wiley, New York 1984.

Supplemented material (Review Questions, Exercises) to Chapter 4 is on enclosed CD.

5

Analysis of variance

The results of observations vary because of changes in the basic factors (both qualitative and quantitative) that control the conditions of the chemical experiment, and also in accidental factors. It is the objective of *analysis of variance (ANOVA)* to investigate the effect of the various factors on the variability of data and to determine which part of the variation in a population is due to systematic reasons (called *factors*) and which is due to random effects. ANOVA has been defined as a statistical technique for analysing measurements that depend on several kinds of effects operating simultaneously, in order to decide which kind of effects are important and to estimate the effects.

The profusion of instrumental techniques in measurements is such that often more than two possible techniques have to be compared. The techniques to be examined may be subject to systematic errors. The choice of a technique is called a *controlled factor.* Moreover, the results are subject to *random errors.* The analysis of variance compares both causes of error, with the purpose of deciding whether or not the controlled factor has a significant effect.

In a practice the analysis of variance often serves

- **(a)** to distinguish between sources of variability between trstiny places, between samples and between replicates, and
- **(b)** to investigate the influence of human factors instrument factors, methodology, concentration or time on the results of measurements.

A survey of ANOVA techniques may be found in the literature [1–5]. We limit ourselves to techniques suitable for evaluation of experimental data mainly.

5.1 Objectives of Analysis of Variance

Let us consider an example from the laboratory. It is desired to examine the influence of different methods of sample homogenization on the result of an analysis. With the use of three different homogenizers Z_1, Z_2 and Z_3, three different samples were prepared and analysed. The observed values are y_{ij}, $i = 1, 2, 3$ and $j = 1, 2, 3$, where y_{ij} denotes the observation for the ith homogenizer and jth sample. The method of homogenization of sample is called a *qualitative factor.* There are also *quantitative factors,* such as, for example, the mean particle size of the homogenized sample or various physico-chemical parameters.

The individual factor Z exists on different levels Z_1, Z_2, Z_3, which are called *treatments.* The treatments are the main sources of variability and may be also be of qualitative or quantitative nature. The model of the response in one-factor ANOVA can be written

$$y_{ij} = \mu_i + \varepsilon_{ij} \tag{5.1}$$

where y_{ij} represents the jth observation ($j = 1, 2, \ldots, n_i$) for the ith treatment ($i = 1, 2 \ldots, k$), μ_i is the true response (mean) at a factor level Z_i, and ε_{ij} is the random error present in the jth observation for the ith treatment. The mean μ_i may be divided into two parts

$$\mu_i = \mu + \alpha_i \tag{5.2}$$

where μ represents a general overall mean and α_i represents the effect of the ith treatment Z_i. The total number of observations is $n = n_1 + n_2 + \ldots + n_k$.

In our example we have $k = 3$ and $n = 9$. We will now test the null hypothesis that there are no differences caused by the method of homogenization, $H_0 : \mu_1 = \mu_2 = \mu_3$ which corresponds to the null hypothesis $H_0 : \alpha_1 = \alpha_2 = \alpha_3 = 0$. If we investigate the differences between just three methods of homogenization, we have a *fixed-effect model.* If the levels (Z_1, Z_2, Z_3) are random samples from the population of all possible methods of homogenization, we have a *random-effect model.*

The types of effects proposed lead to distinctive model assumptions and associated statistical analysis. The analysis of variance can be applied in several distinct forms, according to the structure of the process being investigated. The selection of a particular form usually constitutes a major difficulty in the practical application of the analysis of variance.

The choice between fixed- or random-effect models depends on the *purpose* of the analysis. If we suppose that three homogenizing machines make three different levels of particle size, then instead of considering homogenizing machine to be the factor, we use the mean particle size.

(a) We speak about a fixed-effects model when three homogenizing machines correspond to three different milling finenesses, and we examine whether these homogenizing machines affect significantly the results of the chemical analysis.

(b) We speak about a random-effects model when we test whether the mean particle size has an influence on the results of the analysis. From a population of different particle sizes we randomly select three.

The criteria for choosing between the fixed- and random-effects models are as follows:

(a) Factors with fixed effects are usually a type of treatment, a type of instrument, a testing method, a type of raw material, etc.

(b) Factors with random effects are laboratories, days, people, animals, etc.

One-way ANOVA deals with the influence of a single factor on a single response variable. When that one factor has fixed effects, one-way ANOVA *(fixed-effects one-way ANOVA)* involves comparison of several (two or more) population means. Each population corresponds to one treatment (factor level). Often, the influence of more factors is examined and then we speak about a multi-way ANOVA. In this chapter, we concentrate on one- and two-way ANOVA only.

An example of two factors could be the milling of samples using three grinding mills by two technicians. The second factor is here "technician" with levels L_1 and L_2. The result of chemical analysis y_{ijk}, $i = 1, 2, 3; j = 1, 2$; and $k = 1, 2, 3$ means the result for the ith way of milling mode by the jth technician on the kth sample. The observation is replicated for a given combination of both factors and the corresponding ANOVA model is expressed by

$$y_{ijk} = \mu_{ij} + \varepsilon_{ijk} \tag{5.3}$$

where μ_{ij} is the true, theoretical mean of analysis for the combination of factors $Z_i L_j$ and ε_{ijk} is a random error. The mean μ_{ij} can be written as a sum of effects α_i and β_j of the factors Z_i and L_j, an overall mean μ, and the interaction effect τ_{ij} due to

$$\mu_{ij} = \mu + \alpha_i + \beta_j + \tau_{ij} \tag{5.4}$$

The term τ_{ij} represents the effect of *interaction* of levels Z_i and L_j. It is used when the variability of y_{ijk} cannot be explained by the additive influence of factors. When the effects of both factors are fixed or random, we speak about *models with fixed* or *random effects*. It may happen that, e.g., effects of factor L are random while factor Z has fixed effects. Such models are called *model with mixed effects.*

A combination of levels of individual factors (e.g. $Z_i L_j$) is called a *cell*. When the number of replicate measurement is the same for each, the experiment is termed a *balanced experiment* (or a *balanced plan* of experiments); for different numbers it is called an *unbalanced experiment*.

Treatment of data is easier for balanced experiments. Treatment of data from unbalanced experiments is more complicated and sometimes the ANOVA assumptions (e.g. about normality) are not fulfilled, and this will cause distortion in the ANOVA results.

5.2 One-Way ANOVA

Suppose that some factor A, which we postulate as having some effect on a response variable y, has k levels. We set up an experiment in which n_1 observations are made of the response y at level A_1, n_2 observations at level A_2, and so on, with n_i observations at a given level A_i. The levels A_i are called *treatments*, and there are k treatments in the experimental design. The total number of observations is $n = \sum_{i=1}^{k} n_i$. At each level A_i there are n_i observations $y_{ij}, j = 1, \ldots, n_i$. The layout of a one-way ANOVA experiment with different number of replicates for each treatment is shown in Table 5.1.

Let $\hat{\mu}_i$ denote the mean of the ith (partial) sample, that is, the mean of observations at the ith treatment or level

$$\hat{\mu}_i = \frac{1}{n_i} \sum_{j=1}^{n_i} y_{ij} \qquad (5.5)$$

Table 5.1 Data layout for one-way ANOVA with an unequal number of replications on each treatment

		Factor A levels (treatments)					Overall mean
	I	1	2	...	i	...	k	...
		A_1	A_2	...	A_i		A_k	
Replicate	j							
	1	y_{11}	y_{21}	...	y_{i1}	...	y_{k1}
	2	y_{21}	y_{22}	...	y_{i2}	...	y_{k2}
	...	...	...	...	...	...	...
	...	...	...	...	...	...	...
	n_1	y_{1n_1}	y_{2n_2}	...	y_{in_i}	...	y_{kn_k}

Mean	$\hat{\mu}_1$	$\hat{\mu}_2$	...	$\hat{\mu}_i$	...	$\hat{\mu}_k$	$\hat{\mu}$
Sample size	n_1	n_2	...	n_i	...	n_k	n

The overall (or grand mean) $\hat{\mu}$ of the all samples can be defined as

$$\hat{\mu} = \frac{1}{k}\sum_{i=1}^{k}\hat{\mu}_i = \frac{1}{n}\sum_{i=1}^{k}\sum_{j=1}^{n_i}y_{ij} \tag{5.6}$$

Equations (5.5) and (5.6) determine the estimates of parameters μ_i in Eq. (5.1) or μ in Eq. (5.2). To determine an estimate of effect α_i the following expression is used:

$$\hat{\alpha}_i = \hat{\mu}_i - \hat{\mu} \tag{5.7}$$

To avoid identifiability problems the parameters α_i are constrained by

$$\sum_{i=1}^{k} n_i \alpha_i = 0 \tag{5.8}$$

When the sample size is the same for each treatment (balanced experiment), this condition simplifies to

$$\sum_{i=1}^{k} \alpha_i = 0 \tag{5.9}$$

The estimator $\hat{\alpha}_i$ corresponds to this constraint. The procedure adopted now depends on whether fixed-effects or random-effects ANOVA is considered.

5.2.1 Fixed-efects One-way ANOVA

5.2.1.1 Assumptions

In a *fixed-effects ANOVA model*, all the factor levels being considered are fixed, i.e. the levels of each factor are the only levels of interest [6]. The "effects" referred to in such a model represent measures of the influence (i.e., the effect) that different levels of the factor have on the observed variable. Such measures are often expressed in the form of differences between a at given level and an overall mean. The ith level corresponds to the ith population from which the sample of size n_i can be selected. The effect of the *i*th population is often measured by the amount that the *i*th population mean differs from an overall mean.

The assumptions needed for fixed-effects one-way ANOVA may be stated simply as follows:

(a) Random samples of observations (experimental results, measurements, observations, etc.) are selected from each of k fixed populations or groups. For the *i*th level we have sample $y_{ij}, j = 1, ..., n_i$.

(b) The model of ANOVA defined by Eq. (5.1) is valid.

(c) The observations are normally distributed with constant variance σ^2 in the whole population $y_{ij} = N(\mu_i, \sigma^2)$.

(d) Random errors ε_{ij} are mutually independent random variables normally distributed with a mean equal to zero and variance σ^2, $\varepsilon_{ij} = N(0, \sigma^2)$

In general, classical ANOVA analysis can be applied if none of the assumptions is very badly violated. This is true for more complex ANOVA situations as well as for fixed-effects one-way ANOVA. The term generally used to refer to this property of broad applicability is called *robustness*. We say that a procedure is robust with respect to moderate departures from the basic assumptions.

We must nevertheless be careful to avoid using robustness as an automatic justification for blindly applying the ANOVA model. Certain facts should be borne in mind when the use of ANOVA in a given situation is considered. For example, the normality assumption does not have to be exactly satisfied as long as we are dealing with relatively large samples (e.g., 20 or more observations from each population), although the consequences of large deviations from normality are more severe for random effects than for fixed effects. The assumption of variance homogeneity can also be mildly violated without serious risk, provided that the numbers of observations selected from each population are more or less the same, although, again, the consequences are more severe for random effects.

Violation of the assumption of independence of the observations, however, can lead to very serious errors in both the fixed- and random-effects cases. In general, great care should be taken to ensure that the observations are independent. This concern arises primarily in studies where repeated observations are recorded on the same experimental subjects, since very often the level of response of a subject on one occasion has a decided effect on subsequent responses.

What, then, should be done when one or more of these assumptions are in serious question? One possibility is for the data to be transformed (e.g., by means of a log, square root, or other transformation) so that they more closely satisfy the assumptions.

5.2.1.2 Methodology

The null hypothesis that there is no treatment effect, i.e., the hypothesis of equal population means $H_0 : \mu_1 = \mu_2 = ... = \mu_k = \mu$ is usually tested first.

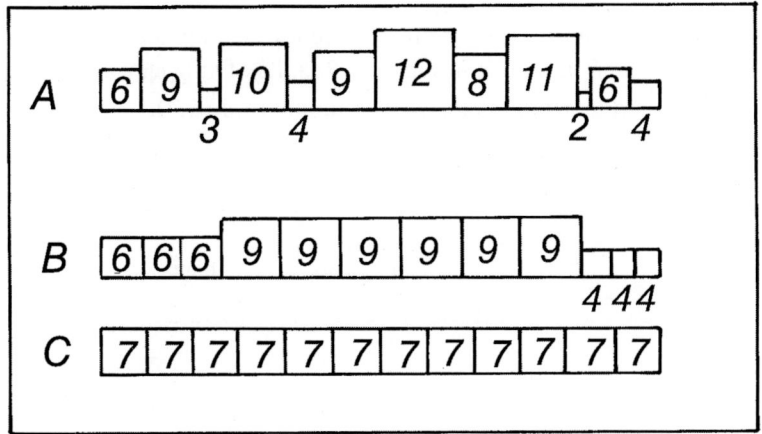

Figure 5.1 Partitioning the sum of squared deviations into members A, B and C, where A are data, B are treatment means and C is the overall mean: $S_C = A - C$, $S_A = B - C$, $S_R = A - B$.

We begin the analysis by partitioning the sum of squared deviations from the overall $\hat{\mu}$ defined by

$$S_C = \sum_{i=1}^{k} \sum_{j=1}^{n_i} (y_{ij} - \hat{\mu})^2 \tag{5.10}$$

into two components, one attributable to the identifiable source of variation (factor A), and denoted by S_A; the other representing the variation due to uncontrolled factors and random errors associated with the response measurement, and denoted by S_R. Rearrangement of Eq. (5.10) leads to

$$S_C = \sum_{i=1}^{k} \sum_{j=1}^{n_i} \left[(y_{ij} - \hat{\mu}_i) + (\hat{\mu}_i - \hat{\mu}) \right]^2 = S_A + S_R \tag{5.11}$$

where
$$S_A = \sum_{i=1}^{k} n_i (\hat{\mu}_i - \hat{\mu})^2 \qquad (5.12)$$

represents the variability *between* individual treatments of a given factor A, and

$$S_R = \sum_{i=1}^{k} \sum_{j=1}^{n_i} (y_{ij} - \hat{\mu}_i)^2 \qquad (5.13)$$

represents the variability *within* all treatments. A graphical interpretation of the partitioning of S_c is shown in Fig. 5.1. Results of an ANOVA procedure are usually presented in a so-called ANOVA table (Table 5.2).

Table 5.2 ANOVA table for one-way fixed-effects model

Source of variation	Degrees of freedom	Mean square	Expected mean
Between treatments S_A	$k-1$	$S_A / (k-1)$	$\sigma^2 + \dfrac{\sum_{i=1}^{k} n_i \alpha_i^2}{k-1}$
Residual (within treatments) S_R	$n-k$	$S_R / (n-k)$	σ^2
Totals	$n-1$	—	—

The last column in Table 5.2 shows the expected mean square. An unbiased estimate of the variance of errors σ^2 is the mean square of residuals defined by

$$\hat{\sigma}^2 = S_R / (n-k) \qquad (5.14)$$

The null hypothesis that the treatments are equal, i.e. insignificance of effect; $H_0 : \alpha_i = 0, i = 1,...,k,$ and the alternativě hypothesis is $H_A : \alpha_i \neq 0, i = 1,...,k,$. The test is based on the fact that S_A / σ^2 has the χ^2-distribution with $(k-1)$ degrees of freedom, and quantity S_R / σ^2 has an independent χ^2-distribution with $(n-k)$ degrees of freedom. Their ratio has the Fisher-Snedecor F-distribution with $(k-1)$ and $(n-k)$ degrees of freedom. The test statistic is calculated by

$$F_e = \frac{S_A(n-k)}{S_R(k-1)} \tag{5.15}$$

When the null hypothesis H_0 is valid, the statistic F_e has the F-distribution with $(k-1)$ and $(n-k)$ degrees of freedom. When F_e is greater than the quantile $F_{1-\alpha}(k-1, n-k)$, the null hypothesis is rejected, and the effect of factor A is taken as significant. If $F_e \leq F_{1-\alpha}(k-1, n-k)$, the effect of factor A should be considered to be insignificant. Then the total variance σ^2 is related only to the uncontrolled (random) factor and may serve as an estimate of the replication variance.

In interpreting the results of an ANOVA, it is important to bear in mind that a very low value of the variance ratio S_A/S_R may be related to the fact that some important uncontrolled factor was not randomized in the course of the experiment. This may lead to an increased variance within the treatments while leaving the variance between the treatments unchanged, resulting in a reduced variance ratio. In these circumstances, the experimental results will not obey the model defined by Eq. (5.1). If, on the other hand, the inequality $F_e > F_{1-\alpha}(k-1, n-k)$ holds, the difference between the two variances is significant, and so is the effect of factor A.

Problem 5.1 *Test of quality of silver nitrate made by various companies*

The bottles containing silver nitrate were manufactured by five different companies. Random samples of silver nitrate taken from the bottles were used for determination of chlorine in organic samples. From each of the five bottles different numbers of random samples are taken to prepare stock solutions of $AgNO_3$: $n_1 = n_3 = 6$; $n_2 = n_5 = 3$, $n_4 = 4$. Test whether the quality of $AgNO_3$ coming from various chemical companies differs.

Data: The percentages of chlorine in a single organic compound determined by using five stock solutions of silver nitrate are listed in Table 5.3.

Table 5.3 Determination of Cl by various stock solutions of $AgNO_3$

Replicate	Source of $AgNO_3$				
	V1	*V2*	*V3*	*V4*	*V5*
1	4.40	4.90	5.55	4.45	5.15
2	4.40	4.95	5.10	5.45	6.25
3	5.20	5.40	5.50	4.65	6.14
4	5.45	–	5.98	4.40	–
5	5.80	–	5.60	–	–
6	5.60	–	5.56	–	–

Program: ADSTAT or QC-Expert: ANOVA-1.

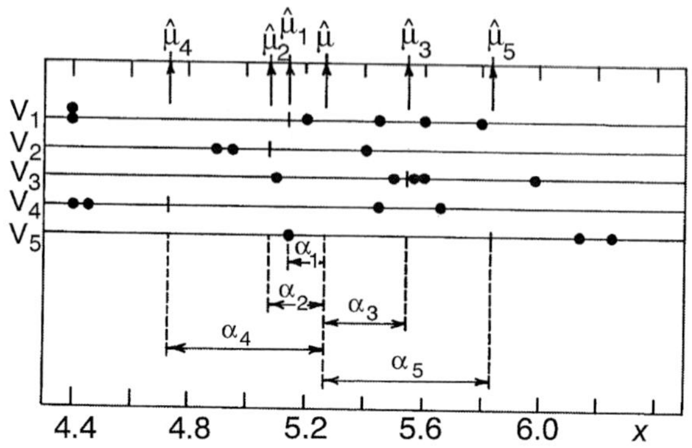

Figure 5.2 Determination of Cl by various stock solutions of AgNO$_3$

Solution: The estimates of the individual means $\hat{\mu}_i$, the overall mean $\hat{\mu}$ and effects $\hat{\alpha}_i$, are:

$\hat{\mu} = 5.2715$; $\hat{\mu}_1 = 5.1417$; $\hat{\mu}_2 = 5.083$; $\hat{\mu}_3 = 5.548$; $\hat{\mu}_4 = 4.738$; $\hat{\mu}_5 = 5.8467$;

$\hat{\alpha}_1 = -0.1298$; $\hat{\alpha}_2 = -0.1885$; $\hat{\alpha}_3 = 0.276$; $\hat{\alpha}_4 = -0.534$; $\hat{\alpha}_5 = 0.575$.

The sums of squared deviations and variance components are summarized in Table 5.4.

Table 5.4 One-way ANOVA table for quality of silver nitrate

Source of variation	Degrees of freedom	Mean square	Fe
Between companies $S_A = 2.7999$	4	0.6999	3.1
Residual $S_R = 3.8322$	17	0.2254	–
Totals $S_c = 6.632$	21	–	–

For significance level $\alpha = 0.05$ the quantile $F_{0.95}(4, 17) = 2.96$.

Conclusion: Since the experimental value F_e is greater than quantile $F_{0.95}$ (4, 17) = 2.96, the null hypothesis $H_0 : \alpha_i = 0, i = 1,...,5$, is rejected, and the quality of AgNO$_3$ coming from different chemical manufacturers significantly differs.

5.2.1.3 Multiple-comparison Procedure

Whenever an ANOVA F-test for simultaneous comparison of several population means is found to be statistically significant, it is of interest to determine which *specific* differences there are among the population means. For example, if four means are being compared (fixed-effects case) and the null hypothesis $H_0 : \mu_1 = \mu_2 = \mu_3 = \mu_4$ is rejected, it is usually desirable to determine which subgroups of means are different by considering some more specific hypothesis such as $H_{01} : \mu_1 = \mu_2, H_{02} : \mu_2 = \mu_3$ or even $H_{04} : (\mu_1 + \mu_2)/2 = (\mu_3 + \mu_3)/2$, which compares the average effect of populations 1 and 2 with the average effect of populations 3 and 4. Such specific comparisons may have been of interest to the investigator *before* the data were collected, or may arise in completely exploratory studies only *after* the data have been examined. In either event, a seemingly reasonable first approach to making inferences about differences among the population means would be to make several t-tests, and to focus on all the tests found to be significant. The justification for all the $Z = {}_k C_2 = k(k-1)/2$ tests being applied simultaneously comes from the Bonferroni inequality. Special quantiles of the t-distribution for level $\alpha/2\,Z$ must then be used. Testing of differences between population means in ANOVA is called the multiple comparisons technique. In Scheffe's multiple comparison procedure the null hypothesis $H_0 : \mu_i = \mu_j$ is rejected for all pairs of (i,j) populations for which

$$\left| \hat{\mu}_i - \hat{\mu}_j \right| \geq \sqrt{(k-1)\sigma^2 \; F_{1-\alpha}(k-1, n-k) \left[1/n_i + 1/n_j \right]} \qquad (5.16)$$

where n is the total number of observations, k is the number of means considered, n_i and n_j are the sizes of the samples selected from the ith and jth populations (treatments), respectively, $\hat{\sigma}^2$ is the estimate of variance calculated by Eq. (5.14). Equation (5.16) is used for all pairs of indices (i,j). In some cases, only selected linear constraints can be proved.

In general, a *linear contrast* is defined to be any linear function of the population means, say

$$L = \sum_{i=1}^{k} c_i \mu_i \qquad (5.17)$$

with known coefficients c_i such that

$$\sum_{i=1}^{k} c_i = 0 \quad \text{and} \quad \sum_{i=1}^{k} c_i^2 > 0 \qquad (5.18a,b)$$

The contrast estimate is defined by

$$\hat{L} = \sum_{i=1}^{k} c_i \hat{\mu}_i \qquad (5.19)$$

When all the observations come from a normal distribution $N(\mu_i, \sigma^2)$, the associated null hypothesis $H_0: L = 0$ may be tested against the alternative H_A: $L \neq 0$ by using the test criterion

$$F_L = \frac{\hat{L}^2}{\left[\hat{\sigma}^2 \sum_{i=1}^{k} c_i^2 / n_i \right]} \qquad (5.20)$$

If the null hypothesis H_0 is valid, the test criterion F_L has a Fisher-Snedecor distribution with 1 and $(n - k)$ degrees of freedom. The null hypothesis is rejected when F_L reaches a value higher than quantile $F_{1-\alpha}(1, n - k)$.

Problem 5.2   *Differences in quality of AgNO$_3$ from two suppliers*

For the data from Problem 5.1 test whether the difference between μ_4 and μ_2 is statistically significant. Is the silver nitrate from supplier No. 4 of better quality than that from supplier No. 2?

Data: Problem 5.1

Program: ADSTAT or QC-Expert: ANOVA-1.

Solution: To test for a difference between μ_4 and μ_2, the linear contrast with coefficients $c_1 = c_3 = c_5 = 0$; $c_2 = 1$, $c_4 = -1$ may be calculated. The estimate of contrast $\hat{L} = \hat{\mu}_2 - \hat{\mu}_4 = 0.345$. From Eq. (5.20), $F_L = 0.119/(0.2254(1/4 + 1/3)) = 0.905$. Because the quantile $F_{0.95}(1, 17) = 4.451$ is greater than F_L, the difference between μ_4 and μ_2 is not statistically significant.

Conclusion: The difference between the quality of silver nitrate from suppliers 4 and 2 is not statistically significant.

5.2.1.4 Regression Model

The procedure of analysis of variance is applicable only when the observations are independent, the errors ε_{ij} have the normal distribution $N(0, \sigma^2)$ with constant variance σ^2. Before use of the ANOVA procedure, all the assumptions should be examined. For this, it is advantageous to convert the ANOVA model into a linear regression model and apply regression diagnostics (from Chapter 6).

Most ANOVA procedures can also be considered in a regression analysis setting; this can be done by defining appropriate dummy variables in a regression model [9]. The ANOVA model, Eq. (5.1), may be expressed as the linear regression model

$$y_{ij} = \mu_1 w_1 + \mu_2 w_2 + ... + \mu_k w_k + \varepsilon_{ij} \tag{5.21}$$

where the w_i are dummy variables which take the following values:

$$w_i = \begin{cases} 1 & \text{for effect } i \\ 0 & \text{otherwise} \end{cases}$$

The means $\mu_1, \mu_2, ..., \mu_k$, are understood as the regression parameters. If all the assumptions about errors are valid, the parameters estimates $\hat{\mu}_i$ can be calculated by the least-squares method—i.e. by minimizing

$$U(\mu) = \sum_{i=1}^{k} \sum_{j=1}^{n_i} (y_{ij} - \sum_{i=1}^{k} w_i \mu_i)^2 \tag{5.22}$$

Analytical minimizing of $U(\mu)$ leads to the system of equations:

$$\frac{\partial U(\mu)}{\partial \mu_i} = 0, \ i = 1, ..., k \tag{5.23}$$

Since $w_j = 1$ only for $i = j$ the solution of Eq. (5.23) has the simple form

$$\hat{\mu}_i = \frac{\sum_{j=1}^{n_i} y_{ij}}{n_i} \tag{5.24}$$

Chapter 6 explains the important role played by the diagonal elements H_{ii} of the projection matrix H in analysis of residuals and leverage points,

$$\mathbf{H} = \begin{bmatrix} (1/n_1)\mathbf{1}_1\mathbf{1}_1^T & 0 & 0 \\ 0 & (1/n_2)\mathbf{1}_2\mathbf{1}_2^T & 0 \\ 0 & 0 & (1/n_k)\mathbf{1}_k\mathbf{1}_k^T \end{bmatrix} \tag{5.25}$$

where I_i is a column unit vector of size $(n_i \times 1)$. Matrix H consists of blocks of size $(n_i \times n_i)$ with values $1/n_i$.

For the same number of observations (replicates) of each treatment (balanced experiments) all the diagonal elements H_{ii} have the same magnitude. It means that estimates $\hat{\mu}_i$ have constant variance.

For different numbers of observation of each treatment, the variances

$$D(\hat{\mu}_i) = \sigma^2 / n_i \qquad (5.26)$$

are not constant. Similarly, the variances of the residuals

$$D(\hat{e}_{ij}) = \sigma^2 (1 - 1/n_i) \qquad (5.27)$$

also are no constant. Residuals $\hat{e}_{ij}$ in ANOVA models are expressed as

$$\hat{e}_{ij} = y_{ij} - \hat{\mu}_i \qquad (5.28)$$

For very different numbers of replicates of the treatments (unbalanced experiments) the residuals will have nonconstant variance even in cases when the errors have constant variance.

From the theory of regression models (Chapter 6) it is evident that for extreme points, the diagonal elements of the projection matrix become larger than $2k/n$. This means that for small sample sizes there is a danger that levels with a particularly small number of observations will have a strong influence on the results of the statistical analysis.

Problem 5.3 *Investigation of the influence of individual suppliers of silver nitrate on ANOVA result*

Examine the influence of the individual suppliers of silver nitrate in Problem 5.1 especially for small sample sizes, and test whether any supplier can be taken as an extreme.

Data: As for Problem 5.1

Solution: For samples from the first and third supplier the diagonal elements of the projection matrix $1/n_i = 1/6 = 0.16$, from the second and the fifth, $1/n_i = 1/3 = 0.33$ and from the fourth $1/n_i = 1/4 = 0.25$. The critical value is $2 \times 5/22 = 0.4545$.

Conclusion: Since all diagonal elements of projection matrix H_{ij}, $i = 1, \ldots,$ 5 have values under the critical limit, all samples can be considered as not to be leverages.

5.2.1.5 Checking for Data Normality

To check the data normality, the rankit plot (Chapter 2) may be used. Examination of standardized residuals is also helpful (see Chapter 6)

$$\hat{e}_{Si} = \frac{\hat{e}_{ij}}{\hat{\sigma}\sqrt{1-1/n_i}} \qquad (5.29)$$

where $\hat{e}_{ij}$ are residuals, $\hat{\sigma}$ is the estimate of the standard deviation and n_i is the number of observations for a treatment. The standardized residuals, in a classical analysis of variance, exhibit approximately a normal distribution with zero mean and unit variance $\hat{e}_{Si} \approx N(0,\sigma^2)$. If the errors are normally distributed, $\varepsilon_{ij} \approx N(0,\sigma^2)$, the rankit plot of the standardized residuals is linear, with zero intercept and unit slope.

Problem 5.4   *Check of data normality*

Check whether the data from Problem 5.1 have a normal distribution, with the use of a rankit plot G12 for standardized residuals $\hat{e}_{Si}$.

Data: Problem 5.1

Program: ADSTAT or QC-Expert: ANOVA-1.

Solution: The rankit plot is shown in Fig. 5.3

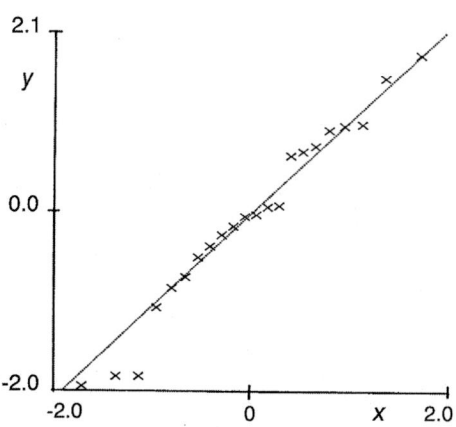

Figure 5.3   The rankit plot G12 for standardized residuals from Problem 5.4.

Conclusion: The rankit plot proves that the data have an approximately normal distribution.

When data do not belong to the normal distribution, some transformation (logarithm, square-root, or other functions) can often be applied. In many practical cases, data are skewed to higher values. Then the logarithmic transformation

$$y^* = \ln(2y + C) \tag{5.30}$$

is suitable. The value C is selected such that

(1) the distribution of residuals is symmetrical with kurtosis near to 3,

(2) a rankit plot of standardized residuals is linear.

In ANOVA models the outliers play an important role (as in regression). Outliers may be detected by Jack-knife residuals $\hat{e}_{J,ij}$ which are defined by

$$\hat{e}_{Jij} = \hat{e}_{Sij} \sqrt{\frac{n-k-1}{n-k-\hat{e}_{Sij}^2}} \tag{5.31}$$

For normally distributed data, these residuals have approximately the Student distribution with $(n-k-1)$ degrees of freedom. Roughly, if $\hat{e}_{J,ij}^2 > 10$, the given value y_{ij} is taken as an outlier. Other diagnostics for finding outliers, described in Chapter 6, may be applied also.

Problem   *5.5 Detection of outliers in data*

Test whether the value 5.15 in the fist sample from the fifth supplier in Problem 5.1 is an outlier.

Data: As for Problem 5.1

Solution: The residual is $\hat{e}_{51} = y_{51} - \hat{\mu}_5 = 5.15 - 5.8467 = -0.6967$ and corresponding standardized residual is $\hat{e}_{S51} = -0.6967 / \sqrt{0.2254(1-1/3)} = -1.7973$. The Jack-knife residual is

$$\hat{e}_{J,ij} = -1.7973 / \sqrt{(22-6)/(22-5-1.797^2)} = -1.932 \quad \text{and} \quad \hat{e}_{J,ij}^2 = 3.74$$

Conclusion: Because $\hat{e}_{J,ij}^2 = 3.74$ is smaller than 10, the value y_{51} is not an outlier.

5.2.1.6 Checking for homoscedasticity

The assumption of homoscedasticity (i.e. constant variance) may be tested by use of the same diagnostics as for the linear regression model. For non-constant numbers of observations on treatments, the heteroscedasticity of classical residuals Eq. (5.27) should be considered. For a sufficient number of observations on a treatment, in addition to the mean $\hat{\mu}_i$ a treatment variance s_i^2 can also be estimated. A test of homoscedasticity may be carried out on the basis of a plot of s_i vs. $\hat{\mu}_i$. If a random pattern of points results, the homoscedasticity in different treatments is accepted. If s_i is related to $\hat{\mu}_i$ by a monotonic function $s = f_2(\hat{\mu}_i)$, the data might be transformed to stabilize the variance. A suitable transformation can be determined from

$$g(y) = \int_y \frac{d\mu}{f_2(\mu)} \tag{5.32}$$

Problem 5.6   *Determination of the transformation for stabilizing a variance*

It was found that a plot of s_i vs. $\hat{\mu}_i$ is linear. Determine a convenient transformation for stabilizing the variance.

Solution: By using Eq. (5.32) for $f_2(\mu) \approx a\,\mu$ the differential equation will be $g(y) = \frac{1}{a}\int_y \frac{d\mu}{\mu}$ with solution $g(y) = (\ln y)/a$. Thus, the optimum transformation will be a logarithmic one, $y_{ij}^* = \ln(y_{ij})$.

Conclusion: When the dependence of s_i vs. $\hat{\mu}_i$ is monotonic, a convenient transformation can easily be found. Such a transformation improves the normality of the data too.

5.2.2 The Random-effects Model for One-way ANOVA

A *random-effect factor* is a factor which has levels that may be regarded as a sample from some large population of levels, whereas a fixed-effect factor is one which has levels that are the only levels of interest. The distinction is important in any ANOVA, since different tests of significance are required for different configurations of random- and fixed-effect factors. For now, it is perhaps useful to give some examples of random and fixed effects factors:

(a) "Subjects" is usually considered to be a random-effects factor, since we ordinarily wish to infer from the subjects used, to a population of potential subjects.

(b) "Observers" is a random-effects factor often considered in the examination of the effect of different observers on the response variable of interest.

(c) "Days", "weeks", and so on are usually considered as random factors in investigation of the effect of time on a response variable observed for different time periods. We usually use many levels for such temporal factors to represent a large number of time periods.

(d) "Sex" is always a fixed-effects factor, since its two levels include all possible levels of interest.

(e) "Location" (e.g., cities, plants, laboratories) may be fixed- or random-effects factors, depending on whether only specific sites are of interest or whether a larger geographical universe is to be considered.

(f) "Treatments", "drugs", "chemicals", "tests" and so on, are usually considered as fixed factors, but they may be considered random if their levels are representative of a much large group of possible levels.

In the random-effects model, the effect α_i is considered as a random variable. For the random-effects model, it is assumed that

(a) α_i are mutually independent random variables with normal distribution $N(0, \sigma_A^2)$;

(b) ε_{ij} are mutually independent random variables having normal distribution $N(0, \sigma_\varepsilon^2)$;

(c) random variables α_i are independent of random variables ε_{ij}. The purpose of ANOVA is to test the differences of theoretical effects $\alpha_i - \alpha_j$ or the variances σ_A^2 and σ_ε^2. In some cases, the overall mean μ or means $\mu_i = \mu + \alpha_i$ for individual treatments are estimated.

For estimates of variances α_A^2 and σ_ε^2 a residual sum of squares S_R and a sum of squares S_A explained by effects are used. Both quantities are calculated as for the fixed-effects model (Table 5.2).

The variances σ_ε^2 and σ_A^2 are calculated from

$$\hat{\sigma}_\varepsilon^2 = \frac{S_R}{n-k} \tag{5.33}$$

$$\hat{\sigma}_A^2 = \frac{n(k-1)\left[S_A/(k-1) - S_R/(n-k)\right]}{n^2 - \sum\limits_{i=1}^{k} n_i^2} \tag{5.34}$$

The estimate σ_ε^2 must be non-negative, i.e. $\hat{\sigma}_A^2 = \max(0, \hat{\sigma}_A^2)$. Estimates $\hat{\mu}$, $\hat{\sigma}_\varepsilon^2$ and $\hat{\sigma}_A^2$ have some useful statistical properties. If the assumptions about errors ε_{ij} and effects α_i are valid, they have from all possible unbiased estimates the minimum variance. The variances $\hat{\sigma}_\varepsilon^2$ and $\sum \hat{\sigma}_A^2$ may be estimated with the use of a maximum likelihood or some other method [5].

The requirement of a zero mean for all α_i is similar in philosophy to the requirement that $\sum_{i=1}^{k} \alpha_i = 0$ for the fixed-effects model. When the ANOVA random model applies, we assume that the average effect of treatments α_i is 0 over the entire population of all treatments α. That is, we assume that $\mu_i = 0$. Because we have required the treatment effects to average out to 0 over the entire population of possible effects, there is only one way to assess whether there are any significant treatment effects at all, and this involves consideration of σ_A^2. If there is no variability (i.e., $\sigma_A^2 = 0$), all treatment effects must be 0. If there is variability (i.e., $\sigma_A^2 > 0$), there are some non-zero effects in the population of treatment effects. Thus, our null hypothesis of no treatment effects should be stated as $H_0 : \sigma_A^2 = 0$. This hypothesis is therefore analogous to the null hypothesis used in the fixed-effects case, although it happens to be stated in terms of a population variance rather than in terms of population means. The F-test criterion is stated for the random-effects model in exactly the same way as that used for the fixed-effects model,

$$F_e = \frac{S_A}{S_R} \frac{(n-k)}{(k-1)} \tag{5.35}$$

If the null hypothesis is valid, this F_e statistic has the Fisher-Snedecor distribution with $(k-1)$ and $(n-k)$ degrees of freedom.

When the means μ_i are estimated, the procedure is the same as in the fixed-effects model [5]. For an estimate of an overall mean μ in balanced experiments, the arithmetic mean $\hat{\mu}$ is used. The variance of $\hat{\mu}$ can be estimated from the equation

$$\hat{\sigma}_{\hat{\mu}}^2 = \frac{S_A}{n^*(k-1)k} \tag{5.36}$$

where $n*$ is number of observations, which should be the same for all treatments. For a significance test of the overall mean μ, the test statistic $t = \hat{\mu} / \hat{\sigma}_{\hat{\mu}}$ may be used. This statistic has, for a valid null hypothesis $H_0 : \mu = 0$, the Student distribution with $(k - 1)$ degrees of freedom.

For unbalanced experiments, in addition to the arithmetic mean $\hat{\mu}_N = \frac{1}{k} \sum_{i=1}^{k} \hat{\mu}_i$ the weighted arithmetic mean

$$\hat{\mu}_w = \frac{1}{n} \sum_{i=1}^{k} n_i \, \hat{\mu}_i \tag{5.37}$$

is also computed. The corresponding variances of $\hat{\mu}_N$ and $\hat{\mu}_w$ are

$$D(\hat{\mu}_w) = \frac{1}{n^2} \sum_{i=1}^{k} n_i^2 \left[\frac{\sigma_\varepsilon^2}{n_i} + \sigma_A^2 \right] \tag{5.38}$$

$$\text{and } D(\hat{\mu}_N) = \frac{1}{k^2} \sum_{i=1}^{k} \left[\frac{\sigma_\varepsilon^2}{n_i} + \sigma_A^2 \right] \tag{5.39}$$

In the numerical calculation of σ_A^2 and σ_ε^2 their estimates are substituted, and from both of these, the mean value taken is the one for which the variance has the lower value.

Problem 5.7   *Evaluation of quality of AgNO₃*

Use the data of Problem 5.1, but suppose that instead of 5 bottles of silver nitrate from known suppliers, five bottles randomly selected from stores were used. Test the quality of $AgNO_3$ available in the stores.

Data: As for Problem 5.1

Program: ADSTAT or QC-Expert: ANOVA-1.

Solution: The residual variance $\sigma_\varepsilon^2 = 0.2254$ from Problem 5.1 is substituted into Eq. (5.34)

$$\sigma_A^2 = \frac{22 \times 4(0.6999 - 0.2254)}{22^2 - 106} = 0.1104$$

Since the test statistic $F_e = 3.10$ computed from Eq. (5.35) is higher than the quantile of the Fisher-Snedecor distribution at a significance level $\alpha = 0.05$, $F_{0.95}(4, 17) = 2.9$, the null hypothesis $H_0 : \sigma_A^2 = 0$ is rejected.

Conclusion: The variability of the quality of silver nitrate in bottles in the stores is significant. The fixed-effects model thus gives a different answer from the random-effects model in interpretation of results.

In random-effects models, the assumption of normality may be violated for variables ε_{ij} and also for α_i. Normality may be checked by the rankit plots for residuals. Rankit plots can also be drawn for means $\hat{\mu}_i$ or $\hat{\alpha}_i$ although the results are not absolutely correct [5].

Like the fixed-effects models, the random-effects models can have the normality of data improved by use of a suitable transformation, but variance estimation can be a problem in the transformed scale. When the assumption of normality is violated, the Jack-knife technique (Chapter 3) can be used for estimation of the variance σ_A^2 and testing its significance. For detection of outliers and heteroscedasticity, the same technique as for the fixed-effects model is adopted.

5.3 Two-way ANOVA

In the previous section we explained the simplest kind of ANOVA problem, that involving a single factor. We now focus on the two-factor case, which is generally referred to as two-way ANOVA. This extension is by no means trivial. We shall describe how the two-factor situation may be classified according to the pattern of the data.

5.3.1 Two-way Data Patterns

Several different types of data patterns for two-way ANOVA are illustrated in Table 5.5. Each of these tables describes a two-factor study with three levels of factor B (the "column" factor) and four levels of factor A (the "row" factor). The combination of level A_i and B_j is called a *cell*. The *y*s in each table correspond to individual observations on a single dependent variable *y*. The number of *y*'s in a given cell is denoted by n_{ij} for the *i*th level of factor A and the *j*th level of factor B. The marginal total for the *i*th row is denoted by n_i and for the *j*th column by n_j. The total number of observations is denoted by n.

Table 5.5 Some two-way data patterns for a 4 × 3 table

(a) Single observation per cell ($n_{ij} = 1$), balanced case:

	Factor B		
Factor A	y y y y	y y y y	y y y y

(b) Equal number of replicates per cell ($n_{ij} = 4$), balanced case:

	Factor B		
Factor A	yyyy yyyy yyyy yyyy	yyyy yyyy yyyy yyyy	yyyy yyyy yyyy yyyy

(c) Equal replications by column, proportional replications by row ($n_{ij} = n_{.j}/4$) unbalanced case:

	Factor B			
Factor A	yyyy	yy	yyy	$n_{1.} = 9$
	yyyy	yy	yyy	$n_{2.} = 9$
	yyyy	yy	yyy	$n_{3.} = 9$
	yyyy	yy	yyy	$n_{4.} = 9$
	$n_{.1} = 16$	$n_{.2} = 8$	$n_{.3} = 12$	

(d) Proportional row and column replications ($n_{ij} = n_{i.} \, n_{.j}/n$), unbalanced case:

	Factor B			
Factor A	yyyy	yy	yyy	$n_{1.} = 9$
	yyyyyyyy	yyyy	yyyyyy	$n_{2.} = 18$
	yyyyyyyyyyyy	yyyyyy	yyyyyyyyy	$n_{3.} = 27$
	yyyyyyyy	yyyy	yyyyyy	$n_{4.} = 18$
	$n_{.1} = 32$	$n_{.2} = 16$	$n_{.3} = 24$	$n = 72$

(e) Nonsystematic replications, unbalanced case:

	Factor B			
	yy	yyy	$yyyyy$	$n_{1.} = 11$
	yyy	$yyyy$	yy	$n_{2.} = 9$
Factor A	y	yyy	$yyyy$	$n_{3.} = 8$
	$yyyyy$	yy	y	$n_{4.} = 8$
	$n_1 = 11$	$n_2 = 12$	$n_3 = 13$	$n = 36$

The simplest two-factor pattern (Table 5.5a), arises when there is a single observation in each cell (i.e., $n_{ij} = 1$ for all i and j).

A second type of pattern (Table 5.5b) occurs when there are equal numbers of observations in each cell. Here, $n_{ij} = 4$ for all i and j. The common property of the last three patterns is that all cells do not have the same number of observations. Unequal cell replications often arise in observational studies in which the levels of certain effects are determined after, rather than before, the data are collected.

For the pattern in Table 5.5c, cells in the same column have the same number of observations, whereas cells in the same row are in the ratio 4 : 2 : 3. For this table each of the four cell frequencies in the jth column is equal to the same fraction of the corresponding total column frequency (i.e., $n_{ij} = n_j/4$ in this case). Note, for example, that $n_1/4 = 16/4$, which is the number of observations in any cell in column 1.

For Table 5.5d the cells in a given column are in the ratio 1 : 2 : 3 : 2, whereas the cells in a given row are in the ratio 4 : 2 : 3. This pattern results because n_{ij} is determined as $n_{ij} = n_{i.} \, n_{.j} / n$, which means that any cell frequency can be obtained by multiplying the corresponding row and column marginal frequencies together and then dividing by the total number of observations. Thus, for cell (1,2) in Table 5.5d, we have $n_{1.} \, n_2 / n = 9(16) / 72 = 2$, which equals n_{12}. Similarly, for cell (4, 3), $n_{4.} \, n_3 / n = 18(24) / 72 = 6$, which equals n_{43}.

There is no mathematical rule for describing the pattern of cell frequencies in Table 5.5e, and so we say that such a pattern is nonsystematic.

5.3.2 Formulation of Various Two-way ANOVA Models

We shall consider the case in which we must set up an experiment to study the effects of two factors A and B on a response variable y. Factor A has N

levels $\alpha_1, \alpha_2, ..., \alpha_N$ whereas factor B has M levels $\beta_1, \beta_2, ..., \beta_M$. For each combination of levels $(\alpha_i \beta_j)$, we measure the response y_{ij} by carrying out n_{ij} observations; The total number of observations is $n = \sum_{i=1}^{N} \sum_{j=1}^{M} n_{ij}$. The model of the response to each treatment may be written

$$y_{ijk} = \mu + \alpha_i + \beta_j + \tau_{ij} + \varepsilon_{ijkq} \tag{5.40}$$

where μ represents an overall mean or a common effect, α_i represents the row effect on the ith level of factor A ($i = 1, 2, ..., N$), β_j represents the column effects on the jth level of factor B ($j = 1, 2,..., M$) and τ_{ij} represents the effect due to the interaction of two factors, A and B.

The interaction term τ_{ij} is the deviation of the mean of observations in the (ij)th set from the sum of the first three terms in the model defined by Eq. (5.40), and $\varepsilon_{ijk}, (k = 1, 2, ..., n_{ij})$ represents the error term.

The simplest model of interaction of rows with columns is the *Tukey model of interaction* defined by

$$\tau_{ij} = C \, \alpha_i \, \beta_j \tag{5.41}$$

where C is a constant. More complicated models of interaction are the *row-linear interaction model* expressed by

$$\tau_{ij} = C_R \, \gamma_i \, \beta_j \tag{5.42}$$

or the *column-linear interaction model* expressed by

$$\tau_{ij} = C_K \, \alpha_i \, \delta_j \tag{5.43}$$

The extended model is the *additive-multiplicative interaction model* expressed by

$$\tau_{ij} = C_W \, \gamma_i \, \delta_j \tag{5.44}$$

These expressions contain, in addition to the column and row constants δ_j and γ_i also the general constants C_R, C_K and C_W.

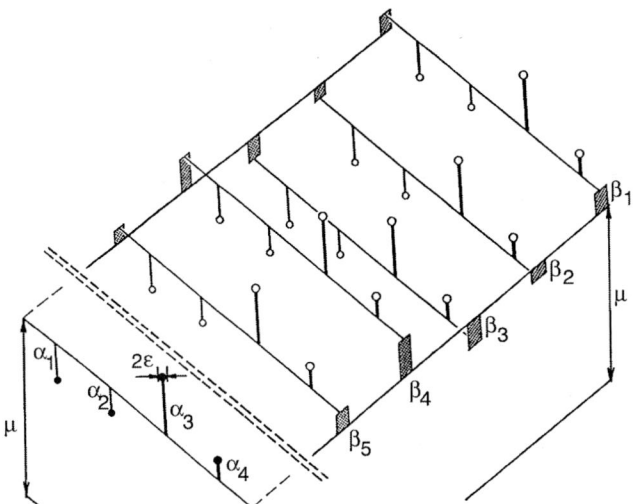

Figure 5.4 Geometric interpretation of two-way ANOVA

Interactions of second and higher orders also exist, and these can express rather complicated structures in data. Analysis of such non-linear interaction models may be found in the literature [10].

We limit ourselves to the simplest (Tukey) model of interaction Eq. (5.41). Since this model contains only one parameter C, it is called the model with *one degree of freedom for non-additivity*. This model expresses approximately the interaction effects in quadratic models for which $\mu_{ij} \approx (\mu + \alpha_i + \beta_j)^2$. After rearrangement of this equation, the interaction term is of the type $2\alpha_i\beta_j$. Use of the Tukey model of interaction is suitable for cases when each cell contains just one observation.

5.3.3 Fixed-effects Two-way ANOVA

This type of two-way screening is the most frequently used. It allows evaluation of the effect of two factors on the results of a chemical analysis. Two-way ANOVA problems can be separated into three groups:

(1) models with a single observation per cell
(2) balanced models
(3) unbalanced models.

For each model, a different computational procedure is required.

5.3.3.1 Models with a single observation per cell

In these models the each cell contains only one observation, and the model is described by Eq. (5.40). The errors ε_{ij} are assumed to be independent identically distributed random variables with zero mean and constant variance. In testing, it is assumed that the error distribution is normal. In the ANOVA model, there are the following constraints

$$\sum_{i=1}^{N} \alpha_i = 0; \quad \sum_{j=1}^{M} \beta_j = 0; \quad \sum_{i=1}^{N} \tau_{ij} = 0; \quad \sum_{j=1}^{M} \tau_{ij} = 0; \qquad (5.45)$$

For pure additive effect of individual factors, $\tau_{ij} = 0$ for all $i = 1, ..., N$ and $j = 1, ..., M$. The estimates of parameters μ, α_i and β_j are in this case calculated from

$$\hat{\mu} = \frac{1}{N\,M} \sum_{i=1}^{N} \sum_{j=1}^{M} y_{ij} \qquad (5.46)$$

$$\hat{\alpha}_i = \frac{1}{M} \sum_{j=1}^{M} y_{ij} - \hat{\mu} \qquad (5.47)$$

$$\hat{\beta}_j = \frac{1}{N} \sum_{i=1}^{N} y_{ij} - \hat{\mu} \qquad (5.48)$$

For residuals

$$\hat{e}_{ij} = y_{ij} - \hat{\mu} - \hat{\alpha}_i - \hat{\beta}_j \qquad (5.49)$$

When Eq. (5.40) is considered as the special linear regression model, the diagonal elements H_{ij} of a projection matrix $\mathbf{H}$ have the same value [7]

$$H_{ii} = \frac{N + M - 1}{N\,M} \qquad (5.50)$$

Off-diagonal elements are not zero, so an outlier in one cell aflfects the estimates of parameters for all cells.

To determine the interaction we use the fact that

$$\tau_{ij} = E(y_{ij}) - \mu - \alpha_i - \beta_j \tag{5.51a}$$

Then, the estimate of interaction is given approximately by

$$\hat{\tau}_{ij} \approx \hat{e}_{ij} \tag{5.51b}$$

by using Eq. (5.51a), the *Tukey model* of interaction [Eq. (5.41)] may be identified. If the plot of $\hat{e}_{ij}$ vs. $\hat{\alpha}_i \hat{\beta}_j$ is linear, the Tukey model of interaction is accepted. The parameter C is calculated from the slope of this straight line

$$C = \left[\sum_{i=1}^{N} \sum_{j=1}^{M} \hat{e}_{ij} \, \hat{\alpha}_i \, \hat{\beta}_j \right] \Big/ \left[\sum_{i=1}^{N} \sum_{j=1}^{M} \hat{\alpha}_i^2 \, \hat{\beta}_j^2 \right] \tag{5.52}$$

The slightly modified plot $\hat{e}_{ij}$ vs. $\hat{\alpha}_i \, \hat{\beta}_j / \hat{\mu}$ is called *the non-additivity graph.* If this plot exhibits a non-random trend, interactions probably exist.

The sums of squares of ANOVA model for a Tukey model of interaction are given in Table 5.6. The quantity S_T is the sum of squared deviations corresponding to the Tukey interaction [3].

$$S_T = \left[\sum_{i=1}^{N} \sum_{j=1}^{M} y_{ij} \, \hat{\alpha}_i \, \hat{\beta}_j \right]^2 \Big/ \left[\sum_{i=1}^{N} \sum_{j=1}^{M} \hat{\alpha}_i^2 \, \hat{\beta}_j^2 \right]. \tag{5.53}$$

and the symbol S_{AB} means a residual sum of squares for the case without interaction

$$S_{AB} = \sum_{i=1}^{N} \sum_{j=1}^{M} (y_{ij} - \hat{\mu} - \hat{\alpha}_i - \hat{\beta}_j)^2 \tag{5.54a}$$

The corresponding mean square M_{AB}

$$M_{AB} = \frac{S_{AB}}{(N-1)(M-1)} \tag{5.54b}$$

represents an unbiased estimate of the variance σ^2.

Table 5.6 Two-way ANOVA model with interaction of Tukey type

Sum of squares for	Degrees of freedom	Mean square	Test criterion F
Factor A, $$S_A = M \sum_{i=1}^{N} \hat{\alpha}_i^2$$	$N-1$	$M_A = S_A/(N-1)$	$F_A = M_A/M_{AB}$
Factor B, $$S_B = N \sum_{j=1}^{M} \hat{\beta}_j^2$$	$M-1$	$M_B = S_B/(M-1)$	$F_B = M_B/M_{AB}$
Interaction (Tukey), S_T $= $ [Eq.(5.53)]	1	$M_T = S_T$	$F_T = M_T/M_E$
Residuals, $$S_R = S_{AB} - S_T$$	$NM - N - M$	$M_E = S_R/(NM - N -M)$	–
Totals, $$S_C = \sum_{i=1}^{N} \sum_{j=1}^{M} (\hat{\mu} - y_{ij})^2$$	$NN - 1$	–	–

Statistical tests based on the Fisher-Snedecor F-criterion may be performed. The null hypothesis H_0: "Tukey interaction is not significant" is tested by the F_T criterion from Table 5.6. If the null hypothesis H_0 is valid, F_T has the Fisher-Snedecor F-distribution with 1 and $(N \times M - N - M)$ degrees of freedom. When this hypothesis is not rejected, a test of the null hypothesis H_0: $\alpha_i = 0$, $i = 1, ..., N$ (the effects of rows or factor A are not significant) using the statistic

F_A, or a test of the null hypothesis H_0: $\beta_j = 0$, $j = 1, ..., $ M (the effects of columns or factor B are not significant) using the statistic F_B may be made. If the null hypothesis H_0 is valid, the F_A statistic has the Fisher-Snedecor F-distribution with $(N-1)$ and $(N-1)(M-1)$ degrees of freedom. F_B has the same distribution with $(M-1)$ and $(N-1)(M-1)$ degrees of freedom. If F_T is higher than corresponding quantile of the Fisher-Snedecor distribution, the effect of Tukey interaction is significant.

Problem 5.8 *Determination of water content in solvents, in various laboratories*

In three samples of solvent A_1, A_2 and A_3, the content of water was determined in four laboratories B_1, B_2, B_3 and B_4. Test whether there are significant differences between the water contents of the three samples and in the results coming from the four laboratories.

Data: N = 3, M = 4

Table 5.7 Water content (%) in solvent found by various laboratories

Sample	Laboratories			
	B_1	B_2	B_3	B_4
A_1	1.35	1.13	1.06	0.98
A_2	1.40	1.23	1.26	1.22
A_3	1.49	1.46	1.40	1.35

Program: ADSTAT or QC-Expert: ANOVA-2P: Two-way, One Observation Per Cell.

Solution: It is found that $\hat{\mu} = 1.277$; $\hat{\alpha}_1 = -0.147$; $\hat{\alpha}_2 = 0$; $\hat{\alpha}_3 = 0.1475$; $\hat{\beta}_1 = 0.1358$; $\hat{\beta}_2 = 0.0042$; $\hat{\beta}_3 = -0.0375$ and $\hat{\beta}_4 = -0.0942$. The slope estimate from Eq. (5.52) is $C = -3.532$. Figure 5.5 shows a non-additivity graph with a slightly significant trend. The sum of square deviations corresponding to Tukey interaction is $ST = 0.0156$. The sum of square is $SAB = 0.02215$ and $MAB = 0.003692$. The results are summarized in the ANOVA Table 5.8.

Table 5.8 ANOVA table of water content in different solvents determined by different laboratories

Sum of squares for	D	Mean square	Test criterion F
Samples A, $S_A =$ 0.174	2	0.087	19.64
Laboratories B, $S_B =$ 0.0862	3	0.0287	* 6.49
Interaction (Tukey), $S_T = 0.0156$	1	0.0156	3.522
Residuals, $S_R =$ 0.0222	5	0.0044	–
Totals, $S_c = 0.2824$	11	0.0257	–

The quantiles of the Fisher-Snedecor distribution are $F_{0.95}$ (1, 5) = 6.61; $F_{0.95}$ (2, 5) = 5.79 and $F_{0.95}$ (3, 5) = 5.41.

Conclusion: The interaction effect is not significant and the additive model of ANOVA can be used: sample effect and laboratory effect are significant. There are non-random difference between laboratory results and samples of solvent.

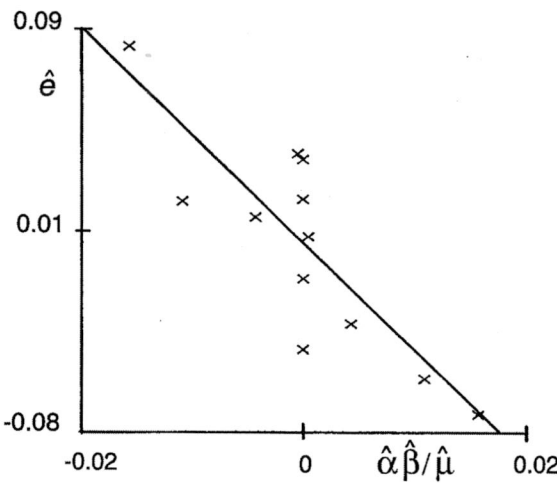

Figure 5.5 Non-additivity graph for water content in different solvents.

In some cases it is convenient to apply the power transformation

$$y_{ij}^* = \begin{cases} \dfrac{(y_{ij} + Q)^\lambda - 1}{\lambda} & \text{for } \lambda \neq 0 \\ \ln(y_{ij} + Q) & \text{for } \lambda = 0 \end{cases} \tag{5.55}$$

where Q is a constant selected to make $(y_{ij} + Q) > 0$ and λ is a parameter of transformation. Details about the power transformation are given in Chapter 2. In ANOVA, the power transformation can be used for eliminating non-additivity. The parameter C is estimated from Eq. (5.52) and it is equal to the slope of the regression straight line divided by $\hat{\mu}$ in a non-additivity graph. The value of λ may then be evaluated from the equation

$$\hat{\lambda} = 1 - \hat{\mu}\, C \tag{5.56}$$

In practice, $\hat{\lambda}$ is rounded to the nearest number from the following series:..., $-1.0; -0.5; -0.3; 0; 0.3; 0.5; 1; 1.5; 2;...$. For the estimate of variance $\hat{\sigma}^2(\hat{\lambda})$ the following expression is used [11]

$$\hat{\sigma}^2(\hat{\lambda}) = \frac{\hat{\mu}^2 \, N \, M \, S_{AB}}{(N-1)(M-1)\sum_{i=1}^{N}\sum_{j=1}^{M}\hat{\alpha}_i^2\hat{\beta}_j^2} \qquad (5.57)$$

For a first guess, transformations in the range $\hat{\lambda} \pm \hat{\sigma}(\hat{\lambda})$ are acceptable. Table 5.6 can be also used for ANOVA without interaction when $S_T = 0$. Since each observation y_{ij} affects parameters in all cells and a projection matrix is not diagonal, the classical analysis of residuals by Eq. (5.49) can lead to wrong conclusions. The non-random trend on a rankit plot indicates an interaction or non-normality. Detection of outliers cannot be performed by residuals $\hat{e}_{ij}$. For identification of outliers, so-called proper tetrads $T_{ij:eg}$ are recommended [8]. These are given by

$$T_{ij:eg} = y_{ij} - y_{ej} - y_{ig} + y_{eg}, \ i \neq e, \ j \neq g \qquad (5.58)$$

Instead of residuals, medians Q_{ij} of all tetrads are calculated which contain observations y_{ij}. Medians Q_{ij} are plotted in rankit graphs.

5.3.3.2 Balanced Models

These models contain $n_{ij} = n^*$ observations in each cell. The ANOVA model is expressed by Eq. (5.3) or (5.4). An estimate of parameter μ_{ij} is represented by the arithmetic mean

$$\hat{\mu}_{ij} = \frac{1}{n^*}\sum_{i=1}^{n} y_{ijk} \qquad (5.59)$$

For estimation of other parameters, the following expressions are used

$$\hat{\mu} = \frac{1}{N \, M}\sum_{i=1}^{N}\sum_{j=1}^{M} \hat{\mu}_{ij} \qquad (5.60)$$

$$\hat{\alpha}_i = \frac{1}{M} \sum_{j=1}^{M} \hat{\mu}_{ij} - \hat{\mu} \tag{5.61}$$

$$\hat{\beta}_j = \frac{1}{N} \sum_{i=1}^{N} \hat{\mu}_{ij} - \hat{\mu} \tag{5.62}$$

Residuals are given by

$$\hat{e}_{ijk} = y_{ijk} - \hat{\mu} - \hat{\alpha}_i - \hat{\beta}_j \tag{5.63}$$

In this case as in Eq. (5.51a), an estimate of interaction is defined

$$\hat{\tau}_{ij} = \hat{\mu}_{ij} - \hat{\mu} - \hat{\alpha}_i - \hat{\beta}_j \tag{5.64}$$

Notice that this expression differs from Eq. (5.51) by the fact that instead of the variable y_{ij} the mean $\hat{\mu}_{ij}$ is used. For a test of non-additivity of Tukey interaction, the plot $\hat{\tau}_{ij}$ vs. $\hat{\alpha}_i \hat{\beta}_j$ can be adopted. A random pattern in this graph indicates additive effects of the two factors. The sums of squares for this type of ANOVA model are given in Table 5.9.

Table 5.9 Two-way ANOVA table for a balanced model

Sum of squares for	Degrees of freedom	Mean square	Test F-criterion
Factor A, $S_A = n^* M \sum_{i=1}^{N} \hat{\alpha}_i^2$	$N-1$	$M_A = \dfrac{S_A}{N-1}$	$F_A = \dfrac{M_A}{M_R}$
Factor B, $S_B = n^* N \sum_{j=1}^{M} \hat{\beta}_i^2$	$M-1$	$M_B = \dfrac{S_B}{M-1}$	$F_B = \dfrac{M_B}{M_R}$
Interaction AB, $S_{AB} = n^* \sum_{i=1}^{N} \sum_{j=1}^{M} \hat{\tau}_i^2$	$(N-1)(M-1)$	$M_{AB} = \dfrac{S_{AB}}{(N-1)(M-1)}$	$F_{AB} = \dfrac{M_{AB}}{M_R}$
Residuals, $S_R = \sum_{i=1}^{N} \sum_{j=1}^{M} \sum_{k=1}^{n^*} (y_{ijk} - \hat{\mu}_{ij})^2$	$MN(n^*-1)$	$M_R = \dfrac{S_R}{MN(n^*-1)}$	
Totals, $S_C = \sum_{i=1}^{N} \sum_{j=1}^{M} \sum_{k=1}^{n^*} (y_{ijk} - \hat{\mu})^2$	MNn^*-1	—	

The corresponding expected values of mean squares are

$$E(M_A) = \sigma^2 + \frac{n^* \, M \sum\limits_{i=1}^{N} \alpha_i^2}{(N-1)\,\sigma^2} = \sigma^2 + n^* \, M \sigma_A^2$$

$$E(M_B) = \sigma^2 + \frac{n^* \, M \sum\limits_{j=1}^{M} \beta_j^2}{(M-1)\,\sigma^2} = \sigma^2 + n^* \, N \sigma_B^2$$

$$E(M_{AB}) = \sigma^2 + \frac{n^* \sum\limits_{i=1}^{N} \sum\limits_{j=1}^{M} \tau_{ij}^2}{(N-1)(M-1)\,\sigma^2} = \sigma^2 + n^* \, \sigma_{AB}^2$$

The expected value $E(M_R) = \sigma^2$ shows that the variance M_R is an unbiased estimate σ^2 of an error variance. Variances σ_A^2, σ_B^2 and σ_{AB}^2 correspond to the effects of rows, columns and interaction. These expressions are used also in calculation of estimates of variance of factors and interaction. Then instead of mean values $E(.)$, the mean squares, and instead of variance σ^2, the residual variance $\hat{\sigma}^2$ (cf. Problem 5.9) are substituted. It is important to note that the mean squares are not estimates of the corresponding variances.

Statistical criteria F_{AB}, F_B and F_A defined in the ANOVA Table 5.9 are used to test whether the interaction effects, column effects and row effects are significant. To test a null hypothesis H_0: $\tau_{ij} = 0$, $i = 1, ..., N$ and $j = 1, ...,$ M, the test criterion F_{AB} is used; if H_0 is valid, this has the Fisher-Snedecor distribution with $(N-1)(M-1)$ and $M\,N(n^* - 1)$ degrees of freedom. To test the significance of row effects (factor A), the null hypothesis H_0 is $\alpha_i = 0$, $i = 1, ..., N$. When H_0 is valid, the test criterion F_A has the Fisher-Snedecor distribution with $(N-1)$ and $M\,N(n^* - 1)$ degrees of freedom. For column effects (factor B), the null hypothesis H_0 is $\beta_j = 0, j =1, ..., M$. When H_0 is valid, the test criterion F_B has the Fisher-Snedecor distribution with $(M-1)$ and $M\,N(n^* - 1)$ degrees of freedom. The unbiased estimator of variance is here M_R.

For balanced models, the diagonal elements H_{ii} of a projection matrix **H** are constant. It may be concluded that the two models, single-observation-per-cell (I) and balanced (II) differ only in replacement of quantities.

An advantage of all balanced models is mutual orthogonality of the individual terms of the ANOVA model, so that the individual partial sums of squares in Tables 5.9 and 5.6 may be added. This may be exploited for simultaneous testing of several hypotheses [17, 18].

Problem 5.9 *Derivation of a test criterion for examining the independence of factor B*

Let us suppose that Table 5.9 is available and we wish to test whether the results depend on a factor B. That is, we test the validity of the two hypotheses, H_{01}: $\tau_{ij} = 0$ and H_{02}: $\beta_j = 0$, simultaneously.

Solution: Since partial sums of squares may be added, we can calculate the sum of squares resulting from factor B simply as $S_{PB} = S_B + S_{AB}$. The corresponding number of degrees of freedom $N(M-1)$ is the sum of degrees of freedom $(M-1)$ and $(N-1)(M-1)$, so that the mean square is $M_{PB} = S_{PB} / [N(M-1)]$. The test criterion $F_{PB} = M_{PB} / M_R$ has the Fisher-Snedecor distribution with $N(M-1)$ and $M N(n^*-1)$ degrees of freedom, if the partial hypotheses H_{01} and H_{02} are valid.

Conclusion: It is possible to test another simultaneous hypotheses, similarly. That is, is the variability of y due to interaction only, which corresponds to H_{01}: $\sigma_i = 0$ and H_{02}: $\beta_j = 0$.

With the use of the procedure shown in Problem 5.9 the validity of various ANOVA submodels may be tested, from the simplest $y_{ijk} = \mu + \varepsilon_{ijk}$ over all partial models (containing only some of the parameters α, β and τ) to the total analysis expressed by Eqs. (5.3) and (5.4). Summation of sums of squares is recommended also in cases when the influence of some factors or interactions is proved to be not significant. Then, a corresponding sum of squares is added to the residual one, and corresponding terms are omitted.

Problem 5.10 *Precision of chromatographic determination of diethyleneglycol in ethyleneglycol*

Three laboratory technicians A_1, A_2 and A_3 can work on three chromatographs B_1, B_2 and B_3 and determine diethyleneglycol (DEG) in ethyleneglycol. Each technician made two determinations with each chromatograph. Besides ANOVA, estimate how much of the variance corresponds to the instrument (the instrumental error) and how much to the technician (the error of the experimenter).

Data: $N = 3$, $M = 3$, $n^* = 2$. Data for percentage of DEG found in ethylene glycol are given in Table 5.10

Table 5.10 Concentration of DEG [%] measured twice by three technicians A_1, A_2 and A_3 on three instruments B_1, B_2 and B_3.

Technician	Instrument		
	B_1	B_2	B_3
A_1	0.110, 0.116	0.101, 0.102	0.108, 0.109
A_2	0.112, 0.111	0.115, 0.106	0.111, 0.109
A_3	0.114, 0.112	0.107, 0.109	0.113, 0.110

Program: ADSTAT or QC-Expert: ANOVA-2B: Two-way balanced.

Solution: The ANOVA table is shown in Table 5.11. For the significance level $\alpha = 0.05$ the quantiles are $F_{0.95}(2, 9) = 4.26$ and $F_{0.95}(4, 9) = 3.63$. From F criteria in Table 5.11, it is evident that the effects of factor A and interaction AB are separately not significant. Let us test whether technicians have an influence on the determination of DEG. Two null hypotheses are formulated, H_{01}: $\alpha_i = 0$ and H_{02}: $\tau_{ij} = 0$. The corresponding mean square is $S_{PA} = (S_A + S_{AB})/(2 + 4) = 1.63 \times 10^{-5}$ and the test criterion $F_{PA} = 1.63 \times 10^{-5}/7.83 \times 10^{-6} = 2.09$. Since the corresponding quantile $F_{0.95}(6, 9) = 3.373$ is higher than the criterion F_{PA}, the technicians have no significant influence on the determination of DEG and the ANOVA model is formulated by the equation $y_{ijk} = \mu + \beta_j + \varepsilon_{ijk}$. Let us test the significance of factor B with H_0: $\beta_j = 0$. In addition, to avoid construction of a new ANOVA table, the sums $S_A + S_{AB}$ will be added to the residual sum of squares S_R. The mean square will then be

$$M_R^* = \frac{S_R + S_A + S_{AB}}{2 + 4 + 9} = \frac{1.688 \times 10^{-5}}{15} = 1.12 \times 10^{-5}.$$

Table 5.11 ANOVA table for DEG determination

Sum of squares	Degrees of freedom	Mean of square	Test criterion F
Technicians A, $S_A = 3.81 \times 10^{-5}$	2	1.906×10^{-5}	2.433
Instrum. B, $S_B = 1.027 \times 10^{-4}$	2	5.139×10^{-5}	6.560
Interaction AB, $S_{AB} = 6.02 \times 10^{-5}$	4	1.506×10^{-5}	1.922
Residual $S_R = 7.05 \times 10^{-5}$	9	7.833×10^{-6}	
Totals $S_c = 2.716 \times 10^{-4}$	17		—

The test criterion F_B is $F_B = M_B / M_R^* = 4.58$ and the corresponding quantile $F_{0.95}(3, 15) = 3.68$ is lower. Thus, the null hypothesis H_0 is rejected, and the influence of the instrument on the determination of DEG has been shown to be significant at significance level $\alpha = 0.05$. The expected value of a mean square is given by $E(M_B) = \sigma^2 + 6\sigma_B^2$ and the estimate of σ^2 is $\hat{\sigma}^2$. The estimate of instrumental error may be calculated from $\hat{\sigma}_B^2 = (M_B - \hat{\sigma}^2)/6 = 6.68 \times 10^{-6}$.

Conclusion: It has been shown that the precision of determination of DEG is affected only by the instrument used. The variability caused by technicians and other random effects is $\hat{\sigma}^2 = 1.12 \times 10^{-5}$, and the variability caused by instruments (instrumental error) is $\hat{\sigma}_B^2 = 6.68 \times 10^{-6}$.

5.3.3.3 Unbalanced Models

For unbalanced models, there are n_{ij} observations in the (ij)th cell. When an experiment is poorly organized, so that differences in n_{ij} values are in tens, the ANOVA is rather complicated. Parameters of Eqs. (5.3) and (5.4) are not orthogonal, and the partitioning of the sum of squares is ambiguous. The analysis of variance is performed with the use of programs for linear regression, with models (5.3) and (5.4) considered as special regression models with dummy variables taking only the values 0 or 1. For practical calculations in chemometrics, an approximate partitioning of overall sum of squares is used. This begins with a calculation of means:

$$\hat{\mu}_{ij} = \frac{1}{n_k} \sum_{i=1}^{n_k} y_{ijk} \tag{5.65}$$

for the cells. From these values the residual sum of squares is estimated

$$S_R = \sum_{i=1}^{N} \sum_{j=1}^{M} \sum_{k=1}^{n_k} (y_{ijk} - \hat{\mu}_{ij})^2 \tag{5.66}$$

For calculation of other components of the partitioned overall sum of squares, the estimates of means $\hat{\mu}_{ij}$ are used, with the assumption that they have been estimated from an equivalent number of observations n_e^* defined by

$$n_e^* = \left[\frac{1}{N\,M} \sum_{i=1}^{N} \sum_{j=1}^{M} (1/n_{ij}) \right]^{-1} \tag{5.67}$$

Analysis of variance is done by the same technique as for balanced models (Table 5.8), and individual sums are defined by

$$S_A = n_e^* \, M \sum_{i=1}^{N} (\hat{\mu}_i - \hat{\mu})^2 \qquad (5.68)$$

with $(N-1)$ degrees of freedom

$$S_B = n_e^* \, N \sum_{j=1}^{M} (\hat{\mu}_j - \hat{\mu})^2 \qquad (5.69)$$

with $(M-1)$ degrees of freedom, and

$$S_{AB} = n_e^* \sum_{i=1}^{N} \sum_{j=1}^{M} (\hat{\mu}_{ij} - \hat{\mu}_i - \hat{\mu}_j - \hat{\mu})^2 \qquad (5.70)$$

with $(N-1)(M-1)$ degrees of freedom. In these expressions, the following notation is used

$$\hat{\mu}_i = \frac{1}{M} \sum_{j=1}^{M} \hat{\mu}_{ij}; \quad \hat{\mu}_j = \frac{1}{N} \sum_{i=1}^{N} \hat{\mu}_{ij}; \quad \hat{\mu} = \frac{1}{N\,M} \sum_{i=1}^{N} \sum_{j=1}^{M} \hat{\mu}_{ij};$$

The sum $S_A + S_B + S_{AB} + S_R$ is not here exactly equal to S_C, but differences are relatively small. Tests for row, column and interaction effects are made as previously for balanced models.

When there are several replicates in individual cells, for each cell the sample variance s_{ij}^2 may be estimated, and a plot s_{ij}^2 vs. $\hat{\mu}_{ij}$ used to check for any variance on the mean (heteroscedasticity).

5.3.4 Mixed Effects Two-way ANOVA

Suppose that some factor A corresponds to fixed effects and another factor B to random effects. The factor B is usually considered as the noisy factor, and it is not usually tested. The ANOVA model is defined by Eqs. (5.3) and (5.4). For fixed effects the constraints are $\alpha_i = 0$, $i = 1, \ldots, N$. For random variables

β_j and ε_{ijk} it is assumed that

(a) β_j are mutually independent random variables with the normal distribution $N(0, \sigma_B^2)$;

(b) ε_{ijk} are mutually independent random variables with the normal distribution $N(0, \sigma_\varepsilon^2)$;

(c) the random variables β_j are independent of the random variables ε_{ijk}

Here is an example of the elucidation of an interaction. Some authors consider that there are random mutually independent variables with the normal distribution $N(0, \sigma_{AB}^2)$. We will consider that τ_{ij} are independent of β_j and identically distributed variables, so that their variance is equal to [5]

$$D(\tau_{ij}) = (1 - 1/N)\sigma_{AB}^2 \qquad (5.71)$$

For equal cell numbers $n_{ij} = n^*$, to analyse a mixed-effects model the ANOVA Table 5.8 may be used. The null hypotheses concerning nonsignificance of interaction, H_0: $\sigma_{AB}^2 = 0$, and nonsignificance of factor B, H_0: $\sigma_B^2 = 0$, are tested with the use of test criteria F_{AB} and F_B from Table 5.8. For a test of significance of fixed effects of factor A, another test criterion is used [5]

$$F_{PA} = M_A/M_{AB} \qquad (5.72)$$

If the null hypothesis H_0: $\alpha_i = 0$ is valid, the test statistic F_{PA} has the Fisher-Snedecor distribution with $(N-1)$ and $(N-1)(M-1)$ degrees of freedom. If there is no interaction, the test criterion F_A from Table 5.8 may be used. This procedure can also be modified for cases where the effects of factor A are random and the effects of factor B fixed.

Problem 5.11   *Influence of the instrument of determination of DEG*

Consider the same task as in Problem 5.12, with the difference that three technicians were chosen randomly from a team of laboratory staff. Factor A now has random effects and factor B fixed. By doing ANOVA for this mixed-factors model, examine the influence of the type of chromatograph.

Solution: In Problem 5.12 it was estimated that $M_B = 5.139 \times 10^{-5}$ and $M_{AB} = 1.506 \times 10^{-5}$. The test criterion F_{PA} (5.72) is now written as $F_{PB} = M_B / M_{AB}$

= 3.412 The corresponding quantile $F_{0.95}(2, 4) = 6.944$ is higher than this F_{PB} value. Therefore, the null hypothesis H_0: $\beta_j = 0$ is rejected and the type of chromatograph does have a significant effect on the determination of DEG.

Conclusion: The ANOVA result demonstrates that a small change in the assumptions about an experiment (here the random selection of technicians) leads to a change in the results of ANOVA.

5.3.5 Random-effects two-way ANOVA

In this case, the effects of both factors are random. The ANOVA model can be defined by Eqs. (5.3) and (5.4). The random components of the model, $\alpha_i, \beta_j, \tau_{ij}$ and ε_{ijk} are characterized by following properties:

(a) α_i are mutually independent random variables with the normal distribution $N(0, \sigma_A^2)$;

(b) β_j are mutually independent random variables with the normal distribution $N(0, \sigma_B^2)$;

(c) τ_{ij} are mutually independent random variables with the normal distribution $N(0, \sigma_{AB}^2)$;

(d) ε_{ijk} are mutually independent random variables with the normal distribution $N(0, \sigma_\varepsilon^2)$;

(e) all these random variables are independent of one another. For interaction τ_{ij} the estimates $\hat{\tau}_{ij}$ are dependent on $\hat{\alpha}_i$ and $\hat{\beta}_j$. For the whole population it is assumed that the interactions are independent for this model [5].

The primary goal of ANOVA here is to find the estimates of the variance components $\sigma_A^2, \sigma_B^2, \sigma_{AB}^2$ and σ_ε^2 test them. We will limit the discussion to balanced models with n^* replication per cell. To find the estimates of the variance components, the statistics from Table 5.8 can be adopted. The following expressions are used for the estimates of the partial variances

$$\sigma_A^2 = \frac{M_A - M_{AB}}{n^* M} \tag{5.73}$$

$$\sigma_B^2 = \frac{M_B - M_{AB}}{n^* M} \tag{5.74}$$

$$\sigma_{AB}^2 = \frac{M_{AB} - M_R}{n^*} \tag{5.75}$$

$$\text{and } \sigma_e^2 = M_R \tag{5.76}$$

The mean squares M_A, M_B and M_{AB} are defined in Table 5.8. For testing the significance of the individual variance components, the F-tests are used as follows:

(a) For a *test of the null hypothesis H_0:* $\sigma_{AB}^2 = 0$ the test criterion F is

$$F_{EAB} = \frac{M_{AB}}{M_R} \tag{5.77}$$

If the null hypothesis H_0 is valid, the statistic F_{EAB} has a Fisher-Snedecor distribution with $(N-1)(M-1)$ and $N M(n^* - 1)$ degrees of freedom.

(b) For a *test of the null hypothesis H_0:* $\sigma_A^2 = 0$ the test criterion F is

$$F_{EA} = \frac{M_A}{M_{AB}} \tag{5.78}$$

If the null hypothesis H_0 is valid the statistic F_{EA} has a Fisher-Snedecor distribution with $(N-1)$ and $(N-1)(M-1)$ degrees of freedom.

(c) For a *test of the null hypothesis H_0:* $\sigma_B^2 = 0$ the test criterion F is

$$F_{EB} = \frac{M_A}{M_{AB}} \tag{5.79}$$

If the null hypothesis is valid the statistic F_{EB} has a Fisher-Snedecor distribution with $(M-1)$ and $(N-1)(M-1)$ degrees of freedom.

When the estimates of parameters are interesting, the estimates $\hat{\mu}_{ij}$ and $\hat{\mu}$ defined for fixed-factors models can be applied. The estimate of variance of parameter $\hat{\mu}$ is defined by [5]

$$\hat{\sigma}_{\hat{\mu}}^2 = \frac{M_A}{N\, M\, n^*} \qquad (5.80)$$

Some alternative procedures for estimating the variance components are discussed in the literature [3–5].

Problem 5.12 *Factors affecting the chromatographic determination of DEG*

Consider the data of Problem 5.10, with the difference that both technicians and instruments were chosen randomly. Moreover, in the laboratory there are more than three technicians and instruments. This represents a random-effects model, and we want to find the effect of the variances of the individual sources on the variability of the final results.

Solution: A test of the null hypothesis H_0: $\sigma_B^2 = 0$ yields the same result as in Problem 5.13. To test the null hypothesis H_0: $\sigma_A^2 = 0$ the criterion is F_{PA} = 1.264. The corresponding quantile $F_{0.95}(2, 4) = 6.944$ is higher, so that the variance σ_A^2 cannot be considered significantly different from zero. For the null hypothesis H_0: $\sigma_{AB}^2 = 0$ in Problem 5.10 it was found that $F_{AB} = F_{EAB} = 1.93$, which is lower than the corresponding quantile $F_{0.95}(4, 9) = 3.63$, so the null hypothesis H_0 cannot be rejected. The estimate of variance corresponding to instruments is $\sigma_B^2 = (M_B - M_{AB})/6 = 6 \times 10^{-6}$.

Conclusion: The variance of any partial source of variance, σ_A^2, σ_B^2 and σ_{AB}^2, cannot be considered to be significantly different from zero, so the result of determination of DEG is loaded by random effects only.

5.4 Summary

The first step of any ANOVA procedure is to recognize, on the basis of the data layout, whether the model is a fixed-, mixed- or random-effects one. For all three models the hypotheses to be tested must be specified, and the parameters to be estimated formulated. It is useful to know whether interaction is likely. The general procedure of ANOVA involves the following steps [20]:

(1) Estimate the parameters of the ANOVA model.
(2) Test the significance of model and construct submodels or models with fixed-effects.

(3) Express the variance components for the random-effects model and test their significance.

(4) Test the assumptions of normality, homogeneity of variance and outliers. Residuals other than the classical ones may be used (Chapter 6).

(5) Interpret the ANOVA results with reference to data and assumptions.

5.5 References

[1] S. R. Searle, *Biometrics,* 1971, 27, 1.

[2] M. S. Bartlett and D. G. Kendall, *J. Roy. Stt. Soc,* 1946, B8, 128.

[3] H. Scheffe, *The Analysis of Variance,* Wiley, New York, 1959.

[4] S. R. Searle, *Linear Models,* Wiley, New York, 1971.

[5] P. G. Miller, *Beyond ANOVA, Basic of Applied Statistics,* Wiley, New York, 1986.

[6] T. P. Speed, *Ann. Statist.* 1987, 15, 885.

[7] J. D. Emerson, D. C. Hoaglin and P. I. Kempthorne, *J. Am. Statist. Assoc.* 1984, 79, 329.

[8] D. Bradu and D. M. Hawkins, *Technometrics,* 1982, 24, 103.

[9] P. Bloomfield and W. Steiger, *Least Absolute Deviations: Theory, Applications and Algorithms.*Birkhauser, Boston, 1983.

[10] K. R. Gabriel, *J. R. Stt. Soc,* 1972, B40, 186.

[11] N. A. C. Cressie, *Biometrics,* 1978, 34, 505.

[12] Potocký R a kol.: *Zbierka úloh z pravdepodobnosti a matematickej štatistiky.* ALFA-SNTL, Bratislava 1986.

[13] Anderson R. L.: *Practical Statistics for Analytical Chemists.* van Nostrand Reinhold Comp., New York 1987.

[14] Miller J. C., Miller J. N.: *Statistics for Analytical Chemistry.* Ellis Horwood, Chichester 1984.

[15] Liteanu C., Rica I.: *Statistical Theory and Methodology of Trace Analysis.* Ellis Horwood, Chichester 1980.

[16] Rice J. A.: *Mathematical Statistics and Data Analysis.* Wadsworth & Brooks, California 1988, s. 397.

[17] Hintze J.: *Number Cruncher Statistical Systems 2000.* Manuál, Kaysville, Utah, October 1998.

[18] Hintze J.: *User's Guide NCSS2000.* Statistical System for Windows, Kaysville, Utah 1999.

[19] MacBerthouex P., Brown L. C.: *Statistics for Environmental Engineers*, Lewis Publishers, London, 2002.

[20] Meloun M., Militký J.: *Statistické zpracování experimentálních dat,* Plus Praha 1994 (1. vydání), EAST PUBLISHING, Praha 1998 (2. vydání), Academia Praha 2004 (3. vydání).

Supplemented material (Review Questions, Exercises, and Results of Exercises) to Chapter 5 is on the enclosed CD.

6
Linear regression models

6.1 Formulation of the Linear Regression Model

In instrumental methods in laboratory, the instrument's response y (output variable) for selected values of the input variables x is often measured. For example, an absorbance A (here, the output variable y) is measured on the scale of a spectrophotometer at

(1) a selected value of wavelength λ (here, the first independent variable $x_{i,1}$),

(2) a concentration of colour-forming solution c (here, the second independent variable $x_{i,2}$),

(3) a value of adjusted pH of solution (here, the third independent variable $x_{i,3}$), and

(4) in kinetic measurements, at an actual time (here, the fourth independent variable $x_{i,4}$).

This results in n observed values of y, measured at four kinds of selected values of independent variables, $m = 4$, written as $\left\{y_i, x_{ij}\right\}$, $i = 1,..., n$, and $j = 1,..., m$

$$\begin{bmatrix} y_1 \\ y_2 \\ . \\ . \\ . \\ y_n \end{bmatrix} \begin{bmatrix} x_{11} & x_{12} & x_{13} & x_{14} \\ x_{21} & x_{22} & x_{23} & x_{24} \\ . & . & . & . \\ . & . & . & . \\ . & . & . & . \\ x_{n1} & x_{n2} & x_{n3} & x_{n4} \end{bmatrix}$$

In matrix notation, this is written as $\{y, X\}$. Vector y has dimensions $(n \times 1)$ and matrix X has dimensions $(n \times m)$.

The statistical analysis is intended to find a relationship between the response (output) variable y and the controllable (independent) variables x. The type of function $y = f(x, \beta)$ depends on the nature of both the variables y and x. There are three possible scenarios.

(a) Variables y and x have no random errors. The function $y = f(x, \beta)$ contains a vector of unknown parameters β of dimension $(m \times 1)$. To estimate them, at least $n = m$ measurements y_i, $i = 1, ..., n$, at adjusted values x_i are necessary to solve a set of n equations of the form

$$y_i = f(x_i, \beta) \tag{6.1}$$

with regard to unknown parameters β. The measured variables y_i are assumed to be measured completely precisely, without any experimental errors. The model function $f(x, \beta)$ is assumed to be correct and to correspond to data y. In the laboratory, none of these assumptions is usually fulfilled.

(b) Variable y is subject to random errors, but variables x are controllable. This case is the *classical regression model*, for which the conditional mean of random variable y at a point x is given by

$$E(y/x) = f(x, \beta) \tag{6.2}$$

. The method of estimation of parameters β depends on the distribution of random variable y. The *additive model of measurement errors* (Chapter 1) is assumed:

$$y_i = f(x_i, \beta) + \varepsilon_i \tag{6.3}$$

where ε_i is a random variable containing the measurement errors $\varepsilon_{M,i}$ and the model errors $\varepsilon_{T,i}$ coming from an approximate model which does not correspond to the true theoretical model $f_T(x_i, \beta)$. Decomposition of the total error ε_i into components $\varepsilon_{M,i}$ and and $\varepsilon_{T,i}$ is illustrated in Fig. 6.1.

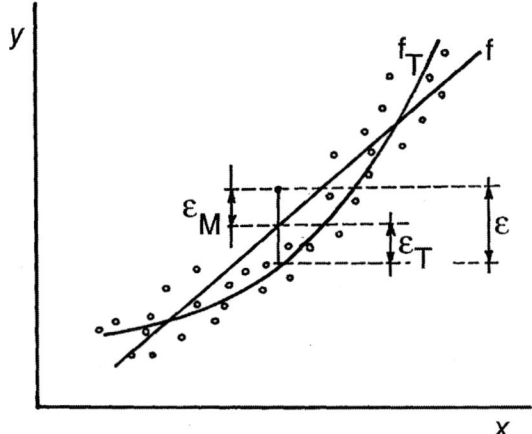

Figure 6.1   Decomposition of the total error ε into two components, the measure error ε_M and the model error ε_T.

Treatment of experimental data by regression analysis involves first the choice of a linear regression model

$$E(y\,/\,\mathbf{x}) = \sum_{j=1}^{m} \beta_j x_j \qquad (6.4)$$

which either can be an approximation of the unknown theoretical function f_T (Fig. 6.1) or can be derived from a knowledge of the experimental system. In Eq. (6.4), instead of variables x_j, their functions which do not contain parameters β may be used. Parameter estimates of model (6.4) may be determined, on the assumption that Eq. (6.3) is valid, either by the *method of maximum likelihood* or the *method of least-squares*.

(c)   Variables y, x are a sample from the random vector (η, ξ^T) with $m + 1$ components. Regression is conditioned by the mean value (6.2) where x represents an actual quantity from the random vector ξ. Unlike regression models, in these "correlation models" the regression function can be derived from a simultaneous probability density frequency function $p(y, x)$ and a conditional probability density function $p(y/x)$. The analysis of correlation models is discussed in Chapter 7.

For either correlation or for regression models, the same expressions are valid, although they differ significantly in meaning.

In this chapter, only linear regression models [Eq. (6.4)] are considered, i.e. models which may be written in the form

$$\begin{bmatrix} y_1 \\ y_2 \\ \cdot \\ \cdot \\ \cdot \\ y_n \end{bmatrix} = \begin{bmatrix} x_{11} & x_{12} & x_{13} \cdots & x_{1m} \\ x_{21} & x_{22} & x_{23} \cdots & x_{2m} \\ \cdot & \cdot & \cdot \cdots & \cdot \\ \cdot & \cdot & \cdot \cdots & \cdot \\ \cdot & \cdot & \cdot \cdots & \cdot \\ x_{n1} & x_{n2} & x_{n3} & x_{nm} \end{bmatrix} \times \begin{bmatrix} \beta_1 \\ \beta_2 \\ \cdot \\ \cdot \\ \beta_m \end{bmatrix} + \begin{bmatrix} \varepsilon_1 \\ \varepsilon_2 \\ \cdot \\ \cdot \\ \cdot \\ \varepsilon_n \end{bmatrix} \tag{6.5a}$$

of dimensions $(n \times 1)$ $(n \times m)$ $(m \times 1)$ $(n \times 1)$ and for m = 4. In matrix notation, Eq. (6.5a) takes the simple form

$$\mathbf{y} = \mathbf{X}\,\boldsymbol{\beta} + \boldsymbol{\varepsilon} \tag{6.5b}$$

Columns $\mathbf{x}_j$ define geometrically the m-dimensional co-ordinate system or the hyperplane L in n-dimensional Euclidean space E^n. The vector $\mathbf{y}$ does not have to lie in this hyperplane L, as shown in Fig. 6.2, which is for two independent variables ($m = 2$).

The vector $\mathbf{X}\boldsymbol{\beta}$ lies in hyperplane L and parameters $\boldsymbol{\beta}$ may be understood as the coefficients of proportionality of the individual components x_j of the co-ordinate system. The regression model is formed by their linear combination. Whatever regression criterion is used for linear regression models, the model function $\mathbf{Xb}$ and the theoretical model $\mathbf{X}\boldsymbol{\beta}$ will lie in an m-dimensional hyperplane L.

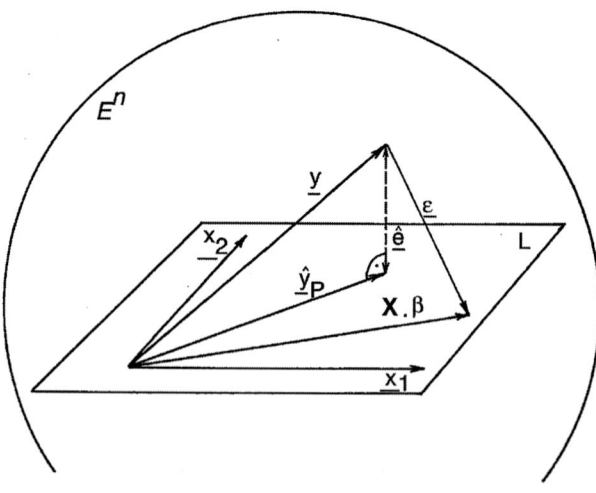

Figure 6.2   Geometric illustration of a linear regression model for two independent variables.

The least-squares method (LS) is the most frequently used method in regression analysis. The geometry of the least-squares method is very simple. For a linear regression (Fig. 6.2), the parameter estimates $\mathbf{b}$ may be found by minimization of distance between the vector $\mathbf{y}$ and the hyperplane L. This is equivalent to finding the minimal length of the residual vector

$$\hat{\mathbf{e}} = \mathbf{y} - \hat{\mathbf{y}}_P \qquad (6.6)$$

where $\hat{\mathbf{y}}_P = \mathbf{Xb}$ is the *prediction vector*. In Euclidean space the length of vector $\hat{\mathbf{e}}$ can be expressed by

$$D = \sqrt{\{\hat{\mathbf{e}}, \hat{\mathbf{e}}\}} = \sqrt{\sum_{i=1}^{n} \hat{e}_i^2} \qquad (6.7)$$

The symbol $\{\mathbf{x}, \mathbf{y}\} = \sum_{i=1}^{n} x_i y_i$ means the scalar product of two vectors. The square of vector $\hat{\mathbf{e}}$ length is consistent with criterion $U(\mathbf{b})$ of the least-squares method $D^2 = U(\mathbf{b})$ so that the estimates of model parameters $\mathbf{b}$ minimize the expression

$$U(\mathbf{b}) = D^2 = \sum_{i=1}^{n} \left[y_i - \sum_{j=1}^{m} x_{ij} b_j \right]^2 = \sum_{i=1}^{n} (y_i - \hat{y}_{P,i})^2 \qquad (6.8)$$

Vectors $\hat{\mathbf{e}}$ and $\hat{\mathbf{y}}_P$ are illustrated in Fig. 6.2. The vector $\hat{\mathbf{y}}_P$ represents a perpendicular projection of vector $\mathbf{y}$ onto hyperplane L. The vector $\hat{\mathbf{e}}$ for which a function D is minimal lies in n–m dimensional hyperplane $L^{\perp}$ that is perpendicular to the hyperplane L and is called the *residual vector*.

The residual vector $\hat{\mathbf{e}}$ is perpendicular to all columns of matrix $\mathbf{X}$ and therefore all corresponding scalar products are zero

$$\{\mathbf{x}_j, \hat{\mathbf{e}}\} = \sum_{i=1}^{n} x_{ij} \hat{e}_i = 0, \quad j = 1, \dots, m \qquad (6.9)$$

This set of equations may be written in matrix notation as

$$\mathbf{X}^{\mathrm{T}} \hat{\mathbf{e}} = 0 \qquad (6.10)$$

Substituting $(\mathbf{y} - \mathbf{Xb})$ for $\hat{\mathbf{e}}$ leads to a set of linear equations in the known vector $\mathbf{b}$

$$\mathbf{X}^T\mathbf{y} = \mathbf{X}^T\mathbf{Xb} \qquad (6.10a)$$

The estimate $\mathbf{b}$ which minimizes the distance $\mathbf{D}$ is then

$$\mathbf{b} = (\mathbf{X}^T\mathbf{X})^{-1}\mathbf{X}^T\mathbf{y} \qquad (6.11)$$

where $\mathbf{A}^{-1}$ represents the inverse of matrix $\mathbf{A}$.

The perpendicular projection of $\mathbf{y}$ into hyperplane L can be made by using projection matrix $\mathbf{H}$ and may be expressed by

$$\hat{\mathbf{y}}_P = \mathbf{Hy} \qquad (6.12)$$

By substitution from Eq. (6.11), Eq. (6.12) may be rewritten as

$$\hat{\mathbf{y}}_P = \mathbf{Xb} = \mathbf{X}(\mathbf{X}^T\mathbf{X})^{-1}\mathbf{X}^T\mathbf{y} \qquad (6.12a)$$

The projection matrix $\mathbf{H} = \mathbf{X}(\mathbf{X}^T\mathbf{X})^{-1}\mathbf{X}^T$ has the property of projecting any vector $\mathbf{V}$ into a plane L. When the vector $\mathbf{V}$ already lies in plane L, $\mathbf{HV} = \mathbf{V}$. However, when vector $\mathbf{V}$ is perpendicular to plane L, $\mathbf{HV} = \mathbf{0}$ where $\mathbf{0}$ is a zero vector.

The projection matrix $\mathbf{P}$ for perpendicular projection into a hyperplane $L^{\perp}$ that is orthogonal to hyperplane L is

$$\mathbf{P} = \mathbf{E} - \mathbf{H} \qquad (6.13)$$

where $\mathbf{E}$ is an $n \times n$ identity matrix. With the use of these two projection matrices the total decomposition of vector $\mathbf{y}$ into two orthogonal components may be written as $\mathbf{y} = \mathbf{Hy} + \mathbf{Py} = \hat{\mathbf{y}}_P + \hat{\mathbf{e}}$. The geometric interpretation is that vector $\mathbf{y}$ is decomposed into two mutually perpendicular vectors (Fig. 6.2). The same expressions can be reached by an analytical minimization, i.e. by a differentiation of Eq. (6.8) and rearrangement.

Problem 6.1 *Parameter estimates of a calibration straight line*

Apply the expressions already derived to a model of a straight calibration line $E(y/x) = \beta_1 x + \beta_2$ where y is the measured quantity and x is usually a concentration. Derive estimates b_1, b_2 and the elements of the projection matrix **H**.

Solution: For this case, we have

$$\mathbf{X}^T\mathbf{X} = \begin{bmatrix} x_1 & \cdots & x_n \\ 1 & \cdots & 1 \end{bmatrix} \begin{bmatrix} x_1 & 1 \\ \cdots & 1 \\ x_n & 1 \end{bmatrix} = \begin{bmatrix} \sum x_i^2 & \sum x_i \\ \sum x_i & n \end{bmatrix}$$

$$\mathbf{X}^T\mathbf{y} = \begin{bmatrix} x_1 & \cdots & x_n \\ 1 & \cdots & 1 \end{bmatrix} \begin{bmatrix} y_1 \\ \cdots \\ y_n \end{bmatrix} = \begin{bmatrix} \sum x_i y_i \\ \sum y_i \end{bmatrix}$$

where all sums are considered for $i = 1$ to n. For determination of the inversion matrix $(\mathbf{X}^T\mathbf{X})^{-1}$ the method based on the adjugate matrix may be used

$$(\mathbf{X}^T\mathbf{X})^{-1} = (1/\det(\mathbf{X}^T\mathbf{X}).adj(\mathbf{X}^T\mathbf{X})) \text{ where } \det(\mathbf{X}^T\mathbf{X}) = n\sum_{i=1}^{n} x_i^2 - \left[\sum_{i=1}^{n} x_i\right]^2 \text{ and}$$

$$adj(\mathbf{X}^T\mathbf{X}) = \begin{bmatrix} n & -\sum x_i \\ -\sum x_i & \sum x_i^2 \end{bmatrix}.$$

Recall that for matrix $\mathbf{A} = \begin{bmatrix} a & b \\ c & d \end{bmatrix}$, $\det(\mathbf{A}) = a \times d - c \times b$ and $adj(\mathbf{A}) = \begin{bmatrix} d & -b \\ -c & a \end{bmatrix}$.

Substitution into Eq. (6.11) leads to the vector of parameter estimates **b**, which is given

$$\mathbf{b} = \begin{bmatrix} b_1 \\ b_2 \end{bmatrix} = \frac{1}{n\sum x_i^2 - \left[\sum x_i\right]^2} \begin{bmatrix} n & -\sum x_i \\ -\sum x_i & \sum x_i^2 \end{bmatrix} \begin{bmatrix} \sum x_i y_i \\ \sum y_i \end{bmatrix} \qquad (6.14)$$

Multiplication yields estimates of the two parameters β_1, β_2 in the closed form

$$b_1 = \frac{n\sum x_i y_i - \sum x_i \sum y_i}{D} \qquad (6.14a)$$

$$b_2 = \frac{\sum x_i^2 \sum y_i - \sum x_i \sum x_i y_i}{D} \qquad (6.14b)$$

$$\text{where } D = n\sum x_i^2 - \left[\sum x_i\right]^2 \qquad (6.14c)$$

Equation (6.12a) allows the diagonal elements of projection matrix **H** to be determined

$$H_{jj} = \frac{1}{D}[x_j \ 1]\begin{bmatrix} n & -\sum x_i \\ -\sum x_i & \sum x_i^2 \end{bmatrix}\begin{bmatrix} x_j \\ 1 \end{bmatrix} = \frac{\sum x_i^2 + nx_j^2 - 2x_j\sum x_i}{D}, \ j,k = 1,...,n$$

and for nondiagonal elements H_{jk}

$$H_{jk} = \frac{1}{D}[x_k \ 1]\begin{bmatrix} n & -\sum x_i \\ -\sum x_i & \sum x_i^2 \end{bmatrix}\begin{bmatrix} x_j \\ 1 \end{bmatrix}$$

$$= \frac{nx_j x_k + \sum x_i^2 - (x_k + x_j)\sum x_i}{D}, \ j,k = 1,...,n$$

Introduction of the arithmetic mean $\bar{x} = 1/n\sum_{i=1}^{n} x_i$ into the expressions for H_{jj} and H_{jk} gives

$$H_{jj} = \frac{1}{n} + \frac{n(x_j - \bar{x})^2}{D}$$

$$\text{and } H_{jk} = \frac{1}{n} + \frac{n(x_j - \bar{x})(x_k - \bar{x})}{D}$$

Conclusion: With the use of simple matrix operations, estimates of the parameters of a straight line and of the elements of the projection matrix may be calculated.

Problem 6.2 *Geometric interpretation of a calibration straight line*

Derive expressions for estimates of parameters b_1 and b_2 for a calibration straight line and make a geometric representation. Use a perpendicular projection of vector y into the plane defined by the columns of matrix $\mathbf{X}$ and also make a geometric representation of the individual projections.

Solution: The model of a calibration straight line is expressed in matrix form by $\mathbf{y} = b_1^* \mathbf{x_C} + b_2^* \mathbf{J} + \hat{\mathbf{e}}$ where $\mathbf{x_C}$ is a centred variable with elements $x_{C,i} = x_i - \bar{x}$ representing the concentration or content of component and $\mathbf{J}$ is an $(n \times 1)$ vector of ones. Parameters b_1^* and b_2^* refer to the model with centred variables. The advantage of the use of a centred variable instead of the original one is that the vectors $\mathbf{x_C}$ and $\mathbf{J}$ are orthogonal and their scalar product is equal to zero,

$$\{\mathbf{x_C}, \mathbf{J}\} = \sum_{i=1}^{n} x_{C,i} \cdot 1 = \sum_{i=1}^{n} (x_i - \bar{x}) = 0$$

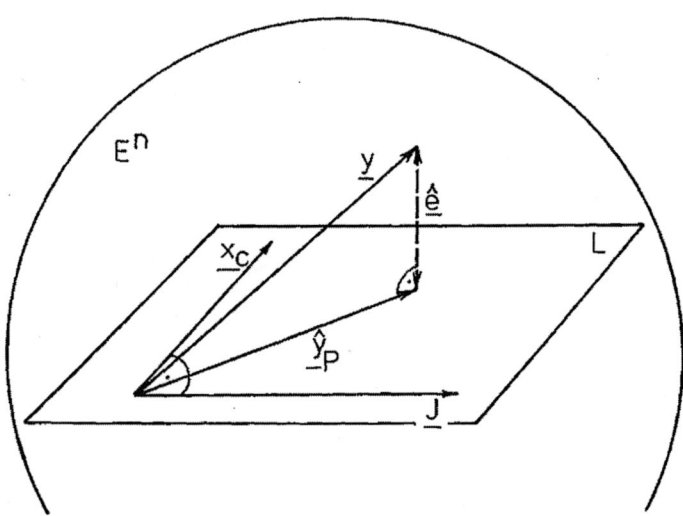

Figure 6.3 Geometrical representation of a calibration straight line (when y is the only random variable) in Euclidean space E^n.

The perpendicular projection $\mathbf{P_L}$ (of vector $\mathbf{y}$ into a plane L) is [owing to the orthogonality of the components] equal to the sum of the projection $\mathbf{P_J}$ (of vector $\mathbf{y}$ on vector $\mathbf{J}$) and the projection $\mathbf{P_X}$ (of vector $\mathbf{y}$ on a vector $\mathbf{x_C}$), $\mathbf{P_L} = \mathbf{P_J} + \mathbf{P_X} = b_1^* \mathbf{x_C} + b_2^* \mathbf{J} = \hat{\mathbf{y}}_P$, (see Fig. 6.4).

Projection $\mathbf{P_X}$ lies on vector $\mathbf{x_C}$ and vector $\mathbf{y} - \mathbf{P_X}$ is on the perpendicular to this vector. Then the corresponding scalar product must be zero,

$$\{y - P_X, x_C\} = \{y - b_1^* x_C, x_C\} = \sum_{i=1}^{n} (y_i - b_1^* x_{C,i}) x_{C,i} = 0$$

The estimate b_1^* will then be

$$b_1^* = \frac{\sum_{i=1}^{n} y_i (x_i - \bar{x})}{\sum_{i=1}^{n} (x_i - \bar{x})^2} = \sum_{i=1}^{n} y_i w_i \qquad (6.14d)$$

where w_i are weight coefficients. Because the vector $(y - P_J)$ is perpendicular to the vector J, the following equality is valid

$$\{y - P_J, J\} = \sum_{i=1}^{n} (y_i - b_2^*) . 1 = 0$$

and therefore the estimate b_2^* is equal to $b_2^* = \bar{y} = \frac{1}{n} \sum_{i=1}^{n} y_i$.

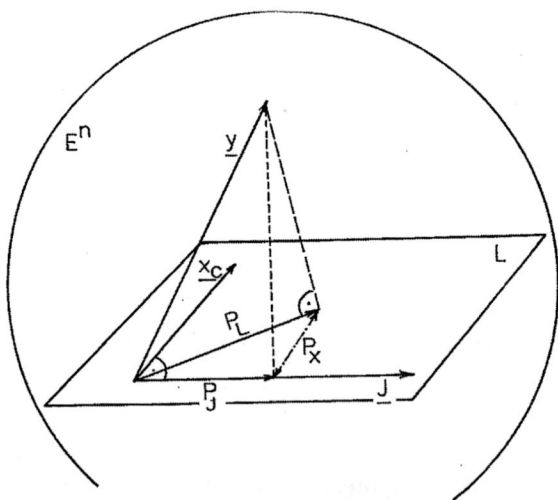

Figure 6.4 Geometrical representation of the individual projections P_L, P_J and P_X.

The estimates b_1 and b_2 for a model with a non-centred variable x, and the estimates b_1^* and b_2^* of a model with centred variable x_C are related in the following ways.

$$b_1 = b_1^*$$

$$b_2 = b_2^* - b_1\bar{x} = \sum_{i=1}^{n}\left(\frac{1}{n} - \bar{x}w_i\right)y_i \qquad (6.14e)$$

where $w_i = [x_i - \bar{x}]/\left[\sum_{i=1}^{n}(x_i - \bar{x})^2\right]$ are the weight coefficients of the individual

values y_i when b_1^* is calculated from Eq. (6.14d). Equations (6.14a) and (6.14d) or (6.14b) and (6.14e) are equivalent. Equations (6.14d, e) show that the parameter estimates are the weighted linear combinations of all y_i. The magnitudes of the weight coefficients depend only on the location of experimental data. This important conclusion tells us that when a value of x_i is far from the mean $\bar{x}$, the weight w_i is large, so the point (x_i, y_i) has great "weight", and is more significant in the estimate b_1.

Conclusion: The estimates of straight line parameters may be found from geometric considerations. These estimates correspond to those found by the least-squares method.

6.2 Conditions for the Least-Squares Method

In determinination of the statistical properties of random vectors $\hat{\mathbf{y}}_P, \hat{\mathbf{e}}$ and $\mathbf{b}$, there are some conditions necessary for the least-squares method (LS) to be valid [1].

(1) The regression parameters $\boldsymbol{\beta}$ can have any value. In chemometric practice, however, there are some restrictions on the parameters, based on physical meaning.

(2) The regression model is linear in the parameters, and an additive model for the measurement errors is valid [Eq. (6.5b)].

(3) The matrix of non-random controllable values of the independent variables X has a column rank equal to m. This means that the two columns $\mathbf{x}_j, \mathbf{x}_k$ are not collinear (i.e. parallel) vectors. This is the same as saying that the matrix X^TX is a symmetric regular invertible matrix with non-zero determinant. That is, plane L is m-dimensional, and vector X b and the parameter estimates b are unambiguously determined.

(4) The mean value of the random errors ε_i is zero; $E(\varepsilon_i) = 0$. This is valid for all correlation models. It may happen that $E(\varepsilon_i) = K$, $i = 1, ..., n$, which means that the model does not contain an intercept

term. If an intercept term is used in such a model, it will be found that
$E(\varepsilon'_i) = 0$ where $\varepsilon'_i = y_i - \hat{y}_{P,i} - K$.

(5) The random errors ε_i have constant and finite variance, $E(\varepsilon_i^2) = \sigma^2$. The conditional variance $D(y/x) = \sigma^2$ is also constant and therefore the data are said to be *homoscedastic*.

(6) The random errors ε_i are uncorrelated and therefore $cov(\varepsilon_i, \varepsilon_j) = E(\varepsilon_i, \varepsilon_j) = 0$. When the errors follow the normal distribution they are also independent. (This corresponds to independence of the measured variables *y*.)

(7) The random errors ε_i have a normal distribution $N(0, \sigma^2)$. The vector y then has a multivariate normal distribution with mean $\mathbf{X}\boldsymbol{\beta}$ and covariance matrix $\sigma^2\mathbf{E}$ where E is the identity matrix.

When first six conditions are met, the parameter estimates **b** found by minimization of a least-squares criterion are best *unbiased linear* estimates of the regression parameters $\boldsymbol{\beta}$ [2].

The term *best* estimates (**b**) means that any linear combination of these estimates has the *smallest* variance of all linear unbiased estimates. That is, the variances of the individual estimates $D(b_j)$ are the smallest from all possible linear unbiased estimates (Gauss-Markov theorem).

It should be noted that there exist *biased estimates,* the variance of which can be smaller than the variance of estimates $D(b_j)$.

The term *unbiased* estimates means that $E(\boldsymbol{\beta} - \mathbf{b}) = 0$ and the mean value of an estimate vector $E(\mathbf{b})$ is equal to a vector of regression parameters $\boldsymbol{\beta}$.

The term *linear* estimates means that they can be written as a linear combination of measurements **y** with weights Q_{ij} which depend only on the locations of variables x_j, $j = 1, ..., m$. If we write Eq. (6.11) $\mathbf{Q} = (\mathbf{X}^T\mathbf{X})^{-1}\mathbf{X}^T$ for the weight matrix, we can then say

$$b_j = \sum_{i=1}^{n} Q_{ij} y_i \qquad (6.15)$$

Each estimate b_j is the weighted sum of all measurements. Also, the estimates **b** have an asymptotic multivariate normal distribution with covariance matrix [2, 4].

$$D(\mathbf{b}) = \sigma^2 (\mathbf{X}^T\mathbf{X})^{-1} \qquad (6.16)$$

When condition (7) is valid, all estimates **b** have a normal distribution, even for finite sample sizes *n*.

Problem 6.3 *Variance of parameter estimates for calibration straight line*

Derive the equations for calculation of the variance of estimates of parameters $D(b_1), D(b_2)$ and the co-variance $cov(b_1, b_2)$ for the calibration straight lines from Problem 6.1.

Solution: From Eq. (6.16) and the answer to Problem 6.1, we can write

$$D(b_1) = \frac{n\sigma^2}{D}, \quad D(b_2) = \frac{\sigma^2 \sum_{i=1}^{n} x_i^2}{D}$$

$$\text{and } cov(b_1, b_2) = \frac{-\sigma^2 \sum_{i=1}^{n} x_i}{D}.$$

The correlation coefficient R_0, expressing the correlation between estimates b_1 and b_2 is calculated from

$$R_0 = \frac{cov(b_1, b_2)}{\sqrt{D(b_1).D(b_2)}} = \frac{-\sum_{i=1}^{n} x_i}{\sqrt{n \sum_{i=1}^{n} x_i^2}}$$

For small values of n and positive values of x, the coefficient R_0 can be near to -1. This means that the estimate of the slope b_1 and the estimate of the intercept b_2 in model $y = \beta_1 x + \beta_2$ are negatively correlated, and the corresponding correlation coefficient can reach very high value.

For the variances $D(b_1^*)$ and $D(b_2^*)$ the equations for estimates b_1^* and b_2^* may be used,

$$D(b_1^*) = \sum_{i=1}^{n} w_i^2 D(y_i) = \sigma^2 \sum_{i=1}^{n} w_i^2 = \frac{\sigma^2}{\sum_{i=1}^{n} (x_i - \bar{x})^2}$$

$$D(b_2^*) = \frac{\sum_{i=1}^{n} D(y_i)}{n^2} = \frac{\sigma^2}{n}$$

Based on Eq. (6.14e), the expression may be written as

$$D(b_2) = \sum_{i=1}^{n} \left(\frac{1}{n} - \bar{x} w_i \right)^2 D(y_i) = \sigma^2 \left[\frac{1}{n} + \frac{\bar{x}^2}{\sum\limits_{i=1}^{n} (x_i - \bar{x})^2} \right]$$

Conclusion: For a calibration straight line, the variance of the estimates of intercept and slope, and the correlation coefficient between the parameters may be calculated from the simple expressions derived. Estimates b_1 and b_2 for positive data $x_i > 0$, $i = 1,..., n$, are always negatively correlated.

6.3 Statistical Properties of the Least-squares Method

When conditions (1) to (7) for the least-squares method are met, some statistical properties of vectors $\mathbf{b}$, $\hat{\mathbf{y}}_P$ and $\hat{\mathbf{e}}$ may be utilized. As the projection matrix $\mathbf{H}$ is non-random, for a *covariance matrix of prediction,* the following expression is valid

$$D(\hat{\mathbf{y}}_P) = \sigma^2 \mathbf{H} \tag{6.17}$$

and for a *covariance matrix of residuals* the expression

$$D(\hat{\mathbf{e}}) = \sigma^2 \mathbf{P} = \sigma^2 (\mathbf{E} - \mathbf{H}) \tag{6.18}$$

Both (6.17) and (6.18) are based on important properties of projection matrices, i.e. idempotence $\mathbf{H} = \mathbf{HH}$ and symmetry $\mathbf{H} = \mathbf{H}^T$.

Variances of the parameter estimates $\mathbf{b}$ are derived from Eq. (6.11) and given by Eq. (6.16). The residual sum of squares *RSS*, denoted also by $U(\mathbf{b})$, may be written as:

$$RSS = U(\mathbf{b}) = \hat{\mathbf{e}}^T \hat{\mathbf{e}} = \mathbf{y}^T (\mathbf{E} - \mathbf{H}) \mathbf{y} = \varepsilon^T \mathbf{P} \varepsilon$$

and the mean residual sum of squares sum is expressed as

$$E(RSS) = \sigma^2 \text{tr}(\mathbf{P}) = \sigma^2 (n - m) \tag{6.19}$$

where tr($\mathbf{P}$) is a trace matrix $\mathbf{P}$. With reference to the idempotence and symmetry of the projection matrix $\mathbf{P}$, the trace of matrix $\mathbf{P}$ is equal to its rank.

An unbiased estimate of the variance of errors σ^2 can be calculated with the use of the variance of the residuals

$$\hat{\sigma}^2 = \frac{U(\mathbf{b})}{n-m} = \frac{\hat{\mathbf{e}}^T\hat{\mathbf{e}}}{n-m} \tag{6.20}$$

Problem 6.4   Estimation of the variances of prediction and residuals for a calibration straight line

Derive expressions for calculation of an estimate of the prediction variance $D(\hat{y}_{P,i})$ and of an estimate of the residual variance $D(\hat{e}_i)$ for the calibration straight line from Problem 6.1.

Solution: From Problem 6.1 and Eq. (6.17) we know that for an estimate of the prediction variance:

$$D(\hat{y}_{P,i}) = \sigma^2\left[\frac{1}{n} + \frac{n(x_i - \overline{x})^2}{D}\right] \quad \text{and from Eq. (6.18) for an estimate of the}$$

residual variance

$$D(\hat{e}_i) = \sigma^2\left[\frac{n-1}{n} - \frac{n(x_i - \overline{x})^2}{D}\right]$$

Thus, the prediction variance and the residual variance are quadratic functions of the distance from $\overline{x}$. At the point $x_i = \overline{x}$, the prediction variance has a minimum and the residual variance a maximum (Fig. 6.5a, b).

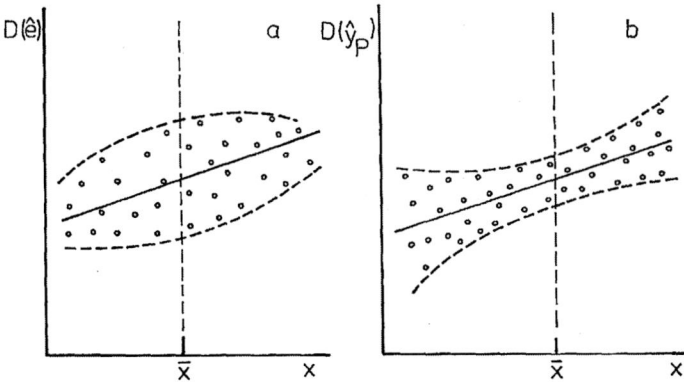

Figure 6.5   Dependence of (a) the residual variance $D(\hat{e}_i)$ and (b) the prediction variance $D(\hat{y}_{P,i})$ on the independent variable x.

Conclusion: At the point $\overline{x}$ the prediction variance is minimal and the residual variance maximal. When a point x_i is far away from $\overline{x}$, the prediction is less precise but an estimate of the residual is more precise.

From Fig. 6.1 it is evident that a square with the length of vector **y** is equal to the sum of squares of the lengths of vectors $\hat{\mathbf{y}}_P$ and $\hat{\mathbf{e}}$,

$$\mathbf{y}^T\mathbf{y} = \hat{\mathbf{y}}_P^T\hat{\mathbf{y}}_P + \hat{\mathbf{e}}_P^T\hat{\mathbf{e}} = \mathbf{y}^T\mathbf{H}\mathbf{y} + \mathbf{y}^T(\mathbf{E} - \mathbf{H})\mathbf{y} \tag{6.21}$$

or in abbreviated notation

$$TSS = SS + RSS \tag{6.21a}$$

Equations (6.21) and (6.21a) may be understood to mean that the total sum of squares *TSS* may be decomposed into two components, the sum of squares *SS* caused by the regression model and the unelucidated residual sum of squares *RSS*. The mean value of the sum-of-squares of a regression model is given by

$$E(SS) = E(\mathbf{y}^T\mathbf{H}\mathbf{y}) = \boldsymbol{\beta}^T\mathbf{X}^T\mathbf{X}\boldsymbol{\beta} + m\sigma^2 \tag{6.22}$$

Instead of the quantities *RSS* and *SS,* their average values are often used. The mean regression sum of squares is defined by

$$MSS = SS / m \tag{6.23a}$$

and its expected value by

$$E(MSS) = \sigma^2 + \frac{1}{m}\boldsymbol{\beta}^T\mathbf{X}^T\mathbf{X}\boldsymbol{\beta} \tag{6.23b}$$

The mean residual sum of squares is defined by

$$MRS = RSS / (n - m) \tag{6.24a}$$

and its expected value by

$$E(RMS) = \sigma^2 \tag{6.24b}$$

If $\boldsymbol{\beta} = 0$, i.e. all parameters of regression model are zero, then

(a) *SS* is independent of *RSS*

(b) $SS = \sigma^2 \chi_m^2$ where χ_m^2 is a random variable with the χ^2-distribution with m degrees of freedom,

(c) $RSS = \sigma^2 \chi_{n-m}^2$ where χ_{n-m}^2 is a random variable with the χ^2 -distribution with $(n - m)$ degrees of freedom.

On the basis of these three facts, the ratio

$$F = MSS / MRS \qquad (6.25)$$

has the Fisher-Snedecor *F*-distribution with m and $(n - m)$ degrees of freedom.

Problem 6.5 *Decomposition of the total sum of squares for the model of a calibration straight line*

For the calibration straight line defined in Problem 6.1 try to decompose the total sum of squares [Eq. (6.21)].

Solution: The full expression for $RSS = \sum_{i=1}^{n} \hat{e}_i^2$ is

$$RSS = \sum_{i=1}^{n} \hat{e}_i (y_i - b_2 - b_1 x_i) = \sum_{i=1}^{n} \hat{e}_i y_i - b_2 \sum_{i=1}^{n} \hat{e}_i - b_1 \sum_{i=1}^{n} \hat{e}_i x_i$$

For models with an intercept, $\sum_{i=1}^{n} \hat{e}_i = 0$ always. The vector $\hat{e}$ is perpendicular to vector **x** so that $\sum_{i=1}^{n} \hat{e}_i x_i = 0$. Therefore, the last two terms in the sum are equal to zero and

$$RSS = \sum_{i=1}^{n} \hat{e}_i y_i = \sum_{i=1}^{n} [y_i - b_2 - b_1 x_i] y_i$$

$$= \sum_{i=1}^{n} y_i^2 - \left[b_2 \sum_{i=1}^{n} y_i + b_1 \sum_{i=1}^{n} x_i y_i \right] = TSS - SS$$

and the regression sum-of-squares is expressed by

$$SS = b_2 \sum_{i=1}^{n} y_i + b_1 \sum_{i=1}^{n} x_i y_i$$

Expressing SS directly from the definition we get

$$SS = \sum_{i=1}^{n} \hat{y}_{P,i}^2 = \sum_{i=1}^{n} (b_2 + b_1 x_i)^2 = nb_2^2 + 2b_1 b_2 \sum_{i=1}^{n} x_i + b_1^2 \sum_{i=1}^{n} x_i^2$$

The resulting equation can be used to express the expected value of SS. From the elemental properties of an expected value, we have

$$E(b_2^2) = \beta_2^2 + D(b_2)$$

$$E(b_1^2) = \beta_1^2 + D(b_1)$$

and $E(b_1, b_2) = \beta_1 \beta_2 + \mathrm{cov}(b_1, b_2) = \beta_1 \beta_2 - \dfrac{\sigma^2 \sum_{i=1}^{n} x_i}{D}$

From these three expressions the expected value of the regression model sum-of-squares is given by

$$E(SS) = 2\sigma^2 + \left[n\beta_2^2 + 2\beta_1 \beta_2 \sum_{i=1}^{n} x_i + \beta_1^2 \sum_{i=1}^{n} x_i^2 \right]$$

This can also be derived by straight substitution into Eq. (6.22).

Conclusion: From estimates b_1 and b_2 it is possible to calculate not only RSS but also SS.

When the model contains an intercept term we will use the linear combination of vectors

$$E(y/x) = \beta_1 \mathbf{x_1} + \ldots + \beta_{m-1} \mathbf{x_{m-1}} + \beta_m \mathbf{J} \tag{6.26}$$

where $\mathbf{J} = (1,1,\ldots, 1)^{\mathrm{T}}$ is the vector of all ones. On introducing centred variables

$$\mathbf{x}_{C,j} = \mathbf{x}_j - \mathbf{J}\bar{x}_j, \qquad j = 1, \ldots, m - 1 \tag{6.27}$$

the scalar products become $\{\mathbf{x}_{C,j} \ \mathbf{J}\} = 0$. This means also that vectors $\mathbf{x}_{C,j}$ and $\mathbf{J}$ are orthogonal. In Eq. (6.27) the symbol $\bar{x}_j = 1/n \sum_{i=1}^{n} x_{ij}$ means the arithmetic mean of the jth controllable independent variable. By using centred variables the regression model will be expressed in matrix form

$$\mathbf{y} = \mathbf{X}_C \boldsymbol{\beta}^* + \beta_m^* \mathbf{J} + \boldsymbol{\varepsilon} \tag{6.28}$$

where $\mathbf{X}_C$ is a matrix of dimension $(n \times (m - 1))$ and β^* is vector of dimension $(m - 1) \times 1$. Because of the orthogonality of the two variables in Eq. (6.28), the estimates $\mathbf{b}^*$ and b_m^* of parameters β^* and β_m^* may be determined independently, from a projection into a plane L_1 as defined by the columns of matrix $\mathbf{X}_C$ or from projection on a vector $\mathbf{J}$ (see Problem 6.2). By projection into a plane L_1 we find

$$\mathbf{X}_C \mathbf{b}^* = \mathbf{X}_C (\mathbf{X}_C^T \mathbf{X}_C)^{-1} \mathbf{X}_C^T \mathbf{y} \tag{6.29a}$$

$$\text{or } \mathbf{b}^* = (\mathbf{X}_C^T \mathbf{X}_C)^{-1} \mathbf{X}_C^T \mathbf{y} \tag{6.29b}$$

By using projection onto a vector $\mathbf{J}$ we find

$$\mathbf{J} b_m^* = \mathbf{J}(\mathbf{J}^T \mathbf{J})^{-1} \mathbf{J}^T \mathbf{y} \tag{6.30a}$$

$$\text{or } b_m^* = (\mathbf{J}^T \mathbf{J})^{-1} \mathbf{J}^T \bar{\mathbf{y}} = \frac{1}{n} \sum_{i=1}^{n} y_i = \bar{y} \tag{6.30b}$$

With the use of Eq. (6.30), the expression for an estimate of intercept term variance may be derived:

$$D(b_m) = \sigma^2 (\mathbf{J}^T \mathbf{J})^{-1} = \frac{\sigma^2}{n} \tag{6.31}$$

By introducing a centred dependent variable

$$\mathbf{y}_C = \mathbf{y} - \mathbf{J}\bar{y} \tag{6.32}$$

the regression model (6.28) is transformed into a model without an intercept term

$$\mathbf{y}_C = \mathbf{X}_C \boldsymbol{\beta}_C + \varepsilon \tag{6.33}$$

For this model, the total sum of squared deviations from the average *TSC* may be decomposed into the sum of squared deviations from the regression model and the sum of squared residuals *RSC* (equal to *RSS*). Then

$$\mathbf{y}_C^T\mathbf{y}_C = \hat{\mathbf{y}}_{PC}^T\hat{\mathbf{y}}_{PC} + \hat{\mathbf{e}}^T\hat{\mathbf{e}} \tag{6.34}$$

$$\text{or } TSC = SSC + RSC \tag{6.35}$$

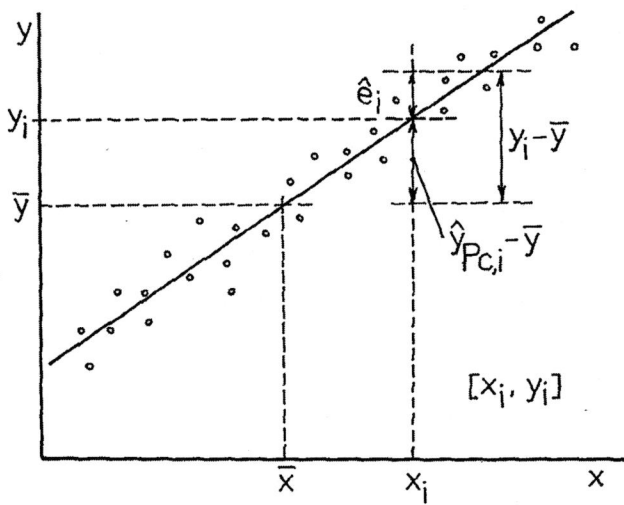

Figure 6.6 Decomposition of the total deviation from the mean $(y_i - \overline{y})$ into an explained part $(\hat{y}_{PC,i} - \overline{y})$ and a part not explained by the regression model, $\hat{e}_i$.

Decomposition of the total variations from an average $(y_i - \overline{y})$ into a part explained $(\hat{y}_{PC,i} - \overline{y})$ and a part explained $\hat{e}_i$ by a regression model is illustrated in Fig. 6.6 for a regression straight line.

We know that

$$TSC = TSS - n\overline{y}^2 \tag{6.36}$$

and also

$$SSC = \mathbf{y}_C^T\mathbf{H}_C\mathbf{y}_C = \mathbf{b}^T\mathbf{X}^T\mathbf{X}\mathbf{b} - n\overline{y}^2 \tag{6.37}$$

where $\mathbf{b}$ is a vector of dimension $(m \times 1)$ containing an intercept term. The cosine of the angle between vectors $\mathbf{y}_C$ and $\hat{\mathbf{y}}_{PC}$ can be calculated by trigonometry (Fig. 6.7):

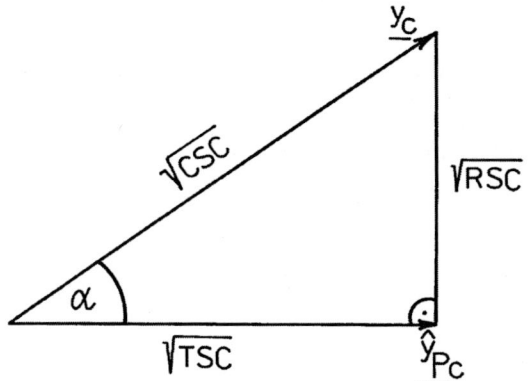

Figure 6.7 Geometry of the determination of $\cos \alpha$.

$$\cos \alpha = \sqrt{\frac{SSC}{TSC}} = \sqrt{1 - \frac{RSC}{TSC}} = \sqrt{1 - \frac{\sum_{i=1}^{n} \hat{e}_i^2}{\sum_{i=1}^{n} (y_i - \bar{y})^2}} \qquad (6.38)$$

The quantity $\cos \alpha$ is numerically equal to the value of the multiple correlation coefficient expressed for correlation models.

The square of the correlation coefficient R is called the *determination coefficient R^2*. For regression models a quantity $R = \cos \alpha$ is interpreted as the measure of relative difference between regression model $M_1 : \hat{\mathbf{y}}_P = \mathbf{Xb}$ and model $M_0 : \hat{\mathbf{y}}_P = \mathbf{J}\bar{\mathbf{y}}$. If R tends to zero, the regression model has all parameters except the intercept term equal to zero and model M_0 is valid. This means that in practical calculations the quantity R cannot be used as a measure of linearity, even for investigation of the quality of a regression model.

From Eq. (6.25) we can calculate the ratio

$$F_R = \frac{(TSC - RSC)(n - m)}{RSC(m - 1)} = \frac{\hat{R}^2 (n - m)}{(1 - \hat{R}^2)(m - 1)} \qquad (6.39)$$

which has the Fisher-Snedecor F-distribution with $(m - 1)$ and $(n - m)$ degrees of freedom. In Eq. (6.39) the quantity $\hat{R}^2$ means an estimate of the

determination coefficient calculated from $R = \cos \alpha$ and the use of estimates
b. With the use of F_R (6.39) the null hypothesis $H_0 : \boldsymbol{\beta}_C = 0$ may be tested; this
is equivalent to the hypothesis H_0: $R^2 = 0$. A test of significance of multiple
correlation coefficient is the same as a test of significance of all regression
coefficients except the intercept term.

Problem 6.6   *Investigation of abrasion resistance and the composition of rubber*

The dependence of the abrasion resistance of rubber, y, on the content of silica
filler x_1 and a binding substance x_2 was studied. Whereas the filler increases
abrasion resistance, the binding substance also increased its resistance
efficiency. Estimate parameters β_1, β_2 and β_3 of this proposed linear
model $E(y/x) = \beta_1 x_1 + \beta_2 x_2 + \beta_3$ and test the statistical significance of the
correlation coefficient.

○ **Data:** $n = 11$, $m = 3$

y	83	113	92	82	100	96	98	95	80	100	92
x_1^*	1	1	-1	-1	0	0	0	0	0	1.5	-1.5
x_2^*	-1	1	1	-1	0	0	0	1.5	-1.5	0	0

Variables x_1^* and x_2^* are transformed from the raw variables x_1 and x_2 as
follows: $x_1 = 6.7 x_1^* + 50$, $x_2 = 2 x_2^* + 4$

Solution: Since the data are the result of a planned experiment, the vectors **J**,
$\mathbf{x}_1^*$ and $\mathbf{x}_2^*$ are orthogonal. The matrix $\mathbf{X}^T\mathbf{X}$ is then diagonal, with the form

$$\mathbf{X}^T\mathbf{X} = \begin{bmatrix} \sum x_{i1}^{*2} & 0 & 0 \\ 0 & \sum x_{i2}^{*2} & 0 \\ 0 & 0 & n \end{bmatrix} = \begin{bmatrix} 8.5 & 0 & 0 \\ 0 & 8.5 & 0 \\ 0 & 0 & 11 \end{bmatrix}$$

The vector $\mathbf{X}^T\mathbf{y}$ has the components

$$\mathbf{X}^T\mathbf{y} = \begin{bmatrix} \sum y_i x_{i1}^* \\ \sum y_i x_{i2}^* \\ \sum y_i \end{bmatrix} = \begin{bmatrix} 34 \\ 62.5 \\ 1031 \end{bmatrix}$$

The following estimates are calculated from Eq. (6.11).

$$b_1^* = \frac{\sum\limits_{i=1}^{n} y_i x_{i1}^*}{\sum\limits_{i=1}^{n} x_{i1}^{*2}} = 4$$

$$b_2^* = \frac{\sum\limits_{i=1}^{n} y_i x_{i2}^*}{\sum\limits_{i=1}^{n} x_{i2}^{*2}} = 7.3529$$

$$b_3^* = \frac{\sum\limits_{i=1}^{n} y_i}{\sum\limits_{i=1}^{n} n} = 93.7273$$

where the stars denote estimates in the transformed variables model. The regression model in the raw variables has the form

$$\hat{y}_P = \left[\frac{x_1 - 50}{6.7}\right] 4 + \left[\frac{x_2 - 4}{2}\right] 7.3529 + 93.7273 = 0.597 x_1 + 3.6765 x_2 + 49.1708$$

The corresponding residuals sum of squares RSC is given by

$$RSC = \sum_{i=1}^{n} (y_i - \hat{y}_{P,i})^2 = 326.6$$

and the estimate of residual standard deviation $(n = 11, m = 3)$

$$\hat{\sigma} = \sqrt{\frac{RSC}{n - m}} = 6.39$$

Because of the diagonality of the matrix $\mathbf{X}^T\mathbf{X}$ we can say that

$$D(\mathbf{b}^*) = \hat{\sigma}^2 \begin{bmatrix} 1/8.5 & 0 & 0 \\ 0 & 1/8.5 & 0 \\ 0 & 0 & 1/11 \end{bmatrix} = \begin{bmatrix} 4.804 & 0 & 0 \\ 0 & 4.804 & 0 \\ 0 & 0 & 3.712 \end{bmatrix}$$

For estimates corresponding to the raw variables, then

$$D(b_1) = \frac{D(b_1^*)}{6.7^2} = 0.107$$

$$D(b_2) = \frac{D(b_2^*)}{2^2} = 1.201$$

$$D(b_3) = \left[\frac{50}{6.7}\right]^2 D(b_1^*) + \left[\frac{4}{2}\right]^2 D(b_2^*) + D(b_3^*) = 290.47$$

The total sum of deviations from the mean TSC is

$$TSC = \sum_{i=1}^{n} y_i^2 - n\bar{y} = 922.1818$$

Introducing this into Eq. (6.38) leads to an estimate of the coefficient of determination:

$$\hat{R}^2 = 1 - \frac{326.6}{922.1818} = 0.6458$$

From Eq. (6.39), the test criterion F_R is

$$F_R = \frac{(922.1818 - 326.6) \times 8}{326.6 \times 8} = 7.2943$$

The value F_R is higher then the corresponding quantile of the Fisher-Snedecor distribution $F_{0.95}(2, 8) = 4.46$, so at the significance level $\alpha = 0.05$, the coefficient of determination is considered to be significantly different from zero.

Conclusion: By using planned experimental data, all the columns of matrix $\mathbf{X}$ are mutually orthogonal, and with use of a suitable transformation of variables, the statistical characteristics of the linear model may be calculated.

6.3.1 Construction of Confidence Intervals

When parameter estimates **b** are determined, it is necessary to remember that **b** represents the point estimates of parameters **β**. These estimates are random quantities and in practice they are less important than the *confidence intervals* in which the true (theoretical) value of parameter $\boldsymbol{\beta}$ lies with some selected probability $(1 - \alpha)$ [57, 60]. As for univariate data samples, the significance level is usually chosen with $\alpha = 0.05$ or 0.01. These levels correspond to the 95% or the 99% confidence intervals.

Confidence intervals are constructed on the assumption that a random quantity $(n-m)\hat{\sigma}^2 / \sigma^2$ has the χ^2-distribution with $(n - m)$ degrees of freedom, and a random quantity $(\mathbf{b} - \boldsymbol{\beta})^{\mathrm{T}} \mathbf{X}^{\mathrm{T}} \mathbf{X} (\mathbf{b} - \boldsymbol{\beta}) / \sigma^2$ has a χ^2-distribution with m degrees of freedom. The corrected ratio of these quantities has a Fisher distribution with m and $n - m$ degrees of freedom. The bounds of the $100(1 - \alpha)\%$ confidence region are described by

$$(\mathbf{b} - \boldsymbol{\beta})^{\mathrm{T}} \mathbf{X}^{\mathrm{T}} \mathbf{X} (\mathbf{b} - \boldsymbol{\beta}) = m\hat{\sigma}^2 F_{1-\alpha}(m, n - m) \qquad (6.40)$$

where $F_{1-\alpha}(m, n-m)$ is the $(1 - \alpha)$ quantile of the Fisher-Snedecor F-distribution with m and $(n - m)$ degrees of freedom. Because the matrix $\mathbf{X}^{\mathrm{T}}\mathbf{X}$ is regular, Eq. (6.40) defines a hyperellipsoid with axes oriented in the directions of the eigenvectors $\mathbf{V}_j$ of the matrix $(\mathbf{X}^{\mathrm{T}}\mathbf{X})^{-1}$. The lengths of the individual half-axes are equal to $p\sqrt{\lambda_j}$ where λ_j are eigenvalues of the matrix $(\mathbf{X}^{\mathrm{T}}\mathbf{X})^{-1}$ and coefficient p is defined by

$$p^2 = m\hat{\sigma}^2 F_{1-\alpha}(m, n - m) \qquad (6.41)$$

Neglecting any correlation between parameter estimates, from Eq. (6.40) the $100(1 - \alpha)\%$ *simple confidence interval* for parameter β_j has the form

$$b_j - t_{1-\alpha/2}(n-m) \times \hat{\sigma}\sqrt{c_{jj}} \le \beta_j \le b_j + t_{1-\alpha/2}(n-m) \times \hat{\sigma}\sqrt{c_{jj}} \qquad (6.42)$$

where c_{jj} is the jth diagonal element of the matrix $(\mathbf{X}^{\mathrm{T}}\mathbf{X})^{-1}$ and $t_{1-\alpha/2}(n-m)$ is the $(1 - \alpha/2)$ quantile of the Student distribution with $(n - m)$ degrees of freedom. Simple confidence intervals are, however, too narrow for correlated estimates $\mathbf{b}$. Therefore we will define the *extreme confidence intervals* as the extremes on the confidence ellipsoid given by

$$b_j - p\sqrt{c_{jj}} \le \beta_j \le b_j + p\sqrt{c_{jj}} \qquad (6.43)$$

In some cases the confidence ellipsoid is created for q regression parameters on the assumption that they are the last q components of vector $\boldsymbol{\beta}$. Then, for the $100(1 - \alpha)\%$ boundary confidence ellipsoid

$$(\mathbf{b}_2 - \boldsymbol{\beta}_2)^\mathrm{T} \mathbf{D}_2^{-1}(\mathbf{b}_2 - \boldsymbol{\beta}_2) = q\hat{\sigma}^2 F_{1-\alpha}(q, n-m) \tag{6.44}$$

where the matrix $\mathbf{D}_2$ of dimension $(q \times q)$ is formed from the matrix $(\mathbf{X}^\mathrm{T}\mathbf{X})^{-1}$ by leaving out the first $(m-q)$ columns and $(m-q)$ rows. The symbol $\boldsymbol{\beta}_2$ denotes the vector of the last q components of vector $\boldsymbol{\beta}$, and vector $\mathbf{b}_2$ is defined analogously.

Similary, the confidence interval for a prediction $\hat{y}_{\mathrm{P},i}$ for point $\mathbf{x}_0 = (x_{01}, \ldots, x_{0m})^\mathrm{T}$ can be calculated. For $100(1-\alpha)\%$ confidence interval of prediction we can write [4]

$$\mathbf{x}_0^\mathrm{T}\mathbf{b} - t_{1-\alpha/2}(n-m) \times \hat{s}_{\mathrm{P},0} \leq \mathbf{x}_0^\mathrm{T}\boldsymbol{\beta} \leq \mathbf{x}_0^\mathrm{T}\mathbf{b} + t_{1-\alpha/2}(n-m) \times \hat{s}_{\mathrm{P},0} \tag{6.45}$$

where $\hat{s}_{\mathrm{P},0}^2$ is the variance of prediction for which [Eq. (6.17)] the following expression may be used

$$D(\hat{y}_{\mathrm{P},0}) \approx \hat{s}_{\mathrm{P},0}^2 = \hat{\sigma}^2 \mathbf{x}_0^\mathrm{T} (\mathbf{X}^\mathrm{T}\mathbf{X})^{-1} \mathbf{x}_0 \tag{6.46}$$

The relationships between the limits of the confidence interval of prediction (6.46) and $\mathbf{x}_0$ form the $100(1-\alpha)\%$ confidence band. The band is narrowest at the centre of gravity of the controllable (independent) variables, $x_{0j} = \bar{x}_j$.

When the confidence bands for all possible values of vector $\mathbf{x} = (x_1, \ldots, x_m)^\mathrm{T}$ are to be calculated, the Schéffe method should be used. With probability $(1-\alpha)$ the theoretical value $\mathbf{x}^\mathrm{T}\boldsymbol{\beta}$ lies in an interval

$$\mathbf{x}^\mathrm{T}\boldsymbol{\beta} \pm \sqrt{mF_{1-\alpha}(m, n-m)\hat{\sigma}^2 \mathbf{x}^\mathrm{T}(\mathbf{X}^\mathrm{T}\mathbf{X})^{-1}\mathbf{x}} \tag{6.45a}$$

Confidence bands constructed from Eq. (6.45a) are called *Working-Hottelling bands* [4] and have the same properties as those constructed on the basis of Eq. (6.45).

Problem 6.7 *Validation of a new laboratory method*

Validate a new laboratory method by comparison of results *(y)* with results obtained by a classical standard method *(x)* for a set of parallel determinations. If both methods lead to same results, the dependence of y on x is linear ($y = \beta_1 x + \beta_2$) with zero intercept $\beta_2 = 0$ and unit slope $\beta_1 = 1$. Estimate the parameters b_1 and b_2 and construct the 95% confidence interval of intercept and slope, and the 95% confidence interval of prediction for a sample with $x_0 = \bar{x}$.

○ **Data:** amount of reagent in mg determined by new laboratory *(y)* and standard *(x)* methods, $n = 24$, $m = 2$

x: 40.2, 43.8, 47.6, 50.7, 56.8, 81.3, 83.3, 97.1, 102.5, 118.7, 129.4, 184.8, 287.5, 295.4, 420.3, 421.3, 427.9, 566.1, 608.5, 640.7, 692.8, 705.2, 714.4, 881.4.

y: 48.9, 39.1, 42.6, 56.9, 70.3, 71.5, 97.6, 99.9, 105.2, 102.3, 106.8, 162.9, 234.0, 303.4, 388.8, 391.1, 369.3, 611.6, 580.2, 643.3, 596.6, 612.6, 633.5, 669.8

Solution: The least-squares estimates of slope is $b_1 = 0.868(\pm 0.030)$ and intercept $b_2 = 14.73(\pm 12.61)$ ($\pm$ standard deviations) were computed. The determination coefficient $\hat{R}^2 = 0.974$ and the estimate of standard deviation of residuals $\hat{\sigma} = 39.54$. From Eq. (6.42)

$$b_2 - t_{1-\alpha/2}(n-m)\sqrt{D(b_2)} \le \beta_2 \le b_2 + t_{1-\alpha/2}(n-m)\sqrt{D(b_2)}$$

whence $14.73 - 2.08 \times 12.61 < \beta_2 < 14.73 + 2.08 \times 12.61$ and $-11.499 < \beta_2 < 40.959$. Since this confidence interval includes zero, the intercept β_2 is not significantly different from zero. The confidence interval for the slope is $0.868 - 2.08 \times 0.0302 < \beta_1 < 0.868 + 2.08 \times 0.0302$ or $0.805 < \beta_1 < 0.930$.

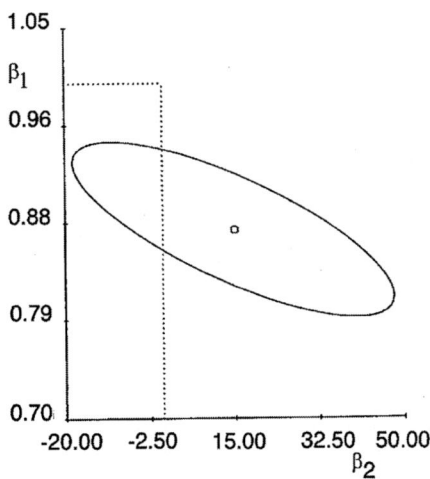

Figure 6.8 Construction of the 95% confidence ellipse and the point $\beta_1 = 1$ and $\beta_2 = 0$.

Because this interval does not include 1.0, the slope can not be considered to be equal to one. Figure 6.8 demonstrates the 95% confidence ellipse for parameters $\beta_1 = 1$ and $\beta_2 = 0$.

For a regression straight line and $x_0 = \bar{x} = 320.7$, according to Eq. (6.46) we may write

$$D(\hat{y}_P) = \hat{s}_{P,0}^2 = \hat{\sigma}^2 \left[\frac{1}{n} + \frac{(x_0 - \bar{x})^2}{D} \right] = \frac{\hat{\sigma}^2}{n}$$

Then introducing numbers into Eq. (6.45) leads to

$$14.73 + 0.868 \times 320.7 - \frac{2.08 \times 39.54}{\sqrt{24}} \leq x_0^T \beta \leq 14.73$$

$$+ 0.868 \times 320.7 + \frac{2.08 \times 39.54}{\sqrt{24}}$$

whence $276.89 \leq x_0^T \beta \leq 309.89$

Despite the small variance of prediction in the point $\bar{x}$, the 95% confidence interval is rather broad. Figure 6.9 shows construction of the 95% confidence bands of the calculated regression straight line together with experimental points.

Conclusion: The confidence intervals of the intercept and the slope indicate that the intercept of regression straight line can be considered to be equal to zero, but the slope significantly differs from unity. Thus the results of the new laboratory method differ from those obtained by the standard method by a multiplicative constant.

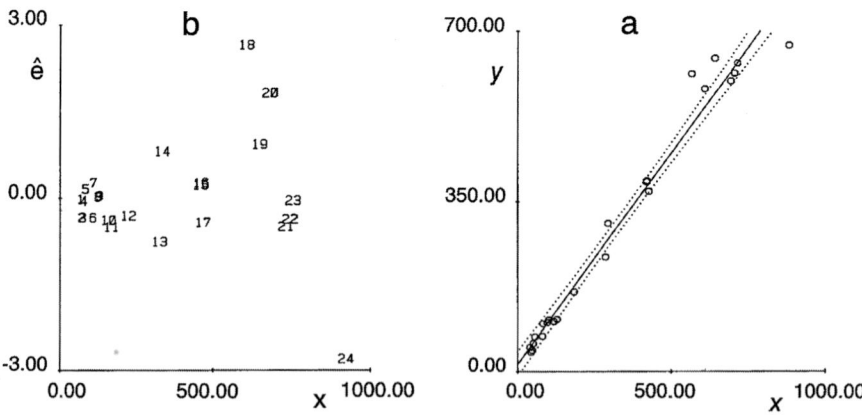

Figure 6.9   (a) Construction of the 95% confidence bands and (b) dependence of residuals $\hat{e} = f(x)$ on variable x.

6.3.2 Testing of Hypotheses

Tests for significance of parameters are closely connected with the construction of confidence intervals. To test the null hypothesis H_0: $\boldsymbol{\beta} = \boldsymbol{\beta}_0$ where $\boldsymbol{\beta}_0$ is the vector of known numbers, against the alternative H_A: $\boldsymbol{\beta} \neq \boldsymbol{\beta}_0$ the test criterion based on Eq. (6.40) may be expressed as

$$F = \frac{(\mathbf{b} - \boldsymbol{\beta}_0)^T \mathbf{X}^T \mathbf{X}(\mathbf{b} - \boldsymbol{\beta}_0)}{m\hat{\sigma}^2} \tag{6.47}$$

which has approximately the Fisher-Snedecor F-distribution with m and $(n - m)$ degrees of freedom. If H_0 is valid. The criteria defined by Eqs. (6.25) and (6.39) are special cases of the test statistic F as defined by Eq. (6.47).

For a test of the simple hypothesis H_0: $\beta_j = \beta_{j,0}$ against the alternative H_A: $\beta_j \neq \beta_{j,0}$ the test criterion based on Eq. (6.42) is

$$T_j = \frac{|b_j - \beta_{j,0}|}{\hat{\sigma}\sqrt{c_{jj}}} \tag{6.48}$$

which has approximately the Student t-distribution with $n - m$ degrees of freedom when H_0 is valid.

Most regression programs perform a Fisher-Snedecor test of significance of the determination coefficient Eq. (6.39) and a Student t-test on the significance of the individual parameters β_j calculated from Eq. (6.48) with $\beta_{j,0} = 0$. The F-test also determines simultaneous significance of all components of vector β except an absolute member. There are four cases:

(1) The F-test is not significant and all t-tests also are not significant. Then the model is considered to be unsuitable because it does not explain the variability of y.

(2) The F-test and all t-tests are significant. Then the model is considered to be suitable to express the variability of y. It does not mean, however, that the model is correct and acceptable.

(3) The F-test is significant but one or more t-test is not significant. Then the model is considered to be suitable, but the controllable variables x_j for which the parameters β_j are not significantly different from zero are rejected.

(4) The F-test is significant but all t-tests for parameters β indicate that all controllable variables are insignificant. Paradoxically, this shows that although the model as a whole is suitable, none of controllable variables is significant. This may result from multicollinearity (Section 6.3.2.1).

It should be noted that a model is considered to be significant when the form of the model is $f(\mathbf{x}, \boldsymbol{\beta}) = \mathbf{X}/\boldsymbol{\beta}$ but not $f(\mathbf{x}, \boldsymbol{\beta}) = J\bar{y}$.

Problem 6.8 *Disadvantages of classical statistical analysis for linear regression*

Anscombe [5] published test data for four simulated samples of size $n = 11$. Test the statistical significance of parameters β_1 and β_2, i.e. H_0: $\beta_1 = 0$ against H_A: $\beta_1 \neq 0$ and H_0: $\beta_2 = 0$ against H_A: $\beta_2 \neq 0$ and compare the results of tests with graphical analysis of residuals.

○ **Data:** To check the efficiency of statistical algorithms for linear regression models, the following test data samples are often used. The four data samples have the same statistical characteristics $b_1 = 0.5$, $b_2 = 3.0$, $D(b_1) = 0.0139$ and $D(b_2) = 1.2656$.

Table 6.1 Test data for linear regression

Data sample	A		B	C	D	
Variable	x	y	y	y	x	y
Point						
1	10	8.04	9.14	7.46	8	6.58
2	8	6.95	8.14	6.77	8	5.76
3	13	7.58	8.74	12.74	8	7.71
4	9	8.81	8.77	7.11	8	8.84
5	11	8.33	9.26	7.81	8	8.47
6	14	9.96	8.10	8.84	8	7.04
7	6	7.24	6.13	6.08	8	5.25
8	4	4.26	3.10	5.39	19	12.50
9	12	10.84	9.13	8.15	8	5.56
10	7	4.82	7.26	6.42	8	7.91
11	5	5.68	4.74	5.73	8	6.89

Solution: When the linear regression model is $E(y/x) = \beta_1 x + \beta_2$, all four sets of data lead to the same parameter estimates i.e. $b_1 = 0.5$ and $b_2 = 3.0$, with parameter variances $D(b_1) = 0.0139$ and $D(b_2) = 1.2656$, and the test criteria are $T_1 = 2.667$ and $T_2 = 4.241$. The test criterion F_R (6.39) has the same value $F_R = 17.97$, the determination coefficient $\hat{R}^2 = 0.66$ and the residual standard deviation $\hat{\sigma} = 1.237$. This leads to the conclusion that both regression parameters β_1 and β_2 are significantly different from zero.

Since the quantile of the F-distribution $F_{0.95}(1, 9) = 5.117$ is less than the calculated F_R, the determination coefficient differs from zero. It would seem that for all four data samples the linear regression model fits quite well. Figures 6.10 to 6.13 show that only data sample A is well characterized by a linear model. A good approximation is also reached for sample C, where just one outlier prevents the data from corresponding to a linear model.

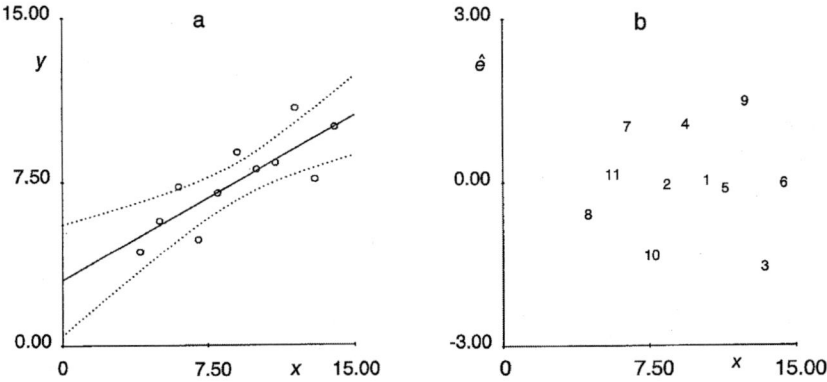

Figure 6.10   (a) Linear regression model $\hat{y}_P = f(x)$ for data sample A, and (b) dependence of residuals $\hat{e} = f(x)$ on variable x.

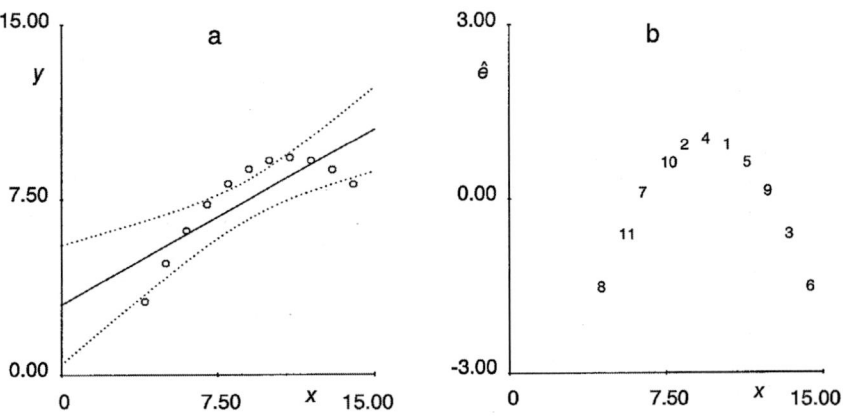

Figure 6.11   (a) Linear regression model $\hat{y}_P = f(x)$ for data sample B, and (b) dependence of residuals $\hat{e} = f(x)$ on variable x.

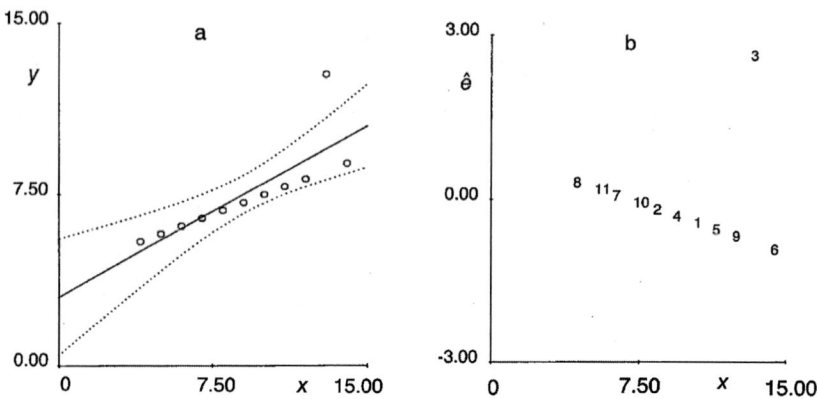

Figure 6.12 (a) Linear regression model $\hat{y}_P = f(x)$ for data sample C, and (b) dependence of residuals $\hat{e} = f(x)$ on variable x.

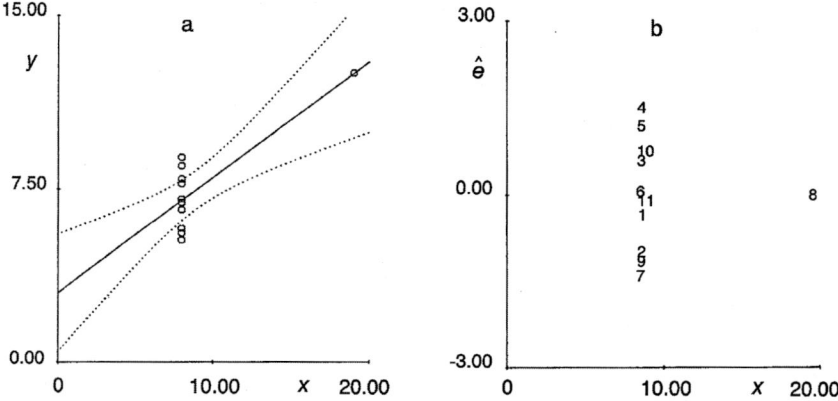

Figure 6.13 (a) Linear regression model $\hat{y}_P = f(x)$ for data sample D, and (b) dependence of residuals $\hat{e} = f(x)$ on variable x.

It may be rather surprising to a user of linear regression analysis that the statistical characteristics do not indicate here either the nonlinear trend of sample B (Fig. 6.11) or the silly data of sample D (Fig. 6.13).

Conclusion: Classical regression analysis may suggest models that do not correspond at all to the data set. Any model must be confirmed by graphical examination of residuals.

6.3.2.1 Testing for Multicollinearity

Paradoxical cases where the *F*-test is significant and all *t*-tests are not significant may result from strong multicollinearity among columns of matrix **X**. In correlation models, this corresponds to a situation when there are high values of paired correlation coefficients between controllable variables. Multicollinearity may be recognized by a finding that vectors $\mathbf{x}_j$ and $\mathbf{x}_k$, $j \neq k$, (which represents columns of matrix $\mathbf{X}$) are approximately parallel.

In the presence of multicollinearity, it is not possible to find the influence of the individual controllable variables x_j. Multicollinearity may exist in models which fit experimental reality quite well. Here, RSC has a small value and the predictions $\hat{y}_{P,i}$ are quite close to the experimental values y_i. Multicollinearity also appears in polynomial models and data which come from unplanned experiments.

Multicollinearity can be removed, for example, by selecting the location of experimental points such that the columns of matrix $\mathbf{X}$ will be mutually orthogonal, i.e. their scalar product will be zero.

$$\{\mathbf{x}_j, \mathbf{x}_k\} = \sum_{i=1}^{n} x_{ij} x_{ik} = 0 \quad \text{for} \quad j \neq k$$

If all columns of matrix $\mathbf{X}$ are mutually orthogonal, the matrix $\mathbf{X}^\mathrm{T}\mathbf{X}$ is diagonal and a solution of Eq. (6.11) can be expressed in the form

$$b_j = \frac{\displaystyle\sum_{i=1}^{n} x_{ij} y_i}{\displaystyle\sum_{i=1}^{n} x_{ij}^2}, \quad j = 1, \ldots, m$$

and the test criterion F_R is

$$F_\mathrm{R} = \frac{\displaystyle\sum_{j=1}^{m-1} T_j^2}{m-1} = T_\mathrm{S}$$

where T_S is an average value of all test statistics T_j^2 defined by Eq. (6.48) for $\beta_{0j} = 0$. It is supposed here that β_m is the intercept term.

To examine the suitability of a proposed linear model with regard to possible multicollinearity, Scott [8] uses a test criterion M_T

$$M_\mathrm{T} = \frac{\dfrac{F_R}{T_S} - 1}{\dfrac{F_R}{T_S} + 1} \tag{6.49}$$

and the following rules for *identification of multicollinearity*.

(a) If $M_T > 0.8$ the model is not suitable because of multicollinearity, so a model correction is necessary.

(b) If $0.33 \le M_T \le 0.8$ the model is poor because of multicollinearity, so some model correction is recommended.

(c) If $M_T < 0.33$, the model has little problem from multicollinearity, so no model correction is necessary.

The M_T criterion is useful in cases where it is necessary to discover all controllable variables which significantly affect the variability of the dependent variable y. When data are approximated by an empirical model, for example, by a polynomial, the M_T values need not be considered.

Problem 6.9 *Approximation of an absorption spectrum by a polynomial*

Find a model which describes the dependence of the molar absorptivity ε on wavelength λ, $\varepsilon = f(\lambda)$. Use a polynomial of the second degree $E(\varepsilon / \lambda) = \beta_1 + \beta_2\lambda + \beta_3\lambda^2$.

○ *Data:* $n = 15$, $m = 3$, ε [mol^{-1}.dm^3.cm^{-1}], λ [nm]:

ε	3	3.4	4.3	5	6	6.8	8.1	9.2	10.7	11.6	12.9	13.6	14.6	15.3	15.5
λ	460	470	480	490	500	510	520	530	540	550	560	570	580	590	600

Solution: Table 6.2 lists the numerical values of the parameter estimates b_1, b_2 and b_3 with their standard deviations and test statistics T_j for $\beta_j = 0$. Since the test criterion $F_R = 696$ is greater than the corresponding quantile of the Fisher-Snedecor F-distribution $F_{0.95}(2, 12) = 3.885$, the proposed model is statistically significant. In contrast, the quantile of the Student t-distribution $t_{0.975}(12) = 2.2$ is greater than both T_2 and T_3, therefore both parameters β_2 and β_3 are insignificant. The test criterion $M_T = 0.989$ (Eq. (6.49)) indicates very strong multicollinearity in the model.

Table 6.2 Parameter estimates and their statistical characteristics

j	Parameter	Estimate b	$\sqrt{D(b_j)}$	T_j
1	β_1	−43.93	19.38	-2.267
2	β_2	0.1018	0.0735	1.386
3	β_3	-2.51×10^{-6}	6.923×10^{-5}	-0.0361

Conclusion: In polynomial models, the significance of individual terms of the equation can not be judged from the result of the Student t-test alone.

Statistical tests are rather insensitive to small deviations of the error distribution from normality. However, in cases of strong non-normality or heteroscedasticity it is necessary to make a correction to the number of degrees of freedom for determination of the quantiles of the Fisher-Snedecor and Student distributions.

6.3.2.2 Test of Significance of the Intercept Term

In laboratory practice, it is important to examine the significance of the intercept term β_m by testing the null hypothesis H_0: $\beta_m = 0$ against the alternative H_A: $\beta_m \neq 0$. An intercept term always exists in correlation models. In regression models the intercept term ensures a zero sum of residuals $\sum_{i=1}^{n} \hat{e}_i = 0$.

In programs for regression analysis the following difficulties exist regarding the intercept term:

(a) The intercept term β_m always exists for centred data.
(b) Because the value $\bar{y} = 0$ is used in its calculation, the determination coefficient (6.38) for models without an intercept, $\hat{R}_B^2$ will be significantly higher than $\hat{R}^2$ for models with an intercept. The residual sum of squares for a model without an intercept, RSC_B, is always higher than or equal to the residual square sum for a model with an intercept, RSC.

Good programs allow calculation for a model with or without an intercept term, and correctly evaluate the determination coefficient because they do not substitute $\bar{y} = 0$. The difficulty can be avoided by introduction of a fictional point (x_{n+1}, y_{n+1}) with the co-ordinates [9]:

$$x_{n+1,j} = n^* \bar{x}_j, \quad j = 1, ..., m-1$$

$$y_{n+1} = n^* \bar{y}$$

where $n^* = 1 + \sqrt{n+1}$ and $\bar{x}_j$ and $\bar{y}$ are arithmetic means calculated from the data. With the use of this extended data set, the classical linear regression with an intercept term leads to the same results as the regression without an intercept term for the original set of n data points.

The influence of an intercept term on regression model may be understood by considering the location of point (x_{n+1}, y_{n+1}) with regard to other points. When this point is an outlier, the model without an intercept term is not suitable. The significance of an intercept term may be evaluated by use of Jack-knife residuals (6.97). If the point (x_{n+1}, y_{n+1}) is far from other points, the data are not in a suitable range to allow testing for presence of an intercept term. The significance of an intercept term may be also examined by the test statistic T_j (6.48), with $\beta_{0j} = 0$.

Problem 6.10   *A Lambert-Beer Law calibration straight line*

Estimate the parameters of the calibration *straight* line for the Lambert-Beer law for the dependence of a measured absorbance A on the concentration c, of p-nitroaniline. Does the straight line pass through the origin?

○ **Data:** $n = 6$, $m = 2$, $d = 1.000$ cm

c, 10^5.mol.dm^{-3}	1.98	2.58	3.42	4.43	5.51	6.58
Absorbance A	0.293	0.374	0.500	0.642	0.804	0.963

Solution: If the Lambert-Beer law holds, the straight line $A = \varepsilon_M \, d \, c$, where ε_M is the molar absorptivity, d is the cell length and c the molar concentration, passes through the origin. We will add the fictional point with co-ordinates $A_{n+1} = (1 + 7)\overline{A} = 2.172$ and $c_{n+1} = (1 + 7)\overline{c} = 14.88$. For this extended data set the model found was $A = (0.146 \pm 0.0003)c + (2.703 \pm 0.206) \times 10^{-5}$. The test statistics $T_1 = 490$ and $T_2 = 0.013$ show that the straight line goes through the origin. To check whether it is correct to neglect the intercept term, the extended data set is plotted in Fig. 6.14.

Conclusion: It is obvious that the point (x_{n+1}, y_{n+1}) does not differ from the others, and the Jack-knife residual (Eq. 6.97) $\hat{e}_{J,n+1} = 0.037$ also indicates that the intercept is insignificant.

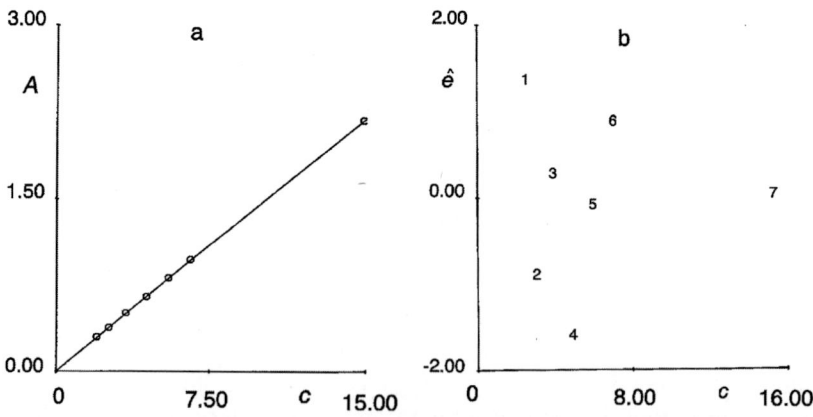

Figure 6.14   Test of significance of the intercept term in a Lambert-Beer law calibration: (a) the straight line for $A = \varepsilon c$ and the fictional point, and (b) the dependence of the residuals $\hat{e}$ on variable c.

Problem 6.11   *Estimates of the parameters of a calibration straight line that passes through the origin*

Demonstrate a procedure for parameter estimation in the case of a calibration straight line which must pass through the origin.

Solution: The regression model $E(y/x) = \beta_1 x$ is shown in Fig. 6.15. The vector $\hat{y}_P$ is the perpendicular projection of vector **y** on vector **x**. The estimate b_1 may be calculated by the simple expression from Problem 6.1, but we use here an analytical derivation of the least-squares criterion $U(\beta) = \sum_{i=1}^{n} (y_i - \beta_1 x_i)^2$,

$$\frac{\delta U(\beta)}{\delta \beta_1} = -2\sum_{i=1}^{n} (y_i - \beta_1 x_i)x_i = 0 \text{ and rewriting } b_1 = \frac{\sum_{i=1}^{n} x_i y_i}{\sum_{i=1}^{n} x_i^2} = \sum_{i=1}^{n} v_i y_i$$

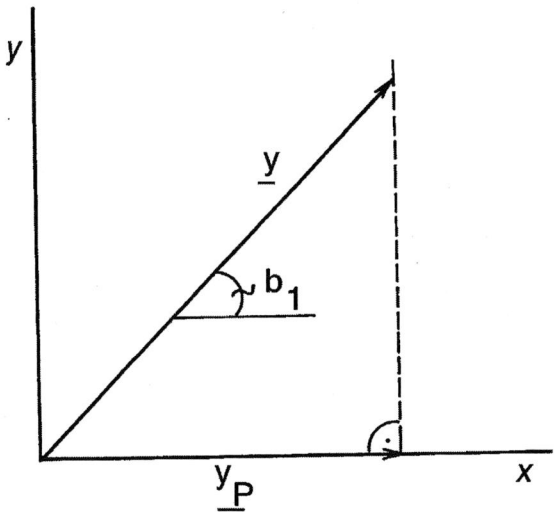

Figure 6.15   Geometrical illustration of the regression model $y = \beta_1 x$, and the projection of vector y on vector x.

The variance of this estimate is calculated from $D(b_1) = \sum_{i=1}^{n} v_i^2 D(y_i) = \dfrac{\sigma^2}{\sum_{i=1}^{n} x_i^2}$

The residual sum of squares *RSC* is given by $RSC = \sum_{i=1}^{n} y_i^2 - b_1 \sum_{i=1}^{n} x_i y_i$

and the theoretical sum of squares *SS* is $SS = b_1 \sum_{i=1}^{n} x_i y_i = b_1^2 \sum_{i=1}^{n} x_i^2$

For the determination coefficient R^2 it holds

$$R^2 = 1 - \frac{\sum\limits_{i=1}^{n} y_i^2 - b_1 \sum\limits_{i=1}^{n} x_i y_i}{\sum\limits_{i=1}^{n} y_i^2 - n\bar{y}^2} = \frac{b_1 \sum\limits_{i=1}^{n} x_i y_i - n\bar{y}^2}{\sum\limits_{i=1}^{n} y_i^2 - n\bar{y}^2}$$

The projection matrix $\mathbf{H}$ contains elements

$$H_{jj} = \frac{x_j^2}{\sum\limits_{i=1}^{n} x_i^2} \quad \text{and} \quad H_{jk} = \frac{x_j x_k}{\sum\limits_{i=1}^{n} x_i^2}, \quad j,k = 1,\ldots,n$$

From Eq. (6.17) we have

$$D(\hat{y}_{\mathrm{P},i}) = \frac{\hat{\sigma}^2 x_j^2}{\sum\limits_{i=1}^{n} x_i^2}$$

and the confidence interval for a model of regression straight line is calculated from Eq. (6.45)

$$x_0 b_1 - t_{1-\alpha/2}(n-1)\frac{\hat{\sigma}.x_0}{\sqrt{\sum\limits_{i=1}^{n} x_i^2}} \le x_0 \beta_1 \le x_0 b_1 + t_{1-\alpha/2}(n-1)\frac{\hat{\sigma}.x_0}{\sqrt{\sum\limits_{i=1}^{n} x_i^2}}$$

The end-points of the two confidence intervals are straight lines going through an origin, whereas for a general model of a regression straight line they are parabolic curves.

6.3.2.3 Simultaneous Test of a Composite Hypothesis

The likelihood ratio test (Section 8.6.2) may be used for testing general parametric hypotheses. In a case where the null hypothesis H_0: $\beta_2 = 0$ is to be tested against the alternative H_A: $\beta_2 \neq 0$, where β_2 represents the last q elements of the vector $\boldsymbol{\beta}$, the regression model is expressed in the divided form

$$y = \begin{bmatrix} X_1 & X_2 \end{bmatrix} \begin{bmatrix} \beta_1 \\ \beta_2 \end{bmatrix} + \varepsilon = X_1\beta_1 + X_2\beta_2 + \varepsilon$$

where X_1 is the matrix of dimension $[n \times (m-q)]$ containing those controllable variables with regression coefficients that are not included in a test vector β_2. Similarly, X_2 is the matrix of dimension $(n \times q)$ containing those controllable variables with regression coefficients that are included in a test vector β_2. When the hypothesis H_0 is valid, it is evident that $\hat{y}_{P,1} = X_1 b_1$ where $b_1 = (X_1^T X_1)^{-1} X_1^T y$ and the corresponding residual sum of squares RSC_1 is $RSC_1 = (y - \hat{y}_{P,1})^T (y - \hat{y}_{P,1})$. When the hypothesis H_A is valid, we have $\hat{y}_P = Xb$ where $b_1 = (X_1^T X_1)^{-1} X_1^T y$ and the corresponding residual sum of squares RSC is $RSC = (y - \hat{y}_P)^T (y - \hat{y}_P)$. The difference $(RSC_1 - RSC)$ corresponds to an increase in the residual sum of squares caused by validity of the null hypothesis H_0. The test criterion has the form

$$F_1 = \frac{(RSC_1 - RSC)(n-m)}{RSCq} \tag{6.50}$$

which if the H_0 hypothesis is valid, has the Fisher-Snedecor F-distribution with q and $(n-m)$ degrees of freedom.

A mistake often made in the application of linear regression in laboratories is a false approach to a choice of test criteria. Instead of the test criterion F_1, the individual test statistics T_j from Eq. (6.48) are calculated, and on their basis, the significance of a composite hypothesis H_0: $\beta_2 = \beta_{2,0}$ against H_A: $\beta_2 \neq \beta_{2,0}$ tested. Here $\beta_{2,0}$ is the vector of known parameters.

For tests of composite hypotheses, the test statistic F_1 should be used, where RSC_1 is the residual sum of squares for the model $\hat{y}_{P,i} = X_1 b_1 + X_2 \beta_{2,0}$ where b_1 is the estimate of parameters β_1 on the assumption that the restriction $\beta_2 = \beta_{2,0}$ is valid.

Problem 6.12 *Simultaneous test of a composite hypothesis for a Lambert-Beer law model*

For the data from Problem 6.10, test the composite null hypothesis H_0: $\beta_2 = 0, \beta_1 = 0.148$ against H_A: $\beta_2 \neq 0, \beta_1 \neq 0.148$. The false approach would be two separate tests of two null hypotheses, H_0: $\beta_2 = 0$ and H_0: $\beta_1 = 0.148$.

Solution: On substitution into Eq. (6.48), we obtain

$$T_2 = \frac{|1.461 \times 10^{-4} - 0|}{0.00398} = 0.037$$

$$T_1 = \frac{|0.1459 - 0.148|}{0.000908} = 2.314$$

Because T_1 and T_2 are less than the quantile of the Student t-distribution, $t_{0.975}(4) = 2.7764$, both tests lead to a conclusion that H_0: $\beta_2 = 0, \beta_1 = 0.148$ should be accepted. This conclusion is, however, *false*.

The more rigorous approach uses a simultaneous test of the composite hypothesis H_0: $\beta_2 = 0$ and $\beta_1 = 0.148$.

The procedure starts with a calculation of $RSC = 5.12 \times 10^{-5}$ for estimates $b_1 = 0.1459$ and $b_2 = 1.461 \times 10^{-4}$. Then, $RSC_1 = 5.3476 \times 10^{-4}$ for parameters $\beta_{2,0} = 0$ and $\beta_{1,0} = 0.148$ is calculated. From Eq. (6.50), the test criterion F_1 is

$$F_1 = \frac{(5.347 \times 10^{-4} - 5.12 \times 10^{-5}) \times 4}{5.12 \times 10^{-5} \times 2} = 18.89$$

Because the quantile of Fisher-Snedecor F-distribution is $F_{0.95}(2, 4) = 6.944$, the null hypothesis H_0: $\beta_2 = 0$ and $\beta_1 = 0.148$ cannot be accepted. The result of this F-test is not in agreement with conclusion of the previous t-tests. Figure 6.16 shows the 95% confidence ellipse of parameters β_1 and β_2, and the point $\beta_{1,0} = 0.148$ and $\beta_{2,0} = 0$ marked by a cross. This point lies outside the 95% confidence interval of the two parameters.

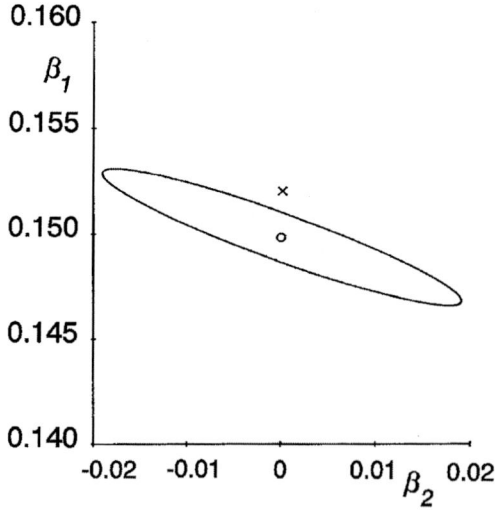

Figure 6.16   The 95% confidence interval for the parameters β_1 and β_2. The point $\beta_1 = 0$, $\beta_2 = 0.148$ is marked by a cross.

Conclusion: It may be concluded that a simultaneous test of the composite hypothesis cannot be replaced by tests of two separate hypotheses. Thus, testing of individual parameters in a vector β_0 can lead to quite false conclusions.

Problem 6.13 *Validation of a new laboratory method by a simultaneous test of a composite hypothesis*

Try to test a composite hypothesis H_0: $\beta_2 = 0$ and $\beta_1 = 1$ in Problem 6.7 against the alternative H_A: $\beta_2 \neq 0$ and $\beta_1 \neq 1$.

○ *Data:* from Problem 6.7

Solution: From the results of Problem 6.7, we have $RSC = 3440$, and when we set $\beta_{1,0} = 1$ and $\beta_{2,0} = 0$, we obtain $RSC_1 = 8221$. On substitution into Eq. (6.50), we find $F_1 = \dfrac{(8220 - 3440) \times 22}{3440 \times 2} = 15.28$ which is greater than the quantile of the Fisher-Snedecor F-distribution $F_{0.95}(2, 22) = 3.44$, so the null hypothesis H_0 cannot be accepted. This conclusion is also in agreement with the partial t-tests and confidence intervals of the two parameters. Figure 6.17 shows the regression straight line $\hat{y}_P = x$ with experimental points and a graphical analysis of residuals.

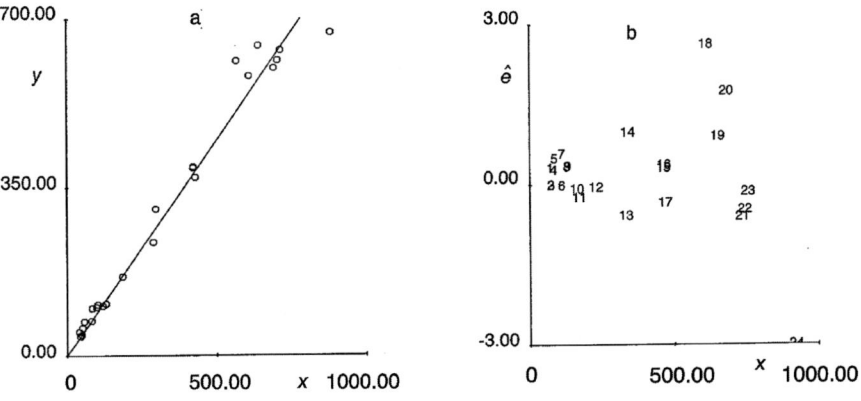

Figure 6.17 (a) Linear regression model of validation of a new laboratory method $\hat{y}_P = x$, and (b) dependence of the residuals on x.

Conclusion: A simultaneous test of the composite hypothesis (H_0: $\beta_{1,0} = 1$ and $\beta_{2,0} = 0$) confirmed that a new laboratory method is not in agreement with the results of a standard one.

6.3.2.4 Test of Agreement of Two Linear Models

The test of a composite hypothesis just described may be re-arranged to allow for testing of agreement of parameters in two linear models

$$\mathbf{y}_1 = \mathbf{X}_1 \boldsymbol{\beta}_1 + \boldsymbol{\varepsilon}_1 \tag{6.51a}$$

$$\mathbf{y}_2 = \mathbf{X}_2 \boldsymbol{\beta}_2 + \boldsymbol{\varepsilon}_2 \tag{6.51b}$$

where $\mathbf{X}_1$ is a matrix of dimension ($n_1 \times m$), $\mathbf{y}_1$ is a vector of dimension ($n_1 \times 1$), $\mathbf{X}_2$ is a matrix of dimension ($n_2 \times m$), and $\mathbf{y}_2$ is a vector of dimension ($n_2 \times 1$). RSC_1 is the residual sum of squares corresponding to model (6.51a), RSC_2 is the residual sum of squares corresponding to model (6.51b) and RSC is the residual sum of squares corresponding to the composite model

$$\begin{bmatrix} \mathbf{y}_1 \\ \mathbf{y}_2 \end{bmatrix} = \begin{bmatrix} \mathbf{X}_1 \\ \mathbf{X}_2 \end{bmatrix} \times \boldsymbol{\beta} + \begin{bmatrix} \boldsymbol{\varepsilon}_1 \\ \boldsymbol{\varepsilon}_2 \end{bmatrix} \tag{6.52}$$

We use the Chow test of the null hypothesis H_0: $\beta_1 = \beta_2$ against the alternative H_A: $\beta_1 \neq \beta_2$ based on the test criterion

$$F_C = \frac{(RSC - RSC_1 - RSC_2)(n - 2m)}{(RSC_1 + RSC_2) \times m} \tag{6.53}$$

where $n = n_1 + n_2$. If the variances of the two samples are the same ($\sigma_1^2 = \sigma_2^2$, homoscedasticity), the test criterion F_C has the Fisher-Snedecor F-distribution with m and $(n - 2m)$ degrees of freedom.

When the variances of the two samples are not the same ($\sigma_1^2 \neq \sigma_2^2$, heteroscedasticity), the Fisher-Snedecor F-distribution may be used with m and r degrees of freedom where

$$r = \frac{\left[(n_1 - m)\sigma_1^2 + (n_2 - m)\sigma_2^2 \right]^2}{(n_1 - m)\sigma_1^4 + (n_2 - m)\sigma_2^4} \tag{6.54}$$

A more accurate version of this equation is given in [6].

Problem 6.14 *Comparison of measurement results from two laboratories*

Determination of the free energy ΔG of the vapour of boron oxide as a function of temperature T was carried out in two laboratories [11]. Compare the results and test whether the values measured in the two laboratories can be considered to be the same.

○ **Data:** $n = 6$, $m = 2$,

T, K	$-\Delta G$, kcal/mol	
	Lab A	Lab B
1409	34.9	34.9
1441	33.8	34.6
1457	33.4	31.9
1492	32.4	33.1
1569	30.3	30.1
1610	29.3	29.1

Solution: If a linear regression model is valid for both data samples, the models are

$$E(\Delta G / T) = \beta_{1,\text{A}}T + \beta_{2,\text{A}}$$

$$E(\Delta G / T) = \beta_{1,\text{B}}T + \beta_{2,\text{B}}$$

We will test the null hypothesis H_0: $\boldsymbol{\beta}_\text{A} = \boldsymbol{\beta}_\text{B}$ against the alternative H_A: $\boldsymbol{\beta}_\text{A} \neq \boldsymbol{\beta}_\text{B}$, where $\boldsymbol{\beta}_\text{A} = (\beta_{1,\text{A}}, \beta_{2,\text{A}})^\text{T}$ and $\boldsymbol{\beta}_\text{B} = (\beta_{1,\text{B}}, \beta_{2,\text{B}})^\text{T}$. We use the Chow test:

$$\text{Laboratory A: } b_{1,\text{A}} = -0.02768(\pm 0.00525)$$

$$b_{2,\text{A}} = 73.73(\pm 7.865)$$

$$\hat{\sigma} = 0.916$$

$$RSC_\text{A} = 3.358$$

$$\text{Laboratory B: } b_{1,\text{B}} = -0.02776(\pm 0.000157)$$

$$b_{2,\text{B}} = 73.82(\pm 0.235)$$

$$\hat{\sigma} = 0.0274$$

$$RSC_B = 0.002992$$

Laboratory A + B: $b_{1,A+B} = -0.02772(\pm 0.00235)$

$$b_{2,A+B} = 73.77(\pm 3.521)$$

$$\hat{\sigma} = 0.58$$

$$RSC_{A+B} = 3.364$$

The standard deviations of the parameters are given in brackets. Substitution into Eq. (6.53) for $RSC = RSC_{A+B}$, $RSC_1 = RSC_A$ and $RSC_2 = RSC_B$, leads to

$$F_C = \frac{(3.364 - 3.358 - 0.002992)(12 - 4)}{(3.358 + 0.002992) \times 2} = 0.0036$$

Because the variances of the samples differ, we calculate the degrees of freedom r from Eq. (6.54).

$$r = \frac{\left[4 \times 0.916^2 + 4 \times 0.0274^2\right]^2}{4 \times 0.916^2 + 4 \times 0.0274^2} = 4.007 \approx 4$$

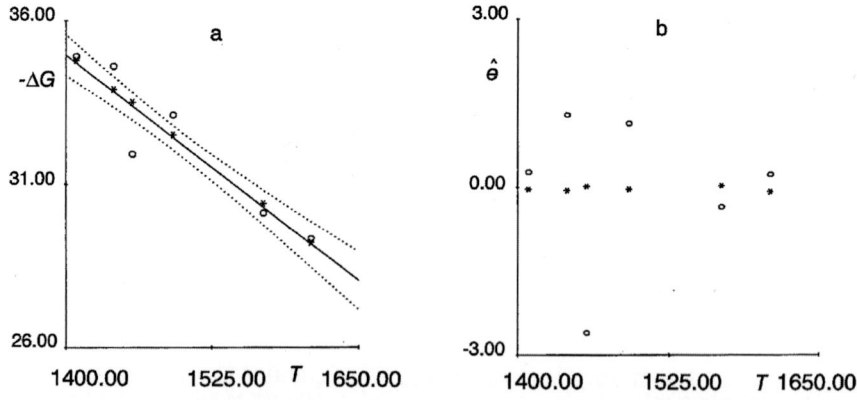

Figure 6.18 (a) Regression straight line for data $(-\Delta G, T)$ for two laboratories. Data for laboratory A (denoted by circle) have more spread; (b) dependence of residuals: lab A (denoted by circle) and lab B (denoted by star).

The quantile of the Fisher-Snedecor F-distribution $F_{0.95}(2, 4) = 6.94$ is greater than F_C, so H_0: $\beta_A = \beta_B$ is accepted. Figure 6.18 shows a graphical interpretation of the laboratory measurements. The data from lab A show much more spread.

Conclusion: On the basis of a Chow test, it may be concluded that results from the two laboratories can be considered to be the same. The data from lab A are less precise.

Problem 6.15 *Comparison of two calibration straight lines*

Two insulin samples, A and B, are compared according to their ability to decrease the level of blood sugar. The sample A was injected into 11 randomly chosen rats and sample B into 9 rats. The decrease of blood sugar level was determined. Compare the efficiency of the two insulin samples.

○ *Data:* Insulin A: x is the amount in μl of insulin A, and y is a decrease in sugar level [%] in blood, $n_A = 11$, $m_A = 2$,

x	120	160	200	240	280	320	360	400	440	480	500
y	17	26	30	27	45	47	48	63	60	69	69

Insulin B: the same for insulin B, $n_B = 9$, $m_B = 2$

x	169	200	240	280	320	360	400	440	480
y	9	18	17	25	39	45	47	57	61

Solution: If the two insulin types have the same effect, the two regression straight lines will not be significantly different. To test the agreement between the lines we use the test criterion F_C (6.53). The statistical characteristics are:

$$\text{Insulin A: } \hat{y}_A = 1.808(\pm 3.504) + 0.1369(\pm 0.0103)x$$

$$RSC_A = 159.6$$

$$\hat{\sigma} = 4.211$$

$$\text{Insulin B: } \hat{y}_B = -18.67(\pm 3.535) + 0.1688(\pm 0.0105)x$$

$$RSC_B = 74.25$$

$$\hat{\sigma} = 3.26$$

$$\text{Insulin } A + B: \quad \hat{y}_{A+B} = -6.397(\pm 4.45) + 0.1481(\pm 0.0132)x$$

$$RSC_{A+B} = 821.1$$

Then from Eq. (6.53) we find:

$$F_C = \frac{(821.1 - 159.6 - 74.25)(20 - 4)}{(159.6 + 74.25) \times 2} = 20.09$$

which is a greater value than the quantile of the Fischer-Snedecor F-distribution $F_{0.95}(2, 16) = 3.63$, so that the null hypothesis H_0 is rejected. From Fig. 6.19 it is evident that, although the straight lines have similar slopes, they differ in intercept.

Conclusion: The insulin samples have significantly different activity.

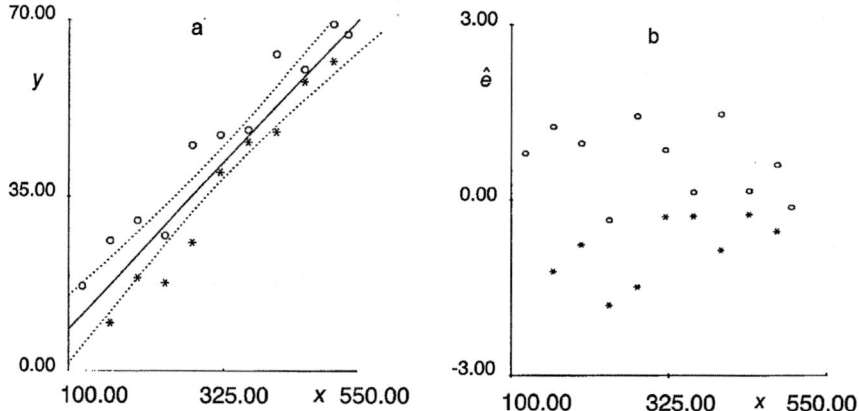

Figure 6.19   (a) Regression straight line for insulin data A and B and (b) dependence of residuals $\hat{e}$ on variable x. Circles mean sample A, stars mean sample B.

6.3.2.5 Acceptance Test for a Proposed Linear Model

Utts [7] has introduced a test of acceptance of a proposed linear regression model $f(\mathbf{x}, \beta) = \mathbf{X}\beta$, based on a ratio of the residual sum of squares. If the regression model $f(\mathbf{x}, \beta)$ is non-linear there exists a group of points n_1 to which a linear model will fit, as shown in Fig. 6.20.

Let us denote by RSC_1 the residual sum of squares corresponding to a linear regression of n_1 points, and by RSC the residual sum of squares corresponding to a linear regression of all n points. Utts criterion for model acceptance is formulated as

$$F_{\mathrm{U}} = \frac{(RSC - RSC_1) \times (n_1 - m)}{RSC_1 \times (n - n_1)} \qquad (6.55)$$

which for the hypothesis H_0: "the linear regression model is valid" has the Fisher-Snedecor F-distribution with $(n - n_1)$ and $(n_1 - m)$ degrees of freedom. Utts recommends choosing $n_1 \approx n/2$ and selecting the points that give the smallest values of the diagonal elements $H_{ii} = \mathbf{x}_i^{\mathrm{T}} (\mathbf{X}^{\mathrm{T}} \mathbf{X})^{-1} \mathbf{x}_i$ of the projection matrix $\mathbf{H}$. Such selected points lie close to the centre of gravity of the controllable independent variables. If the calculated F_{U} is smaller than the corresponding quantile of F-distribution, the linear regression model can be accepted.

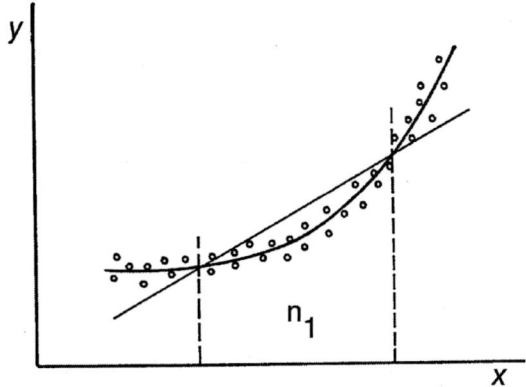

Figure 6.20 Principle of the Utts test.

Another group of tests of acceptance of a proposed regression model [7] is based on an application of extended regression model. The proposed linear model is usually extended to include higher powers of the independent variables and eventually also their interactions. If original linear model was correct, the parameters for higher members of the new model will be statistically insignificant.

For a model of a regression straight line, the assumption of linearity may be checked by a test of significance of parameter β_2 in an extended model

$$E(y/x) = \beta_1 x + \beta_2 x^2 + \beta_3$$

Here, the t-test could be used, but as any multicollinearity would change the results of the t-test, the Fisher-Snedecor F-test is applied. Let RSC_Q be the residual sum of squares for a quadratic model and RSC_L the residual sum of squares for a linear model. The test criterion of linearity will have the form

$$F_{\mathrm{L}} = \frac{(RSC_L - RSC_Q)(n-3)}{RSC_Q \times 1} \tag{6.56}$$

If the null hypothesis H_0: $\beta_2 = 0$ is valid, the F_{L} criterion has the Fisher-Snedecor F-distribution with 1 and $(n-3)$ degrees of freedom. The statistic $\sqrt{F_{\mathrm{L}}}$ has the Student distribution with $(n-3)$ degrees of freedom.

Instead of the various test criteria for testing linearity, the statistical characteristics for comparison of different models may also be used. One of these is the *mean quadratic error of prediction* defined by:

$$MEP = \frac{\sum_{i=1}^{n}(y_i - \mathbf{x}_i^{\mathrm{T}}\mathbf{b}_{(i)})^2}{n} \tag{6.57a}$$

where $\mathbf{b}_{(i)}$ is the estimate of regression parameters when all points except the ith one were used and $\mathbf{x}_i$ is ith row of matrix $\mathbf{X}$. The statistic MEP uses a prediction $\hat{y}_{\mathrm{P},i}$ from an estimate constructed without including the ith point. Another mathematical expression for MEP is:

$$MEP = \sum_{i=1}^{n} \frac{\hat{e}_i^2}{(1-H_{ii})^2 n} \tag{6.57b}$$

For large sample sizes n the element H_{ii} tends to zero $(H_{ii} \approx 0)$ and then

$$MEP = RSC/n \tag{6.57c}$$

If MEP is used instead of RSC in the equation for the determination coefficient (6.38), the resulting statistic $\hat{R}_{\mathrm{P}}^2$ is called the *predicted determination coefficient*

$$\hat{R}_{\mathrm{P}}^2 = 1 - \frac{n \times MEP}{\sum_{i=1}^{n} y_i^2 - n \times \bar{y}^2} \tag{6.58}$$

Another statistical characteristic has quite general use and is derived from information theory and entropy [12] and known as the *Akaike information criterion, AIC*

$$AIC = n\ln\left(\frac{RSC}{n}\right) + 2m \tag{6.59}$$

The most suitable model is the one which gives the lowest value of the Akaike information criterion.

Problem 6.16 *Selection from three polynomial models*

For the data from Problem 6.9, do a regression analysis, and test whether the data sample should be fitted by a polynomial of the third or fifth degree.

○ *Data:* Problem 6.9

Solution: Table 6.3 lists the statistical characteristics $MEP, \hat{R}_P^2, \hat{R}^2$ and AIC for the hypotheses that the regression model is expressed by a polynomial of the second, third and fifth degree.

Table 6.3 Selection of best model according to the statistics $MEP, \hat{R}_P^2, \hat{R}^2$ and AIC

Polynomial degree		MEP	$\hat{R}_P^2$	$\hat{R}^2$	AIC
2	✓	0.3502	0.9905	0.9915	−21.65
3		0.0283	0.9992	0.9992	−56.02
5		0.0613	0.9985	0.9997	−55.04

Because of the good precision of the experimental data, the statistics MEP and $\hat{R}_P^2$ do not indicate that the polynomial of third degree is the most suitable. Only statistic AIC indicates that the best model is the polynomial of third degree:

$$y_P = 860.2(\pm\, 85.17) - 5.057(\pm 0.485)x + 0.00977(\pm 0.00092)x^2 - 6.146 \times$$

$$10^{-6}(\pm 5.78 \times 10^{-7})x^3$$

and estimates of all three parameters are statistically significant. In the case of the polynomial of fifth degree, all the parameter estimates except β_3 are statistically insignificant, as a consequence of multicollinearity.

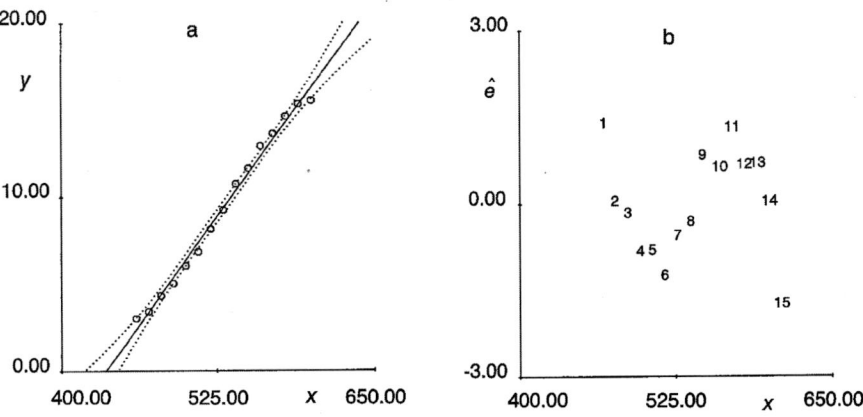

Figure 6.21 (a) Curve fitting of data from Problem 6.9 by a polynomial of the second degree, and (b) graphical examination of residuals.

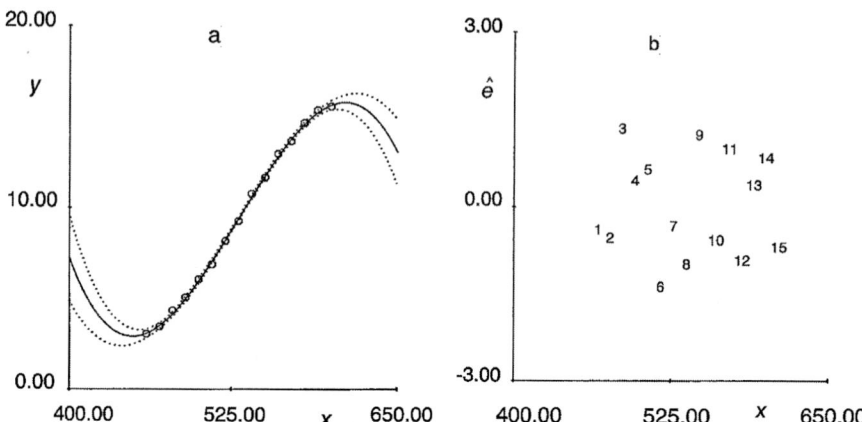

Figure 6.22 (a) Curve fitting of data from Problem 6.9 by a polynomial of the third degree, and (b) graphical examination of residuals.

Figure 6.21 shows the curve fitting of the data by a polynomial of the second degree and Fig. 6.22 by a polynomial of the third degree. The numerical statistics $\hat{R}_P^2$ and $\hat{R}^2$ are not able to distinguish between these two polynomials, but the graphical analysis of residuals is a more efficient tool for deciding among several plausible models.

Conclusion: There are cases when statistical characteristics fail, and graphical examination of residuals gives more satisfactory results for model specification.

Problem 6.17 *Examination of linearity of four samples of test data*

Examine the linearity of the four data samples from Problem 6.8. These four samples have the same values of statistical characteristics, but only sample A may be considered as an acceptable straight line.

○ **Data:** Problem 6.8

Solution: To test the linearity of the data, the linear model $E(y/x) = \beta_1 x + \beta_2$ is compared with the quadratic model $E(y/x) = \beta_1 x + \beta_2 x^2 + \beta_3$. The table lists the statistical characteristics RSC, MEP, $\hat{R}_P^2$, AIC, T_2 for a test of H_0: $\beta_2 = 0$ in the quadratic model and F_L. Nonlinearity is clearly indicated by F_L and by AIC. MEP and $\hat{R}_P^2$ show that there is an improvement of fit with the use of quadratic model for sample B, which exhibits a nonlinear curve.

	Proposed linear model				Proposed quadratic model				Tests	
	RSC	MEP	$\hat{R}_P^2$	AIC	RSC	MEP	$\hat{R}_P^2$	AIC	T_2	F_L
A	13.76	1.871	0.708	6.46	12.9	1.955	0.976	31.74	0.72*	0.53*
B	13.76	2.204	0.642	6.46	2.23×10^{-5}	3.11×10^{-6}	1	−114	2219	49
C	13.76	2.147	0.653	6.46	13.0	3.107	0.961	31.82	0.68*	0.467*
D	13.76	9×10^{99}	0.767	6.46	13.8	9×10^{99}	−	32.43	4	0

Conclusion: For examination of linearity, the F_L test criterion for comparing the linear with the quadratic model seems to be the most reliable. Other statistical characteristics, i.e. *AIC, MEP* and $\hat{R}_P^2$ can be used, but the rejection of the nonlinearity assumption does not lead to automatic acceptance of the linear model (see samples C and D).

There are many criteria for examination of linearity in regression models. Suitability of a proposed model can be easily checked if information about the measurement variance or the errors variance, σ^2, is available. When a proposed model is correct, *RSC* will be approximately equal to $(n - m)\,\sigma^2$. When the model is incorrect, the residual sum of squares *RSC* can be decomposed into the sum of squares *SSE* corresponding to "pure" errors and the sum of squares *SSL* corresponding to the poor choice of model, the so-called lack of fit:

$$RSC = SSE + SSL \tag{6.59a}$$

To estimate a variance σ^2 that is independent of a proposed linear regression model, the method of repeated measurements can be used. At *M* different values of vector $\mathbf{x}_i$, $i = 1,..., M$, there are always n_i repeated measurements y_{ij}, $i = 1,..., M$, $j = 1, ..., n_i$. The regression model is then expressed by

$$y_{ij} = \sum_{k=1}^{m} \beta_k x_{ik} + \varepsilon_{ij}, \quad j = 1,..., n_i \tag{6.60}$$

This model represents a linear regression model for $n = \sum_{i=1}^{M} n_i$ measurements. The matrix $\mathbf{X}$ has dimension $(n \times m)$ and the vector $\mathbf{y}$ has dimension $(n \times 1)$. On substitution into Eq. (6.11) the estimate $\mathbf{b}$ is obtained and the residual sum of squares is

$$RSC = \mathbf{y}^T\mathbf{y} - \mathbf{b}^T\mathbf{X}\mathbf{y} = \sum_{i=1}^{n} \hat{e}_i^2 \tag{6.61}$$

An independent estimate of the sum of squares due to pure errors SSE is calculated from

$$SSE = \sum_{i=1}^{M} \sum_{j=1}^{n_i} (y_{ij} - \bar{y}_i)^2 \tag{6.62}$$

where $\bar{y}_i = \left[\sum_{i=1}^{n_i} y_{ij}\right] / n_i$ is the arithmetic mean of repeated y_{ij} values for a given x_i. The sum of squares corresponding to the lack of fit is $SSL = RSC - SSE$. The independent estimate of variance is here $SSE/(n - M)$ and the mean value of SSL is $SSL/(M - m)$. The test criterion

$$F_N = \frac{(RSC - SSE)(n - M)}{SSE(M - m)} \tag{6.63}$$

has, for a correct model, the Fisher-Snedecor F-distribution with $(M - m)$ and $(n - M)$ degrees of freedom. When $F_N < F_{1-\alpha}(M - m, n - M)$, the proposed model is correct and the sum of squares SSL is not significantly different from zero.

The quantity $SSE/(n - M)$ represents an unbiased estimate of the variance σ^2, whether the proposed model is correct or not.

Problem 6.18 *Examination of a kinetic model of recombination of the Bromocresol Green anion*

The kinetic cosntant has been determined for the recombination reaction of the anion of Bromocresol Green (BCG) with a proton in a solution of glycerol and water [13]. For different concentrations $\mathbf{x}$ of BCG the reciprocal values of the relaxation times y_{ij}, $i = 1,..., n$ were measured. The kinetic model $y_{ij} = -k_D + k_R$ $x_i + \varepsilon_{ij}$ is proposed, where k_D is the kinetic constant of recombination, and k_R is the kinetic constant of dissociation. Determine the two kinetic constants and examine the proposed model.

○ *Data:* $n = 24$, $n_i = 2$, $i = 1, ..., M$, $M = 12$, $m = 2$

x_i [10^{-6}mol dm^{-3}]	7.98	8.96	10.37	12.08	16.81	24.22	29.5	36.75	37.69	65.32	87.32	145.5
y_{i1} [10^6 s^{-1}]	0.44	0.36	0.37	0.43	0.79	0.8	1.04	1.04	1.22	2.06	2.32	3.70
y_{i2} [10^6 s^{-1}]	0.35	0.40	0.46	0.51	0.59	0.9	1.04	0.94	1.27	1.73	2.08	3.66

Solution: With the use of the least-squares method, the estimates $\hat{k}_D = 0.232$ (± 0.033) and $\hat{k}_R = 0.0238$ (± 0.0006) were obtained (standard deviations in brackets). The determination coefficient $\hat{R}^2 = 0.987$ and the test criterion $F_R = 1616$ (Eq. (6.39)) prove that the linear model is valid.

The residual sum of squares $RSC = 0.285$, the residual standard deviation $\hat{\sigma} = 0.1139$ and an independent estimate of sum of squares corresponding to pure errors SSE (6.62) is for $M = 12$ and $n_i = 2$ equal to $SSE = 0.1274$. The independent estimate of the residual standard deviation is then

$$\hat{\sigma}_I = \sqrt{0.1274 / (24 - 12)} = 0.103.$$

On substituting into Eq. (6.63), we find the test criterion is

$$F_N = \frac{(0.285 - 0.1274)(24 - 12)}{0.1274(12 - 2)} = 1.484$$

As the quantile $F_{0.95}(10, 12) = 2.75$ is greater than F_N, the proposed kinetic model is correct and the residual sum of squares corresponding to lack of fit, $SSL = 0.285 - 0.1274 = 0.1575$ may be considered as insignificant. Figure 6.23 shows the regression model with the 95% confidence interval.

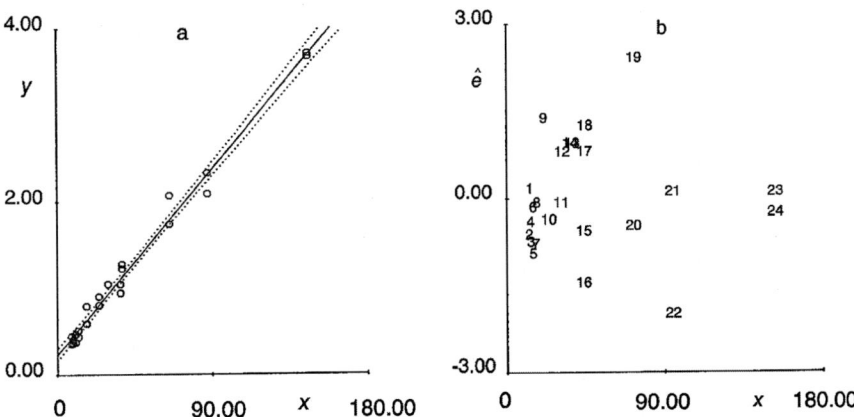

Figure 6.23 (a) Degree of fit of straight line for proposed kinetic model, and (b) graphical examination of residuals.

Conclusion: When replicate results are available, it is relatively easy to test the validity of a regression model.

When replicate measurements for all i are not available, the data may be divided into groups of approximately the same x values [14].

6.3.3 Comparison of Regression Lines

Often in data treatment we need to compare M proposed regression models

$$y_{ij} = \beta_{2j} + \beta_{1j} x_{ij} + \varepsilon_{ij}, \quad j = 1,..., M, \ i = 1,..., n_j \tag{6.64}$$

for M groups of experimental data $[(x_{ij}, y_{ij}), i = 1, ..., n_j], j = 1, ..., M$. Typical examples are Lambert-Beer law calibration lines, i.e. the dependence of absorbance on concentration at M different wavelengths. We want to know:

(a) if the regression lines have the same intercept;
(b) if the regression lines have the same slope;
(c) if the regression lines are identical.

The first step of the statistical analysis is always estimation of the parameters b_{2j}, b_{1j} and $\hat{\sigma}_j^2$ for each set of data, individually, by the least-squares method.

The second step involves examination of homoscedasticity, i.e. constancy of variance $\hat{\sigma}_j^2$ because testing hypotheses (a), (b) and (c) requires constant and identical variance in all groups.

The Bartlett test for homoscedasticity is a commonly used test. In this test, we compare M independent variance estimates $\hat{\sigma}_j^2, j = 1, ..., M$, with v_j degrees of freedom. The null hypothesis H_0: $\sigma_j^2 = \sigma^2, j = 1, ..., M$, is tested. For models of a regression straight line, the degrees of freedom are $v_j = n_j - 2$. We define:

$$V = \sum_{j=1}^{M} v_j \tag{6.65a}$$

$$\hat{\sigma}_C^2 = \frac{\sum_{j=1}^{M} v_j \times \hat{\sigma}_j^2}{V} \tag{6.65b}$$

$$\text{and } L = 1 + \frac{\sum_{j=1}^{M} v_j^{-1} - V^{-1}}{3M - 3} \tag{6.66}$$

The test criterion of the Bartlett test of homoscedasticity is given by

$$B = \frac{V.\ln\hat{\sigma}_C^2 - \sum_{j=1}^{M} v_j.\ln\hat{\sigma}_j^2}{L} \qquad (6.67)$$

which, if the null hypothesis is valid, has the χ^2-distribution with $(M-1)$ degrees of freedom. If $B < \chi_{1-\alpha}^2(M-1)$ [where $\chi_{1-\alpha}^2(M-1)$ is the $100(1-\alpha)\%$ quantile of the χ^2-distribution], the null hypothesis is accepted and the estimate of constant variance σ^2 is called the *pooled variance* $\hat{\sigma}_C^2$ (Eq. 6.65b). The Bartlett test, is, however, sensitive to deviations of the residuals from normality.

To compare two groups of points, $M = 2$, the identity of two variances H_0: $\sigma_1^2 = \sigma_2^2$ may be tested by the test criterion

$$F_2 = \frac{\max(\hat{\sigma}_1^2, \hat{\sigma}_2^2)}{\min(\hat{\sigma}_1^2, \hat{\sigma}_2^2)} \qquad (6.68)$$

which, if the null hypothesis is valid, has the Fisher-Snedecor F-distribution with $(n_1 - 2)$ and $(n_2 - 2)$ degrees of freedom when $\hat{\sigma}_1^2 > \hat{\sigma}_2^2$. Generally, the degrees of freedom used in calculation of $\hat{\sigma}_i^2$, $i = 1, 2$ are also used here.

6.3.3.1 Test for Homogeneity of Intercepts

When the null hypothesis H_0: $\beta_{21} = \beta_{22} = \ldots \beta_{2j} = \ldots \beta_{2M} = \beta_{2C}$ is valid, the pooled estimate of the overall intercept β_{2C} as a weighted combination of the estimates of the individual intercepts b_{2j} may be obtained from

$$b_{2C} = \frac{\sum_{j=1}^{M} w_{Bj}.b_{2j}}{\sum_{j=1}^{M} w_{Bj}} \qquad (6.69)$$

where the jth weight coefficient w_{Bj} corresponding to the estimate of the jth straight line is given by

$$w_{Bj} = \frac{n_j \sum_{i=1}^{n_j} (x_{ij} - \overline{x}_j)^2}{\sum_{i=1}^{n_j} x_{ij}^2} \qquad (6.70)$$

For testing the estimate of errors, variance σ^2 is calculated from the variance of individual parameter estimates b_{2j} around their weighted average b_{2C} and from a combination of the variability of all points around the regression line inside the invididual data groups. The test criterion is

$$F_I = \frac{\displaystyle\sum_{j=1}^{M} w_{Bj}(b_{2j} - b_{2C})^2 / (M-1)}{\displaystyle\sum_{j=1}^{M}\sum_{i=1}^{n_j} \hat{e}_{ij}^2 / (n-2M)} \tag{6.71}$$

where $n = \displaystyle\sum_{j=1}^{M} n_j$. When the null hypothesis H_0 is valid the test criterion F_I has the Fisher-Snedecor distribution with $(M-1)$ and $(n-2M)$ degrees of freedom. The residuals $\hat{e}_{ij}$ are calculated for the individual regression lines. We can write $\displaystyle\sum_{j=1}^{M}\sum_{i=1}^{n_j} \hat{e}_{ij}^2 = \sum_{j=1}^{M} RSC_j$ where RSC_j is the residual sum of squares for the jth group.

When $F_I < F_{1-a}(M-1, n-2M)$, then all straight lines have, at significance level α, the same intercept; and its estimate is given by Eq. (6.69). The variance of this intercept is calculated from

$$D(b_{2C}) = \frac{\hat{\sigma}^2}{\displaystyle\sum_{j=1}^{M} w_{Bj}} = \frac{\displaystyle\sum_{j=1}^{M}\sum_{i=1}^{n_j} \frac{\hat{e}_{ij}^2}{n-2M}}{\displaystyle\sum_{j=1}^{M} w_{Bj}} \tag{6.72}$$

The intercept estimate has an asymptotically normal distribution and represents an unbiased estimate of parameter β_{2C}.

6.3.3.2 Test for Homogeneity of Slopes

The test of homogeneity of slopes is known as a test of parallelism of regression straight lines. If the null hypothesis H_0: $\beta_{11} = \beta_{12} = \dots \beta_{1j} = \dots \beta_{1M} = \beta_{1C}$ is valid, the pooled estimate of overall (common) slope β_{1C} as a weighted combination of individual slope estimates b_{1j} may be calculated from

$$b_{1C} = \frac{\displaystyle\sum_{j=1}^{M} w_{Sj} \times b_{1j}}{\displaystyle\sum_{j=1}^{M} w_{Sj}} \tag{6.73}$$

$$\text{where } w_{Sj} = \sum_{i=1}^{n_j} (x_{ij} - \overline{x})^2 \tag{6.74}$$

As in the test for homogeneity of intercepts, a test criterion may be derived

$$F_S = \frac{\sum_{j=1}^{M} w_{Sj}(b_{1j} - b_{1C})^2 / (M-1)}{\sum_{j=1}^{M} \sum_{i=1}^{n_j} \hat{e}_{ij}^2 / (n-2M)} \tag{6.75}$$

which, when the null hypothesis is valid, has the Fisher-Snedecor F-distribution with $(M-1)$ and $(n-2M)$ degrees of freedom. When $F_S < F_{1-\alpha}(M-1, n-2M)$, all the regression straight lines can, at significance level α, be considered as parallel. The best estimate of overall slope is b_{1C}, from Eq. (6.73), and its variance estimate may be calculated from

$$D(b_{1C}) = \frac{\sum_{j=1}^{M} \sum_{i=1}^{n_j} \hat{e}_{ij}^2 / (n-2M)}{\sum_{j=1}^{M} w_{Sj}} \tag{6.76}$$

When the null hypothesis H_0 is valid, the slope estimate b_{1C} has a normal distribution and represents an unbiased estimate of parameter β_{1C}.

6.3.3.3 Test for Coincidence of Regression Lines

The test for coincidence of regression lines H_0: $\beta_{2j} = \beta_{2C}$, $\beta_{1j} = \beta_{1C}$, $j = 1,...,M$ is a combination of the two previous tests F_I and F_S (Sections 6.3.3.1 and 6.3.3.2). The test compares two residual sums of squares, RSC_K with RSC_C. RSC_K was obtained after fitting all M groups of data by a single common straight line with estimates b_{1K} and b_{2K} and RSC_C is calculated from the individual groups of data separately, $RSC_C = \sum_{j=1}^{M} RSC_j$. The test criterion is

$$F_A = \frac{(RSC_K - RSC_C)(2M-2)}{RSC_C / (n-2M)} \tag{6.77}$$

When the null hypothesis H_0 is valid, the test criterion F_A has the Fisher-Snedecor F-distribution with $(2M-2)$ and $(n-2M)$ degrees of freedom.

When $F_A < F_{1-\alpha}(M-2), (n-2M)$, then all regression straight lines may be considered as identical, with slope b_{1K} and intercept b_{2K}. Individual groups of data are then collected into a single common sample of size n. When the null hypothesis is not accepted, it is usually possible to find subgroups of data which are homogeneous enough.

Problem 6.19   *Comparison of three methods for determination of gold in jewellery alloys*

The classical method for determination of gold in jewellery alloys is rather tedious and time-consuming. Three variants of the X-ray fluorescence (XRF) method for gold have been proposed. Ten samples were measured by the classical method (values x) and the new methods (values y_1, y_2 and y_3). Test whether the results from the three variants differ from one another and if these results are in agreement with those measured by the classical method.

○ **Data:** $M = 4$, $n_{ij} = 10$

i	x	y_1	y_2	y_3
1	59.0	63.8	72.2	64.3
2	59.0	62.6	71.1	63.1
3	59.0	64.0	69.9	64.3
4	59.3	58.9	63.5	59.2
5	62.7	60.2	64.8	60.5
6	64.8	62.9	70.8	63.4
7	72.3	70.0	68.7	70.0
8	75.2	77.4	76.7	77.4
9	75.2	78.7	85.2	79.1
10	75.2	80.7	78.5	80.6

Solution: The parameter estimates b_{2j}, b_{1j}, RSC_j and the residual standard deviations for all three variants of determination of gold are listed in Table 6.4. All three variants have the same sample size $n_j = 10$ and have a common x-axis. Therefore the weights w_{Bj} and w_{Sj} are independent of the j value, so that:

$$w_{Bj} = \frac{10\sum_{i=1}^{10}(x_i - \bar{x})^2}{\sum_{i=1}^{10} x_i^2} = 0.1124 \quad \text{and} \quad w_{Sj} = 10\sum_{i=1}^{10}(x_i - \bar{x})^2 = 497.59.$$

Before testing the agreement of residuals, the variances must be examined. On introducing numbers into Eqs. (6.64–6.67), we find $V = 24$, $\hat{\sigma}_C^2 = 15.804$, $L = 1.0556$ and $B = 4.9087$. As the quantile $\chi_{0.95}^2(2) = 5.99$ is greater than B, the null hypothesis about homoscedasticity is accepted. The variance for the second X-ray fluorescence method is significantly greater than for the first and third one.

Table 6.4 Regression analysis of gold determination by three X-ray fluorescence methods.

Variant j	b_{2j}	b_{1j}	RSC_j	$\hat{\sigma}_j$
1	1.087	1.01	92.98	3.409
2	31.55	0.61	192.8	4.91
3	2.74	0.989	93.44	3.418
1+2+3	11.79	0.871	540.4	4.393

To test the homogeneity of intercepts, the pooled estimate of the intercept in Eq. (6.69), $b_{2C} = 11.79$, is calculated, and $\sum_{j=1}^{3} RSC_j = 379.22$. The test criterion F_I is calculated from Eq. (6.71):

$$F_I = \frac{65.9692 / 2}{379.22 / 24} = 2.088$$

Since the quantile $F_{0.95}(2, 24) = 3.4$ is greater than F_I the three intercepts can be considered to be identical and equal to the estimate $b_{2C} = 11.792$ with variance $D(b_{2C}) = 35.151$ (Eq. (6.72)).

If the three methods are not systematically biased, the null hypothesis H_0: $\beta_{2C} = 0$ should be valid. From Eq. (6.48), $T_2 = 1.989$ is smaller than the quantile $t_{0.975}(8) = 2.3$, so the null hypothesis H_0 is accepted, and all three variants do not lead to systematically biased results.

To test the homogeneity of the slopes, the pooled estimate of the slope is calculated (6.73), $b_{1C} = 0.870$, and from Eq. (6.75) the test criterion $F_s = 1.596$, which is a lower value than the quantile $F_{0.95}(2, 24) = 3.4$. Therefore all slopes can be considered to be identical and equal to $b_{1C} = 0.870$ with variance [Eq. (6.76)] $D(b_{1C}) = 0.0079$.

When all three variants of the XRF method (y_1, y_2, y_3) give the same results as the standard method (x), the null hypothesis H_0: $\beta_{1C} = 1$ should be valid. From Eq. (6.48), $T_1 = 1.459$, which is smaller than the quantile $t_{0.975}(8) = 2.3$, so the null hypothesis is accepted and all results y_1, y_2, y_3 are identical with x.

In the test for identity of the three regression lines by Eq. (6.77), the test criterion

$$F_A = \frac{(540.4 - 379.22)/4}{379.22/24} = 2.55$$

is smaller than the quantile $F_{0.95}(4, 24) = 2.776$, so all three XRF methods can be considered to be identical.

To test the homoscedasticity of the three variants, three repetitive measurements at $x = 59.0$ and $x = 75.2$ are used. Table 6.5 lists independent estimates of variance, calculated from these repetitive measurements.

Table 6.5 Variance estimates calculated from repeated measurements, and the F-test criterion (Eq. (6.65)).

Variant	$\hat{\sigma}_1^2$	$\hat{\sigma}_2^2$	$\hat{\sigma}_K^2$	$F_2 = \dfrac{\hat{\sigma}_1^2}{\hat{\sigma}^2}$	$F_j = \dfrac{\hat{\sigma}_j^2}{\hat{\sigma}_K^2}$
	(x=59.0)	(x=75.2)	(both x)		
1	0.573	2.763	1.668	4.882	6.97
2	1.323	20.06	10.692	15.163	1.59
3	0.480	2.564	1.522	5.341	7.67

The null hypothesis H_0: $\sigma_1^2 = \sigma_2^2$ is tested by the Fisher-Snedecor F-test, and Table 6.5 lists the F_2 values. The second null hypothesis H_0: $\sigma_j^2 = \sigma_K^2$ says that any jth variance is the same as the variance of all measurements. When F_2 values are compared with the quantile $F_{0.95}(2, 2) = 19$, the null hypothesis H_0 (i.e. an assumption about homoscedasticity) is accepted. When F_j values are compared with the quantile $F_{0.95}(24, 4) = 5.77$, it may be concluded that the residual variances of the first and the third variants are higher than the variance of measurement. The conclusion is here affected by the fact that the replicate measurements were carried out only for two different levels of x.

Conclusion: Examination of three XRF variants proved that they do not differ significantly and lead to the same results as the standard method. The first variant seems to be the most precise one.

6.4 Numerical Problems in the Computer Calculation of Linear Regression

The determination of parameter estimates of a linear model [Eq. (6.11)] seems to be a simple task. When subprograms for matrix operations are available in a package of algorithms, the formal solution of Eq. (6.11) is quite easy. Some difficulties arise when the matrix $\mathbf{X}^T\mathbf{X}$ appears to be singular, from the point of view of the machine precision and the algorithm. In some cases, especially with polynomial models, the parameter estimates may be without physical meaning. The regression curve goes quite close to experimental points but oscillates among them (for polynomials of higher degree) or is systematically shifted.

The reasons for numerical difficulties in the computer evaluation of parameter estimates **b** are as follows:

(1) Neglect of the limited precision of computer in building the matrix $\mathbf{X}^T\mathbf{X}$.

(2) Inconvenient procedures for matrix inversion or solving the set of linear equations.

(3) Multicollinearity leading to the ill-conditioning of matrix $\mathbf{X}^T\mathbf{X}$.

(4) Linear dependence of some columns of matrix $\mathbf{X}^T\mathbf{X}$, leading to its non-invertability because of a singularity.

Good linear-regression programs overcome these difficulties and always give correct solutions. Among the most effective programs are algorithms which do not build matrix $\mathbf{X}^T\mathbf{X}$ but instead solve the overdetermined set of n linear equations of m unknowns $\mathbf{y} = \mathbf{X}\,\mathbf{b}$. For example, the algorithm SVD (Singular value decomposition [16]) works even on computer with poor data precision.

Problem 6.20 *Examination of the quality of a regression algorithm (LS)*

Many test examples are available for examining the quality and effectiveness of linear regression algorithms. An example suitable for numerical control of quality of regression programs comes from the linear model $E(y\,/\,x) = \beta_1 x_1 + \beta_2 x_2$. Calculate the estimates b_1 and b_2 by the least-squares method.

○ *Data:* The numerical constant ε in algorithms examination is selected such that the condition $\varepsilon < 10^{-(d+1)/2}$ is fulfilled, where d is the number of valid digits used in the actual computer.

i	y	x_1	x_2
1	3	1	1
2	ε	ε	0
3	2ε	0	ε

Solution:

(a) *The analytical approach:*

$$\mathbf{X}^\mathrm{T}\mathbf{X} = \begin{bmatrix} 1+\varepsilon^2 & -1 \\ -1 & 1+\varepsilon^2 \end{bmatrix}$$

$$\mathbf{X}^\mathrm{T}\mathbf{y} = \begin{bmatrix} 3+\varepsilon^2 \\ 3+2\varepsilon^2 \end{bmatrix}$$

According to Problem 6.1 the inversion matrix is determined as

$$(\mathbf{X}^\mathrm{T}\mathbf{X})^{-1} = \frac{1}{(1+\varepsilon^2)^2 - 1}\begin{bmatrix} 1+\varepsilon^2 & 1 \\ 1 & 1+\varepsilon^2 \end{bmatrix}$$

and on substituting into Eq. (6.11), the estimates of the parameters are found to be

$$b_1 = \frac{(3+\varepsilon^2)(1+\varepsilon^2)}{(1+\varepsilon^2)^2 - 1} - \frac{(3+2\varepsilon^2)}{(1+\varepsilon^2)^2 - 1} = 1$$

$$b_2 = \frac{-(3+\varepsilon^2)}{(1+\varepsilon^2)^2 - 1} + \frac{(1+\varepsilon^2)(3+2\varepsilon^2)}{(1+\varepsilon^2)^2 - 1} = 2$$

The estimates b_1 and b_2 do not depend on the magnitude of ε, and moreover $RSC = 0$.

(b) *The numerical approach, by computer:* If the condition $\varepsilon < 10^{-(d+1)/2}$ is valid, then $1 + \varepsilon^2 = 1$. All elements of the matrix $\mathbf{X}^\mathrm{T}\mathbf{X}$ will be ones, and its inversion will be impossible because $\det(\mathbf{X}^\mathrm{T}\mathbf{X}) = 0$. If the computer works with a precision of 11 digits, the choice of $\varepsilon = 10^{-6}$ will cause the computation to fail.

Conclusion: Because of limited precision of computers and possible ill-conditioning of the normal equations $\mathbf{X}^\mathrm{T}\mathbf{X}$, even simple tasks may cause numerical difficulties.

To make statistical analysis easier, most programs work with matrix $\mathbf{X}^\mathrm{T}\mathbf{X}$. To avoid difficulties with large differences its centred or normalized version is used. The variables are expressed as deviations from the arithmetic mean

$$x_{Cij} = x_{ij} - \overline{x}_j \quad \text{and} \quad y_{Ci} = y_i - \overline{y} \ .$$

The resulting centred variables then form the elements of the matrix $\mathbf{X}_C$ or the vector $\mathbf{y}_C$. Centring all variables results in the intercept term cancelling out.

The matrix $(\mathbf{X}_C^T \mathbf{X}_C)^{-1}$ is a submatrix of $(\mathbf{X}^T \mathbf{X})^{-1}$, so that after its inversion all elements of matrix $(\mathbf{X}^T \mathbf{X})^{-1}$ may be calculated. These elements are necessary for the statistical analysis. Then

$$(\mathbf{X}^T\mathbf{X})^{-1} = \begin{bmatrix} c_0 & \mathbf{c}^T \\ \mathbf{c} & (\mathbf{X}_C^T\mathbf{X}_C)^{-1} \end{bmatrix}$$

If we define the averages vector $\mathbf{x}_P = (\overline{x}_1, ..., \overline{x}_{m-1})^T$, we may write

$$c_0 = \frac{1}{n} - \mathbf{c}^T\mathbf{x}_P \quad \text{and} \quad \mathbf{c}^T = (-1)\mathbf{x}_P^T(\mathbf{X}_C^T\mathbf{X}_C)^{-1}$$

With normalized variables, the standard deviations $\hat{\sigma}(x_j)$ and $\hat{\sigma}(y)$ are used. If we introduce normalized variables

$$Z_{ij} = \frac{x_{Cij}}{\hat{\sigma}(x_j)\sqrt{n-1}} \quad \text{and} \quad q_i = \frac{y_{Ci}}{\hat{\sigma}(y)\sqrt{n-1}}$$

the matrix $\mathbf{R} = \mathbf{Z}^T\mathbf{Z}$ is formally identical with the correlation matrix of controllable variables and the vector $\mathbf{r} = \mathbf{Z}^T\mathbf{q}$ formally contains the correlation coefficients of all controllable variables with the response variable. In correlation models there are real correlation coefficients. In the least-squares method, the parameter estimates $\mathbf{b}_N = \mathbf{R}^{-1}\mathbf{r}$ are found, for which

$$b_{Nj} = b_j \frac{\hat{\sigma}(x_j)}{\hat{\sigma}(y)}$$

The advantage of normalized variables is that the elements of matrix $\mathbf{R}$ are the numbers in the interval from -1 to $+1$. A disadvantage is the possible distortion of the calculated matrix $\mathbf{R}$, e.g. by use of the "for pocket-calculator" modified expression $\sum_{i=1}^{n} x_i^2 - n\overline{x}^2$ instead of $\sum_{i=1}^{n} (x_i - \overline{x})^2$. With the modified expression, the final result is the difference between two large numbers, with a result close to zero. The limited precision of computer or calculator can result in a sum of squares of residuals from the mean with a value that is zero or even negative.

From many techniques of numerical solution of the least-squares problem, we select here the following two cases:

(a) the *method of orthogonal functions,* which is simple and convenient for polynomial models; and

(b) the *method of rational ranks,* used in the program ADSTAT and QC-Expert. Another algorithm is described by Lawson and Hanson [18].

6.4.1 The Method of Orthogonal Functions

Orthogonal functions are frequently used because they result in considerable simplification of the statistical analysis. For the linear regression model

$$E(y/\mathbf{x}) = \sum_{j=1}^{m} \beta_j f_j(\mathbf{x}) \tag{6.78}$$

where $f_j(\mathbf{x})$ are any functions of the input variables $\mathbf{x}$ which do not contain regression parameters. Equation (6.11) is used to estimate the parameters β. For this case, Eq. (6.11) contains the matrix $\mathbf{F}$ of dimension $(n \times m)$ with elements $f_j(x_i), j = 1, \ldots, m, i = 1, \ldots, n,$ instead of matrix $\mathbf{X}$. For further analysis it is convenient for the matrix $\mathbf{F}^T\mathbf{F}$ to be diagonal. Therefore the scalar products of all pairs of columns of $\mathbf{F}$ must be equal to zero. For $\mathbf{F}^T\mathbf{F}$ to be diagonal,

$$\sum_{i=1}^{n} f_j(\mathbf{x}_i) f_k(\mathbf{x}_i) \rightarrow \begin{cases} 0 & \text{for } k \neq j \\ \sum_{i=1}^{n} f_j^2(\mathbf{x}_i) & \text{for } k = j \end{cases} \tag{6.79}$$

From Eq. (6.79) it follows that the diagonality of $\mathbf{F}^T\mathbf{F}$ may be achieved

(a) by the adjustment of values x_i for a given function f_j, $i = 1, \ldots, n$. This is the case in designed experiments;

(b) by the special choice of functions f_j for given locations x_i, $i = 1, \ldots,$ n. That is, the construction of orthogonal functions $g_j(\mathbf{x})$ from the original ones $f_j(x_j)$.

Orthogonal functions are generated by use of the recurrent relation

$$g_j(\mathbf{x}) = f_j(\mathbf{x}) + \sum_{L=j-1}^{1} Q_{jL} g_L(\mathbf{x}) \tag{6.80}$$

The coefficients Q_{jL} may be found from the conditions of orthogonality. A set of j linear equations is formed, and each equation contains just one unknown Q_{jL} for which:

$$Q_{jL} = \frac{-\sum\limits_{i=1}^{n} f_j(\mathbf{x}_i)\, g_L(\mathbf{x}_i)}{\sum\limits_{i=1}^{n} g_L^2(\mathbf{x}_i)} \tag{6.81}$$

With the orthogonal functions $g_j(\mathbf{x})$ generated from the original ones $f_j(\mathbf{x})$ by Eqs. (6.80) and (6.81), the linear regression model may be expressed in the form

$$E(y/\mathbf{x}) = \sum_{j=1}^{m} c_j g_j(\mathbf{x}) \tag{6.82}$$

Since $g_j(\mathbf{x})$ are orthogonal, the estimates of parameters c_j may be obtained by straight substitution into

$$c_j = \frac{\sum\limits_{i=1}^{n} g_j(\mathbf{x}_i) \times y_i}{\sum\limits_{i=1}^{n} g_j^2(\mathbf{x}_i)} \qquad j = 1,\dots,m \tag{6.83}$$

The variances of the parameter estimates are given by

$$D(c_j) = \frac{\hat{\sigma}^2}{\sum\limits_{i=1}^{n} g_j^2(\mathbf{x}_i)} \tag{6.84}$$

Thus, when the orthogonal functions are known, the data evaluation for a linear regression requires only substitution into simple expressions.

Problem 6.21 *Examination of the quality of the orthogonal functions method*

Estimate parameters β_1 and β_2 in Problem 6.20 by the method of orthogonal functions.

○ **Data:** from Problem 6.20

Solution: The original functions are $f_1(x) = x_1$, $f_2(x) = x_2$. On substituting into Eq. (6.80) we get

$$g_1(x) = x_1$$

$$g_2(x) = x_2 + Q_{21}$$

$$g_1(x) = x_2 + Q_{21}x_1$$

$$g_2(x) = x_2 + Q_{21}g_1(x) = x_2 + Q_{21}x_1$$

For functions $g_1(x)$ and $g_2(x)$ to be orthogonal for the given values x_i, $i = 1, ..., n$, their scalar product must be equal to zero, that is

$$\sum_{i=1}^{n} g_1(x_i)g_2(x_i) = \sum_{i=1}^{n} x_{1i}(x_{2i} + Q_{21}x_{1i}) = 0$$

From this equation, the term Q_{21} is given by $Q_{21} = \dfrac{-\sum\limits_{i=1}^{n} x_{1i}x_{2i}}{\sum\limits_{i=1}^{n} x_{1i}^2}$. For the data in the problem

$$\sum_{i=1}^{n} x_{1i}x_{2i} = 1 + 0 + 0 = 1, \quad \sum_{i=1}^{n} x_{1i}^2 = 1 + \varepsilon^2 \quad \text{and} \quad \text{therefore} \quad Q_{21} = \dfrac{-1}{1 + \varepsilon^2}.$$

Then, on substituting into Eq. (6.83), we have $c_1 = \dfrac{\sum\limits_{i=1}^{n} x_{1i}y_i}{\sum\limits_{i=1}^{n} x_{1i}^2} = \dfrac{3 + \varepsilon^2}{1 + \varepsilon^2}$ and

$$c_2 = \dfrac{\sum\limits_{i=1}^{n} \left[x_{2i} - x_{1i}/(1+\varepsilon^2) \right] \times y_i}{\sum\limits_{i=1}^{n} \left[x_{2i} - x_{1i}/(1+\varepsilon^2) \right]^2} = \dfrac{4 + 6\varepsilon^2 + 2\varepsilon^4}{2 + 3\varepsilon^2 + \varepsilon^4} = 2 .$$ The regression model of

type (6.82) then has the form $E(y/x) = \dfrac{3+\varepsilon^2}{1+\varepsilon^2}g_1(x) + 2g_2(x)$. After substituting

for $g_1(x)$ and $g_1(x)$, we obtain $E(y/x) = \dfrac{3+\varepsilon^2}{1+\varepsilon^2}x_1 + 2\left(x_2 - \dfrac{x_1}{1+\varepsilon^2}\right) = 1x_1 + 2x_2$.

Thus, the estimates of the parameters are $b_1 = 1$ and $b_2 = 2$.

Conclusion: The method of orthogonal functions can find estimates of parameters for linear models. The quality of the parameter estimates achieved by computer is determined by the precision of calculation of the individual orthogonal functions.

This example demonstrates that the application of orthogonal functions is quite simple. The errors caused by limited computer precision accumulate according to Eq. (6.80) and increase with the number of orthogonal functions m. The use of orthogonal functions is nearly equivalent to the use of adequate methods for the matrix inversion.

The advantage of orthogonal functions is that when some functions $g_e(x)$ are omitted, the coefficients C_j for the remaining functions $g_j(x)$ will be unchanged. The method may be used to search for the optimum combination of polynomial terms (degrees of polynomial model). The disadvantage of orthogonal functions is the rather complicated manipulation. It is useful to know that all types of orthogonal polynomials may be expressed by three-term recurrent expressions [19].

6.4.2 The Method of Rational Ranks (Generalised Principal Components Regression, GPCR)

To detect ill-conditioning of $\mathbf{X}^T\mathbf{X}$ or the $\mathbf{R}$, the matrices are decomposed into eigenvalues and eigenvectors. Since the matrix $\mathbf{R}$ is symmetrical it may be expressed by eigenvalues $\lambda_1 \leq \lambda_2 \leq ... \leq \lambda_m$, and corresponding eigenvectors $\mathbf{P}_j$, $j = 1, ..., m$, in the form of the sum

$$\mathbf{R} = \sum_{j=1}^{m} \lambda_j \mathbf{P}_j \mathbf{P}_j^T \qquad (6.85)$$

The inverse matrix $\mathbf{R}^{-1}$ may be expressed in the form

$$\mathbf{R}^{-1} = \sum_{j=1}^{m} \lambda_j^{-1} \mathbf{P}_j \mathbf{P}_j^T \qquad (6.86)$$

With the use of Eq. (6.86) Eq. (6.11) can be rewritten in the form

$$\mathbf{b}_N = \sum_{j=\omega}^{m} \left[\lambda_j^{-1} \mathbf{P}_j \mathbf{P}_j^T \right] \mathbf{r} \tag{6.87}$$

The covariance matrix of normalized estimates $\mathbf{b}_N$ may be rewritten in form

$$D(\mathbf{b}_N) = \hat{\sigma}_N^2 \sum_{j=\omega}^{m} \left[\lambda_j^{-1} \mathbf{P}_j \mathbf{P}_j^T \right] \tag{6.88}$$

In the case of the least-squares in Eqs. (6.86) and (6.88) the parameter ω is set to $\omega = 1$. From both equations it follows that when the eigenvalues λ_j, are small, the estimates $\mathbf{b}_N$ and their variances are rather high. According to the magnitude of the eigenvalues λ_j, regression problems can be divided into three groups:

(1) All eigenvalues are significantly higher than zero. The use of the least-squares method does not cause any problems.

(2) Some eigenvalues are close to zero. This is a typical example of multicoUinearity when some common methods fail.

(3) Some eigenvalues are equal to zero. Then the matrix $\mathbf{X}^T\mathbf{X}$ or R is singular and cannot be inverted.

The only way of avoiding difficulties with groups (2) and (3) is the use of the method of rational ranks. Here, the terms (or parts of them) with small values of eigenvalues λ_j are neglected [20]. The criterion for omitting terms corresponding to small eigenvalues has the form

$$\text{abs}\left[\frac{\sum_{j=1}^{\omega} \lambda_j}{\sum_{j=1}^{m} \lambda_j} \right] = P$$

where P is the chosen precision (usually 10^{-5}). The value ω determines the lower limit from which, in Eqs. (6.87) and (6.88), the summation is carried out. Let us define $W = \sum_{j=1}^{\omega} \lambda_j$ and $E = \sum_{j=1}^{m} \lambda_j$. When the condition $\frac{W}{E} > P$ is valid, i.e. the value ω is not an integer, the summation is made from $\omega - 1$ and the eigenvalue $\lambda_{\omega-1}$ is "weighted" by the factor

$$u = \frac{W - EP}{\lambda_\omega} \qquad (6.90)$$

Therefore, the length of estimates $\|\mathbf{b}_N\|$ with their variances may be continuously decreased as a function of increasing precision P. However, it is followed by an increase of the estimate bias and a decrease in the multiple correlation coefficient. The bias of estimates is here caused by neglecting terms in Eqs. (6.87) and (6.88) at $\omega > 1$. It has been proposed [20] that the squared bias $h_V^2(\mathbf{b}_N) = [\beta - E(\mathbf{b})]^2$ achieved by the method of rational ranks is equal to

$$h_V^2(\mathbf{b}_N) = \beta_N^T \left[\sum_{j=1}^{\omega} \mathbf{P}_j \mathbf{P}_j^T \right] \beta_N \qquad (6.91)$$

The optimum magnitude of P may be determined by finding a minimum of the mean quadratic error of prediction MEP, Eq. (6.57). In program ADSTAT or QC-Expert the user chooses the value of precision P, or it takes the default value $P = 10^{-35}$.

Problem 6.22 *Examination of the method of rational ranks*

Determine the parameter estimates b_1 and b_2 for the data from problem 6.20 by the method of rational ranks, with $P = 5 \times 10^{-9}$ and $\varepsilon = 10^{-4}$.

○ *Data:* from Problem 6.20

Solution: By decomposition of matrix $\mathbf{X}^T\mathbf{X}$ into eigenvalues and eigenvectors it is found that $\lambda_1 = \varepsilon^2$ and $\lambda_2 = 2 + \varepsilon^2$ and that $\mathbf{P}_1^T = (-\sqrt{0.5}, \sqrt{0.5})$, $\mathbf{P}_2^T = (\sqrt{0.5}, -\sqrt{0.5})$. Because

$$\frac{\lambda_1}{\lambda_1 + \lambda_2} = \frac{10^{-8}}{2 + 2 \times 10^{-8}} = 5.0 \times 10^{-9} \quad \omega = 1 \text{ and in Eq. (6.87) only the second}$$

term is used. Because the decomposition of the matrix $\mathbf{X}^T\mathbf{X}$ (and not of $\mathbf{R}$) was made the parameter estimates from Eq. (6.87) are calculated

$$\mathbf{b} = \begin{bmatrix} b_1 \\ b_2 \end{bmatrix} = \frac{1}{2 + \varepsilon^2} \begin{bmatrix} \sqrt{0.5} \\ \sqrt{0.5} \end{bmatrix} \begin{bmatrix} \sqrt{0.5} & \sqrt{0.5} \end{bmatrix} \begin{bmatrix} 3 + \varepsilon^2 \\ 3 + 2\varepsilon^2 \end{bmatrix} = \begin{bmatrix} 1.5(2 + \varepsilon^2)/(2 + \varepsilon^2) \\ 1.5(2 + \varepsilon^2)/(2 + \varepsilon^2) \end{bmatrix} = \begin{bmatrix} 1.5 \\ 1.5 \end{bmatrix}$$

The fact was used here that $\mathbf{X}^T\mathbf{y} = [3 + \varepsilon^2 \quad 3 + 2\varepsilon^2]^T$. The parameter estimates found differ from the true values $\beta_1 = 1$ and $\beta_2 = 2$, but the residual sum of squares $RSC = 0.5\varepsilon^2 = 5 \times 10^{-9}$ is rather small. By using the classical least-squares method, and a computer working with precision of 7 valid digits the problem is not at all solved. It is evident that, even with smaller values of ε, these biased estimates remain the same, and "numerical underflowing" may occur when ε^2 is smaller than the computer precision.

Conclusion: The method of rational ranks enables biased parameter estimates to be found, and for singular or ill-conditioned matrices X^TX are more suitable than estimates found by the least-squares method, which are always unbiased.

For the ill-conditioned matrix X^TX, the biased parameter estimates are shorter and smaller than least-squares estimates. They are "more precise" because they have smaller parameter variances. Moreover, these estimates exist even for a singular X^TX matrix, when the least-squares method always fails.

Problem 6.23 *Approximation of a convex increasing function by a polynomial*

Many problems in data treatment concern approximation of instrumental data of convex (or concave) increasing (or decreasing) values by a polynomial of any degree, so that the polynomial represents the shape of a data curve. Use the method of rational ranks for approximation of convex increasing data. For approximation, choose a polynomial of the sixth degree $E(y/x) = \sum_{j=1}^{6} b_j x^j + b_7$.

Calculate the value of the dependent variable y_0 at the origin, i.e. parameter estimate β_7.

○ **Data:** $n = 10$

x	25	35	45	55	65	75	85	95	105	115
y	150	160	170	190	210	230	270	310	370	450

Solution: Table 6.6 lists parameter estimates found by the classical least-squares method (LS), with $P = 10^{-30}$, and by the method of rational ranks (RV), with $P = 3.5 \times 10^{-4}$ for which the statistic *MEP* was smallest.

Figure 6.24a shows the regression model, with the 95% confidence interval for the LS method and Fig. 6.24b the regression model for the RV method, for $P = 3.5 \times 10^{-4}$. From the figures and Table 6.6 it is evident that the parameter estimates for the LS method do not match the data very well.

Table 6.6 The parameter estimates for different P values

Method	P	MEP	b_7	b_1	b_2	b_3	b_4	b_5	b_6
LS	10^{-30}	160.8	195.5	−5.92	0.258	−0.0049	5.33×10^{-5}	-2.9×10^{-7}	6.98×10^{-10}
RV	3.5×10^{-4}	8.59	134.7	0.35	0.0092	3.2×10^{-5}	5.3×10^{-8}	3.9×10^{-9}	4.5×10^{-11}

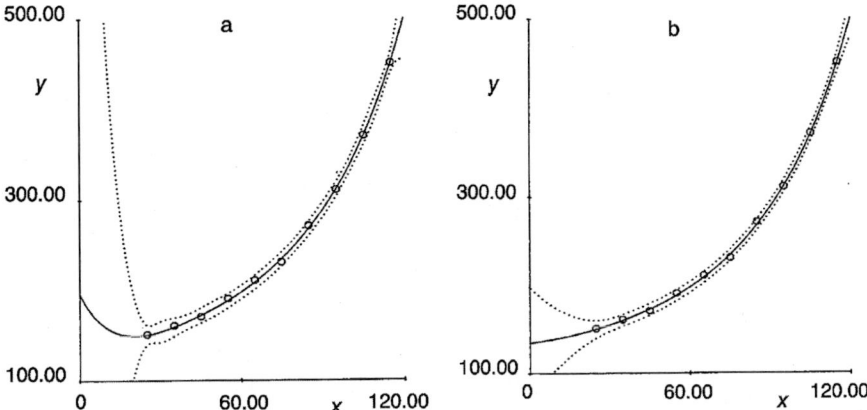

Figure 6.24   The curve fitting of the model proposed, with the 95% confidence interval and the experimental *data:* (a) by the LS method ($P = 10^{-30}$), and (b) by the RV method ($P = 3.5 \times 10^{-4}$).

The estimate of parameter β_7 is larger than the value of y_1 so the proposed model equation has a minimum between the origin and the point (x_1, y_1). The confidence intervals are rather broad, and do not permit prediction of y outside the measurement interval.

The estimates determined by the rational ranks method are biased. The parameter b_7 is, however, smaller than y_1 and the confidence intervals show some possibility of prediction, even outside the measurement interval. As the polynomial degree was known and had physical meaning, no corrections based on the statistical analysis will be attempted.

Conclusion: The method of rational ranks allows us to find parameter estimates which give a curve shape corresponding to the data trend, with no extra extremes or inflections. With classical LS, such problems can be solved only with the introduction of some restrictions on parameters. Moreover, it is found that, for practical purposes, the biased estimates are not unsatisfactory.

6.5 Regression Diagnostics

In linear and nonlinear regression analysis, the method of least-squares is often used. This method, however, does not ensure that the model is fully acceptable from statistical and physical point of view. A source of problems may be found in components of a regression triplet [data, model and method of estimation]. The least-squares method provides accurate estimates only when all assumptions about data and about a regression model are fulfilled. When some assumptions are not fulfilled, the least-squares method is inconvenient. Regression diagnostics represent the procedures for identification of

(a) the data quality for a proposed model,

(b) the model quality for a given set of data, and

(c) fulfilment of all least-squares assumptions.

In the literature [21] the term regression diagnostics refers to methods for identification of influential points and multicollinearity. Atkinson [22] also includes as part of regression diagnostics, methods for proposing an actual regression model, perhaps with use of transformation of variable(s). Weisberg [23] includes as regression diagnostics

(1) the examination of all assumptions for parameter estimation,

(2) the statistical analysis of parameters, i.e. testing of the model,

(3) the identification of influential points, i.e. critical examination of data.

In this book we understand by regression diagnostics

(1) methods of exploratory data analysis of individual variables (Chapter 2),

(2) methods for analysis of influential points, and

(3) methods for identification of violations of the conditions for least-squares (Section 6.2).

The main difference between the use of regression diagnostics and classical statistical tests is that there is no necessity for an alternative hypothesis, but all types of deviations from an ideal regression triplet are discovered. Our concept of exploratory regression analysis is based on the fact that "the computer user knows more about the data than the computer". The personal computer serves us as an efficient tool for interactive diagnosis of data, model, and estimation method. The procedure of model building with the help of a personal computer involves interactive co-operation between the user and the computer program. Therefore, formal models that do not have physical meaning should not be proposed and analysed.

6.5.1 Exploratory Regression Analysis

Methods of exploratory data analysis have been described in Chapter 2. In exploratory regression analysis we will use these methods for (a) determination of statistical peculiarities of individual variables or residuals, (b) examination of assumptions regarding the distribution of variables and residuals.

In some cases, simply plotting the measured variable y_i against an index i may uncover a latent variable, often related to time or order of measurement [25].

The first view into the relationship between individual variables comes from an x-y scatter plot of y against x. Some information about multicollinearity can be obtained by plotting pairs of controllable variables, x_j against x_k, $j \neq k$. An approximately linear dependence indicates strong multicollinearity. However, a plot of the response y against variable x_j, $j = 1, ..., m$, may suggest nonlinearity of a model which is, in fact, of linear nature.

Problem 6.24 *Danger of false conclusions from inappropriate application of the scatter plot*

Draw the scatter plot of response y against variable x_1 and y against x_2 for a linear regression model $y_i = 10 - 6x_{1i} + 0.5x_{2i}$, $i = 1,..., 10$. Choose values of independent variable $x_i = i - 5$ and $x_{2i} = x_{1i}^2$. Draw conclusions from these two plots.

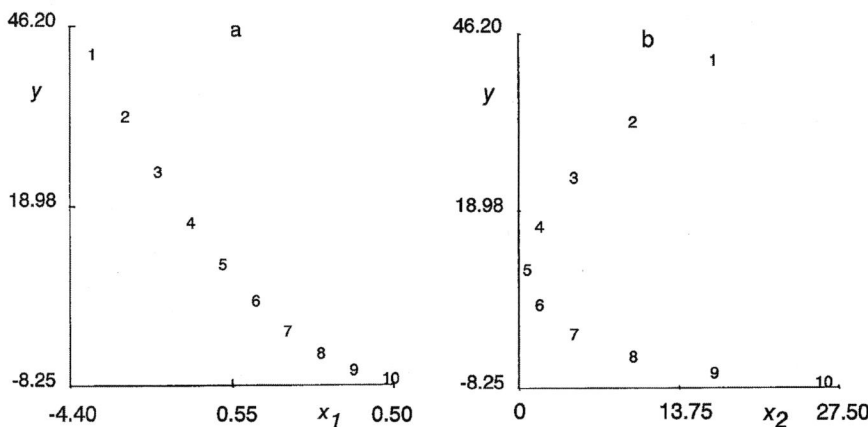

Figure 6.25   The scatter plots for (a) response y as a function of variable x_1 and (b) response y as a function of x_2.

Solution: Figure 6.25 shows two scatter plots indicating quite strong nonlinearity. This could lead to a hypothesis that y is a nonlinear function of variables x_1 and x_2. The apparent nonlinear nature is caused by a particular choice of controllable values x_{2i} and by quite strong multicollinearity between the independent variables x_1 and x_2.

Conclusion: For multivariate cases, scatter plots of y against x_j often are not helpful in identification of the regression model type.

Data normality is examined by the quantile-quantile (Q-Q) plot (Chapter 2). The principle methods of exploratory regression analysis include the determination of data range, data variability and the presence of outliers. All the graphical diagnostic tools described in Chapter 2 may be applied here. The EDA techniques allow identification of situations where

(1) the range of measured data is too restricted,

(2) the proposed model is false because there are some latent variables,

(3) multicollinearity exists,

(4) data do not have the normal distribution when the controllable variables are random numbers.

6.5.2 Examination of Data Quality

Data quality has a strong influence on any proposed regression model. Examination of data quality involves detection of the *influential points* (*IP*), which cause many problems in regression analysis by shifting the parameter estimates or increasing the variance of the parameters. Influential points may be classified into three groups:

(a) *Gross errors*, caused by outliers in the measured variable or by the leverage points (extremes in the controllable variables).

(b) *Golden points* are special chosen points which have been very precisely measured to extend the prediction capability of model.

(c) *Latently influential points* are the consequence of a poor regression model.

Influential points may instead be classified according to data location:

(1) *Outliers* differ from the other points in value on the *y*-axis;

(2) *Leverage points* differ from the other points in values on the *x*-axis or in a combination of these quantities (in the case of multicollinearity).

There are also points, however, which are outliers and leverage points together. Outliers are identified by examination of the residuals. Leverage points are found from the diagonal elements H_{ii} of the projection hat matrix.

6.5.2.1 Statistical Analysis of Residuals

(1) *Classical residuals*

Residuals $\hat{e}_i$ are defined by the expression $\hat{e}_i = y_i - \mathbf{x}_i \mathbf{b}$ where $\mathbf{x}_i$ is the *i*th row of matrix $\mathbf{X}$. Classical analysis is based on the assumption that residuals are estimates of errors ε_i. With the use of residuals the properties of errors are examined. This assumption, however, is not quite correct and can sometimes lead to false results. The false assumptions about residuals include the following:

(a) the distribution of residuals is the same as the error distribution and the statistical properties of the residuals are identical with those of the errors, and

(b) if the residual value is large, a large effect is caused by the corresponding point, so the point should be excluded from the data.

Let us note the differences between errors ε_i and residuals $\hat{e}_i$. The geometric illustration (Fig. 6.2) shows that the residuals $\hat{e}_i$ *are not independent* even when the errors ε_i are independent. The residuals $\hat{e}_i$ are a projection of vector **y** into a subspace of dimension $(n - m)$. By using projection matrix **P** we can write

$$\hat{\mathbf{e}} = \mathbf{Py} = \mathbf{P}(\mathbf{X\beta} + \varepsilon) = \mathbf{P\varepsilon} = (\mathbf{E} - \mathbf{H})\varepsilon \qquad (6.92)$$

To rearrange Eq. (6.92), we use the fact the vector **Xβ** lies in the plane perpendicular to the projection plane, so that a zero vector results. For the *i*th residual

$$\hat{e}_i = (1 - H_{ii})y_i - \sum_{j \neq i}^{n} H_{ij}y_j = (1 - H_{ii})\varepsilon_i - \sum_{j \neq i}^{n} H_{ij}\varepsilon_j \qquad (6.93)$$

Each residual $\hat{e}_i$ is a linear combination of all errors ε_i. The distribution of residuals depends on

(a) the error distribution,

(b) the elements of the projection matrix **H**,

(c) the sample size n.

Because the residual $\hat{e}_i$ represents a sum of random quantities with bounded variance, the *supernormality effect* appears for small sample sizes. Even when the errors ε_i do not have a normal distribution, the distribution of residuals is close to normal. In small samples, the elements of the projection matrix **H** are large and the main role of an actual point is to influence the sum of terms $H_{ij}\varepsilon_j$. The distribution of this sum is closer to a normal one than the distribution of errors ε_i. For large sample sizes, where $1/n \approx 0$, we find that $\hat{e}_i \approx \varepsilon_i$ and analysis of the residual distribution gives direct information about the distribution of errors.

Equation (6.18) may be used to calculate the residual variance

$$D(\hat{e}_i) = (1 - H_{ii})\hat{\sigma}^2 \qquad (6.94)$$

The variance of residuals $D(\hat{e}_i)$ is not constant even when the variance of errors is constant. According to Eq. (6.18), the paired correlation coefficient r_{ij} between two residuals $\hat{e}_i$ and $\hat{e}_j$ is given by

$$r_{ij} = \frac{-H_{ij}}{\sqrt{(1-H_{ii})(1-H_{jj})}} \tag{6.95}$$

which shows that residuals are correlated even when errors ε_i and ε_j are independent.

For strong leverage points (extremes), the diagonal elements $H_{ii} \to 1$ while non-diagonal elements $H_{ij} \approx 0$. From Eq. (6.93), it may be concluded that an equation $\hat{e}_i = 0$ is valid, whatever the magnitude of y_i. The residuals do not always indicate correctly some strongly deviant values.

When a regression analysis is carried out by the least-squares method, for a model with an intercept term it is true that $\dfrac{\sum\limits_{i=1}^{n} \hat{e}_i}{n} = 0$ which corresponds to saying that the mean value of errors is equal to zero, $E(\varepsilon) = 0$.

Classical residuals are always associated with non-constant variance; they sum to be more normal and may not indicate strongly deviant points. The common practice of programs for statistical analysis of residuals is to use for examination some statistical characteristics of residuals such as the mean, the variance, the skewness and the kurtosis.

It should be particularly noted that in the case of small sample sizes the estimates of skewness and kurtosis are rather distorted and cannot reliably indicate the correctness of a proposed model.

Problem 6.25 *Inappropriate application of some simple statistics for residual analysis*

To illustrate the overestimated approach to examination of the reliability of parameter estimates, the following residual statistics are calculated: the mean of absolute values of residuals $|\bar{e}|$, the estimate of the standard deviation of residuals $\hat{\sigma}(\hat{e})$, the estimate of residual skewness $\hat{g}_1(\hat{e})$, the estimate of residual kurtosis $\hat{g}_2(\hat{e})$. Calculate these statistics for the four data samples from Problem 6.8.

○ **Data:** from Problem 6.8

Solution: The numerical values of some statistical characteristics for the four data samples from Problem 6.8 are listed in Table 6.7.

Table 6.7 Statistical characteristics for the four data samples of Problem 6.8

Sample	$\lvert\overline{e}\rvert$	$\hat{\sigma}(\hat{e})$	$\hat{g}_1(\hat{e})$	$\hat{g}_2(\hat{e})$	$L(\hat{e})$
A	0.837	1.237	0.13	2.24	0.29
B	0.967	1.237	0.63	1.94	1.25
C	0.716	1.237	−2.28	6.57	15.39
D	0.903	1.237	−0.011	1.93	0.52

The last column of Table 6.7 lists values of the criterion of the Jarque-Berra test, which combines both skewness and kurtosis (Section 6.5.4). When $L(\hat{e}) > \chi^2_{1-\alpha}(2) = 5.99$, the normality of the data distribution is not proved.

From Table 6.7 it is evident that the statistics $\lvert\overline{e}\rvert$ and $\hat{\sigma}(\hat{e})$ do not indicate the model quality. Only sample C, which has one strong outlier, exhibits significant deviation in skewness and kurtosis, and the Jarque-Berra test does not prove the normality of the sample distribution. Neither the nonlinear dependence (sample B) nor the spurious data with one strong outlier (sample D) are correctly detected by the four statistics of residuals.

Conclusion: The statistical characteristics $\lvert\overline{e}\rvert$, $\hat{\sigma}(\hat{e})$, $\hat{g}_1(\hat{e})$, $\hat{g}_2(\hat{e})$ often do not give a correct indication of the quality of a model.

(2) *Normalized residuals*

In statistical data treatment the normalized residuals $\hat{e}_{Ni}$ defined by $\hat{e}_{Ni} = \dfrac{\hat{e}_i}{\hat{\sigma}}$ are often recommended. It is often assumed that these residuals are normally distributed quantities with zero mean and variance equal to one, $\hat{e}_{Ni} \sim N(0,1)$. When normalized residuals are used, the rule of 3σ is classically recommended: quantities with $\hat{e}_{Ni}$ of magnitude greater than $\pm 3\sigma$ are classified as the outliers. For a normal distribution, only 0.3% of all values lie outside the interval $x \pm 3\hat{\sigma}$. Such assumptions about normalized residuals are misleading.

From Eq. (6.94) it is obvious that the variance $D(\hat{e}_{Ni}) = (1 - H_{ii})$ is not constant, and also not equal to one. For strong leverage points, $\hat{e}_i \approx 0$, so application of 3σ rule could lead to exclusion of correct points but retention of erroneous values.

Problem 6.26 *Inappropriate application of normalized residuals for identification of influential points*

For a dependence $y_i = x_i^2$, $i = 1,\dots, 12$, the data contain one gross error, the number x_{12} is replaced by y_{12}. Estimate parameters and with the use of normalized residuals try to locate the false leverage point 12.

○ *Data:*

i	1	2	3	4	5	6	7	8	9	10	11	12
x_i	1	2	3	4	5	6	7	8	9	10	11	144
y_i	1	4	9	16	25	36	49	64	81	100	121	12

Solution: By the least-squares method for the model $E(y/x) = \beta_1 x^2$, the parameter estimate was found $b_1 = 0.000671$. Point 12 has $\hat{e}_{N12} = -0.03$ and point 1 has the highest value, $\hat{e}_{N1} = 1.94$. With the use of the standardized residual, Eq. (6.96), the maximum for the 12th point was indicated, $\hat{e}_{S12} = -3.16$.

Conclusion: Normalized residuals are not able to indicate leverage points. Such points may be discovered, for example, by standardized residuals.

(3) *Standardized residuals*

The standardized residuals $\hat{e}_{Si}$, defined by

$$\hat{e}_{Si} = \frac{\hat{e}_i}{\hat{\sigma}\sqrt{1 - H_{ii}}} \tag{6.96}$$

exhibit constant variance. The statistical properties of standardized residuals are the same as those of classical residuals. Standardized residuals are, apart from the multiplicative constant $1/\sqrt{n-m}$, equal to cosine θ_i the angle between the vector $\hat{\mathbf{e}}$ and the vector $\mathbf{i}^{\perp}$ (which is the projection of the ith column of matrix $\mathbf{E}$ onto hyperplane $L^{\perp}$) so that $\cos \theta_i = \dfrac{\hat{e}_{Si}}{\sqrt{n-m}}$. The maximum value of $\hat{e}_{Si}$ is bounded by $\sqrt{n-m}$. The variable $\hat{e}_{Si}/(n-m)$ has the beta distribution $Be[0.5, (n-m-1)/2]$.

(4) *Jack-knife residuals*

If, in Eq. (6.96), instead of the standard deviation we use the estimate of standard deviation $\hat{\sigma}_{(-i)}$ obtained by leaving out the ith point, we obtain the Jack-knife or *fully studentized residuals*

$$\hat{e}_{Ji} = \hat{e}_{Si}\sqrt{\frac{n-m-1}{n-m-\hat{e}_{Si}^2}} = \sqrt{n-m}\,\cot g\theta_i \tag{6.97}$$

which, with an assumption of normality of errors, have the Student distribution with $(n - m - 1)$ degrees of freedom. Jack-knife residuals correspond to the criterion of a *t*-test of the null hypothesis H_0: $C = 0$ in the model of a simple shift

$$y = X\beta + Ci + \varepsilon \tag{6.98}$$

where i is the identity vector with the ith element equal to one and other elements equal to zero. The model [Eq. (6.98)] expresses the case of an outlier where C is directly equal to the value of deviation, but also the case of a leverage point $C = d_i^T\beta$ where d_i^T is the vector of the deviation of the individual x-components of the ith point. Jack-knife residuals are often used instead of classical residuals $\hat{e}_i$ for identification of outliers. In the case of leverage points, these residuals do not give a reliable indication.

(5) *Predicted residuals*

The estimate of parameter C in Eq. (6.98) is represented by the predicted residuals

$$\hat{e}_{Pi} = y_i - x_i b_{(i)} = \frac{\hat{e}_i}{1 - H_{ii}} \tag{6.99}$$

where $b_{(i)}$ is a vector of the parameter estimates obtained by the least-squares method with the ith point omitted. Predicted residuals sensitively monitor the magnitude of shift C.

(6) *Recursive residuals*

All the residuals already mentioned are correlated. To find uncorrelated residuals, the recursive least-squares method can be used. The resulting recursive residuals are very useful diagnostically as they allow identification of any instability in a model, for example, instability in time. Recursive residuals are defined

$$\hat{e}_{Ri} = 0, \quad i = 1,...,m \tag{6.100a}$$

$$\hat{e}_{Ri} = \frac{(y_i - x_i b_{i-1})}{\sqrt{1 + x_i(X_{i-1}^T X_{i-1})^{-1} x_i^T}}, \quad i = m+1,...,n \tag{6.100b}$$

where b_{i-1} are estimates obtained from the first $(i - 1)$ points. The matrix X_{i-1} contains the first $(i - 1)$ rows of matrix X. These recursive residuals are independent and have constant variance. They are often used in normality tests or in tests of stability of regression coefficients.

Problem 6.27 *Identification of influential points by various types of residuals*

The outlier in sample C and the leverage point in sample D from Problem 6.8 can be identified by use of some types of residuals.

○ *Data:* from Problem 6.8

Solution: The classical residuals $\hat{e}_i$ standardized residuals $\hat{e}_{Si}$ and Jack-knife residuals $\hat{e}_{Ji}$ are used to detect the outlier in sample C ($x_3 = 13$, $y_3 = 12.74$) and one leverage point ($x_8 = 19$, $y_8 = 12.5$). The results are shown in Table 6.8. The outlier in sample C is most effectively detected by its Jack-knife residual. The leverage point (8 in sample D) is not detected by any residual.

Table 6.8 Various types of residuals for samples C and D

Sample	i	x_i	y_i	$\hat{e}_i$	$\hat{e}_{Si}$	$\hat{e}_{Ji}$
C	3	13	12.74	3.24	3	1203.54
D	8	19	12.5	0	0	0

Conclusion: Neither standardized nor Jack-knife residuals are always suitable for identification of influential points.

The various types of residuals differ in suitability for diagnostic purposes.

(1) The standardized residuals $\hat{e}_{Si}$ serve for identification of heteroscedasticity.

(2) The Jack-knife residuals $\hat{e}_{Ji}$ or the predicted residuals $\hat{e}_{Pi}$ are suitable for identification of outliers.

(3) Recursive residuals $\hat{e}_{Ri}$ are used for identification of autocorrelation.

For analysis of residuals a variety of plots are used. Three principal types of plots can indicate inaccuracy of a proposed model, some trends, heteroscedasticity or influential points in data.

Plot type I (the index sequence plot) is a plot of residuals $\hat{e}_i$ against the index i.

Plot type II (the plot against the independent variables) is a plot of residuals $\hat{e}_i$ vs. the independent variable x_j.

Plot type III (the plot against the prediction) is represented by a plot of residuals $\hat{e}_i$ against the predicted value y_i.

Figure 6.26 shows possible graph shapes which can occur in plots of residuals. If the graph shape is a random pattern (Fig. 6.26a), the least-squares assumption is correct. Some systematic pattern indicates that the approach is incorrect in some way. A sector pattern in graph types I, II and III indicates heteroscedasticity in data (Fig. 6.26b). A band pattern in graph types I and II indicates some error in calculation or absence of x_j in model (type II). The band pattern may be also caused by outlying points or in type III by a missing intercept term in the regression model.

It should be noted that the plot of $\hat{e}_i$ against the dependent variable y_i is not recommended, because the two quantities are strongly correlated. The smaller the correlation coefficient, the more linear is this plot.

A nonlinear pattern in all three graph types I, II and III indicates that the model proposed is incorrect.

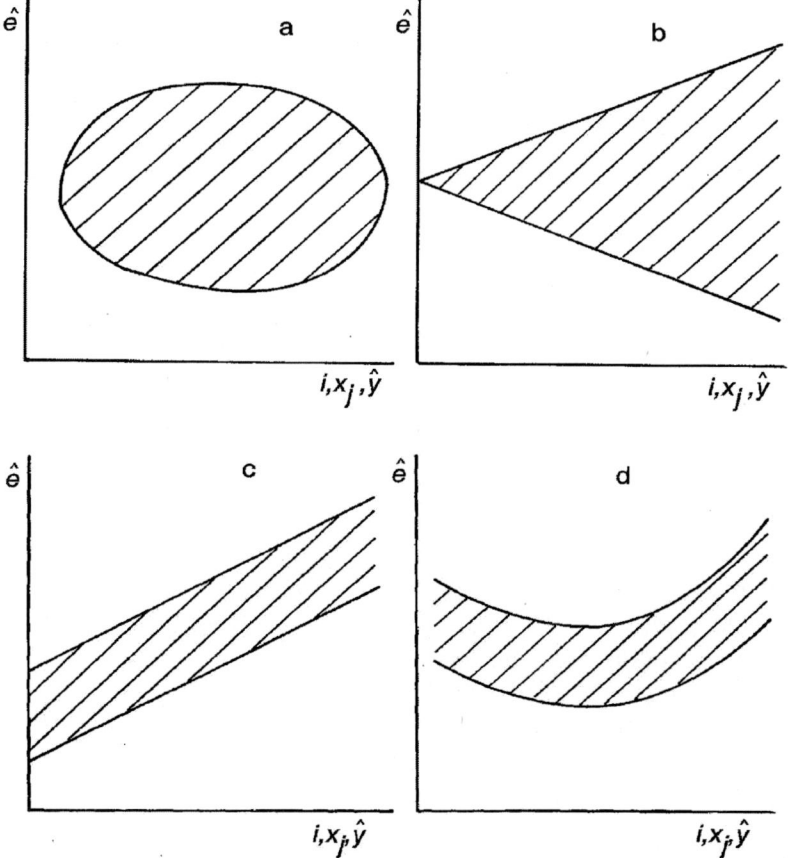

Figure 6.26 Possible shapes of residual plots: (a) random pattern shape, (b) sector pattern shape, (c) band shape, (d) nonlinear curved band shape.

6.5.2.2 Analysis of Projection Matrix Elements

Analysis of elements of the projection hat matrix plays an important role in regression diagnostics because the diagonal elements of this matrix $H_{ii} = x_i (X^T X)^{-1} x_i^T$ indicate the presence of leverage points which are not detected by analysis of residuals. Diagonal elements (denoted in literature as "leverage") have some properties which come from the symmetry and idempotency of matrix H. Among the properties of matrix H are:

(1) The condition for the diagonal elements of a projection matrix is $0 < H_{ii} < 1$ and for nondiagonal elements $-1 < H_{ij} < 1$. When a model also contains an intercept term and the rank of matrix X is m, another condition for diagonal elements is valid, $1/n < H_{ii} < 1/C$, where C is the number of replicate measurements at each value of the controllable variable.

(2) For a model with an intercept term and the full rank of matrix, X:

$$\sum_{i=1}^{n} H_{ii} = m \quad \text{and} \quad \sum_{i=1}^{n} H_{ij} = 1.$$ The mean value of the diagonal element

is $H_{ii} = m / n$.

(3) From the idempotency of matrix H it follows that $H_{ii} = H_{ii}^2 + \sum_{j \neq i}^{n} H_{ij}^2$

$= \sum_{j=1}^{n} H_{ij}^2$. From this equation two important properties of diagonal elements H_{ii} follow:

(a) If the diagonal elements are close to zero, $H_{ii} \to 0$, all nondiagonal elements are also close to zero, $H_{ij} \to 0$, for $j = 1,..., n$;

(b) If the diagonal elements are close to 1, $H_{ii} \to 1$, all nondiagonal elements are close to zero, $H_{ij} \to 0$, for $j = 1, ..., n$.

(4) If the matrix X comes from the multivariate normal distribution, the quantity $F = (n - m)[H_{ii} - 1/n][(1 - H_{ii})(m - 1)]$ has the Fisher-Snedecor distribution $F(m - 1, n - m)$.

(5) The larger the diagonal elements H_{ii}, the more the ith point of prediction $\hat{y}_i$, is affected. If the H_{ii} elements are close to 1 ($H_{ii} \to 1$, and $\hat{y}_i = y_i$) then all of the variability in x_i is explained by the regression model.

(6) The diagonal elements $H_{ii} = \partial \hat{y}_i / \partial y_i$ express the sensitivity of the prediction $\hat{y}_i$ to any change in variable y_i. A zero value, $H_{ii} = 0$, indicates a point which has no influence on prediction.

(7) The diagonal elements H_{ii} are a nondecreasing function of the controllable variables m, and a nonincreasing function of the number of points n.

(8) The further point x_i lies from the centre of gravity of all points, the more it is likely to be a leverage point, and the more the value of diagonal elements H_{ii} will increase.

(9) If the controllable variables $\mathbf{x}$ have the normal distribution, for large sample sizes $n(nH_{ii} - 1)$ has approximately the $\chi^2_m(2)$ distribution.

For more complex analysis, it is useful to form the extension of matrix $\mathbf{X}$ by a vector $\mathbf{y}$ to give matrix $\mathbf{X}^* = (\mathbf{X}|\mathbf{y})$. This matrix corresponds to the projection matrix

$$\mathbf{H}^* = \mathbf{H} + \frac{\hat{\mathbf{e}}\hat{\mathbf{e}}^T}{\hat{\mathbf{e}}^T\hat{\mathbf{e}}} \qquad (6.101)$$

Since the matrix $\mathbf{H}^*$ contains information about all variables it can be used as the total measure of influential points. Diagonal elements of this matrix are given by

$$H^*_{ii} = H_{ii} + \frac{\hat{e}^2_i}{(n-m)\hat{\sigma}^2} \qquad (6.102)$$

To look at elements of the projection matrix, the *index graph* of H_{ii} elements against the index i is used.

Problem 6.28 *Identification of influential points from elements of the projection matrix*

The outlying point in sample C and the leverage point in sample D from Problem 6.8 may be used to test the identification of influential points by elements H_{ii} and H^*_{ii} of projection matrix $\mathbf{H}$.

○ *Data:* from Problem 6.8

Solution: The calculated diagonal elements H_{ii} and H^*_{ii} of the projection matrix $\mathbf{H}$ are listed in Table 6.9

Table 6.9 Elements of the projection matrix H_{ii} and the extended projection matrix H^*_{ii} for samples C and D

Sample	x_i	y_i	H_{ii}	H^*_{ii}
C	13	12.75	0.236	1
D	19	12.5	1	1

The diagonal elements of the extended projection matrix indicate a strong influential point in both samples. The leverage point in sample D is indicated even by the diagonal element H_{ii} of the original projection matrix.

Conclusion: The diagonal elements of an extended projection matrix are useful for detecting outlier and leverage points in data. The leverage point was not detected by any type of residuals (Problem 6.27).

6.5.2.3 Plots for Identification of Influential Points

For identification of different types of influential points, various types of residuals are combined with the diagonal elements of the projection matrix **H**.

(1) *Graph of predicted residuals (GPR)*

(x-axis: the predicted residuals $\hat{e}_{P,i}$; y-axis: the classical residuals $\hat{e}_i$)

This graph is one of the simplest graphs. The leverage points are easy detected by their location as they lie outside the line $y = x$, and they are located quite far from this line. The outliers are located on the line $y = x$ but far from its central pattern (Fig. 6.27).

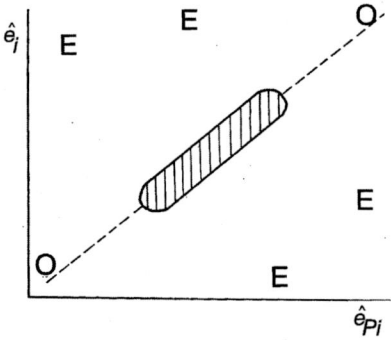

Figure 6.27   Graph of predicted residuals (GPR): E is a high leverage point and O is an outlier.

(2) *Williams graph (WG)*

(x-axis: the diagonal elements H_{ii}; y-axis: the Jack-knife residuals $\hat{e}_{J,i}$)

In this graph two boundary lines are drawn. The first line is for outliers, $y = t_{0.95}(n - m - 1)$ and the second line is for leverage points, $x = 2m/n$ (Fig. 6.28). Denote that $y = t_{0.95}(n - m - 1)$ is the 95% quantile of the Student distribution with $(n - m - 1)$ degrees of freedom.

(3) *Pregibon graph (PG)*

(x-axis: the diagonal elements H_{ii}, y-axis: the normalized residuals $\hat{e}_{N,i}^2$)

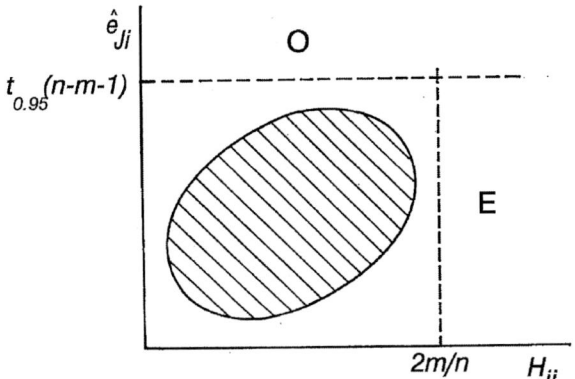

Figure 6.28 Williams graph (WG): E is the leverage point and O is the outlier.

Since the expression $E(H_{ii} + \hat{e}_{Ni}^2) = (m+1)/n$ is valid for this graph, two different constraining lines can be drawn, $y = -x + 2(m+1)/n$ and $y = -x + 3(m+1)/n$. To distinguish among influential points the following rules are used:

(a) a point is strongly influential if it is located above the upper line;

(b) a point is influential if it is located between the two lines. The influential point can be either an outlier or a leverage point.

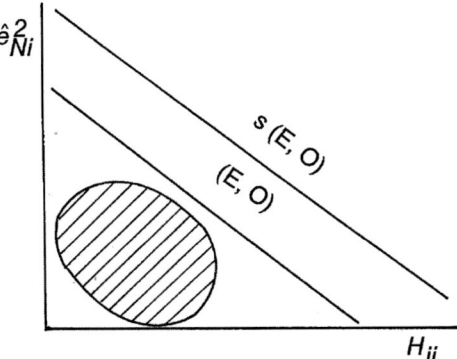

Figure 6.29 Pregibon graph (PG): (E, O) are influential points, and s(E, O) are strongly influential points.

(4) *McCulloh and Meeter graph (MMG)*

(x-axis: $\ln\left[H_{ii}/(m(1-H_{ii}))\right]$; y-axis: the standardized residuals $\hat{e}_{S,i}^2$)

In this plot the solid line drawn represents the locus of points with identical influence, with slope -1. The 90% confidence line is defined by

$y = -x - \ln F_{0.9}(n - m, m)$. The boundary line for leverage points is defined as $x = \ln [2/(n - 2m)]$. The boundary line for outliers is defined by $y = \ln [(n - m)(t^2_{0.95}(n - m)]$ where $t_{0.95}(n - m)$ is the 95% quantile of the Student distribution with $(n - m - 1)$ degrees of freedom.

(5) *Index graph (IG)*

(x-axis: the index i; y-axis: the residuals $\hat{e}_i, \hat{e}_{Si}, \hat{e}_{Ni}, \hat{e}_{Pi}, \hat{e}_{Ji}, \hat{e}_{Ri}$, or the diagonal elements H_{ii} or H^*_{ii} or estimates b_j)

The x-axis always contains the order index i, but the y-axis can be a residual or the diagonal elements of the projection matrix. Sometimes also the parameter estimates b_i are on this axis.

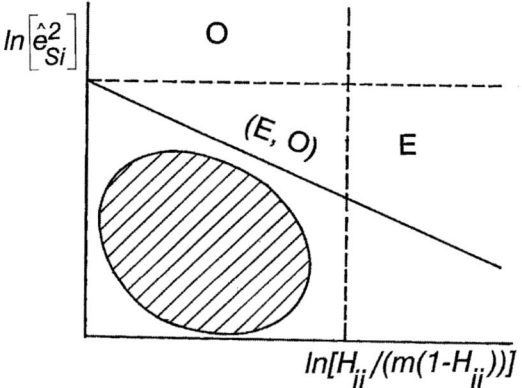

Figure 6.30 McCulloh and Meeter graph (MMG): E is a leverage point, O is an outlier and (E, O) is an influential point.

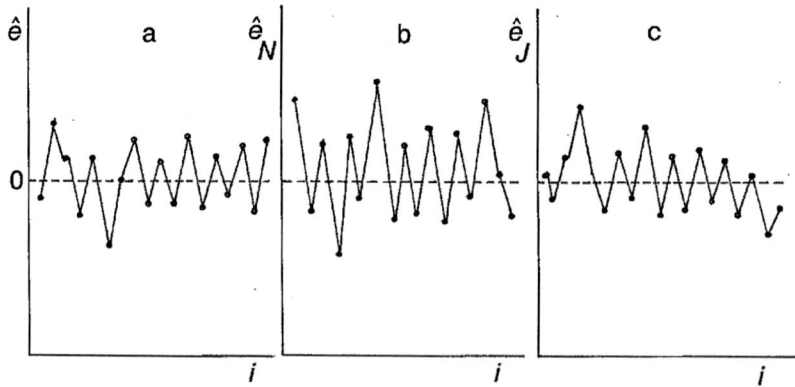

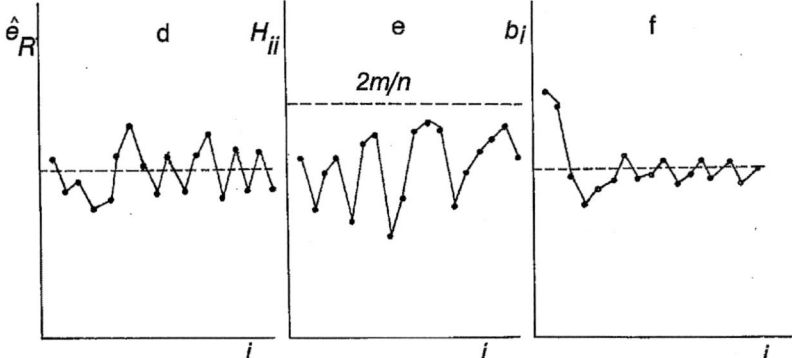

Figure 6.31 Various types of index graph: (a) $\hat{e}_i$ vs. i, (b) $\hat{e}_{Ni}$ vs. i, (c) $\hat{e}_{Ji}$ vs. i, (d) $\hat{e}_{Ri}$ vs. i, (e) H_{ii} vs. i, (f) $\mathbf{b}_i$ vs. i

(6) *Rankit graph (Q-Q plot)*

(*x*-axis: the quantile of the standardized normal distribution u_{P_i} ; *y*-axis: the ordered residuals $\hat{e}_{(i)}, \hat{e}_{S(i)}, \hat{e}_{N(i)}, \hat{e}_{P(i)}, \hat{e}_{J(i)}, \hat{e}_{R(i)}$,)

On the *x*-axis are quantiles of the standardized normal distribution u_{P_i} for $P_i = i/(n+1)$ and on the *y*-axis the order statistics of the residuals, i.e. increasing ordered values of various types of residuals.

Problem 6.29 *Identification of influential points by graphical analysis of residuals and examination of elements of projection matrix*

Use a variety of methods of graphical analysis of residuals and elements of projection matrix to identify influential points in the data from Problem 6.7. Compare the efficiency of the various graphical tools for detecting outliers and leverage points.

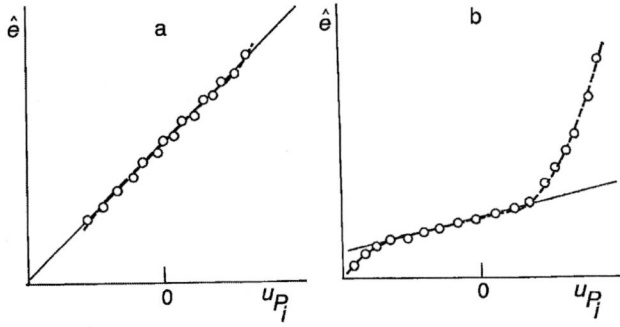

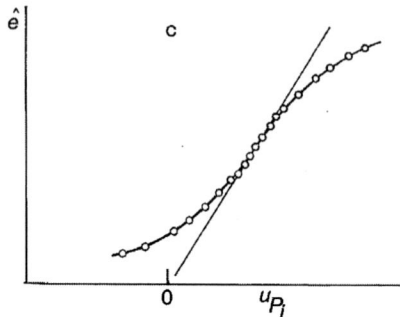

Figure 6.32 Possible variations of the rankit (Q-Q) graph of residuals: (a) no outliers, (b) presence of outliers or residuals distribution with heavy tails, (c) flat distribution of residuals.

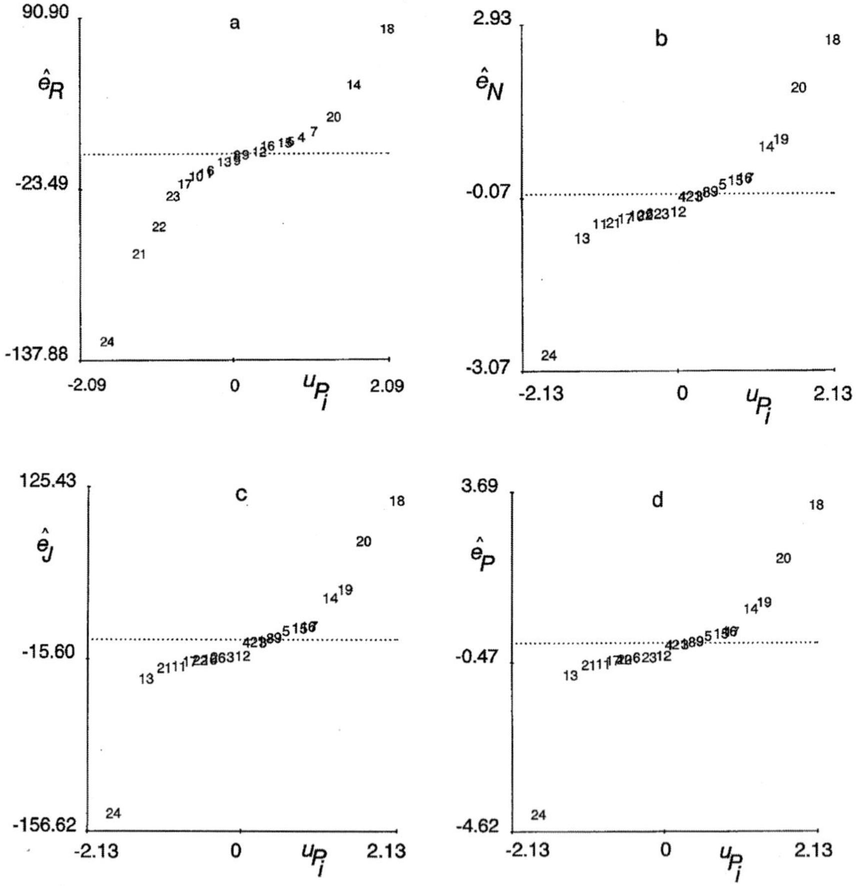

Figure 6.33 Four types of rankit graph: (a) recursive residuals, (b) normalized residuals, (c) Jack-knife residuals, and (d) predicted residuals.

○ *Data:* from Problem 6.7

Solution: Figure 6.33 shows four types of rankit graph and Fig. 6.34 shows four types of index graph. The graph of the third type of residuals $\hat{e}_i$ against prediction $\hat{y}_{Pi}$ indicates that the model proposed is incorrect; some trends and heteroscedasticity in data are also seen (Fig. 6.35).

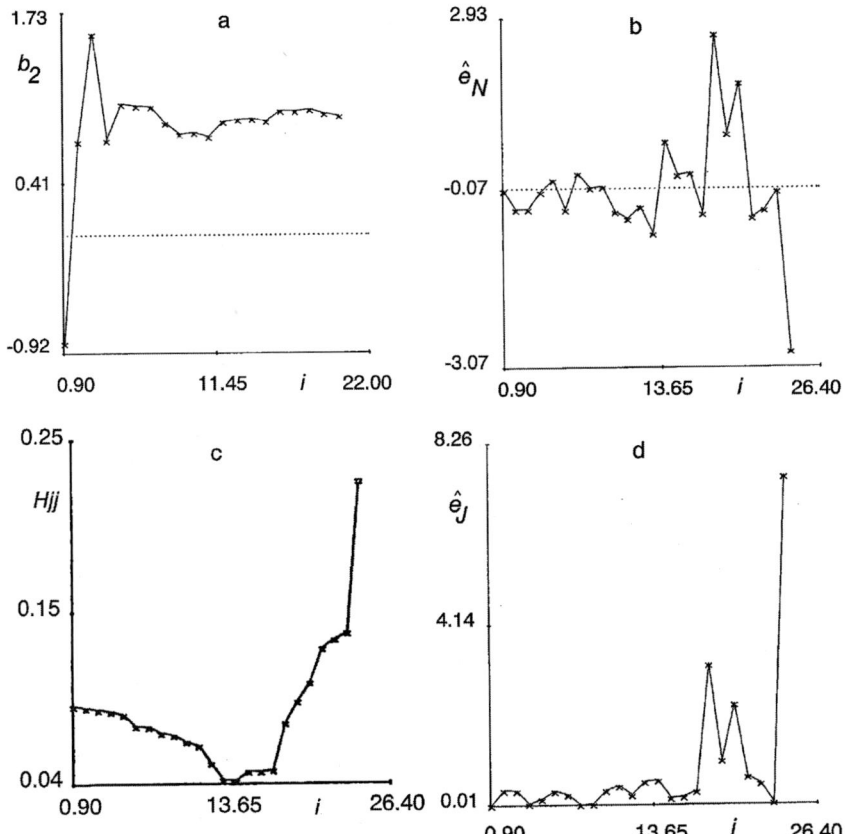

Figure 6.34 Four types of index graph: (a) recursive estimates of slope b_{2i} (b) normalized residuals, (c) elements of projection matrix, and (d) Jack-knife residuals.

For identification of influential points we can use the information from four graphs. The graph of predicted residuals (GPR) on Fig. 6.36a discovers three outliers, points 24, 20 and 18, which, although they lie on the axis $y = x$, are rather far from other points. The Williams graph (WG) on Fig. 6.36b also discovers three outliers, points 18, 20 and 24.

The Pregibon graph (PG) in Fig. 6.36c shows two outlying points above the upper limiting line, detecting that these two points are strongly influential.

Point 20, lying between the two parallel limiting lines, is the only influential point. This point can be either an outlier or a leverage point.

The McCulloh and Meeter graph (MMG) in Fig. 6.36d includes the line with slope −1, connecting points which are of the same influence, and two boundary lines. This plot also discovers three strongly influential points, 24, 20 and 18.

Conclusion: Graphical methods of analysis of residuals are rather simple and illustrative. They enable quick identification of influential points, i.e. here points 18, 20 and 24. To distinguish between outliers and leverage points, some boundary lines must be constructed.

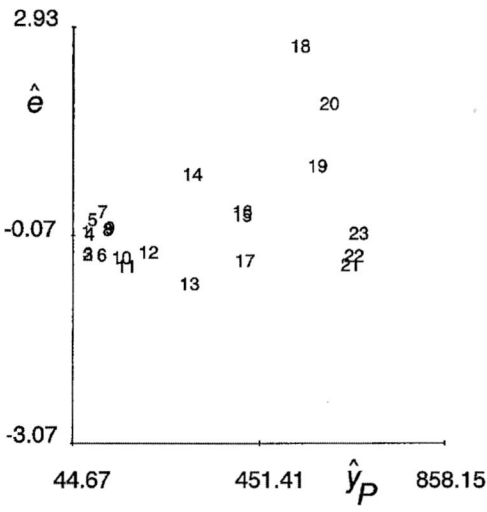

Figure 6.35   Graph of type III, of classical residuals $\hat{e}_i$ against prediction, $\hat{y}_{\mathrm{P}i}$.

6.5.2.4 Other Additional Diagnostics of Influential Points

In the classification of influential points, it is important to remember that they can affect the various regression characteristics differently. Points affecting the prediction $\hat{y}_i$, for example, may not affect the parameter variance. The degree of influence of individual points can be classified according to the characteristics that are affected. For identification of influential points, there are many additional diagnostics which may be divided according to two principal approaches.

The first is based on the examination of changes which occur when certain points are omitted.

The second approach concerns the validity of the linear regression model (6.5b) when the variance of errors is abnormal. For the ith error, the normal distribution $N(0,\sigma^2 / w_i)$ is valid, but the other errors $\varepsilon_j, j \neq i$, have the normal distribution of constant variance σ, i.e. $N(0, \sigma^2)$. The weight parameter lies in the interval $0 < w_i < 1$. This second approach leads to the *model of inflated variance.*

For $w_i = 1$ this assumption leads to the classical least-squares method. If we write $b(w_i)$ for the parameter estimate calculated according to Eq. (6.5b) when the variance of the ith error is just equal to σ^2 / w_i then the following expression is valid

$$\mathbf{b}(1) - \mathbf{b}(w_i) = \frac{(\mathbf{X}^T\mathbf{X})^{-1}\mathbf{x}_i(1 - w_i)\hat{e}_i}{1 - (1 - w_i)H_{ii}} \tag{6.103}$$

where $\mathbf{x}_i$ is the ith row of matrix $\mathbf{X}$ which contains x components of the ith point.

For $w_i = 0$, Eq. (6.103) leads to $\mathbf{b}(1) - \mathbf{b}(0) = \mathbf{b} - \mathbf{b}_{(i)}$ where $\mathbf{b}_{(i)}$ is the estimate reached by the least-squares method by using all points except the ith one. Leaving out the ith point is therefore the same as the case when this point has unbounded infinite variance. To express the sensitivity of parameter estimates to the perturbance parameter w_p, the sensitivity function $\partial\mathbf{b}(w_i)/\partial w_i$ can be used

$$\frac{\partial\mathbf{b}(w_i)}{\partial w_i} = (\mathbf{X}^T\mathbf{X})^{-1}\mathbf{x}_i\hat{e}_i \frac{s_A + (1 - w_i)H_{ii}}{s_A^2}. \tag{6.104}$$

where $s_A = 1 - (1 - w_i)H_{ii}$. The following types of sensitivity function of parameter estimates are possible.

(1) *The Jack-knife influence function*

The sensitivity function of parameter estimates defined by Eq. (6.104) at the value $w_i = 0$, is given by

$$\frac{\partial\mathbf{b}(w_i)}{\partial w_i}\bigg|_{w_i=0} = (\mathbf{X}^T\mathbf{X})^{-1}\mathbf{x}_i \frac{\hat{e}_i}{(1 - H_{ii})^2} = \frac{JC_i}{n - 1} \tag{6.105}$$

The term JC_i is the Jack-knife influence function. It is related to the sensitivity function of parameter estimates for the case when the ith point is omitted, because $\mathbf{b}(0) = \mathbf{b}_{(i)}$.

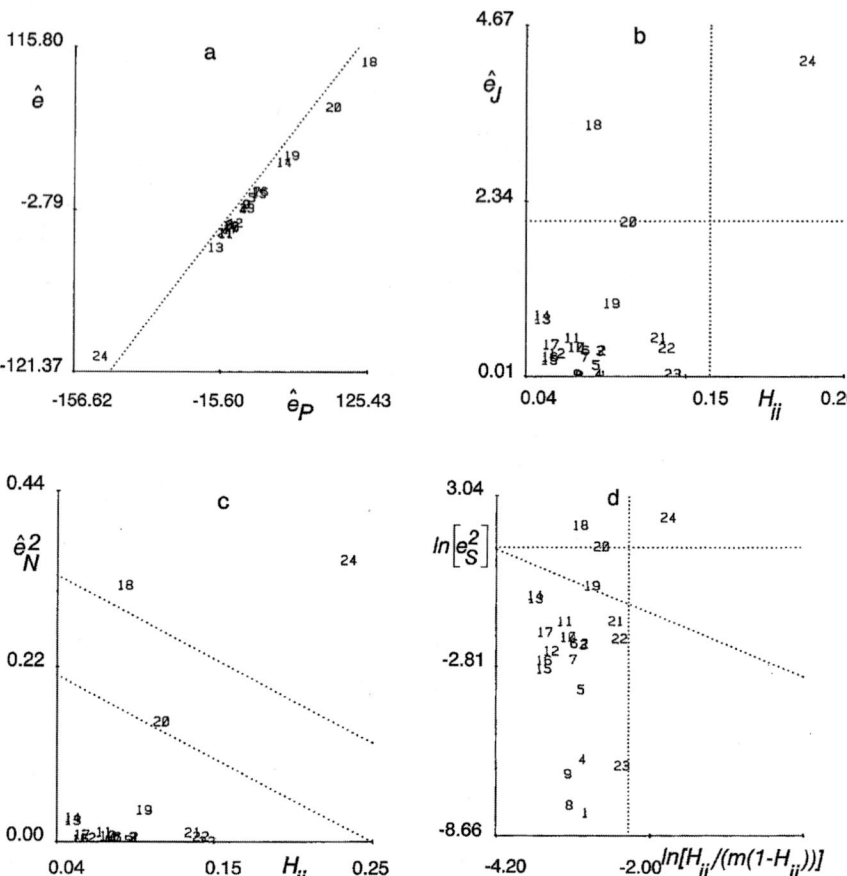

Figure 6.36 Graphs for identification of outliers and leverage points: (a) GPR, (b) WG, (c) PG, and (d) MMG.

(2) *The empirical influence function* EC_i

The sensitivity function of parameter estimates [Eq. (6.104)] at the value of $w = 1$ is given by

$$\left.\frac{\partial \mathbf{b}(w_i)}{\partial w_i}\right|_{w_i=1} = (\mathbf{X}^T\mathbf{X})^{-1}\mathbf{x}_i\hat{e}_i = \frac{EC_i}{n-1} \tag{6.106}$$

The term EC_i is the empirical influence function. It is related to the sensitivity function of parameter estimates at the location of parameter estimates, $\mathbf{b}$, by the least-squares method.

(3) *The sample influence function SC_i*

The sample influence function is proportional to the change in the vector of parameter estimates when the ith point is left out. With the use of Eq. (6.103) we can write

$$SC_i = n(\mathbf{b} - \mathbf{b}_{(i)}) = n(\mathbf{X}^T\mathbf{X})^{-1}\mathbf{x}_i \frac{\hat{e}_i}{1 - H_{ii}} \qquad (6.107)$$

All three influence functions differ only in a single term $(1 - H_{ii})$ so they are not identically sensitive to the presence of leverage points, for which $H_{ii} \to 1$. The disadvantage of all these influence functions is the fact that they are m-dimensional vectors. Their components define the influence of the ith point on the estimate of the jth parameter. Therefore, normalization of these vectors is used [26] to obtain scalar measures corresponding to distances which express the relative influence of the given point on all parameter estimates.

A popular scalar measure of the relative influence of the ith point on all parameter estimates is the *Cook distance D_i*. This is derived by normalization of the sample influence function SC_i. The resulting D_i has the form

$$D_i = \frac{(\mathbf{b} - \mathbf{b}_{(i)})^T \mathbf{X}^T\mathbf{X}(\mathbf{b} - \mathbf{b}_{(i)})}{m\,\hat{\sigma}^2} = \frac{(\hat{\mathbf{y}} - \hat{\mathbf{y}}_{(i)})^T (\hat{\mathbf{y}} - \hat{\mathbf{y}}_{(i)})}{m\,\hat{\sigma}^2} = \frac{\hat{e}_{Si}}{m} \frac{H_{ii}}{1 - H_{ii}} \qquad (6.108)$$

The Cook distance is related to the confidence ellipsoid of the estimates, and it also permits comparison with the quantiles of the Fisher-Snedecor F-distribution. However, the shift of estimates appears here when the ith point is left out. It is approximately true that when $D_i > 1$, the shift is greater than the 50% confidence region, so the relevant point is rather influential. Another interpretation of the Cook distance D_i is based on the Euclidean distance between the prediction vector $\hat{\mathbf{y}}$ estimated by the least-squares method and the prediction vector $\hat{\mathbf{y}}_{(i)}$ estimated by the least-squares method when the ith point is left out. The Cook distance D_i expresses the influence of the ith point on the parameter estimate $\mathbf{b}$ only.

When the ith point does not affect the parameter estimates $\mathbf{b}$ significantly, the value of the Cook distance D_i is low. Such a point, however, can strongly affect the estimate of the residual variance $\hat{\sigma}^2$.

The relative changes in the parameter estimates caused by leaving out the ith point may be expressed by the standardized deviations of the jth estimate

b_j of that parameter estimate $b_{(i)j}$ which has been obtained by leaving out the ith point. The corresponding diagnostic is defined by

$$DS_{ij} = \frac{b_j - b_{(i)j}}{\hat{\sigma}_{(i)}\sqrt{C_{ii}}} \qquad (6.109)$$

where C_{ii} is the diagonal element of matrix $\mathbf{X}^T\mathbf{X}$. The influence of the ith point on the estimate of the yth regression parameter is significant when $DS > 2/\sqrt{n}$.

Problem 6.30   *The change in the estimate of the slope and intercept of a calibration straight line, caused by an outlier*

Determine the change of estimate value for the slope and intercept of the regression straight line $E(y/x) = \beta_1(x - \bar{x}) + \beta_2^*$ in the presence of one outlier.

Solution: From Eq. (6.103) for $w_i = 0$, the expression for the change in parameters $\Delta = \mathbf{b}_i - \mathbf{b}_{(i)}$ is given by

$$\Delta = \begin{bmatrix} b_1 - b_{(i)1} \\ b_2 - b_{(i)2} \end{bmatrix} = \begin{bmatrix} \sum\limits_{i=1}^{n} (x_i - \bar{x})^{-2} & 0 \\ 0 & n^{-1} \end{bmatrix} \begin{bmatrix} x_i - \bar{x} \\ 1 \end{bmatrix} \hat{e}_i \left[1 - n^{-1} - \frac{(x_i - \bar{x})^2}{\sum\limits_{i=1}^{n}(x_i - \bar{x})^2} \right]$$

For the slope change $\Delta_1 = b_1 - b_{(i)1}$ it is $\Delta_1 = \dfrac{n \times \hat{e}_i(x_i - \bar{x})}{(n-1)\sum\limits_{j=1}^{n}(x_j - \bar{x})^2 - n(x_i - \bar{x})^2}$

and for the intercept change $\Delta_2 = b_2^* - b_{(i)2}^*$ it is

$$\Delta_2 = \frac{\hat{e}_i}{(n-1)\sum\limits_{j=1}^{n}(x_j - \bar{x})^2 - n(x_i - \bar{x})^2}.$$

From these expressions it may be concluded that for $x_i = \bar{x}$, $\Delta_1 = 0$ regardless of the magnitude y_i. The slope of the regression straight line will not change whether the point located at $x_i = \bar{x}$ is an outlier or not. The estimate of the intercept will change, however, in dependence on the magnitude of $\hat{e}_i$.

Conclusion: The points of a calibration straight line located on the x-axis far from the mean $\bar{x}$ have the most significant effect on the slope. A point having a negligible effect on the slope may have a strong influence on the intercept estimate.

To express the sensitivity of distance measures to influential points, the *Atkinson distance* A_i is used

$$A_i = DF_i \sqrt{\frac{n-m}{m}} = |\hat{e}_{Ji}| \sqrt{\frac{n-m}{m} \frac{H_{ii}}{1-H_{ii}}} \tag{6.110}$$

which is also convenient for graphical interpretation. With designed experiments, usually $H_{ii} = m/n$ and the Atkinson distance A_i is numerically equal to the Jack-knife residual $\hat{e}_{Ji}$.

By normalizing the sample influence function and using the variance estimate $\sigma^2_{(i)}$ obtained from estimates $\mathbf{b}_{(i)}$, we obtain the characteristic DF_i defined by

$$DF_i^2 = \frac{(\hat{y}_i - \hat{y}_{(i)})^2}{\hat{\sigma}^2_{(i)} H_{ii}} = \hat{e}^2_{Ji} \frac{H_{ii}}{1-H_{ii}} \tag{6.111}$$

The ith point is considered to be significantly influential when $DF_i > 2\sqrt{m/n}$. The characteristic DF_i was recommended by Belsey, Kuh and Welsch [21] as the basic diagnostic characterizing the influence of individual points on prediction $\hat{y}$. The term $H_{ii}/(1-H_{ii})$ in Eqs. (6.108–6.111) is equal to the ratio of variances $D(\hat{y}_i)/D(\hat{e}_i)$, and gives a measure of the sensitivity of regression to the location of the ith point.

There are many regression diagnostics indicating influential points which are based on the approach of leaving out the ith point. In addition to DS_{ij} and DF_i several other characteristics may be useful [26].

The *Anders-Pregibon diagnostic* AP_i expresses the influence of the ith point on the volume of the confidence ellipsoid

$$AP_i = \frac{\det(\mathbf{X}^{*\mathrm{T}}_{(i)}\mathbf{X}^*_{(i)})}{\det(\mathbf{X}^{*\mathrm{T}}\mathbf{X}^*)} \tag{6.112}$$

where $\mathbf{X}^* = (\mathbf{X} \mid \mathbf{y})$ is the matrix extended by the vector $\mathbf{y}$. The diagnostic AP_i is related to the elements of the extended projection matrix $\mathbf{H}^*$ by the expression

$$AP_i = 1 - H_{ii} - \hat{e}^2_{Ni} = 1 - H^*_{ii} \tag{6.113}$$

A point is considered to be influential if $H_{ii}^* = 1 - AP_i > \dfrac{2(m+1)}{n}$. To unify some of the expressions for identification of influential points, Cook and Weisberg [24] have recommended the use of a general diagnostic called the *likelihood distance LD* defined by

$$LD_i = 2\left[L(\hat{\theta}) - L(\hat{\theta}_{(i)})\right] \qquad (6.114)$$

where $L(\hat{\theta})$ is the maximum of the logarithm of the likelihood function when all points are used and $L(\hat{\theta}_{(i)})$ is corresponding value when the ith point is omitted. The parametric vector θ contains either the parameter **b** or the variance estimate $\hat{\sigma}^2$. For strongly influential points $LD_i > \chi^2_{1-\alpha}(m+1)$ where $\chi^2_{1-\alpha}(m+1)$ is the quantile of the χ^2-distribution.

With the use of different variants of LD_i it is possible to examine the influence of the ith point on the parameter estimates or on the variance estimate or on both [26].

(a) To examine the influence of individual points on the parameter estimates b the likelihood distance $LD_i(\mathbf{b})$ is expressed by

$$LD_i(\mathbf{b}) = n\ \ln\left[\frac{d_i\,H_{ii}}{1-H_{ii}} + 1\right] \qquad (6.115)$$

where $d_i = \hat{e}_{si}^2 / (n-m)$.

(b) To examine the influence of individual points on the residual variance estimate, the likelihood distance $LD_i(\hat{\sigma}^2)$ has the form

$$LD_i(\hat{\sigma}^2) = n\ \ln\left[\frac{n}{n-1}\right] + n\ln(1-d_i) + \frac{(n-1)d_i}{(1-d_i)} - 1 \qquad (6.116)$$

(c) To examine the influence of individual points on the parameters b and variance $\hat{\sigma}^2$ together, the likelihood distance $LD_i(\mathbf{b},\hat{\sigma}^2)$ has the form

$$LD_i(\mathbf{b},\hat{\sigma}^2) = n\ \ln\left[\frac{n}{n-1}\right] + n\ \ln(1-d_i) + \frac{(n-1)d_i}{(1-d_i)(1-H_{ii})} - 1 \qquad (6.117)$$

Investigation of the three variants of the likelihood distance leads to the following conclusions [26].

(a) The diagnostic $LD_i(\mathbf{b})$ is a monotonic function of the Cook distance D_i (6.108) and has no advantage over the diagnostic D_i.

(b) The diagnostic $LD_i(\hat{\sigma}^2)$ does not depend on H_{ii} and therefore it is not affected by high leverage points.

(c) The diagnostic $LD_i(\mathbf{b}, \hat{\sigma}^2)$ expresses the influence of individual points on $\mathbf{b}$ and $\hat{\sigma}^2$. It is more useful than both diagnostics A_i and DF_i, especially for models without intercept [26]. It seems to be enough to examine just this diagnostic. Generally the LD_i measures are not quite universal, and for estimation of influential points, many diagnostics must be combined.

Another test for influential points [27] is based on the influence of individual points on the sum of mean quadratic errors of estimates, of the mean quadratic errors of prediction and on the integral mean quadratic error of prediction. To test the influence of the ith point on all these characteristics, the Jack-knife residual $\hat{e}_{Ji}$ may be used as test criterion. This is suitable either for models of simple shift, Eq. (6.98), or models of inflated variance $D(\varepsilon_i) = \sigma^2 / w_i$.

When more points are examined simultaneously for the model of simple shift, the validity of condition

$$\hat{e}_{Ji}^2 \leq F_{1-\alpha/n}(1, n-m-1, 0.5) \qquad (6.118)$$

means that no influential points are present in data. Here, $F_{1-\alpha/n}(1, n-m-1, 0.5)$ means the $100(1 - \alpha/n)\%$ quantile of the non-central F-distribution with non-centrality parameter 0.5 and 1, and $(n - m - 1)$ degrees of freedom. For the model of inflated variance, analogously the validity of the condition

$$\hat{e}_{Ji}^2 \leq 2\, F_{1-\alpha/2}(1, n-m-1) \qquad (6.119)$$

means that influential points are absent. Here $F_{1-\alpha/n}(1, n-m-1)$ means the $100(1-\alpha/n)\%$ quantile of the central F-distribution with 1 and $(n-m-1)$ degrees of freedom. On the basis of these two tests, an approximate rule may be formulated: strongly influential points have squared Jack-knife residuals $\hat{e}_{Ji}^2$ greater than 10.

Problem 6.31 *Comparison of various diagnostics for identification of influential points*

For the outlier from sample C and for the high leverage point from sample D (Problem 6.8) calculate the following five diagnostics: DF_i, D_i, $LD_i(\mathbf{b})$, $LD_i(\hat{\sigma}^2)$ and $LD_i(\mathbf{b},\hat{\sigma}^2)$.

◯ *Data:* from Problem 6.8.

Solution: The calculated diagnostics for identification of influential points are listed in Table 6.10. It can be seen that the leverage point in sample D leads to the indefinite relation 0/0 for D_i, and DF_i and a computer interpreted it as zero. Even the characteristics LD_i do not indicate the leverage point in sample D.

Table 6.10 Comparison of five diagnostics for identification of influential points

Sample	x_i	y_i	D_i	DF_i	$LD_i(\mathbf{b})$	$LD_i(\hat{\sigma}^2)$	$LD_i(\mathbf{b},\hat{\sigma}^2)$
C	13	12.75	1.39	670	2.97	1.81×10^6	2.37×10^6
D	19	12.5	0	0	0	4.84×10^{-2}	4.84×10^{-2}

Conclusion: If the influential points are leverage points, then $\hat{e}_i = 0$ and $H_{ii} = 1$. Detection of these points depends on calculation of indefinite relations by a computer.

To test for influential points, some diagnostic graphs may be used:

(1) The index graph (IG) shows the characteristics of influential points as a function of index i of the point. These graphs may also be plotted for elements of the projection matrix, H_{ir} etc.

(2) The L-R graph introduced by Gray [28] has on the y-axis the squared residuals $\hat{e}^2_{Ni} = \hat{e}^2_i / RSC$ and on the x-axis the elements H_{ir}. All the points will lie under the hypotenuse of the triangle with a 90° angle in the origin of the two axes and the hypotenuse defined by the limiting equality $H_{ii} + \hat{e}^2_{Ni} = 1$.

Most of the characteristics of influential points may be expressed in the form $K(m,n) \times f(H_{ii},\hat{e}^2_{Ni})$, where $K(m, n)$ is a constant depending only on m and n. Therefore the characteristic DF_i [Eq. (6.111)] can be rewritten as

$$DF_i = \sqrt{n-m-1}\sqrt{\frac{H_{ii}\hat{e}^2_{Ni}}{(1-H_{ii})(1-H_{ii}-\hat{e}^2_{Ni})}} \tag{6.120}$$

In the L-R graph, contours of the same critical influence are plotted, and the locations of individual points are compared with them. It may be determined from Eq. (6.120) that for the characteristics DF_i the contours are hyperbolic, as described by the equation $y = \dfrac{(2x - x^2 - 1)}{x(K - 1) - 1}$ where $K = n$ $(n - m - 1)/(c^2 m)$ and c is a constant. For $c = 2$, the constant K corresponds to the limit $2 / \sqrt{m / n}$. The constant c is usually equal to 2, 4 or 8. L-R graphs for other characteristics of influential points may also be drawn.

Problem 6.32   *Examination of infuential points in the validation of a new laboratory method*

Use the L-R graph for DF_i to examine the influential points in Problem 6.7 (validation of a new laboratory method by a comparison with a standard one).

○ *Data:* from Problem 6.7

Solution: Figure 6.37a shows the L-R graph for DF_i. This indicates that points 18, 20 and 24 are strongly influential. With these three points omitted, the regression equation $y = 9.413\ (\pm 5.67) + 0.876\ (\pm 0.016)\ x$ is estimated, with determination coefficient $\hat{R}^2 = 0.994$. The standard deviations of parameter estimates are given in brackets. When these results are compared with those of Problem 6.7, it may be concluded that elimination of influential points will not significantly affect the parameter estimates, but does affect their variances.

The standard deviation of the residuals for the original model with $n = 24$ points, $\hat{\sigma} = 39.54$, decreased on elimination of points 18, 20 and 24, to the value $\hat{\sigma} = 17.24$. Omitting three influential points caused a significant decrease in the quadratic error of prediction from $MEP = 1942$ to $MEP = 333.6$.

Figure 6.37b shows the regression model with the 95% confidence bands. If these are compared with those on Fig. 6.9, the confidence band can be seen to have narrowed.

Conclusion: The L-R graph permits easy identification of influential points. Eliminating influential points causes an improvement in the interval estimates, and this also affects the results of statistical tests.

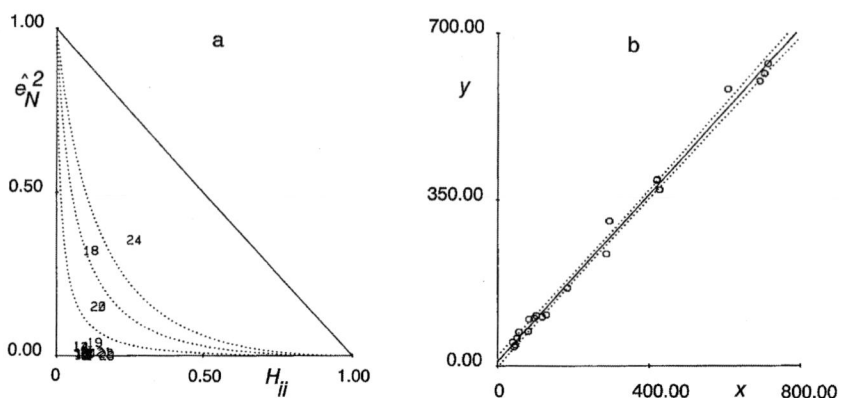

Figure 6.37   (a) The L-R graph for the diagnostic DF_{it} and (b) comparing a new laboratory method with the standard one, with three influential points omitted.

6.5.3 Examination of a Proposed Regression Model

The quality of a proposed model can be considered in case of one controllable variable x directly from the scatter plot of y vs. x. In the case of more controllable variables, scatter plots can falsely indicate nonlinearity in a linear model (cf. Problem 6.24). There are many various plots for considering y on x_j but we limit the choice here to (a) partial regression leverage plots, and (b) partial residual plots. Both plots are augmented here by the graph of residual $\hat{e}$ vs. prediction $\hat{y}$, which can indicate a false model when the points form a nonlinear pattern.

6.5.3.1 Partial Regression Leverage Plots

Belsey [21] named these graphs partial regression leverage plots (PRL plots) and considers them as the basic computer tools for interactive analysis of regression models. They permit classification of the quality of a regression model proposed and also indicate the presence of an influential point and lack of fulfillment of the assumptions of the classical least-squares method. They show the dependence between $\mathbf{y}$ and a selected controllable variable x_j when the other controllable variables forming columns in the matrix $\mathbf{X}_{(j)}$ are kept constant. By the symbol $\mathbf{X}_{(j)}$ we mean a matrix formed by leaving out the jth column $\mathbf{x}_j$.

To discuss the properties of these plots, we assume the regression model (6.5b) expressed in the form

$$\mathbf{y} = \mathbf{X}_{(j)}\boldsymbol{\beta}^* + \mathbf{x}_j c + \varepsilon \tag{6.121}$$

where $\boldsymbol{\beta}^*$ is of dimension $(m-1) \times 1$ and c is the regression parameter of the jth variable. On projecting both sides of Eq. (6.121) into a space orthogonal to the space spanned by the columns of matrix $\mathbf{X}_{(j)}$, we obtain

$$\mathbf{P}_{(j)}\mathbf{y} = \mathbf{P}_{(j)}\mathbf{x}_j c + \mathbf{P}_{(j)}\boldsymbol{\varepsilon} \qquad (6.122)$$

In Eq. (6.122), the product $\mathbf{P}_{(j)}\mathbf{X}_{(j)}$ is equal to zero. The projection matrix $\mathbf{P}_{(j)} = \mathbf{E} - \mathbf{H}_{(j)}$ leads to a projection into the space of the residuals. From Eq. (6.92) it follows that

(a) the term $\hat{\mathbf{v}}_j = \mathbf{P}_{(j)}\mathbf{x}_j$ is the residual vector of regression of one variable $\mathbf{x}_j$ on the other variables which form columns of the matrix $\mathbf{X}_{(j)}$;

(b) the term $\hat{\mathbf{u}}_j = \mathbf{P}_{(j)}\mathbf{y}$ is the residual vector of regression of variable y on other variables which form columns of the matrix $\mathbf{X}_{(j)}$.

The mean value $E(\hat{\mathbf{u}}_j)$ is then given by

$$E(\hat{\mathbf{u}}_j) = cE(\hat{\mathbf{v}}_j) \qquad (6.123)$$

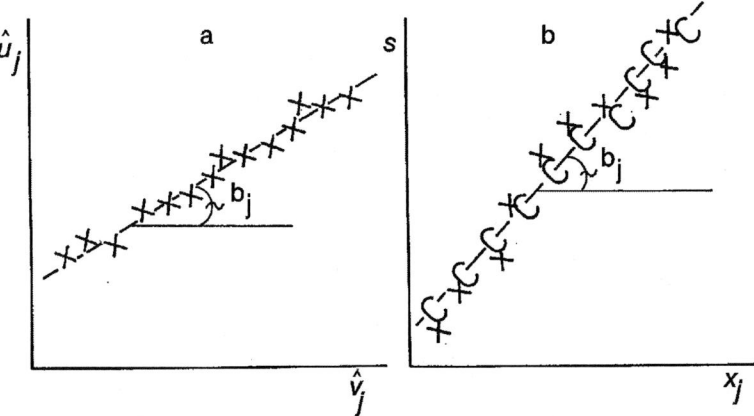

Figure 6.38 (a) A partial regression leverage plot, and (b) a partial residual plot.

The dependences of $\hat{\mathbf{u}}_j$ on $\hat{\mathbf{v}}_j$ form the *partial regression leverage plots.*

When Eq. (6.121) is valid, Eq. (6.123) is linear with zero intercept. The slope estimate obtained by the least-squares method is calculated from

$$\hat{c} = \frac{\hat{\mathbf{u}}_j^T \hat{\mathbf{u}}_j}{\hat{\mathbf{v}}_j^T \hat{\mathbf{v}}_j} = \frac{\mathbf{x}_j^T \mathbf{P}_{(j)} \mathbf{y}}{\mathbf{x}_j^T \mathbf{P}_{(j)} \mathbf{x}_j} \qquad (6.124)$$

After some rearrangements it may be shown that the slope estimate $\hat{c}$ is identical with the estimate b_j determined by the classical least-squares method for *unpartitioned model* $E(y \,/\, \mathbf{x}) = \mathbf{X}\boldsymbol{\beta}$. Moreover, an important equality

$$\hat{\mathbf{e}} = \hat{\mathbf{u}}_j - \hat{\mathbf{v}}\hat{c} \tag{6.125}$$

shows how the residual $\hat{\mathbf{e}}$ from the least-squares method (6.92) is connected with the partial residuals $\hat{\mathbf{u}}_j$ and $\hat{\mathbf{v}}_j$.

The partial regression leverage plots have the following properties:

(a) The slope $\hat{c}$ in the PRL plot is identical with the estimate b_j in an unpartitioned model and the intercept is equal to zero. This linear dependence is valid only when the proposed model [Eq. (6.121)] is correct.

(b) The correlation coefficient between $\hat{\mathbf{v}}_j$ and $\hat{\mathbf{u}}_j$ corresponds to the partial correlation coefficient $R_{yxj}(\mathbf{x})$.

(c) Residuals corresponding to a regression straight line in the PRL plot are identical with residuals for an unpartitioned model.

(d) In the PRL plot the influential points stand out, and also any violation of the assumptions for the least-squares method, for example, about homoscedasticity.

Partial regression leverage plots have also some disadvantages:

(a) On the x-axis, the co-ordinates $\hat{\mathbf{v}}_j$ are not in the original scale of variable $\mathbf{x}_j$. If there is, for example, some scatter in the residuals of functions $\mathbf{x}_j$, the PRL plots may not indicate it.

(b) If individual controllable variables (in the columns of matrix $\mathbf{X}$) are strongly correlated, the PRL plot may not indicate correctly the non-linearity, so a false hypothesis of the model may be proposed (6.121).

The partial regression leverage plots are in the standard output of the regression module of ADSTAT or QC-Expert, because they correctly indicate various types of influential points.

Problem 6.33   *Application of partial regression leverage plots*

Construct PRL plots for a linear regression model with the simulated data from Problem 6.24.

○ *Data:* Generated from Problem 6.24

Solution: The PRL plots for variables x_1 and x_2 are shown in Fig. 6.39. The linear course and the zero residuals show that the data are in accord with a linear function of x_1 and x_2. The strong multicollinearity between the variables does not influence their course.

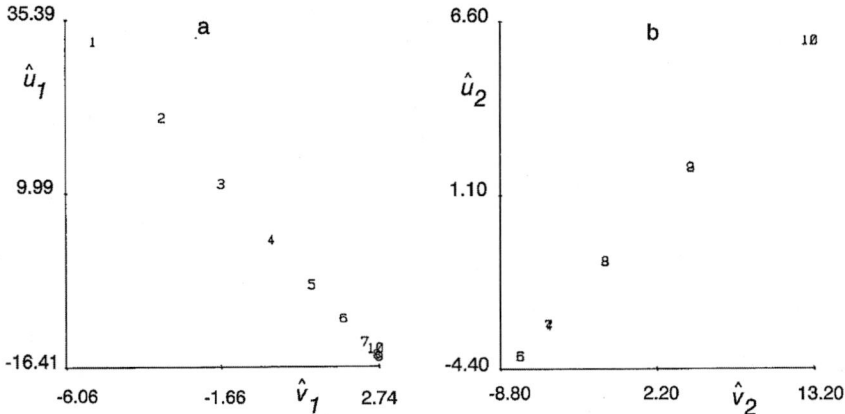

Figure 6.39 The partial regression leverage plots for (a) variable x_1 and (b) variable x_2.

Conclusion: The linearity of all partial regression leverage plots proves the correctness of a proposed regression model. The quality of estimates may be classified according to the spread of points around the regression straight line in partial regression leverage plots.

6.5.3.2 Partial Residual Plots

Partial residual plots are also termed "component + residual" plots. Rewriting Eq. (6.125) in the form

$$\hat{\mathbf{u}}_j = \hat{\mathbf{e}} + b_j(\mathbf{E} - \mathbf{H}_{(j)})\mathbf{x}_j \qquad (6.126)$$

gives the partial regression leverage plot expressed as a dependence of $\hat{\mathbf{e}} + b_j(\mathbf{E} - \mathbf{H}_{(j)})\mathbf{x}_j$ on $(\mathbf{E} - \mathbf{H}_{(j)})\mathbf{x}_j$. The *partial residual plot* is the special case $\mathbf{H}_{(j)} = 0$. It is, in fact, a dependence of partial residuals $\mathbf{s}$ on variable $\mathbf{x}_j$. For variable $\mathbf{s}$ we have

$$\mathbf{s} = \hat{\mathbf{e}} + \mathbf{b}\mathbf{x}_j = y - \sum_{k \neq j}^{m} \mathbf{x}_k b_k \qquad (6.127)$$

where $\mathbf{x}_k$ is the kth column of matrix $\mathbf{X}$.

When the regression model contains an intercept, the modified partial residuals may be used

$$s_i^* = \hat{e}_i + (x_{ij} - \overline{x}_j)b_j + \overline{y} \qquad (6.128)$$

where $\overline{x}_j, \overline{y}_j$ are the arithmetic averages of variables x_j and y.

In "component + residual" plots a deterministic component is plotted separately.

$$c_{ij} = (x_{ij} - \bar{x}_j)b_j, \quad i = 1,...,n \qquad (6.129)$$

which is usually marked on a plot by the letter "C". The partial residual $s_i = c_{ij} + \hat{e}_i$, $i = 1,...,n$, are in this plot marked by crosses. If $\mathbf{x}_j$ is orthogonal to all the columns of matrix $\mathbf{X}_{(i)}$, then $\hat{\mathbf{v}}_j = \mathbf{x}_j$ and the partial regression leverage plot would be identical to the partial residual plot. The partial residual plots provide rather different information from the partial regression leverage plots. Partial residual plots have the following properties:

(a) the slope of **s** vs. $\mathbf{x}_j$ is equal to b_j and the intercept is zero. The linear dependence shows the suitability of proposed variable $\mathbf{x}_j$ in the model;

(b) the residuals of these regression lines are the residuals $\hat{e}_i$ for the unpartitioned model;

(c) if the angle between $\mathbf{x}_j$ and some columns of matrix $\mathbf{X}_{(j)}$ is small (multicolinearity) the partial residual plot has falsely small scatter around the regression line $b_j\mathbf{x}_j$ and the effect of influential points is suppressed.

Partial residual plots are recommended for indication of different types of nonlinearity in the case of a poorly proposed regression model.

Problem 6.34 *Building partial residual plots*

Construct the partial residual plots for the linear regression model and simulated data from Problem 6.24.

○ *Data:* from Problem 6.24.

Solution: The partial residual plots for variables x_1 and x_2 are drawn in Fig. 6.40.

The linear course together with the zero residuals $\hat{e}_i$ show again the linearity with respect to variables x_1 and x_2. Since the x-axes are not transformed, the magnitude of slopes may be considered. This magnitude should correspond to parameter estimate b_j in the regression model.

Conclusion: The linearity in all partial residual plots shows the correctness of the regression model proposed. It is recommended to combine examination of the partial regression leverage plots with the partial residual plots.

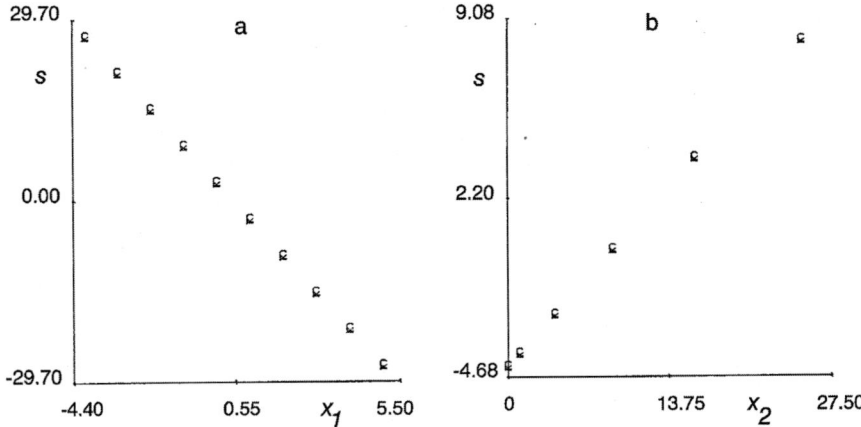

Figure 6.40   Partial residual plots for (a) variable x_1 and (b) variable x_2.

Problem 6.35 *Examination of the model for the relationship between rubber composition and its abrasion resistance*

Make a graphical examination of a proposed linear model expressing the relationship between the composition of rubber and its abrasion resistance (Problem 6.6). Identify any influential points.

○ **Data:** from Problem 6.6

Solution: Figure 6.41 shows the partial regression leverage plots and Fig. 6.42 the partial residual plots for x_1 and x_2.

Because of orthogonality of the two variables, the plots in Figs. 6.41 and 6.42 are nearly the same. The variable x_1 is not significantly affected, as in both plots 6.41a and 6.42a the points form a random pattern. Variable x_2 shows a distinct trend which may result either from nonlinearity or from outliers in the data, and particularly point 8. In Fig. 6.43 the plot of residuals $\hat{e}_i$ vs. the prediction $\hat{y}$ shows a random pattern of points, proving that the proposed model is suitable, despite two points, 8 and 2, seeming to be outliers.

Both graphs in Fig. 6.44 show significant influence from point 8 and also from points 3 and 1. The values of the Jack-knife residuals $\hat{e}_{Ji}$ do not indicate strongly influential points, because the maximum value $\hat{e}_{Ji}$ for $i = 8$ is $\hat{e}_{J8} = -2.404$.

Conclusion: For a small sample size it is not possible to consider whether the model has been correctly proposed. From the graphs, it can be concluded that the data may contain the outliers. Repetition of experiments proved that the linear model is not correct.

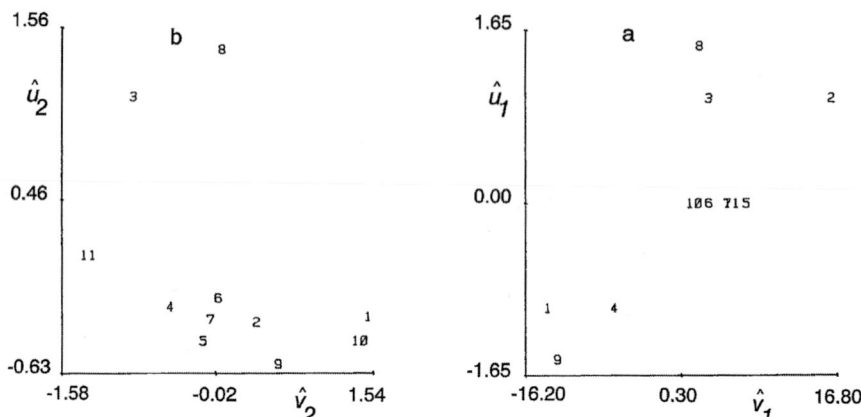

Figure 6.41   Partial regression leverage plot for (a) the variable x_1 and (b) the variable x_2.

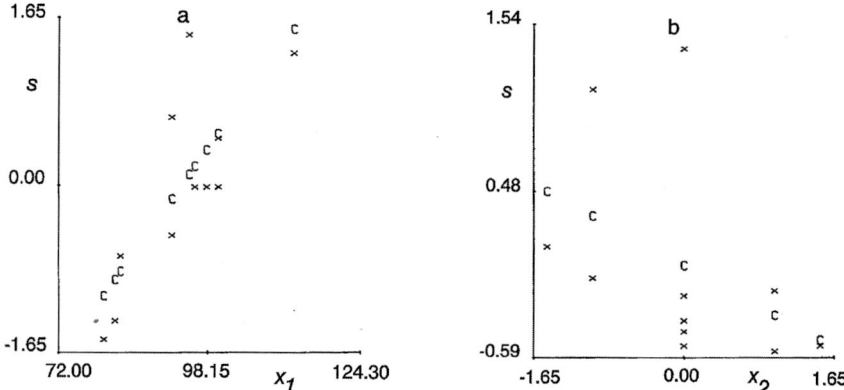

Figure 6.42   Partial residual plot for (a) the variable x_1 and (b) the variable x_2.

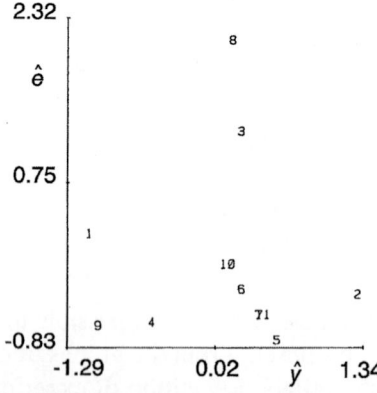

Figure 6.43   Plot of residuals $\hat{e}_i$ vs. prediction $\hat{y}$.

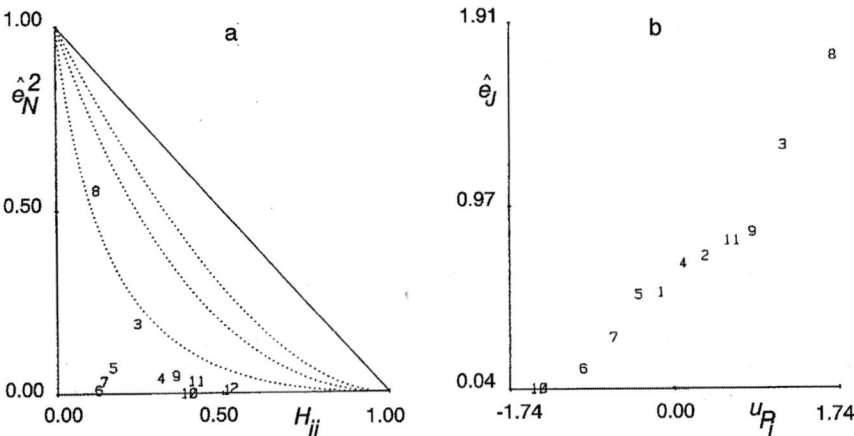

Figure 6.44 (a) The L-R graph for DF_i and (b) the Q-Q plot for the Atkinson distance A_r

6.5.3.3 Sign Test for Model Specification

To check a proposed regression model with reference to the data, all tests of specification (linearity) from Section 6.3 may be applied. A simple test based on the residuals $\hat{e}_i$ is the *sign test*. Incorrectness of a proposed model causes non-randomness of residuals, and this non-randomness may be tested by a sign test. The number of sequences n_U of the same sign of residuals is estimated, e.g. for residuals $-1, -1, 1, -1, 1, 2, 1$ the number of sequences is equal to 4, $\hat{n}_U = 4$. Then the number of residuals with positive sign $(n+)$ and negative sign $(n-)$ is determined. For medium sample sizes the theoretical number of sequences n_t and its variance D_t are defined by

$$n_t = 1 + \frac{2n_+ n_-}{n_+ + n_-} \approx 1 + \frac{n}{2} \tag{6.130}$$

$$D_t = \frac{2n_+ n_- (2n_+ n_- - n_+ + n_-)}{(n_+ + n_-)^2 (n_+ + n_- - 1)} \approx \frac{n}{4}$$

When $n_U < n_t - 2\sqrt{D_t}$, there is a trend in the residuals and the model is incorrect.

Problem 6.36 *Examination of a proposed model by the sign test*

For samples A, B, C and D from Problem 6.8, test for correctness of the proposed model of a regression straight line.

○ **Data:** from Problem 6.8

Solution: Table 6.11 lists the numbers of sequences for samples A, B, C and D.

From the table it is evident that small values of $\hat{n}_U$ (less than $n_t - 2\sqrt{D_t} \approx 4.84$) for samples B and C correspond to non-randomness of residuals and also the incorrectness of the proposed straight line model.

Table 6.11 The number of sequences for samples A, B, C and D

Data sample	A	B	C	D
The number of sequences $\hat{n}_U$	7	3	4	7

Conclusion: The sign test can test for non-randomness of residuals, caused either by a false model (sample B) or by outliers (sample C).

6.5.4 Examination of Conditions for the Least-squares Method

The violation of the basic conditions for the least-squares method is discussed in Section 6.6. In this section the graphical diagnostics for indication of heteroscedasticity, autocorrelation and non-normality of errors ε are described.

6.5.4.1 Heteroscedasticity

Heteroscedasticity often appears in instrumental data measured in the chemical laboratory. The variance of measurement is usually an increasing function of variable y because the relative precision of the measurement is constant. This type of heteroscedasticity may be detected by the plot of $\hat{e}_i^2$ vs. $\hat{y}_i$ which gives a pattern of typically linear or nonlinear shape. If the measurement variance is dependent on x_j, the plot $\hat{e}_i^2$ vs. x_{ij} leads also to a linear or nonlinear pattern. The heteroscedasticity may be detected by plots of residuals or by partial regression leverage plots.

Identification of heteroscedasticity in data is based on the idea that the variance of a measured quantity at the ith point is an exponential function of the variable $x_i\beta$ of the type $\sigma_i^2 = \sigma^2 \exp(\lambda x_i\beta)$ where x_i is the ith row of matrix **X**. The test for homoscedasticity is carried out by checking the null hypothesis H_0: $\lambda = 0$. Cook and Weisberg [30] introduced the test criterion

$$S_f = \frac{\left[\sum_{i=1}^{n} (\hat{y}_i - \hat{y}_P)\hat{e}_i^2 \right]^2}{2\hat{\sigma}^4 \sum_{i=1}^{n} (\hat{y}_i - \hat{y}_P)^2} \tag{6.131}$$

where $\bar{y}_P = \left(\sum_{i=1}^{n} \hat{y}_i \right) / n$. When the null hypothsis is valid, the test statistic S_f

has approximately the $\chi^2(1)$ distribution with one degree of freedom.

The corresponding diagnostic plot has the squares of standardized residuals $\hat{e}_{Si}^2$ on the y-axis and $(1 - H_{ii})\hat{y}_i$ on the x-axis. If heteroscedasticity is not present in the data, a random pattern of points appears. When heteroscedasticity is present, a wedge-shaped pattern appears and most of the points are located in this part of plot.

Problem 6.37 *Examination of the homoscedasticity assumption in validation of a new laboratory method*

Figure 6.9 shows that for larger values of x the variance of points round the regression line increases. Are the data from Problem 6.7 homoscedastic or heteroscedastic?

○ *Data:* from Problem 6.7

Solution: The test criterion $S_f = 119.45$ [Eq. (6.131)] has a higher value than the quantile $\chi^2_{0.975}(1) = 5.02$ and the null hypothesis $H_0: \lambda = 0$ is rejected. The data exhibit heteroscedasticity. The plot of $\hat{e}_i^2$ against $\hat{y}_i$, Fig. 6.45a shows a recognizable systematic trend indicating heteroscedasticity. Points 18, 20 and 24 are influential points. The typical wedge-shaped pattern of points in the plot of $\hat{e}_{Si}^2$ on the y-axis and $(1 - H_{ii})\hat{y}_i$ on the x-axis in Fig. 6.45b also proves heteroscedasticity. When points 18, 20 and 24 are left out, the test criterion $S_f = 2.75$ is lower than $\chi^2_{0.975}(1) = 5.02$ and heteroscedasticity is not proved.

Conclusion: Plots indicating heteroscedasticity can also detect whether the heteroscedasticity is caused by the presence of influential points.

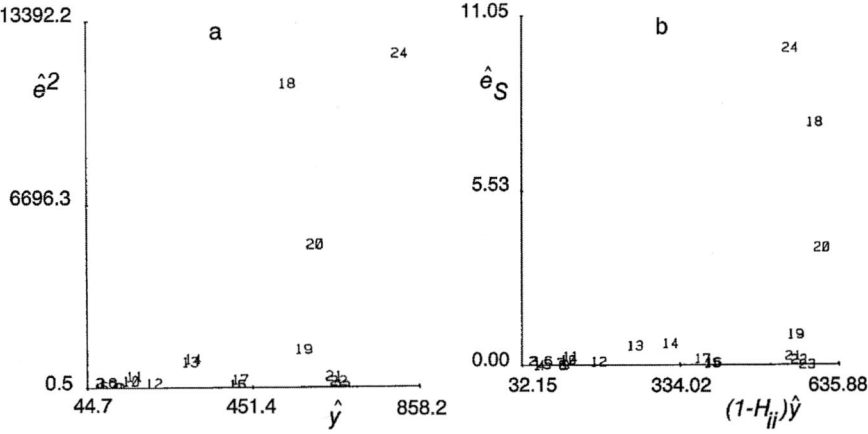

Figure 6.45 Plots for testing for heteroscedasticity (a) $\hat{e}_i^2$ vs. $\hat{y}_i$, and (b) $\hat{e}_{Si}^2$ on the y-axis and $(1 - H_{ii})\hat{y}_i$ on the x-axis.

6.5.4.2 Autocorrelation

When data are a time series, the errors ε are not independent but are correlated with one another. We will discuss only the most frequent case of autocorrelation of the first order, the *autoregressive process of the first order* AR(1), described by the expression

$$\varepsilon_i = \rho_1 \varepsilon_{i-1} + u_i \qquad (6.132)$$

where $u_i \approx N(0, \sigma^2)$ is an independent, random variable with constant variance and $\rho_1 \leq 1$ is the *autocorrelation coefficient of the first order.* For $\rho_1 = 1$, Eq. (6.132) defines a case of cumulative errors, which appears quite often in laboratory. When the model $\mathbf{X}\boldsymbol{\beta}$ does not contain all the significant variables and is falsely proposed, the mean values of the residuals correspond to an AR(1) process, with a positive autocorrelation coefficient of the first order, ρ_1. Tests of autocorrelation can be understood as tests of accuracy of a proposed model, with reference to the number of controllable variables. From Eq. (6.132) it may be concluded that for an AR(1) process, the dependence of ε_i on ε_{i-1} is linear, with slope ρ_1. To test for autocorrelation, the graph of $\hat{e}_i$ against $\hat{e}_{i-1}$ is plotted, and an approximately linear trend proves significant autocorrelation.

Classical residuals are, however, correlated even in cases when the errors ε_i are not correlated. For small sample sizes, this may lead to a false finding of linearity of the dependence $\hat{e}_i$ against $\hat{e}_{i-1}$. The use of recursive residuals $\hat{e}_{Ri}$ is more convenient.

Problem 6.38   *Autocorrelation test for kinetic data*

Kinetic data for inversion of a saccharide in 1M HCl at 30°C were measured. Find out whether the autocorrelation effect in the data is caused by the method of taking samples. Use graphical tests.

○ *Data:* x is time in minutes; y is the logarithm of the fraction of saccharide remaining unreacted in the reaction mixture, multiplied by 10.

x	0	10	20	30	40	50	60	70	80
y	1	0.954	0.895	0.843	0.791	0.735	0.685	0.628	0.581

Solution: The plot of $\hat{e}_i$ against $\hat{e}_{i-1}$ in Fig. 6.46a and the plot of $\hat{e}_{Ri}$ vs. $\hat{e}_{Ri-1}$ in Fig. 6.46b show significant negative autocorrelation. Since both plots are similar in nature, classical residuals may be used.

Conclusion: Plots for examination of autocorrelation of residuals allow the sign and the magnitude of the autocorrelation coefficient of the first order ρ_1 to be estimated. Here, the classical $\hat{e}_i$ and recursive residuals $\hat{e}_{Ri}$ give similar results.

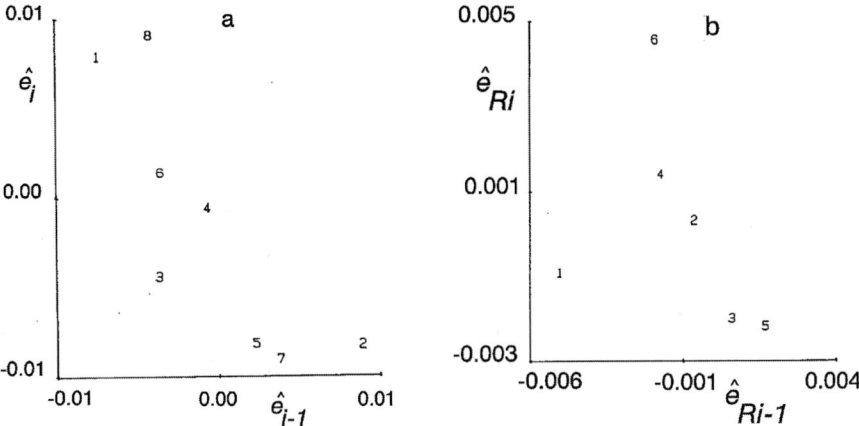

Figure 6.46 Plots for examination of autocorrelation of the first order (a) $\hat{e}_i$ vs. $\hat{e}_{i-1}$ and (b) $\hat{e}_{Ri}$ vs. $\hat{e}_{Ri-1}$

6.5.4.3 Normality of Errors

The normality of errors is examined by a Q-Q plot containing the order statistics of classical residuals $\hat{e}_{(i)}$ in dependence on the quantile of the normalized normal distribution u_{P_i} for $P_i = i / (n+1)$. Since small samples exhibit a supernormality effect, independent recursive residuals $\hat{e}_{Ri}$ are used instead of classical residuals, because this effect then does not exist.

To test the normality of residuals, some tests from Chapter 3 may be used. The most convenient test seems to be the *Jarque-Berra test of combined skewness and kurtosis* [32] which is based on the criterion

$$L(\hat{e}) = n\left[\frac{\hat{u}_3^2}{6\hat{u}_2^3} + \frac{(\hat{u}_4 / \hat{u}_2^2) - 3}{24}\right] + n\left[\frac{3\hat{u}_1^2}{2\hat{u}_2} + \frac{\hat{u}_3\,\hat{u}_1}{\hat{u}_2^2}\right] \qquad (6.133)$$

where the symbol $\hat{u}_j$ denotes the jth general moment of the sample residuals, and is defined by

$$\hat{u}_j = \frac{1}{n}\sum_{i=1}^{n}\hat{e}_i^j \qquad (6.133a)$$

When the errors have a normal distribution, the test statistic $L(\hat{e})$ has asymptotically the $\chi_{1-\alpha}^2(2)$ distribution. When $L(\hat{e}) > \chi_{1-\alpha}^2(2) = 5.99$, the null hypothesis H_0 about the error normality is rejected. In this test, the supernormality effect of small samples may again disturb statistical testing.

For linear models with an intercept term, $E(\hat{e}) = \hat{u}_1 = 0$ and the Jarque-Berra criterion can be simplified to the form

$$L(\hat{e}) = n\left[\frac{\hat{g}_1}{6} + \frac{(\hat{g}_2 - 3)}{24}\right]$$
(6.134)

Where $\hat{g}_1 = \hat{u}_3^2 / \hat{u}_2^3$ and $\hat{g}_2 = \hat{u}_4 / \hat{u}_2^2$. This procedure however, is not convenient for small samples, because of the supernormality effect, and moreover the distribution of $L(\hat{e})$ differs from the asymptotic $\chi_{1-\alpha}^2(2) = 5.99$. For small samples it is more convenient to determine the distribution of $L(\hat{e})$ from a simulation calculation for the given matrix $\mathbf{X}$. As $L(\hat{e})$ is dependent of the error variance σ^2, the errors ε_i may be generated from the normalized normal distribution $N(0, 1)$.

Problem 6.39 *Examination of normality for four samples*

For samples A, B, C and D in Problem 6.8, use the Jarque-Berra test criterion $L(\hat{e})$ to test for normality.

○ *Data:* from Problem 6.8

Solution: For four samples the Jarque-Berra test criterion $L(\hat{e})$ is the last column in Table 6.7 (Problem 6.25). This test disproved normality only for sample C. The other samples A, B and D exhibit normality of residuals.

Conclusion: Tests of normality of residuals do not prove incorrectness of a proposed regression model or unsuitability of data. When normality is not proved, the presence of outliers is often the cause; the Q-Q plot is then recommended for detection of influential points.

6.6 Procedures when Conditions for Least-squares are Violated

In Section 6.2, seven conditions were mentioned which must be met if the least-squares method is to give the best unbiased linear estimates of parameters. The construction of confidence intervals and hypothesis tests also depend on these conditions being satisfied. In the laboratory some of the conditions, however, are not met. In this chapter we give our attention to regression procedures when

(1) some restrictions are placed on the parameters;
(2) the covariance matrix of errors is not diagonal and data do not exhibit the same variance;

(3) the matrix $\mathbf{X^T X}$ is ill-conditioned because of multicollinearity;

(4) the distribution of data is not normal and some influential points exist in data;

(5) the independent variables x are also subject to random errors.

The most important diagnostic procedures for identification of violations of the least-squares conditions are described in Section 6.5. This section gives a modified procedure for parameter estimation and some special tests.

6.6.1 Restrictions Placed on the Parameters

In many laboratory problems some restrictions are placed on parameters because of their physical meaning and interpretation. Positive values, for example, are often requested for most chemical or physical parameters. The regression procedure with a restriction depends on whether the restrictions are *precise* (deterministic as they are fixed numbers) or *statistical* (they are random numbers). The restrictions can be stated in form of equalities, or inequalities when they concern restricted intervals.

The most frequent request in laboratory problems is that the regression line should fit the data and also pass through the origin. This last request can be fulfilled by omitting the intercept term. We will discuss cases when the parameter restrictions are given as an equality, and when parameters should be numerically greater than a given limit.

A restriction in the form of an equality

This group of restrictions includes the following requests about parameters,

(a) some parameters should reach specified values;

(b) some parameters should have a specified mutual ratio;

(c) the sums or differences of some parameters should be equal to a given number; and

(d) the regression model should also fit certain points specified by co-ordinates.

To satisfy these four requests, the condition of linearity may be formulated as

$$P_{11}\beta_1 + P_{12}\beta_2 + ... + P_{1m}\beta_m = p_1$$
$$P_{21}\beta_1 + P_{22}\beta_2 + ... + P_{2m}\beta_m = p_2$$
$$\cdots\cdots\cdots\cdots\cdots\cdots\cdots\cdots\cdots \qquad (6.135)$$
$$P_{k1}\beta_1 + P_{k2}\beta_2 + ... + P_{km}\beta_m = p_k$$

or written in a matrix, notation

$$\mathbf{P}\boldsymbol{\beta} = \mathbf{p} \tag{6.135a}$$

where $\mathbf{P}$ is the matrix of dimension $(k \times m)$ of known coefficients and $\mathbf{p}$ is the vector of dimension $(k \times 1)$ of known components, estimated on the basis of the requested restrictions. The mathematical condition for solution is that the rank of matrix $\mathbf{P}$ should be equal to k and also $k < m$. This means that rows of matrix $\mathbf{P}$ are linearly independent.

To estimate parameters $\mathbf{b}_R$ which fulfil the condition of the minimum of the least-squares method with a restriction (6.135a), the technique of Lagrange multipliers is used. This method involves minimization of the conditioned sum of squares

$$U_R = (\mathbf{y} - \mathbf{X}\mathbf{b}_R)^T (\mathbf{y} - \mathbf{X}\mathbf{b}_R) + \boldsymbol{\lambda}^T (\mathbf{p} - \mathbf{P}\mathbf{b}_R) \tag{6.136}$$

where $\boldsymbol{\lambda}$ is the vector of Lagrange multipliers of dimension $(k \times 1)$; its estimate is also sought. The method is called *the conditioned least-squares method* (CLS). As in the classical least-squares method (LS), the estimates $\mathbf{b}_R$ and $\boldsymbol{\lambda}$ may be found with the use of the first derivative of function U_R according to these parameters,

$$\frac{\partial U_R}{\partial \mathbf{b}} = -2\mathbf{X}^T\mathbf{y} + 2\mathbf{X}^T\mathbf{X}\mathbf{b}_R - \mathbf{P}^T\boldsymbol{\lambda} = 0 \tag{6.137a}$$

$$\frac{\delta U_R}{\delta \boldsymbol{\lambda}} = \mathbf{p} - \mathbf{P}\mathbf{b}_R = 0 \tag{6.137b}$$

This equation defines $(n + k)$ linear equations according to parameters $\boldsymbol{\lambda}$ and $\mathbf{b}_R$. After rewriting we obtain the estimate $\boldsymbol{\lambda}$ in the form

$$\boldsymbol{\lambda} = 2\left[\mathbf{P}(\mathbf{X}^T\mathbf{X})^{-1}\mathbf{P}^T\right]^{-1}(\mathbf{p} - \mathbf{P}\mathbf{b}) \tag{6.138}$$

where $\mathbf{b} = (\mathbf{X}^T\mathbf{X})^{-1}\mathbf{X}^T\mathbf{y}$ is the parameter estimate found by classical least-squares. The estimate of restricted parameters is calculated from

$$\mathbf{b}_R = \mathbf{b} - (\mathbf{X}^T\mathbf{X})^{-1}\mathbf{P}^T\left[\mathbf{P}(\mathbf{X}^T\mathbf{X})^{-1}\mathbf{P}^T\right]^{-1}(\mathbf{P}\mathbf{b} - \mathbf{p}) \tag{6.139}$$

When a given parameter restriction is valid

(a) the estimate $\mathbf{b}_R$ is unbiased;

(b) its covariance matrix is given by

$$D(\mathbf{b}_R) = D(\mathbf{b})\left[\mathbf{E} - \mathbf{P}^T\mathbf{S}\mathbf{P}(\mathbf{X}^T\mathbf{X})^{-1}\right] \qquad (6.140)$$

where $D(\mathbf{b})$ is a covariance matrix for estimates $\mathbf{b}$ by the classical least-squares method (6.16), $\mathbf{E}$ is the unit matrix and $\mathbf{S} = [\mathbf{P}(\mathbf{X}^T\mathbf{X})^{-1}\mathbf{P}^T]^{-1}$; and

(c) the unbiased estimate of the residual variance σ^2 is calculated from

$$\hat{\sigma}_R^2 = \frac{(\mathbf{y} - \mathbf{X}\mathbf{b}_R)^T(\mathbf{y} - \mathbf{X}\mathbf{b}_R)}{n - m + k} \qquad (6.141)$$

If the errors have a normal distribution, confidence intervals and tests of significance may be constructed as in Section 6.3. From Eq. (6.140) it may be concluded that the variance $D(b_{Rj})$ for restricted parameter estimates are always smaller than for $D(b_j)$. The main task here is to check the validity of Eq. (6.135a), i.e. H_0: $\mathbf{P\beta} - \mathbf{p} = 0$ against H_A: $\mathbf{P\beta} - \mathbf{p} \neq 0$. The hypothesis H_0 may be tested by the classical Fisher-Snedecor F-test with the test criterion

$$F_0 = \frac{(\mathbf{Pb} - \mathbf{p})^T\mathbf{S}(\mathbf{Pb} - \mathbf{p})/k}{(\mathbf{y} - \mathbf{Xb})^T(\mathbf{y} - \mathbf{Xb})/(n-m)} \qquad (6.142)$$

which, if H_0 is valid, has the Fisher-Snedecor F-distribution with k and $(n - m)$ degrees of freedom [73]. When $F_0 > F_{0.95}(k, n - m)$ the parameter restrictions are not suitable for the given data, and the estimate $\mathbf{b}_R$ is biased. For a small bias, the variances often decrease, so that the estimate $\mathbf{b}_R$ seems to be better than the estimate $\mathbf{b}$ [33].

From the computational point of view, it is more convenient to express the test criterion (6.142) in terms of the residual sum of squares (Section 6.3). If we write $RSC = (\mathbf{y} - \mathbf{Xb})^T(\mathbf{y} - \mathbf{Xb})$ and $RSC_0 = (\mathbf{y} - \mathbf{Xb}_R)^T(\mathbf{y} - \mathbf{Xb}_R)$ the Eq. (6.142) may be expressed in an equivalent form as

$$F_0 = \frac{(RSC_0 - RSC)/k}{RSC/(n-m)} \qquad (6.143)$$

To use Eq. (6.143) both estimates $\mathbf{b}$ and $\mathbf{b}_R$ must be calculated and RSC and RSC_0 evaluated.

Problem 6.40   *The conditional least-squares method in the case of a single parameter restriction*

Derive equations for estimation of parameters $\mathbf{b}_R$ in a case where only one restriction is given:

$$P_1\beta_1 + P_2\beta_2 + ... + P_m\beta_m = p \quad \text{or} \quad \mathbf{P}\boldsymbol{\beta} = p \text{ where } \mathbf{P} \text{ is the row vector.}$$

Solution: Since the matrix $\mathbf{S}$ contains just one element we will speak about the scalar S. The matrix $\mathbf{M} = (\mathbf{X}^\mathsf{T}\mathbf{X})^{-1}\mathbf{P}^\mathsf{T}$ becomes the column vector $\mathbf{M}$. Let us introduce the matrix $\mathbf{C}$ as it is $\mathbf{C} = (\mathbf{X}^\mathsf{T}\mathbf{X})^{-1}$ with elements C_{jk}. Then

$$S = \sum_{j=1}^{m}\sum_{k=1}^{m} P_j C_{jk} P_k \text{ and the vector } \mathbf{M} \text{ has the following elements } M_j = \sum_{k=1}^{m} C_{jk} P_k$$

. Then, from Eq. (6.139) we have $\mathbf{b}_R = \mathbf{b} - \mathbf{M}\sum_{j=1}^{m}(P_j b_j - p)/S$.

Conclusion: Since the matrix $\mathbf{S}$ becomes a scalar S, the expressions for the covariance matrix of estimates and the test criterion F_0 may also be simplified.

Problem 6.41   *Finding the relationship between the surface under a chromatographic peak and the ethanol concentration, by linear regression with restriction*

The relationship between the concentration of ethanol in water (x) and the corresponding area of the chromatographic peak (y) was examined, and the model $E(y\,/\,x) = \beta_1 x + \beta_2 x^2 + \beta_3$ was proposed. To have physical meaning, this curve should go through the two points with co-ordinates $(0,0)$ and $(100,100)$, corresponding to limits, the first for pure water and the second for pure ethanol. Estimate the model parameters and test whether the restrictions correspond to the given data set.

○ *Data:* x is the volume percentage of ethanol in water and y is the relative area of the chromatographic peak as a percentage.

x	10	20	30	40	50	60	70	80	90
y	8.16	15.9	22.7	31.5	39.8	49.4	59.7	70.6	83.6

Solution: Because the first restriction requires the regression curve to go through the origin, the intercept term β_3 should be equal to zero, $\beta_3 = 0$. The second restriction leads to the equation $100 = \beta_1 \times 100 + \beta_2 \times 100^2$ which can be rewritten as $1 = \beta_1 + \beta_2 \times 100$. On simplifying this equation by elmination of β_1 we obtain the following equation (which is linear with respect to β_2)

$E(y / x) = (1 - 100\beta_2)x + \beta_2 x^2 = x + \beta_2(x^2 - 100x)$. By using the least-squares

criterion and the analytical derivative we obtain $b_{R2} = \dfrac{\displaystyle\sum_{i=1}^{n}(y_i - x_i)(x_i^2 - 100x_i)}{\displaystyle\sum_{i=1}^{n}(x_i^2 - 100x_i)^2}$

and find that $b_{R2} = 4.1199 \times 10^{-3}$ and $RSC_0 = 31.82$. The classical least-squares estimates for the regression model without restriction has the form $y = 2.724(\pm 0.648) + 0.557(\pm0.0297)\ x + 3.726(\pm0.29)\times10^{-3}x^2$. The standard deviations of the parameter estimates are given in brackets. The corresponding value of RSC is 1.555, and

$$F_0 = \frac{(31.82 - 1.555) / 2}{1.555 / (9 - 3)} = 58.39$$

.

Since F_0 is significantly greater than $F_{0.95}(2, 6) = 5.14$, the given restrictions do not correspond to the data. The two models are compared in Fig. 6.47.

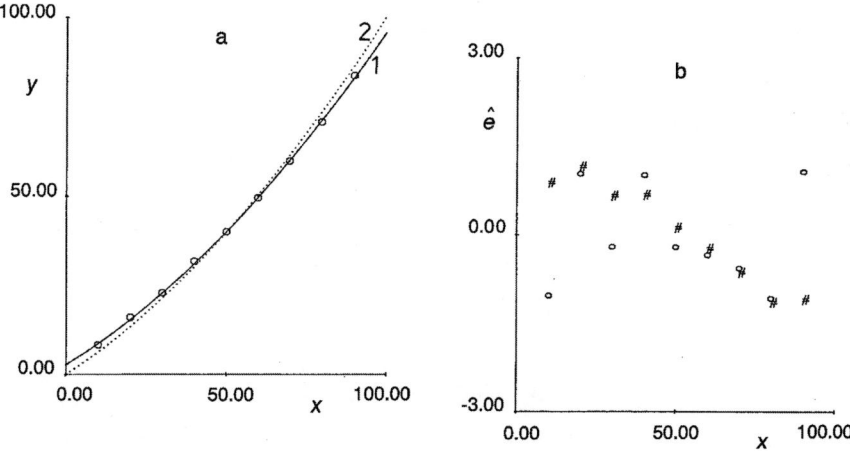

Figure 6.47 (a) Regression model without restrictions (full curve 1) and with restrictions (dash curve 2), fitted through the experimental points. The restriction requests the curve to go through two points, (0, 0) and (100, 100); (b) the graphical analysis of residuals: o model without and # model with restrictions.

Conclusion: In some cases, the restrictions given enable parameters to be derived in such a way that the method of Lagrange multipliers is not required. The statistical criterion F_0 examines whether the data are in agreement with the given restrictions.

6.6.2 The Method of Generalized Least-squares (GLS)

In the analysis of instrumental data in the laboratory, it is often found that the errors are often not independent or that they do not exhibit the same variance. The covariance matrix of errors $D(\mathbf{\mathring{a}}) = \mathbf{C}_\varepsilon$ is then not equal to $\sigma^2 \times \mathbf{E}$ and the more generalized relationship should be used

$$D(\mathbf{\varepsilon}) = \mathbf{C}_\varepsilon = \sigma^2 \mathbf{K} \tag{6.144}$$

As the matrix $\mathbf{K}$ is, apart from the multiplicative constant σ^2, the same as the covariance matrix, it should be the case that it is symmetric and $K_{ij} = K_{ji}$. Moreover, the inequality $\left|K_{ij}\right| \le \sqrt{K_{ii}K_{jj}}$ which makes certain that matrix $\mathbf{K}$ is positive definite. These properties allow the inversion matrix $\mathbf{K}^{-1}$ to be expressed as the product of weight matrices $\mathbf{V}$ in the form $\mathbf{K}^{-1} = \mathbf{V}^{\mathrm{T}}\mathbf{V}$. When all other conditions for the least-squares method are met, the parameter estimates may be obtained by minimization of the *generalized least-squares criterion* in the form

$$U(\mathbf{b}) = (\mathbf{y} - \mathbf{Xb}_G)^{\mathrm{T}} \mathbf{K}^{-1} (\mathbf{y} - \mathbf{Xb}_G) \tag{6.145}$$

By using an analytical minimization of Eq. (6.145), the expression for the estimate $\mathbf{b}_G$ may be derived in the form

$$\mathbf{b}_G = (\mathbf{X}^{\mathrm{T}}\mathbf{K}^{-1}\mathbf{X})^{-1} \mathbf{X}^{\mathrm{T}}\mathbf{K}^{-1}\mathbf{y} \tag{6.146}$$

where the index G denotes characteristics of the *method of generalized least-squares* (GLS). The estimate $\mathbf{b}_G$ is called the *Aitken estimate*. When the weight matrix $\mathbf{V}$ is introduced into Eq. (6.146) the resulting estimate by the GLS method is

$$\mathbf{b}_G = (\mathbf{Z}^{\mathrm{T}}\mathbf{Z})^{-1} \mathbf{Z}^{\mathrm{T}}\mathbf{w} \tag{6.147}$$

where $\mathbf{Z} = \mathbf{VX}$ and $\mathbf{w} = \mathbf{Vy}$. Equation (6.147) shows how the parameter estimate $\mathbf{b}_G$ by the GLS method can easily be transformed into the estimate $\mathbf{b}$ by the LS method with the use of simple multiplication by the weight matrices. When we multiply the Eq. (6.5b) by the weight matrix $\mathbf{V}$, we obtain

$$\mathbf{Vy} = \mathbf{VX\beta} + \mathbf{V\varepsilon} \tag{6.148a}$$

$$\text{or } \mathbf{w} = \mathbf{Z\beta} + \mathbf{V\varepsilon} \tag{6.148b}$$

The mean value $E(V\varepsilon)$ is equal to zero, and the covariance matrix is calculated from

$$D(V\varepsilon) = V^T V K \sigma^2 = E\sigma^2 \qquad (6.149)$$

This means that the transformed errors $V\varepsilon$ already satisfy the conditions for the classical LS method, and Eq. (6.147) may be used. With the use of variables Z and w, the expressions valid for the LS method, including the interval estimates and hypothesis tests, may be used. For example, the covariance matrix of estimates b_G may be written, by using Eq. (6.16), as

$$D(b_G) = \sigma^2 (Z^T Z)^{-1} = \sigma^2 (X^T K^{-1} X)^{-1} \qquad (6.150)$$

The estimates b_G are unbiased and best in the class of all linear unbiased estimates. When the errors ε have the normal distribution $N(0, \sigma^2 K)$, the estimates b_G are normally distributed with mean value $E(b_G) = \beta$ and the covariance matrix is defined by Eq. (6.150). The estimate of the residual variance $\hat{\sigma}_G^2$ is calculated from

$$\hat{\sigma}_G^2 = \frac{(y - X b_G)^T K^{-1} (y - X b_G)}{n - m} \qquad (6.151)$$

When the classical LS method is used for a case when Eq. (6.144) is valid, instead of the more correct approach by the GLS method it is true that,

(a) the parameter estimates b remain unbiased;

(b) the covariance matrix $D(b)$ does not correspond to the correct covariance matrix $D(b_G)$, so that the estimates are not already best as they have greater variances;

(c) the estimate of the residual variance $\hat{\sigma}^2$ will be biased.

For these reasons the interval estimates and statistical tests will give quite false results.

A special case of the GLS method is the method of weighted least-squares (WLS). The matrix K is diagonal, condition 6 is valid and errors ε are independent in this case.

If K_{ii} are the diagonal elements of matrix K, we can write for the diagonal element of a matrix V that $V_{ii} = \sqrt{1 / K_{ii}}$. When this is introduced into Eq. (6.145), we obtain:

$$U(b) = \sum_{i=1}^{n} K_{ii}^{-1} \left(y_i - \sum_{j=1}^{m} x_{ij} b_j \right)^2 = \sum_{i=1}^{n} \left[y_i V_{ii} - \sum_{j=1}^{m} V_{ii} x_{ij} b_j \right]^2 \qquad (6.152)$$

When all variables are multiplied by the corresponding weights, the same conditions apply as for the classical LS method. Linear regression programs based on the least-squares method can readily be extended for weighted least-squares.

6.6.2.1 Heteroscedasticity

Heteroscedasticity in data means that condition 5, about constancy of variance, is violated. In laboratory problems, nonconstancy of variance in measured data is common.

The operation of some instruments is such that there is a constant relative error, so that

$$\sigma_i^2 = \sigma_0^2 \left[E(y/\mathbf{x}) \right]^2 \tag{6.153}$$

In other cases, the variance σ_i^2 may be estimated from error propagation for all operations, chemicals, glassware, procedures, etc. When the values of individual variances σ_i^2 can be exactly specified, then setting $V_{ii} = 1/\sigma_i$ permits application of the WLS method.

A large group of laboratory problems in which there is heteroscedasticity in the data arises from the use of transformed variables. The variable y is often transformed to give a linear relationship (the linearization method). When the original values y_i have constant variance $D(y_i) = \sigma^2$, nonlinear transformation $g(y_i)$ causes the variance $D(g(y_i))$ to be non-constant, and for the first approximation the following expression will hold

$$\sigma_i^2 = D(g(y_i)) = \left[\frac{\partial g(y_i)}{\partial y_i} \right]^2 D(y_i) \tag{6.154}$$

The variances can be equalized by introduction of weights given by

$$V_{ii} = \left[\frac{\partial g(y)}{\partial y} \right]_{y=y_i}^{-1} \tag{6.155}$$

with the use of the WLS method. The method of weighted least-squares with weights V_{ii} defined by Eq. (6.155) is called *quasilinear regression*. It is generally true that each transformation distorts the error distribution, so it is always better to use the method of nonlinear regression.

Problem 6.42 *Examination of the dependence of the solubility of Na$_2$SO$_3$ on temperature*

The dependence of the solubility of Na$_2$SO$_3$ (y) on temperature (x) may be described by the empirical expression $y = \exp(\beta_1 + \beta_2 x)$. Estimate parameters β_1 and β_2 with the use of the linearization, quasilinearization and non-linear methods of least-squares.

○ *Data:* solubility y, %, and temperature x, °C.

x	0	10	20	30	40	50	60	70	80
y	33.5	37.0	41.2	46.1	50.0	52.0	56.3	64.3	69.9

Solution: The expression $y = \exp(\beta_1 + \beta_2 x)$ can be transformed into linear form $\ln y = \beta_1 + \beta_2 x$. The weights V_{ii} will eliminate the heteroscedasticity and can be expressed by $V_{ii} = y_i$ [Eq. (6.155)]. Table 6.12 lists the parameter estimates by the three least-squares methods.

Table 6.12 Comparison of parameter estimates found by the LS, WLS and NLS methods

Method	Transformation	b_1	b_2	RSC*
LS	linearization	3.532	8812	10.22
WLS	quasilinearization	3.535	8756	10.13
NLS	nonlinear regression	3.537	8720	10.11

*in transformed variables

The estimates achieved by the quasilinearization (WLS) are in quite good agreement with those found by nonlinear regression (NLS). The precision of prediction should be considered in the original and not in the transformed variables.

Conclusion: Application of statistical weights V_{ii} from Eq. (6.155) in the WLS method increases the accuracy of parameter estimates.

To solve laboratory problems with heteroscedasticity in the data, the procedure is usually as follows:

(1) Identification of the presence of heteroscedasticity in the data with the use of tests from section 6.5.4.1.

(2) Identification of the actual type of heteroscedasticity, which determines the effect of the errors variance on the variables of the regression model.

(3) Determine parametric estimates for the known type of heteroscedasticity.

Step 1: Identification of heteroscedasticity

Instead of a sample diagnostic test (Section 6.5.4) or the various plots, there are also *nonconstructive tests,* which do not require knowledge of the heteroscedasticity model, and *constructive tests,* which require knowledge of the heteroscedasticity model. A common test is the *test of residual trend,* which has the test criterion

$$D = \sum_{i=1}^{n} \left[P(|\hat{e}_i|) - i \right]^2 \tag{6.156}$$

where $P(|\hat{e}_i|)$ stands for the order of absolute value of the ith residual. This criterion D is connected with the Spearman correlation coefficient ρ_s by the expression

$$\rho_s = 1 - \frac{6D}{n^3 - n} \tag{6.157}$$

The heteroscedasticity test therefore becomes the test of a null hypothesis H_0: $\rho_s = 0$ (i.e. homoscedasticity) against an alternative H_A: $\rho_s \neq 0$ (i.e. heteroscedasticity). For larger sample sizes, $n > 10$, another test-criterion can be also used

$$t_s = \sqrt{\frac{\hat{\rho}_s^2 (n-2)}{1 - \hat{\rho}_s^2}}$$

which, when the null hypothesis is valid (i.e. homoscedasticity) has the Student t-distribution with $n - 2$ degrees of freedom.

The *Szroeter test* requires the data to be rearranged in ascending order of variance, $\sigma_{i-1}^2 \leq \sigma_i^2, i = 2, ..., n$; in order to examine the values of the variable which is a monotonie function of the variances. If Eq. (6.153) is valid, the ordering is made according to the magnitude of the y-values (or prediction $\hat{y}_i$, respectively) in ascending order. The null hypothesis of homoscedasticity H_0: $\sigma_i^2 = \sigma_{i-1}^2, i = 2, ..., n$ is tested against the alternative H_A: $\sigma_i^2 > \sigma_{i-1}^2, i = 2, ..., n$. The test criterion of the Szroeter test is defined by

$$Q_T = \sqrt{\frac{6n}{n^2 - 1}} \left(Q - \frac{n+1}{2} \right) \tag{6.159}$$

$$\text{where } Q = \frac{\sum_{k=1}^{n} k\hat{e}_k^2}{\sum_{k=1}^{n} \hat{e}_k^2} \text{ and residuals } \hat{e}_k \text{ correspond to ordered data. The statistic}$$

Q_T has, asymptotically, the standardized normal distribution $N(0, 1)$. When Q_T > 1.645, heteroscedasticity is proved at the significance level $\alpha = 0.05$.

The constructive tests are based on the known model of heteroscedasticity and on significance tests.

Step 2: Identification of the type of heteroscedasticity

When the matrix $\sigma^2 \mathbf{K}$ is not known, it is necessary to estimate its diagonal elements, which correspond to variances σ_i^2. For large sample sizes, the variance estimates σ_i^2 can be replaced by the squared residuals $\hat{e}_i^2$ obtained by the classical LS method. This procedure is usually used in seeking parametric models of heteroscedasticity.

Horn [35] suggested application of so-called *AUE estimates of variances*, defined by $\hat{\sigma}_i^2 = \dfrac{\hat{e}_i^2}{1 - H_{ii}}$. The estimates of variance $\hat{\sigma}_i^2$ may be used directly in the method of weighted least-squares (WLS) where $V_{ii} = 1/\hat{\sigma}_i$ or for examination of various types of heteroscedasticity. Three principal models of heteroscedasticity are distinguished.

(a) The *multiplicative model of heteroscedasticity* is expressed by

$$\sigma_i^2 = \sigma_0^2 \exp(\delta x_{ij}) \tag{6.160a}$$

$$\text{or } \sigma_i^2 = \sigma_0^2 |x_{ij}|^{\delta} \tag{6.160b}$$

where δ is a parameter.

Instead of variable x_j in these two models, the theoretical value $E(y/x_i) = \eta_i$ may be used. The multiplicative model is valid when the dependence of $\ln(\hat{e}_i^2)$ on x_{ij} or $\hat{y}_i$ is approximately linear. The significance test of the slope δ here corresponds to the test for the multiplicative model of heteroscedasticity.

(b) The *additive model of heteroscedasticity* is expressed by

$$\sigma_i^2 = \sigma_0^2 (1 + \delta x_{ij})^2 \tag{6.161}$$

where instead of x_j, $E(y/x_i) = \eta_i$ may be used. The additive model is valid for cases when the dependence of $|\hat{e}_i|$ on x_{ij} or $\hat{y}_i$ is approximately linear. The significance test of the slope δ corresponds to the test for the additive model of heteroscedasticity.

(c) The *mixed model of heteroscedasticity* is expressed by

$$\sigma_i^2 = \delta_0 + \delta_1 x_{ij} \qquad (6.162)$$

where instead of x_j, $E(y/x_i) = \eta_i$ may be used. The mixed model is valid for cases when the dependence of $\hat{e}_i^2$ on x_{ij} or $\hat{y}_i$ is approximately linear. The significance test of this slope δ_1 shows the presence of the mixed type of heteroscedasticity.

In laboratory practice, the most common model seems to be the model of constant relative error (6.153) which corresponds to the multiplicative type of heteroscedasticity.

Step 3: Estimation of parameters

There are many methods of parameter estimation for linear models with heteroscedasticity in the data, but we restrict ourselves to the simplest one, i.e. the method of weighted least-squares (WLS), which is possible with most linear regression programs. The general procedure consists of the following steps:

(1) Estimation of parameters β in the linear model by the classical LS method and estimation of the residuals $\hat{e}_i$.

(2) Estimation of the parameters of the chosen heteroscedasticity type, with $\hat{\sigma}_i^2 = \hat{e}_i^2$.

(3) Estimation of weights from $V_{ii} = 1/\hat{\sigma}_i^*$ where $\hat{\sigma}_i^*$ is the estimate of the standard deviation determined from the parametric model of heteroscedasticity.

The main problem of parameter estimation for heteroscedastic models lies in the transformation of the squared residuals $\hat{e}_i^2$. This problem can be solved partly by use of quasilinear regression.

For a case defined by Eq. (6.153) the weights may be chosen such that $V_{ii} = 1/|y_i|$, or with the use of predicted values calculated by the classical least-squares method, $V_{ii} = 1/|\hat{y}_i|$.

Problem 6.43　*Tests for heteroscedasticity in the validation of a new laboratory method*

Data from Problem 6.7, on the validation of a new laboratory method by comparison with a standard one, were examined in Problem 6.37 and heteroscedasticity was proved. For the multiplicative model of heteroscedasticity [Eq. (6.153)], estimate the unknown parameters by the weighted LS method.

○ **Data:** from Problem 6.7

Solution: Data are examined for two assumptions:

(a) *Assumption of constant relative error of measurement.* With the use of the WLS method and weight $V_{ii} = 1/|y_i|$ the regression equation is found to be $y = 8.23(\pm 5.177) + 0.879(\pm 0.0249)x$, with determination coefficient $\hat{R}^2 = 0.983$ and quadratic error of prediction $MEP = 20560$. Figure 6.48b illustrates "reverse" heteroscedasticity with regard to variable x: the variance decreases with increasing values of x.

(b) *Assumption of multiplicative heteroscedasticity.* The results of Problems 6.7 and 6.37 suggest that the multiplicative model of heteroscedasticity is applicable. Figure 6.49 shows the plot of $\ln \hat{e}_i^2$ vs. x_i, with the straight line $\ln \hat{e}_i^2 = 3.239 + 0.005098x$.

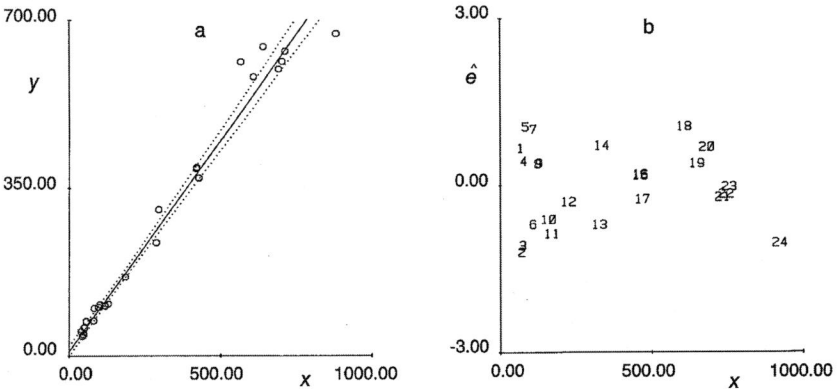

Figure 6.48 (a) Regression model with 95% confidence interval of prediction, and (b) the graphical examination of residuals $\hat{\mathbf{e}}$. The weight $V_{ii} = 1/|y_i|$ is used.

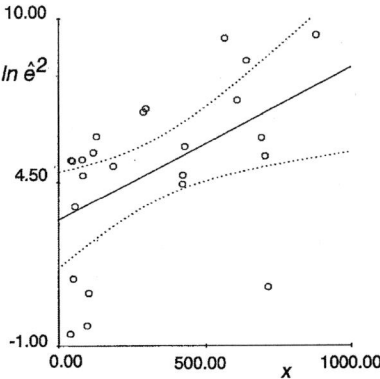

Figure 6.49 The plot of $\ln \hat{e}_i^2$ vs. x_i indicates a multiplicative model of heteroscedasticity.

For parameter estimation, the WLS method was used with weights $V_{ii} = 1/\sqrt{\exp(3.239 + 0.005098x_i)}$ and the regression equation was estimated as: $\hat{y} = 7.937(\pm 6.898) + 0.895(\pm 0.0259)x$ with determination coefficient $\hat{R}^2 = 0.982$ and the mean quadratic error $MEP = 1410$. In Fig. 6.50, the residuals form a random pattern, and therefore the heteroscedasticity has been removed.

Conclusion: It was found that application of weights according to $V_{ii} = 1/|y_i|$ is not always the best solution. It is better to determine the actual type of heteroscedasticity.

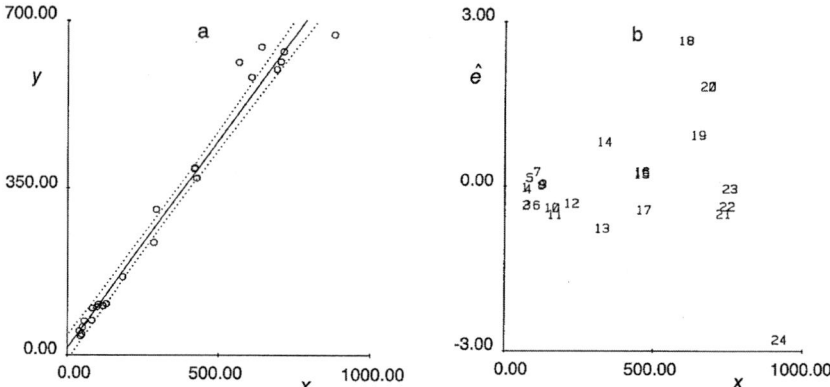

Figure 6.50   (a) Regression model with the 95% confidence intervals and (b) graphical examination of residuals. Weights V_{ii} were found from a multiplicative model of heteroscedasticity.

6.6.2.2 Autocorrelation

Autocorrelation in data represents a violation of condition 6 for least-squares methods, concerning *independence of measurement errors*. Autocorrelation may be found in laboratory problems involving data concerned with time dependencies, for example, the data from the kinetics of a reaction. The covariance matrix of errors $\mathbf{C}_\varepsilon$ contains off-diagonal elements.

In laboratory problems, we are often faced with cumulative errors which are the consequence of a sampling technique used when all experiments are carried out on a single solution. For example, investigation of the kinetics of a chemical reaction is performed by measurement of the concentration of initial substances or resulting reaction products in a single experiment. The *process error* ε_t at time t is, in an ideal case, given by

$$\varepsilon_t = \sum_{j=1}^{t} u_j \qquad (6.163)$$

where u_j are independent random variables of the normal distribution $N(0, \sigma^2)$. This equation shows that the process error ε_t is a sum of all the random effects which have affected the process throughout the experiment. The model [Eq. (6.163)] is a special case of the *autoregressive model of the first order* AR(1), for which Eq. (6.132) is valid. Other eventualities leading to a non-diagonal matrix $\mathbf{C}_\varepsilon$ rarely appear in laboratory. In the case of model AR(1), Eq. (6.132) may be expressed in matrix form as

$$\varepsilon = \mathbf{A}_1 \mathbf{u} \tag{6.164}$$

where $\mathbf{A}_1$ is the lower triangular matrix

$$\mathbf{A}_1 = \begin{bmatrix} 1 & 0 & 0 & .. & .. & .. & 0 \\ \rho_1 & 1 & 0 & .. & .. & .. & 0 \\ \rho_1^2 & \rho_1 & 1 & .. & .. & .. & 0 \\ .. & .. & .. & .. & .. & .. & .. \\ .. & .. & .. & .. & .. & .. & .. \\ .. & .. & .. & .. & .. & .. & .. \\ \rho_1^{n-1} & \rho_1^{n-2} & \rho_1^{n-3} & .. & .. & .. & 1 \end{bmatrix} \tag{6.165}$$

and moreover it is valid that $E(\varepsilon) = 0$ and the variance of the ith error is given by

$$D(\varepsilon_t) = \sigma^2 \sum_{i=0}^{t-1} \rho_1^{2i} \approx \frac{\sigma^2}{1-\rho_1^2} \tag{6.166}$$

The last term in this equation is valid on the assumption that t has a sufficiently high value, or that the autoregressive process started at $t = -\infty$.

For an autoregressive process of the first order, simple expressions for covariance of errors, $E(\varepsilon_t \ \ \varepsilon_s)$ may be found and the stationary covariance matrix of errors formed

$$\mathbf{C}_\varepsilon = \frac{\sigma^2}{1-\rho_1^2} \begin{bmatrix} 1 & \rho_1 & \rho_1^2 & .. & .. & .. & \rho_1^{n-1} \\ \rho_1 & 1 & \rho_1 & .. & .. & .. & \rho_1^{n-2} \\ \rho_1^2 & \rho_1 & 1 & .. & .. & .. & \rho_1^{n-3} \\ .. & .. & .. & .. & .. & .. & .. \\ .. & .. & .. & .. & .. & .. & .. \\ .. & .. & .. & .. & .. & .. & .. \\ \rho_1^{n-1} & \rho_1^{n-2} & \rho_1^{n-3} & .. & .. & .. & 1 \end{bmatrix} \tag{6.167}$$

with a general element $C_{ij} = \rho_1^{|i-j|}$. The inverse matrix has the simple structure

$$
\mathbf{C}_\varepsilon^{-1} = \frac{1}{\sigma^2}
\begin{bmatrix}
1 & -\rho_1 & 0 & 0 & .. & .. & 0 \\
-\rho_1 & 1+\rho_1^2 & -\rho_1 & 0 & .. & .. & 0 \\
0 & -\rho_1 & 1+\rho_1^2 & -\rho_1 & .. & .. & 0 \\
.. & & .. & & .. & .. & .. \\
.. & & .. & & .. & .. & .. \\
.. & & .. & & .. & .. & .. \\
0 & 0 & 0 & & .. & .. & 1
\end{bmatrix}
\qquad (6.168)
$$

For the autoregressive model of the first order, the covariance matrix of errors may be determined by substitution into Eq. (8.21a) (Chapter 8). In calculation of the inverse matrix $\mathbf{C}_\varepsilon^{-1}$ the matrix $\mathbf{A}_1^{-1}$ should be known

$$
\mathbf{A}_1^{-1} =
\begin{bmatrix}
1 & 0 & 0 & .. & .. & .. & 0 \\
-\rho_1 & 1 & 0 & .. & .. & .. & 0 \\
0 & -\rho_1 & 1 & .. & .. & .. & 0 \\
.. & & .. & & .. & .. & .. \\
.. & & .. & & .. & .. & .. \\
.. & & .. & & .. & .. & .. \\
0 & 0 & 0 & .. & .. & .. & 1
\end{bmatrix}
$$

This matrix is composed from unit diagonal elements and one underdiagonal band of identical elements $-\rho_1$. The corresponding matrix $\mathbf{C}_\varepsilon^{-1}$ differs from the matrix of Eq. (6.168) only in that the first element on the main diagonal is not equal to 1 but to $(1 + \rho_1^2)$. For the case of cumulative errors in Eq. (6.169), $\rho_1 = 1$. When ρ_1 is unknown, the GLS method with a weight matrix $\mathbf{V} = \mathbf{A}_1^{-1}$ is used.

Let us derive the equation for the transformed vector $\mathbf{w}$ and the matrix $\mathbf{Z}$ used in Eq. (6.147). By a straight multiplication we determine that $w_1 = y_1$ and $w_j = y_j - \rho_1 y_{j-1}$ for $j = 2, ..., n$. The first row in the matrix $\mathbf{Z}$ is $\mathbf{x}_1$. The general element Z_{ij} of other rows of this matrix is given by

$$
Z_{ij} = x_{ij} - \rho_1 x_{i-1,j}, \quad i = 2,...,n, \quad j = 1,...,m \qquad (6.169)
$$

When the first experimental point is omitted and the first row in matrix $\mathbf{Z}$ is neglected, we obtain the *Cochrane-Orcutt estimate*, which corresponds to the minimum of the LS criterion for the first differences

$$U_0 = \sum_{i=2}^{n} \left[(y_i - \rho_1 y_{i-1}) - (\mathbf{x}_i - \rho_1 \mathbf{x}_{i-1}) \mathbf{b}_Z \right]^2 \tag{6.170}$$

where $\mathbf{x}_i$ stands for the ith row in matrix $\mathbf{X}$. However, it is better to use all experimental points, and the regression criterion

$$U_K = U_0 + (y_1 - \mathbf{x}_1 \mathbf{b}_Z)^2 \tag{6.171}$$

When the autocorrelation coefficient ρ_1 is not known it can be estimated by

$$\hat{\rho}_1 = \frac{\sum\limits_{i=2}^{n} \hat{e}_i \, \hat{e}_{i-1}}{\sum\limits_{i=2}^{n} \hat{e}_{i-1}^2} \tag{6.172}$$

Equation (6.172) represents the slope of the regression straight line of the plot of $\hat{e}_i$ vs. $\hat{e}_{i-1}$, estimated by the classical LS method. As the residuals do not have constant variance, it is more convenient, in Eq. (6.172), to use the standardized residuals $\hat{e}_{Si} = \hat{e}_i / \sqrt{1 - H_{ii}}$. By substituting for ρ_1 from Eq. (6.172), U_0 or U_K can be minimized and the estimates $\mathbf{b}_Z$ of parameters $\boldsymbol{\beta}$ may be found. These estimates are biased, because the estimate of the autocorrelation coefficient of the first order, $\hat{\rho}_1$ from Eq. (6.172) is not ideal for small sample sizes. A significant improvement can be achieved by iterative refinement, as follows:

(1) For a given $\hat{\rho}_1$ the estimates $\mathbf{b}_Z$ and residuals $\hat{\mathbf{e}}$ are evaluated;

(2) With the use of the residuals $\hat{\mathbf{e}}$, estimate $\hat{\rho}_1$ is refined, then step (1) is repeated.

The iteration process terminates when the estimates for $\hat{\rho}_1$ in two successive steps do not differ.

It is permissible to make a simultaneous search for both estimates $\mathbf{b}_Z$ and ρ_1 by minimization of U_K by nonlinear regression, because of ρ_1, even though the model is linear in $\boldsymbol{\beta}$.

Problem 6.44 *Estimates of the parameters of a regression straight line in data with cumulative errors*

Write expressions for the parameter estimates of the calibration straight line $E(y \,/\, x) = \beta_1 x + \beta_2$ when the experimental arrangement produces data with cumulative errors.

Solution: Because the case of cumulative errors is a special case of model AR(1) for $\rho_1 = 1$, we start with the more general solution which is valid for any ρ_1. For a regression straight line, $E(y/x) = \beta_1 x + \beta_2$ the regression criterion U_K is $U_K = (y_1 - b_{1Z}x_1 - b_{2Z})^2 + \sum_{i=2}^{n}\left[(y_i - \rho_1 y_{i-1}) - b_{1Z}(x_i - \rho_1 x_{i-1}) - b_{2Z}(1 - \rho_1)\right]^2$.

From both derivatives $\delta U_K / \delta b_{1Z}$ and $\delta U_K / \delta b_{2Z}$, estimates may be found which minimize U_K. Then a set of two linear equations is formed

$$\begin{bmatrix} y_1 x_1 + \sum_{i=2}^{n}(y_i - \rho_1 y_{i-1})(x_i - \rho_1 x_{i-1}) \\ y_1 + (1-\rho_1)\sum_{i=2}^{n}(y_i - \rho_1 y_{i-1}) \end{bmatrix} =$$

$$\begin{bmatrix} x_1^2 + \sum_{i=2}^{n}(x_i - \rho_1 x_{i-1}) & x_1 + (1-\rho_1)\sum_{i=2}^{n}(x_i - \rho_1 x_{i-1}) \\ x_1 + (1-\rho_1)\sum_{i=2}^{n}(x_i - \rho_1 x_{i-1}) & 1 + (1-\rho_1)^2 \end{bmatrix}\begin{bmatrix} b_{1Z} \\ b_{2Z} \end{bmatrix}$$

from which the estimates b_{1Z} and b_{2Z} are calculated. For a case of cumulative errors, when $\rho_1 = 1$, the formulation is simpler

$$\begin{bmatrix} y_1 x_1 + \sum_{i=2}^{n}(y_i - y_{i-1})(x_i - x_{i-1}) \\ y_1 \end{bmatrix} = \begin{bmatrix} x_1^2 + \sum_{i=2}^{n}(x_i - x_{i-1}) & x_1 \\ x_1 & 1 \end{bmatrix}\begin{bmatrix} b_{1Z} \\ b_{2Z} \end{bmatrix} \qquad (6.173)$$

From this, the following estimates are calculated

$$b_{2Z} = y_1 - b_{1Z}x_1 \qquad (6.174)$$

and $b_{1Z} = \dfrac{\sum_{i=2}^{n}(y_i - y_{i-1})(x_i - x_{i-1})}{\sum_{i=2}^{n}(x_i - x_{i-1})^2}$ \qquad (6.175)

Equation (6.175) corresponds to the minimum of U_K, and also of the simplified criterion U_0 of the LS method in first differences. From the set of normal equations, the estimates of variance may be derived

$$D(b_{1Z}) = \frac{\sigma^2}{\sum\limits_{i=2}^{n}(x_i - x_{i-1})^2} \quad \text{and} \quad D(b_{2Z}) = \sigma^2 \left[1 + \frac{x_1^2}{\sum\limits_{i=2}^{n}(x_i - x_{i-1})^2} \right]. \quad \text{From Eq}.$$

(6.175), we can also prove that for the case of constant difference between the location of experimental points $\Delta = x_i - x_{i-1}$, $i = 2, ..., n$, a simple expression can be derived for the slope

$$b_{1Z} = \frac{y_n - y_1}{\Delta \times (n-1)} \tag{6.176}$$

Conclusion: For the case of cumulative errors, the estimates of the calibration straight line parameters and their variances can be found from Eqs. (6.174) and (6.175). The estimate of a slope is not affected by the use of the simple criterion of the LS method.

Problem 6.45   *Parameters of the kinetics of sugar inversion, with consideration of various errors in the data*

Problem 6.38 presented kinetic data for the inversion of sugar. Because samples were taken from a single sugar solution after different time intervals, it can be expected that the data contain cumulative errors. Estimate parameters β_1 and β_2 of the regression straight line $E(y/x) = \beta_1 x + \beta_2$ with the use of (a) the classical LS method, (b) the generalized LS method for the case of cumulative errors, and (c) the generalized LS method for the AR(1) model of errors.

○ ***Data:*** from Problem 6.38

Solution: (a) *The LS method*:

By use of the classical LS method, the regression equation found was y = 1.002($\pm$0.0017) $-$ 0.005303($\pm$ 0.0000357)x with the residual standard deviation $\hat{\sigma}$ = 0.00276. The sign test confirmed a trend in residuals. The number of sequences n_U = 8 is significantly higher than the expected mean value $E(n_U) \approx 4$ for independent residuals. The estimate of the autocorrelation coefficient $\hat{\rho}_1$ = -0.715 shows that the assumption of cumulative errors is not quite correct. The value $\hat{\rho}_1$ is strongly affected by the small number of data points.

(b) *The GLS method for cumulative errors*:

Substitution into Eqs. (6.173) and (6.174) and the corresponding expressions for variances yields the regression equation y = 1.000($\pm$2.19 $\times$ 10^{-5}) $-$ 0.005238($\pm$0.000166)x with residual standard deviation $\hat{\sigma}$ = 0.00469.

(c) *The GLS method for the AR(1) model of errors*:

With the use of estimates from Eq. (6.173) the best estimate of ρ_1 was refined iteratively, to $\hat{\rho}_1 = -0.864$. The calculated regression equation is then $y = 1.003(\pm 6.8 \times 10^{-4}) - 0.00532(\pm 1.49 \times 10^{-5})x$ with residual standard deviation $\hat{\sigma} = 0.00189$.

Conclusion: The large number of sequences of residuals n_U in comparison with the mean value $E(n_U) \approx n/2$ shows that the model has structure AR(1). With a model of cumulative errors, it may happen that the results are much worse than those from the LS method. The general expression (6.173), with iterative refinement of $\hat{\rho}_1$ is more convenient to use. This method gave decreased residual standard deviation and variances of parameter estimates.

Many varied tests may be used to test the significance of the autocorrelation coefficient ρ_1. The *Wald test* is a simple one which examines the null hypothesis H_0: $\rho_1 = 0$ against the alternative one H_A: $\rho_1 \neq 0$ by using the *Wald test criterion*

$$W_a = \frac{n\hat{\rho}_1^2}{1 - \hat{\rho}_1^2} \qquad (6.177)$$

When H_0 is valid, the test statistic W_a has approximately the $\chi^2(1)$ distribution with one degree of freedom. The *Durbin-Watson test* is based on the test criterion

$$D_W = \frac{\sum\limits_{i=2}^{n} (\hat{e}_i - \hat{e}_{i-1})^2}{\sum\limits_{i=1}^{n} \hat{e}_i^2} \qquad (6.178)$$

and $D_W \approx 2 - 2\hat{\rho}_1$. The range of rejection of a null hypothesis H_0: $\rho_1 = 0$ depends not only on the selected significance level α but also on the location of experimental points x_i. For positive autocorrelation, $0 < D_W < 2$, while for negative autocorrelation $2 < D_W < 4$. If $D_W \approx 2$ then the autocorrelation coefficient is not significant. If $D_W \approx 0$ or $D_W \approx 4$, respectively, the null hypothesis H_0 is rejected and ρ_1 is significantly different from zero. In statistical tables both critical limits, the lower, d_L, and the upper, d_U, for a given significance level α and number of controllable variables m, may be found. When $\rho_1 > 0$, for $D_W > d_U$ the null hypothesis H_0 is accepted, and for

$D_W < d_L$ it is rejected. When $d_L < D_W < d_U$, the test is not conclusive. When the value of the autocorrelation coefficient ρ_1 is very high, the proposed regression model may be false, and a significant variable may have been excluded from the model.

Problem 6.46 *Examination of the autocorrelation coefficient for the kinetics of inversion of sugar*

Examine the significance of the autocorrelation coefficient ρ_1. In Problem 6.45, a value $\hat{\rho}_1 = -0.715$ was found for its estimate.

○ *Data:* $\hat{\rho}_1 = -0.715$, $n = 9$.

Solution: For the Wald test (6.177), the test criterion is

$$W_a = \frac{9(-0.715^2)}{(1-0.715^2)} = 9.413$$

Since W_a is greater then the quantile $\chi^2_{0.95}(1) = 3.84$, the null hypothesis H_0: $\rho_1 = 0$ is rejected and the autocorrelation coefficient ρ_1 may be considered to be significantly different from zero.

Conclusion: Examination of the autocorrelation coefficient confirmed the conclusion of Problem 6.45. By using the iterative method of refining the autocorrelation coefficient, the refined estimate is $\hat{\rho}_1 = -0.864$.

6.6.3 Multicollinearity

Multicollinearity does not mean a violation of the conditions for the least-squares methods. It concerns an assumption about positive definite matrix X^TX and therefore the solution of Eq. (6.11).

According to Section 6.1, we understand the columns of matrix X as the column vectors which define the hyperplane L in n-dimensional Euclidean space E^n (Fig. 6.2). According to the angle θ_{jk} between two vectors x_j and x_k (or between columns of matrix X) two limiting cases may be distinguished:

(1) *Orthogonality* is found when the cosine of angle θ_{jk} is zero

$$\cos\theta_{jk} = \frac{\{x_j, x_k\}}{\|x_j\| \times \|x_k\|} \qquad (6.179)$$

and also the scalar product $\{x_j, x_k\} = 0$ where the symbol $\|x_j\| = \sqrt{\{x_j, x_k\}}$ means the length of vector x_j. If all the columns of matrix X are mutually orthogonal, then the matrix X^TX is diagonal and the regression analysis simplifies (Section 6.4.1).

(2) *Collinearity* is found when the cosine of angle θ_{jk} is equal to 1, cos $\theta_{jk} = 1$, because the angle between vectors x_j and x_k is zero, $\theta_{jk} = 0$, and the two vectors x_j and x_k are parallel and linearly dependent, and the following expression holds for them

$$c_j x_j + c_k x_k = 0 \qquad (6.180)$$

where c_j and c_k are nonzero constants. When Eq. (6.180) holds for q pairs of columns of matrix X, its rank is equal to $m - q$ and the matrix X^TX is singular.

Equation (6.180) can be valid for more vectors yet, when one of the columns x_j is the result of a linear combination of several other columns. This situation is called *perfect multicollinearity*. The term multicollinearity, however, can include other cases when some columns of matrix X have nearly zero angle and are therefore approximately linearly dependent.

$$\sum_{j=1}^{m} c_j x_j = \delta \qquad (6.181)$$

where δ is the vector with components near to zero, and the vector c with elements c_j is nonzero, $\|c\| \gg \|\delta\|$. The multicollinearity causes ill-conditioning of the matrix X^TX, and this has two consequences:

(a) the determinant of matrix X^TX is close to zero;

(b) some eigenvalues of matrix X^TX are close to zero.

Multicollinearity causes many difficulties in inversion of matrix X^TX and also numerical errors, depending on the machine precision of the computer used. As well as numerical difficulties, multicollinearity causes statistical difficulties. From Eqs. (6.88) and (6.87), it is evident that for k values near to zero, the parameter estimates and their variance will be abnormally high. The special difficulties are caused by the sensitivity of the parameter estimates b to small changes in data, such as adding another point to the data.

Figure 6.51 shows a geometric interpretation of the LS method for two nearly collinear controllable variables. Figure 6.51a shows vector y projected into a segment of angle θ_{12}, and Fig. 6.51b shows a case when a small change in vector **y** causes its perpendicular projection to lie out of this segment.

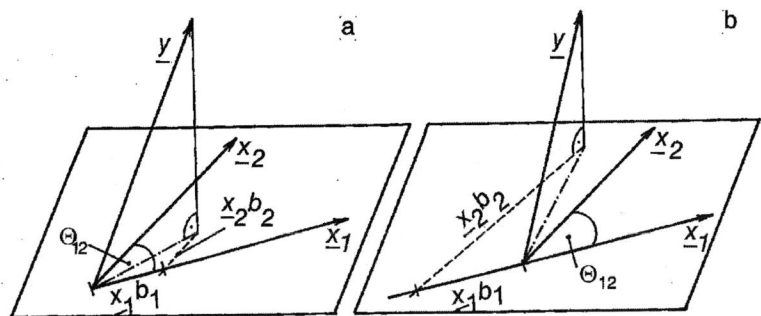

Figure 6.51 Geometric interpretation of the sensitivity of the estimates in the case of multicollinearity: (a) the estimates b_1 and b_2 are small and positive, and (b) the estimate b_1 is negative while b_2 is large and positive.

Multicollinearity causes the following statistical difficulties:

(a) *Non-stability of estimates* is caused by great sensitivity of parameter estimates to small changes in the data. The estimates often have the wrong sign, and this damages their physical interpretation.

(b) *Large variances $D(b_j)$* of individual estimates cause t-tests to indicate that parameter β_j (cf. Section 6.3) is statistically insignificant.

(c) *Strong correlation* between elements of the estimates vector **b** means that they cannot be interpreted separately.

On the other hand, in cases of multicollinearity the determination coefficient is always high and the regression model may fit the data quite well. For data approximation and data smoothing by regression, multicollinearity does not cause difficulties apart from numerical ones related to the ill-conditioning of matrix $\mathbf{X}^T\mathbf{X}$.

When data are measured according to an experimental design the problem of multicollinearity is removed. The plan of designed experiments leads to orthogonality of the columns of matrix **X**.

When data are not measured according to a designed experiment multicollinearity always exists to some extent. However, strong multicollinearity causes the parameter estimates and hypotheses tests to be affected more by linear connections between the columns of matrix **X** than by the regression model itself. In the chemical laboratory, the values of the controllable variables may be adjusted freely, so the problem of multicollinearity may be avoided by an appropriate data measurement step.

With reference to multicollinearity in data, we can identify three cases of interest.

(a) The *over-estimated regression model* contains too many controllable variables expressing the same basic factors. An example is a structure properties model in which properties of substances are described by various measurable changeable structures.

(b) *Inappropriate location of experimental points* causes multicollinearity to form "artificially" because of the choice of location of points. Often the values of significantly important variables oscillate in a small range and seem to be nearly constant, and they are collinear with the vector corresponding to the intercept term.

(c) *Physical constraints in model or data* refers to limits on the values of the controllable variables derived from the chemistry of the system. An example is an investigation of multicomponent mixtures where the controllable variables are represented by the content of each component. Because the sum of all relative concentrations should be equal to 100%, in a q-component mixture there will be $(q - 1)$ independent components. In a model, only $(q - 1)$ variables are assumed: for a two-component mixture there is only one variable, for a three-component mixture, only two, etc. Similar restriction may apply to stoichiometric ratios, etc.

From knowledge about the controllable variables, and their significance and restrictions, multicollinearity can be completely removed from the data. In the case of polynomial models, the multicollinearity is defined by the model structure. If the experimental strategy cannot be changed, other techniques for decreasing the influence of multicollinearity should be used, despite the fact that the parameter estimates are then biased, as in the case of the method of rational ranks (section 6.4.2).

Multicollinearity can be detected from scatter plots for x_j and x_k when the approximate linear dependence proves the strong multicollinearity. The multicollinearity may be exposed or masked by the presence of influential points and especially by high leverage points. For diagnostic purposes, the residuals v_j, of regression variable x_j on the remaining controllable variables in a matrix $X_{(j)}$ which does not contain the column x_j can be used. Let us use $H_{(j)}$ to denote the projection matrix which corresponds to the projection into a subspace of columns of matrix $X_{(j)}$. For diagnosis of influential points from the point of view of multicollinearity, the plot of $v_{ji}^2 / (v_j^T v_j)$ against $H_{(j)ii}$ is used, where v_{ij} is the ith component of vector v_j and $H_{(j)ii}$ is the ith diagonal element of matrix $H_{(j)}$.

In Fig. 6.52, the points strongly affected by multicollinearity are located in the bottom right-hand corner and the top left-hand corner of the graph. The points located in the top left corner cause multicollinearity only when variable x_j is included in the model. The points located in the bottom right corner are strongly influential only when variable x_j is not included in model.

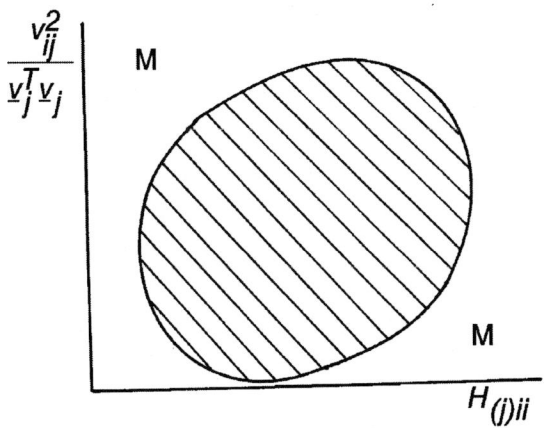

Figure 6.52   Identification of multicollinearity in *data:* M denotes multicollinearity.

The presence of multicollinearity can be identified on the basis of numerical and statistical criteria. Instead of the matrix $\mathbf{X}^T\mathbf{X}$, its normalized version $\mathbf{R}$ is used. Matrix $\mathbf{R}$ is formally identical with the correlation matrix of controllable variables. The following numerical criteria are commonly used.

(a)   The *determinant of matrix* $\mathbf{R}$ is calculated from $\det(\mathbf{R}) = \prod\limits_{j=1}^{m} \lambda_j$ where the λ_j are eigenvalues of the matrix $\mathbf{R}$. If $\det(\mathbf{R})$ is small and less than 10^{-3}, multicollinearity is detected.

(b)   The *conditioning number* K is calculated from

$$K = \lambda_{max} / \lambda_{min} \qquad (6.182)$$

where λ_{max} and λ_{min} are the maximal and minimal eigenvalues of a matrix $\mathbf{R}$. If $K > 10^3$, strong multicollinearity is detected.

(c)   The main statistical criterion used is the *VIF factor,* defined as the ratio of the variance of they jth regression coefficient to the same variance for orthogonal variables when $\mathbf{R}$ is the unit matrix. It is given by

$$VIF_j = \tilde{R}_{jj} \qquad (6.183)$$

where $\tilde{R}_{jj}$ is the jth diagonal element of matrix $\mathbf{R}^{-1}$. The *VIF* factors are related to the determination coefficient $\hat{R}_{xj}^2$ of regression $\mathbf{x}_j$ on $\mathbf{X}_{(j)}$ when $\mathbf{x}_j$ is expressed as a combination of other controllable variables. Then

$$VIF_j = \frac{1}{1 - \hat{R}_{xj}^2} \qquad (6.184)$$

If $VIF_j > 10$, strong multicollinearity is detected.

Problem 6.47   *Testing for multicollinearity in the dependence of the mean activity coefficient on temperature*

The dependence of the logarithm of the mean activity coefficient on the temperature, $\ln \gamma_{\pm} = f(T)$ can be expressed by a polynomial of the third degree. Consider the extent of multicollinearity and use the method of rational rank to decrease the multicollinearity level.

○ *Data:* measured for $m_{HCl} = 0.1$

$T°C$	0	10	20	30	40	50	60	70	80	90
$\ln y_{\pm}$	0.8067	0.8038	0.8000	0.7964	0.7927	0.7867	0.7828	0.7750	0.7690	0.7650

Solution: For the proposed model, $\ln \gamma_{\pm} = \beta_1 T + \beta_2 T^2 + \beta_3 T^3 + \beta_4$, the regression equation is found to be

$$\ln y_{\pm} = 0.807(\pm 1.06 \times 10^{-3}) - 2.654 \times 10^{-4}(\pm 1.07 \times 10^{-4}) \, T - 3.13 \times 10^{-6}$$
$$(\pm 2.87 \times 10^{-6}) \, T^2 + 9.44 \times 10^{-9}(\pm 2.09 \times 10^{-8}) \, T^3$$

by the classical least-squares method (estimated standard deviations of parameter estimates are given in brackets). The determination coefficient $\hat{R}^2$ = 0.9957, the quadratic error of prediction $MEP = 3.507 \times 10^{-6}$ and the Akaike criterion $AIC = -132.26$. Table 6.13 lists the eigenvalues and the *VIF* factors. From these numbers, $\det(\mathbf{R}) = 3.97 \times 10^{-4}$ and the conditioning number K = 1989.73 are calculated. From t-tests at the significance level $\alpha = 0.05$, parameters β_2 and β_3 are statistically insignificant.

Table 6.13 Characteristics detecting multicollinearity

P	Characteristic	$j = 1$	$j = 2$	$j = 3$
10^{-35}	VIF_j	70.42	439.1	184
	λ_j	0.00146	0.0935	2.905
0.05	VIF_j	6.204	0.260	4.373

With the use of the method of rational ranks, the regression equation with precision $P = 0.05$ is estimated in the form $\ln y_{\pm} = 0.807(\pm8.72 \times 10^{-4}) - 3.22 \times 10^{-4}(\pm3.28 \times 10^{-5})T - 1.476 \times 10^{-6}(\pm7.18 \times 10^{-8})T^2 - 2.837 \times 10^{-9}(\pm3.314 \times 10^{-9})T^3$ with the determination coefficient $\hat{R}^2 = 0.9955$, the mean quadratic error of prediction $MEP = 2.264 \times 10^{-6}$ and Akaike criterion $AIC = -131.7$. Table 6.13 gives the VIF factors. The matrix $\mathbf{R}^{-1}$ constructed according to Eq. (6.86), and replacing) $j = 1$ with $j = \omega$, removed multicollinearity, and t-tests at significance level $\alpha = 0.05$ showed that the parameter β_3 is statistically insignificant.

Figure 6.53 shows the regression model found by the method of least-squares, with the 95% confidence intervals, and Fig. 6.54 shows the model found by the method of rational ranks ($P = 0.05$). From comparison of these figures, it is obvious that elimination of multicollinearity leads to narrower confidence bands.

Conclusion: Significant multicollinearity, as indicated by the VIF criterion having a value higher than 10, causes an increase in the estimates of variance, and hence an increase in width of the confidence bands. Elimination of multicollinearity leads to a decrease in goodness-of-fit (a decrease of $\hat{R}^2$) but to an improvement in the prediction ability of model (the criterion MEP), in addition to the decrease in the variance of estimates and narrowing of confidence bands. Elimination of multicollinearity is rather important in calibration in the instrumental methods of laboratory chemistry.

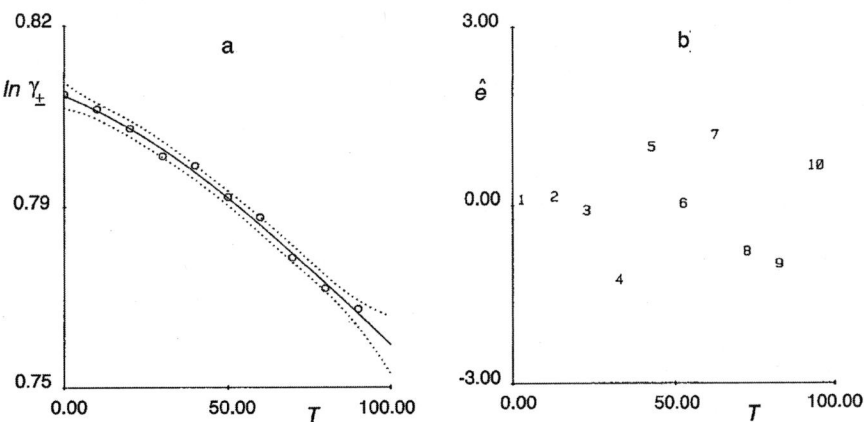

Figure 6.53   (a) Regression model estimated by the LS method, and (b) graphical examination of residuals.

6.6.4 Variables Subject to Random Errors

In laboratory problems, both the dependent variable y and controllable variables $\mathbf{x}$ are measured quantities. The variances of $\mathbf{x}$ are usually significantly smaller than the variance of y, and also smaller than the differences between the locations of individual points. Under such conditions the assumption about the deterministic matrix $\mathbf{X}$ may be abandoned. In some cases it is necessary to suppose that instead of variables $\mathbf{x}_j$ we measure experimental values t_j given by

$$t_{ij} = x_{ij} + \kappa_{ij} \qquad\qquad (6.185)$$

where κ_{ij} are errors of measurement of the jth independent variable at the ith point. The result of measurement is the set of n points $\{y_i, t_{ij};\ j = 1,...,m\}$, $i = 1,..., n$. If the x_{ij} are deterministic quantities we speak about *functional models*, but if the x_{ij} are random quantities we speak about *structural models*.

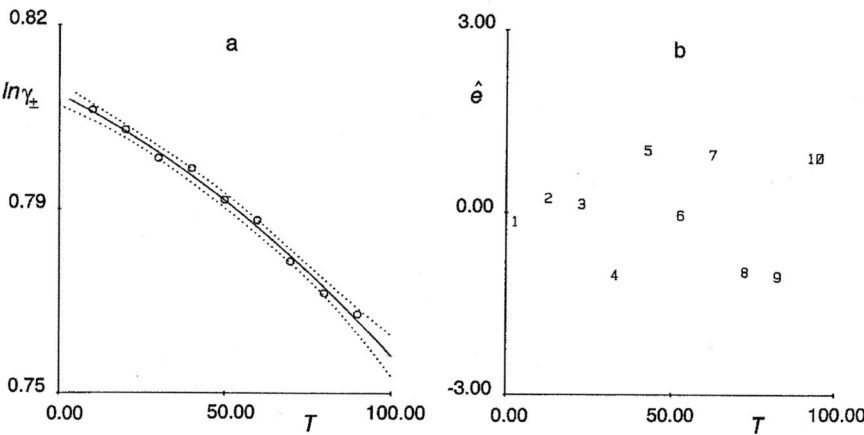

Figure 6.54   (a) Regression model estimated by the rational rank method and (b) graphical examination of residuals

The errors ε_i and κ_{ij} have the properties:

(a) Both errors, ε_i and κ_{ij}, have zero means.

(b) The variance of errors $D(\varepsilon_i^2) = \sigma^2$ and $D(\kappa_{ij}^2) = \tau_j^2$ are at all n points constant (homoscedasticity).

(c) The errors ε_i and κ_{ij} at different points (i.e. measurements) are uncorrelated, so that

$$E(\varepsilon_i\ \ \varepsilon_k) = 0, i \neq j \ \text{ and } \ E(\kappa_{ij}\ \ \kappa_{kj}) = 0, i \neq k .$$

(d) The errors ε_i and κ_{ij} are mutually uncorrelated so that $E(\varepsilon_i\, \kappa_{ij}) = 0$, $j = 1,...,m$.

If errors ε have a normal distribution $N(0,\sigma^2)$ and errors κ_j also have a normal distribution $N(0,\tau_j^{\,2})$, then according to the maximal likelihood method, the criterion of the *extended least-squares method* (ELS) can be expressed as:

$$U_E(\mathbf{b}, \mathbf{X}) = \sum_{i=1}^{n} \left[\frac{1}{\sigma^2} \left(y_i - \sum_{j=1}^{m} b_j x_{ij} \right)^2 + \sum_{j=1}^{m} \frac{1}{\tau_j^2} (t_{ij} - x_{ij})^2 \right] \qquad (6.186)$$

By minimizing the function $U_E(\mathbf{b}, \mathbf{X})$ with respect to $\mathbf{b}$ and to $\mathbf{x}_p$ we find the extended estimates $\mathbf{b}_E$ and also correct quantities $\hat{x}_{ij}$ of controllable variables x_{ij}. With some simplifying assumptions, Eq. (6.186) can be expressed as

$$U_E(\mathbf{b}) = \frac{\sum_{i=1}^{n} \left(y_i - \sum_{j=1}^{m} b_j t_{ij} \right)^2}{\sigma^2 + \sum_{j=1}^{m-1} b_j^2 \tau_j^2} \qquad (6.187)$$

where b_m is the intercept term. For a regression straight line, Eq. (6.187) is simplified to:

$$U_E(b_1, b_2) = \frac{\sum_{i=1}^{n} (y_i - b_1 t_i - b_2)^2}{\sigma^2 + \tau^2 b_1} \qquad (6.188)$$

Figure 6.55 shows that for $\sigma^2 = \tau^2$, the criterion $U_E(b_1, b_2)$ leads to minimization of the squares of perpendicular distances between the regression function and the experimental points.

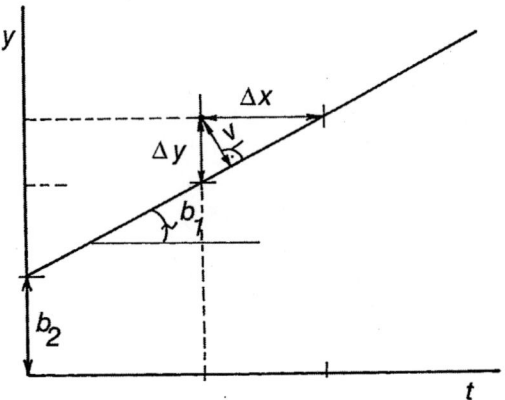

Figure 6.55   Illustration of the criterion of squares of perpendicular distances

$$v^2 = \Delta y^2 / (1 + (\Delta y / \Delta x)^2) \text{ for a straight line } y = b_1 t + b_2.$$

The estimate of parameters **b** in Eq. (6.187) may be achieved by the iterative procedure of the weighted least-squares method with weight values given by

$$P_i = \left(\sigma^2 + \sum_{j=1}^{m-1} \tilde{b}_{Ej} \tau_j^2 \right)^{-1/2}$$ where $\tilde{b}_{Ej}$ are the parameter values estimated in the

previous iteration. However, the procedure requires knowledge of variances σ^2 and τ_j^2. From the structure of Eqs. (6.187) and (6.188), it may be concluded that a knowledge of the ratios of variances, $K_j = \sigma^2 / \tau_j^2, j = 1,...,m$, should be useful in application of the iterative procedure.

We restrict ourselves now to the simplest linear case, i.e. the straight line. From Eq. (6.185) we can write

$$y_i = b_1(t_i - \kappa_i) + b_2 + \varepsilon_i = b_1 t_i + b_2 + \varepsilon_i^* \tag{6.189}$$

where $\varepsilon_i^* = \varepsilon_i - b_1 \kappa_i$ represents errors related to a magnitude b_1. To express the variance of real x values, the mean quadratic deviation σ_x^2 is used,

$\sigma_x^2 = \dfrac{\sum_{i=1}^{n}(x_i - \bar{x})^2}{n}$. The ratio $K_x = \dfrac{\sigma_x^2}{\sigma_t^2} = \dfrac{\sigma_x^2}{\sigma_x^2 + \tau^2}$ is called the *reliability ratio*.

If the true values of x_i and the errors κ_i are not correlated, the mean value $E(b_1)$ estimated by least-squares with neglect of an error ε_i^* structure is given by

$$E(b_1) = \beta_1 K_x \tag{6.190}$$

The corresponding determination coefficient is expressed by

$$R_{yt}^2 = R_{yx}^2 K_x \tag{6.191}$$

If the classical least-squares method is used, measurement errors cause a decrease in slope estimate b_1 and in the correlation coefficient. The magnitude of this decrease depends on the reliability ratio K_x or on the ratio of σ_x^2 to σ_t^2. If K_x is significantly lower than 1, Eq. (6.188) is used to estimate slope b_{1E}.

To remove the intercept b_2 we introduce the centred variables $(y_i - \bar{y})$ and $(t_i - \bar{t})$. If we know the variance ratio $K = \sigma^2 / \tau^2$, Eq. (6.188) may be expressed in the form

$$U_E(b_1) = \frac{\sum_{i=1}^{n}\left[(y_i - \bar{y}) - b_1(t_i - \bar{t}) \right]^2}{K + b_1} \tag{6.192}$$

After analytical minimization of $U_E(b_1)$ we obtain

$$b_{1E} = L + \text{sign}(S_{yt})\sqrt{K + L^2} \qquad (6.193)$$

where $L = \dfrac{S_y - K S_t}{2S_t}$ and $\text{sign}(S_{yt})$ gives the sign of variable S_{yt}. Symbol S represents the sums of squares

$$S_y = \sum_{i=1}^{n}(y_i - \bar{y})^2, \quad S_t = \sum_{i=1}^{n}(t_i - \bar{t})^2, \quad S_{yt} = \sum_{i=1}^{n}(y_i - \bar{y})(t_i - \bar{t}) \ .$$

When the slope b_{1E} is known, the intercept b_{2E} of the regression straight line may be calculated from

$$b_{2E} = \bar{y} - b_{1E}\bar{x} \qquad (6.194)$$

The influence of the magnitude of K on the set of regression straight lines is evident from Eq. (6.193).

(a) For $K \to \infty$, the regression line corresponds to the *LS method*.

(b) For $K = 1$, the regression line minimizes the perpendicular distances from experimental points. This is called *orthogonal regression*.

(c) For $K \to 0$ the regression line is an *inverse regression* i.e. a linear dependence of t on y.

Unsuitable selection of the magnitude of K leads, however, to an increase in variances, so the techniques for simultaneous estimation of parameters and variance ratio are used. Some other procedures of regression analysis, for the case when all variables are subject to random errors, are described by Fuller [36].

We will write an expression for a structural model for which the random variables x, ε and κ have independent normal distributions with variances σ_x^2, σ^2 and τ^2, and for which the ratio $K = \sigma^2 / \tau^2$ is also known. The variance estimates are then given by

$$\hat{\sigma}_x^2 = \frac{S_{yt}}{K(n-1)}\left[\sqrt{K + L^2} - L\right] \qquad (6.195)$$

and $$\hat{\tau}^2 = \frac{1}{2K(n-1)}\left[S_y + KS_t - 2S_{yt}\sqrt{K + L^2}\right] \qquad (6.196a)$$

and $$\hat{\sigma}^2 = \hat{\tau}^2 K \qquad (6.196b)$$

The variance of the slope of the regression straight line is given by

$$D(b_{1E}) = \frac{1}{(n-1)\hat{\sigma}_x^4} \left[\hat{\sigma}_x^2 S_V + \hat{\tau}^2 S_V - b_{1E}^2 \hat{\tau}^4 \right] \qquad (6.197)$$

where $S_V = \dfrac{n-1}{n-2}(K + b_{1E}^2)\hat{\tau}^2$. To test hypotheses about parameter b_{1E}, the test criterion

$$T_E = \frac{|b_{1E} - b_1^*|}{\sqrt{D(b_{1E})}} \qquad (6.198)$$

is used. If the null hypothesis H_0: $\beta_1 = b_1^*$ is accepted, this criterion has approximately the Student t-distribution with $(n - 2)$ degrees of freedom. The variance of the intercept of the regression straight line is estimated by

$$D(b_{2E}) = \frac{S_V}{n} + \overline{t}^2 D(b_{1E}) \qquad (6.199)$$

$$\text{where } \overline{t} = \frac{\sum\limits_{i=1}^{n} t_i}{n}.$$

Problem 6.48   Validation of a new laboratory method when both variables are subject to random errors

In Problem 6.7 the results of new laboratory method (y) are compared with the standard one (x). Estimate both parameters of regression straight line $y = \beta_1 x + \beta_2$ when both variables x and y are loaded by experimental error and the variances of both methods are same, $K = 1$. Test the null hypothesis H_0: $\beta_1 = 1$.

○ **Data:** From Problem 6.7

Solution: Sum of squares $S_y = 1.327 \times 10^6$, $S_x = 1.714 \times 10^6$, $S_{yt} = 1.489 \times 10^6$ substituted into Eq. (6.193) give the estimate of slope $b_{1E} = 0.8784$ and into Eq. (6.194) the estimate of intercept $b_{2E} = 11.521$. As the estimates of variances are $\hat{\sigma}_x^2 = 7.367 \times 10^4$ and $\hat{\tau}^2 = 848.53 = \hat{\sigma}^2$, the estimate of variance of the slope from Eq. (6.197) will be $D(b_1) = 9.337 \times 10^{-4}$ and of the intercept from Eq. (6.199) $D(b_2) = 161.53$. When we test the null hypothesis H_0: $\beta_1 = 1$ with Eq. (6.198), we find that the test criterion $T_E = |0.8784 - 1| / \sqrt{9.337 \times 10^{-4}} = 3.9795$ is higher than the quantile $t_{0.975}(24 - 2) = 2.074 = 2.074$ and therefore the null hypothesis H_0 is rejected and the slope β_1 differs significantly from 1. Figure

6.56 demonstrates the regression straight line which minimizes perpendicular distances from experimental points.

Conclusion: Correctly recognizing that both variables are subject to random errors does not cause any difficulties in estimation of the parameters of the regression straight line. A set of regression straight lines arranged according to the precision of individual variables (the magnitude *K*) may be calculated. For *K* = 1 the useful criterion of perpendicular distances from experimental points is obtained.

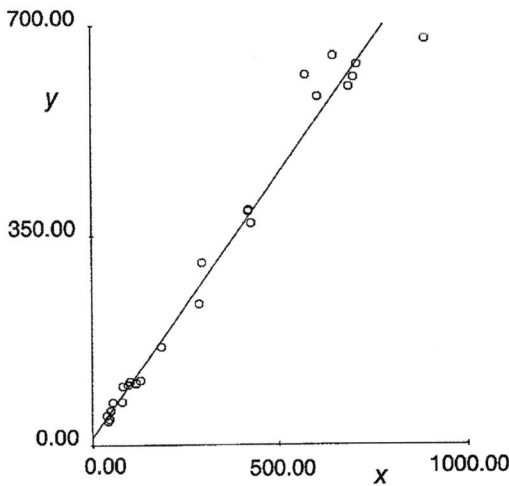

Figure 6.56 The regression straight line minimizing the perpendicular distances from experimental points.

6.6.5 Other Error Distributions of the Dependent Variable

6.6.5.1 The M-estimates Method

When the distribution of the errors in the dependent variable *y* is not normal (violation of condition 7 for the LS method, Section 6.2) the parameter estimates obtained by the LS method are not the best possible estimates. In such a case, instead of the least-squares criterion some other *robust* criterion can be used, that is not so sensitive to violation of the condition about the error distribution, and not sensitive to influential points. The most convenient robust criteria seem to be the group of *M*-estimates. The *M*-estimates are maximal likelihood estimates for a given probability density function of errors $p(\varepsilon)$. All *M*-estimates are related to the minimization criterion

$$U_M = \sum_{i=1}^{n} \rho(e_i / \sigma) = \sum_{i=1}^{n} \rho\left[(y_i - \mathbf{x}_i \mathbf{b}_M) / \sigma\right] \qquad (6.200)$$

where x_i is the ith row of matrix $\mathbf{X}$, σ is the parameter of spread and $\rho(.)$ is a convenient function determined from the probability density $p(\varepsilon)$. By analytical minimization of U_M (6.200) a set of normal equations is obtained:

$$\sum_{i=1}^{n} \psi(e_i / \sigma)x_{ij} = 0, \quad j = 1,...,m \tag{6.201}$$

where the function $\psi(x) = \dfrac{\partial \rho(x)}{\partial x}$ represents the derivative of function $\rho(x)$ with respect to x. Then, if $r_i = e_i / \sigma$, Eq. (6.201) may be expressed in a form which corresponds to the weighted least-squares method

$$\sum_{i=1}^{n} w_i(r)y_i x_{ij} = \sum_{i=1}^{n}\sum_{k=1}^{m} w_i(r)x_{ij}x_{ik}b_k, \quad j = 1,...,m \tag{6.202}$$

where $w_i(r) = \psi(r_i)/r_i$. The parameters are estimated by the *iterative method of re-weighted least-squares* (IRWLS), by using the following procedure:

(1) Select $w_i(r) = l$, $i = 1,...,n$ and set $l = 1$.

(2) Estimate the residuals $r_i = \hat{e}_i / \hat{\sigma}_l$ by the classical least-squares method. In order to reach convergence, corrected least-squares estimates are used [40].

(3) Calculate the weights $w_i(r_l)$ from Eq. (6.203), for $l = l + 1$.

(4) Use the reweighted least-squares to estimate b_1 and the residuals r_r

(5) If the estimates b_l and b_{l+1} are not close enough, go to step 3, otherwise $b_l = b_{M}$

It should be noted that in the jth iteration the weights used have been calculated from residuals $\hat{e}_{l-1}$ in the $(l-1)$th iteration. By applying this method, the robust estimate of parameter σ can be evaluated. An independent estimate $\hat{\sigma}_l$ from the residuals $\hat{e}_{l-1}$ determined in the previous iteration seems to be most convenient. A useful expression is

$$\hat{\sigma} = \frac{med\left(\left|\hat{e}_l - med(\hat{e}_l)\right|\right)}{0.6745} \tag{6.203}$$

where $med(\hat{e}_l)$ is the median calculated from all residuals and for sake of simplicity, the indices $(l - 1)$ denoting the actual iteration used for residual estimation, are omitted. The constant 0.6745 for large sample size fixes the value $\hat{\sigma}$ to be equal to the residual standard deviation $\hat{\sigma}$ but for a normal error distribution. A simpler option is

$$\hat{\sigma} = 2.1 \times \text{med}\left(\left|\hat{e}_l\right|\right) \tag{6.204a}$$

Hill and Holland [37] recommended the expression

$$\hat{\sigma} = \frac{\text{med}\left(\text{largest}\left[n - m\right]\left|\hat{e}_l\right|\right)}{0.6745} \tag{6.204b}$$

Huber [38] recommends a procedure of simultaneous estimation of $\mathbf{b}_j$ and $\hat{\sigma}_j$ in every iteration. Some variants of IRWLS method are described in a paper by Li [39].

It can be difficult to make the initial guess of the parameters to be estimated. Application of the classical least-squares method can cause difficulties from non-convergence of the estimates. The simple procedure of the corrected least-squares method was suggested by Phillip and Eyring [40]. It starts with estimates $\mathbf{b}$ determined by the classical least-squares method. From residuals $\hat{\mathbf{e}}$ the robust parameter of scale is estimated

$$S = \text{med}_i\left(\left|\hat{e}_i\right|\right) \tag{6.205}$$

and the winsorized residuals are calculated by the rule

$$e_i^w = \left\{\begin{array}{ll} -1.5S & \text{for } \hat{e}_i < -1.5S \\ \hat{e}_i & \text{for } \left|\hat{e}_i\right| \leq 1.5S \\ 1.5S & \text{for } \hat{e}_i > 1.5S \end{array}\right\}.$$ The vector of correction $\hat{\mathbf{q}} = (q_1,...,q_m)^T$ is

calculated as the vector of regression coefficients $\mathbf{e}^w$ on $\mathbf{X}$ from $\hat{\mathbf{q}} = (\mathbf{X}^T\mathbf{X})^{-1} \mathbf{X}^T\mathbf{e}^w$. To calculate the quantities r_1 and $w_i(r_i)$, the following corrected parameters values are taken as initial values $\mathbf{b}^w = \mathbf{b} + \hat{\mathbf{q}}$. This procedure does not require much computer time, since the matrix $(\mathbf{X}^T\mathbf{X})^{-1}$ is already evaluated.

The statistical analysis of M-estimates is based on fact that estimates $\mathbf{b}_M$ have an asymptotically normal distribution with mean $\boldsymbol{\beta}$ and covariance matrix

$$D(\mathbf{b}_M) = \tau^2 (\mathbf{X}^T\mathbf{X})^{-1} \tag{6.206}$$

where $\tau^2 = \dfrac{E(\psi^2)}{[E(\psi)]^2}$. Estimate $\hat{\tau}^2$ can be found from the expression

$$\hat{\tau}^2 = K_e \frac{\displaystyle\sum_{i=1}^{n} \frac{\psi^2(r_i)}{n-m}}{\left[\displaystyle\sum_{i=1}^{n} \frac{\psi'(r_i)}{n}\right]^2} \tag{6.207}$$

The constant K_e is the correction for finite samples; it may be set equal to one (according to Li [39]) or calculated from an expression suggested by Huber [38].

The advantage of the IRWLS method is the fact that after termination of iterative refinement of parameter estimates the covariance matrix of the LS method is already the estimate $D(\mathbf{b}_M)$.

To examine robustness, functions such as $\rho(r)$ should be selected in order to get the derivative $\psi(r)$ bounded. From Fig. 6.57b it is obvious that for the LS criterion the function $\psi(r)$ is not bounded, because it increases with an increase of r.

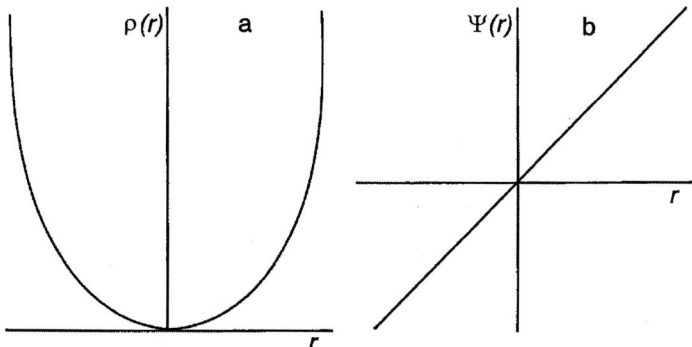

Figure 6.57   (a) Function $\rho(r)$, and (b) function $\psi(r)$ for the least-squares method.

Of the robust methods, we restrict ourselves here to the L_1 method and the method of combined procedure with "Biweights", i.e. *regression with limited influence*. The second method is robust for all types of influential points.

6.6.5.2 The L_1 approximation method

The method of L_1 approximation is also called the *method of least absolute residuals*. The criterion is in the form

$$L_1(\mathbf{b}) = \sum_{i=1}^{n} \left| y_i - \sum_{j=1}^{m} x_{ij} b_j \right| \tag{6.208}$$

This is a special case of M-estimates for $\rho(r) = |r|$ and $\psi(r) = \mathrm{sign}(r)$. Both are shown in Fig. 6.58.

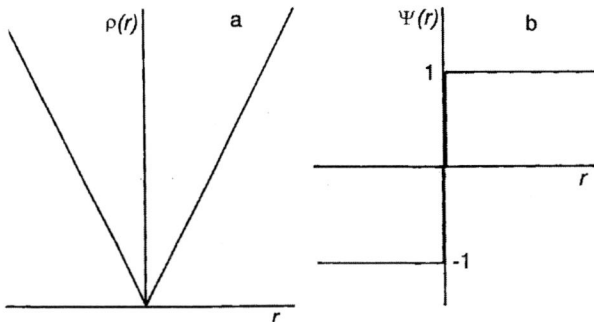

Figure 6.58 (a) The function $\rho(r)$, and (b) function $\psi(r)$ for the L_1 approximation method.

It can be seen from Fig. 6.58b that the function $\psi(r)$ is bounded for all r by the value ± 1. This means that the criterion (6.208) is *robust* for all residuals. The estimates $\mathbf{b}_L$ achieved by minimization of the criterion $L_1(\mathbf{b})$ are maximum likelihood estimates when the errors ε have the Laplace distribution. For a symmetric distribution of errors with kurtosis greater than 3, the estimates $\mathbf{b}_M$ are more effective, i.e. they have smaller variances than the estimates $\mathbf{b}$ from the classical LS method. From Eq. (6.208) it arises that the $L_1(\mathbf{b})$ criterion consists of several linear segments. Figure 6.59 shows the dependence of $L_1(\mathbf{b})$ on b for the case of $m = 1$, for a regression straight line passing through the origin.

From Fig. 6.59b it is evident that many different estimates may exist that correspond to a minimum of $L_1(\mathbf{b})$. Minimization of the criterion $L_1(\mathbf{b})$ is a linear programming problem, i.e. to search for the minimum of $\left[\sum_{i=1}^{n}\left(e_i^+ + e_i^-\right)\right]$ when $\mathbf{Xb}_L + \mathbf{e}^+ + \mathbf{e}^- = \mathbf{y}$ where $\mathbf{e}^+, \mathbf{e}^- > 0$ are vertical deviations from the regression plane $\mathbf{Xb}_L$ (Fig. 6.60). Estimates for parameters b_L may be obtained by the program IRWLS in ADSTAT by using the weights $w_i(r) = 1/|r_i|$ [Eq. (6.202)].

For simple regression models such as the equation of a straight line, we can use the condition that the regression function corresponding to a minimum of the criterion $L_1(\mathbf{b})$ *must* go through just m experimental points. This fact can be used in writing an algorithm which, for all combinations of m points, determines the parameter, estimates by solving linear equations for m unknowns, and the values for the estimates $\mathbf{b}_M$ are those for which the criterion $L_1(\mathbf{b})$ has a minimum. For a regression straight line, and a small number of experimental points, this algorithm is rather simple i.e. it formulates a search for the slope and intercept of a straight line going through two points.

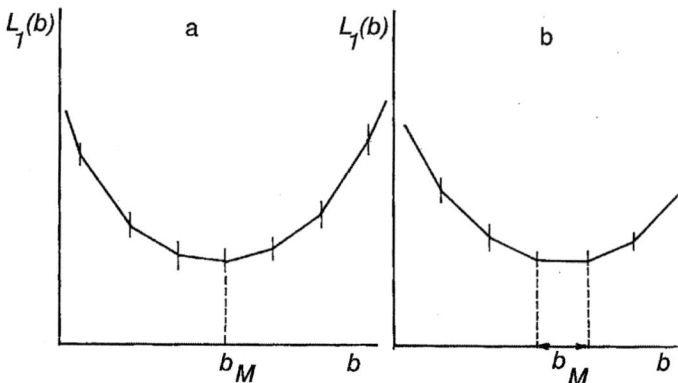

Figure 6.59 Two possible shapes of the criterion function $L_1(b)$: (a) with an obvious minimum, (b) the minimum covers an interval.

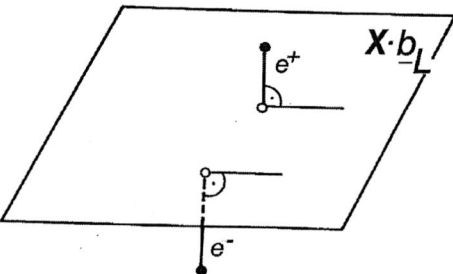

Figure 6.60 Representation of the vertical deviations e^+ and e^- from the regression plane Xb_L.

Problem 6.49 *Examination of the relationship between the change of surface energy of adsorption and the effective specific surface of a sorbent*

Nikulichev and Kanchenko [41] studied the adsorption of stearic acid from decane solution onto various sorbents including stearates and 12-oxystearates. The effective specific surface S_e of these sorbents and the change of surface energy as a consequence of adsorption, $-\Delta G$, were measured. Compare the L_1 approximation and LS methods and, construct a linear model between the variables $-\Delta G$ and S_e.

○ **Data:** S_e in m².kg⁻¹, $-\Delta G$ in kJ.mol⁻¹:

S_e	2.6	3.3	4.4	4.2	6.2	6.5
$-\Delta G$	17.8	18.6	16.2	17.3	15.8	15.2

Solution: The LS method estimates the regression equation as $-\Delta G = -0.7524$ $S_e + 20.23$ with the residual sum of squares $RSS = 0.266$ and the mean absolute deviation $A = 0.44$. By determination of all possible straight lines going through two points and substituting into the $L_1(\mathbf{b})$ criterion, the following model was found: $-\Delta G = -0.5556\ S_e + 19.24$ with the residual sum of squares $RSS = 0.352$ and the mean absolute deviation $A = 0.435$. The two straight lines are compared in Fig. 6.61.

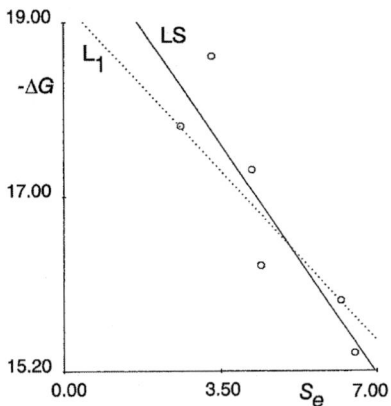

Figure 6.61   Comparison of the regression straight lines found by the LS method (LS) and the L_1 approximation (L$_1$).

Conclusion: For a regression straight line, application of the L_1 approximation is simple. This method is robust enough to cope with the outlying point number 2.

Statistical analysis of the results of the L_1 approximation depends on the asymptotic normality of estimates $\mathbf{b}_L$. The covariance matrix $D(\mathbf{b}_L)$ is calculated from Eq. (6.206), and the variance τ^2 is estimated from $\tau = \dfrac{1}{2p(\tilde{e})}$ where $p(\tilde{e})$ is the probability density function of errors at the median. It is approximately valid that $p(\tilde{e}) = 0.5(\tilde{e}_{0.75} - \tilde{e}_{0.25})$ where $\tilde{e}_{0.75}$ is the upper and $\tilde{e}_{0.25}$ is the lower quartile of the residuals (Chapter 2). Statistical analysis is similar to the LS method, but instead of $\hat{\sigma}^2$ the quantity τ^2 is used. Many authors describe the L_1 approximation as a generally robust method, but this method is robust only with reference to outlying points and not to leverages.

Problem 6.50   *Comparison of the robustness of the LS method with the L_t approximation, in the presence of one influential point*

To illustrate the efficiency of robustness of the two methods, the LS and L_1 approximation, six data points are used. The first data set (A) contains one outlier (y is equal to 10 instead of the correct value, 1) and the second data set (B) contains one leverage (x is equal to 10 instead of the correct value 1). If the regression method is robust enough it should estimate both parameters $\beta_1 = 1$ and $\beta_2 = 0$ in the model $E(y\,/\,x) = \beta_1 x + \beta_2$. Estimate b_1 and b_2 by the LS and L_1 approximation methods, and compare the results.

○ **Data:**

Data set A	x	1	2	3	4	5	6
	y	10	2	3	4	5	6
Data set B	x	10	2	3	4	5	6
	y	1	2	3	4	5	6

Solution: The estimates of parameters β_1 and β_2 by the LS and L_1 approximation method are listed in Table 6.14, and the regression straight lines are shown in Fig. 6.62.

Table 6.14 Comparison of parameter estimates b_1 and b_2 found by the LS and L_1-approximation methods for the model $y = 1 \times x + 0$.

	LS method		L_1-approximation	
	b_1	b_2	b_1	b_2
Data set A	−0.2857	6	1	0
Data set B	−1.25	4.125	−0.2857	3.857

The poor robustness of the LS method leads to a change of sign of the slope of the straight line. The L_1 approximation is robust enough towards the outlying point (set A) but not towards the leverage point (outlying in x value) (set B).

Conclusion: The L_1 approximation method is not generally robust enough to cope with all types of influential points. One influential point can be enough for both methods to give a false estimate of slope; the difference may be big enough to result in a change of sign of the slope.

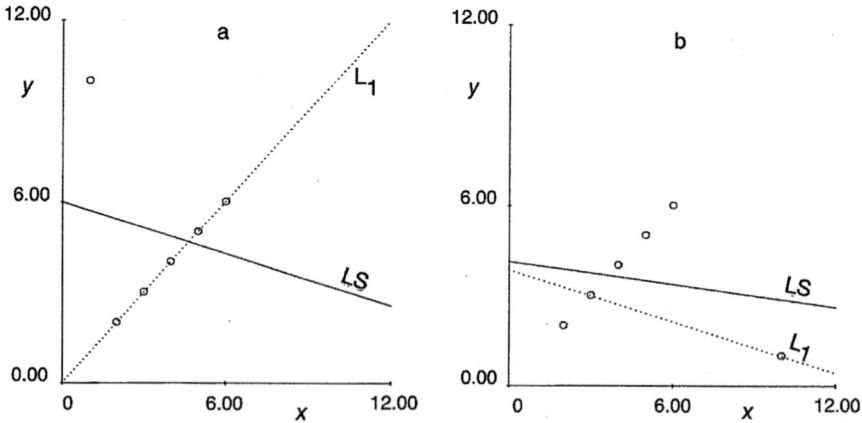

Figure 6.62 Robustness of the LS method and the L_1 approximation for (a) set A (one outlier only) and (b) set B (one leverage and one outlier together).

6.6.5.3 Robust Estimates with Bounded Influence

By using a convenient choice of function $\rho(r)$ or $w(r)$, the robust M-estimates may be found by the iterative reweighted least-squares method. Table 6.15 lists the most frequently used types of M-estimates for $\rho(r)$ and $w(r)$ with numerical values of the constant term.

Table 6.15 Functions $\rho(r)$ and $w(r)$ for five selected robust methods

Author of method	$\rho(r)$	$w(r)$	Range	Constant
Andrews	$\dfrac{A^2\left(1-\cos\left(r/A\right)\right)}{2A^2}$	$(A/r)\sin\left(r/A\right)$	$\|r\|\le A$ $\|r\|>A$	$A=1.339$
Tukey	$\dfrac{\left(B^2/2\right)\left(1-\left(1-\left(r/B\right)^2\right)^3\right)}{B^2/2}$	$\left(1-\left(r/B\right)^2\right)^2$	$\|r\|\le B$ $\|r\|>B$	$B=4.865$
Huber	$\dfrac{r^2/2}{k\|r\|-k^2/2}$	$\dfrac{1}{k/\|r\|}$	$\|r\|\le k$ $\|r\|>k$	$k=1.345$
Talwar	$\dfrac{r^2/2}{T^2/2}$	$\dfrac{1}{0}$	$\|r\|\le T$ $\|r\|>T$	$T=2.795$
Welsch	$\left(W^2/2\right)\left(1-\exp\left(-\left(r/W\right)^2\right)\right)$	$\exp\left(-\left(r/W\right)^2\right)$	–	$W=2.985$

For analysis of laboratory problems the Tukey "biweight" is recommended. It is suitable for calculation of $r_i = e_i/\sigma$, with $\sigma = S$ from Eq. (6.205). For the normal distribution of errors this estimate is equal to 0.67σ.

The estimates of parameters determined by methods from Table 6.15 are robust only on outliers but not on leverages (outlying in x values). To ensure robustness against all types of influential points, estimates with bounded influence are constructed. In the simplest way the set of equations (6.201) is modified by introducing new weights $V(\mathbf{x}_i)$ into the expression $\sum_{i=1}^{n}\psi(e_i/\sigma)x_{ij}V(\mathbf{x}_i)=0$. The weights $V(\mathbf{x}_i)$ eliminate the influence of leverages and are proportional to the magnitudes H_{ii} of the diagonal elements of the projection matrix $\mathbf{H}$.

Krasker and Welsch [43] recommend selecting weights by using the expression

$$V\left(\mathbf{x}_i\right)=\frac{1-H_{ii}}{\sqrt{H_{ii}}} \tag{6.209}$$

Effective procedures for a construction of estimates with bounded influence may be found in the work of Hettmansperger [44]. Introduction of the weights $V\left(\mathbf{x}_i\right)$ from Eq. (6.209) into computer programs does not cause any difficulty. In the IRWLS method the weights $w_i\left(r\right)$ are replaced by weights $V\left(\mathbf{x}_i\right)w_i\left(r\right)$.

Problem 6.51 *Examination of the robustness of estimates with bounded influence*

Estimate the parameters of the regression straight line for data from Problem 6.50 by using a combination of Welsch weights $V\left(\mathbf{x}_i\right)$ (Table 6.15) in Eq. (6.209).

○ *Data:* from Problem 6.50

Solution: Table 6.16 lists estimates of the parameters for the regression straight line for both data sets. Although both sets contain one strongly influential point, the slope estimates are always equal to 1.

Table 6.16 Estimates b_1 and b_2 of parameters $\beta_1 = 1$ and $\beta_2 = 0$ determined by the use of Welsch weights

	b_1	b_2
Set A	1	1.87×10^{-6}
Set B	0.995	0.0196

Conclusion: Estimates with limited influence are robust against all types of influential points.

Some other global robust methods exist. Strong robust methods are methods in which, instead of the sum of squared residuals, the median of squared residuals is sought. These robust methods may be used to locate groups of influential points.

Problem 6.52 *Operation of a plant for the oxidation of ammonia to nitric acid*

The operation of a plant for the oxidation of ammonia to nitric acid was studied [45], and a set of data from 21 days of operation was collected. The dependent variable y represents the percentage of the input ammonia that is lost by escaping as unabsorbed nitric oxides. This is an inverse measure of the yield of nitric acid for the plant. Three independent variables are x_1 the rate of operation, x_2 the temperature of the cooling water in the coils of the absorption tower for the nitric acid, and x_3 the concentration of nitric acid in the absorbing liquid. Investigation of plant operations indicates that the following sets of runs can be considered as replicates: (1, 2), (4, 5, 6), (7, 8), (11,12), and (18,19). While the runs in each set are not exact replicates, the points are sufficiently close to each other in x-space for them to be used as such. Suppose the linear model is $E(y / \mathbf{x}) = \beta_1 x_1 + \beta_2 x_2 + \beta_3 x_3 + \beta_4$ and apply the LS method and the L_1-approximation method to indicate influential points.

○ **Data:** (*denotes the strongly influential point in output)

Run No.	Stack Loss %, y	Air Flow, x_1	Temperature, x_2	[HNO$_3$], x_3	Residual, $\hat{e}_{L1}$
1	42	80	27	89	2.53 *
2	37	80	27	88	1.85×10^{-6}
3	37	75	25	90	2.715*
4	28	62	24	87	3.814*
5	18	62	22	87	−0.61
6	18	62	23	87	−0.89
7	19	62	24	93	0.50
8	20	62	24	93	0
9	15	58	23	87	−0.73
10	14	58	18	80	-2.8×10^{-6}
11	14	58	18	89	0.279
12	13	58	17	88	0.039
13	11	58	18	82	−1.43
14	12	58	19	93	−0.887
15	8	58	18	89	0.601
16	7	50	18	86	0.008
17	8	50	19	72	−0.217
18	8	50	19	79	9.1×10^{-5}
19	9	50	20	80	0.241
20	15	56	20	82	0.812
21	15	70	20	91	−4.722*

Solution: The classical LS method finds the regression equation $\hat{y} = -37.68 + 0.7336x_1 + 1.3883x_2 - 0.2164x_3$ with determination coefficient $\hat{R}^2 = 0.913$ and residual standard deviation $\hat{\sigma} = 3.243$. The partial regression graphs for the independent variable x_1 and x_2 indicate that points 21 and 4 are outliers. The L-R graph indicates that point 21 is a strongly influential point and points 1, 3 and 4 are less influential. With the use of the L_1 approximation, the regression equation takes the form $\hat{y} = -39.65 + 0.83x_1 + 0.581x_2 - 0.0621x_3$ with mean absolute deviation $A_p = 2.004$. The last column in the Table shows the residuals $\hat{e}_{L1} = \hat{e} / A_p$ which indicate that points 21 and also 1, 3 and 4 are influential.

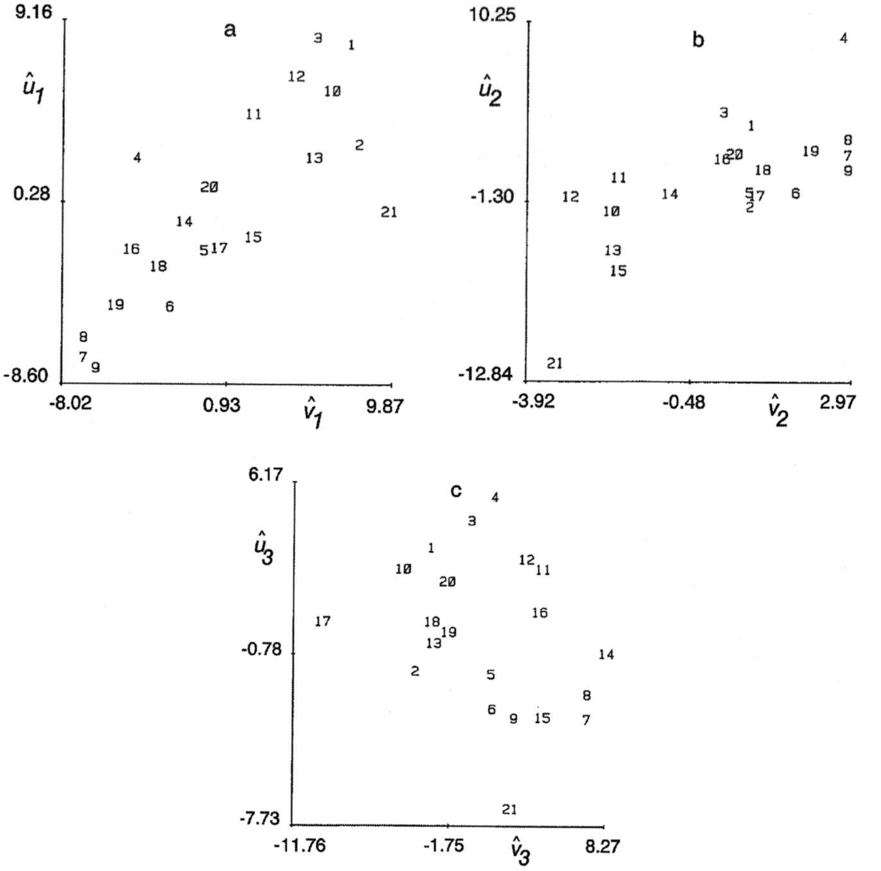

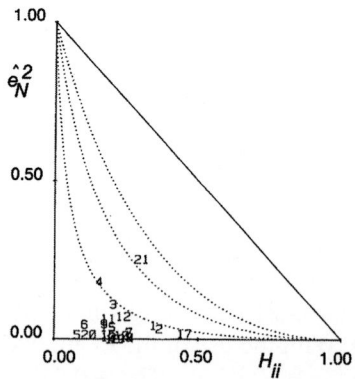

Figure 6.63 Partial regression graph for independent variables (a) x_1, (b) x_2, and (c) x_3.

Figure 6.64 The L-R graph for the diagnostic DF_t indicating outliers and leverages

Conclusion: The robust methods can be useful for identification of influential points.

Problem 6.53 *Examination of the effect of three different factors on the amount of ozone in air*

The dependence of the amount of ozone in the air (y) on the intensity of the sun's radiation for the range of wavelengths 400–700 nm (x_1), the mean velocity of wind (x_2) and the highest daytime temperature (x_3) was studied [45]. The linear model $E(y/\mathbf{x}) = \beta_1 x_1 + \beta_2 x_2 + \beta_3 x_3 + \beta_4$ was proposed. Compare the robust estimates of the parameters with the estimates found after strongly influential points are rejected.

○ *Data:*

Measurement	y [ppm]	x_1	x_2 [miles/h]	x_3 [°F]
1	41	190	74	67
2	36	118	8.0	72
3	12	149	12.6	74
4	18	313	11.5	62
5	23	299	8.6	65
6	19	99	13.8	59
7	8	19	20.1	61
8	16	256	9.7	69
9	11	290	9.2	66
10	14	274	10.9	68
11	18	65	13.2	58
12	14	334	11.5	64
13	34	307	12.0	66
14	6	78	18.4	57
15	30	322	11.5	68
17	1	8	9.7	59
18	11	320	16.6	73
19	4	25	9.7	61
20	32	92	12.0	61
21	23	13	12.0	67
22	45	252	14.9	81
23	115	223	5.7	79

Solution: The classical LS method leads to the regression equation

$$\hat{y} = -79.99 - 0.01868x_1 - 1.996x_2 + 1.963x_3.$$

The partial regression graphs and the L-R graph indicate that there is only one influential point, number 23. From these graphs it is also evident that the dependence on the chosen independent variables x_1, x_2 and x_3 is not strong. When point 23 is omitted, the regression equation becomes

$$\hat{y} = -37.52 + 0.00559x_1 - 0.7488x_2 + 0.9935x_3$$

The L_1 approximation gives the equation

$$\hat{y} = -75.36 + 0.00665x_1 - 0.3391x_2 + 1.527x_3.$$

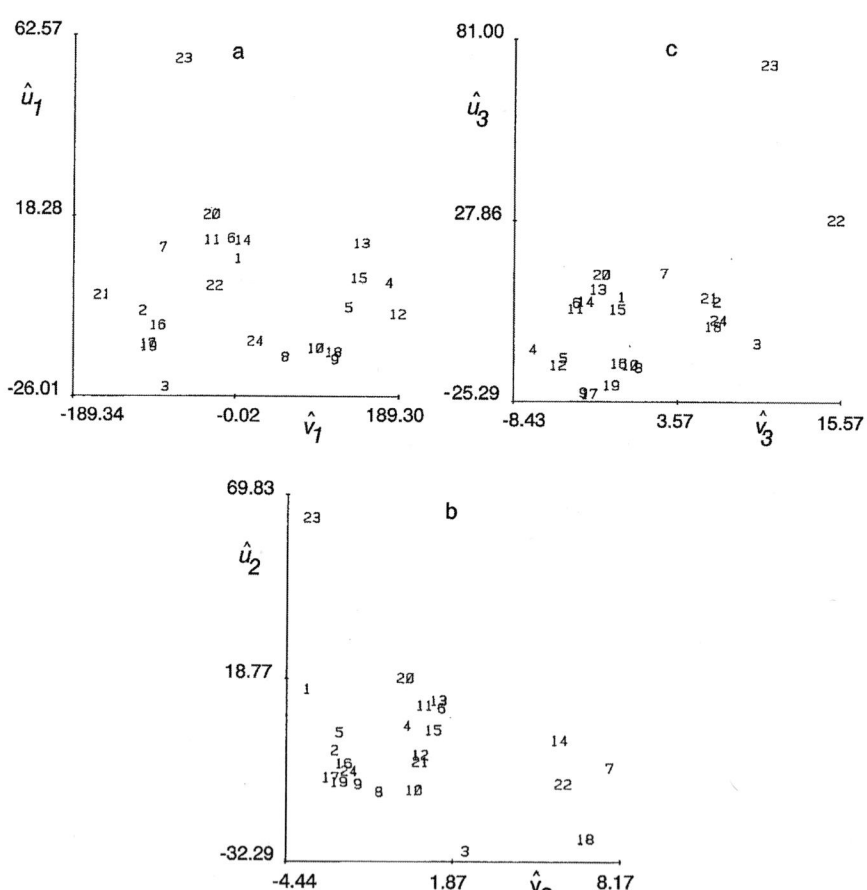

Figure 6.65 Partial regression graph for independent variable (a) x_1, (b) x_2, and (c) x_3.

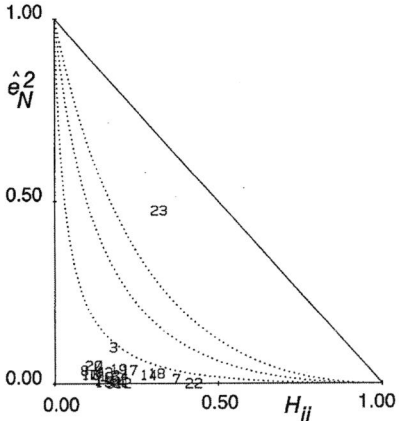

Figure 6.66   The L-R graph for diagnostic DF_i. indicating outliers and leverages

It is interesting that when the LS method is used, some predicted points (amount of ozone) have a negative sign: this does not happen with the alternative regression methods. From the physical point of view, predicted values of $\hat{y}$ should always be positive. For example, for the point 7, by the LS method $\hat{y}_7 = -0.7$, by the LS method without point 23, $\hat{y}_7 = 8.14$, and by the L_1 approximation $\hat{y}_7 = 11.13$.

The presence of a single influential point caused the model to be unsuitable for prediction. Omitting the outlier (23) has a more beneficial effect than the use of the L_1 approximation.

Conclusion: Interactive data analysis based on identification of influential points often leads to better and more correct parameter estimates than robust regression.

One of greatest disadvantages of robust methods is a preference for the regression model proposed. If the proposed model is unsuitable, robust methods lead to suppression of the influence of individual points and therefore also to a suppression of the detection of unsuitable proposed models. Therefore, robust methods should be applied only with careful regard to the peculiarities of the model and data.

Sometimes it is falsely believed that the effects of influential points (outliers and leverages) are suppressed in large samples. Let us illustrate the effect of a single influential point (outlier O or leverage L) in a set of 50 points. Samples of 50 points containing one outlier O or one high-leverage point L, were generated, and the influence on the estimate b_1 of the slope of the regression straight line was examined. Data x_i were simulated by generation from a rectangular distribution $R[0, 1]$, and then linearly transformed into the interval $(10, 20)$. The variable y_i

was calculated using the relation $y_i = \beta_1 x_1 + \beta_2 + N(0,1)$ where $\beta_1 = 1$, $\beta_2 = 1$ and $N(0, 1)$ is a random number with standardized normal distribution. Into this data, for point 37, either an outlier O or a leverage L was introduced. Table 6.17 lists values of the slope estimate b_1 of the regression straight line $y = x + 1$, when point 37 has magnitudes 40, 70, and 80. Regression analysis was performed by classical LS (LS), by the M-estimate with Welsch weights from Table 6.15 (WR) and by the estimate with bounded influence (KWR).

Table 6.17 Estimation of slope b_1 by three different regression methods: LS, WR and KWR for a data set of 50 points with one influential point (O = outlier, L = leverage point). $\beta_1 = 1$.

Value of influential point	LS		WR		KWR	
	L	O	L	O	L	O
40	0.46	1.180	0.926	0.973	0.958	0.962
70	0.17	1.490	0.203	0.969	0.965	0.959
80	0.13	1.602	0.140	0.969	0.965	0.959

Table 6.17 illustrates that one outlier or leverage point in a set of 50 points causes the classical LS method or the M-estimate method to determine a totally false estimate of the slope b_1.

The method of slope estimate with bounded influence (KWR) is robust and found a true estimate of parameter β_1. This example of the influence of a single outlier or high leverage point indicates that without an analysis of influential points in interactive co-operation with the computer, routine data treatment may be totally invalidated by false and meaningless estimates. Just one decimal point falsely writtten may cause totally erroneous parameter estimates.

6.7 Calibration

Calibration is one of the most important applications in the chemical laboratory for regression analysis. Calibration consists of two steps:

(1) building a calibration model;

(2) application of the calibration model.

Building a calibration model is identical with the task of building a regression model. The second step of calibration involves inversion of the first step, i.e. for a measured response y^* the corresponding value x^* and its statistical characteristics are calculated. The main attention in this section is paid to calibration straight lines.

6.7.1 Types of calibration and calibration models

Calibration tasks have been classified according to different criteria by Rossenblatt and Spiegelman [46].

(1) *Absolute calibration* is the most frequently used procedure in lab oratory instrumentation. In the construction of a calibration model, the measured quantity η, called the signal (potential, EMF, electric resistance, pH, absorbance, etc.) is related to the quantity ξ which describes a state or a property of the system (composition, concentration, temperature, time, etc.). An example of an absolute calibration is the dependence of the absorbance of a solution (η) on its concentration (ξ).

In a calibration experiment for n samples with known (or precisely measured) values of variable ξ, the corresponding quantities η are measured. Frequently, both variables are monitored instrumentally, and there will be n points $\{x_i y_i\}$, $i = 1,..., n$, where

$$y_i = \eta_i + \varepsilon_i \tag{6.210a}$$

$$x_i = \xi_i + \delta_i \tag{6.210b}$$

where ε_i and δ_i are experimental errors. If the variable ξ_i is measured precisely, or exactly defined standards are used, $\delta_i = 0$, $i = 1, ..., n$. The quantity η_i is replaced by a calibration model $f(x, \beta)$, and data treatment leads to estimation of parameter β.

In the second phase, there are M repeated measured values of an laboratory signal $\{y_j^*\}$, $j = 1,..., M$ from which the mean value of property $\hat{x}^*$ with its confidence interval is estimated. An example of a signal that depends on concentration is illustrated in Fig. 6.67, where symbols L_L and L_U denote the lower and upper limits of the confidence interval of concentration. In the rest of this chapter we will consider only absolute calibration.

(2) *Comparative calibration* is a procedure in which one instrument is calibrated agianst a second one, and either may be used as the standard. An example is the determination of concentration with the use of absorbance (Lambert-Beer law) as the first method, and potentiometric titration as the second method. Absorbance values are compared with volumes of titrant added. The errors δ_i are not negligible, and to construct the calibration model the regression analysis for the case when both variables are subject to experimental errors must be used.

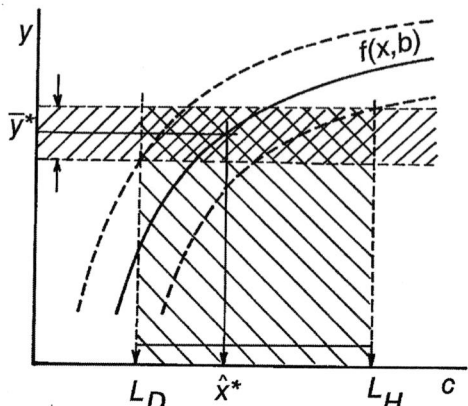

Figure 6.67   Absolute calibration and a procedure for determination of concentration $\hat{x}^*$
for the mean value of laboratory signal $\overline{y}^*$. L_L and L_U are the lower and
upper limits of the confidence interval of concentration.

With reference to the application of the calibration model, the following
cases may be distinguished:

(a) single application of the calibration model: the calibration model is
constructed from n measured points $\{x_i, y_i\}$, $i = 1,...,n$, and then
one estimate $\hat{x}^*$ with its confidence interval is calculated from one y^*
value;

(b) mutliple application of the calibration model: from the calibration
model, several estimates $\hat{x}^*$ are determined from values of the
laboratory signal.

(c) single or multiple application in combination with other measurements:
the result of the second phase of calibration is used together with
other variables and constants for determination of a quantity which is
a function of more variables. Here, any bias in the estimates $\hat{x}^*$ which
will be included in the final systematic error of the result.

The difficulty of the calibration task depends on the model used. For linear
regression models, the confidence bands around the model may be expressed
by Eq. (6.45) or for all possible values by Eq. (6.45a). The components of
vector **x** are functions of a measured property (i.e. usually concentration),
and when polynomial models are considered, the individual components

correspond to powers of this measured property. To find a value of $\hat{x}^*$, a root
of a polynomial must be found.

For nonlinear regression models the solution is sought in the form

$$\hat{x}^* = f^{-1}(y^*) \tag{6.211}$$

On the base of the Taylor series for this function, the approximate formula for the variance $D(\hat{x}^*)$ may be found in the form [47]

$$D(\hat{x}^*) \approx \left[\frac{\delta f(x,\mathbf{b})}{\delta x}\right]^{-2}\left[\frac{D(y^*)}{M} + D(f(x,\mathbf{b}))\right] \qquad (6.212)$$

where $D(y^*)$ is the variance of y^* values, usually equal to σ^2 and $D(f(x, \mathbf{b})) = D(\hat{y})$ is the variance of prediction, estimated from the Taylor series of function $f(x, \mathbf{b})$. For the linear regression model the variance of prediction is given by

$$D(\hat{y}) = \sigma^2\left[\frac{1}{n} + \frac{(x^* - \bar{x})^2}{\sum\limits_{i=1}^{n}(x_i - \bar{x})^2}\right] = \sigma^2\left[\frac{1}{n} + \frac{(y^* - \bar{y})^2}{b_1^2\sum\limits_{i=1}^{n}(x_i - \bar{x})^2}\right]$$

where b_1 is the estimate of the slope of the regression line. On substituting into Eq. (6.212) we obtain

$$D(\hat{x}^*) \approx \frac{\sigma^2}{b_1^2}\left[\frac{1}{M} + \frac{1}{n} + \frac{(y^* - \bar{y})^2}{b_1^2\sum\limits_{i=1}^{n}(x_i - \bar{x})^2}\right] \qquad (6.212a)$$

Difficulties are caused by the generally non-symmetric distribution of quantity x^*. Only in the case of a calibration straight line and small residual variance can the distribution of x^* be assumed to be approximately normal [48].

If both y and y^* are random variables with normal distribution, the difference $\Delta = \bar{y}^* - f(x^*, \mathbf{b})$ will also have the normal distribution. The standardized random variable $\Delta / \sqrt{D(\Delta)}$ has the Student distribution with the number of degrees of freedom used for determination of $D(\Delta)$. To find the $100(1 - \alpha)\%$ confidence interval of the quantities $\hat{x}^*$ defined in Eq. (6.211) it is necessary to solve the equation [51]

$$(\bar{y}^* - f(\hat{x}^*, \mathbf{b}))^2 = F_{1-\alpha}(1, r) \times D(\bar{y}^* - f(\hat{x}^*, \mathbf{b}))$$

where $r = n - 2$. The variance $D(\Delta) = D(\bar{y}^* - f(\hat{x}^*, \mathbf{b}))$ may be estimated by the Taylor series expansion of function $f(\hat{x}^*, \mathbf{b})$. It is approximately valid that

$$D(\Delta) = D(\bar{y}^*) + \sum_{j=1}^{m} \left[\frac{\delta f(\hat{x}^*, \mathbf{b})}{\delta b_j} \right]^2 D(b_j) + 2\sum_{i=1}^{m-1} \sum_{j>1}^{m} \frac{\delta f(\hat{x}^*, \mathbf{b})}{\delta b_i} \times \frac{\delta f(\hat{x}^*, \mathbf{b})}{\delta b_j} \operatorname{cov}(b_i, b_j)$$

$$(6.213)$$

here m is the number of regression parameters. The special case represented here is the model of the calibration straight line

$$y = b_1(x - \bar{x}) + \bar{y} \qquad (6.213a)$$

for which, after substitution into Eq. (6.213), we obtain

$$D(\Delta) = \sigma^2 \left[\frac{1}{M} + \frac{1}{n} + \frac{(\hat{x}^* - \bar{x})^2}{\sum_{i=1}^{n}(x_i - \bar{x})^2} \right] \qquad (6.213b)$$

When $D(\Delta)$ is known, the two limits of the confidence interval of $\hat{x}^*$ may be estimated. This involves finding roots of the quadratic equation

$$\bar{y}^* - \bar{y} - b_1(\hat{x}^* - \bar{x}) = F_{1-\alpha}(1, n-2) \times D(\Delta) \qquad (6.213c)$$

with respect to the variable $\hat{x}^*$.

6.7.2 Calibration straight line

A straight line is the usual calibration model in a chemical laboratory. Usually, it is supposed that this model fits all the measured points for a given set of variables x and y. For example, the Lambert-Beer law $A = \varepsilon\, d\, c$ expresses a linear relationship between absorbance and concentration c where ε is the molar absorptivity and d is the path length of the cell.

In some cases, however, the straight-line model is valid only in a limited interval, and above a limiting point $\{x_A, y_A\}$ there is a significant departure from linearity. For example, the Kubelka-Munk relationship between the remission function $(1 - R^2)/2R$ and the concentration c is valid only for low concentrations. The Lambert-Beer law too is valid only up to some limiting concentration, above which curvature occurs.

For statistical data treatment, the model in the form of Eq. (6.213a) may be used, or some other equivalent expression such as

$$y_i = \beta_1 x + \beta_2 + \varepsilon_i, \quad i = 1, \dots, n$$

$$\text{or } y_i^* = \beta_1 \kappa + \beta_2 + \varepsilon_i^*, \quad j = 1, \dots, M.$$

The task of calibration is to find an estimate of parameter x^*, the primary parameter, and of parameters β_1 and β_2, the supplementary parameters. The estimation assumes normality of the errors ε_i and ε_j^*. The estimate $\hat{x}^*$ and its confidence interval may be calculated by several procedures.

By substituting into Eq. (6.211) from Eq. (6.213a) we obtain the straight estimate of parameter κ in the form

$$\hat{x}^* = \bar{x} + \frac{(y^* - \bar{y})}{b_1} \tag{6.214}$$

where y^* is the measured signal (or the average $\bar{y}^*$ for $M > 1$ repeated measurements, respectively) and b_1 is the estimate of the slope. This estimate is generally biased and a correction is made by Naszodi's modified estimates [49]

$$\hat{x}_B^* = \bar{x} + (y^* - \bar{y}) \frac{b_1}{b_1^2 + \dfrac{\sigma^2}{\displaystyle\sum_{i=1}^{n}(x_i - \bar{x})^2}} \tag{6.215}$$

Krutchkoff [50] proposed the inversion estimate

$$\hat{x}_I^* = \bar{x} + (y^* - \bar{y}) \frac{\displaystyle\sum_{i=1}^{n}(x_i - \bar{x})(y_i - \bar{y})}{\displaystyle\sum_{i=1}^{n}(y_i - \bar{y})^2} \tag{6.216}$$

which refers to the inversion regression model $E(x / y) = \alpha_1 (y - \bar{y}) + \alpha_2$. From analysis of the estimate $\hat{x}_I^*$ it was found that it too is a biased estimate which is not better than the straight estimate $\hat{x}^*$. Moreover, in the estimation of parameters α_1 and α_2 it is falsely supposed that the y values are measured with negligible errors in comparison with the x-values. Schwartz [51] proposed the nonlinear estimate given by

$$\hat{x}_N^* = \frac{\sum_{i=1}^n x_i \exp\left[\dfrac{-(y^* - b_2 - b_1 x_i)^2}{2\hat{\sigma}^2}\right]}{\sum_{i=1}^n \exp\left[\dfrac{-(y^* - b_2 - b_1 x_i)^2}{2\hat{\sigma}^2}\right]} \tag{6.217}$$

which, however, assumes normality of residuals.

Problem 6.54 *Point estimates of concentration from an AAS calibration straight line*

The atomic absorbances of solutions of various concentration of lithium were measured. Determine the calibration straight line and from it then the concentration of lithium for measured absorbance values $A_1 = 0.0002$, $A_2 = 0.5$ and $A_3 = 1.0$.

○ *Data: n = 16*, Concentration C, g of Li in 25 ml, Absorbance A:

C, A:						
	2.5 0.063	5.0 0.120	7.5 0.189	10.0 0.251	12.5 0.316	15.0 0.393
17.5 0.442	20.0 0.502	22.5 0.568	25.0 0.639	27.5 0.694	30.0 0.749	32.5 0.821	35.0 0.884
37.5 0.947	40.0 1.010					

Solution: The classical method of least-squares leads to the regression equation $A = 0.02525(\pm 0.00011)C + 0.0002(\pm 0.0028)$ with correlation coefficient $\hat{R} = 0.9999$. Table 6.18 lists the estimates $\hat{x}^*$, $\hat{x}_B^*$, $\hat{x}_I^*$ and $\hat{x}_N^*$ for $A = 0.0002$, 0.5 and 1.

Table 6.18 Concentration estimates by various methods

Absorbance	$\hat{x}^*$	$\hat{x}_B^*$	$\hat{x}_I^*$	$\hat{x}_N^*$
0.0002	0	4.32×10^{-4}	6.05×10^{-3}	2.5
0.5	19.795	19.795	19.795	20
1.0	39.597	39.597	39.592	40

Within experimental error, all estimates except the nonlinear one lead to the same result.

Conclusion: For sufficiently precise data with small spread around the regression straight line, the classical estimate $\hat{x}^*$ is satisfactory.

In the construction of confidence intervals of the estimates $\hat{x}^*$ and $\hat{x}^*_B$ for more scattered data, the simplest is the determination of $D(\hat{x}^*)$ and to use Eq. (6.213) with an assumption of normality. The limits of the 95% confidence interval are calculated by

$$L_L = \hat{x}^* - 1.96\sqrt{D(\hat{x}^*)} \qquad (6.218a)$$

$$L_U = \hat{x}^* + 1.96\sqrt{D(\hat{x}^*)} \qquad (6.218b)$$

To construct the confidence interval, the ratio

$$\frac{\left[(b_2 + b_1)(\hat{x}^* - y^*)\right]^2}{\sigma^2 \left[\dfrac{1}{n} + \dfrac{(x^* - \overline{x})^2}{\displaystyle\sum_{i=1}^{n}(x_i - \overline{x})^2}\right]} \approx F_{1-\alpha}(1, n-2) \qquad (6.219)$$

is often used. This ratio exhibits the Fisher-Snedecor distribution with 1 and $(n-2)$ degrees of freedom. The corresponding $100(1-\alpha)\%$ confidence interval of parameter x is calculated from

$$L_{L,U} = \overline{x} + \frac{(y^* - \overline{y}) \pm \hat{\sigma}\sqrt{F_{1-\alpha}(1, n-2)}\left[\dfrac{1+\lambda}{n} + \dfrac{(y^* - \overline{y})^2}{b_1^2 \displaystyle\sum_{i=1}^{n}(x_i - \overline{x})^2}\right]}{b_1(1-\lambda)} \qquad (6.220)$$

Parameter λ is given by $\lambda = \dfrac{\hat{\sigma}^2 F_{1-\alpha}(1, n-2)}{b_1^2 \displaystyle\sum_{i=1}^{n}(x_i - \overline{x})^2}$ and is the variation coefficient

of slope β_1. When this ratio is smaller than 0.1 the slope estimate is sufficiently precise for the approximate confidence interval for paramater κ to be used in the form

$$L_{L,U} = x^* \pm t_{1-\alpha/2}(n-2) \times \frac{\hat{\sigma}}{|b_1|}\sqrt{\dfrac{1}{n} + \dfrac{(y^* - \overline{y})^2}{b_1^2 \displaystyle\sum_{i=1}^{n}(x_i - \overline{x})^2}} \qquad (6.221)$$

If we want information about the whole regression (calibration) line, we replace in Eq. (6.220) the term $\sqrt{F_{1-\alpha}(1, n-2)}$ by the term $\sqrt{2F_{1-\alpha}(2, n-2)}$. The Scheffe's confidence interval of one predicted value y^* at $\hat{x}^*$ is calculated by

$$L_{L,U} = y^* \pm \hat{\sigma}\sqrt{2F_{1-\alpha}(2, n-2)}\sqrt{1 + \frac{1}{n} + \frac{(\hat{x}^* - \bar{x})^2}{\sum_{i=1}^{n}(x_i - \bar{x})^2}} \qquad (6.222)$$

This confidence interval is larger by the variance σ^2 because the variable y^* is used instead of its mean value $E(y^*)$. By rearrangement of Eq. (6.222) we find the $100(1 - \alpha)\%$ confidence interval of variable κ in the form

$$L_{L,U} = \hat{x}^* \pm \frac{\hat{\sigma}\sqrt{2F_{1-\alpha}(2, n-2)}}{\lambda_1}\sqrt{\lambda_1 + \frac{\lambda_1}{n} + \frac{(y^* - \bar{y})^2}{\sum_{i=1}^{n}(x_i - \bar{x})^2}} \qquad (6.223)$$

where $\lambda_1 = \dfrac{b_1^2 - \sqrt{2F_{1-\alpha}(2, n-2)}}{\sum_{i=1}^{n}(x_i - \bar{x})^2}$. When the arithmetic mean $\bar{y}^*$ is used, the term 1 is replaced by term $1/M$ in the brackets of Eq. (6.222). An analogous adjustment can be made to give Eq. (6.223), which corresponds to Eq. (6.213c). The graphical interpretation of the confidence interval of parameter x is shown in Fig. 6.68.

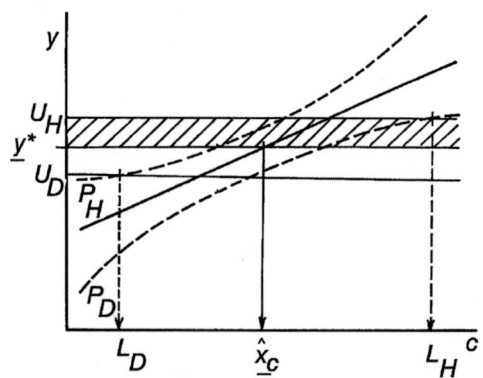

Figure 6.68 Determination of the confidence interval of parameter x for a calibration straight line. The confidence interval of the signal is indicated by the hatched area.

When there are replicate values of y, and $\bar{y}^*$ has been determined, the confidence straight lines U_L and U_U should be calculated. The intersection of straight line U_U with the lower confidence parabola P_L of the calibration straight line leads to point L_U and the intersection of straight line U_L with the upper confidence parabola P_U leads to point L_L.

If the variance of measurement, σ^2, is known it is easy to define the $100(1 - \alpha)\%$ confidence interval of signal $\bar{y}^*$ in the form $U_{L,U} = \bar{y}^* \pm u_{1-\alpha/2}\sigma$ where $u_{1-\alpha/2}$ is the quantile of the normalized normal distribution. If σ^2 is unknown, the inequality

$$\sigma^2 < \frac{(n-2)\hat{\sigma}^2}{\chi^2_{\alpha/2}(n-2)\,M} \tag{6.224}$$

may be used, where $\chi^2_{\alpha/2}$ is the lower $100\,\alpha/2\%$ quantile of the χ^2-distribution. The confidence interval of signal $U_{L,U}$ is then calculated from

$$U_{L,U} = \bar{y}^* \pm u_{1-\alpha/2}\,\frac{\hat{\sigma}}{\sqrt{M}}\sqrt{\frac{n-2}{\chi^2_{\alpha/2}(n-2)}} \tag{6.225}$$

Instead of the quantile $u_{1-\alpha/2}$ in this equation, for $M = 1$ the more convenient quantile of the Student distribution $t_{1-\alpha/2}(n-2)$ is used and the variance σ^2 is replaced by its estimate $\hat{\sigma}^2$.

From Eq. (6.45a) the limiting $100(1 - \alpha)\%$ confidence parabola are given by

$$P_{L,U} = b_1 x + b_2 \pm \sigma \left\{ 2F_{1-\alpha}(2, n-2)\left[\frac{1}{n} + \frac{(x-\bar{x})^2}{\sum\limits_{i=1}^{n}(x_i - \bar{x})^2}\right]\right\}^{1/2} \tag{6.226}$$

The limiting value L_U represents the solution of the equation

$$U_U = P_L \tag{6.227a}$$

with respect to variable x. The limiting value L_L is the solution of the equation

$$U_L = P_U \tag{6.227b}$$

Both equations are quadratic with respect to variable x. From Fig. 6.68 it can be seen that in some cases the intersection of a straight line with the parabola does not exist, but in other cases the confidence straight line of the signal may intersect the parabola of calibration line at two points. This indicates that the spread of data is too large when the slope of the calibration straight line is small, and such a calibration straight line is a poor model.

Problem 6.55 *Various confidence intervals of concentration from a photometric calibration straight line*

For the photometric calibration data from Problem 6.54, calculate the bounds of the 95% confidence interval of concentration, L_L and L_U, for the following absorbance values: $A_1^* = 0.0002$, $A_2^* = 0.5$, $A_3^* = 1$, $\overline{A} = 0.51$ (the mean of two measured values 0.50 and 0.52; $M = 2$) and $\overline{A_4^*} = 0.977$ (the mean of measured values 0.95, 0.98, 1.00; $M = 3$) by using Eqs. (6.218), (6.220), (6.221), (6.223) and (6.227).

○ **Data:** from Problem 6.54

Solution: The calculated limits $L_{L,U}$ for the 95% confidence intervals of concentration are listed in Table 6.19.

Table 6.19 The 95% confidence interval of concentration calculated by various expressions

M	A^*	(6.218) lower L_L; upper L_U	(6.220) lower L_L; upper L_U	(6.221) lower L_L; upper L_U	(6.223) lower L_L; upper L_U	(6.227) lower L_L; upper L_U
1	0.0002	−0.46; 0.46	−0.302; 0.295	−0.299; 0.299	−0.646; 0.639	−0.398; 0.417
1	0.50	19.37; 20.22	19.65; 19.94	19.65; 19.94	19.20; 19.94	19.39; 19.20
2	(0.50; 0.52)	19.89; 20.50	20.05; 20.33	20.05; 20.33	19.76; 20.62	19.90; 20.48
1	1.00	39.15; 40.05	39.33; 39.87	39.33; 39.87	38.97; 40.23	39.19; 40.01
3	(0.95; 0.98; 1.00)	38.37; 38.97	38.42; 38.93	38.42; 38.93	38.26; 39.09	38.43; 38.91

By using Eq. (6.223) instead of $\lambda_1(1+1/n)$ when $M > 1$, the term $\lambda_1(1/M + 1/n)$ is used.

The confidence intervals from Eqs. (6.220) and (6.221) do not reflect a higher precision of determination of $\overline{y}^*$. The confidence limits (6.227a, b)

were evaluated by a simplified expression for the confidence straight line of the signal by

$$U_{L,U} = \bar{y}^* \pm u_{1-\alpha/2} \, \sigma / \sqrt{M} \qquad (6.228)$$

with $\alpha = 0.05$. From Table 6.19 it is seen that for sufficient precision of data, Eq. (6.220) and its approximation (6.221) lead to the same results. The other confidence intervals are, however, rather different. The approximation (6.218) leads to values $L_{L,U}$ which are close to the values calculated from Eq. (6.227).

Conclusion: For data with a small spread around the regression straight line, the simpler approximation (6.218) should be used. For replicate signal measurements the expressions (6.220) and (6.221) are *not* suitable. The quality of the confidence interval around the parameter x is improved by

(1) repeating the signal measurement y^*, i.e. increasing the number of measurements M. For a sufficient number of replicates, M, the estimate $U_{L,U}$ can be calculated from Eq. (6.228), with σ^2 replaced by the variance σ_y^{2*} and the quantile $u_{1-\alpha/2}$ replaced by the quantile of the Student distribution $t_{1-\alpha/2}$.

(2) The confidence parabola may be narrowed by elimination of influential points. In polynomial calibration models the confidence bands may be narrowed by the use of biased estimates calculated by the method of the rational ranks.

(3) Decreasing the residual variance $\hat{\sigma}^2$ and so increasing the precision of measurement, or by the use of a correct calibration model.

6.7.3 The Precision of Calibration

To express the precision of a calibration, limiting values of the concentration for which the measurement signal is still statistically significantly different from the noise are usually defined. To express precision and sensitivity of calibration methods, three levels of signal are identified:

(1) The *critical level* y_c represents the upper limit of the $100(1 - \alpha)\%$ confidence interval of the predicted signal from the calibration model for the concentration equal to zero, i.e. the *blank measurement.* By replacing $\sqrt{2F_{1-\alpha}(2, n-2)}$ by the quantile $t_{1-\alpha/2}(n-2)$ in Eq. (6.222) and setting $x = 0$, we obtain an expression for the critical level y_C in the form

$$y_C = \bar{y} - b_1\bar{x} + \hat{\sigma}\, t_{1-\alpha/2}(n-2) \sqrt{1 + \frac{1}{n} + \frac{\bar{x}^2}{\sum\limits_{i=1}^{n}(x_i - \bar{x})^2}} \qquad (6.229)$$

Signals above this critical level y_C are significantly different from the noise. The concentration x_C corresponding to this critical level y_C is determined from the calibration model from

$$x_C = \frac{y_C - \bar{y}}{b_1} + \bar{x} \qquad (6.229a)$$

(2) The *detection limit* y_D corresponds to the concentration for which the lower $100(1 - \alpha)\%$ confidence interval of signal prediction from the calibration model is equal to y_C. The detection limit y_D and its corresponding concentration x_D are illustrated on Fig. 6.69. For the linear calibration model we have

$$y_D = y_C + \hat{\sigma}\, t_{1-\alpha/2}(n-2)\, \sqrt{1 + \frac{1}{n} + \frac{(x_D - \bar{x})^2}{\sum\limits_{i=1}^{n}(x_i - \bar{x})^2}} \qquad (6.230)$$

Oppenhelmer [52] proposed the following approximation

$$y_D = y_C + \hat{\sigma}\, t_{1-\alpha/2}(n-2)\, \sqrt{1 + \frac{1}{n} + \frac{\bar{x}^2}{\sum\limits_{i=1}^{n}(x_i - \bar{x})^2}} \qquad (6.231)$$

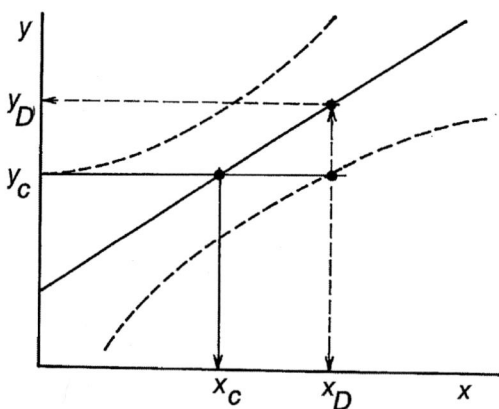

Figure 6.69   Definition of the critical level y_C, the detection limit y_D and their corresponding concentrations x_C and x_D.

The corresponding concentration x_D is calculated from

$$x_D = \frac{y_D - \bar{y}}{b_1} \qquad (6.231a)$$

The detection limit gives the lowest true signal level which still permits detection. The quantity x_D gives the minimum concentration which can be distinguished from zero with probability $(1 - \alpha)$.

(3) The *quantification (determination) limit* y_Q is the smallest signal level for which the relative standard deviation of prediction from the calibration model is sufficiently small and equal to the number C, where $C = 0.1$, usually. If the predicted value at point x_Q is given by $y(x_Q) = \bar{y} + b_1(x_Q - \bar{x})$ and the condition of determination y_Q is then equal to

$$\frac{\sqrt{D(y(x_Q))}}{\hat{y}(x_Q)} = C \tag{6.232}$$

Substitution and rearrangement leads to the expression

$$y_Q = \frac{\hat{\sigma}}{C}\sqrt{1 + \frac{1}{n} + \frac{(x_Q - \bar{x})^2}{\sum\limits_{i=1}^{n}(x_i - \bar{x})^2}} \tag{6.233}$$

In practice, in the chemical laboratory, an approximation is used, as follows.

$$y_Q \approx \frac{\hat{\sigma}}{C}\sqrt{1 + \frac{1}{n} + \frac{\bar{x}^2}{\sum\limits_{i=1}^{n}(x_i - \bar{x})^2}} \tag{6.234}$$

The corresponding concentration x_Q is given by

$$x_Q = \frac{y_Q - \bar{y}}{b_1} + \bar{x} \tag{6.234a}$$

For nonlinear calibration models, Schwartz [53] recommends that the upper L_U and lower L_L limits of the confidence interval of concentration which correspond to different signal levels y^* are determined. Instead of the relative standard deviation of prediction from the calibration model, Schwartz uses the effective relative standard deviation

$$C(x') = \frac{L_U - L_L}{2x' \, t_{1-\alpha/2}(n-2)} \tag{6.235}$$

(4) The *modified determination limit* y'_Q is the value of x' for which $C(x')$ $= C$. This y'_Q limit is found graphically by plotting $C(x')$ against x and substituting in the calibration model. Equation (6.235) may be used for linear models as well as nonlinear ones.

All four definitions may be simply used to calculate the detection limit y_D and the determination limit y_Q for nonlinear calibration models, and for data for which the variance of measurement is not constant (heteroscedasticity) [52]. Generally, it is valid that $y_C < y_D < y_Q$. Ebel and Kamm [54] have described an alternative procedure of determination of the detection limit y_D, and this is illustrated in Fig. 6.70.

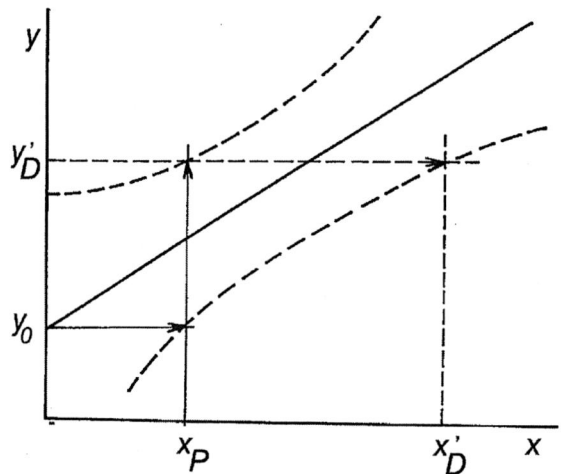

Figure 6.70 Illustration of the procedure for determination of the detection limit y_D according to Ebel and Kamm [54].

Even for this case for linear calibration models it is easy to determine x_P by the use of L_L from Eqs. (6.220) and (6.221). Substitution of x_P into the expression for L_U leads to $L_U = y'_D$

Problem 6.56   *Precision and sensitivity of a photometric calibration model*

For the photometric calibration from Problem 6.54 calculate the critical level y_C, the detection level y_D and the determination limit y_Q for the relative standard deviation $C = 0.1$.

○ **Data:** from Problem 6.54

Solution: The values calculated are $\bar{x} = 21.25$, $\bar{y} = 0.5368$, $\sum_{i=1}^{16}(x_i - 21.25)^2$ $= 2125$, $K = 0.0252$, $\hat{\sigma}^2 = 1.722 \times 10^{-6}$ and $t_{0.975}(14) = 2.14$. The limiting levels of absorbance and the corresponding concentrations are determined to be: y_C = 0.0129 and x_C = 0.504; y_D = 0.0257 and x_D = 1.008; and y_Q = 0.0593 and x_Q = 2.339. To calculate y_D and y_Q, the approximate expressions (6.231) and (6.234) were used. Figure 6.71 shows the dependence of $C(x')$ on x' for the interval $0.1 \le x' \le 1.0$. Here, $x'_Q \approx 0.6$ and $y_Q = \bar{y} + (0.6 - \bar{x})b_1 = 0.0164$, for $C(x') = 0.1$.

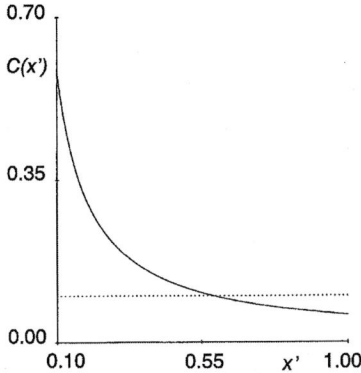

Figure 6.71 Plot of $C(x')$ vs. x' in a search for the determination limit. Dotted line corresponds to $C = 0.1$.

Conclusion: For linear calibration models, all three limits are easy to calculate.

A modified determination limit x'_Q, y'_Q may be determined by use of the expression for variance $D(x)$, Eq. (6.212). Setting $C = 0.1$, the quantity x'_Q is the root of the equation

$$0.01x_Q^2 = D(x')$$
(6.236)

For linear calibration models with $M = 1$, the variance $D(x')$ is defined by Eq. (6.212a). By a simple rearrangement, Eq. (6.236) can be transformed into a quadratic equation with the following root

$$x'_Q = \frac{-\bar{x} + \sqrt{\bar{x}^2 + AD}}{A}$$
(6.237)

where $A = \dfrac{b_1 \sum\limits_{i=1}^{n} (x_i - \bar{x})^2}{100\hat{\sigma}^2} - 1$ and $D = \bar{x}^2 + \dfrac{(n+1)\sum\limits_{i=1}^{n} (x_i - \bar{x})^2}{n}$. When x'_Q is known, the value y'_Q is determined by substitution of the value into the calibration model.

Problem 6.57 *Modified determination limit for the photometric calibration model*

Calculate the modified determination limit for photometric calibration model from Problem 6.54.

○ **Data:** from Problem 6.54

Solution: From Eqs. (6.236) and (6.237) the following numerical values are calculated: $A = 7936.5$, $D = 2709.375$ and $x'_Q = 0.582$, which corresponds to $y'_Q = 0.01596$.

Conclusion: For the linear calibration model the modified determination limit is readily calculated from Eq. (6.237).

6.8 Procedure for Linear Regression Analysis

The procedure for examination and construction of a linear regression model consists of following steps [71, 72].

(1) *Proposal of a model*

The procedure should always start from the simplest model, with individual independent controllable variables not raised to powers other than the first, and with no interaction terms of the type $x_j x_k$ included. Only in cases when it is known that the model contains functions of the controllable variables is an exception made.

(2) *Exploratory data analysis in regression*

The scatter of individual variables and all possible pair combinations are examined. The scatter plots of x_j vs. x_k or the index plots x_j vs. j are often used here. In this step of a regression analysis, the significance of individual variables with reference to scatter and the presence of multicollinearity is examined. An approximately linear relationship between variables in scatter plots of x_j vs. x_k indicates multicollinearity. Also, in this step the influential points causing multicollinearity are detected.

(3) *Parameter estimation*

The parameters of the proposed regression model and the corresponding basic statistical characteristics of this model are determined by the classical least-squares method (LS). Individual parameters are tested for significance by using the Student t-test, the determination coefficient $\hat{R}^2$ and the predicted determination coefficient $\hat{R}_p^2$. Other statistical characteristics calculated are the total F-test, the model significance test, the model complexity test, the mean quadratic error of prediction MEP and the Akaike information criterion AIC, to examine the linearity of model.

(4) *Analysis of regression diagnostics*

The statistical analysis of classical residuals leads to estimates of residual variance $\hat{\sigma}^2(\hat{e})$, residual standard deviation $s(\hat{e})$, residual skewness $\hat{g}_1(\hat{e})$, residual kurtosis $\hat{g}_2(\hat{e})$, the Pearson χ^2-test of residual normality and the Jarque-Berra normality test. Different diagnostic graphs are used to examine the regression diagnostics for identification of influential points, and to test the conditions for the least-squares method, namely homoscedasticity, absence of autocorrelation, and normality of error distribution. If influential points are found, it has to be decided whether these points should be eliminated from data. If points are eliminated, the whole data treatment must be repeated. When there are several controllable variables, the significance of each variable and its function is examined by partial-regression graphs and by the partial-residual graph.

(5) *Construction of a more accurate model*

According to the test for fulfilment of the conditions for the least-squares method, and the result of regression diagnostics, a more accurate regression model is constructed as follows.

(a) When heteroscedasticity is found in the data, the weighted least-squares method (WLS) is used.

(b) When autocorrelation is found in the data, the generalized least-squares method (GLS) is used.

(c) When some restrictions apply to the parameters, the conditioned least-squares method (CLS) is used.

(d) When multicollinearity is found in the data, the method of rational ranks (MRV) is used.

(e) When all variables are subject to random errors, the extended least-squares method (ELS) is used.

(f) When the data have an error distribution other than normal, or the data contain outliers or high leverage points, some robust methods are used.

(6) *Evaluation of the quality of the model proposed*

With the use of classical tests, regression diagnostics and some supplementary information about the "model + data + method", the quality of the proposed linear regression model is evaluated.

(7) *Analysis of calibration models*

For a calibration model proposed for the given signal value y^*, the quality of the independent variable x^* together with its confidence interval is estimated. Before application of the calibration model, the detection limit and the determination limit should be estimated. These limits determine the allowable lower limit of the calibration model.

(8) *Statistical hypothesis testing*

In some cases, to compare several straight lines, statistical hypothesis testing is performed.

6.9 References

[1] N. R. Draper and H. Smith, *Applied Regression Analysis*, 2nd Ed., Wiley, New York, 1981.

[2] G. A. F. Seber, *Linear Regression Analysis,* Wiley, New York, 1977.

[3] I. Guttman, *Linear Models—An Introduction,* Wiley, New York, 1982.

[4] S. R. Searle, *Linear Models,* Wiley, New York, 1971.

[5] F. J. Anscombe, *Am. Statist.,* 1973, 27, 17.

[6] J. Utts, *Commun. Statist.,* 1982, 11, 2801.

[7] W. Kramer and H. Sonnberger, *The Linear Regression Model under Tests,* Physica Verlag, Heidelberg, 1986.

[8] J. R. Scott, *Appl. Statist.,* 1975, 24, 42.

[9] J. Cassela, *Am. Statist.,* 1983, 37, 147.

[10] R. Suich, G. C. Derringer, *Technometrics,* 1977, 19, 213.

[11] A. N. Kornilov and L. B. Smenina, *Zh. Fiz. Khim.,* 1970, 44, 1932.

[12] J. Militky, *Proc. Conf European Simulation Conference 87,* Prague, September, 1987.

[13] G. R. Phillip, J. M. Harris and E. M. Eyring, *Anal. Chem.,* 1982, 54, 2053.

[14] J. W. Neil and D. E. Johnson, *Commun. Stat.,* 1984, 13, 485.

[15] J. R. Green and D. Margerison, *Statistical Treatment of Experimental Data,* Elsevier, Amsterdam, 1978.

[16] J. C. Nash, *Compact Numerical Algorithms for Computer,* A. Hilger, Bristol, 1979.

[17] A. M. Antila and M. L. Sikvonen, *Z. Anal. Chem.,* 1987, 327, 799.

[18] Ch. Lawson and R. Hanson, *Solving Least-Squares Problems,* Englewood Cliffs, New Jersey, 1974.

[19] A. G. Dahlquist and A. Bjorck, *Numerical Methods,* Prentice-Hall, Englewood Cliffs, 1974.

[20] D. M. Marquardt, *Technometrics,* 1970, 12, 591.

[21] D. A. Belsey, E. Kuh and R. E. Welsch, *Regression Diagnostics,* Wiley, New York, 1980.

[22] A. C. Atkinson, *Plots, Transformations and Regression,* Clarendon Press, Oxford, 1985.

[23] S. Weisberg, *Technometrics,* 1983, 25, 219.

[24] R. D. Cook and S. Weisberg, *Residuals and Influence in Regression,* Chapman and Hall, New York, 1982.

[25] B. Joiner, *Am. Statist.,* 1981, 35, 227.

[26] S. Chatterjee and A. S. Hadi, *Statist. Sci.,* 1986, 1, 379.

[27] M. A. O'Gorman and R. M. Myers, *Commun. Statist.,* 1987, 16, 771.

[28] J. B. Gray, *Proc. Statist. Comput. Sect.,* p. 159, ASA, Washington 1983.

[29] C. L. Mallows, *Technometrics,* 1986, 28, 313.

[30] R. D. Cook and S. Weisberg, *Biometrika,* 1983, 70, 1.

[31] N. Querry, *Technometrics,* 1964, 6, 225.

[32] C. M. Jarque and A. K. Bera, *Int. Stat. Rev.,* 1987, 55, 163.

[33] G. G. Judge and M. E. Bock, *Statistical Implications of Pre-test and Stein Rule Estimators in Econometrics,* North Holland, Amsterdam, 1978.

[34] J. J. Leary and E. B. Messick, *Anal. Chem.,* 1985, 57, 956.

[35] S. D. Horn, R. A. Horn and D. B. Duncan, *J. Am. Statist. Assoc,* 1975, 70, 380.

[36] W. A. Fuller, *Measurement Error Models,* Wiley, New York, 1987.

[37] R. W. Hill and D. W. Holland, *J. Am. Statist. Assoc,* 1977, 72, 828.

[38] D. J. Huber, *Robust Statistics,* Wiley, New York, 1981.

[39] Li G., in Hoaglin D. C. *et al,* eds., *Exploring Data Tables, Trends and Shapes,* Wiley, New York, 1985.

[40] G. R. Phillip and E. M. Eyring, *Anal. Chem.,* 1983, 55, 1134.

[41] Yu. G. Nikulichev, Yu. A. Kanchenko, A. E. Mysak and I. F. Kobilinskaya, *Kolloidn. Zh.,* 1988, 50, 473.

[42] J. Mala and I. Slama, *Chem. Pap.,* 1988, 42, 319.

[43] W. S. Krasker and R. E. Welsch, *J. Am. Statist. Assoc,* 1982, 77, 595.

[44] T. P. Hettmansperger, *Aust. J. Statist.,* 1987, 29, 1.

[45] P. J. Rousseeuw and A. M. Leroy, *Robust Regression and Outliers Detection,* Wiley, New York, 1987.

[46] J. R. Rosenblatt and C. H. Spiegelman, *Technometrics*, 1981, 23, 329.

[47] S. Ebel and U. Becht, Z. *Anal. Chem.,* 1987, 327, 157.

[48] L. M. Schwartz, *Anal. Chem.,* 1976, 48, 2287.

[49] L. J. Naszodi, *Technometrics,* 1978, 20, 201.

[50] R. G. Krutchkoff, *Technometrics,* 1967, 9, 425.

[51] L. M. Schwartz, *Anal. Chem.,* 1977, 49, 2062.

[52] L. Oppenhelmer, T. P. Capizzi, R. M. Weppelman and H. Mehta, *Anal. Chem.,* 1983, 55, 638.

[53] L. M. Schwartz, *Anal. Chem.,* 1983, 55, 1424.

[54] S. Ebel and U. Kamm, Z. *Anal. Chem.,* 1984, 318, 293.

[55] S. Ebel and R. Brockmeyer, Z. *Anal. Chem.,* 1970, 326, 770.

[56] D. Himmelblau, *Process Analysis by Statistical Methods,* Wiley, New York, 1969.

[57] Swed F., Eisenhart C.: Annal of Math. Statist. 14, 66 (1943).

[58] Liteanu C., Rica I.: Statistical Theory and Methodology of Trace Analysis. Ellis Horwood, Chichester 1980.

[59] Davídek J. a kol.: Laboratorní příručka analýzy potravin. SNTL, Praha 1981.

[60] Kraft G., Dosch H.: Z. Anal. Chem. 271, 264 (1974).

[61] Truxová I.: Diplomová práce, VŠCHT Pardubice 1991.

[62] Rice J. A.: Mathematical Statistics and Data Analysis. Wadsworth & Brooks, California 1988.

[63] Cyhelský L. a kol.: Úlohy k základům statistiky. SNTL, Praha 1988.

[64] Potocký R. a kol.: Zbierka úloh z pravdepodobnosti a matematickej štatistiky. ALFA, Bratislava 1986.

[65] Kleinbaum D. G. a kol.: Applied Regression Analysis and Other Multivariate Methods. PWS-KENT Publishing Comp., Boston, 1988.

[66] Ebel S., G. Herold: Z. Anal. Chem. 270, 20 (1974).

[67] Anderson R. L.: Practical Statistics for Analytical Chemists. van Nostrand Reinhold Company, New York 1987.

[68] Gottschalk G.: Z. Anal. Chem. 282, 1 (1976).

[69] Miller J. C., Miller J. N.: Statistics for Analytical Chemistry. Ellis Horwood, Chichester, 1984.

[70] Graybill F. A., Iyer H. K.: Regression Analysis: Concepts and Applications. Duxbury Press, International Thomson Publishing 1994.

[71] Neter J., Kutner M. H., Nachtsheim CH.J., Wasserman W.: Applied Linear Statistical Models. RICHARD D. IRWIN, Chicago 1990.

[72] Meloun M., Militký J.: Statistické zpracování experimentálních dat, Plus Praha 1994 (1. vydání), EAST PUBLISHING, Praha 1998 (2. vydání), Academia Praha 2004 (3. vydání).

[73] Beyer, W., editor (1981), CRC Standard Mathematical Tables, 25th edition, CRC Press, Boca Raton, Florida, str. 144 a str. 435. Lord, John (1995), Sizes: The Illustrated Encyclopedia, Harper Collins, New York, New York, str. 117.

Supplemented material (Review Questions, Exercises, Results of Exercises) to Chapter 6 is on enclosed CD.

Chapter 4 considers the characteristics and procedures of multivariate data analysis, and Chapter 6 describes the construction of linear regression models. In this chapter we describe relationships expressing dependencies among the components $\xi_1, ..., \xi_m$ of an m-dimensional random vector ξ by using regression. The difference from construction of linear regression models in Chapter 6 is that here the data form a random sample from the m-dimensional distribution of ξ. There is no consideration about which component ξ_j of the vector ξ is the response (in the linear model, the dependent variable) and which components of vector ξ are controllable (in the linear model, the independent variable).

The random sample $\{x_i\}$, $i = 1, ..., n$ of size n represents an $(n \times m)$ array of data

$$
\begin{bmatrix}
x_{11} & x_{12} & \cdots & x_{1m} \\
x_{21} & x_{22} & \cdots & x_{2m} \\
\cdots & \cdots & \cdots & \cdots \\
x_{n1} & x_{n2} & \cdots & x_{nm}
\end{bmatrix}
$$

where the number of rows n (i.e., the number of m-dimensional "points" x_i) is larger than the number of columns m (i.e., the number of "variables" or components of vector x). The characteristic fact is that all components of the data vector are random (measured) and not controllable by the experimenter.

In the regression models of Chapter 6, some independent variables such as temperature, concentration, etc. are also measured (and therefore random) variables, but the experimenter could adjust and control their magnitude.

Although in statistical practice, correlation problems do not often require detailed analysis, we find that problems such as (a) comparison of various testing methods on different samples or (b) searching for relationships among various properties or characteristics of compounds, are more problems of correlation than regression.

7.1 Correlation Models

As for univariate random variables, the components of random vectors can be characterized by use of means and variances. A *measure of intensity* of the dependence between components ξ_i and ξ_j, $i \neq j$ is given by the second central mixed moment $\text{cov}(\xi_i, \xi_j)$, denoted as the *covariance*. The *standardized covariance* or *correlation coefficient* $\rho(\xi_i, \xi_j)$ is more useful.

Covariance and correlation coefficients and methods for their estimation are described in Chapter 4. Here we use the covariance matrix $\mathbf{C}$ with elements formed by individual covariances or the correlation matrix $\mathbf{R}$ with elements formed by individual correlation coefficients. The covariance matrix $\mathbf{C}$ has the variances on the diagonal while the correlation matrix has ones.

A random vector is characterized by the vector of mean values $\mathbf{u} = (u_1, ..., u_m)^T$ where $u_j = E(\xi_j)$, and by the covariance matrix $\mathbf{C}$. This information is generally not sufficient. Analogously to the mean values $E(\xi_j)$, conditional means or conditional variances can also be defined. We will define these characteristics for a case of two random quantities ξ_1 and ξ_2, and then for the general random vector ξ.

7.1.1 Correlation Models for two Random Variables

For two random variables ξ_1 and ξ_2 the conditional means are given by

$$E(\xi_1 / x_2) = \int_{-\infty}^{+\infty} x_1 f(x_1 / x_2) dx_1 \tag{7.1a}$$

$$E(\xi_2 / x_1) = \int_{-\infty}^{+\infty} x_2 f(x_2 / x_1) dx_2 \tag{7.1b}$$

where $f(x_2/x_1)$ and $f(x_1/x_2)$ are the conditional probability densities (cf. Chapter 4). From Eq. (7.1a,b) it is evident that the conditional mean value $E(\xi_1 / x_2)$ is in fact a mean value of the random variable ξ_1, with condition that the random variable ξ_2 lies in the infinitely small interval around the value x_2. The conditional mean value $E(\xi_2 / x_1)$ is defined similarly. Because they are conditioned by a random variable, the conditional mean values are *random variables* which may be characterized by the means and variances. The means of the conditional mean values do not provide any new information because $E(E(\xi_1 / x_2)) = E(\xi_1)$ and $E(E(\xi_2 / x_1)) = E(\xi_2)$.

By introducing the variances of the conditional mean values $D(E(\xi_1 / x_2))$ and $D(E(\xi_2 / x_1))$ the total variances $D(\xi_1)$ and $D(\xi_2)$ may be decomposed into the components

$$D(\xi_1) = E(D(\xi_1 / x_2)) + D(E(\xi_1 / x_2)) \qquad (7.2a)$$

$$D(\xi_2) = E(D(\xi_2 / x_1)) + D(E(\xi_2 / x_1)) \qquad (7.2b)$$

The first terms on the right hand sides represent the mean values of the conditional variances which may be defined in a similar way to the conditional mean values (Eq. 7.1), with the use of conditional probability densities.

The conditional mean values have the same properties as the unconditional.

For the conditional mean value $E(\xi_2 / x_1)$ it is also true that

(1) For any x_1 of random variable ξ_1, the values $E(\xi_2 / x_1)$ exist if $E(\xi_2)$ $<\infty$.

(2) If a random variable ξ_1 does not depend on the random variable ξ_2, the conditional mean value is independent of the condition and $E(\xi_2 / x_1) = E(\xi_2)$.

(3) If $\xi_2 = g(\xi_1)$ where $g(.)$ is a function notation, then $E(\xi_2 / x_1) = g(x_1)$.

(4) The conditional mean value is a not symmetric function of the arguments, so that $E(\xi_2 / x_1) \neq E(\xi_1 / x_2)$.

Property (3) shows that the conditional mean value is a function of quantity x_1 of condition ξ_1 and therefore it is denoted as the regression of variable ξ_2 on the variable ξ_1.

Generally, two types of regression are distinguished [1]:

(1) The *theoretical regression* is a conditional mean value derived from knowledge of a conditional probability density $f(x_2/x_1)$ or the knowledge of a joint probability density $f(x_1, x_2)$ and both marginal densities $f(x_1)$ and $f(x_2)$. It is valid that for all elliptic conditional distributions, including the normal one, the theoretical regression is a linear one [2]. For some conditional distributions, however, the theoretical regression may be nonlinear [3].

(2) The empirical regression is any conveniently selected function $g(x_1)$ which approximates the behaviour of the conditional mean value $E(\xi_2 / x_1)$. To find the function $g(.)$ and the parameters estimates, the methods in Chapter 4 may be used.

We will now deal with the theoretical regression when all the components of vector ξ have the normal distribution, and also the joint distribution of vector ξ is normal.

Problem 7.1   *Deriving a theoretical regression for the normalized normally distributed random variables*

Let us assume that the random variables ξ_1 and ξ_2 have the normalized normal distribution $N(0, 1)$ with zero mean and variance equal to one. Let the joint distribution of the variables is also normal with the probability density function

$$f(x_1, x_2) = \frac{1}{2\pi\sqrt{1-\rho^2}} \exp\left[-\frac{x_1^2 - 2\rho x_1 x_2 + x_2^2}{2(1-\rho^2)}\right]$$

where $\rho = \rho(\xi_1, \xi_2)$ is the correlation coefficient between the random variables ξ_1 and ξ_2. Derive the theoretical regression $E(\xi_2 / x_1)$.

Solution: In the first step, the conditional probability density $f(x_2/x_1)$ should be calculated $f(x_2 / x_1) = \dfrac{f(x_1, x_2)}{f(x_1)}$. On substitution and rearrangement, we get

$$f(x_2 / x_1) = \frac{1}{\sqrt{2\pi(1-\rho^2)}} \exp\left[-\frac{(x_2 - \rho x_1)^2}{2(1-\rho^2)}\right] \qquad (7.3)$$

By substituting into Eq. (7.1) and analytical integration we get

$$E(\xi_2 / x_1) = \int_{-\infty}^{+\infty} \frac{x_2}{\sqrt{2\pi(1-\rho^2)}} \exp\left[-\frac{(x_2 - \rho x_1)^2}{2(1-\rho^2)}\right] dx_2 = \rho\, x_1 \qquad (7.4)$$

Conclusion: For this case of random variables, the theoretical regression is linear with zero intercept, and the slope corresponds to the correlation coefficient ρ.

The conditional variances are the characteristics of variability of conditional distributions. The conditional variance $D(\xi_2 / x_1)$ expresses the variability of the random variable ξ_2 around the theoretical regression $E(\xi_2 / x_1)$, on condition that ξ_1 has a realization x_1 where function x_1 is called the *scedastic function*. If $D(\xi_2 / x_1)$ is a constant independent of ξ_1 (or x_1) it is a *heteroscedastic function*. The homoscedastic and heteroscedastic functions are illustrated in Fig. 7.1.

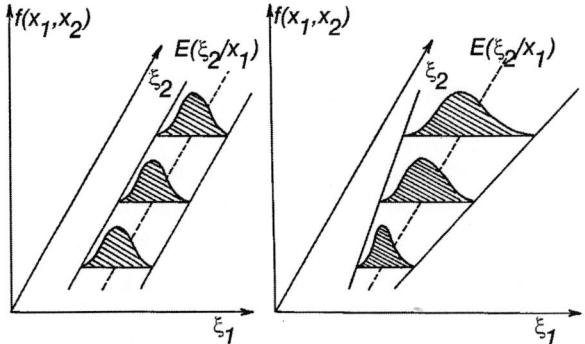

Figure 7.1 (a) Homoscedastic and (b) heteroscedastic relationship between two linearly dependent random variables

For independent random variables the following expressions are valid: $E(\xi_2 / x_1) = E(\xi_2)$ and $D(\xi_2 / x_1) = D(\xi_2)$. The theoretical regression $E(\xi_2 / x_1)$ and $E(\xi_1 / x_2)$ then represents two mutually perpendicular straight lines, parallel with the co-ordinates when the scedastic functions are constant. For dependent random quantities, either the conditional mean value (Fig. 7.1) or conditional variance, or both, is/are non-constant.

Problem 7.2   *Determination of a scedastic function*

Determine the scedastic function $D(\xi_2 / x_1)$ for the variables ξ_1 and ξ_2 defined in Problem 7.1.

Solution: From the definition of variance we can write

$$D(\xi_2 / x_1) = \int_{-\infty}^{+\infty} (x_2 - E(\xi_2 / x_1))^2 f(x_2 / x_1) dx_2.$$ After substitution from Eq.(7.3)

and analytical integration, we find

$$D(\xi_2 / x_1) = (1 - \rho^2) \qquad (7.5)$$

Conclusion: For normalized normally distributed variables, when their joint distribution is also normal the scedastic function is constant.

Conditional variances are also random variables (dependent on condition) which may characterized by the mean values and variances. In the linear regression of ξ_2 on ξ_1 the mean conditional variance is

$$E(D(\xi_2 / x_1)) = D(\xi_2)\left[1 - \rho^2\right] \qquad (7.6)$$

and the variance of conditional mean value is

$$D(E(\xi_2 / x_1)) = D(\xi_2)\rho^2 \qquad (7.7)$$

For homoscedastic functions, the mean conditional variance is equal to the conditional variance which does not depend on the conditional values. Conditional variances have all the properties of unconditional variances.

The mean conditional variances generally characterize a stochastic dependence between random variables which can be nonlinear.

If $E(D(\xi_2 / x_1)) = D(\xi_2)$, ξ_1 and ξ_2 are independent.

If $E(D(\xi_2 / x_1)) \leq D(\xi_2)$, there is a stochastic relationship between the variables.

From Eq. (7.2b) and the definition of regression it follows that the variance of the conditional mean is that part of the total variance concerned with "the theoretical regression" caused by the influence of variable ξ_1 on the variability of variable ξ_2. The mean value of the conditional variance expresses the influence of all other (not considered) variables which cause variability in output variable ξ_2.

For a measure of regression quality, we use the ratio R_R^2 which is defined as $R_R^2 = \dfrac{D(E(\xi_2 / x_1))}{D(\xi_2)}$ and which determines the part of the variability of random variable ξ_2 explained by theoretical regression. From Eq. (7.7) it follows that R_R^2 for linear regression is equal to the square of correlation coefficient or to the determination coefficient.

Let us mention the theoretical regression and conditional variances for two random variables, ξ_1 with distribution $N(\mu_1, \sigma_1^2)$ and ξ_2 with distribution $N(\mu_2, \sigma_2^2)$ when their joint distribution is also normal. On the basis of Problem 7.1, the theoretical regression $E(\xi_2 / x_1)$ may be expressed in the form

$$E(\xi_2 / x_1) = \mu_2 + \rho \frac{\sigma_2}{\sigma_1}(x_1 - \mu_1) \qquad (7.8)$$

This is a straight line with slope $b_1 = \rho\sigma_2 / \sigma_1$ and intercept $b_2 = \mu_2 - b_1\mu_1$ which passes through the centre of gravity of co-ordinates $[\mu_1, \mu_2]$.

Similarly, on the basis of the results of Problem 7.2, the conditional variance may be determined as

$$D(\xi_2 / x_1) = \sigma_2^2(1 - \rho^2) \tag{7.9}$$

From the definition it follows that

$$D(\xi_2 / x_1) = E\left[\xi_2 - E(\xi_2 / x_1)\right]^2 \tag{7.10}$$

and the conditional variance is equal to the mean value of the square of deviations of random quantity ξ_2. In a similar way, the linear expression for the regression $E(\xi_1 / x_2)$ may be found

$$E(\xi_1 / x_2) = \mu_1 + \rho\frac{\sigma_1}{\sigma_2}(x_2 - \mu_2) \tag{7.11}$$

The conditional variance

$$D(\xi_1 / x_2) = \sigma_1^2(1 - \rho^2) \tag{7.12}$$

It is obvious that both theoretical regressions go through the same point. The product of their slopes is equal to the square of the correlation coefficient. If the correlation coefficient $\rho = 1$, both slopes of theoretical regressions will be equal to one and both regressions will be identical. If $\rho = 0$, both slopes will be equal to zero and the regressions will be parallel with the axis of the co-ordinate system and will have an angle of 90°. The angle φ between the theoretical regressions gives a measure of the linear dependence between the random quantities ξ_1 and ξ_2. For this angle

$$\tan\varphi = \frac{\sigma_1\sigma_2(1 - \rho^2)}{\rho(\sigma_1^2 + \sigma_2^2)} \tag{7.13}$$

Figure 7.2 shows the relationship between the two theoretical regressions.

In cases when the correlation coefficient is not equal to zero or one, there exist two different regressions. Often it is possible to determine which variable is a response and which is the controllable one, and according to that, to select a suitable type of regression. When it is not possible to determine the type of variables, to determine the linear relationship between ξ_1 and ξ_2 we can use:

(1) the *principal axis* which corresponds to the minimum of the squares of the perpendicular distances of points from the regression straight line.

(2) the *reduced principal axis* of the ellipses of constant densities, corresponding to the minimum of the products of deviations in the two variables. The slope of the corresponding reduced principal axis is directly equal to the variance ratio $d = \sigma_2 / \sigma_1$ and the regression goes through the centre of gravity $[\mu_1, \mu_2]$.

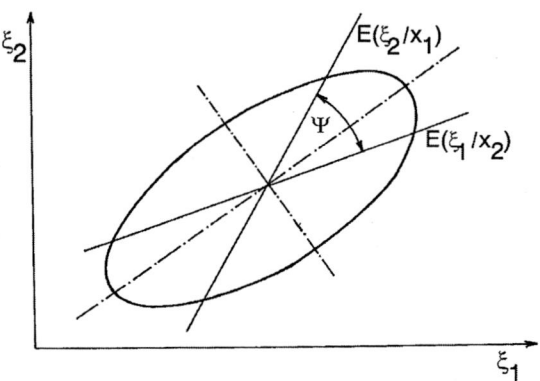

Figure 7.2 The relationship between theoretical regressions

These expressions may be used for practical purposes. The means μ_1 and μ_2 may be replaced by the arithmetic averages $\bar{x}_1$ and $\bar{x}_2$ the variances σ_1^2 and σ_2^2 by the sample variance estimates s_1^2 and s_2^2, and the correlation coefficient ρ by the sample estimate of the correlation coefficient

$$R = \frac{\sum_{i=1}^{n}(x_{1i} - \bar{x}_1)(x_{2i} - \bar{x}_2)}{\sqrt{\sum_{i=1}^{n}(x_{1i} - \bar{x}_1)^2 \ \sum_{i=1}^{n}(x_{2i} - \bar{x}_2)^2}} \qquad (7.14)$$

The slope b_1 and the intercept b_2 of the regression $E(\xi_2 / x_1)$ correspond to estimates found by the least-squares method, and $D(\xi_2 / x_1)$ corresponds to the residual sum of squares.

Problem 7.3 *Influence of solvent type on the degree of polymerization of cotton*

For 17 differently degraded samples of cotton the relative viscosity was determined in (a) a solution of the ethylenediamine complex of copper

(CUEN), and (b) an alkaline solution of copper tetra-ammine hydroxide (CUOXAN). From viscosity values, the degree of polymerization values DP was calculated. Examine the relationship between the degree of polymerization DP_1 in solution CUEN and DP_2 in solution CUOXAN. Estimate the regression straight line corresponding to the reduced principal axis for the data.

Data: $n = 17$

$DP_1 \times 1000$	5.913	1.837	5.732	3.792	5.823	2.837	4.382	2.169	2.218
$DP_2 \times 1000$	2.341	1.740	2.863	1.648	2.608	1.391	3.269	1.694	1.726

$DP_1 \times 1000$	1.813	1.642	1.629	3.680	3.601	3.617	1.317	1.286
$DP_2 \times 1000$	1.354	1.451	1.295	4.110	2.300	4.065	1.480	1.096

Solution: From the data, the following statistics are calculated: For $DP_1 \times 1000$: $\bar{x}_1 = 3.1346, s_1 = 1.5995$ and for $DP_2 \times 1000$: $\bar{x}_2 = 2.143, s_2 = 0.9465$ and the correlation coefficient $R = 0.6142$. For regression of DP_2 on DP_1,

$$DP_2 = 2.143 + 0.6142\frac{0.9465}{1.5999}(DP_1 - 3.1346) = 1.0037 + 0.3634DP_1.$$

The residual variance corresponding to this regression is equal to $\hat{\sigma}^2 \approx D(\xi_2 / x_1) \approx s_2^2\left[1 - R^2\right] = 0.558$. For straight line corresponding to the reduced principal axis: $DP_2 = \bar{x}_2 + \frac{s_2}{s_1}(DP_1 - \bar{x}_1) = 0.5917DP_1 + 0.288$.

Figure 7.3 shows the regression straight line $E(\xi_2 / x_1)$ and the straight line corresponding to the reduced principal axis (RPA) fitted through experimental data.

Conclusion: If the basic statistical characteristics of sample are known, the regression coefficients may be estimated. To test for a theoretical regression, the normality of DP_1, DP_2 and also the joint distribution of these quantities should be examined. The straight line RPA expresses a relationship between components DP_1 and DP_2 of a two-dimensional random vector. The calculation of RPA is quite simple.

7.1.2 The Correlation Model for Many Random Variables

In the case of a multi-dimensional random vector ξ, we can define different conditional values according to the number of variables we include in a condition. Let us limit ourselves to a case where the component ξ_1 is a response and the other components $\xi_2, \ldots, \xi_m$ forming the vector ξ^* are explanatory variables. The regression ξ_1 on ξ^* is then defined as the conditional mean value $E(\xi_1/x^*)$ where

$\mathbf{x}^*$ is the actual realization of vector ξ^*. Also, we again suppose that vector ξ and all its components have the normal distribution.

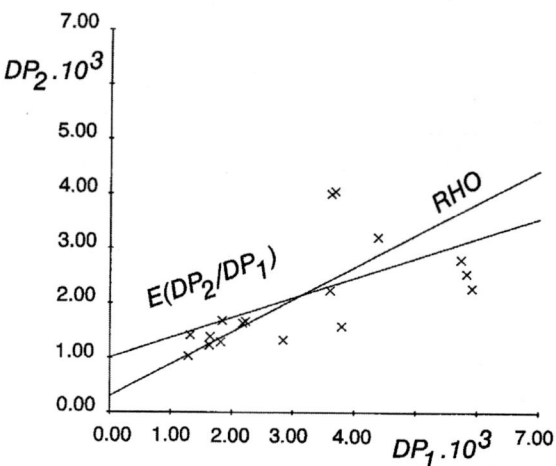

Figure 7.3   The straight line corresponding to the reduced principal axis (RPA) and to the regression $E(\xi_2/x_1)$.

Let us separate the vector of mean values μ (and the covariance matrix $\mathbf{C}$) into components

$$\mu = \begin{bmatrix} \mu_1 \\ \mu^* \end{bmatrix}, \quad \mathbf{C} = \begin{bmatrix} \sigma_1^2 & \mathbf{c}_1^T \\ \mathbf{c}_1 & \mathbf{C}^* \end{bmatrix} \text{ where } \mathbf{c}_1^T = \left[\text{cov}(\xi_1, \xi_2), ..., \text{cov}(\xi_1, \xi_m) \right] \text{ is the}$$

vector containing the covariance between the response variable ξ_1 and the explanatory variables ξ^*. The symbol $\mathbf{C}^*$ represents the covariance matrix of the explanatory variables.

As for the case of two variables, the conditioned probability density $f(x_1/\mathbf{x}^*)$ may be determined, and this has a normal distribution, so the conditional mean value is

$$E(\xi_1 / \mathbf{x}^*) = \mu_1 + \mathbf{c}_1^T \mathbf{C}^{*-1}(\mathbf{x}^* - \mu^*) \tag{7.15}$$

Let us introduce the vector $\mathbf{a} = (a_1, ..., a_{m-1})^T$ expressed as

$$\mathbf{a} = \mathbf{C}^{*-1}\mathbf{c}_1 \tag{7.16}$$

Then Eq. (7.15) represents the linear function of variables $\mathbf{x}^*$ in the form

$$E(\xi_1 / \mathbf{x}^*) = \mu_1 + \mathbf{a}^T(\mathbf{x}^* - \boldsymbol{\mu}^*) = \mu_1 + \sum_{i=1}^{m-1} a_i(x_{i+1} - \mu_{i+1}) \qquad (7.17)$$

If the joint distribution of the random vector is normal and the distribution of all ξ_j components is also normal, the resulting theoretical regression is linear.

The vector of regression coefficients $\mathbf{a}$ is estimated here as a solution of the set of $(m-1)$ linear equations

$$\mathbf{a}C^* = \mathbf{c}_1 \qquad (7.18)$$

where C^* and $\mathbf{c}_1$ contain individual covariances. The corresponding conditional variance is given by the relation

$$D(\xi_1 / \mathbf{x}^*) = \sigma_1^2 - \mathbf{c}_1^T \mathbf{C}^{*-1} \mathbf{c}_1 = \sigma_1^2 - \sum_{i=1}^{m} c_i^1 C_{ij}^{22} c_j^1 \qquad (7.19)$$

where c_i^1, c_j^1 are elements of vector $\mathbf{c}_1$, and C_{ij}^{22} are elements of matrix $\mathbf{C}^{*-1}$. If all components of the random vector ξ^* are mutually independent, matrix C is diagonal with variances σ_j^2 on the diagonal. For individual regression coefficients, then

$$a_{j-1} = \frac{\mathrm{cov}(\xi_1, \xi_j)}{\sigma_j^2} = \rho(\xi_1, \xi_j) \frac{\sigma_1}{\sigma_j}, \quad j = 2, \ldots, m \qquad (7.20)$$

Similarly, the expression for the conditional variance will simplify to

$$D(\xi_1 / \mathbf{x}^*) = \sigma_1^2 - \sum_{j=2}^{m} \rho^2(\xi_1, \xi_j)\sigma_1^2 = \sigma_1^2 \left[1 - R_{1(2,\ldots,m)}^2\right] \qquad (7.21)$$

where $R_1(2, \ldots m)$ is the multiple correlation coefficient between ξ_1 and the vector ξ^*. For this correlation coefficient

$$R_{1(2,\ldots,m)} = \sqrt{\frac{D(E(\xi_1 \mathbf{x}^*))}{\sigma_1^2}} = \sqrt{1 - \frac{\det(\mathbf{R})}{\det(\mathbf{R}_{11})}} \qquad (7.22)$$

where $\mathbf{R}_{ij}$ is the matrix formed by leaving out the ith row and the jth column of the correlation matrix $\mathbf{R}$.

Basic properties of the multiple correlation coefficient are

(1) $0 \leq R_{1(2,\ldots,m)} \leq 1$;

(2) if $R_{1(2,\ldots,m)} = 1$, the random quantity ξ_1 is exactly a linear combination of quantities $\xi_2, \ldots, \xi_m$;

(3) if $R_{1(2,\ldots,m)} = 0$, all pairwise correlation coefficients $\rho(\xi_1, \xi_2) = 0$, $j = 2, \ldots, m$;

(4) for the case of a single explanatory variable, the multiple correlation coefficient is identical with the absolute value of the paired correlation coefficient, $R_{1(2)} = |\rho(\xi_1, \xi_2)|$;

(5) as the number of explanatory variables increases, the multiple correlation coefficient never decreases: $R_{1(2)}^2 \leq R_{1(2,3)}^2 \leq R_{1(2,2,3,4)}^2 \leq \ldots \leq R_{1(2,\ldots,m)}^2$

Problem 7.4 *Multiple correlation coefficient for two explanatory variables*

Estimate the multiple correlation coefficient $R_{1(2,3)}$ between a variable ξ_1 and two variables ξ_2 and ξ_3.

Solution: For correlation matrices $\mathbf{R}$ and $\mathbf{R}_{11}$ we can write $\mathbf{R} = \begin{bmatrix} 1 & R_{12} & R_{13} \\ R_{12} & 1 & R_{23} \\ R_{13} & R_{23} & 1 \end{bmatrix}$

and $\mathbf{R}_{11} = \begin{bmatrix} 1 & R_{23} \\ R_{23} & 1 \end{bmatrix}$. In these expressions, the symmetry $R_{ij} = R_{ji}$ of paired correlation coefficients is used. After substitution into Eq. (7.22) and some rearrangement we get

$$R_{1(2,3)} = \sqrt{\frac{R_{12}^2 + R_{13}^2 - 2R_{12} R_{13} R_{23}}{1 - R_{23}^2}} \tag{7.23}$$

Equation (7.23) shows that paired correlation coefficients can not reach any value in the range $-1 \leq R_{ij} \leq 1$, but they are mutually bounded by the condition $R_{1(2,3)} \leq 1$.

If $R_{23} = 0$, the explanatory variables are mutually uncorrelated, and

$$R_{1(2,3)}^2 = R_{12}^2 = R_{13}^2 \tag{7.24a}$$

Conclusion: The multiple correlation coefficient may be estimated as the function of paired correlation coefficients. When the explanatory variables

$\xi_2,...,\xi_m$ are mutually uncorrelated, the square of the multiple correlation coefficient is equal to the sum of squares of paired correlation coefficients.

In some cases the centred random variables or normalized random variables are used. For centred random variables $\xi_{c,j} = \xi_j - \mu_j$, $j = 1,...,m$ and for normalized random variables $\xi_{N,j} = \dfrac{\xi_j - \mu_j}{\sigma_j}$, $j = 1,...,m$. The regression defined by Eq. (7.17) may be expressed with the use of centred random variables in the form

$$E(\xi_1 / \mathbf{x}_c^*) = \mathbf{c}_1^T \mathbf{C}^{*-1} \mathbf{x}_c^* = \sum_{i=1}^{m-1} a_i x_{c,i+1} \qquad (7.24b)$$

It can be seen that centring does not change the estimates of the regression coefficients, but the intercept term is equal to zero. With normalized random variables, the regression $E(\xi_1 / \mathbf{x}_c^*)$ takes the form

$$E(\xi_{N1} / \mathbf{x}_N^*) = \mathbf{R}^T \mathbf{R}^{*-1} \mathbf{x}_N^* = \frac{\mathbf{a}^T \mathbf{D} \mathbf{x}_c^*}{\sigma_1} = \mathbf{b}^T \mathbf{x}_N^* \qquad (7.25)$$

where $\mathbf{R}$ is a vector of size $((m-1) \times 1)$ containing paired correlation coefficients $\rho(\xi_1, \xi_j)$, $j = 2, ..., m$, and $\mathbf{R}^*$ is the correlation matrix of the vector of explanatory variables of size $(m-1) \times (m-1)$, $\mathbf{D}$ denotes a diagonal transformation matrix with elements σ_j, $j = 2,... m$, on the main diagonal. The coefficients $\mathbf{b}_j = \mathbf{R}^{*-1}\mathbf{R}$ are called the normalized regression coefficients. From Eq. (7.25) it follows that a relationship exists between non-normalized (a_j) and normalized (b_j) regression coefficients

$$a_{j-1} = b_{j-1} \frac{\sigma_1}{\sigma_j}, \qquad j = 2,...,m \qquad (7.26)$$

The normalization changes the magnitude of the regression coefficients. The advantage of normalized regression coefficients is the fact that they concern directly the paired correlation coefficients and are easier to interpret.

Problem 7.5 *Regression for two explanatory variables*

Estimate coefficients $\mathbf{a}$ and $\mathbf{b}$ of the linear regression of one response variable ξ_1 and two explanatory variables ξ_2, ξ_3.

Solution: The vector $\mathbf{R}$ is (R_{12}, R_{13}). Matrix $\mathbf{R}^*$ is identical to the matrix $\mathbf{R}_{11}$ from Problem 7.4. For the matrix $\mathbf{R}^{*-1}$ it is valid $\mathbf{R}^{*-1} = (1 - R_{23}^2)^{-1} \begin{bmatrix} 1 & -R_{23} \\ -R_{23} & 1 \end{bmatrix}$. The normalized regression coefficients then are

$$\begin{bmatrix} b_1 \\ b_2 \end{bmatrix} = \frac{1}{1 - R_{23}^2} \begin{bmatrix} 1 & -R_{23} \\ -R_{23} & 1 \end{bmatrix} \begin{bmatrix} R_{12} \\ R_{13} \end{bmatrix}$$

leading to the expressions

$$b_1 = \frac{R_{12} - R_{23} R_{13}}{1 - R_{23}^2} \tag{7.27a}$$

$$b_2 = \frac{R_{13} - R_{23} R_{12}}{1 - R_{23}^2} \tag{7.27b}$$

Non-normalized regression coefficients are expressed by $a_1 = b_1 \dfrac{\sigma_1}{\sigma_2}$ and $a_2 = b_2 \dfrac{\sigma_1}{\sigma_3}$. If the explanatory variables are mutually uncorrelated, $R_{23} = 0$ and the normalized regression coefficients correspond to the paired correlation coefficients

$$b_1 = R_{12} \tag{7.28a}$$

and

$$b_2 = R_{13} \tag{7.28b}$$

Conclusion: It is obvious that the regression coefficients are functions of the only paired correlation coefficients. When the explanatory variables are mutually uncorrelated, the normalized regression coefficients are directly equal to the paired correlation coefficients between the jth response and the explanatory random variable.

The regression coefficients $\mathbf{a}$ and $\mathbf{b}$ are such that the correlation between random variables ξ_1 and $\hat{\xi}_1 = (\xi_1 / \mathbf{x}^*)$ is maximal. Random variable $\hat{\xi}_1$ is the linear combination of components $\xi_1, \dots, \xi_m$ of random vector ξ for realization $\mathbf{x}^*$. It is also valid that

$$\rho(\xi_1, \hat{\xi}_1) = R_{1(2,3,\dots,m)} \tag{7.29}$$

In regression analysis $\hat{\xi}_1$ is called the *prediction*. Equation (7.29) shows that the multiple correlation coefficient is, in fact, the paired correlation coefficient between the vector ξ_1 and vector $\hat{\xi}_1$.

The random variable $\varepsilon = \xi_1 - \hat{\xi}_1$ is called the residual in point $\mathbf{x}^*$. The residuals are uncorrelated with individual explanatory variables because $cov(\varepsilon, \xi_j) = E(\varepsilon \xi_j) = E(\xi_1 \ \xi_j - \hat{\xi}_1 \ \xi_j) = 0$. The covariance of residuals with a controllable variable ξ_1 is equal to the conditioned variance

$$cov(\varepsilon, \xi_1) = D(\xi_1 / \mathbf{x}^*) \tag{7.30}$$

which is also the residual variance $D(\varepsilon)$. As for the case of one exploratory variable (Section 7.1.1) the estimates of the regression parameters and other random variables may be obtained on the basis of the sample means $\overline{x}_1, ..., \overline{x}_m$, and the sample covariance, or correlation matrix, respectively. It is also valid that these estimates are identical with the estimates obtained by the least-squares method.

Problem 7.6 *The effect of inorganic and organic nitrogen in soil on the nitrogen content of corn*

The effect of the concentration of inorganic nitrogen (x_2) and organic nitrogen (x_3) in the soil on the content of nitrogen in corn (x_1) has been studied [4]. Estimate the linear regression model for the regression of x_1 on x_2 and x_3, and calculate the multiple correlation coefficient.

Data:

x_1	64	60	71	61	54	77	81	93	93	51
x_2	0.4	0.4	3.1	0.6	4.7	1.7	9.4	10.1	11.6	12.6
x_3	53	23	19	34	24	65	44	31	29	58

76	96	77	93	95	54	168	99
10.6	23.1	23.1	21.6	23.1	1.9	26.8	29.9
37	46	50	44	56	36	59	51

Solution: From the expressions for the mean, variance and pair correlation coefficient we estimate

$$\bar{x}_1 = 81.28 \quad \bar{x}_2 = 11.94 \quad \bar{x}_3 = 42.11$$
$$s_1^2 = 728.8 \quad s_2^2 = 103.6 \quad s_3^2 = 185.6$$

and $\mathbf{R} = \begin{bmatrix} 1 & 0.6934 & 0.3545 \\ 0.6934 & 1 & 0.4616 \\ 0.3545 & 0.4616 & 1 \end{bmatrix}$. On substitution into Eq. (7.23) we obtain

$$R_{1(2,3)} = \sqrt{\frac{0.6934^2 + 0.3545^2 - 2 \times 0.6934 \times 0.3545 \times 0.4616}{1 - 0.4616^2}} = 0.6945.$$

From Eq. (7.29a) we obtain

$$b_1 = \frac{0.6934 - 0.4616 \times 0.3545}{1 - 0.4616^2} = 0.6732 \text{ and from Eq. (7.29b) } b_2 = 0.04375.$$

For non-normalized regression coefficients $a_1 = \sqrt{\frac{728.8}{103.6}} \, 0.6732 = 1.7855$

and $a_2 = 0.08669$ For the intercept term, from Eq. (7.17), we have $a_0 = \bar{x}_1 - a_1\bar{x}_2 - a_2\bar{x}_3 = 56.31$. The linear regression model has the form $x_1 = 56.31 + 1.7855x_2 + 0.08669x_3$.

The multiple correlation coefficients are the same, and the coefficients a_0, a_1, a_2 correspond to the estimates by the least-squares method.

Conclusion: The coefficients of linear regression models and the multiple correlation coefficient may be calculated directly from the definitions.

From a practical point of view, it is convenient to use computer programs for linear regression. However, these expressions show that regression models can be directly derived from the random vector, and moreover they often aid the interpretation of the statistical characteristics.

In some cases it is useful to examine a relationship between two components ξ_1 and ξ_j of a random vector, when the other components of vector ξ are excluded. To express the intensity of this dependence, the *partial correlation coefficients* of various orders are used. The simplest are the partial correlation coefficients of zero order, which correspond to the paired correlation coefficients.

The partial correlation coefficient of the first order $R_{1,3(2)}$ corresponds to the paired correlation coefficient between the residuals $\varepsilon_2 = \xi_1 - E(\xi_1 / x_2)$ and the residuals $\kappa_2 = \xi_3 - E(\xi_3 / x_2)$. Then

$$R_{1,3(2)} = \frac{R_{13} - R_{12}\, R_{23}}{\sqrt{(1 - R_{12}^2)(1 - R_{23}^2)}} \qquad (7.31)$$

Similarly, other partial correlation coefficients $R_{1i(j)}$ of the first order can be defined from the paired correlation coefficients between residuals $\varepsilon_j = \xi_1 - E(\xi_1 / x_j)$ and residuals

$$\kappa_j = \xi_i - E(\xi_i / x_j) \qquad (7.32)$$

Then, it can be shown that $R_{1,i(j)} = \dfrac{R_{1i} - R_{1j}\, R_{ij}}{\sqrt{(1 - R_{1i}^2)(1 - R_{ij}^2)}}$. The partial

correlation coefficients of the second order $R_{1,i(j,k)}$ are the same as the paired correlation coefficients of residuals $\varepsilon_{j,k} = \xi_1 - E(\xi_1 / (x_j, x_k))$ and residuals $\kappa_{j,k} = \xi_i - E(\xi_i / (x_j, x_k))$. To estimate these, an equation analogous to Eq. (7.32) may be used, where instead of the paired correlation coefficients the partial correlation coefficients of the first order are used

$$R_{1,i(j,k)} = \frac{R_{1i(j)} - R_{1j(k)}\, R_{ij(k)}}{\sqrt{(1 - R_{1j(k)}^2)(1 - R_{ij(k)}^2)}} \qquad (7.33)$$

The partial correlation coefficient of the $(m-1)$th order $R_{1,i(2,3,\dots,m)}$ corresponds to the paired correlation coefficient between residuals $\varepsilon_{2,\dots,m} = \xi_1 - E(\xi_1 / \mathbf{x}^*)$ and residuals $\kappa_{2,\dots,m} = \xi_i - E(\xi_i / \mathbf{x}^*)$ where the vector $\mathbf{x}^*$ contains the components $x_2, x_3, \dots, x_{i-1}\, x_{i+1}, \dots, x_m$.

Generally, the partial correlation coefficients of the higher orders are estimated according to a recursive formula

$$R_{1,j(2,3,\dots,j-1)} = \frac{A - B\, C}{\sqrt{(1 - B^2)(1 - C^2)}} \qquad (7.34)$$

where $A = R_{1,j(2,3,\dots,j-2)}$, $B = R_{1,j-1(2,3,\dots,j-2)}$, and $C = R_{j,j-1(2,3,\dots,j-2)}$.

When individual partial correlation coefficients of all orders are known, the multiple correlation coefficient can be estimated from

$$R_{1(2,\dots,m)}^2 = 1 - (1 - R_{1,2}^2)(1 - R_{1,3(2)}^2)(1 - R_{1,4(2,3)}^2)\dots(1 - R_{1,m(2,3,\dots,(m-1))}^2) \qquad (7.35)$$

All these expressions can be evaluated on a pocket calculator.

In the computer estimation of the partial correlation coefficients, matrix notation is convenient.

$$R_{1,i(2,3,...,m)} = \frac{(-1)^i \det(\mathbf{R}_{1,i})}{\sqrt{\det(\mathbf{R}_{1,1}) \det(\mathbf{R}_{i,i})}} \tag{7.36}$$

where $\mathbf{R}$ is the correlation matrix corresponding to the vector ξ and $\mathbf{R}_{i,j}$ is the matrix formed by leaving out the ith row and the jth column of matrix $\mathbf{R}$.

Problem 7.7 *Partial correlation coefficients of the first order*

For the random variables ξ_1, ξ_2, ξ_3 estimate with the use of Eq. (7.36) the partial correlation coefficients $R_{1,2(3)}$ and $R_{1,3(2)}$.

Solution: For calculation of the correlation coefficients the matrices $\mathbf{R}, \mathbf{R}_{11}, \mathbf{R}_{12}, \mathbf{R}_{22}, \mathbf{R}_{13}$ and $\mathbf{R}_{33}$ are necessary. We determined

$$\mathbf{R} = \begin{bmatrix} 1 & R_{12} & R_{13} \\ R_{12} & 1 & R_{23} \\ R_{13} & R_{23} & 1 \end{bmatrix} \qquad \mathbf{R}_{11} = \begin{bmatrix} 1 & R_{23} \\ R_{23} & 1 \end{bmatrix}$$

$$\mathbf{R}_{12} = \begin{bmatrix} R_{12} & R_{23} \\ R_{13} & 1 \end{bmatrix} \qquad \mathbf{R}_{33} = \begin{bmatrix} 1 & R_{12} \\ R_{12} & 1 \end{bmatrix}$$

$$\mathbf{R}_{13} = \begin{bmatrix} R_{12} & 1 \\ R_{13} & R_{23} \end{bmatrix} \qquad \mathbf{R}_{22} = \begin{bmatrix} 1 & R_{13} \\ R_{13} & 1 \end{bmatrix}$$

On substitution into Eq. (7.36) we find

$$R_{1,2(3)} = \frac{R_{12} - R_{23} R_{13}}{\sqrt{(1 - R_{23}^2)(1 - R_{13}^2)}} \tag{7.37}$$

$$\text{or } R_{1,3(2)} = \frac{R_{13} - R_{12} R_{23}}{\sqrt{(1 - R_{23}^2)(1 - R_{12}^2)}} \tag{7.38}$$

Conclusion: The partial correlation coefficients may be estimated directly from Eq. (7.36).

It is interesting that by use of the partial correlation coefficients, the normalized regression coefficients may also be found. The intensity of the mutual relationship between components of a random vector may be better estimated by the partial correlation coefficients than by the paired correlation coefficients.

Problem 7.8   *Partial correlation between the nitrogen content in corn and in soil*

For the data from Problem 7.6, calculate the partial correlation coefficients between the nitrogen content in corn and (a) the content if inorganic nitrogen in soil $R_{1,2(3)}$ and (b) the content of organic nitrogen in soil, $R_{1,3(2)}$.

Data: from Problem 7.6

Solution: By direct substitution into Eq. (7.37), we find

$$R_{1,2(3)} = \frac{0.6934 - 0.4616 \times 0.3545}{\sqrt{(1 - 0.4616^2)(1 - 0.3545^2)}} = 0.6386 \text{ and from Eq. (7.38)}$$

$$R_{1,3(2)} = 0.05325.$$

Conclusion: The nearly zero value of the partial correlation coefficient $R_{1,3(2)}$ shows that the influence of organic nitrogen in soil on the nitrogen content in corn is negligible. The relatively high value of the paired correlation coefficient $R_{13} = 0.3545$ is strongly affected by the correlation $R_{23} = 0.4616$ between the organic and inorganic nitrogen in soil. Detailed data analysis shows that point 17 is an outlier, so the analysis should be repeated with that point omitted.

7.2 Correlation Coefficients

Correlation coefficients serve as basic measures for expressing "closeness" of the linear stochastic dependence between components of the random vector ξ. In the literature [5] many other characteristics are used which can also cover nonlinear stochastic dependences. We restrict ourselves here to a description of the distributions of the sample correlation coefficients, and some selected tests.

7.2.1 Paired Correlation Coefficient

The paired correlation coefficient $\rho(\xi_i, \xi_j) = R_{ij}$ is a measure of the linear stochastic dependence between the random variables ξ_i and ξ_j. For sample size n, the same correlation coefficient may be estimated from Eq. (7.14). For simplicity, we denote the paired correlation coefficient by the letter ρ and the sample paired correlation coefficient by R.

At first we restrict discussion to a case when the joint distribution of quantities ξ_1 and ξ_2 is normal and $\rho = 0$. Then the probability density of random quantity R is symmetrical around zero and has the shape [6] expressed by

$$f(R) = \frac{2^{n-3}}{\pi} \frac{\left[\Gamma\left[\frac{n-1}{2}\right] \right]^2}{\Gamma[n]} (1 - R^2)^{(n-4)/2} \tag{7.39}$$

where $\Gamma(.)$ is the Gamma function. For $n = 5, 9$ and 51, the courses $f(R)$ are illustrated in Fig. 7.4.

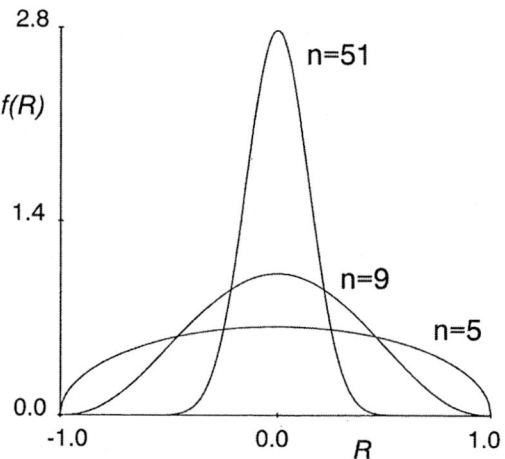

Figure 7.4   The probability density of the sample correlation coefficient for $\rho = 0$ and for sample size $n = 5$, 9 and 51

In construction of significance tests the following test criterion is used

$$t = \frac{R\sqrt{n-2}}{\sqrt{1 - R^2}} \tag{7.40}$$

which for $\rho = 0$ has the Student distribution with $(n - 2)$ degrees of freedom. This may be used for testing the independence between a pair of random variables. If their distribution is two-dimensionally normal, lack of correlation is identical to independence.

The null hypothesis H_0: $\rho = 0$ is tested *vs.* various alternatives. If the criterion $|t|$ from Eq. (7.40) is larger than the corresponding quantile of the

Student distribution, the null hypothesis is rejected and the random variables are correlated. This test is strongly non-robust and is valid only in the case of two-dimensional normality of ξ_1 and ξ_2. To speed up the convergence of $f(R)$ to the normal distribution, various transformations are used. The simple Ruben transformation has the form

$$R(R) = \frac{R\sqrt{n-2.5}}{\sqrt{1-0.5R^2}} \tag{7.41}$$

The random variable $R(R)$ has, even for small sample sizes, the normalized normal distribution $N(0, 1)$.

Problem 7.9 *Significance of the degree of polymerization of cotton in two solutions*

Determine the significance of the correlation coefficient between the degrees of polymerization of cotton determined in solution CUEN and CUOXAN (Problem 7.3) when the sample correlation coefficient $R = 0.6142$ was estimated from 17 data values.

Data: from Problem 7.3

Solution: We select the significance level $\alpha = 0.05$. Since there must be a positive linear relationship between the degrees of polymerization in the two solutions, we know that $\rho \geq 0$. We examine the null hypothesis H_0: $\rho = 0$ vs. H_A: $\rho \geq 0$.

(a) By substituting into Eq. (7.40), we calculate the test criterion $t = 3.104$. This value is higher than the quantile of the Student distribution $t_{0.95}(15) = 1.753$, so the inequality $\rho > 0$ is accepted.

(b) By substituting into Eq. (7.41) we calculate the test criterion of the Ruben transformation $R(R) = 2.596$. This value is higher than the quantile of the normalized normal distribution $u_{0.95} = 1.645$, so the null hypothesis H_0 is rejected.

Conclusion: Both tests prove that the population correlation coefficient is, with 95% probability, positive, and therefore correlation exists between the two degrees of polymerization of the two solutions.

A common case is a simultaneous distribution of two random variables that are two-dimensionally normal with $\rho \neq 0$. For $n > 3$ the probability density function of the sample correlation coefficient may be expressed in the form [6]

$$f(R/\rho) = \frac{2^{n-3}}{\pi(n-3)!}(1-\rho^2)^{(n-1)/2}(1-R^2)^{(n-4)/2}\sum_{j=0}^{\infty}\frac{(2\rho R)^j}{j!}\left[\Gamma\left[\frac{n+j-1}{2}\right]\right]^2$$

for $-1 \le R \le +1$, and $f(R/\rho) = 0$ elsewhere.   (7.42)

The probability density function $f(R/\rho)$ is rather asymmetrical for small sample sizes (see Fig. 7.5). For sufficiently large sample sizes ($n > 500$), the distribution of $f(R/\rho)$ can be approximated by a normal distribution with mean $E(R) = \rho$ and variance $D(R) = (1-\rho^2)^2 / (n-1)$.

If random variables ξ_1 and ξ_2 have a two-dimensional *elliptic* distribution with correlation coefficient ρ and kurtosis g_2, the random variable

$$u_n = \frac{|R-\rho|\sqrt{n-1}}{1-\rho^2}$$ (7.43)

has an asymptotically normal distribution with zero mean value and variance equal to $(1+g_2)$.

Problem 7.10   *The confidence interval of the correlation coefficient*

For 600 random samples, the content of iron was determined by two analytical methods with correlation coefficient $R = 0.85$. Estimate the 95% confidence interval of the correlation coefficient ρ. Examine the null hypothesis $H_0: \rho = 0.9$ against the alternative $H_A: \rho \ne 0.9$.

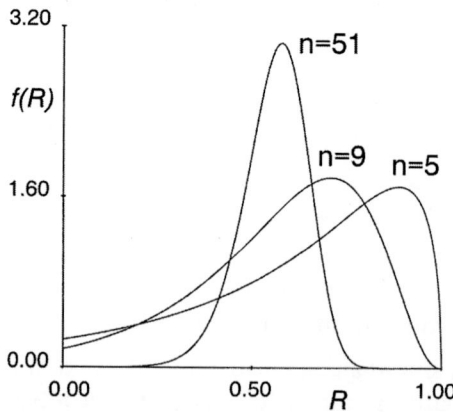

Figure 7.5 The probability density function of the sample correlation coefficient (for $\rho = 0.6$) for sample sizes $n = 5$, 9 and 51

Solution: We make use of the asymptotic normality of the distribution of the correlation coefficient. It is valid that $R - u_{1-\alpha/2}\dfrac{(1-R^2)}{\sqrt{n-1}} \leq \rho \leq R + u_{1-\alpha/2}\dfrac{(1-R^2)}{\sqrt{n-1}}$ where $u_{1-\alpha/2}$ is the quantile of the normalized normal distribution. On substituting, we get $0.828 \leq \rho \leq 0.872$. To test the H_0 hypothesis, we use the test criterion u_n [Eq. (7.43)], $u_n = \dfrac{|0.85 - 0.9|\sqrt{600-1}}{1 - 0.9^2} = 6.44$ which is higher than the quantile $u_{0.975} = 1.96$, so we reject the null hypothesis $H_0 : \rho = 0.9$ at the significance level $\alpha = 0.05$. The correlation between the two analytical methods for iron determination differs significantly from the value 0.9.

Conclusion: For large sample sizes the asymptomatic normality can be utilized to permit a test of paired correlation coefficients.

If the sample correlation coefficient is estimated from the sample size n, some measures of location, spread and distribution shape may be approximated [6]. The mean may be expressed in the form

$$E(R) = \rho + (1-\rho^2)\left[-\frac{\rho}{2(n-1)} + \frac{\rho - 9\rho^3}{8(n-1)^2} + \frac{\rho + 42\rho^3 - 75\rho^5}{16(n-1)^3} + \ldots\right] \quad (7.44)$$

the variance

$$D(R) = \frac{1}{n}(1-\rho^2)^2\left[1 + \frac{11\rho^2}{2(n-1)} + \frac{-24\rho^2 + 75\rho^4}{2(n-1)^2} + \ldots\right] \quad (7.45)$$

the skewness

$$g_1(R) = \frac{6\rho}{n-1}\left[1 + \frac{-30 + 70\rho^2}{12(n-1)} + \ldots\right] \quad (7.46)$$

the kurtosis

$$g_2(R) \approx \frac{6(12\rho^2 - 1)}{n-1} \quad (7.47)$$

The bias $E(R) - \rho$ is, as an initial approximation, equal to

$$E(R) - \rho = \frac{-\rho(1-\rho^2)}{2(n-1)}$$

and the estimate R calculated by Eq. (7.14) is rather underestimated for $\rho > 0$. For very small sample size ($n < 15$), the corrected correlation coefficient is used for practical calculations. This is given by

$$R^* = R\left[1 + \frac{1 - R^2}{2(n - 4)}\right] \qquad (7.48)$$

The square of the correlation coefficient is strongly *overestimated* in cases when the sample is not random. For larger intervals of sample values and more uniform scaling, the R^2 value is more overestimated.

To improve the statistical properties of a distribution of the sample correlation coefficient, many transformations which speed up convergence to normality are used. The best known is the Fisher transformation [7] which takes the form

$$Z(R) = \arctan(R) = 0.5\ln\left[\frac{1 + R}{1 - R}\right] \qquad (7.49)$$

This transformation stabilizes the variance. For $n > 50$, the distribution of quantity $Z(R)$ is approximately normal, with mean value $E(Z)$ and variance $D(Z)$ calculated from

$$E(Z) = Z(\rho) \qquad (7.50a)$$

$$D(Z) = \frac{1}{n - 3} \qquad (7.50b)$$

More exact estimates of the mean, variance, skewness and kurtosis are given by

$$E(Z) = Z(\rho) + \frac{0.5\rho}{n - 1} + \frac{\rho(5 + \rho^2)}{8(n - 1)^2} + \frac{\rho(11 + 2\rho^2 + 3\rho^4)}{16(n - 1)^3} + \ldots \qquad (7.51)$$

the variance $D(Z) = \frac{1}{n - 1}\left[1 + \frac{0.5(4 - \rho^2)}{n - 1} + \ldots\right] \approx \frac{1}{n - 3} - \frac{\rho^2}{2(n - 3)^2} \qquad (7.52)$

$$\text{the skewness } g_1(Z) = \frac{\rho^3}{(n-1)^{2/3}} + \dots \tag{7.53}$$

$$\text{the kurtosis } g_2(Z) = \frac{2}{n+1} + \dots \tag{7.54}$$

For small sample sizes, the Sammiunddin transformation [8] is recommended

$$S = \frac{(R-\rho)\sqrt{n-2}}{\sqrt{(1-R^2)(1-\rho^2)}} \tag{7.55}$$

The quantity S has approximately the Student distribution with $n - 2$ degrees of freedom for $\rho \neq 0$.

Kraemer [9] replaces the correlation coefficient ρ in Eq. (7.55) with the median $\tilde{\rho}$ of a distribution of the sample correlation coefficient, for which

$$\tilde{\rho} = \rho + (1-\rho^2)\rho\left[\frac{2}{n-1} + \frac{(-7\rho^2+15)}{24(n-1)^2} + \dots\right] \tag{7.56}$$

There are many other transformations [6] which are recommended for small or large sample sizes.

Problem 7.11 *Examination of the correlation coefficient between the degrees of polymerization of cotton in two solutions*

Suppose that if the correlation coefficient between the degrees of polymerization of cotton in CUEN and CUOXAN solutions is not smaller than $\rho_0 = 0.85$, a significant linear association between the results exists. Examine the null hypothesis H_0: $\rho = 0.85$ against alternative H_A: $\rho < 0.85$, with the use of various transformations.

Solution: (a) *Fisher transformation* (7.49) leads to the test criterion we get $u_F = |Z(R) - Z(\rho_0)|\sqrt{n-3}$ with approximate distribution $N(0, 1)$. From Eq. (7.49) we get $Z(R) = 0.5\ln\left[\frac{1+0.6142}{1-0.6142}\right] = 0.7156$ and $Z(\rho_0) = Z(0.85) = 1.256$

Then $u_F = |0.7156 - 1.256| \sqrt{17 - 3} = 2.021$ is higher than the quantile $u_{0.95} = 1.64$ and therefore the null hypothesis H_0: $\rho = 0.85$ cannot be accepted.

(b) The *Sammiunddin transformation* (7.55) leads to the test criterion

$$S = \frac{|0.6142 - 0.85| \sqrt{17 - 2}}{\sqrt{(1 - 0.614^2)(1 - 0.85^2)}} = 2.197 \quad \text{is higher than the quantile}$$

$t_{0.95}(15) = 1.725$, so the null hypothesis H_0: $\rho = 0.85$ is rejected.

(c) *Kraemer modification* (7.56) leads to the median $\tilde{\rho} \approx 0.85 + (1 - 0.85^2)$

$$0.85 \left[\frac{2}{16} + ... \right] = 0.879. \text{ The test criterion } S = \frac{|0.6142 - 0.879| \sqrt{17 - 2}}{\sqrt{(1 - 0.879^2)(1 - 0.6142^2)}}$$

$= 2.726$ is higher than the quantile $t_{0.95}(15) = 1.725$ and therefore the null hypothesis H_0: $P = 0.85$ is rejected.

Conclusion: All three tests used show that the correlation coefficient ρ is significantly lower than the value 0.85 and therefore the stochastic dependence between the degrees of polymerization is not very strong.

When the joint distribution of random variables is not normal and the sample contains strong outliers, the normalized transformation is not valid and the correlation coefficient is not suitable for expressing a stochastic association. We can then use various robust estimates of correlation coefficients, which apply robust estimates of parameters of location, spread and covariance. Some techniques have been described in [10].

The correlation coefficients should be interpreted very carefully. As a general rule, a significant paired correlation is not the proof of a causal dependence. Sometimes false correlations are formed when either ξ_1 or ξ_2 strongly correlate with some unconsidered random variable ξ_3, and a high value of $\rho(\xi_1, \xi_2)$ is the consequence of high values of $\rho(\xi_1, \xi_3)$ and $\rho(\xi_2, \xi_3)$. In the interpretation of correlation coefficients, the partial correlation coefficients should also be considered.

7.2.2 Partial Correlation Coefficients

For calculation of partial correlation coefficients either the recursive formulae [Eq. (7.34)] or the matrix method [Eq. (7.36)] can be used.

For statistical testing and building of the confidence interval, we use a rule that a distribution of the partial correlation coefficient of the order $(m - 1)$ is identical to the distribution of the paired correlation coefficients for sample size $(n - m + 1)$. Thus, techniques described in Section 7.2.1, with modified sample size, may also be used.

Problem 7.12 *Significance of the dependence between the organic nitrogen in soil and the content of nitrogen in corn*

For the data from Problem 7.8, examine the significance of the correlation coefficient $R_{1,3(2)}$ as an expression of the association between organically bound nitrogen in soil and content of nitrogen in corn.

Data: from Problem 7.8

Solution: To examine the significance of the null hypothesis H_0: $R_{1,3(2)} = 0$ against H_A: $R_{1,3(2)} \neq 0$ we use the above relationship. Because the partial correlation coefficient is of the first order, we have $m - 1 = 1$ and the reduced sample size is $n - 1$. From Eq. (7.40) the test criterion $t_p = \dfrac{\hat{R}_{1,3(2)}\sqrt{n-3}}{\sqrt{1 - \hat{R}_{1,3(2)}^2}} = 0.227$

is

smaller than the quantile $t_{0.975}(15) = 2.13$, so the null hypothesis H_0: $R_{1,3(2)} = 0$ is accepted. In calculation of t_p, the partial correlation coefficient $R_{1,3(2)} = 0.05858$ for $n = 18$ was used.

Conclusion: On the basis of the test of the partial correlation coefficient, it is concluded that there is no significant correlation between the organic nitrogen content in soil and in corn.

Partial correlation coefficients can be used to elucidate some false correlations. Consider a case when the paired correlation coefficient between ξ_1 and ξ_2 is $R_{12} = H$ where $H \rightarrow 1$. Suppose there is a random variable ξ_3 which strongly correlates with ξ_1 and ξ_2, so that $R_{13} = H^2$ and $R_{23} = H$. Then the multiple correlation coefficient $R_{1(2,3)} = H$ may be estimated, and the partial correlation coefficients are equal to $R_{1,3(2)} = 0$,

$$R_{1,2(3)} = \frac{H}{\sqrt{1 + H^2}}.$$

Despite the high value of the paired correlation coefficient ($R_{13} = H^2$ is close to one) the quantity ξ_3 does not contribute to the explanation of the variability of ξ_1 and ξ_2. It is a typical *parasite variable*. When $R_{23} = R_{13} / R_{12}$, the variable ξ_3 is a parasite.

This situation can arise from the neglect of a significant variable such as, for example, time or temperature. For example, at various time values during a degradation process, the mechanical or optical properties of the materials will be different. If time is ignored, a significant "false" correlation among these properties appears. When time is included as a variable, the optical properties do not contribute to explaining the variability of the mechanical properties. A high value of the paired correlation coefficient is not always a guarantee of a significant association between variables.

Similarly, there are cases when a low value of the paired correlation coefficient leads to high partial correlation coefficients and a high multiple correlation coefficient. When $R_{13} = 0$ and $R_{12} = \varepsilon$ $(\varepsilon \to 0)$, but variables ξ_2 and ξ_3 are strongly correlated i.e. $R_{23} = \sqrt{(1 - \varepsilon^2)}$ then $R_{12(3)} = 1, R_{13(2)} = -1$ and $R_{1(2,3)} = 1$. From this it follows that a zero paired correlation coefficient does not mean automatically that a given random variable is insignificant or parasite and may be excluded.

Moreover it is not valid that strongly correlated random variables are always redundant. These examples demonstrate that often no conclusions can be made from the paired correlation coefficients. It is because other variables are not considered that a "false" correlation may be concluded.

7.2.3 Multiple Correlation Coefficient

The multiple correlation coefficient, denoted as $R_{1(2,...,m)}$, is a measure of the overall linear stochastic association of one random variable ξ_1 with the best linear combination of the other components $\xi_2, ..., \xi_m$, of the random vector ξ. The sample correlation coefficient $R_{1(2,...,m)}$ may be readily calculated from Eq. (7.22) by replacing the correlation matrix $\mathbf{R}$ by the sample correlation matrix $\hat{\mathbf{R}}$. For the sake of simplicity, we refer to the multiple correlation coefficient of the population as R_m and its sample estimate as $\hat{R}_m$.

Let us suppose that the vector ξ has an m-multidimensional normal distribution and that all its components have a normal distribution.

Case $R_m = 0$: The probability density of random variable $\hat{R}_m^2$ is given by

$$f(\hat{R}_m^2) = K_m \ (\hat{R}_m^2)^{(m-3)/2} \ (1 - \hat{R}_m^2)^{(n-m-2)/2} \ . \tag{7.57}$$

where K_m is a constant dependent on m and n. The distribution defined by Eq. (7.57) is a beta-distribution $Be[(m - 1)/2, (n - m)/2]$. Then the random variable given by

$$F_R = \frac{(n - m) \ \hat{R}_m^2}{(m - 1)(1 - \hat{R}_m^2)} \tag{7.58}$$

has the F-distribution with $(m-1)$ and $(n-m)$ degrees of freedom. For large sample sizes, the distribution of

$$C_R = (n-1)\,\hat{R}_m^2 \qquad (7.59)$$

is χ^2 with $(m-1)$ degrees of freedom.

For the mean value of the sample squared multiple correlation coefficient, we have

$$E(\hat{R}_m^2) = \frac{m-1}{n-1} \qquad (7.60)$$

and for the variances

$$D(\hat{R}_m^2) = \frac{2(n-m+2)(m-1)}{(n^2-1)(n-1)} \approx \frac{2(m-1)}{n^2} \qquad (7.61)$$

Equation (7.60) shows that for a small number of measurements and a larger number of explanatory variables, the quantity $\hat{R}_m^2$ will be significantly different from zero even in cases when the population multiple correlation coefficient R_m^2 is equal to zero. For example, if $n = 12$ and $m = 9$, then $E(\hat{R}_m^2) = 0.727$ even though $R_m^2 = 0$. This negative effect may be removed by decreasing m or increasing n. A sample size higher than $n_{\min} = (1+100\ m)$ ensures that $\hat{R}_m^2 \approx 0.1$ for uncorrected random variables.

In the case of a multi-variable normal distribution, the null hypothesis H_0: $R_m = 0$ against the alternative H_A: $R_m \neq 0$, with use of criterion F_R, is suitable as a test for independence.

Problem 7.13 *Significance of the relationship between the nitrogen content in soil and in corn*

In Problem 7.6 the multiple correlation coefficient expressing the relationship between the nitrogen content in corn and a linear combination of organically bound nitrogen and inorganically bound nitrogen in soil is equal to $\hat{R}_{1(2,3)}^2 = 0.6945$. Examine the null hypothesis H_0: $R_{1(2,3)} = 0$.

Solution: According to Eq. (7.58), the test criterion $F_R = \dfrac{(18-3)\ 0.6945^2}{(3-1)\ (1-0.6945^2)} =$

6.988 is higher than the quantile of the Fisher-Snedecor distribution $F_{0.95}(2, 15) = 3.682$, and therefore the null hypothesis H_0: $R_{1(2,3)} = 0$ is rejected at significance level $\alpha = 0.05$.

Conclusion: The content of nitrogen in soil significantly affects the content of nitrogen in corn. Inorganically bound nitrogen contributes predominantly.

Case $R_m > 0$: To calculate the sample multiple correlation coefficient, $\hat{R}_m^2$ the complicated exact expression or a convenient approximation may be used. Gurland [6] has proposed a relatively precise approximation

$$\frac{\hat{R}_m^2}{1-\hat{R}_m^2} \approx \frac{\dfrac{(n-1)\hat{R}_m^2}{1-\hat{R}_m^2}+(m-1)}{n-m} \times F_{r,n-m} \tag{7.62}$$

where the quantity $F_{r,n-m}$ has the F-distribution with r and $(n-m)$ degrees of freedom. Then

$$r = \left[K(n-1)+(m-1)\right]/Z \tag{7.62a}$$

where

$$Z = \frac{(n-1)K(K+2)+(m-1)}{(n-1)K+(m-1)} \tag{7.62b}$$

$$\text{and } K = \frac{\hat{R}_m^2}{1-\hat{R}_m^2} \tag{7.62c}$$

For large sample sizes, the square of the multiple correlation coefficient reaches approximately a normal distribution with the mean value $E(\hat{R}_m^2) = R_m^2$ and variance $D(\hat{R}_m^2) = \dfrac{4R_m^2(1-R_m^2)^2}{n-1}$. The random variable

$$u_R = \frac{\sqrt{n-1}(\hat{R}_m^2 - R_m^2)}{2R_m(1-R_m^2)} \tag{7.63}$$

has the normalized normal distribution. Also, the Fisher and other transformations for speeding up convergence to normality can be used.

For the mean value of the squared multiple correlation coefficient

$$E(\hat{R}_m^2) = R_m^2 + \frac{m-1}{n-1}(1-R_m^2) - \frac{2(n-m)}{n^2-1}R_m^2(1-R_m^2) + \dots \tag{7.64}$$

The variance is given by

$$D(\hat{R}_m^2) = \frac{4R_m^2(1-R_m^2)^2(n-m)^2}{(n^2-1)(n+3)} \approx \frac{4R_m^2(1-R_m^2)^2}{n} \tag{7.65}$$

For smaller sample sizes, the estimate $\hat{R}_m^2$ is overestimated. The corrected multiple correlation coefficient is expressed by

$$\hat{R}_m^{*2} = \hat{R}_m^2 - \frac{m-3}{n-m}(1-\hat{R}_m^2) - \left[\frac{2(n-3)}{(n-m)^2}(1-\hat{R}_m^2) + \dots\right] \tag{7.66}$$

It can be seen that $\hat{R}_m^{*2} < \hat{R}_m^2$. For small values of $\hat{R}_m^2$, the corrected $\hat{R}_m^{*2}$ can be even be negative and therefore it should be restricted to the interval $[0,1]$.

7.2.4 Rank Correlation

In some cases, the classical paired correlation coefficient can be replaced by the rank correlation coefficient, which is not very sensitive to the presence of outliers.

The rank of the ith element of a sample is equal to the index of the order statistic. Let us write the sample ranks for variable ξ_1 as x_{1Si} and sample ranks regarding to the variable ξ_2 as x_{2Si}. The Spearman rank correlation coefficient is then expressed by

$$\tilde{\rho}_S = 1 - \frac{6}{n(n^2-1)}\sum_{i=1}^{n}(x_{1Si} - x_{2Si})^2 \tag{7.67}$$

For $\rho_S = 0$, the distribution of $\hat{\rho}_S$ is symmetric with mean value $E(\hat{\rho}_S) = 0$ and variance $D(\hat{\rho}_S) = 1/(n-1)$.

For $n > 10$, the quantity

$$t_S = \frac{|\hat{\rho}_S|\sqrt{n-2}}{\sqrt{1-\hat{\rho}_S^2}} \tag{7.68}$$

has the Student distribution, asymptotically, with $n-2$ degrees of freedom, if the theoretical coefficient $\rho_S = 0$.

Problem 7.14 *Correlation between the effective specific surface energy and the change of surface energy of adsorption*

For six different stearates, the effective surface ξ_1 and the change of surface energy of adsorption ξ_2 were evaluated. Estimate the Spearman correlation coefficient and examine its significance.

Data: $\{x, y\}$: 2.6 17.8, 3.3 18.6, 4.4 16.2, 4.2 17.3, 6.2 15.8, 6.5 15.2.

Solution: Table 7.1 lists the ranks of x_{1Si} and x_{2Si}.

Table 7.1 The order of quantities of x_{1Si} and x_{2Si}.

x_{1Si}	1	2	3	4	5	6
x_{2Si}	5	6	3	4	2	1

From Eq. (7.67), we can calculate $\hat{\rho}_S = 1 - \dfrac{6}{6(6^2 - 1)}(4^2 + 4^2 + 1 + 1 + 3^2 + 5^2) =$ -0.943. Then substitution into Eq. (7.68) leads to $t_S = \dfrac{0.943\sqrt{4}}{\sqrt{1 - 0.943^2}} = 5.66$.

Because the quantile $t_{0.975}(4) = 2.776$ is lower than 5.66, the null hypothesis $H_0 : \rho_S = 0$ is rejected.

Conclusion: The nonparametric test used showed significant negative correlation between the effective specific surface and the change of surface energy of adsorption. For small sample sizes, the conclusion is of little consequence.

In practical problems, often several elements of a sample have the same rank. In this case, these elements have the same mean rank as if they had different values, and the Spearman correlation coefficient is then estimated from

$$\hat{\rho}_S = \frac{\dfrac{n(n^2 - 1)}{6} - \displaystyle\sum_{i=1}^{n}(x_{1Si} - x_{2Si})^2 - a - b}{\left\{\left[\dfrac{n(n^2 - 1)}{6} - 2a\right]\left[\dfrac{n(n^2 - 1)}{6} - 2b\right]\right\}^{1/2}} \tag{7.69}$$

where a and b are correcting coefficients for rank, expressed by

$$a = \frac{1}{12}\sum_j (a_j^3 - a_j) \qquad\qquad (7.69a)$$

$$b = \sum_k (b_k^3 - b_k) \qquad\qquad (7.69b)$$

where j is the number of clusters of the same rank for x_1 and a_j is the number of values of the same rank in the jth cluster. The definitions for k and b_k are similar.

The rank correlation coefficient ρ_S lies in the interval $-1 < \rho_S < 1$. If the sample comes from a two-dimensional normal distribution and $n \geq 30$, then

$$R_{12} = \rho(\xi_1, \xi_2) = 2\sin\left(\frac{\pi}{6}\rho_S\right) \qquad\qquad (7.70)$$

When rank correlation coefficients are used, it should be remembered that transforming data from x_{1i} and x_{2i} into x_{1Si} and x_{2Si} always causes loss of information. Robustness and a decrease in sensitivity to deviations from normality are the compensation.

7.3 Procedure for Correlation Analysis

The procedure of correlation analysis assumes some mutual relationships (bounds) among the components of the random vector. Besides a pair correlation coefficient, a partial correlation coefficient should also be computed, to enable deeper analysis of mutual bounds. Interpretation should be made carefully, especially when the sample size is not large.

Problem 7.15 *Procedure for examining correlation*

As an example we will use Exercise **B7.05** *Content of tar, nicotine, and CO in cigarettes* The US Federal Trade Commission assesses domestic cigarettes according to their content of tar x_1 [mg], nicotine x_2 [mg] and weight x_3 [g] and finally also the content of carbon monoxide CO x_4 [mg] in the cigarette smoke released. This is because the US surgeon general considers factors x_1, x_2 and x_4 to be highly dangerous to human health. The most recent studies have shown that increased tar and nicotine content result in increased carbon monoxide levels. Examine at a level of significance of $\alpha = 0.05$ whether a correlation exists between variables (a) x_1 and x_4, as well as (b) x_2 and x_4, and (c) x_3 and x_4.

Solution:

  1.   Proposed model: Correlation of x_4 with others.

Variable	Pearson's corr. coefficient	level of significance	Spearman's corr. level of coefficient	significance
x_1	0.9575	0.0000	0.9448	0.0000
x_2	0.9259	0.0000	0.8778	0.0000
x_3	0.4640	0.0195	0.2170	0.2975
x_4	1.0000	-----	1.0000	-----

  Multiple correlation coefficient r 0.95843

  Coefficient of determination 100 % D : 91.859

  Predicted coefficient of determination D_p : 0.91326

  Pearson's correlation coefficient x_4 vs. x_1, x_4 vs. x_2, x_4 vs. x_3 shows a high correlation, two independent variables x_1 and x_2 have a strong linear relationship to dependent variable x_4, while the final independent variable x_3 is weakly related. Analogous conclusions are drawn from Spearman's rank correlation coefficient. *Pearson's multiple correlation coefficient r* indicates that the proposed linear regression model is statistically significant and its second power, the *coefficient of determination $D = r^2$* shows that 91.86% of the points correspond well with the model.

  2.   Pearson's pair correlation: Pearson's parametric correlation coefficient and Cronbach's correlation coefficient γ are calculated.

Pearson's pair correlation coefficients between pairs of descriptive variables		Calculated level of significance
x_1 versus x_2:	0.97661	0.0000
x_1 versus x_3:	0.49077	0.0127
x_1 versus x_4:	0.95749	0.0000
x_2 versus x_3:	0.50018	0.0109
x_2 versus x_4:	0.92595	0.0000
x_3 versus x_4:	0.46396	0.0195
Cronbach's correlation coefficient γ : 0.6939		
Standardised Cronbach's correlation coefficient γ : 0.9111		

Pearson's pair correlation coefficient x_4 vs. x_1, x_4 vs. x_2, x_4 vs. x_3 shows a high correlation, two independent variables x_1 and x_2 have a strong linear relationship to dependent variable x_4, while the final independent variable x_3 is weakly related

3. **Spearman's pair correlation**: Spearman's non-parametric rank correlation coefficient is calculated.

Spearman's rank correlation coefficients between pairs of descriptive variables		Calculated level of significance
x_1 versus x_2:	0.92843	0.0000
x_1 versus x_3:	0.15539	0.4583
x_1 versus x_4:	0.94480	0.0000
x_2 versus x_3:	0.19623	0.3472
x_2 versus x_4:	0.87781	0.0000
x_3 versus x_4:	0.21697	0.2975

Analogous to Pearson's correlation coefficient, *Spearman's rank correlation coefficient between two independent variables* shows a strong correlation between the first two independent variables x_1 vs. x_2. There is a much weaker linear relationship between x_1 vs. x_3 and x_2 vs. x_3.

4. **Matrix of differences between Pearson's and Spearman's correlation coefficients**: Some software provides a comparison of these two types of correlation matrices by creating a matrix of their calculated differences.

Matrix of differences between Pearson's and Spearman's correlation coefficients:		
x_1 versus x_2: 0.04817		
x_1 versus x_3: 0.33538	x_2 versus x_3: 0.30395	
x_1 versus x_4: 0.01269	x_2 versus x_4 0.04813	x_3 versus x_4 0.24699

Certain authors recommend identifying which variable pairs require more in-depth examination. In such cases however, we feel that a far more useful tool is the regression diagnostics used frequently in this book, and the graphs of leverage points which easily and clearly reveal outlying values.

7.4 References

[1]   V. V. Gubarev, *Algoritmy statisteceskich izmerenij*, Energoatomizdat, Moskva, 1986.

[2]   I. Nimo-Smith, *Biometrika*, 1979, 66, 390.

[3]   Ch. J. Kovalski, *Am. Statist.*, 1973, 27, 103.

[4]   P. Prescott, *Technometrics*, 1975, 17, 129.

[5]   G. J. Mirskij, *Charakteristiki stochasticeskoj vzaimnosvzaji i ich izmerenije*, Energoizdat, Moskva, 1982.

[6]   M. Siotani, T. Hyakawa and Y. Fujikoshi, *Modern Multivariate Statistical Analysis*, American Science Press, 1985.

[7]   R. A. Fisher, *Metron*, 1921, 1, 1.

[8]   M. S. Srivastava, *Commun. Statist.*, 1983, **A12**, 125.

[9]   H. Ch. Kraemer, *J. Am. Statist. Assoc*, 1973, 68, 1004.

[10] S. J. Devlin, R. Gnanadesikan and J. R. Kettenring, *Biometrika*, 1975, 62, 531.

[11] M. Meloun, J. Militký: Statistická analýza experimentálních dat, Academia Praha, 2004.

Supplemented material (Review Questions, Exercises, Results of Exercises) to Chapter 7 is on the enclosed CD.

8
Nonlinear Regression Models

Nonlinear models are often used in the chemical laboratory. There are three main ways in which nonlinear models are utilized.

(1) Construction of *calibration models* when the measured variable y is a nonlinear function of the independent (adjustable, controllable) variable x.

(2) Construction of *chemical models* describing the stoichiometry, concentration and equilibrium constants of all the products of chemical reactions at equilibrium, or the kinetics of chemical reactions.

(3) Construction of *empirical models* based on a study of the nonlinear dependence between the dependent variable y and independent explanatory variables x.

According to the actual type of task, an approach to building the regression model $f(\mathbf{x}, \boldsymbol{\beta})$ is chosen. The regression model $f(\mathbf{x}, \boldsymbol{\beta})$ is a function of a vector of controllable independent variables x and of a vector of unknown parameters $\boldsymbol{\beta}$ of dimension $(m \times 1)$, $\boldsymbol{\beta} = \{\beta_1, ..., \beta_m\}^T$. Nonlinear regression considers the set of points $\{y_i, \mathbf{x}_i^T\}$, $i = 1, ..., n$, where y represents the response (dependent) variable.

The dimension of vector $\mathbf{x}_{i,}$ does not affect the dimension of vector $\boldsymbol{\beta}$. The regression problem is formulated with regard to a regression triplet:

(1) the data set,

(2) a proposed model, and

(3) a regression criterion.

The regression problem consists of a search for the best model $f(\mathbf{x}, \boldsymbol{\beta})$ on a basis of the data set $\{y_i, x_i\}$, $i = 1, ..., n$, such that the model sufficiently fulfils the given regression criterion.

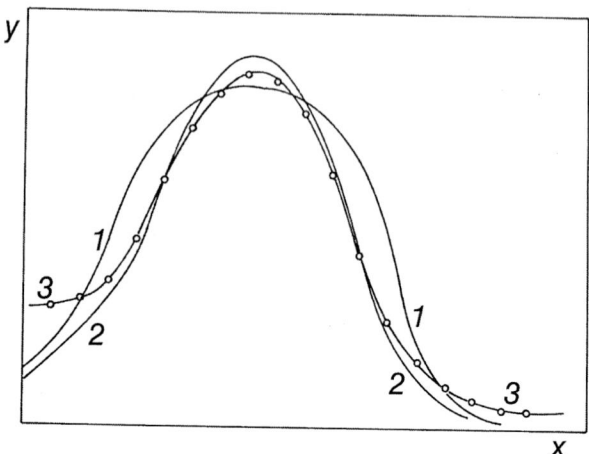

Figure 8.1   Regression model, $y = f(x, \boldsymbol{\beta})$ at three stages of a search for unknown parameters $\boldsymbol{\beta}$: (1) for an initial guess of parameters $\mathbf{b}^{(0)}$, (2) in the kth iteration of parameters refinement $\mathbf{b}^{(t)}$, (3) for the best estimates $\mathbf{b}$.

In chemometrics, the model $f(\mathbf{x}, \boldsymbol{\beta})$ is usually known, so the regression problem consists of searching for the best estimates of unknown parameters $\boldsymbol{\beta}$. In contrast to linear regression models, the parameters $\boldsymbol{\beta}$ play a very important role in nonlinear models. In linear regression models, the regression parameters have no physical meaning but are just numerical coefficients; the parameters in a nonlinear model can have a specific physical meaning. Finding the numerical values is often the main purpose of the regression analysis. Examples are equilibrium constants (dissociation constants, stability constants, solubility products) of reactions, rate constants in kinetic models, or unknown concentrations in titration curves. In the interpretation of estimates of model parameters, it must be remembered that they are *random variables* which have variance, and which are often strongly correlated.

Problem 8.1   *Formulation of the parameters and variables of a regression model*

The dependence of a rate constant for a chemical reaction, k, on temperature T is described by the Arrhenius equation

$$k = k_0 \exp(-E / RT) \tag{8.1}$$

where k_0 is the activation entropy of the chemical reaction, E is the activation energy, and R is the universal gas constant. Formulate the model parameters and examine their correlation.

Solution: The rate constant k (response variable, y) was measured at various temperatures, T (explanatory variable, x). The unknown parameters in the regression model [Eq. (8.1)] are $\beta_1 = k_0$ and $\beta_2 = -E/R$. If the additive errors model is used

$$y_i = f(x_i, \boldsymbol{\beta}) + \varepsilon_i = \beta_1 + \exp(\beta_2 / x_i) + \varepsilon_i \tag{8.2}$$

Here, b_1 and b_2 are estimates of β_1 and β_2 and are determined from experimental data $\{k_i, T_i\}$, $i = 1, ..., n$, based on a regression criterion. When errors ε_i in Eq. (8.2) are independent (values of y_i are mutually independent) random variables of the same distribution, and have constant variance $\sigma^2(y)$.

The regression criterion corresponding to the least-squares method may be used. That is

$$U(\boldsymbol{\beta}) = \sum_{i=1}^{n} (y_i - f(x_i, \boldsymbol{\beta})) \tag{8.3}$$

The estimates **b** then minimize the criterion $U(\boldsymbol{\beta})$.

It can be shown here that a strong correlation exists between estimates ln b_1 and b_2. This correlation may be expressed by the paired correlation coefficient

$$r = \sqrt{c^2 + c} \ \ln(1 + 1/c) \tag{8.4}$$

where $c = T_1/(T_n - T_1)$ *is* related to minimum (T_1) and maximum (T_n) temperatures. When $T_1 = 300$ K and $T_2 = 360$ K, then $r = 0.9986$ and ln b_1 and b_2 are nearly linearly dependent. This means that the ratio $(\ln b_1)/b_2$ is constant and hence, individual parameters cannot be estimated independently. When parameters are correlated, unfortunately a change in the first parameter is often compensated for by a change in the second one. There may be several different pairs of parameter estimates $(\ln b_1, b_2)$ which give nearly same values of the least squares criterion $U(\mathbf{b})$. The parameter estimates achieved by various regression programs may differ by some orders of magnitude, but nevertheless apparently a "best" fit to the experimental data, and low values of $U(\mathbf{b})$ are reached.

Conclusion: Even a simple nonlinear model may lead to difficulties in the accuracy of the parameter estimates and also in their interpretation.

Often, attempts are made to apply nonlinear regression models in situations which are totally inappropriate. Models are often applied outside the range of their validity, and generally, it is supposed that they can substitute for missing data. In chemical kinetics, for example, attempts may be made to estimate parameters, from data far from equilibrium. The calculated parameters then differ significantly from the true equilibrium parameters, and should be interpreted as model parameters only.

The result of nonlinear regression depends on the quality of the regression triplet, i.e. (1) the data, (2) the model, and (3) the regression criterion. Correct formulation leads to parameter estimates which have meaning not only formally but also physically.

In this chapter, we solve problems for which a regression model is known. For solving calibration problems or searching for empirical models, the regression model is appropriate. Some procedures are mentioned in Chapter 9.

8.1 Formulation of a Nonlinear Regression Model

A *linear* regression model is a model which is formed by a linear combination of model parameters. This means that linear regression models can, with reference to the model functions, be nonlinear. For example, the model $f(x, \beta) = \beta_1 + \beta_2 \times \sin x$ is sinusoidal, but with regards to parameters it is a linear model. For linear regression models, the following condition is valid

$$g_j = \frac{\partial f(x, \beta)}{\partial \beta_j} = \text{constant}, \ j = 1, ..., m \qquad (8.5)$$

If for any parameter, β_j the partial derivative is not a constant, we say that the regression model is nonlinear. Nonlinear regression models may be divided into the following groups:

(1) *Non-separable models*, when condition (8.5) is not valid for any parameter. For example, in the model $f(x, \beta) = \exp(\beta_1 x) + \exp(\beta_2 x)$.

(2) *Separable models*, when condition (8.5) is valid for one model parameter. For example, the model $f(x, \beta) = \beta_1 + \beta_2 \exp(\beta_3 x)$ is nonlinear only with regards to the parameter β_3.

(3) *Intrinsically linear models* are nonlinear, but by using a correct transformation they can be transformed into linear regression models.

For example, the model $f(x, \beta) = \beta^2 x$ is nonlinear in parameter β, but the shape of the model is a straight line. With the use of the reparameterization $\gamma = \beta^2$ the nonlinear model is transformed into a linear one.

Reparameterization means transformation of parameters β into parameters y which are related to the original ones by a function

$$\gamma = g(\beta) \tag{8.6}$$

By reparameterization, many numerical and statistical difficulties of regression may be avoided or removed and non-separable models transformed into separable models. The model of the Arrhenius equation (8.2) is separable, i.e. linear with regard to β_1 and by the reparameterization, $f(x,\gamma) = \exp(\gamma_1 + \gamma_2 / x)$ is transformed into a non-separable model, where $\gamma_1 = \ln \beta_1$ and $\gamma_2 = \beta_2$. Each regression model may be reparameterized in many ways, one of which is described in Section 8.5.

In chemometrics, we often distinguish models that are linearly transformable, which can, by use of an appropriate transformation, be transformed into linear regression models. For example, the Arrhenius regression model (8.2), may be transformed into the form (if random errors e are neglected) $\ln y = \gamma_1 + \gamma_2 z$ where $\gamma_1 = \ln \beta_1, \gamma_2 = \beta_2$ and $z = 1/x$. The resulting model is a linear model with respect to parameters y. For finite errors ε, however, this transformation is not correct, and causes heteroscedasticity. When the measured rate constants k_i have constant variance $\sigma^2(k_i)$, then the quantities $\ln k_i$ have non-constant variance $\sigma^2(\ln k) = \sigma^2(k_i) / (k_i)^2$, i.e. constant relative error. The linear transformation is useful for simplification of the search for parameters, but it leads to biased estimates and is therefore used to find a guess for initial estimates of unknown parameters (Section 8.5). The derivatives g_j in Eq. (8.5) are *sensitivity measures* of parameter β_j in model $f(x, \beta)$.

From the sensitivity measures of individual parameters, a preliminary analysis of nonlinear regression models can be made, classifying their quality and identifying any redundancy caused by an excessive number of parameters. A model should not contain excessive parameters and its parameters may be unambiguously estimated if the sensitivity measures, g_j, for given data are found to be linearly independent. This means that it is not possible to determine non-zero coefficients $v_j, j = 1,..., m$, such that the Eq. (8.7) is fulfilled

$$\sum_{j=1}^{m} g_j v_j = 0 \tag{8.7}$$

However, if at least one non-zero coefficient, $v_j \neq 0$, exists for which Eq. (8.7) is fulfilled, the regression model is redundant and should be simplified by *excluding* some parameters. If Eq. (8.7) is valid, all parameters may not be individually estimable.

Problem 8.2 *Examination of redundancy of a regression model*

Test for redundant parameters in the regression model $f(x,\boldsymbol{\beta}) = \beta_1 \exp(\beta_2 + \beta_3 x)$. Apply the sensitivity measures, g_j.

Solution: We first compute sensitivity measures,

$$g_1 = \exp(\beta_2 + \beta_3 x)$$
$$g_2 = \beta_1 \exp(\beta_2 + \beta_3 x)$$

and $g_3 = \beta_1 x \exp(\beta_2 + \beta_3 x)$. On substituting into Eq. (8.7), we get $(v_1 + v_2 \beta_1 + v_3 \beta_1 x) \exp(\beta_2 + \beta_3 x) = 0$. For $v_1 = -\beta_1, v_2 = 1$ and $v_3 = 0$, Eq. (8.7) is fulfilled, so the model contains redundant parameters.

To confirm the redundant parameters, reparameterization of the model may be used i.e.

$f(x, \boldsymbol{\gamma}) = \exp(\gamma_1 + \gamma_2 x)$ or $f(x, \boldsymbol{\delta}) = \delta_1 \exp(\delta_2 x)$ where $\gamma_1 = \ln \beta_1 + \beta_2, \gamma_2 = \beta_3$ or $\delta_1 = \beta_1 \exp(\beta_2)$ and $\delta_2 = \beta_3$.

Conclusion: Parameters β_1 and β_2 cannot be estimated separately. Only their functions γ_1, γ_2 or δ_1, δ_2 may be estimated.

Examination for redundancy in regression models should be part of the investigation of any regression model. Some models exhibit redundancy for only some combinations of parameters β. The model cannot be simplified without knowledge of preliminary estimates of some parameters.

Problem 8.3 *Influence of the magnitude of parameters on redundancy in a regression model*

Examine the redundancy in the regression model $f(x,\boldsymbol{\beta}) = \exp(\beta_1 x) + \exp(\beta_2 x)$ with regard to the magnitude of its parameters β_1 and β_2.

Solution: The sensitivity measures $g_1 = x \exp(\beta_1 x)$ and $g_2 = x \exp(\beta_2 x)$ are substituted into Eq. (8.7), resulting in the expression $x(v_1 \exp(\beta_1 x) + v_2 (\beta_2 x)) = 0$. There are no values for v_1 and v_2 that satisfy the equation, unless $\beta_1 = \beta_2$, in which case $v_1 = 1$ and $v_2 = -1$ fulfils the condition.

Conclusion: When in a search for the best estimates of parameters β_1 and β_2, estimate b_1 is nearly equal to b_2, the model is ill-conditioned, and if $b_1 = b_2$, the model is redundant.

There are models which exhibit local redundancy for selected points or values of the independent variable x.

Problem 8.4 *Examination of local redundancy of parameters of the Arrhenius equation*

Find the conditions under which the model of the Arrhenius equation (8.2) is redundant.

Solution: The sensitivity measures substituted into Eq. (8.7) lead to $(v_1 + v_2\beta_1 / x)\exp(\beta_2 / x) = 0$. For $v_1 = -\beta_1 / x$ and $v_2 = 1$ this equation is satisfied. Local redundancy occurs when the value β_1 is of the same magnitude as some of the experimental quantities x_i, y_i, $i = 1,..., n$.

Conclusion: Redundant parameters in the Arrhenius equation occur when $\beta_1 \approx x_i$.

Redundancy always leads to singularity of matrix $\mathbf{J}^T\mathbf{J}$ (cf. Section 8.4). This means that algorithms for the inversion of this matrix by classical procedures will fail (Section 8.5). The local redundancy of parameters may be avoided by using pseudoinversion of matrix $\mathbf{J}^T\mathbf{J}$.

Ill-conditioned nonlinear models cause problems when Eq. (8.7) is fulfilled only approximately. It is analogous to multicollinearity in linear regression models. Although parameter estimates may be found when $\mathbf{J}^T\mathbf{J}$ is ill-conditioned, some numerical difficulties appear during its inversion. If we know the approximate magnitude of the parameter estimates $\mathbf{b}^{(0)}$, we may construct the matrix $\mathbf{L} = n^{-1}(\mathbf{J}^T\mathbf{J})$ with elements

$$L_{jk} = \frac{1}{n}\sum_{i=1}^{n} \frac{\partial f(x_i,\mathbf{b})}{\partial b_j} \frac{\partial f(x_i,\mathbf{b})}{\partial b_k}\bigg|_{\mathbf{b}=\mathbf{b}^{(0)}} \tag{8.8}$$

Matrix $\mathbf{L}$ corresponds to the matrix $(1/n)\mathbf{X}^T\mathbf{X}$ for linear regression models. To estimate the ill-conditioning, matrix $\mathbf{L}$ is transformed into the standardized form $\mathbf{L}^*$ with elements

$$L_{ij}^* = \frac{L_{ij}}{\sqrt{L_{ii}L_{jj}}} \tag{8.9}$$

The conditioning of matrix $\mathbf{L}^*$ gives a guide to the conditioning of parameters $\mathbf{b}^{(0)}$ in a given model for a given experimental data set.

A simple measure of ill-conditioning is the determinant of matrix $\mathbf{L}^*$, $\det(\mathbf{L}^*)$. When the determinant is less than 0.01, i.e. $\det(\mathbf{L}^*) < 0.01$, the nonlinear model is ill-conditioned and hence has to be simplified [1].

In many regression programs, the inversion of matrix $(\mathbf{J}^T\mathbf{J})$ involves its eigenvalues, $\lambda_1 \geq \lambda_2 \geq ... \geq \lambda_m$. (An indication of redundancy is the zero values of some eigenvalues.) For a measure of ill-conditioning the ratio $\lambda_p = \lambda_1 / \lambda_m$ may be used. If $\lambda_p > 900$, the regression model is ill-conditioned [2].

Problem 8.5 *Examination of ill-conditioning of the Arrhenius equation for a chemical reaction in the solid phase*

Examine the conditioning of a model of the Arrhenius equation (8.2) for a simulated data set [3] of a chemical reaction in the solid phase. Guessed values of initial estimate are $\beta_1^{(0)} = 10^7$ min^{-1}, $\beta_2^{(0)} = -15047$.

Data:

k, min^{-1}	0.0112	0.0120	0.0325	0.0535
T, K	730	750	770	790

Solution: From Eq. (8.9), the elements of matrix $\mathbf{L}$ are estimated by

$$L_{11} = \frac{1}{4}\sum_{i=1}^{4}\exp(2\beta_2^{(0)}/T_i) = 1.105458 \times 10^{-17}$$

$$L_{12} = L_{21} = \frac{\beta_1^{(0)2}}{4}\sum_{i=1}^{4}\frac{1}{T_i^2}\exp(2\beta_2^{(0)}/T_i) = 1.417623 \times 10^{-13}$$

$$L_{22} = \frac{\beta_1^{(0)2}}{4}\sum_{i=1}^{4}\frac{1}{T_i^2}\exp(2\beta_2^{(0)}/T_i) = 1.818692 \times 10^{-9}$$

By using a standardization procedure and Eq. (8.9), we get

$$\mathbf{L}^* = \begin{bmatrix} 1 & 0.9979 \\ 0.9979 & 1 \end{bmatrix}$$

The determinant of this matrix, $\det(\mathbf{L}^*) = 4.1182 \times 10^{-4}$, shows significant ill-conditioning.

Conclusion: Ill-conditioning is caused by the small range of experimental temperatures.

8.2 Models of Measurement Errors

Suppose that the experimental data $\{\mathbf{x}_i^T, y_i\}$, $i = 1, ..., n$, and the regression model are known. The response variable y is the variable measured and subject to various kinds of errors. Common errors include measurement errors ε_M, errors of model formulation ε_T, errors of adjusting the independent controllable variable x, ε_x and the random errors of the experiment ε_N. The total error, ε of the dependent variable y, is the sum of the individual errors. It is assumed that the total error of measurement, for all values of y_i, $i = 1, ..., n$, has a mean value equal to zero, i.e. $E(\varepsilon_i) = 0$. When $E(\varepsilon_i) = $ constant, the intercept term in model is missing and when $E(\varepsilon_i) \neq$ constant, the model is falsely proposed. The general regression model can be expressed in the form

$$y_i = Z_i(\mathbf{x}_i, \varepsilon_i, \boldsymbol{\beta}), \quad i = 1, ..., n \tag{8.10}$$

where the function Z_i depends on the type of errors and on the form of the regression function.

When the data represent the results of experimental measurement, the *additive model* of measurement errors is usually assumed

$$Z_i = f(\mathbf{x}_i, \boldsymbol{\beta}) + \varepsilon_i \tag{8.11}$$

In many experiments, there are some restrictions on the measured variable y_i, $i = 1, ..., n$. For example, y_i may take only positive values, with non-constant variance, $\sigma^2(y_i)$, but with constant relative error, $\sigma^2(y)/y$. Such conditions are valid in the *multiplicative model* of measurement errors

$$Z_i = f(\mathbf{x}_i, \boldsymbol{\beta}) \exp(\varepsilon_i) \tag{8.12}$$

In chemical practice, the *combined model* of measurement errors

$$Z_i = f(\mathbf{x}_i, \boldsymbol{\beta}) \exp(v_i) + \varepsilon_i \tag{8.13}$$

is also used. The errors v_i and ε_i in Eq. (8.13) are assumed to be independent.

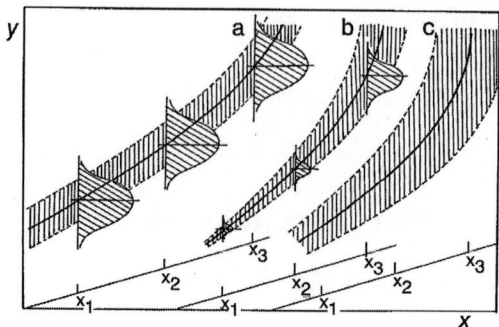

Figure 8.2 Three models of measurement errors: (a) the additive model, (b) the multiplicative model, and (c) the combined model.

In a chemical laboratory the measurement is usually made on just one experimental system. For example, in the investigation of the equilibria of reaction products, the voltage of the glass electrode cell (or absorbance) is monitored during a titration after each addition of a volume of titrant. Cumulative errors can appear in such an experimental procedure. Instrumental measurements are often subject to a constant relative error

$$v(y) = \frac{\sigma(y)}{f(x,\boldsymbol{\beta})}$$

so that the variance of measured variable y is proportional to the square of the value of function $f(x,\boldsymbol{\beta})$,

$$\sigma^2(y) \approx f^2(x,\boldsymbol{\beta}).$$

The *total error* ε_i is then expressed by

$$\varepsilon_i = \sum_{j=1}^{i} u_j + v_i \tag{8.14}$$

where v_i represents the *measurement error* and u_j is the *process error*. Process errors v_i are caused by fluctuations in experimental conditions such as temperature, pressure, purity of reagents, etc., and they are cumulative. The total error ε_i expressed by Eq. (8.14) is therefore additive.

In order to find a proper criterion for regression and to make a statistical analysis, the distribution of the random quantities y_i must be determined. This distribution is closely related to the distribution of errors ε_i given by the probability density function $p_\varepsilon(\varepsilon)$. This function depends on distribution parameters such as the variance σ^2, etc.

In chemometric problems, the error distribution is assumed to be unimodal and symmetrical, with the maximum at $E(\varepsilon) = 0$. It is often assumed, but falsely, that the measurement errors ε_i are mutually independent. The point probability density function $p_\varepsilon(\varepsilon)$ is then given by the product of the *marginal* densities $p_\varepsilon(\varepsilon_i)$

$$p_\varepsilon(\varepsilon) = \prod_{i=1}^{n} p_\varepsilon(\varepsilon_i) \tag{8.15}$$

Several distributions, including the normal, rectangular, Laplace and trapezoidal ones may be expressed by the probability density function

$$p_\varepsilon(\varepsilon_i) = Q_N \exp(-|\varepsilon_i|^p / \alpha) \tag{8.16}$$

where Q_N is the normalizing constant and α is a parameter proportional to the variance. If $p = 1$, the resulting distribution is Laplace. When $p = 2$, the distribution is normal and when $p \to \infty$, rectangular. The disadvantage of describing distribution $p_\varepsilon(\varepsilon_i)$ by Eq. (8.16) is that for $p < 2$, in the neighbourhood of origin, the distribution is not locally quadratic. Therefore, alternative probability density functions, such as the generalized Student distribution, are used [4].

In some cases, the errors are not independent but are characterized by a covariance matrix of errors C_ε. When the errors ε_i come from a symmetric and unimodal distribution with the mean $E(\varepsilon) = 0$, the probability density function is chosen from a class of elliptic distributions

$$p_\varepsilon(\varepsilon) = Q_N \sqrt{\det(\mathbf{B})}\ h(\sqrt{\varepsilon^T \mathbf{B} \varepsilon}) \tag{8.17}$$

where Q_N is the normalizing coefficient, $\mathbf{B}$ is the covariance matrix of errors C_ε and $h(\bullet)$ is a positive function defined on the interval $[0, \infty]$ with finite moments up to $(n + 1)$. The most widely used distribution is the multivariate normal distribution, $N(0, C_\varepsilon)$, which for $h(x) = \exp(-0.5x^2)$ gives the probability density function

$$p_\varepsilon(\varepsilon) = (2\pi)^{-n/2} (\det C_\varepsilon)^{-1/2} \exp(-0.5 \varepsilon^T C_\varepsilon^{-1} \varepsilon) \tag{8.18}$$

It is also possible to use the multivariate Laplace, the Student or other distributions [4].

The form of the covariance matrix of errors C_ε depends on the type of error dependence. A simple example is the case of heteroscedasticity, when errors are mutually independent but have non-constant variance $E(\varepsilon_i^2) = \sigma_i^2$. The matrix C_ε is then diagonal with the elements σ_i^2 on a diagonal and the probability density function (8.17) is transformed into Eq. (8.15). For other types of autocorrelation, the matrix C_ε is not diagonal and their *off-diagonal* elements C_{ij} correspond to the covariance between ε_i and ε_j, $C_{ij} = E(\varepsilon_i \; \varepsilon_j)$.

Problem 8.6   *Covariance matrix of errors for a combination of measurement errors and process errors*

Derive the covariance matrix of errors for a case when the errors ε_i result from errors of measurement v_i and process errors u_i according to Eq. (8.14). Make the following assumptions:

(a) the process errors u_j and errors of measurement v_i are mutually independent, $E(u_j v_i) = 0$;

(b) the process errors u_j are independent, $E(u_j u_i) = 0$ for $j \neq i$ and have constant variance, $E(u_i^2) = \sigma^2$,

(c) the measurement errors v_i are independent, $E(v_i v_j) = 0$ for $j \neq i$ and have non-constant variance, $E(v_i^2) = \sigma_0^2 f^2(\mathbf{x}, \boldsymbol{\beta})$.

Solution: Equation (8.14) is rewritten as

$$\varepsilon_i = \varepsilon_{i-1} + u_i + v_i = \varepsilon_{i-1} + w_i \tag{8.19a}$$

where w_i is the total error on changing from the $(i-1)$th state to the ith state. This error has zero mean $E(w_i) = 0$ but non-constant variance, $\tau_i^2 = \sigma^2 + \sigma_0^2 f^2(x_i, \boldsymbol{\beta})$. From Eq. (8.19a), the individual errors ε_i may be written in the form

$$\varepsilon_1 = w_1$$

$$\varepsilon_1 = w_1 + w_2$$

$$\cdots$$

$$\varepsilon_n = w_1 + w_2 + \ldots + w_n$$

and in matrix notation

$$\varepsilon = Aw \tag{8.19b}$$

where A is the lower triangular matrix of ones, on and under the main diagonal

$$A = \begin{bmatrix} 1 & 0 & 0 & \cdots & 0 \\ 1 & 1 & 0 & \cdots & 0 \\ 1 & 1 & 1 & \cdots & 0 \\ \cdots & \cdots & \cdots & \cdots & \cdots \\ 1 & 1 & 1 & \cdots & 1 \end{bmatrix}$$

The covariance matrix of errors C_ε is given by

$$C_\varepsilon = (\varepsilon \, \varepsilon^T) = E(A \, w \, w^T \, A^T) = AE(ww^T)A^T = AVA^T \tag{8.20}$$

where $E(w \, w^T) = V$ is the covariance matrix of errors w.

With the given assumptions, the errors w_i are independent so that V is the diagonal matrix with elements on the diagonal $V_{ii} = \tau_i^2$. Substitution into Eq. (8.20) results in

$$C_\varepsilon = \begin{bmatrix} \tau_1^2 & \tau_1^2 & \cdots & \tau_1^2 \\ \tau_1^2 & \tau_1^2 + \tau_2^2 & \cdots & \tau_1^2 + \tau_2^2 \\ \cdots & \cdots & \cdots & \cdots \\ \tau_1^2 & \tau_1^2 + \tau_2^2 & \cdots & \sum_{i=1}^{n} \tau_i^2 \end{bmatrix} \tag{8.21}$$

with the general element of this covariance matrix $C_{ij} = \sum_{j=1}^{k} \tau_j^2$ where $k = \min(i,j)$.

Conclusion: Knowledge of error composition is important in covariance matrix building.

If we know the point probability density function of the measurement errors $p_\varepsilon(\varepsilon)$ or the marginal densities $p_\varepsilon(\varepsilon_i)$ we can determine the probability density

$p(\mathbf{y})$ or $p(y_i)$ from the expression for the probability density for a function of a random variable.

In the case of independent random errors ε

$$p(y_i) = p_\varepsilon \left[Z_i^{-1}(\mathbf{x}_i, y_i, \varepsilon) \right] \left| \frac{\partial Z_i^{-1}(.)}{\partial y_i} \right| \qquad (8.22)$$

where $Z_i^{-1}(.)$ denotes the inverse of the function $Z(.)$. For the additive model of measurements [Eq. (8.11)] the following function may be written $Z_i^{-1}(.) = y_i - f(x_i, \boldsymbol{\beta})$ with the derivative $\left| \dfrac{\partial Z_i^{-1}(.)}{\partial y_i} \right| = 1$. Substitution into Eq. (8.22) gives

$$p(y_i) = p_\varepsilon (y_i - f(x_i, \boldsymbol{\beta})) \qquad (8.23)$$

Hence, it may be concluded that the additive model does not cause any deformations of the distribution of the measured quantities with regard to the error distribution. In the case of the multiplicative model of measurements [Eq. (8.12)], the equation obtained is $Z_i^{-1}(.) = \ln y_i - \ln f(x_i, \boldsymbol{\beta})$ with the derivative $\left| \dfrac{\delta Z_i^{-1}(.)}{\delta y_i} \right| = \dfrac{1}{y_i}$ where only positive values of the measured variable y are allowed. Substitution into Eq. (8.22) gives

$$p(y_i) = \frac{1}{y_i} p_\varepsilon \left(\ln y_i - \ln f(x_i, \boldsymbol{\beta}) \right) \qquad (8.24)$$

The probability density obtained does not correspond to the probability density of the errors $p_\varepsilon(.)$.

Problem 8.7 *Distribution of the variable y for combined errors ε*

Determine the distribution of the vector of measured variables $\mathbf{y}$, for a case of combined errors (8.14) assuming that the errors ε have multivariate normal distribution $N(0, \mathbf{C}_\varepsilon)$ and the additive model of measurements is valid.
Solution: According to Eq. (8.23), the probability density function $p_\varepsilon(\varepsilon)$ is defined for a general covariance matrix of errors $\mathbf{C}_\varepsilon$ in Eq. (8.18). It is necessary

to evaluate $\det(\mathbf{C}_\varepsilon)$ and $\mathbf{C}_\varepsilon^{-1}$ for $\mathbf{C}_\varepsilon$ defined by Eq. (8.21). If $\mathbf{C}_\varepsilon = \mathbf{AVA}^{\mathrm{T}}$, then its inverse is

$$\mathbf{C}_\varepsilon^{-1} = (\mathbf{A}^{-1})^{\mathrm{T}} \mathbf{V}^{-1} (\mathbf{A}^{-1}) \tag{8.25}$$

From Eq. (8.20), the matrix $\mathbf{A}^{-1}$ is

$$\mathbf{A}^{-1} = \begin{bmatrix} 1 & 0 & \cdots & 0 & 0 \\ -1 & 1 & \cdots & 0 & 0 \\ 0 & -1 & \cdots & 0 & 0 \\ \cdots & \cdots & \cdots & \cdots & \cdots \\ 0 & 0 & \cdots & -1 & 1 \end{bmatrix}$$

The bidiagonal matrix $\mathbf{A}^{-1}$ is a matrix with one diagonal and one underdiagonal band. Since the matrix $\mathbf{V}$ is diagonal, the matrix $\mathbf{V}^{-1}$ is also diagonal with elements τ_i^{-2} on a diagonal. Substituting into Eq. (8.25) leads to

$$\mathbf{C}_\varepsilon^{-1} = \begin{bmatrix} (\tau_1^{-2} + \tau_2^{-2}) & -\tau_2^{-2} & \cdots & 0 & 0 \\ -\tau_2^{-2} & (\tau_2^{-2} + \tau_3^{-2}) & -\tau_3^{-2} & \cdots & \cdots \\ 0 & -\tau_3^{-2} & (\tau_3^{-2} + \tau_4^{-2}) & \cdots & \cdots \\ \cdots & \cdots & \cdots & (\tau_{n-1}^{-2} + \tau_n^{-2}) & \tau_n^{-2} \\ 0 & 0 & \cdots & \tau_n^{-2} & \tau_n^{-2} \end{bmatrix} \tag{8.26}$$

This matrix is a tridiagonal matrix. Its determinant $\det(\mathbf{C}_\varepsilon)$ is calculated from Eq. (8.21)

$$\det(\mathbf{C}_\varepsilon) = \det(\mathbf{A})\det(\mathbf{V})\det(\mathbf{A}^{\mathrm{T}}) = \det(\mathbf{V}) = \prod_{i=1}^{n} \tau_i^2$$

Conclusion: The joint probability density function of a vector $\mathbf{y}$ is, according to Eqs. (8.18) and (8.23), given by the expression

$$p(\mathbf{y}) = (2\pi)^{-n/2} \left[\prod_{i=1}^{n} \tau_i^2 \right]^{-1/2} \exp\left[-0.5(\mathbf{y} - \mathbf{f})^{\mathrm{T}} (\mathbf{A}^{-1})^{\mathrm{T}} \mathbf{V}^{-1} \mathbf{A}^{-1} (\mathbf{y} - \mathbf{f}) \right] \tag{8.27}$$

where the vector **f** contains the elements $f(x_i, \beta)$, $i = 1,..., n$. The variable y also has a multivariate normal distribution with the same covariance matrix of errors as $\mathbf{C}_e$.

From a survey of error models, it follows from experimental conditions and assumptions about various types of errors, that the distribution of the measured variable y can be derived. In the measurements made in a chemical laboratory, most of the observed errors have the normal distribution, and follow the additive model of errors. Any differences are characterized by the covariance matrix $\mathbf{C}_e$, which may contain only diagonal elements, or off-diagonal elements in addition.

8.3   Formulation of the Regression Criterion

For the vector of measured values $\mathbf{y} = \{y_1,..., y_n\}^T$, the joint probability density function is denoted by the likelihood function $L(\boldsymbol{\theta})$. This function depends on the vector of parameters, $\boldsymbol{\theta}$ which contains the model parameters, β and distribution parameters, σ. The maximum likelihood estimates of parameters, $\hat{\boldsymbol{\theta}}$, are determined by maximization of the logarithm of the function

$$\ln L(\hat{\boldsymbol{\theta}}) = \ln p(\mathbf{y}) = \sum_{i=1}^{n} \ln p(y_i) \tag{8.28}$$

The maximum likelihood estimates $\hat{\boldsymbol{\theta}}$ have an asymptotic variance equal to the inverse of the expected *Fisher information matrix*

$$D(\hat{\boldsymbol{\theta}}) = \mathbf{I}^{-1}(\hat{\boldsymbol{\theta}}) \tag{8.29}$$

The elements of matrix $I(\boldsymbol{\theta})$ are given by

$$I_{ij} = -E\left[\frac{\partial^2 \ln L(\theta)}{\partial \theta_i\, \partial \theta_j}\right] \tag{8.30}$$

For practical purposes, the Fisher information matrix $\mathbf{I}(\theta)$ is replaced by the *estimated information matrix* $\hat{\mathbf{I}}(\theta)$ with elements

$$\hat{I}_{ij} = -\left[\frac{\partial^2 \ln L(\boldsymbol{\theta})}{\partial \theta_i\, \partial \theta_j}\right]_{\theta=\hat{\theta}} \tag{8.31}$$

The estimated information matrix can be used to construct confidence intervals more conveniently. For maximum likelihood estimates, some important properties may be derived:

(1) The estimates $\hat{\theta}$ are asymptotically $(n \to \infty)$ unbiased. Therefore the bias

$$\mathbf{h} = \mathbf{\theta} - E(\hat{\mathbf{\theta}}) = 0 \qquad (8.32)$$

is the zero vector. For a finite sample size n, the estimates $\hat{\theta}$ are biased and the magnitude $\mathbf{h}$ depends on the degree of non-linearity of the regression model.

(2) The estimates $\hat{\theta}$ are asymptotically efficient and the variance estimates are minimal of all unbiased estimates. The covariance matrix $D(\hat{\theta})$ lies on the lower limit of the Cramer-Rao inequality [5]. For finite samples, this property is generally not fulfilled.

(3) The random vector $\sqrt{n}(\hat{\mathbf{\theta}} - \mathbf{\theta})$ has, asymptotically, the normal distribution $N(0, \mathbf{I}^{-1})$ with zero mean and variance equal to the inverse of the Fisher information matrix. When the error distribution is approximately normal, the normality of estimates is valid for finite samples.

For sufficiently large sample sizes, many interesting properties of the estimates $\hat{\theta}$ may be used. For finite sample sizes, some difficulties arise from the bias estimates $\hat{\theta}$. If the probability density function $p(\mathbf{y})$ is known, the maximum likelihood estimates or a criterion for their determination (the *regression criterion*) may be found.

Problem 8.8   *Regression criterion for additive errors*

Derive the regression criterion for the case when measurements errors are independent, with zero mean, constant variance, and the normal distribution $N(0, \sigma^2 \mathbf{E})$; and with the assumption that the additive model of measurement errors (8.11) is valid.

Solution: Let $f_i = f(x_i; \beta)$ and $\theta^T = (\beta^T, \sigma^2)$. If the distribution of measured variable y_i is normal, $N(f_i, \sigma^2)$, then

$$p(y_i) = \frac{1}{\sqrt{2\pi\sigma^2}} \exp\left[\frac{-(y_i - f_i)^2}{2\sigma^2}\right] \qquad (8.33)$$

The logarithm of the likelihood function, $\ln L(\theta)$, has the form

$$\ln L(\theta) = \sum_{i=1}^{n} \ln p(y_i) = \frac{-n}{2} \ln(2\pi\sigma^2) - \frac{2}{(2\sigma)^2} U(\beta) \tag{8.34}$$

where $U(\beta)$ is the *least-squares criterion* or the residual sum of squares of deviations defined as

$$U(\beta) = \sum_{i=1}^{n} (y_i - f_i)^2 \tag{8.35}$$

Analytical maximization of $\ln L(\theta)$ according to σ^2 leads to

$$\frac{\partial \ln L(\theta)}{\partial \sigma^2} = -\frac{n}{2\sigma^2} + \frac{1}{2\sigma^4} U(\beta) = 0 \tag{8.36a}$$

and therefore

$$\hat{\sigma}^2 = \frac{U(\beta)}{n} \tag{8.36b}$$

The estimate σ^2 is biased for a small number of measurements. The unbiased form is

$$\hat{\sigma}^2 = \frac{U(\beta)}{n - m} \tag{8.37}$$

On substituting from Eq. (8.36b) into Eq. (8.34), the concentrated likelihood function $\ln L(\beta)$ is formulated as

$$\ln L(\beta) = -\frac{n}{2}(1 + \ln(2\pi)) - 0.5 \ln U(\beta) \tag{8.38}$$

The maximum of $\ln L(\beta)$ corresponds to the minimum of the regression criterion $U(\beta)$ which is, in fact, a condition for the least-squares method (LS). That is, the method of maximum likelihood is identical to the least-squares method.

On the basis of Eq. (8.29), the covariance matrix of estimates $D(\hat{\theta})$ is given by

$$D(\hat{\theta}) = \begin{bmatrix} \sigma^2 (\mathbf{J}^T\mathbf{J})^{-1} & 0 \\ 0 & 2\sigma^4 / n \end{bmatrix} \tag{8.39}$$

where $\mathbf{J}$ is the $(n \times m)$ Jacobi matrix of the first derivatives of the model, with elements

$$J_{ij} = \frac{\partial f(x_i, \beta)}{\partial \beta_j} \tag{8.40}$$

From Eq. (8.39), it follows that the estimates $\hat{\sigma}^2$ and $\mathbf{b}$ are independent and the parameter covariance matrix is $D(\mathbf{b}) = \sigma^2 (\mathbf{J}^T\mathbf{J})^{-1}$.

Conclusion: The maximum likelihood method enables either the formulation of a regression criterion or the determination of the covariance matrix of estimates. It may be concluded that with the use of the properties of the maximum likelihood method, we can simplify the construction of the confidence intervals and carry out statistical hypothesis testing.

Maximization of the likelihood function leads to the problem of *nonlinear optimization*. When the covariance matrix of errors $\mathbf{C}_\varepsilon$ is known, we can for the additive model of measurement errors and normal error distribution, $\varepsilon \approx N(0, \mathbf{C}_\varepsilon)$, find the maximum likelihood estimates $\mathbf{b}$ of parameters β by minimizing the criterion of the generalized least-squares

$$U(\beta) = (\mathbf{y} - \mathbf{f})^T \mathbf{C}_\varepsilon^{-1} (\mathbf{y} - \mathbf{f}) = \mathrm{Tr}\left[\mathbf{C}_\varepsilon^{-1} \ \hat{\mathbf{e}} \ \hat{\mathbf{e}}^T \right] \tag{8.41a}$$

where $\hat{\mathbf{e}} = \mathbf{y} - \mathbf{f}$ is the deviation or residual and the symbol $\mathrm{Tr}(\mathbf{A})$ denotes the trace of matrix $\mathbf{A}$. If the matrix $\mathbf{C}_\varepsilon$ is diagonal, the situation is much simpler. The least-squares criterion [Eq. (8.41a)] transforms into the relationship

$$U(\beta) = \sum_{i=1}^{n} w_i^2 (y_i - f_i)^2 = \sum_{i=1}^{n} w_i^2 (y_i - f(x_i, \beta))^2 \tag{8.41b}$$

where $w_i^2 = 1/C_{ii}$ is the weight equal to the reciprocal value of the elements of covariance matrix. If the variables $y_i^* = w_i y_i$ and $f_i^* = w_i f(x_i, \beta)$ are introduced, Eq. (8.41b) takes the form of the *classical least-squares method*

$$U(\boldsymbol{\beta}) = \sum_{i=1}^{n} (y_i^* - f_i^*)^2 \tag{8.41c}$$

When the weights are known, a weighted least-squares problem can be converted into a classical least-squares problem with modified variables. This procedure can also be used for the unknown matrix $\mathbf{C}_\varepsilon$ when its elements are estimated separately, for example, because of heteroscedasticity.

For an unknown matrix $\mathbf{C}_\varepsilon$, the technique of consecutive maximization is used. First, the estimate $\mathbf{C}_\varepsilon$ is computed and substituted into the likelihood function. The resulting concentrated likelihood function contains only the parameters $\boldsymbol{\beta}$. Bard [6] derived the following derivatives

$$\frac{\partial \ln \det(\mathbf{C}_\varepsilon)}{\partial \mathbf{C}_\varepsilon} = (\mathbf{C}_\varepsilon)^{-1} \tag{8.42a}$$

and

$$\frac{\partial \mathrm{Tr}\left[\mathbf{C}_\varepsilon^{-1}\, \hat{\mathbf{e}}\, \hat{\mathbf{e}}^{\mathrm{T}}\right]}{\partial \mathbf{C}_\varepsilon} = -(\mathbf{C}_\varepsilon)^{-1}\hat{\mathbf{e}}\, \hat{\mathbf{e}}^{\mathrm{T}}(\mathbf{C}_\varepsilon^{\mathrm{T}})^{-1} \tag{8.42b}$$

The concentrated likelihood function is

$$\ln L = -\frac{n}{2}\ln 2\pi - \frac{1}{2}\ln \det(\mathbf{C}_\varepsilon) - \frac{1}{2}\hat{\mathbf{e}}^{\mathrm{T}}\mathbf{C}_\varepsilon^{-1}\hat{\mathbf{e}} \tag{8.43a}$$

and since

$$\mathbf{C}_\varepsilon = \hat{\mathbf{e}}\hat{\mathbf{e}}^{\mathrm{T}} \tag{8.43b}$$

Then

$$\mathrm{Tr}\left[\hat{\mathbf{e}}\, \hat{\mathbf{e}}^{\mathrm{T}}\right] = \mathrm{Tr}(\mathbf{E}) = n \tag{8.43c}$$

and the concentrated likelihood function has the form

$$\ln L(\boldsymbol{\beta}) = -\frac{n}{2}\left[\ln 2\pi + 1\right] - \frac{1}{2}\ln \det(\hat{\mathbf{e}}\ \hat{\mathbf{e}}^{\mathrm{T}}) \tag{8.44}$$

Maximization of the function $\ln L(\boldsymbol{\beta})$ is the same as minimization of the criterion $U_D(\boldsymbol{\beta})$ where

$$U_D(\boldsymbol{\beta}) = \det(\mathbf{e} \quad \mathbf{e}^{\mathrm{T}}) = \det\left[(\mathbf{y}-\mathbf{f})(\mathbf{y}-\mathbf{f})^{\mathrm{T}}\right] \tag{8.45a}$$

When matrix $\mathbf{C}_\varepsilon$ is diagonal, matrix $\hat{\mathbf{C}}_\varepsilon$ is also diagonal and Eq. (8.45a) converts into the form

$$U(\boldsymbol{\beta}) = \sum_{i=1}^{n} \frac{1}{\hat{C}_{ii}} (y_i - f(x_i,\boldsymbol{\beta}))^2 \tag{8.45b}$$

Hence, in the case of heteroscedasticity, application of the classical least-squares method leads to unbiased estimates **b** but the estimates of the covariance matrix $\hat{\mathbf{C}}_\varepsilon$ are biased.

Problem 8.9 *Regression criterion for combined errors*

Derive a regression criterion for a case of combined errors [Eq. (8.14)] assuming that errors ε have the multivariate normal distribution and that

(1) the measurement errors v_i are negligible in comparison to the process errors u_j,

(2) the process errors u_j are negligible with respect to the errors of measurement v_i.

Solution: The joint probability density function $p(\mathbf{y})$ is expressed by Eq. (8.27). The logarithm of the likelihood function may be expressed as

$$\ln L(\boldsymbol{\theta}) = \frac{n}{2}\ln(2\pi) - \frac{1}{2}\sum_{i=1}^{n}\ln \tau_i^2 - \frac{1}{2}(\mathbf{y}-\mathbf{f})^{\mathrm{T}}(\mathbf{A}^{-1})^{\mathrm{T}}\mathbf{V}^{-1}(\mathbf{y}-\mathbf{f}) \tag{8.46}$$

The last term in this equation can be expressed as

$$U_c(\boldsymbol{\beta}) = \sum_{i=1}^{n}\tau_i^{-2}\left[(y_i - y_{i-1}) - (f(x_i,\boldsymbol{\beta}) - f(x_{i-1},\boldsymbol{\beta}))\right]^2 = \sum_{i=1}^{n}\tau_i^{-2}\left[L_i - K_i(\boldsymbol{\beta})\right]^2$$

$$\tag{8.47}$$

where $y_0 = 0$ and $f(x_0, \beta) = 0$. Moreover, $L_i = y_i - y_{i-1}$ and $K_i(\beta) = f(x_i, \beta)$ $- f(x_{i-1}, \beta)$. Equation (8.47) corresponds to the weighted least-squares method for first differences.

The maximum of the function $\ln L(\theta)$ generally corresponds to a minimum of

$$Q = \sum_{i=1}^{n} \ln\left[\sigma^2 + \sigma_0^2 f(x_i, \beta)\right] + U_c(\beta) \tag{8.48}$$

Maximization of Q in terms of σ^2, σ_0^2 and β may be achieved by general minimization methods:

(1) Small measurement errors. For $\sigma^2 \gg \sigma_0^2 f(x_i, \beta), \tau_i^2 = \sigma^2 = $ constant and we have a *model of pure process errors*. On substituting into the likelihood function (8.46) and differentiating with respect to σ^2, we get

$$\frac{\partial \ln L(\beta)}{\partial \sigma^2} = -\frac{0.5n}{\sigma^2} + \frac{0.5}{\sigma^4} \sum_{i=1}^{n} \left[L_i - K_i(\beta)\right]^2 = 0 \tag{8.49}$$

On rearrangement, we find

$$\hat{\sigma}^2 = \frac{U_P(\beta)}{n} \tag{8.50}$$

Here

$$U_P(\beta) = \sum_{i=1}^{n} \left[L_i - K_i(\beta)\right]^2 \tag{8.51}$$

is the regression criterion for the minimum of the sum of the squared first differences. If estimate $\hat{\sigma}^2$ [Eq. (8.50)] is substituted into the likelihood function (8.46), the concentrated likelihood function is obtained

$$\ln L(\beta) = -\frac{n}{2}[1 + \ln 2\pi] - \frac{1}{2} \ln U_P(\beta) \tag{8.52}$$

The maximum of the function $\ln L(\beta)$ corresponds to a minimum of $U_p(\beta)$. The function $U_p(\beta)$ [Eq. (8.51)] may be minimized by many nonlinear regression programs, after a simple rearrangement of y_i and $f(x_i, \beta)$ into variables L_i and $K_i(\beta)$.

(2) When the process errors (fluctuations of the system) are small in comparison to errors of measurement, we have $\sigma^2 \ll \sigma_0^2 f(x_i, \beta)$, and the variance $\tau^2 \approx \sigma_0^2 f(x_i, \beta)$. On substituting into the likelihood function [Eq. (8.46)] and differentiating with respect to σ_0^2, we get

$$\frac{\partial \ln L(\theta)}{\partial \sigma_0^2} = -\frac{n}{2\sigma_0^2} + \frac{0.5}{\sigma_0^4} \sum_{i=1}^{n} f^{-2}(x_i, \beta)(L_i - K_i(\beta))^2 = 0 \qquad (8.53a)$$

On rearrangement, we get

$$\hat{\sigma}_0^2 = \frac{1}{n} \sum_{i=1}^{n} f^{-2}(x_i, \beta)(L_i - K_i(\beta))^2 \qquad (8.53b)$$

and on substituting $\hat{\sigma}_0^2$ into the likelihood function [Eq. (8.46)], the concentrated likelihood function becomes

$$\ln L(\beta) = -\frac{n}{2}[1 + \ln n + \ln 2\pi] - \sum_{i=1}^{n} \ln f(x_i, \beta) - \frac{1}{2}\ln U_W(\beta) \qquad (8.54)$$

where $U_W(\beta)$ is the criterion of the *weighted least-squares method* for first differences

$$U_W(\beta) = \sum_{i=1}^{n} \left[\frac{L_i - K_i(\beta)}{f(x_i, \beta)}\right]^2 \qquad (8.55)$$

The maximum of the function $\ln L(\beta)$ [Eq. (8.54)] corresponds to the minimum of function $U_C(\beta)$

$$U_C(\beta) = U_W(\beta) \prod_{i=1}^{n} f^2(x_i, \beta) \qquad (8.56)$$

The criterion $U_C(\boldsymbol{\beta})$ may be minimized either by general algorithms or by the *iterative method of weighted least-squares*.

It is obvious that this situation is more complicated than the case of small measurement errors. When the accumulated process errors are negligible then $\varepsilon_i = v_i$ in Eq. (8.44) and the problem of heteroscedasticity has to be solved. The corresponding regression criterion will have the form (8.56) but the function $U_W(\boldsymbol{\beta})$ is expressed in terms of variables $y_i, f(x_i, \boldsymbol{\beta})$ instead of $L_i, K_i(\boldsymbol{\beta})$.

Conclusion: When the errors are complicated, a suitable regression criterion may be derived by using the maximum likelihood method.

8.4 Geometry of Nonlinear Regression

Although some laboratory problems lead to criteria different from the classical least-squares method (LS), the LS method is still the most commonly used method in experimental practice. In Problem 8.8, it was shown that the LS method is really a special case of the maximum likelihood method for an additive model of measurement errors and the normal distribution of independent errors s with zero mean and constant variance. For the purpose of geometric interpretation, the least-squares criterion $U(\boldsymbol{\beta})$ in Eq. (8.35) is rewritten in vector notation as

$$U(\boldsymbol{\beta}) = \|\mathbf{y} - \mathbf{f}\|^2 \qquad (8.57)$$

where $\mathbf{y} = (y_1, ..., y_n)^T, \mathbf{f} = (f(x_1, \boldsymbol{\beta}), ..., f(x_n, \boldsymbol{\beta}))^T$ and the symbol $\|\mathbf{x}\| = \sqrt{\mathbf{x}^T \mathbf{x}}$ means the Euclidean norm.

Examination of the shape of the criterion function $U(\boldsymbol{\beta})$ in the space of the estimators helps to explain why the search for the function minimum is so difficult. In this $(m + 1)$-dimensional space, values of criterion $U(\boldsymbol{\beta})$ are plotted against the parameters $\beta_1, ..., \beta_m$.

For linear regression models, the criterion function $U(\boldsymbol{\beta})$ is an elliptic hyperparaboloid with its centre at $[\mathbf{b}, U(\mathbf{b})]$, the place where $U(\boldsymbol{\beta})$ reaches a minimum (the "pit point"). For linear models, the criterion function $U(\boldsymbol{\beta})$ has a quadratic form of type $\boldsymbol{\beta}^T (\mathbf{X}^T \mathbf{X}) \boldsymbol{\beta}$, and the matrix $\mathbf{X}^T \mathbf{X}$ is positive-definite.

In some cases, the parameter space is used for interpretation of $U(\boldsymbol{\beta})$. Parameter space is an m-dimensional space with the components of vector $\boldsymbol{\beta}$ on the axes. The value $U(\boldsymbol{\beta})$ is a perpendicular projection of an $(m + 1)$-dimensional object into this m-dimensional space. For the two estimated parameters β_1 and β_2 where $m = 2$, the criterion function $U(\boldsymbol{\beta})$ for a linear model is drawn in the $(m + 1)$-dimensional (i.e. 3-dimensional) space, in Fig. 8.3.

Rather complicated shapes can occur in nonlinear models, as a result of the nonlinear function $f(\mathbf{x}, \boldsymbol{\beta})$; there may be a number of extremes and saddle points. Figure 8.4 is an illustration of a criterion function $U(\boldsymbol{\beta})$ with two minima and one saddle point.

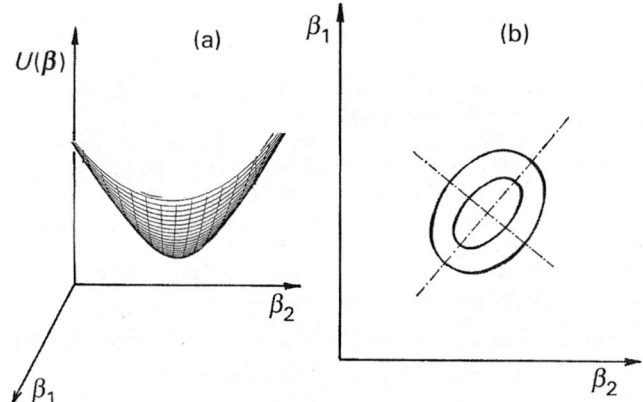

Figure 8.3 Interpretation of the criterion function $U(\boldsymbol{\beta})$ for (a) a linear model $(m = 2)$ forming the elliptic hyperparaboloid in the $(m + 1)$-dimensional space of estimators, and (b) the concentric ellipses as contours in the m-dimensional parameter space.

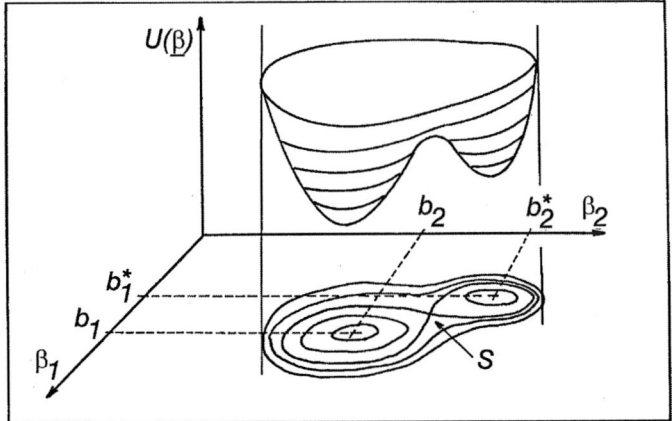

Figure 8.4 A criterion function $U(\boldsymbol{\beta})$ with two local minima $\boldsymbol{\beta}$ and $\boldsymbol{\beta}^*$ and one saddle point S.

Quantitative information on the local behaviour of the criterion function $U(\boldsymbol{\beta})$ in the vicinity of any point β_j may be obtained from a Taylor series expansion up to quadratic terms:

$$U(\boldsymbol{\beta}) = U(\boldsymbol{\beta}_j) + \Delta\boldsymbol{\beta}_j^T \mathbf{g}_j + \frac{1}{2}\Delta\boldsymbol{\beta}_j^T \mathbf{H}\Delta\boldsymbol{\beta}_j \qquad (8.58)$$

where $\Delta\boldsymbol{\beta}_j = \boldsymbol{\beta} - \boldsymbol{\beta}_j$ and $\mathbf{g}_j$ is the gradient vector of a criterion function containing the components

$$\mathbf{g}_k = \frac{\partial U(\boldsymbol{\beta})}{\partial \beta_k}, \quad k = 1,\dots,m \qquad (8.59)$$

The matrix $\mathbf{H}_j$ of dimension $(m \times m)$, is the symmetric *Hessian matrix* defined by the second derivative of the criterion function $U(\boldsymbol{\beta})$ with components

$$H_{lk} = \frac{\partial U(\boldsymbol{\beta})}{\partial \beta_l \, \partial \beta_k}, \quad l,k = 1,\dots,m \qquad (8.60)$$

Equation (8.59a) is valid for any criterion function $U(\boldsymbol{\beta})$. In the least-squares method the gradient of the criterion function $U(\boldsymbol{\beta})$ from Eq. (8.57) has the form

$$\mathbf{g}_j = -2\mathbf{J}^T \mathbf{e} \qquad (8.59b)$$

where $\mathbf{e}$ is the difference vector with elements $e_i = y_i - f(x_i, \boldsymbol{\beta})$, $i = 1,\dots,m$.

The matrix $\mathbf{J}$ of dimension $(n \times m)$ is called the *Jacobian matrix* with elements corresponding to the first derivative of the regression model in terms of the individual parameters at given points. These elements have the form

$$J_{ik} = \frac{\partial f(x_i, \boldsymbol{\beta})}{\partial \beta_k}, \quad i = 1,\dots,n, \ k = 1,\dots,m \qquad (8.61)$$

With the least-squares method, a similar relationship involving the Hessian matrix may be derived:

$$\mathbf{H}_j = 2\left[\mathbf{J}^T\mathbf{J} + \mathbf{B}\right] \tag{8.62}$$

where $\mathbf{B}$ is a matrix containing the second derivatives of the regression function with elements

$$B_{kj} = \sum_{i=1}^{n} e_i \frac{\partial^2 f(x_i,\beta)}{\partial\beta_k \partial\beta_j}, \quad k,j = 1,...,m \tag{8.63}$$

In the vicinity of local minima $\mathbf{b}$, the gradient $\mathbf{g}$ is approximately equal to zero. This means that

(1) the error vector $\hat{\mathbf{e}}$ is perpendicular to the columns of a matrix $\mathbf{J}$ in -dimensional space;

(2) the criterion function $U(\beta)$ is proportional to the quadratic form $\Delta\beta_i^T \mathbf{\varsigma}_i \Delta\beta_i$.

The type of local extreme is distinguished by a matrix $\mathbf{H}$. When the matrix $\mathbf{H}$ is

(a) positive-definite, the extreme is a minimum and $U(\beta)$ approximates to an elliptic hyperparaboloid;

(b) negative-definite, the extreme is a maximum;

(c) indefinite, no extreme is present.

Definiteness of matrices is examined by the Sylvester conditions. For practical calculation, it is necessary that the positive-definite matrix is regular, has rank m and has eigenvalues that are all positive. It is useful to compare the Taylor series expansion of criterion function $U(\beta)$ by Eq. (8.58) with the criterion function $U(\beta)$ into which the Taylor series expansion of a model function is substituted. With the use of Taylor series expansion, the function $f(x_i,\beta)$ in the vicinity of the point β_j may be approximated by

$$f(x_i,\beta) = f(x_i,\beta_j) + \mathbf{J}_i(\beta - \beta_j) + \frac{1}{2}(\beta-\beta_j)^T \mathbf{G}_i(\beta- \beta_j) \tag{8.64}$$

where $\mathbf{G}_i$ is the matrix of second derivatives of a model function $f(x_i,\beta)$ with elements

$$G_{jk} = \frac{\partial^2 f(x_i,\beta)}{\partial\beta_j \partial\beta_k}, \quad j,k = 1,...,m \tag{8.65}$$

and $\mathbf{J}_i$ is the ith row of the matrix $\mathbf{J}$. Generally, a vector $\mathbf{f}$ may be approximated by the Taylor series expansion into quadratic terms

$$\mathbf{f} \approx \mathbf{f}(\boldsymbol{\beta}_j) + \mathbf{J}\Delta\boldsymbol{\beta}_j + \frac{1}{2}\Delta\boldsymbol{\beta}_j^{\mathrm{T}} \mathbf{G}_i \Delta\boldsymbol{\beta}_j \tag{8.66}$$

where $\mathbf{G}_i$ is an $(n \times m \times n)$-dimensional array with layers formed by the matrices $\mathbf{G}_i$. Usually, a linearization of the function $\mathbf{f}$ is used:

$$\mathbf{f} \approx \mathbf{f}(\boldsymbol{\beta}_j) + \mathbf{J}\Delta\boldsymbol{\beta}_j \tag{8.66a}$$

Substituting Eq. (8.66a) into (8.57) we get the criterion for the "linearly-transformed" least-squares method [7]

$$U_L(\boldsymbol{\beta}) = \mathbf{e}^{\mathrm{T}}\mathbf{e} - 2\Delta\boldsymbol{\beta}\mathbf{J}^{\mathrm{T}}\mathbf{e} + \Delta\boldsymbol{\beta}^{\mathrm{T}}(\mathbf{J}^{\mathrm{T}}\mathbf{J})\Delta\boldsymbol{\beta} \tag{8.67}$$

The first term of this equation is equal to $U(\boldsymbol{\beta}_j)$

$$\mathbf{e}^{\mathrm{T}}\mathbf{e} = U(\boldsymbol{\beta}_j) \tag{8.67a}$$

and the second one to $\Delta\boldsymbol{\beta}_j^{\mathrm{T}}\mathbf{g}$

$$-2\Delta\boldsymbol{\beta}\mathbf{J}^{\mathrm{T}}\mathbf{e} = \Delta\boldsymbol{\beta}_j^{\mathrm{T}}\mathbf{g} \tag{8.67b}$$

Equation (8.67) differs from (8.58) only in the third term containing matrix $2\mathbf{J}^{\mathrm{T}}\mathbf{J}$ instead of matrix $\mathbf{H}$. It follows from Eqs. (8.62) and (8.63) that for small error values e_i the matrix $\mathbf{B}$ may be neglected, making Eqs. (8.67) and (8.58) identical. This means that the linearization of the regression model corresponds to the Taylor series expansion of the criterion function $U(\boldsymbol{\beta})$ into quadratic terms, assuming that matrix $\mathbf{B}$ is negligible. From Eq. (8.67), it also follows that for nonlinear regression models the matrix $\mathbf{J}^{\mathrm{T}}\mathbf{J}$ corresponds to $\mathbf{X}^{\mathrm{T}}\mathbf{X}$ in linear models. If the linearization (8.67) is sufficiently precise, the statistical analysis may be performed in a similar way to that used for linear regression models.

(A) *The geometry of linear least-squares*

For the interpretation of the geometry of linear regression, n-dimensional sample space is used. The vector of observations $\mathbf{y} = (y_1,...,y_n)^{\mathrm{T}}$ defines

a line OY from the origin O to the point Y with co-ordinates $(y_1,...,y_n)$. The
X matrix has m column vectors x_i, $i = 1,..., n$, each containing n elements. The
elements of the jth column define the co-ordinates $(x_{j1}, x_{j2},..., x_{jn})$ of a point
$\mathbf{X}_j$ in the sample space, and the jth column vector of matrix **X** defines the vector
OX_j in sample space. The m vectors OX_1, OX_2,..., OX_m define a subspace of
m dimensions called the *estimation space* which is contained within the sample
space. Any point in this subspace can be represented by the termination of a
vector which is a linear combination of the vectors defining the space—that is,
a linear combination of the columns of **X**, such as **Xβ** where $\boldsymbol{\beta} = (\beta_1,..., \beta_m)$ is
an $m \times 1$ vector. Suppose the vector **Xβ** defines the point T. Then the squared
distance YT^2 is given by

$$(\mathbf{y} - \mathbf{X}\boldsymbol{\beta})^{\mathrm{T}}(\mathbf{y} - \mathbf{X}\boldsymbol{\beta}) = U(\boldsymbol{\beta})$$

as defined in Chapter 6. Thus the sum of squares $U(\boldsymbol{\beta})$ represents, in the sample
space, the squared distance of Y from a general point T in the estimation space.
Minimization of $U(\boldsymbol{\beta})$ with respect to $\boldsymbol{\beta}$ implies finding that value of $\boldsymbol{\beta}$, say **b**,
which provides a point P (defined by the vector $\hat{\mathbf{y}} = \mathbf{Xb}$) in the estimation space
closest to the point Y. Then, geometrically, P must be the foot of the perpendicular
from Y to the estimation space, that is, the foot of a line passing through Y and
the orthogonal to all the columns of matrix **X**. In terms of the vectors from the
origin, we can write

$$\mathbf{y} = \hat{\mathbf{y}} + (\mathbf{y} - \hat{\mathbf{y}}) = \mathbf{y} + \hat{\mathbf{e}}$$

where **e** is the *vector of residuals*. The vector **y** is thus divided into two orthogonal
components:

(1) $\hat{\mathbf{y}}$, which lies entirely in the *estimation space*, and
(2) $\hat{\mathbf{e}}$, the vector of residuals, which lies in the *residual space*. The residual
space is defined as the $(n \times m)$-dimensional subspace, which is the
remainder of the full n-dimensional space, after the m-dimensional
estimation space has been defined. The estimation and residual spaces
are thus orthogonal.

If T is a general point in the estimation space and YP is orthogonal to the
space, then $\mathrm{YT}^2 = \mathrm{YP}^2 + \mathrm{PT}^2$ or $U(\boldsymbol{\beta}) = U(\mathbf{b}) + \mathrm{PT}^2$. Thus, the contours for
which $U(\boldsymbol{\beta}) = $ constant must be such that $\mathrm{PT}^2 = U(\boldsymbol{\beta}) - U(\mathbf{b}) = $ constant. In the
sample space, then, the contours defined by $U(\boldsymbol{\beta}) = $ constant consist of all points
T such that $\mathrm{PT}^2 = $ constant; that is, points in the estimation space with the form
Xβ which lie on an m-dimensional sphere centered at the point P defined by **Xb**.
The radius of this sphere is $\sqrt{U(\boldsymbol{\beta}) - U(\mathbf{b})}$.

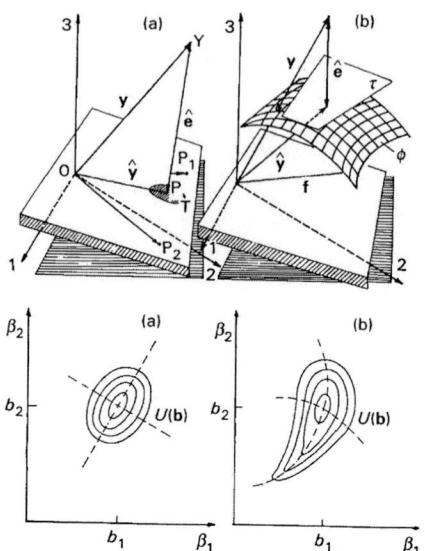

Figure 8.5 Geometrical representation of (a) linear least-squares, and (b) nonlinear least-squares. Upper diagram shows the sample space and the lower diagram illustrates the parameter space.

In order to illustrate the geometry of the linear least-squares, we will look at a sample space with $n = 3$ (Fig. 8.5a) and $m = 2$. That is, there are three components (y_1, y_2, y_3) of the vector $\mathbf{y}$, two parameters β_1 and β_2 of the parametric vector $\boldsymbol{\beta}$ and a three by two matrix $\mathbf{X}$ of the form

$$\mathbf{X} = \begin{bmatrix} x_{11} & x_{21} \\ x_{12} & x_{22} \\ x_{13} & x_{23} \end{bmatrix}$$

The columns of $\mathbf{X}$ define two points P_1 and P_2 with co-ordinates (x_{11}, x_{12}, x_{13}) and (x_{21}, x_{22}, x_{23}) respectively, and the vectors OP_1 and OP_2 define a plane which represents the 2-dimensional estimation space in which vector $\hat{\mathbf{y}} = \mathbf{Xb}$ must lie. The point Y lies above this plane and the perpendicular YP from Y to the plane OP_1P_2 meets the plane at P. Thus YP is the shortest distance from Y to any point in the estimation space, P, and is defined by $\hat{\mathbf{y}} = \mathbf{Xb}$ and $U(\boldsymbol{\beta}) = YP^2$. Then, from Pythagoras's theorem:

$$OY^2 = OP^2 + YP^2$$

If $\mathbf{y} = \hat{\mathbf{y}} + \mathbf{e}$, then the geometrical vector equation is OY = OP + YP. We recall that, in general, contours with constant $U(\boldsymbol{\beta})$ are represented by

m-dimensional spheres in the *estimation space*. Here, the contours must be circles on the plane OP_1P_2. It is evident that if T is a general point $\mathbf{Xb}$ on the plane, $U(\boldsymbol{\beta}) = $ constant means that $YT^2 = $ constant, so that $PT^2 = YT^2 - YP^2 = $ constant. Hence, we obtain circles about P as shown in Fig. 8.5a.

The parameter space is an m-dimensional space in which a set of parameter values $(\beta_1, ..., \beta_m)$ defines a point. The minimum value of $U(\boldsymbol{\beta})$ is attained at the point $\mathbf{b} = (b_1, ..., b_m)$. We recall that

$$U(\boldsymbol{\beta}) - U(\mathbf{b}) = (\boldsymbol{\beta} - \mathbf{b})^T \mathbf{X}^T \mathbf{X}(\boldsymbol{\beta} - \mathbf{b}).$$

All values of $\boldsymbol{\beta}$ which satisfy $U(\boldsymbol{\beta}) = $ constant $= K$ are given by $(\boldsymbol{\beta} - \mathbf{b})^T \mathbf{X}^T \mathbf{X}(\boldsymbol{\beta} - \mathbf{b}) = K - U(\mathbf{b})$. It can be shown that this is the equation of a closed ellipsoidal contour surrounding the point $\mathbf{b}$. When $K_1 > K_2$, the contour $U(\boldsymbol{\beta}) = K_1$ completely encloses the contour $U(\boldsymbol{\beta}) = K_2$ and $\mathbf{b}$ lies in the centre of these nested m-dimensional "eggs". A $100(1 - \alpha)\%$ confidence region for the true (but unknown) value of $\boldsymbol{\beta}$ is enclosed by the contour given by

$$\frac{[U(\boldsymbol{\beta}) - U(\mathbf{b})]/m}{U(\mathbf{b})/(n-m)} F_{1-\alpha}(m, n-m)$$

only if errors are normally distributed, i.e. ε comes from $N(0, \sigma^2)$. The equation can be rearranged as

$$U(\boldsymbol{\beta}) = U(\mathbf{b})\left[1 + \frac{m}{n-m} F_{1-\alpha}(m, n-m)\right]$$

where the expression on the right-hand side is a constant value that defines the contour. The outer contour shown in Fig. 8.5a is labelled as the $100(1 - \alpha)\%$ confidence contour, defined by the above equation. In the 2-dimensional space (β_1, β_2), the contours are concentric ellipses about the point (b_1, b_2). Note that contours of this type are obtained, irrespective of the value of n (the number of observations), since the dimension of the parameter space depends on m alone.

(B) *The geometry of nonlinear least-squares*

When the model is nonlinear there is no $\mathbf{X}$ matrix as in the linear model. Although there is still an estimation space, it is not one that is defined by a set of vectors and it may be very complex. This estimation space is called the *solution locus* and it consists of all points with co-ordinates of the form

$\{f(x_1,\beta), f(x_2,\beta),...f(x_n,\beta),\}$. Since the sum of squares $U(\beta)$ still represents the square of the distance from the point of the solution locus, minimization of $U(\beta)$ still corresponds geometrically to finding the point P of the solution locus which is nearest to Y.

Figure 8.5b shows the sample space for an example involving $n = 3$ observations, y_1, y_2 and y_3, taken at x_1, x_2 and x_3 respectively, and two parameters β_1 and β_2. The curved lines $f(\beta_j), j = 1,2$, also called the *estimation space curves*, indicate the co-ordinate system of parameters on the estimation space or solutions locus. It consists of all points of the form $\{f(x_1,\beta_1,\beta_2), f(x_2,\beta_1,\beta_2), f(x_3,\beta_1,\beta_2),\}$ as β_1 and β_2 vary with x_1, x_2 and x_3 fixed. Generally, this co-ordinate system is formed by all possible combinations of parameters values β in vector f, where f denotes $f(\beta_j)$ in which all parameters β_k, for $k \neq j$, are constant. The co-ordinate system of the estimation space curves forms the *estimation surface* Φ (the hatched part in Fig. 8.5b) of all possible solutions. From Fig. 8.5b, it is obvious that the termination points of all vectors f lie on this estimation surface. When a solution lies on the estimation surface, a vector of parameter estimates b exists for which $y = f(b)$ i.e. $U(\beta) = 0$ and the regression model goes through all experimental points. The tangent plane τ in a location b denoted by the unhatched region, is expressed by Eq. (8.66a) when β_j is replaced by b_j. From the geometry shown in Fig. 8.5b, the following conclusions are drawn:

(1) The minimum $U(\beta)$ corresponds to the minimum distance between a vector y and the estimation surface.

(2) When the tangent plane τ sufficiently approximates the estimation surface Φ in the vicinity of the point b, the vector $f = [f(x_i,b),..., f(x_n,b),]$, called the *prediction*, is a perpendicular projection of a vector y onto the tangent plane. The corresponding projection matrix has the form

$$P = J(J^T J)J^T \tag{8.68}$$

(3) The *residual vector* $\hat{e}$ with components $\hat{e}_i = y_i - f(x_i,b)$ is perpendicular to the tangent plane τ. Therefore a condition for the existence of a minimum of $U(\beta)$ is the validity of the expression $J^T\hat{e} = 0$. It is then important to determine how precisely the tangent plane approximates the estimation surface.

In the linear model, the contours of constant $U(\beta)$ in parameter space consist of concentric ellipses. When the model is nonlinear, the contours are sometimes banana-shaped, often elongated. Sometimes the contours stretch to infinity and do not close, or they may have multiple loops surrounding a number of stationary values. When several stationary values exist they may have different levels or provide alternative minima for $U(\boldsymbol{\beta})$.

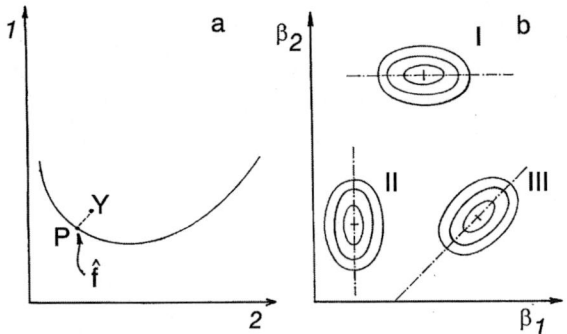

Figure 8.6   (a) The solution locus; (b) the parameter space for $m = 2$ and $n = 3$: (I) β_2 well determined but not β_1; (II) β_1 well determined but not β_2; (III) relationship between β_1 and β_2.

It is convenient to draw both the solution locus and the parameter space, simultaneously (Fig. 8.6). From m-dimensional parameter space a projection is made onto the estimation surface. Estimation of the parameters **b** requires a matrix inversion:

$$\mathbf{b} = \mathbf{f}^{-1}(\mathbf{y}) \qquad (8.69)$$

A condition of unambiguousness is that each **y** value in sample space always corresponds to just one point in parameter space. When the estimation surface is nonlinear, the arbitrary straight line in the parameter space through $\boldsymbol{\beta}^{(0)}$ given by

$$\boldsymbol{\beta} = \boldsymbol{\beta}^{(0)} + l\mathbf{h} \qquad (8.70)$$

where $\mathbf{h} = \left(h_1, ..., h_m\right)^{\mathsf{T}}$ is any nonzero vector, generates a curve or "lifted line" on the solution locus given by

$$\mathbf{f}_h = \mathbf{f}(\boldsymbol{\beta}^{(0)} + l\mathbf{h}) \qquad (8.71)$$

In both Eqs. (8.70) and (8.71), **h** represents the direction vector and l is the parameter of the straight line, and

$$\Delta\boldsymbol{\beta} = l\mathbf{h}$$

The tangent $\mathbf{f}'_h$ to this curve f_h at $\boldsymbol{\beta}^{(0)}$, is found from Eq. (8.66a), to be

$$\mathbf{f}'_h = \mathbf{Jh} \qquad (8.72)$$

The set of all such linear combinations is referred to as the tangent plane τ at $\boldsymbol{\beta}^{(0)}$. To express the curvature of the estimation surface, the vector of second partial derivatives, known as the acceleration of the lifted line f_h, may be shown to be

$$\mathbf{f}''_h = \mathbf{h}^{\mathrm{T}}\mathbf{Gh} \qquad (8.73)$$

Each element of $\mathbf{f}''_h$ has the form $\mathbf{h}^{\mathrm{T}}\mathbf{G}_i\mathbf{h}$ where $\mathbf{G}_i$ is the ith plane of the $\mathbf{G}$ array. The acceleration vector $\mathbf{f}''_h$ comprises three components: the first component $\mathbf{f}''^{\mathrm{N}}_h$ determines the change in direction of the instantaneous velocity vector $\mathbf{f}'_h$ normal to the tangent plane. The second and the third components, which can be added together to give $\mathbf{f}''^{\mathrm{P}}_h$, determine the change in direction of $\mathbf{f}'_h$ parallel to the tangent plane and the change in speed of the moving point respectively. From the projection matrix $\mathbf{P}$ defined by Eq. (8.68) it follows that

$$\mathbf{f}''^{\mathrm{N}}_h = \mathbf{P}\mathbf{f}''_h \qquad (8.74a)$$

$$\mathbf{f}''^{\mathrm{P}}_h = (\mathbf{E} - \mathbf{P})\mathbf{f}''_h \qquad (8.74b)$$

The acceleration components may be converted into curvatures, namely the *intrinsic curvature*

$$K^{\mathrm{N}}_h = \frac{\left\| \mathbf{f}''^{\mathrm{N}}_h \right\|}{\left\| \mathbf{f}'_h \right\|} \qquad (8.75a)$$

and the *parameter-effects curvature*

$$K^{\mathrm{P}}_h = \frac{\left\| \mathbf{f}''^{\mathrm{P}}_h \right\|}{\left\| \mathbf{f}'_h \right\|^2} \qquad (8.75b)$$

Only the latter depends on the particular parameterization chosen.

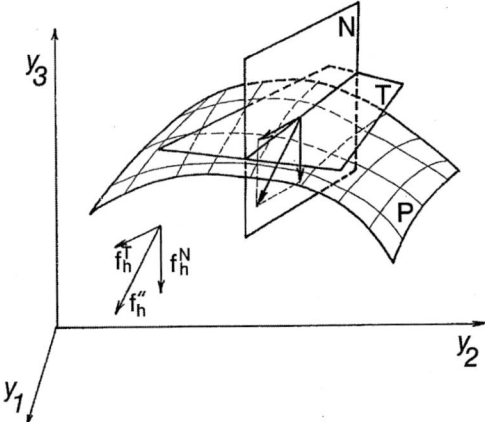

Figure 8.7   Geometrical illustration of curvature; decomposition of the acceleration vector $\mathbf{f}_h''$ into the components $\mathbf{f}_h''^{\mathrm{T}}$ and $\mathbf{f}_h''^{\mathrm{N}}$.

Interpreted geometrically, the intrinsic curvature K_h^{N} represents the reciprocal of the radius of a circle which approximates the estimation surface in the direction $\mathbf{h}$. This curvature depends on the actual type of regression model and on the data used. It is not affected by reparameterization.

The parameter-effects curvature K_h^{P} corresponds to the nonparallelity of curves formed by the projection of uniformly spaced points on parallel straight lines from the parameter space into non-uniformly spaced points on the estimation surface. This curvature may be removed by reparameterization.

For characterization of nonlinear behaviour of regression models, we look for a value of vector $\mathbf{h}$ such that the value of K_h^{N} and K_h^{P} have maximum values. Thus, both curvatures may be converted into response-invariant standardized relative curvatures Γ^{N} and Γ^{P} respectively. Multiplication by the standard radius $\rho = \hat{\sigma}\sqrt{m}$, where $\hat{\sigma}$ is the square root of the estimated residual variance $\hat{\sigma}^2$, results in the *maximum intrinsic curvature*

$$\Gamma^{\mathrm{N}} = \hat{\sigma}\sqrt{m}\,\max(K_h^{\mathrm{N}}) \tag{8.75c}$$

and the *maximum parameter-effects curvature*

$$\Gamma^{\mathrm{P}} = \hat{\sigma}\sqrt{m}\,\max(K_h^{\mathrm{P}}) \tag{8.75d}$$

8.5 Numerical Procedure for Parameter Estimation

If a regression model $f(\mathbf{x}, \boldsymbol{\beta})$ is nonlinear in at least one model parameter β_r, substitution into the criterion function [Eq. (8.57)] leads to a task of *nonlinear minimization*. The application of maximum likelihood (Section 8.3) leads to the task of *nonlinear maximization*. The application of any regression criterion leads to the problem of finding an extreme, where the regression parameters $\boldsymbol{\beta}$ are "variables".

This task can be solved by the application of general optimization methods to search for a free extreme (if no restrictions are placed on the regression parameters) or a search for a constrained extreme if the regression parameters are subject to certain restrictions.

Owing to the great variability of regression models, regression criteria and data, ideal algorithms that can achieve convergence to a global extreme sufficiently fast cannot be found. Most algorithms for many numerical methods often fail, i.e. they converge very slowly or diverge. The more complicated procedures for complicated problems are rather slow and require a large amount of computer memory.

Any program for solving nonlinear regression problems should contain procedures for

(1) searching for extremes in a given direction (one-dimensional optimization);

(2) inversion of matrices;

(3) numerical differentiation (in derivative methods);

(4) methods of overcoming local areas of divergence.

Programs that use the same algorithms may differ in practical applicability. Comparison of individual programs requires special problems [8] that allow testing under approximately similar conditions. Some typical programs for nonlinear regression have been compared [9].

Even with recommended algorithms, the correct result may not be reached. Kuesters and Mize [10] and Wolfe [11] have proposed schemes for solving partial problems such as numerical differentiation, a search in a given direction, etc. Some practical recommendations for the construction of optimization programs have been suggested by Gill, Murray and Wright [12]; and Schmidt [13] has proposed a program for minimization of the least-squares criterion.

For nonlinear regression, it has been recommended that a package of various regression algorithms is applied either individually or in combination [14, 34].

Nonlinear regression algorithms may be classified into the following principal groups:

(1) Derivative-free optimization methods;

(2) Derivative methods for the least-squares method (LS);

(3) General derivative methods;

(4) Algorithms for special cases.

The selection of a particular group depends on many factors. Generally, when a criterion function cannot be differentiated, the derivative-free methods should be used. The derivative methds use a special form of the least-squares criterion (LS) which are based on a quadratic approximation of the regression criterion. The general derivative methods enable solution of the task of maximization of likelihood function, for any regression model.

In this chapter we concentrate on the procedures of the first two groups. The general derivative methods are the most commonly used today [10, 12]. The algorithms for special cases are determined either by other regression procedures, such as the robust or L_p approximation methods, or by sums of exponential, etc. From our experience, the first two groups of regression programs can solve most types of experimental laboratory problems.

8.5.1 Supplemented Material (Non-derivative Optimization Procedures) to Chapter 8 is on Enclosed CD.

8.5.2 Derivative Procedures for the Least-squares Method

Algorithms of this group are very commonly used, not only because the least-squares method is a frequent regression criterion but also to provide information necessary for subsequent statistical analysis of the regression results [64].

These algorithms are useful for all model functions which are twice differentiable. They have the disadvantage that the local convergence depends on the choice of the initial guess $\beta^{(0)}$. All algorithms of this group are of iterative nature. In the ith iteration the procedure starts from the estimates $\beta^{(i)}$ to which a suitable increment vector Δ_i is added by $\beta^{(i+1)} = \beta^{(i)} + \Delta_i$.

Generally, the procedure which searches for a minimum of $U(\beta)$ consists of four steps:

(1) Determination of an initial guess $\beta^{(0)}$.

(2) A search for a convenient directional vector V_i.

(3) Determination of scalar α_i satisfying the condition, $\mathbf{\Delta}_i = \alpha_i \mathbf{V}_i$.

(4) Examination of the minimum obtained.

The vector $\mathbf{\Delta}_i$ is usually considered to be acceptable if

$$U(\boldsymbol{\beta}^{(i)} + \mathbf{\Delta}_i) < U(\boldsymbol{\beta}^{(i)}) \tag{8.113}$$

Some algorithms also allow equality of $U(\boldsymbol{\beta}^{(i+1)})$ and $U(\boldsymbol{\beta}^{(i)})$, or even a small increase. Individual algorithms differ in the realization of steps (2) and (3). Let us discuss each of the four steps:

(1) *Determination of the initial guess*

For many algorithms, this step is decisive for success of the minimization procedure. With a good initial guess, $\boldsymbol{\beta}^{(0)}$, even simple unsophisticated methods usually converge. With a poor initial guess, either a minimum can not be found at all or the minimum obtained is a local one. When the regression model can be linearly transformed, the initial guess $\boldsymbol{\beta}^{(0)}$ may be found by the linear least-squares method. In some cases, the initial guess may be obtained from physical or geometrical characteristics.

A transformation into stable parameters expressing geometrically defined characteristics of a regression model may be made, with, for example, function values or their derivatives at selected points, etc. With a personal computer, the path of the function $f(x, \mathbf{b}^{(0)})$ with given data may readily be tested and therefore the quality of the initial guess examined.

(2) *Determination of a directional vector*

The derivative of a criterion function $U(\boldsymbol{\beta})$ at a point $\mathbf{\Delta} = \boldsymbol{\beta} + \alpha \mathbf{V}$, has the form

$$\frac{\partial U(\boldsymbol{\beta})}{\partial \alpha} = \left[\frac{\partial U(\boldsymbol{\beta})}{\partial \boldsymbol{\beta}}\right]^{\mathrm{T}} \frac{\partial \boldsymbol{\beta}}{\partial \alpha} \tag{8.114}$$

For $\alpha \to 0$ we get, from Eq. (8.114), the *directional derivative*,

$$S_D = \frac{\partial U(\boldsymbol{\beta})}{\partial \alpha}\bigg|_{\alpha \to 0} = \mathbf{g}^{\mathrm{T}} \mathbf{V} \tag{8.115}$$

where $\mathbf{g}$ is the gradient for which Eqs. (8.59a) and (8.59b) are valid. From Fig. 8.15, it is evident that the gradient vector is the vector perpendicular to the tangent vector $\mathbf{t}$. All directional derivatives S_D of the hatched area are acceptable because they do not cause an increase in $U(\beta)$.

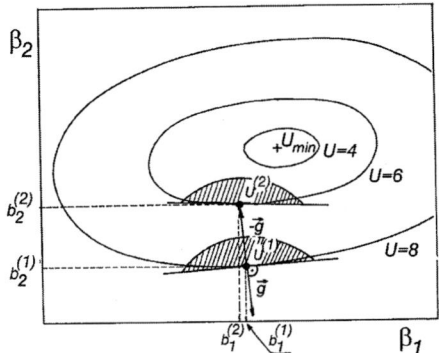

Figure. 8.15 Geometrical illustration of the gradient of a criterion function $U(\beta)$ for $m = 2$. Hatched areas denote admissible directional vectors.

The steepest decrease of the criterion function is in the direction $-\mathbf{g}$. The condition of acceptability of a given directional vector $\mathbf{V}$ requires the directional derivative not to be positive. Any direction for which $\mathbf{g}^T\mathbf{V} > 0$ is unsuitable.

Moreover if the directional vector $\mathbf{V}$ is acceptable a positive definite matrix $\mathbf{R}$ exists such that $\mathbf{V} = -\mathbf{Rg}$. The directional derivative S_D is then

$$S_D = -\mathbf{g}^T\mathbf{Rg} \qquad (8.116)$$

For a positive definite matrix $\mathbf{R}$, the quadratic forms are positive so that S_D in Eq. (8.115) is negative.

(3) *Calculation of an optimum increment*

In searching for an optimal increment $\alpha\mathbf{V}$ in the direction $\mathbf{V}$, an approximation of $U(\beta)$ in this direction by the Taylor series up to the second order can be used. This leads to the form

$$U(\beta + \alpha\mathbf{V}) \approx U(\beta) + \alpha\mathbf{g}^T\mathbf{V} + \frac{\alpha^2}{2}\mathbf{V}^T\mathbf{HV} \qquad (8.117)$$

where $\mathbf{H}$ is the Hessian matrix defined by Eq. (8.62). Equation (8.117) is approximately quadratic with respect to α, so that an optimal α may be estimated by setting the first derivative with respect to α, $U(\boldsymbol{\beta} + \alpha \mathbf{V})$, equal to zero. Hence, we get

$$\alpha^* = -\frac{\partial U(\mathbf{b})}{\partial \alpha} \Big/ \frac{\partial^2 U(\mathbf{b})}{\partial \alpha^2} = -\mathbf{g}^{\mathrm{T}} \mathbf{V} \left[\mathbf{V}^{\mathrm{T}} \mathbf{H} \mathbf{V} \right]^{-1} \tag{8.118}$$

After substitution from Eq. (116) we obtain the *Raleigh coefficient*

$$\alpha^* = \mathbf{g}^{\mathrm{T}} \mathbf{R} \mathbf{g} \left[\mathbf{g}^{\mathrm{T}} \mathbf{R}^{\mathrm{T}} \mathbf{H} \mathbf{R} \mathbf{g} \right]^{-1} \tag{8.119}$$

The Raleigh coefficient, α^*, is restricted to a region in which an approximation of type (8.117) can be used.

Another possibility in the search for an optimal α_i value in the direction $\mathbf{V}_i$ is the one-dimensional minimization of the function $U(\boldsymbol{\beta} + \alpha_i \mathbf{V}_i)$.

(4) *Termination of the iteration process*

The natural criterion for an optimum $\mathbf{b}$ is a zero value of the gradient $\mathbf{g}$ of the criterion function. Many methods terminate the iterative process searching for a minimum when the norm of the gradient $\|\mathbf{g}\|^2 = \sum_{j=1}^{m} g_j^2$ is sufficiently small. It is possible to select a critical value e.g. 10^{-4}, at which the point $\mathbf{b}^{(i)}$ is considered to be an extreme $\mathbf{b}$. Often, an iteration is terminated when the changes in the parameter estimates are very small. None of these criteria lead to termination at a true minimum.

Minimization may terminate less heuristically if the residual vector $\hat{\mathbf{e}}$ is approximately perpendicular to the columns of the matrix $\mathbf{J}$. From Fig. 8.5, it follows that $\mathbf{J}^{\mathrm{T}} \hat{\mathbf{e}} = \mathbf{0}$. The angle, α_i, between the residual vector $\hat{\mathbf{e}}$ and the jth column $\mathbf{J}_j$, of matrix $\mathbf{J}$, is given by the following expression:

$$\cos \alpha_j = \hat{\mathbf{e}}^{\mathrm{T}} \mathbf{J}_j \left[\mathbf{J}_j^{\mathrm{T}} \mathbf{J}_j \hat{\mathbf{e}}^{\mathrm{T}} \hat{\mathbf{e}} \right]^{-1/2} \tag{8.120}$$

When the maximum value of $\cos \alpha_j$ is sufficiently small (e.g. smaller than 10^{-9}) it is assumed that a minimum of $U(\boldsymbol{\beta})$ is found. Many other termination criteria have been proposed [33].

We concentrate our attention on the following derivative algorithms for the least-squares method:

(a) Gauss-Newton methods;

(b) Marquardt methods;

(c) the dog-leg method.

There is a wide spectrum of different improvements and modifications to these methods (see e.g. [71 – 91]), but we restrict ourselves here to some simple and efficient techniques.

8.5.2.1 Gauss-Newton Methods

To determine a convenient directional vector V_i, the quadratic approximation of a criterion function $U(\beta)$ from Eq. (8.58) may be used, and this also corresponds to Eq. (8.117) for $\alpha = 1$. From $\dfrac{\partial U(\beta + V)}{\partial V} = 0$ the optimum direction sector $V_i = N_i$ can be computed. The result takes the form

$$N_i = -H^{-1}g = (J^T J + B)^{-1} J^T \hat{e} \qquad (8.121)$$

On substituting into Eq. (8.119), we estimate that $\alpha^* = 1$ and N_i is directly an increment vector Δ_i. This method is called the *Newton method*. When the criterion $U(\beta)$ is a quadratic function, (an elliptic paraboloid), the minimum **b** will be found in one step. However, for other forms of criterion function $U(\beta)$ and estimates $\beta^{(0)}$ far from **b**, this method does not converge sufficiently fast. Moreover it requires knowledge of the matrix of second derivatives G_i for determination of the matrix **B** in Eq. (8.63). Neglecting matrix **B**, which is equivalent to the linearization of the regression model, is theoretically acceptable for a case when the residual vector $\hat{e}$ is small. The corresponding direction vector L_i has the form

$$L_i = \left(J^T J\right)^{-1} J^T \hat{e} \qquad (8.122)$$

Methods applying the directional vector L_i are called *Gauss-Newton methods*. The methods are simple, and are the most frequently used procedure of nonlinear regression. Substituting $H \approx (J^T J)$ into Eq. (8.119) leads to $\alpha^* = 1$. From a practical point, the Gauss-Newton method works well if some of following conditions are fulfilled:

(1) The residuals $\hat{e}_i = y_i - f(x_i, \beta)$ are small.

(2) The model function $f(x_i, \beta)$ is nearly linear, i.e. the Hessian matrix, **H**, has a small norm and its elements are nearly zero.

(3) The residuals $\hat{e}_i$ have alternating signs so that **B** is an approximate zero matrix. This condition is valid in the vicinity of the optimum **b**.

It is possible to use other methods to extend the region of convergence of this very simple method. The principle ones include:

(a) Inversion of the matrix $\mathbf{J^T J}$ and solution of the set of linear equations

$$\left(\mathbf{J^T J}\right)\mathbf{L} = \mathbf{J^T}\hat{\mathbf{e}} \tag{8.123}$$

(b) Improving the matrix $(\mathbf{J^T J})$ in order to be close to the Hessian matrix **H**.
(c) Choice of the optimal length of the step α.

We shall describe some successful methods which, on combination, lead to more effective modification of the original Gauss-Newton method.

(a) *Inversion of the matrix* $(\mathbf{J^T J})$

When the matrix $(\mathbf{J^T J})$ is well conditioned, the columns of the matrix **J** are linearly independent. Then, for a solution to the set of linear equations (8.123), various procedures may be applied. One of the simplest techniques, with minimal requirement on computer memory, is *Choleski decomposition*. In many practical problems (involving exponential and other nonlinear models), matrix **J** has some nearly collinear columns and therefore the matrix $(\mathbf{J^T J})$ is ill-conditioned. The length of vector **L** estimated from Eq. (8.123) is usually too large and its components have "inconvenient" signs. When some columns of a matrix **J** are linearly dependent, the matrix $(\mathbf{J^T J})$ is singular. These problems may be eliminated by pseudoinversion of matrix $\mathbf{J^T J}$ or generalized inversion of matrix **J** by the algorithm SVD.

Jennrich and Sampson [35] solved the set of linear equations in Eq. (8.123) by stepwise regression. Components of a vector **L** significantly decreasing the function $U(\boldsymbol{\beta})$ are found.

In our programs, the matrix $\mathbf{J^T J}$ is decomposed into eigenvalues and eigenvectors. To invert matrix $(\mathbf{J^T J})$, we use the technique of rational ranks described in Chapter 6.

One of the most advanced procedures for searching for a suitable vector **L** was devised by Schmidt [36]. Instead of the matrix **J**, Schmidt constructs matrix **M** containing only those columns of a matrix **J** for which it is valid that

- they are not linearly dependent,
- they cause the largest decrease in $U(\boldsymbol{\beta})$,
- at least one of them is not orthogonal to vector $\hat{\mathbf{e}}$.

This procedure protects the minimization process against difficulties arising from ill-conditioning

(b) *Improvement of the Hessian matrix*

This group includes methods of variable metric also known as the quasi-Newton methods. Here, each step treats the matrix $(\mathbf{J}^{\mathrm{T}}\mathbf{J})$ so that it approximates the Hessian matrix $\mathbf{H}$. These methods are suitable for cases with large residuals, that is when $U(\boldsymbol{\beta}) \gg 0$. The main idea is simple, and comes from the fact that the Hessian matrix is the derivative of the gradient with respect to a parameter vector, and this derivative is approximated by the difference

$$\mathbf{H}_{i+1} \approx \frac{\mathbf{g}_{i+1} - \mathbf{g}_i}{\boldsymbol{\beta}^{(i+1)} - \boldsymbol{\beta}^{(i)}} = \Delta\mathbf{g}_i\mathbf{S}_i^{-1} = \mathbf{B}_{i+1} \tag{8.124}$$

where $\mathbf{S}_i$ is the increment vector and index i denotes the ith iteration. The matrix $\mathbf{B}_{i+1}$ is an approximation of the matrix $\mathbf{H}_{i+1}$ calculated only from information about gradients and values of the vector $\boldsymbol{\beta}^{(i)}$. The course of the procedure is that, instead of $\mathbf{B}_{i+1}$, the increment $\Delta\mathbf{B}_i = \mathbf{B}_{i+1} - \mathbf{B}_i$ is calculated in the individual iterations. In many cases $\Delta\mathbf{B}_i$ is calculated from [40]

$$\Delta\mathbf{B}_i = \frac{(\Delta\mathbf{g}_i - \mathbf{B}_i\mathbf{S}_i)\mathbf{C}_i^{\mathrm{T}} + \mathbf{C}_i(\Delta\mathbf{g}_i - \mathbf{B}_i\mathbf{S}_i)^{\mathrm{T}}}{\mathbf{C}_i^{\mathrm{T}}\mathbf{S}_i} - \frac{\mathbf{S}_i^{\mathrm{T}}(\Delta\mathbf{g}_i - \mathbf{B}_i\mathbf{S}_i)\mathbf{C}_i\mathbf{C}_i^{\mathrm{T}}}{(\mathbf{C}_i^{\mathrm{T}}\mathbf{S}_i)^2} \tag{8.125}$$

where $\Delta\mathbf{g}_i = \mathbf{g}_{i+1} - \mathbf{g}_i$ and the vector $\mathbf{C}_i$ allows the choice of various strategies of improving the Hessian matrix. From a theoretical point of view, the best option is $\mathbf{C}_i = \mathbf{g}_i$. The process of improvement starts with the zero matrix $\mathbf{B}$.

Instead of approximating all components of the Hessian matrix $\mathbf{H}$, it is possible to improve only the part $\mathbf{B}$ containing second derivatives. The matrix $\mathbf{B}_{i+1}$ has the form

$$\mathbf{B}_{i+1} = \mathbf{J}_{i+1}^{\mathrm{T}}\mathbf{J}_{i+1} + \mathbf{K}_{i+1} \tag{8.126}$$

The matrix $\mathbf{K}_{i+1}$ is symmetrical and corresponds to the condition

$$\mathbf{K}_{i+1}\mathbf{S}_i = \mathbf{V}_i \tag{8.127}$$

The choice of $\mathbf{V}_i$ corresponds to the various variable metric strategies. The matrix $\mathbf{K}_{i+1}$ is again improved in each iteration according to the expression

$$\mathbf{K}_{i+1} = \mathbf{K}_i + \Delta\mathbf{K}_i \tag{8.128}$$

For determination of the matrix $\Delta\mathbf{K}_i$, Eq. (8.125) may be used, but with $\mathbf{B}_i$, replaced by $\mathbf{K}_i$ and $\Delta\mathbf{g}_i$ replaced by $\mathbf{V}_i$. The vector $\mathbf{V}_i$ may be computed from the *Broyden-Dennis formula* [37] where

$$\mathbf{V}_i^{(\mathrm{BD})} = \Delta\mathbf{g}_i - \mathbf{J}_{i+1}^{\mathrm{T}}\mathbf{J}_{i+1}\mathbf{S}_i \tag{8.129}$$

or the *Betts formula*

$$\mathbf{V}_i^{(\mathrm{B})} = \Delta\mathbf{g}_i - \mathbf{J}_i^{\mathrm{T}}\mathbf{J}_i\mathbf{S}_i \tag{8.130}$$

It is possible to use a linear combination of these two formulae or to use more complicated procedures of adaptive improvement of the matrix $\mathbf{H}$, described in detail in the literature [40].

The adaptive improvement of the Hessian matrix, by applying Eq. (8.126), does not automatically result in positive-definiteness of matrix $\mathbf{B}_{i+1}$. Therefore, this technique should be combined with procedures of pseudoinversion.

A mixed strategy is sometimes used when, according to parameter α, a direction $\mathbf{V}_S$ is selected between the linearization $\mathbf{L}$ and approximately Newton direction $\mathbf{N}$.

Gill and Murray [38] propose calculation of an approximation of the Hessian matrix by the difference formulae and then application of the SVD procedure for determination of significant components of the gradient. Many authors [39] recommend the method of variable metrics as a standard part of a library of programs for the minimization of the criterion function $U(\boldsymbol{\beta})$. It is best to restart the calculation of matrix $\mathbf{B}_{i+i}$ when matrix $\mathbf{H}_i$ becomes unsuitable because of cumulated errors.

(c) *Selection of step length*

Many variants of the Gauss-Newton method use, for a selection of the optimal step α^*, the quadratic approximation $U(\mathbf{b} + \alpha\mathbf{L})$ in the direction $\mathbf{L}$. With values $U(\boldsymbol{\beta}^{(i)}) = U_i$, and $U(\boldsymbol{\beta}^{(i)} + \mathbf{L}_i) = U_{i+1}$ and the direction derivative S_D

$$\mathbf{g}_i^{\mathrm{T}}\mathbf{L}_i = -2\mathbf{e}\mathbf{J}(\mathbf{J}^{\mathrm{T}}\mathbf{J})^{-1}\mathbf{J}^{\mathrm{T}}\mathbf{e}^{\mathrm{T}} = S_\mathrm{D}$$

the optimal step length of α^* is estimated from

$$\alpha_i^* = \frac{-S_D}{2(U_{i+1} - S_D - U_i)} \qquad (8.131)$$

If $|S_D|$ is small, the value α_i^* is also small. Therefore $\alpha_i^* = \max\left(0.25, \alpha_i^*\right)$ may be selected. There are various heuristic strategies for selection of convenient α values, and these can speed up an iterative search for a minimum.

8.5.2.2   Marquardt-type Methods

The obvious selection of a directional vector $\mathbf{V}_i$ is the direction of steepest descent, -$\mathbf{g}$. From Eq. (8.119), an optimum coefficient α^* is given by

$$\alpha^* = \mathbf{g}^T\mathbf{g}\left[\mathbf{g}^T\mathbf{Hg}\right]^{-1} \approx \mathbf{g}^T\mathbf{g}\left[\mathbf{g}^T\left(\mathbf{J}^T\mathbf{J}\right)^{-1}\mathbf{g}\right]^{-1} \qquad (8.132)$$

The increment vector $\mathbf{\Delta}_i = -\alpha^*\mathbf{g}$ corresponds to the gradient method.

The gradient method converges slowly in the vicinity of an optimum. On the other hand, in cases when $\mathbf{\beta}^{(i)}$ is far from $\mathbf{b}$, the direction leading to a minimum can be found. It is effective to use a combination of directions of the Newton method $\mathbf{N}_i$ or the direction of linearization $\mathbf{L}_i$ together with the direction -$\mathbf{g}$, to construct a more robust algorithm. These procedures are called *hybrid procedures*. The best known example is the Marquardt method, which calculates the directional vector $\mathbf{V}_i(\lambda)$ from

$$\mathbf{V}_i(\lambda) = \left(\mathbf{J}^T\mathbf{J} + \lambda\mathbf{D}_i^T\mathbf{D}\right)^{-1}\mathbf{J}^T\hat{\mathbf{e}} \qquad (8.133)$$

where λ is the parameter and $\mathbf{D}_i$ is the diagonal matrix which eliminates the influence of various magnitudes of the components of the matrix $\mathbf{J}$. Usually the diagonal elements D_{ii} are equal to diagonal elements of matrix $(\mathbf{J}^T\mathbf{J})$. According to the magnitude of λ, the vector $\mathbf{V}_i(\lambda)$ has following properties:

(a) The length $\|\mathbf{V}_i(\lambda)\|$ is a decreasing function of λ. For $\lambda \to \infty$, $\|\mathbf{V}_i(\lambda)\| \to 0$. The parameter λ operates similarly to the parameter α, i.e. it enables a change in the length of the increment vector.

(b) The cosine of the angle between the vector $\mathbf{V}_i(\lambda)$ and the negative gradient -$\mathbf{g}$, increases as a function of λ. As $\lambda \to \infty$ it approaches a value of one. It then follows that for large λ values, the directional vector $V_i(\lambda)$ approaches the directional vector of the gradient method.

(c) The cosine of the angle between the direction of linearization L_{j}, and the vector $V_i(\lambda)$ is a decreasing function of λ. When $\lambda = 0$, it reaches value 1 and $V_i(\lambda)$ is identical with the direction of the Gauss-Newton method.

The curve $V_i(\lambda)$ in parameter space begins at the point $\boldsymbol{\beta}^{(i)} + L_i$ and ends at the point $\boldsymbol{\beta}^{(i)}$ where it has direction -**g**. However, the space curve does not lie in the plane of vectors L_j, and -**g**. Appropriate selection of parameter λ ensures:

(1) positive definiteness of the matrix $R = (J^TJ + \lambda D^TD)$, which ensures that its inverse can be found;

(2) a shortening step $V_i(\lambda)$ moving from the direction of linearization L_j;

(3) the possibility of choosing between the direction L_i and approximate direction -**g**;

(4) restriction of the magnitude of the incremental vector V_i to the "admissible" region in the vicinity of $\boldsymbol{\beta}^{(i)}$.

The necessity of repeated matrix inversion for each λ is a disadvantage of this procedure, as it is rather time-consuming. Moreover, it may occur that a large λ results in a very small magnitude of V_i. Therefore, the use of the maximum value of the magnitude of λ is limited. Individual modifications of the Marquardt method differ, especially in the strategy of the adaptive setting of λ. The original algorithm begins with $\lambda_0 = 0.01$. After each successful step, the calculation $\lambda_{i+1} = \lambda_i / 10$ is performed, and after an unsuccessful step, $\lambda_{i+1} = 10\lambda_i$.

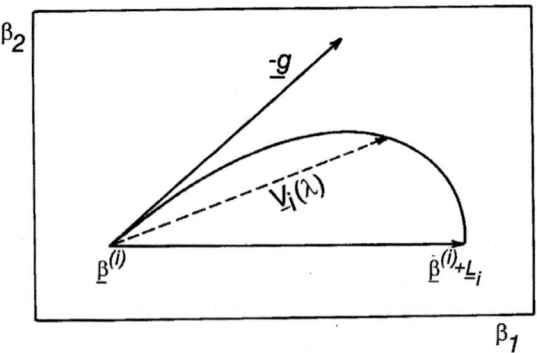

Figure. 8.16 Schematic path of function $v_i(\lambda)$ for two parameters ($m = 2$).

In the Nash algorithm [40], with minimal need for memory for solving a set of linear equations (i.e. determination of $V_i(\lambda)$), the Choleski decomposition is used. The adaptive adjustment of λ starts from $\lambda = 10^{-4}$ and after a successful step $\lambda_{i+1} = \max(10^{-6}, \lambda_i/10)$ is chosen. If a step is unsuccessful, $\lambda_{i+1} = \max(10^6, 4\lambda_i)$ is selected.

In another procedure [41], λ_i is selected according to the magnitude of the maximal diagonal element of matrix $(J^T J)$. Let us denote this element, J_{max}. In each iteration, the procedure begins with $\lambda_1 = C_{min}J_{max}$. If there is no decrease in $U(\beta)$, this coefficient is increased according to

$$\lambda_{i+1} = \lambda_i(C_{max}/C_{min})^{1/4}.$$

This process of increasing λ is performed until a decrease in $U(\beta)$ occurs or until λ is equal to J_{max}. Then another search method is used, where $C_{max} = 1$ and $C_{min} = 0.01$ are chosen.

A very good and effective method was proposed by More [42]; this forms part of the program NL2SOL.

Some authors recommend choosing the optimal value of λ_{opt} as the one which leads to a maximal decrease in $U(\beta)$. This local optimal strategy, in cases of narrow curved-valley shape of $U(\beta)$, causes a global deceleration of convergence.

One strategy for changing λ, which ensures a global speeding up of convergence, concerns three typical situations which may appear in the course of minimization of $U(\beta)$, [43].

(a) It is possible to use the direction of linearization L, i.e. $\lambda = 0$.

(b) When the criterion function $U(\beta)$ has a curved-valley shape, the smallest eigenvalue C_1 of matrix $(J^T J)^{-1}$ is significantly smaller than the second smallest eigenvalue C_2, e.g. $10C_1 < C_2$. The direction of this valley is determined by the eigenvector k_1 corresponding to the smallest eigenvalue. In this case, an increment Δ_i is chosen so that the criterion function $U(\beta)$ remains approximately unchanged. A search is then carried out on the hill-side of the valley [43].

(c) In other cases, the curved valley is not so distorted and the quadratic approximation is inappropriate. For these cases, a suitable strategy for the selection of Δ_i has been described [43].

Meyer and Roth [44] have proposed the method MDLS (modified damped least-squares) for finding the vector $V(\lambda)$, which envolves a one-way minimization in a specific direction.

Generally, Marquardt-type methods are included in standard program libraries because of their robustness.

8.5.2.3 Dog-leg Type Procedures

The main disadvantages of Marquardt methods include:

(a) the need for an inversion after every change of parameter λ;

(b) the small length of vector $\mathbf{V}(\lambda)$ for a large λ.

Both these disadvantages are removed in the next hybrid methods, in which the optimal directional vector $\mathbf{V}(\mu)$, a convex combination of vectors $\mathbf{L}$ and $-\alpha^*\mathbf{g}_i$ is searched for. Here α^* is estimated from Eq. (8.132) and $0 < \mu < 1$. It follows that

$$\mathbf{V}(\mu) = \boldsymbol{\beta}^{(i)} + (1-\mu)\mathbf{L}_i\alpha_1 - \mu\alpha^*\mathbf{g}_i \qquad (8.134)$$

The function $\mathbf{V}(\mu)$ for $\alpha_1 = 1$ and $\alpha_1 < 1$ is shown in Fig. 8.17 as hypotenuses of right angle triangles. The dotted line represents $\alpha_1 < 1$ and the solid line $\alpha_1 = 1$. The classical strategy of the *Powell dog-leg method* estimates an optimal vector $\mathbf{V}_i(\mu)$ on the abscissa, TB, of a triangle defined by the vertices $\mathbf{O} = \mathbf{b}^{(i)}$, $\mathbf{T} = \mathbf{b}^{(i)} + \mathbf{L}_i$ and $\mathbf{B} = \mathbf{b}^{(i)} - \alpha^*\mathbf{g}_i$ where α^* is defined by Eq. (8.132).

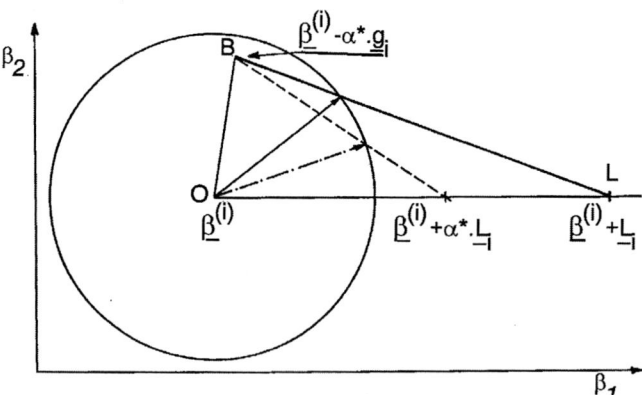

Figure. 8.17 Geometrical illustration of the dog-leg strategy. The circle shows the admissible range of increments. The solid hypotenuse is $\mathbf{V}(\mu)$ for $\alpha_1 = 1$ and the dotted hypotenuse is $\mathbf{V}(\mu)$ for $\alpha_1 < 1$.

It can be seen that for $\mu = 0$, the vector $\mathbf{V}(\mu)$ is identical to the linearization direction $\mathbf{L}$, and for $\mu = 1$, the vector $\mathbf{V}(\mu)$ is identical to the direction of the negative gradient $-\mathbf{g}$. The magnitude of the total increment in direction $-\mathbf{g}$ corresponds to the optimal value α^*.

Dennis and Mei [45] use the *double dog-leg* strategy in which, instead of the vector $\mathbf{L}_i$ the "shorter" vector $\alpha_1 \mathbf{L}_i$ is used. The parameter α_1 is determined in the linearization direction so that the increment corresponds approximately to the Cauchly point [Eq. (8.119)]. Therefore, it can be shown [45] that $\alpha_1 = 0.2 + 0.8 \|\mathbf{g}_i\|^4 \left[\mathbf{g}_i^T (\mathbf{J}^T\mathbf{J})^{-1} \mathbf{g}_i \mathbf{g}_i^T (\mathbf{J}^T\mathbf{J}) \mathbf{g}_i \right]^{-1}$. From Fig. 8.17, it is evident that shortening $\alpha_1 \mathbf{L}_i$ leads to a directional vector $\mathbf{V}_1^*(\mu)$ which is nearer to the linearization direction than vector $\mathbf{V}(\mu)$, calculated from Eq. (8.134) with for $\alpha_1 = 1$. The actual strategy of these techniques differs in the admissable range adopted, in the method of inversion for finding $\mathbf{L}_i$, and in improvement of the Hessian matrix by variable metric methods.

The program MINOPT in our ADSTAT program, described later in this chapter, is based on the Dennis and Mei procedure [46].

8.5.3 Complications in Nonlinear Regression

In nonlinear regressions, many complications arise that are not found in linear regression models.

(a) A minimum in $U(\boldsymbol{\beta})$ exists for some regression models only.
(b) There may be local minima and saddle points in $U(\boldsymbol{\beta})$.
(c) Parameters may be inestimable.
(d) Parameters may be ill-conditioned.

8.5.3.1 Parameter Estimability

Complications (c) and (d) may be identified by analysing the sensitivity coefficients $\mathbf{g}_i$ defined by Eq. (8.5). For practical purposes the normalized sensitivity coefficients [47]

$$C_{j(i)} = \beta_j \frac{\partial f(x_i, \boldsymbol{\beta})}{\partial \beta_j}, \quad j = 1, \ldots, m, \ i = 1, \ldots, n. \tag{8.135}$$

are recommended. The ill-conditioning of parameters β_j and β_h is a consequence of the approximate multicollinearity between parameters β_j and β_h. For a visual interpretation of the examination of the conditioning of parameters in a model, the *sensitivity graph* is used. The sensitivity graph is a plot of $C_{j(i)}$ and $C_{h(j)}$ vs.

x_i, $i = 1, ..., n$. The dependence of the normalized sensitivity coefficients on the index i may also be plotted.

Figure 8.18a shows several possible ill-conditioned models, where the sensitivity coefficients C_j and C_h are linearly dependent. Figure 8.18b, on the other hand, shows linearly independent sensitivity coefficients. More details of similar cases have been presented in the literature [47]. From Fig. 8.18, it follows that for the examination of the linear dependence of the sensitivity coefficients the location of points for $C = 0$ is important. The situation is more complicated when there are more parameters in the model [47].

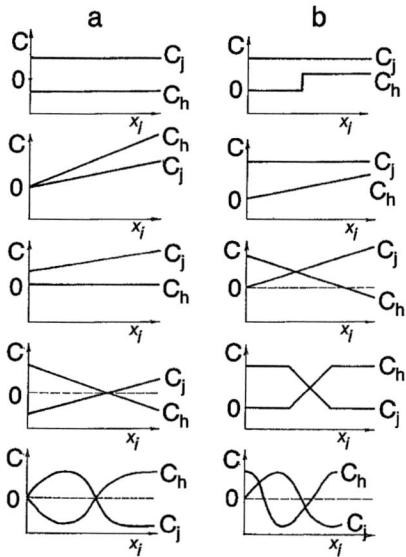

Figure. 8.18 The sensitivity graphs for the sensitivity coefficients where they are (a) linearly dependent, and (b) linearly independent.

To express the sensitivity of a regression model in terms of a change in parameter β_j, the *total sensitivity function* [1] C_{cj} may be used. This function is defined by

$$C_{cj} = \frac{1}{n} \sum_{i=1}^{n} \left[\frac{\partial f(x_i, \beta)}{\partial \beta_j} \right]^2 \tag{8.136}$$

This sensitivity function has meaning only for the nonlinear parameters β_j in the regression model $f(x, \beta)$. For the linear parameters β_h in the regression model $f(x,\beta)$, the total sensitivity function C_{ch} reaches a constant value. A graphical illustration shows the *sensitivity graph of parameters*, which plots the dependence of C_{cj} on β_j in the vicinity of $\beta_j^{(0)}$ or b_j. If the sensitivity graph of parameters is nearly constant, then the regression model has a low sensitivity to changes in the jth parameter, or the regression model $f(x,\beta)$ is linear with respect to the parameter β_j.

8.5.3.2 Existence of a Minimum of $U(\beta_j)$

If, in the vicinity of a minimum, the model function $f(x,\beta)$ reaches an infinite value, then $U(\beta) \to \infty$. For example, for the model $f(\mathbf{x},\beta) = \dfrac{\beta_0 + \beta_1 x_{1i}}{\beta_2 x_{1i} + \beta_3 x_{2i}}$

with two independent variables x_{1i} and x_{2i} such that $\beta_2 x_{2i} = -\beta_3 x_{1i}$ then $U(\beta) \to \infty$.

Gallant [48] has given an example when, with a simulated data set and an exponential model, for one parameter $U(\beta)$ does not have a minimum, but a maximum. Demidenko [49] has proposed a procedure for testing for the existence of a least-squares estimate, and therefore a minimum in $U(\beta)$, in which the regression criterion is restricted (from below) to a limit value. Generally, the existence of a minimum is connected with the problem of identifying the parameters $\mathbf{b}$ of the regression model $f(\mathbf{x},\beta)$[50].

A given regression model is *globally unidentified on a region* Ω, if, for any parameter vector $\mathbf{b} \in \Omega$, we can find another vector $\mathbf{b}^* \in \Omega$ for which

$$f(\mathbf{x},\mathbf{b}^*) = f(\mathbf{x},\mathbf{b}) \tag{8.137}$$

where the symbol $f(\mathbf{x},\beta)$ means the vector with elements $f(x_i,\beta)$.

When the columns of the Jacobian matrix $\mathbf{J}$ are independent, the unidentifiability is *structural* in nature, i.e. independent of the actual numerical values of parameters. The cause of unidentifiability is symmetry of the parameters [51]. This means that the model $f(\mathbf{x},\beta)$ is invariant to a transformation of points in the parametric space. The condition of invariance of this function, with respect to the continuous transformation corresponding to Lie group, is given by [51]

$$\frac{\partial f(\mathbf{x},\beta)}{\partial \beta} \mathbf{h}(\beta) = 0 \tag{8.138}$$

where $\mathbf{h}(\boldsymbol{\beta})$ is the tangent vector which unambiguously defines continuous transformations of parameters, and $\dfrac{\delta f(\mathbf{x}, \boldsymbol{\beta})}{\delta \boldsymbol{\beta}}$ is the vector with components $\dfrac{\delta f(\mathbf{x}, \boldsymbol{\beta})}{\delta \beta_j}, j = 1, ..., m$. If we compare Eq. (8.138) with Eq. (8.5), we find that they express the same condition. This means that for unidentified models the sensitivity coefficients are linearly dependent.

8.5.3.3 Existence of Local Minima

The existence of local minima is characteristic of overdetermined models. Local minima exist in various models formed as a sum of partial nonlinear terms, sums of exponentials, etc. Let us suppose, for example, that for the model $f(\mathbf{x}, \boldsymbol{\beta}) = \beta_1 \exp(\beta_2 x)$ a global minimum $U(\boldsymbol{\beta}^*)$ with estimates, β_1^* and β_2^* were found by the least-squares method. If, for the same data set, we use the model $f(\mathbf{x}, \boldsymbol{\beta}) = \beta_1 \exp(\beta_2 x) + \beta_3 \exp(\beta_4 x)$ we will notice that behind a global minimum many local stationary points will satisfy the condition, $\delta U(\boldsymbol{\beta}) / \delta \boldsymbol{\beta}$. There are points for which

(a) $\beta_1 = \beta_3 = 0$ or $\beta_2 = \beta_4 = 0$ i.e. the model is reduced to a simplified form;

(b) $\beta_4 = \beta_3 = \beta_2^*$ and all combinations of β_2, β_1 for which $\beta_1 + \beta_2 = \beta_1^*$, once again simplifying the model.

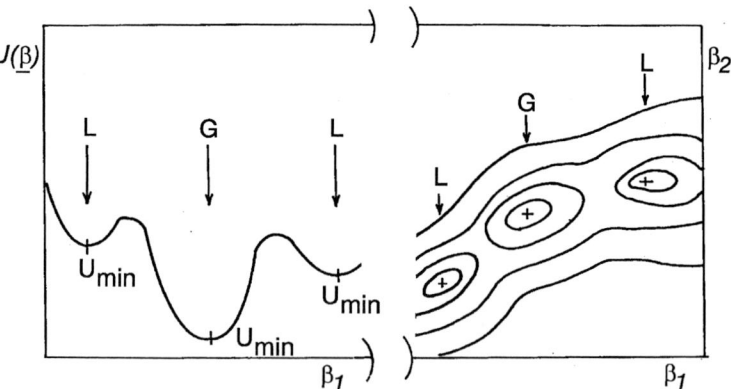

Fig. 8.19   Existence of local minima (L) beside the global one (G).

For higher numbers of exponential terms, the number of local minima increases sharply. Owing to the influence of measurement errors, the global minimum can be determined for a model with fewer terms [52].

A decision about whether a minimum found is global or local may be made on the basis of the Hessian matrix of the criterion function $U(\beta)$. If the Hessian matrix $\mathbf{H}$ is positive-definite at point $\mathbf{b}$, a global minimum has been found [49]. For practical purposes, instead of the matrix $\mathbf{H}$, we examine the positive-definiteness of matrix $(\mathbf{J}^T\mathbf{J})$. If both matrices $\mathbf{H}$ or $(\mathbf{J}^T\mathbf{J})$, are positive-definite, a global minimum has been found. Positive semi-definiteness indicates an overestimated or unidentified model.

8.5.3.4 Ill-conditioning of Parameters

Ill-conditioning of parameters in a model which causes approximately linear dependence in the sensitivity graphs, depends not only on the type of regression model but also on the location and the range of experimental data. Ill-conditioning of parameters is indicated as ill-conditioning of the matrix $(\mathbf{J}^T\mathbf{J})$. Algorithms used for matrix inversion, such as pseudoinversion or the Marquardt type methods, are resistant against this difficulty. Difficulties arise when Gauss-Newton type methods are used. In many cases, ill-conditioning appears on application of numerical differentiation instead of analytical, because there is a loss of precision in the construction of matrix $(\mathbf{J}^T\mathbf{J})$. Numerical differentiation may cause even good algorithms to fail. For numerical differentiation, the method of forward difference is often used.

$$\frac{\delta f(x_i, \beta)}{\delta \beta_j} \approx \frac{1}{h_j}\left[f\left(x_i, h_j \mathbf{I} + \beta\right) - f(x_i, \beta) \right] \qquad (8.139)$$

where $\mathbf{I}$ is a unit vector, with the jth component equal to 1, and other components equal to zero. The increment, h_j, in many cases determines the quality of numerical differentiation. If can be selected as

$$h_j = \left|\beta_j\right| + \sqrt{EP}$$

where EP is the computer precision. For nonlinear regression, it is convenient to use the algorithm written by Brown and Dennis [52], where h_j is evaluated by

$$h_j = \min(U(\beta), \delta_j)$$

and the parameter δ_j is estimated by

$$\delta_j = \begin{cases} 10^{-9} \\ 10^{-3}|\beta_j| \end{cases} \quad \begin{aligned} &\text{if } |\beta_j| < 10^{-6} \\ &\text{if } |\beta_j| \geq 10^{-6} \end{aligned}$$

This technique is also used in our algorithm MINOPT in ADSTAT program [46].

Reparameterization [53] may cause significant improvement in the shape of the criterion function $U(\boldsymbol{\beta})$ and hence improve the conditioning of the matrix $(\mathbf{J}^T\mathbf{J})$. Suitable reparameterization procedures for some nonlinear models are described by Ratkowsky [53].

8.5.3.5 Small Range of Experimental Data

In many practical problems, difficulties with overdetermination and ill-conditioning of model parameters are partly the consequence of a small range of experimental data.

The application of the least-squares method leads then to a model which gives a good description of experimental dependence, but the parameters have no physical meaning. In these situations, it is possible

(a) to collect more experimental data;
(b) to examine the possibility of model simplification;
(c) to investigate the possibility of estimating some parameters on the basis of other supplementary experiments, previous knowledge, experience, theory, etc.
(d) to select suitable restrictions to be placed on the parameters so that their estimates have physical meaning.

The actual procedure depends on the problem in question and on the experience of the experimenter.

Problem 8.12   *Search for a model of the Kohlrausch equation*

The tension relaxation after a jump change of deformation on the value $\varepsilon = 0.4$ was studied for laboratory-made fibres PADt-G. The RETEST apparatus monitored the dependence between tension N_t (MPa) and time t (sec) in range 0-500 sec; and 21 points were recorded. The proposed model was

$$N_t = \beta_1 + (\beta_1 - \beta_2)\exp\left[-(\beta_4 t)^{\beta_3}\right] \tag{8.140}$$

where β_1 is the initial tension, β_2 is the equilibrium tension and β_3, β_4 are empirical constants. A graphical illustration of Eq. (8.140) is shown in Fig. 8.20

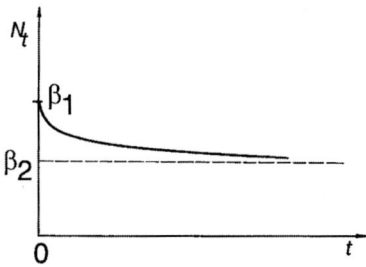

Fig. 8.20 The parameters of the Kohlrausch model.

The task is to estimate the parameters, by using five variants of calculation strategy.

Variant I: All parameters are estimated simultaneously.

Variant II: The model is simplified by assuming that $\beta_2 = 0$.

Variant III: The value for parameter $\beta_1 = 69$ MPa is fixed from previous knowledge.

Variant IV: The value for parameter $\beta_2 = 12$ MPa is fixed from previous knowledge.

Variant V: The values of parameters $\beta_1 = 69$ MPa and $\beta_2 = 12$ MPa are fixed.

Data: $n = 21$

t, sec	N_t, MPa	t, sec	N_t, MPa	t, sec	N_t, MPa
1	62.22	85	52.22	220	49.26
5	58.52	100	52.22	250	48.15
10	58.15	115	51.11	310	47.78
25	55.56	130	50.74	340	47.04
40	54.07	145	50.74	400	46.66
55	53.33	160	50.37	430	46.29
70	52.59	190	49.63	490	45.93

Solution: Variant I: From the initial guesses of the parameters $\beta_1^{(0)} = 65, \beta_2^{(0)} = 35, \beta_3^{(0)} = 0.118, \beta_4^{(0)} = 0.811$, refined values were found (Table 8.1). Parameter β_2 is negative and therefore has no physical meaning. Another initial guess of parameters leads to the refined estimates in column Ib of Table 8.1. The approximate singularity of matrix $\mathbf{J}^{\mathrm{T}}\mathbf{J}$ causes problems.

Table 8.1 Parameter estimates **b** and $U(\boldsymbol{\beta})$ with $\hat{\sigma}$ for the various variants of the model

Parameter estimates	Model variants					
	Ia	Ib	II	III	IV	V
b_1 Mpa	71.32	72.11	58.80	69*	58.27	69*
b_2, Mpa	−811.2	−4414	0*	−38.69	−12*	12*
b_3	0.163	0.152	0.216	0.202	0.231	0.222
b_4	0.0106	0.0227	0.106	0.0691	0.121	0.131
$U(\mathbf{b})$	1.665	1.775	1.750	1.718	1.792	1.818
σ	0.09797	0.104	0.0972	0.0954	0.0966	0.957

*constant values

Variant II: The refined parameters are in column II of Table 8.1. The simplification of the model leads to an insignificant increase in the minimum of the criterion function $U(\boldsymbol{\beta})$. However, the corresponding residual variance $\hat{\sigma}^2$ shows that this model is better than variant I. The matrix $\mathbf{J}^{\mathrm{T}}\mathbf{J}$ is regular and well-conditioned.

Variant III: Even with the improvement in the model (lower $\hat{\sigma}$), the estimate of parameter β_2 has no physical meaning so that knowledge of parameter β_1 does not help here.

Variant IV: Knowledge of parameter β_2 leads to acceptable estimates of other parameters and the degree of fit is still quite good.

Variant V: With two parameters, β_1 and β_2, known, the model becomes much simpler.

Conclusion: The goodness-of-fit for various variants of the model was compared by the residual standard deviation $\hat{\sigma} = \sqrt{U(\mathbf{b})/(n-m)}$. For a given data set, the individual variants of the model do not differ significantly. Moreover, it may be concluded that

(a) The Kohlrausch equation is inappropriate for practical purposes, because the quality of the results depends on the time units. After modification of the exponential term, $\exp\left[-\left(\beta_4 t\right)^{\beta_3}\right]$, the parameter β_3 is dimensionless and β_4 has dimensions of reciprocal time. This modification does not solve the problem of insufficient sample size.

(b) With regard to the physical meaning of model and knowledge about the data, the best method is a reduced model with $\beta_2 = 0$. Also, other physically acceptable variants IV and V have small differences in the estimates of β_3 and β_4.

(c) The estimate of parameter β_1 varies markedly depending on the choice of model variant.

From the results, it can be seen that without previous knowledge of the model system, it is not possible to estimate the equilibrium tension β_2. Also, estimate β_1 becomes loaded by quite a high uncertainty which comes from the variants used.

The aim of Problem 8.11 was to demonstrate that numerical application of linear regression does not guarantee finding a useful model. Statistical analysis of this Problem is then necessary and is detailed in Section 8.6.

8.5.4 Examination of the Reliability of the Regression Algorithm

In the literature, many regression algorithms and packages of programs for nonlinear regression are described [54, 55]. To examine the reliability of a regression algorithm, various test models are used. A good reliable algorithm should estimate correct values of the regression parameters.

For six test models, with their typical data sets, the final results depend on the initial guesses of the parameters. In comparison of the numerical results of these models, no restart or other technique of repeated determination of new initial guesses of parameters (if divergence occured) was used.

Six test models:

Model I	$y = \beta_1 + \beta_2 \exp(\beta_3 x)$
Model II	$y = \exp(\beta_1 x) + \exp(\beta_2 x)$
Model III	$y = \beta_1 \exp[\beta_2 / (\beta_3 + x)]$
Model IV	$y = \beta_1 \exp(\beta_3 x) + \beta_2 \exp(\beta_4 x)$

Model V	$y = \beta_1 x^{\beta_3} + \beta_2 x^{\beta_4}$
Model VI	$y = \beta_1[\exp(-\beta_2 x_1) + \exp(\beta_3 x_2)]$

Test data sets $\{x, y\}$ for 3

Model I: $n = 10$

x	1	5	10	20	25	30	35	40	45	50
y	16.7	16.8	16.9	17.1	17.2	17.4	17.6	17.9	18.1	18.7

Model II: $n = 10$

x	1	2	3	4	5	6	7	8	9	10
y	4	6	8	10	12	14	16	18	20	22

Model III: $n = 16$

x	50	55	60	65	70	75	80
y	34780	28610	23650	19630	16370	13720	11540
x	95	100	105	110	115	120	125
y	7030	6005	5147	4427	3820	3307	2872

Model IV: $n = 16$

x	7.448	7.448	7.969	8.176	9.284	9.439	7.552
y	57.544	53.546	19.498	16.444	4.305	3.006	45.290

x	7.877	8.522	9.314	7.607	7.847	8.176	8.523
y	27.952	11.803	4.764	51.286	31.623	21.777	13.996

x	8.903	9.314
y	7.727	4.999

Model V: $n = 12$

x	12	13	14	15	16	17	18	19	20	21	22	23
y	7.31	7.55	7.80	8.05	8.31	8.57	8.84	9.12	9.40	9.69	9.99	10.30

Model VI: $n = 23$

x_1	0	0.6	0.6	1.4	2.6	3.2	0.8	1.6	2.6	4.0	1.2	2.0
x_2	0	0.4	1.0	1.4	1.4	1.6	2.0	2.2	2.2	2.2	2.6	2.6

y	40	10	5.0	2.5	2.5	2.0	1.0	0.7	0.8	0.7	0.4	0.4
x_1	4.6	3.2	1.6	4.2	4.2	3.2	2.8	4.2	5.4	5.6	3.2	
x_2	2.8	3.0	3.2	3.4	3.4	3.8	4.2	4.2	4.4	4.8	5.0	
y	0.3	0.22	0.22	0.1	0.05	0.07	0.03	0.03	0.03	0.02	0.01	

Table 8.2 Initial guess of parameters estimated for the six test models

Model	$\beta_1^{(0)}$	$\beta_2^{(0)}$	$\beta_3^{(0)}$	$\beta_4^{(0)}$	$U(\beta^{(0)})$
I	20	2	0.5	---	2×10^{23}
II	0.3	0.4	---	---	4×10^3
III	0.02	4000	250	---	1.7×10^9
IV	10^5	10^5	-1.679	-1.31	1.12×10^4
V	100	0.1	2	10	2.68×10^3
VI	12	1.0	25	---	226.9

Table 8.3 Best estimates of parameters of the six test models

Model	b_1	b_2	b_3	b_4	$U(b)$
I	15.67	0.9994	0.0222	---	5.98×10^{-3}
II	0.2578	0.2578	---	---	124.34
III	0.005618	6180	345.2	---	87.9
IV	8.315×10^7	5.088×10^3	-1.95	-0.7786	134
V	3.802	4.141×10^{-3}	0.223	2.061	2.98×10^{-5}
VI	31.5	1.51	19.9	---	1.25

Three regression methods, the method of modified simplex (MSM), the Gauss-Newton method (GN) and program MINOPT were tested. Tables 8.2 and 8.3 show the initial and final parameter estimates, respectively. Table 8.4 shows that program MINOPT works very reliably.

Table 8.4. Number of iterations necessary to reach a minimum $U(\beta)$ for the six test models (F means convergence to a false solution, S means a slow convergence)

Model	MSM	GN	MINOPT
I	F	55	22
II	179	S	10
III	2452	15	32
IV	F	S	42
V	F	F	65
VI	387	12	16

Models I-V were used to compare the ability of well known statistical packages to solve these problems. The packages tested were:

BMDP SOLO version 3.1	abbreviation SOLO
BMDP version 1987	abbreviation BMDP
SAS version 6.03	abbreviation SAS
SPSS PC+ version 3.1	abbreviation SPSS
STATGRAPHICS version 4.2	abbreviation STATGR
ASYSTANT + version 1.0	abbreviation ASYST
SYSTAT version 4.0	abbreviation SYSTAT
ADSTAT 1.1 (TRILOBYTE Ltd.)	abbreviation ADSTAT
MINSQ 3.12 (MICROMATH)	abbreviation MINSQ

For overall comparison, the performance index, PI, is defined as $PI = 100*(\text{number of correct results})/(r*\text{number of methods})$ was computed. Here T is number of tests used. The greater PI, the better the package at solving nonlinear regression problems. 'Number of methods' means the number of optimization methods available in the packages. The possibility of combining methods (as in MINSQ) was not tested. The results are summarized in Table 8.4a. The best results were obtained with ADSTAT and SPSS. Program MINSQ is very quick and can use a combination of methods. Other packages were not very satisfactory for these test problems.

Table 8.4b Comparison of various packages for nonlinear regression

Package	PI	Number of methods
BMDP	25	2
SAS	25	4
SYSTAT	37.5	2

STATGRAPHICS	50	1
ASYSTANT	8.3	4
SPSS	100	1
ADSTAT	100	1
SOLO	20	1
MINSQ	80	1

This comparison will disappoint many users of the standard statistical packages, showing, as it does, that errors due to false optimum location can cause failure of the whole regression analysis.

8.6 Statistical Analysis of Nonlinear Regression

Statistical analysis in nonlinear regression depends on the model used, the model of measurement errors and the criterion function. Let us limit ourselves to the method of maximum likelihood when the estimates **b** minimize the logarithm of maximum likelihood, $l(\boldsymbol{\beta}) = \ln L(\boldsymbol{\beta})$, as defined by Eq. (8.28).

In the construction of confidence intervals for parameters $\boldsymbol{\beta}$ or in the testing of statistical hypotheses, three main approaches are used.

(a) *The method of linearization* is based on the asymptotic normality of the vector $\sqrt{n}(\mathbf{b} - \boldsymbol{\beta})$ and an estimate of the variance D(**b**), defined by Eq. (8.29). As was shown in Problem 8.8, for additive and normally distributed errors:

$$D(\mathbf{b}) = \sigma^2 (\mathbf{J}^\mathsf{T} \mathbf{J})^{-1}$$

This expression corresponds to a *linearization* of the regression model [Eq. (8.66a)]. Here, the same expression is used as in a linear regression but matrix **J** replaces matrix **X**.

(b) *The method of Lagrange multipliers* is based on the fact that the quadratic form

$$QF = \mathbf{U}^\mathsf{T} \mathbf{I}^{-1} \mathbf{U} \tag{8.141}$$

has asymptotically the $\chi^2(m)$ -distribution, and the matrix **I** is defined either by Eq. (8.30) or Eq. (8.31). The vector **U**, of dimension $(m \times 1)$, has components $\delta l(\boldsymbol{\beta}) / \delta \beta_j$.

(c) *The method of the maximum likelihood ratio* is based on:

$$P(\beta) = \frac{L(\beta)}{L(b)} \tag{8.142}$$

The random variable, $\{-2\ln[P(\beta)]\}$, has asymptotically the $\chi^2(m)$ -distribution.

The method of linearization is, in practice, the most widely used one as it does not involve additional calculation. The method of Lagrange multipliers and the maximum likelihood ratio require a numerical search for the roots of nonlinear functions, but lead to more accurate and more useful results.

Statistical properties of the least-squares method (LS)

The LS method is a special case of the maximum likelihood method for an additive model of measurement and an independent normal distribution of measurement errors. Gallant [5] derived the following equations:

$$\mathbf{b} = \boldsymbol{\beta}^* + (\mathbf{J}_t^T \mathbf{J}_t)^{-1} \mathbf{J}_t^T \boldsymbol{\varepsilon} + \text{term}\,(1/\sqrt{n}) \tag{8.143}$$

$$\sigma^2 = \frac{\boldsymbol{\varepsilon}^T (\mathbf{E} - \mathbf{J}_t (\mathbf{J}_t^T \mathbf{J}_t)^{-1} \mathbf{J}_t^T) \boldsymbol{\varepsilon}}{n - m} + \text{term}(1/n) \tag{8.144}$$

Here $\boldsymbol{\beta}^*$ is the true value of the parameters in the model, $\mathbf{J}_t$ is the Jacobian matrix evaluated at the theoretical point $\boldsymbol{\beta}^*$ and term $\left(1/\sqrt{n}\right)$ denotes a random quantity with its mean value equal to $\left(1/\sqrt{n}\right)$. If in Eqs. (8.143) and (8.144), the higher terms are neglected, then from the theory of linear regression, it follows that the distribution of the random quantity $\mathbf{b}$ is m-dimensionally normal

$$\mathbf{b} \approx N\left[\boldsymbol{\beta}, \sigma^2 (\mathbf{J}^T \mathbf{J})^{-1}\right] \tag{8.145}$$

The random variable, $(n-m)\hat{\sigma}^2 / \sigma$, has a $\chi^2(n-m)$ -distribution and $\mathbf{b}$ and $\hat{\sigma}^2$ are independent. In practice, the matrix $\mathbf{J}_t$, is replaced by matrix $\mathbf{J}$, evaluated at point $\mathbf{b}$. The asymptotic normality of estimates $\mathbf{b}$, determined by the least-squares method, does not require normality of errors $\boldsymbol{\varepsilon}$, [5]. Application of Eq. (8.143) requires the following conditions to be fulfilled:

(1) The regression model must be twice differentiable.

(2) The regression model must be identifiable; that is, the function

$$U(\boldsymbol{\beta}^{(k)}) = \lim_{n \to \infty} \frac{1}{n} \sum_{i=1}^{n} \left[f(x_i, \boldsymbol{\beta}^{(k)}) - f(x_i, \boldsymbol{\beta}) \right]^2$$

should have an unambiguous minimum at the point $\boldsymbol{\beta}^{(k)} = \boldsymbol{\beta}$.

(3) The matrix $\mathbf{Q} = \lim_{n \to \infty} (1/n) \mathbf{J}_t^T \mathbf{J}_t$, must be asymptotically regular.

Practical examination of all these conditions is rather complicated [5].

For real experimental data, the estimates **b** and other statistical characteristics are *biased*, and therefore the application of Eq. (8.145) is limited.

The usefulness of the statistical analysis of nonlinear regression models by least-squares methods depends on the magnitude of the bias, and this depends on the degree of nonlinearity in the regression model.

The covariance matrix of parameter estimates

From Eq. (8.145), it follows that the asymptotic covariance matrix of estimates **b** obtained by the LS method is expressed by

$$D(\mathbf{b}) = \sigma^2 (\mathbf{J}^T \mathbf{J})^{-1} \tag{8.146}$$

When errors are independent and identically distributed with constant variance, it is possible to find a more accurate approximation, based on a linearization of estimate $\mathbf{b}(\varepsilon)$ as a function of ε:

$$D(\mathbf{b}) = 4\sigma^2 \mathbf{H}^{-1} (\mathbf{J}^T \mathbf{J}) \mathbf{H}^{-1} \tag{8.147}$$

When $(\mathbf{J}^T \mathbf{J}) \approx 0.5 \mathbf{H}$ is simplified, a less accurate approximation is obtained:

$$D(\mathbf{b}) = 2\sigma^2 \mathbf{H}^{-1} \tag{8.148}$$

When the residuals $\hat{\mathbf{e}}$ are small or the elements of matrix **B** in Eq. (8.62) are approximately zero, then $\mathbf{H}^{-1} = 0.5(\mathbf{J}^T \mathbf{J})^{-1}$. The effect of the application of Eqs. (8.146), (8.147) and (8.148) on the accuracy of the estimate of the confidence regions of parameters **b** has been studied [57], and it was concluded that the two more accurate relations, Eqs. (8.147) and (8.148), are not significantly better. For practical calculations, the asymptotic formula [Eq. (8.146)] is obviously acceptable.

With a knowledge of the covariance matrix $D(\mathbf{b})$, either the variance of individual parameters $D(b_j)$ or the correlation coefficients r_{ij} between estimates b_i and b_j, may be estimated. From Eq. (8.146), we can write

$$D(b_j) = \sigma^2 V_{jj} \qquad (8.149)$$

where V_{jj} are the diagonal elements of the matrix $\mathbf{V} = (\mathbf{J}^T\mathbf{J})^{-1}$. Similarly, the correlation coefficient between the parameters b_i and b_j is

$$r_{ij} = \frac{V_{ij}}{\sqrt{V_{jj}V_{ii}}} \qquad (8.150)$$

If the value of r_{ij} is close to unity, the estimates b_i and b_j are linearly dependent and the model is overdetermined or ill-conditioned with respect to parameters b_i and b_j.

8.6.1 Degree of Nonlinearity of a Regression Model

For characterization of nonlinear behaviour in regression models, the intrinsic curvature K_h^N [Eq. (8.75a)], the parameter-effects curvature K_h^P [Eq. (8.75b)] or the maximum intrinsic curvature Γ^N [Eq. (8.75c)] and the maximum parameter-effects curvature Γ^P [Eq. (8.75d)] can be adopted. If Γ^N and Γ^P are sufficiently small [56] for statistical analysis and for construction of confidence intervals, the *linearization of regression model* [Eq. (8.145)] may be used. From many practical experiments, it has been concluded that

(a) the influence of nonlinearity may be, in many cases, removed by a suitable reparameterization when Γ^N is small and Γ^P high;

(b) the efficiency of reparameterization also depends on the data;

(c) in some cases, even for high values of Γ^N and Γ^P, linearization may be applied. There are cases, however, when even for small values of Γ^N and Γ^P the confidence intervals estimated after linearization are not suitable.

The measures of nonlinearity based on a curvature are not universal.

8.6.1.1 Bias of Parameter Estimates

For expressing the bias of parameter estimates,

$$\mathbf{h} = E(\mathbf{b} - \boldsymbol{\beta}^*)$$

many approximations exist in the literature. For simplicity, we use the following definition of parameter bias [58]

$$\mathbf{h} = (\mathbf{J}^T\mathbf{J})^{-1}\mathbf{J}^T\mathbf{d} \tag{8.151}$$

where $\mathbf{d}$ is the $(n \times 1)$ vector with the components

$$d_i = \frac{-\sigma^2 \mathrm{tr}\left[\left(\mathbf{J}^T\mathbf{J}\right)^{-1}\mathbf{G}_i\right]}{2} \tag{8.152}$$

where $\mathrm{tr}(\bullet)$ denotes the trace of a matrix and $\mathbf{G}_i$ is the matrix of second derivatives of the model function [Eq. (8.65)]. The vector $\mathbf{d}$ is the expected value of the difference between the linear approximation [Eq. (8.66a)] and the quadratic approximation of the model function. Equation (8.151) enables the bias $\mathbf{h}$ to be found from the coefficients (parameters) of the linear regression model,

$$\mathbf{d} = \mathbf{Jh} \tag{8.153}$$

It is obvious that the bias $\mathbf{h}$ will be small if

(a) the vector $\mathbf{d}$ is perpendicular to the tangent hyperplane, defined by columns of matrix $\mathbf{J}$, such that $\mathbf{J}^T\mathbf{d} = 0$.

(b) the vector elements $\mathbf{d}$ are small, i.e. the increment of the quadratic term will be insignificant and the model $f(\mathbf{x}, \boldsymbol{\beta})$ will be well linearized.

Similarly, the bias of residuals is given by $\hat{e}_i = y_i - f(x_i, \mathbf{b})$. The mean value of the vector of residuals,

$$\mathbf{M}_r = E(\hat{\mathbf{e}}) \tag{8.154a}$$

can be rewritten as

$$\mathbf{M}_r = (\mathbf{E} - \mathbf{P})\mathbf{d} \tag{8.154b}$$

where $\mathbf{P}$ is the projection matrix [Eq. (8.68)]. The mean value of the residuals is called the *residuals bias* because it is assumed that $E(\varepsilon) = 0$.

For practical calculation, we often use the relative bias of the parameter estimates defined by

$$h_{R,j} = \frac{h_j}{b_j} \, 100 \, [\%] \tag{8.154c}$$

The bias of estimates is considered significant if $h_{R,j} > 1\%$. For such biased estimates, the statistical analysis based on linearization of regression model cannot be legitimately used.

Problem 8.13 *Calculation of the relative bias of parameters*

Calculate the relative bias of parameters h_{R_j} of individual variants of the Kohlrausch model from Problem 8.12.

Data: from Problem 8.12

Solution: The values calculated for the relative bias of parameter estimates for all the variants of Problem 8.11 are given in Table 8.5.

Table 8.5 Relative bias h_{R_j} of the parameters of the Kohlrausch model

Relative bias, %	Variant of model					
	I_a	I_b	II	III	IV	V
$h_{R,1}$	-1.0×10^{-4}	-1.8×10^6	0.18	_*	0.163	_*
$h_{R,2}$	-4.1×10^6	-6.0×10^8	_*	379.3	_*	_*
$h_{R,3}$	8.62×10^4	1.46×10^7	0.058	0.44	0.058	0.007
$h_{R,4}$	4.12×10^6	3.52×10^9	1.49	-22.5	1.402	1.42

* Constant value of parameter

Conclusion: The original model (Ia, Ib) is very ill-conditioned and has high values of bias, indicating the inadequacy of the proposed model. For variants II, IV and V, the parameter estimates have physical meaning and the bias values are sufficiently small. Therefore for these three variants, the statistical analysis based on linearization of the model may be used.

To express the *total bias* of parameter estimates, Box [59] proposed the scalar characteristic

$$\hat{M} = \frac{\mathbf{h}^T (\mathbf{J}^T \mathbf{J}) \mathbf{h}}{m \hat{\sigma}^2} \tag{8.155}$$

and proved that the scalar $\hat{M}$ is related to the maximum parameter-effect curvature caused by parameters, by following the inequality $\hat{M} \leq \dfrac{(\Gamma^P)^2}{4}$. Similarly, for the norm of the residual bias, the following inequality applies $\|\mathbf{E}\| \leq 0.5 \Gamma^N \hat{\sigma} \sqrt{m}$. The bias of parameters is related to the curvature caused by parameters K_h^P and the residual bias is related to the intrinsic curvature K_h^N.

Problem 8.14 *Kinetic parameters of dyeing*

A modified polyester fibre, Tesil 32, was dyed by a disperse dye, C.I. Dispersion Red 54, and the kinetics of the process were studied. The relative concentration of dye C_i, on the fibre was measured as a function of time, t_i. The kinetics of the isothermal dyeing may be expressed by the Cegarra-Puente model $C_i = \beta_1 \sqrt{1 - \exp(-\beta_2 t_i)}$, where β_1 is the relative equilibrium concentration of the dye on a fibre and β_2 is the rate constant. Estimate b_1 and b_2. Calculate the correlation r_{12} between the parameters β_1 and β_2, the total bias of parameters $\hat{M}$ and the norm of vector, $\|\mathbf{E}\|$.

Data: $n = 6$

t, min	20	40	60	80	100	120
C,%	43.5	53.6	64.1	66.5	72.0	76.5

Solution: From the initial guesses of parameters $\beta_1^{(0)} = 100$, $\beta_2^{(0)} = 10^{-4}$ with $U(\beta^{(0)}) = 12300$, the minimum found was $U(\mathbf{b}) = 12.88$, and the best estimates were $b_1 = 82.37\%$ and $b_2 = 0.0147$ min^{-1}. The correlation coefficient $r_{12} = -0.963$ indicates strong multicollinearity. The values of the relative bias of parameters $h_{R,1} = 0.46\%$ and $h_{R2} = 0.5\%$ indicate low insignficant bias. The scalar value of the total bias of parameters $\hat{M} = 0.1316$ and the norm of vector $\|\mathbf{E}\| = 3.58$ shows low bias also.

Conclusion: Linearization may be used for the statistical analysis of this model.

Because the bias of parameters is related to the curvature caused by parameters, it may be affected by reparameterization. Let us suppose that, instead of parameters $\boldsymbol{\beta}$, the transformed parameters γ defined by Eq. (8.6) are used. Each new parameter, γ_j, is a function of all components of the vector $\boldsymbol{\beta}$ such that $\gamma_j = l(\boldsymbol{\beta})$. The bias of the parameter estimates C_j of parameters γ_j, is given by

$$h_j(\mathbf{C}) = \mathbf{l}^\mathrm{T}\mathbf{h}(\mathbf{b}) + \frac{\sigma^2}{2}\mathrm{tr}\left[\mathbf{M}(\mathbf{J}^\mathrm{T}\mathbf{J})^{-1}\right] \tag{8.156}$$

where is the vector comprising first derivatives of the transformation with elements

$$l_j = \frac{\partial l(\boldsymbol{\beta})}{\partial \beta_j} \tag{8.156a}$$

and the matrix $\mathbf{M}$ contains the second derivatives of the transformation with elements

$$M_{jk} = \frac{\partial^2 l(\boldsymbol{\beta})}{\partial \beta_j \partial \beta_k} \tag{8.156b}$$

The symbol $\mathbf{h}(\mathbf{b})$ denotes the bias, $\mathbf{h}$, of the parameter estimates, $\mathbf{b}$, calculated from Eq. (8.151). If the actual reparameterization is known, the change of bias of the parameters may be predicted. Another task is to select the reparameterization that leads to the smallest bias.

Problem 8.15 *Change of bias of parameters after model reparameterization*

Determine the change in the bias of estimate b_1 in the model $f(x,\boldsymbol{\beta}) = \beta_1 \exp(\beta_2 x)$ by reparameterization, $\gamma_1 = \ln \beta_1$.

Solution: The model reparameterization of $f(x,\boldsymbol{\beta}) = \beta_1 \exp(\beta_2 x)$ leads to the function $f(x,\gamma_1,\beta_2) = \exp(\gamma_1 + \beta_2 x)$. The vector $\mathbf{1}$ has elements, $\mathbf{1} = \begin{bmatrix} 1/b_1 & 0 \end{bmatrix}$ and matrix $\mathbf{M}$ has the form $\mathbf{M} = \begin{bmatrix} 1/b_1^2 & 0 \\ 0 & 0 \end{bmatrix}$. On substitution into Eq. (8.156), we get $h_1(\mathbf{C}) = \dfrac{h_1(b_1)}{b_1} - \dfrac{\sigma^2 V_{11}}{2b_1^2} = \dfrac{h_1(b_1) - 0.5D(b_1)/b_1}{b_1}$ where V_{11} is the first

diagonal element of the matrix $(\mathbf{J}^T\mathbf{J})^{-1}$ and $D(b_1)$ is the variance b_1.

Conclusion: The decrease of bias, $h_1(b_1) - h_1(\mathbf{C})$, is bigger for big variances of estimate b_1, and smaller for small values of b_1. It is assumed that b_1 is always positive.

8.6.1.2 Asymmetry of Parameter Estimates

Nonlinearity of the regression model results in an asymmetric distribution of estimates **b**. The measure of nonlinearity is a measure of the asymmetry of the estimates. Ratkowsky [53] used n points generated as identically distributed quantities, ε_i^*, with mean value equal to zero and variance $\hat{\sigma}^2$. Then, the random dependent variables y_i^+ and y_i^- are generated by

$$y_i^+ = f(x_i, \mathbf{b}) + \varepsilon_i^* \qquad (8.157a)$$

$$y_i^- = f(x_i, \mathbf{b}) - \varepsilon_i^* \qquad (8.157b)$$

Parameter estimates obtained by the least-squares method when using $\mathbf{y}^+$, instead of $\mathbf{y}$, are denoted $\mathbf{b}^+$ and similarly, $\mathbf{b}^-$ when using $\mathbf{y}^-$, instead of $\mathbf{y}$. Linear models are symmetrical, in that $(b_i^+ - \beta_i^*) = -(b_i^- - \beta_i^*)$. For nonlinear models, there is no such symmetry. A convenient measure of asymmetry is the expression,

$$\psi_i = \frac{1}{2}\left[(b_i^+ - \beta_i^*) + (b_i^- - \beta_i^*)\right] \qquad (8.158)$$

The mean value, $E(\psi_i)$, is equal to the bias, h_i and the variance $D(\psi_i)$ is given by

$$D(\psi_i) = 0.25D(b_i^+) + 0.25D(b_i^-) + 0.25\,\mathrm{cov}(b_i^+, b_i^-) \qquad (8.159)$$

Nonlinearity is indicated by the ratio $\lambda_{Ni} = \dfrac{D(\psi_i)}{D(b_i)}$. When $\lambda_{Ni} < 0.01,$, the distribution of parameter estimates is nearly symmetric. For $\lambda_{Ni} > 0.01$, the distribution of estimate b_i is strongly asymmetric and the model, with respect to this parameter, is strongly nonlinear.

Morton [60] published expressions relating the statistical measure, λ_{Ni} to the bias and other measures of nonlinearity. The great advantage of the measure λ_{Ni} is that it is based on statistical arguments. For the determination of λ_{Ni}, Ratkowsky [53] generated many estimates $\mathbf{b}^+$ and $\mathbf{b}^-$ from various vectors of errors, ε^*.

8.6.2   Interval Estimates of Parameters

Point estimates $\mathbf{b}$ of regression parameters $\boldsymbol{\beta}$ are, from a statistical point of view, worthless as they do not mention the region in which a true value $\boldsymbol{\beta}^*$ may be expected. The estimates $\mathbf{b}$ are random quantities estimated from a sample of size n, $\{x_i, y_i\}$, $i = 1, ..., n$. The confidence regions, simultaneous confidence regions and confidence intervals for multivariate samples are constructed similarly. For their elucidation, the same rules as for univariate data are applied (Chapter 3).

In nonlinear regression models for construction of confidence regions and intervals, the linearization often used has elliptic confidence regions. However, linearization is useful only when the model is *not* strongly nonlinear and the nonlinearity measures of asymmetry and bias are small. More accurate confidence regions can be found by using Lagrange multipliers or the likelihood ratio; these are non-elliptic and do not have to be continuous.

8.6.2.1   Confidence Regions of Parameters

From the normality of estimates $\mathbf{b}$, it follows that the quadratic form

$$Q = (\boldsymbol{\beta}^* - \mathbf{b})^{\mathrm{T}} D(\mathbf{b})^{-1} (\boldsymbol{\beta}^* - \mathbf{b}) \qquad (8.159a)$$

has the $\chi^2(m)$-distribution. The corresponding $100(1-\alpha)\%$ confidence region of parameters $\boldsymbol{\beta}^*$ is the m-dimensional ellipsoid with boundaries expressed by

$$(\boldsymbol{\beta}^* - \mathbf{b})^{\mathrm{T}} D(\mathbf{b})^{-1} (\boldsymbol{\beta}^* - \mathbf{b}) = \chi^2_{1-\alpha}(m) \qquad (8.160)$$

where $\chi^2_{1-\alpha}(m)$ is the $100(1-\alpha)\%$ quantile of $\chi^2(m)$ with m degrees of freedom. The centre of this ellipsoid is at the point $\mathbf{b}$. For the least-squares method $D(\mathbf{b}) = \dfrac{U(\mathbf{b})(\mathbf{J}^{\mathrm{T}}\mathbf{J})^{-1}}{n-m}$. After substitution into Eq. (8.159), the quadratic form can be formulated as $\left[\dfrac{\Delta\boldsymbol{\beta}^{\mathrm{T}}(\mathbf{J}^{\mathrm{T}}\mathbf{J})\Delta\boldsymbol{\beta}}{\sigma^2 m}\right]\left[\dfrac{U(\mathbf{b})}{\hat{\sigma}^2(n-m)}\right]^{-1} = \dfrac{\Delta\boldsymbol{\beta}^{\mathrm{T}}(\mathbf{J}^{\mathrm{T}}\mathbf{J})\Delta\boldsymbol{\beta}}{\hat{\sigma}^2 m}$

where $\Delta\boldsymbol{\beta} = \boldsymbol{\beta}^* - \mathbf{b}$. This quadratic form has the distribution $\chi^2(m)/\chi^2(n-m)$ $= F(m, n-m)$, i.e. the Fisher-Snedecor distribution with m and $(n-m)$ degrees of freedom. The confidence ellipsoid then has the boundary

$$\Delta\boldsymbol{\beta}^{\mathrm{T}}(\mathbf{J}^{\mathrm{T}}\mathbf{J})^{-1}\Delta\boldsymbol{\beta} = m\hat{\sigma}^2 F_{1-\alpha}(m, n-m) \tag{8.161b}$$

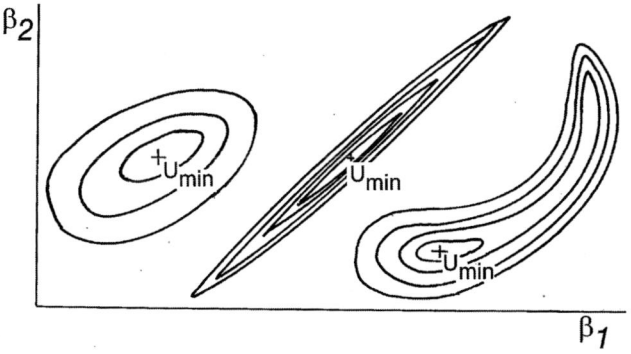

Figure 8.21 Contours for various hyperparaboloid shapes in the vicinity of the minimum (pit).

In some chemometrics programs, instead of confidence ellipsoids, the boundary of the last contour of the least squares criterion is used (Fig. 8.21).

When the bias of parameters, $\mathbf{h}$, is calculated, the correction $\Delta\boldsymbol{\beta}^* = \mathbf{b} - \mathbf{h} - \boldsymbol{\beta}^*$ may be used instead of $\Delta\boldsymbol{\beta}$. To express the geometry of the confidence ellipsoids, the decomposition of the matrix $(\mathbf{J}^{\mathrm{T}}\mathbf{J})^{-1}$, the eigenvalues λ_i and eigenvectors $\mathbf{V}_i$ may be introduced such that

$$(\mathbf{J}^{\mathrm{T}}\mathbf{J})^{-1} = \mathbf{V}\boldsymbol{\lambda}\mathbf{V}^{\mathrm{T}} \tag{8.162a}$$

where $\mathbf{V}$ is the matrix containing the eigenvectors in its columns and the diagonal matrix λ contains eigenvalues $\lambda_1 \geq \lambda_2 \geq \ldots \lambda_m$ on the diagonal. For the corresponding decomposition of the matrix $(\mathbf{J}^{\mathrm{T}}\mathbf{J})$, we have

$$(\mathbf{J}^{\mathrm{T}}\mathbf{J}) = \mathbf{V}^{\mathrm{T}}\boldsymbol{\lambda}^{-1}\mathbf{V} \tag{8.162b}$$

The matrix λ^{-1} is, once again, a diagonal matrix with reciprocal values λ_j^{-1}, $j = 1, \ldots, m$ as elements. After substitution from Eq. (8.162a) into Eq. (8.161), we have

$$\Delta\boldsymbol{\beta}^T\mathbf{V}^T\lambda^{-1}\mathbf{V}\Delta\boldsymbol{\beta} = \mathbf{Y}^T\lambda^{-1}\mathbf{Y} = \sum_{i=1}^{m}\frac{1}{\lambda_i}Y_i^2 \qquad (8.163)$$

Here $\mathbf{Y} = \mathbf{V}\Delta\boldsymbol{\beta}$, which is the new orthogonal set of co-ordinates having the important property that the axes of the confidence ellipsoid are identical with the axes of the co-ordinate system. If we introduce the notation $p^2 = m\hat{\sigma}^2 F_{1-\alpha}(m, n-m)$ the confidence ellipsoid can be expressed by the simple formula

$$\sum_{i=1}^{m}\frac{Y_i^2}{\lambda_i} = p^2 \qquad (8.164)$$

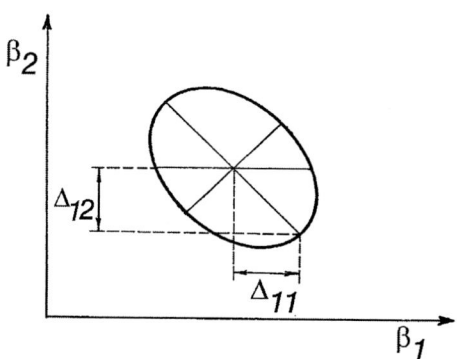

Figure 8.22   Graphical illustrations of the projections Δ_{11} and Δ_{12}.

The lengths of the half-axes of the ellipsoid are equal to $p\sqrt{\lambda_i}$. The projection Δ_{jk} of the jth half-axis into the axis of parameter β_k, is given by

$$\Delta_{jk} = p\left|V_{kj}\sqrt{\lambda_j}\right| \qquad (8.165)$$

where V_{kj} is the kth element of the vector $\mathbf{V}_j$ which is the jth column of matrix $\mathbf{V}$.

Problem 8.16   *Confidence ellipsoid for the parameters of the Cegarra-Puente model*

Estimate the 95% confidence ellipsoid for parameters β_1 and β_2 of the Cegarra-Puente model given in Problem 8.13.

Data: from Problem 8.13

Solution: With use of Eq. (8.164) the co-ordinates of the confidence region in systems Y_1 and Y_2 were calculated, then, by reverse transformation, β_1 and β_2. The results are shown in Fig. 8.23.

Conclusion: The elongated shape of the confidence ellipsoid and its orientation prove strong negative correlation between parameters β_1 and β_2.

When the dimension, m, of the parameter vector is greater than 2, the partial confidence ellipsoid is constructed for only two of the model parameters. Let us assume that the vector $\boldsymbol{\beta}^* = (\boldsymbol{\beta}_1^*, \boldsymbol{\beta}_2^*)^T$ is known, and that the confidence region for parameters $\boldsymbol{\beta}_2^*$ is to be constructed. As for linear regression models, the boundaries of the $100(1-\alpha)\%$ confidence ellipsoid are found from

$$(\boldsymbol{\beta}_2^* - \mathbf{b}_2)^T \mathbf{D}_2^{-1} (\boldsymbol{\beta}_2^* - \mathbf{b}_2) = q\hat{\sigma}^2 F_{1-\alpha}(q, n-m) \qquad (8.166)$$

where q is the dimension of vector $\boldsymbol{\beta}_2^*$. The matrix $\mathbf{D}_2$ is of dimension $(q \times q)$ and is formed from the matrix $(\mathbf{J}^T\mathbf{J})^{-1}$ by omitting $(m-q)$ rows and $(m-q)$ columns. Before constructing the confidence region, the parameters must be renumbered to make the parameters for which the ellipsoid is constructed the last ones.

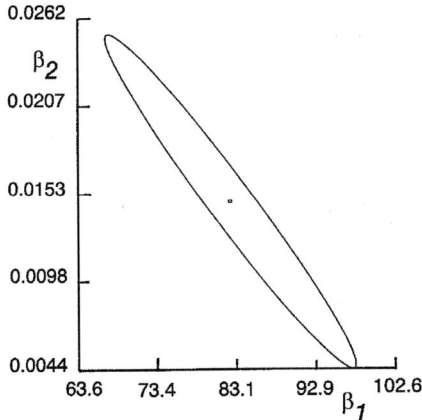

Figure 8.23   The simultaneous confidence region of parameters β_1 and β_2 for the Cegarra-Puente model.

For the method of Lagrange multipliers, the quadratic QF from Eq. (8.141) is directly applied. In the general case of the maximal likelihood method, the boundary of the $100(1-\alpha)\%$ confidence region is defined by

$$\mathbf{U}^T \mathbf{I}^{-1} \mathbf{U} = \chi^2_{1-\alpha}(m) \qquad (8.167a)$$

Various formulae can be derived, depending on the actual likelihood function. For example, in the case of the least-squares method, the matrix $I^{-1} \approx \sigma^2 \left(J^T J\right)^{-1}$ and for vector U is given by

$$U = J^T \hat{e} \qquad (8.167b)$$

where $\hat{e}$ is the vector with elements $\hat{e}_i = y_i - f(x_i, \boldsymbol{\beta}^*)$. When σ^2 is estimated, the boundary of the $100(1 - \alpha)\%$ confidence region has the form

$$\hat{e}^T J (J^T J)^{-1} J \hat{e} = m\hat{\sigma}^2 F_{1-\alpha}(m, n-m) \qquad (8.167c)$$

The boundary of the confidence region is determined as a set of vectors of parameters $\boldsymbol{\beta}$ which fulfil Eq. (8.167c). These confidence regions [Eq. (8.167c)] do not always have to be elliptic.

According to the *likelihood ratio* method, the boundaries of the $100(1 - \alpha)\%$ confidence region may be defined as

$$\ln L(\mathbf{b}) - \ln L(\boldsymbol{\beta}^*) = \chi^2_{1-\alpha}(m) / 2 \qquad (8.168a)$$

In the least-squares method, when both the parameter vector $\mathbf{b}$ and the variance $\hat{\sigma}^2$ are estimated, Eq. (8.168a) may be expressed in the form

$$U(\boldsymbol{\beta}^*) - U(\mathbf{b}) = p^2 \qquad (8.168b)$$

where p^2 is defined by Eq. (8.164). For determination of the confidence region boundaries, a numerical method should be used.

Problem 8.17   *Confidence regions for a model describing the activity of sea-weed as a function of temperature*

To express the empirical dependence of the activity P, of sea-weed on the temperature T at illumination level of 96 W. m^{-2}, the model proposed was

$$P = \beta_2 \left[\frac{\beta_3 - T}{\beta_3 - \beta_1}\right]^Z \exp\left[Z\left[1 - \frac{\beta_3 - T}{\beta_3 - \beta_1}\right]\right] \qquad (8.169)$$

where $Z = B^2 \left(1 + \sqrt{1 + 40/R}\right)^2 / 400$, $B = \ln(\beta_4)(\beta_3 - \beta_1)$, β_1 corresponds to the temperature with the maximum activity of sea-weed, $P_{\max} = \beta_2$, β_3 is the temperature at which the activity of the sea-weed is zero and β_4 is the shape factor. A graphical illustration of the proposed model $P = f(T, \beta_1, \beta_2, \beta_3, \beta_4)$ is shown in Fig. 8.24. Estimate all four parameters and determine the confidence regions for pairs of parameters, $\beta_1^* - \beta_2^*, \beta_2^* - \beta_4^*$ and $\beta_1^* - \beta_3^*$.

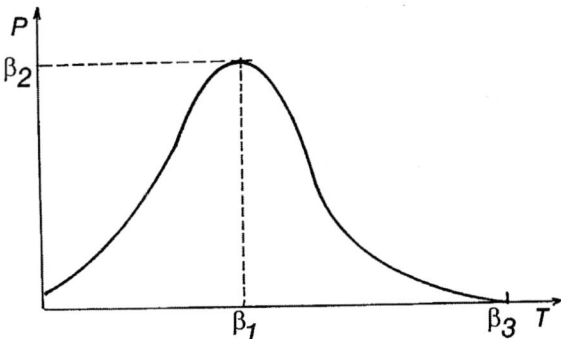

Figure 8.24   Graphical illustration of a model [Eq. (8.169)] and the individual parameters.

Data: $n = 1$

T, °C	5	10	15	20	25	30	35
P	2.18	3.16	4.11	6.67	9.04	5.83	1.13

Solution: From the initial guesses of parameters $\beta_1^{(0)} = 25$, $\beta_2^{(0)} = 9$, $\beta_3^{(0)} = 36$ and $\beta_4^{(0)} = 1.8$, the corresponding value of the sum of squares function $U(\boldsymbol{\beta}^{(0)})$ $= 10.52$. The minimization process terminated at $U(\mathbf{b}) = 1.782$. The parameter estimates $\mathbf{b}$ with their relative bias are listed in Table 8.6.

Table 8.6   Parameter estimates and their relative bias, for the model in Eq. (8.169)

Parameter	Best estimate b_j	Relative bias $h_{R,j}$, %
β_1	25.06	0.0114
β_2	8.296	0.499
β_3	37.23	2.551
β_4	2.201	2.284

Figures 8.25, 8.26 and 8.27 show the 95% confidence region of pairs of parameters $\beta_1^* - \beta_2^*, \beta_1^* - \beta_3^*$ and $\beta_2^* - \beta_4^*$. The solid curve refers to Eq. (8.169) and the dotted curve to Eq. (8.168).

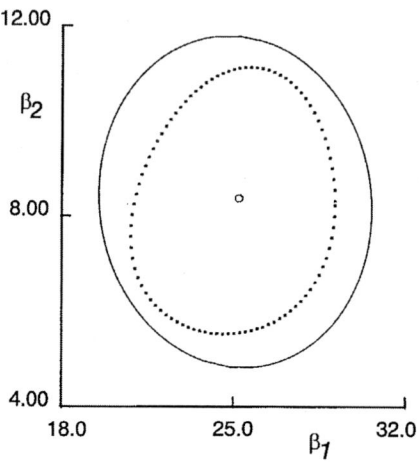

Figure 8.25 The confidence region for parameters $\beta_1^* - \beta_2^*$ for the proposed model, Eq. (8.169), (solid curve) and with the use of Eq. (8.168) (dotted curve).

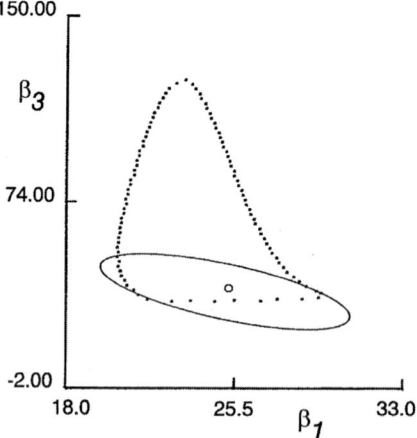

Figure 8.26 The confidence region for parameters $\beta_1^* - \beta_3^*$ for the proposed model, Eq. (8.169), (solid curve) and with the use of Eq. (8.168) (dotted curve).

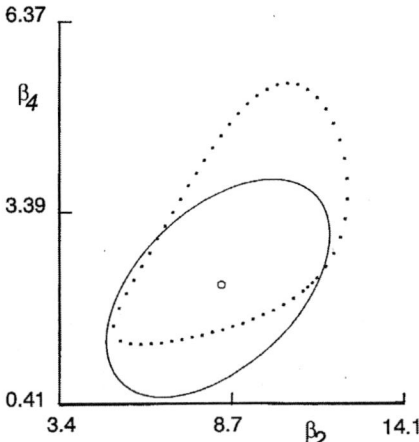

Figure 8.27   The confidence region for parameters $\beta_2^* - \beta_4^*$ for the proposed model, Eq. (8.169), (solid curve) and with the use of Eq. (8.168) (dotted curve).

Conclusion: From the figures, it is obvious that for even a small bias, the differences between the confidence ellipsoids and the more accurate confidence regions are highly significant. The smallest difference is for the pair of parameters, $\beta_1^* - \beta_2^*$ where the values of the relative bias are smaller than 1%.

For the construction of the confidence regions, a reparameterization limiting the bias followed by a reverse parameterization, may also be used [61]. If a nonparameteric technique is required, the Jackknife or the Bootstrap methods are often used. The principle involved is the same as for univariate samples (Chapter 3).

8.6.2.2   Confidence Intervals of Parameters

With the use of Eq. (8.161), the $100(1 - \alpha)\%$ confidence interval of parameter β_j in the form

$$b_j - \hat{\sigma}\sqrt{V_{jj}}\, t_{1-\alpha/2}(n - m) \leq \beta_j^* \leq b_j + \hat{\sigma}\sqrt{V_{jj}}\, t_{1-\alpha/2}(n - m) \qquad (8.170)$$

is a direct analogy of the confidence intervals of the parameters of linear models. The influence of other parameters is neglected. When all off-diagonal elements of the matrix $(\mathbf{J}^T\mathbf{J})^{-1}$ are zero, Equation (8.170) may be used. However, the elements of the vector **b** are often mutually correlated, so that the intervals of Eq. (8.170) are underestimated, i.e. they are too narrow.

A more suitable determination of the confidence interval of parameter β_k^* is on the basis of the maximal length Δ_k of the projection $\Delta_{k,j}$ onto the parameter axis β_k. In the program LETAGROP, and the related system ABLET in ADSTAT, the estimate of the standard deviation of the kth parameter, β_k^*, is calculated from

$$\Delta_k = \max_j(\Delta_{kj})$$

(8.171a)

and the confidence interval of the parameter β_k is estimated from

$$b_k - \Delta_k \le \beta_k^* \le b_k + \Delta_k$$

(8.171b)

Instead of projections it is simpler to search directly for the co-ordinates of the extreme points on the confidence ellipsoid in the directions of the individual parameter axes [62]. The confidence interval of the parameter β_k^* is given by

$$b_k - p\sqrt{V_{kk}} \le \beta_k^* \le b_k + p\sqrt{V_{kk}}$$

(8.172)

For $m = 1$, these confidence intervals are identical. When the number of regression parameters m is increased, the confidence intervals in Eqs. (8.171) and (8.172) become broader than those in Eq. (8.170). All confidence intervals are *symmetrical*.

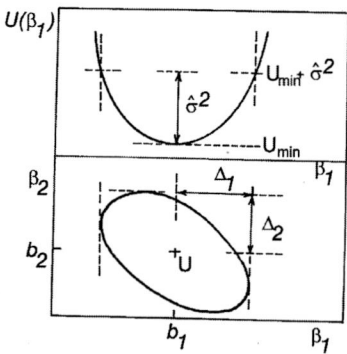

Figure 8.28 The estimates of the parameter standard deviations Δ_1 and Δ_2 by programs LETAGROP and ABLET.

Problem 8.18   *The confidence intervals for the sea-weed activity model*

Estimate the half-length of the 95% confidence intervals for four parameters of the proposed model for the dependence of activity of sea-weed on temperature, from Problem 8.17.

Data: from Problem 8.17

Solution: The half-length of the confidence intervals of the four parameters of the model (8.169) calculated by three different approaches are shown in Table 8.7.

Table 8.7 Half-lengths of the 95% confidence interval of four parameters of model (8.169)

Parameter	b_k	Δ_k (8.170)	Δ_k (8.171)	Δ_k (8.172)
β_1	25.06	2.931	3.922	5.541
β_2	8.296	1.845	3.466	3.488
β_3	37.23	8.199	15.470	15.500
β_4	2.201	0.8984	1.160	1.690

Conclusion: Equation (8.170) leads to the confidence intervals that are false and too narrow. The broad confidence intervals are the consequence of too small a sample size, $n = 7$, relative to the number of unknown parameters, $m = 4$.

Asymmetrical confidence intervals of parameter estimates may be obtained when Eqs. (8.167) and (8.168) are solved numerically with respect to parameter β_k^*, when estimates **b** are supplied for the other components of vector $\boldsymbol{\beta}^*$.

8.6.2.3   Confidence Intervals of Prediction

If the regression model can be linearized, the $100(1-\alpha)\%$ confidence interval of a prediction $f(x^*, \mathbf{b})$ at the point x^* may be calculated. It then follows that

$$f(x^*,\mathbf{b}) - t_{1-\alpha/2}(n-m)\,\hat{\sigma}_P(x^*) \le f(x^*,\boldsymbol{\beta}) \le f(x^*,\mathbf{b}) + t_{1-\alpha/2}(n-m)\,\hat{\sigma}_P(x^*)$$

$$(8.173)$$

where $\hat{\sigma}_P^2(x^*)$ is the estimate of the prediction variance for which

$$\hat{\sigma}_P^2(x^*) = \mathbf{J}^T D(\mathbf{b})\mathbf{J} \qquad (8.173a)$$

The symbol **J** denotes the vector of derivatives of a model function at the point x^* with elements

$$J_j = \frac{\partial f(x^*,\boldsymbol{\beta})}{\partial \beta_j} \qquad (8.173b)$$

The confidence intervals of prediction calculated for the whole range of the independent variable x (if scalar) form the *confidence bands*. Accurate confidence bands may be constructed with the aid of a suitable reparameterization [61].

Problem 8.19   *Confidence bands of prediction for the model of the effect of temperature on sea-weed activity*

Calculate the 95% confidence bands of prediction for the model of activity of sea-weed *vs.* temperature, Eq. (8.169).

Data: from Problem 8.17

Solution: Figure 8.29 shows the regression line (solid curve) and the 95% bands of prediction (dotted curves).

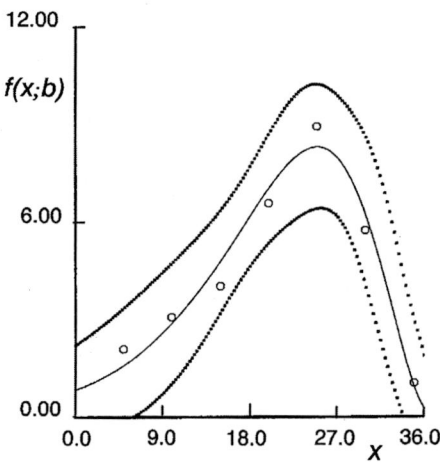

Figure 8.29   Confidence bands of prediction (dotted curve) with calculated regression line for the model, Eq. (8.169) (solid curve) with experimental points.

Conclusion: Here, the width of the confidence band is negatively affected by the small sample size.

8.6.3   Hypothesis Tests about Parameter Estimates

Hypothesis testing is closely related to construction of confidence bands. If parameters β_0 lie in the 95% confidence range around **b**, the differences $(\beta^* - \beta_0)$ may be considered as statistically insignificant at the significance level $\alpha = 0.05$. (The principle of testing is described in Chapter 3.) We restrict ourselves here to the main tests.

To examine the regression parameters, the null hypothesis, H_0: $\boldsymbol{\beta} = \boldsymbol{\beta}_0$ is often tested against the alternative H_A: $\boldsymbol{\beta} \neq \boldsymbol{\beta}_0$, where $\boldsymbol{\beta}_0$ is a given parameter vector. If a regression model can be linearized, Eq. (8.161) leads to the test statistic

$$T = \frac{(\mathbf{b} - \boldsymbol{\beta}_0)^T (\mathbf{J}^T \mathbf{J})(\mathbf{b} - \boldsymbol{\beta}_0)}{m\hat{\sigma}^2} \tag{8.174}$$

which, if the null hypothesis is valid, has the Fisher-Snedecor F-distribution with m and $(n - m)$ degrees of freedom. If $T > F_{1-\alpha}(n - m)$ the null hypothesis H_0, about the equality of $\boldsymbol{\beta}$ and $\boldsymbol{\beta}_0$, is rejected.

When the null hypothesis about one parameter, H_0: $\beta_j = \beta_0$, is tested against the alternative, H_A: $\beta_j \neq \beta_0$, the criterion

$$T_j = \frac{|b_j - \beta_0|}{\hat{\sigma}\sqrt{V_{jj}}} \tag{8.175}$$

may be used. When the null hypothesis is valid, T_j has the Student distribution with $(n - m)$ degrees of freedom. If $T_j > t_{1-\alpha/2}(n - m)$ the null hypothesis about a parameter identity is rejected at significance level α.

When test criteria T and T_j are used, the same restrictions as for the confidence regions hold. If the error distribution ε does not differ from normality, the distribution of the test statistic T_j is Student, even for strongly nonlinear models. As in the construction of confidence regions for parameter subsets $\boldsymbol{\beta}_2$, tests can be constructed for parameter subsets $\boldsymbol{\beta}_2$. When $\boldsymbol{\beta}_0 = 0$ is selected, the classical tests of significance of the regression parameters result.

Problem 8.20   Tests of parameters

For the kinetic model in Problem 8.13, examine the following null hypotheses:

(1) H_0: $\beta = 0$ (which implies that the model is insignificant) against H_A: $\beta \neq 0$;

(2) H_0: $\beta_1 = 80$ against H_A: $\beta_1 \neq 80$.

Data: from Problem 8.13

Solution: The parameter estimates obtained by the least-squares method are $b_1 =$ 82.36, $b_2 = 0.0148$, the variance estimate $\hat{\sigma}^2 = 3.22$. The matrix $(\mathbf{J^T J})$ has the

form $\mathbf{J^T J} = \begin{bmatrix} 3.59 & 5330 \\ 5330 & 8.55 \times 10^6 \end{bmatrix}$ and the matrix $\hat{\sigma}^2 (\mathbf{J^T J})^{-1} = D(\mathbf{b})$ is equal to

$D(\mathbf{b}) = \begin{bmatrix} 12.18 & -0.00758 \\ -0.00758 & 5.11 \times 10^6 \end{bmatrix}$. For testing, we select the significance level

$\alpha = 0.05$.

(1) From Eq. (8.174), we estimate $\mathbf{b^T (J^T J) b} = 3.91 \times 10^4$. The test criterion $T = 5081.9$ is significantly higher than the quantile $F_{0.95}(2, 4)$ = 6.94 and therefore the null hypothesis about model insignificance, $\beta = 0$, is rejected.

(2) From Eq. (8.175), the test criterion is $T_1 = \dfrac{|82.36 - 80|}{\sqrt{12.18}} = 0.0675$.

In comparison with the quantile $t_{0.975}(4) = 2.776$, the T_1 criterion is significantly lower, and therefore the null hypothesis H_0: $\beta_1 = 80$ cannot be rejected.

Conclusion: The statistical tests described are quite simple and do not require complicated calculations.

For testing general parametric hypotheses, the tests of the likelihood ratio and the Lagrange multipliers may be used. Any parametric hypothesis may be expressed as H_0: $\boldsymbol{\beta} \in \omega$ against H_A: $\boldsymbol{\beta} \in \Omega - \omega$, where Ω is an admissible parameter space and ω is its subspace. Often the null hypothesis, expressing q relationships between regression parameters, H_0: $f_1(\boldsymbol{\beta}) = 0, f_2(\boldsymbol{\beta}) = 0,..., f_q(\boldsymbol{\beta}) = 0$, is tested. Then ω is given by restriction conditions of the type $f_j(\boldsymbol{\beta}) = 0, j = 1, ..., q$.

For a test of the null hypothesis H_0, the likelihood ratio has the form

$$l(\mathbf{b}_\omega) = \frac{\underset{\beta \to \omega}{\max} L(\boldsymbol{\beta})}{\underset{\beta \to \Omega}{\max} L(\boldsymbol{\beta})} = \frac{L(\mathbf{b}_\omega)}{L(\mathbf{b})} \tag{8.176}$$

where $\mathbf{b}_\omega$ is the maximum likelihood estimate of parameters $\boldsymbol{\beta}$, with the restriction that it must be in a range ω. The test uses the fact that $-2 \ln l(\mathbf{b}_\omega)$ has the $\chi^2(q)$ distribution. Especially in the case of the least-squares method, when the residual variance is calculated from Eq. (8.176), the test criterion has the form

$$TL = \frac{[U(\mathbf{b}_\omega) - U(\mathbf{b})](n - m)}{qU(\mathbf{b})} \tag{8.177}$$

This statistic has, if the null hypothesis is valid, the Fisher-Snedecor F-distribution with q and $(n - m)$ degrees of freedom.

Problem 8.21 *Tests of parametric hypotheses*

For the Cegarra-Puente kinetic model from Problem 8.13, examine the following hypotheses:

(a) $H_0: \beta_1 = 80$ vs. $H_A: \beta_1 \neq 80$,

(b) $H_0: \beta_1 = 80$ and $\beta_2 = 0.01$ vs. $H_A: \beta_1 \neq 80$ and $\beta_2 \neq 0.01$. Use the *TL* test criterion.

Data: from Problem 8.13

Solution: (1) For the model, $y = 80\sqrt{(1 - \exp(-\beta_2 x))}$, the minimum $U(\beta_2)$ is reached when $U(\beta_2) = 14.51$. In Problem 8.13, $U(\boldsymbol{\beta}) = 12.88$ was achieved. Putting these values into the test criterion, we get $TL = (14.51 - 12.88) \times 4/12.88 = 0.506$, which is smaller than the quantile $F_{0.95}(1, 4) = 7.7$. Therefore the null hypothesis $H_0: \beta_1 = 80$ cannot be rejected at the significance level $\alpha = 0.05$.

(2) The criterion function $U(\boldsymbol{\beta}_\omega) = 469.3$ for $\beta_1 = 80$ and $\beta_2 = 0.01$, for the model $y = 80\sqrt{(1 - \exp(-0.01x))}$. Equation (8.177) then gives the test statistic $TL = 70.87$. This value is significantly higher than the quantile $F_{0.95}(2, 4) = 5.94$, and therefore the null hypothesis $H_0: \beta_1 = 80$ and $\beta_2 = 0.01$ is rejected.

Conclusion: We have shown that application of the *TL* criterion requires two minimizations, except when all values of parameters are known or assumed.

In some cases of a search for a minimum, overdetermined models are formed. Galant [5] has proposed a special procedure for these.

8.6.4 Goodness-of-fit Tests

The examination of residuals is useful, not only for the linear regression model (Section 6.5.2.1), but also for nonlinear regression models and analysis of variance models. Residuals are defined as the differences

$$\hat{e}_i = y_i - \hat{y}_{P,i} = y_i - f(x_i, \mathbf{b}), \quad i = 1,...,n \qquad (8.178)$$

where y_i is an observation and $\hat{y}_{P,i} = f(x_i, \mathbf{b})$ is the calculated value, a "prediction" i.e. the value found from the equation fitted.

We now use graphical and analytical methods for examining the residuals, in order to check the quality of nonlinear models.

8.6.4.1 Graphical Analysis of Residuals

The principle ways of plotting the residuals $\hat{e}_i$ have already been described in regression diagnostics for linear regression models. The following plots are often used in the examination of nonlinear models:

(1) The *overall diagram* gives an initial impression of the residuals. If the model is correct, the residuals should resemble observations from a normal distribution with zero mean.

(2) Plot type I (the *index plot*) is a plot of residuals $\hat{e}_i$ against the index i in time order.

(3) Plot type II (the plot *vs.* the independent variable) is a plot of residuals $\hat{e}_i$ against the independent variable $x_j, j = 1, ..., m$.

(4) Plot type III (the plot *vs.* the prediction) is a plot of residuals against the predicted value $\hat{y}_{P,i}$.

Plot type II is usually adopted as the standard plot (Fig. 8.30). If the proposed model represents the data adequately, the residuals should form a random pattern. Systematic departures from randomness indicate that the model is not satisfactory. To examine the normality of a residual distribution, the rankit plot, used in regression diagnostics for linear models, may be applied.

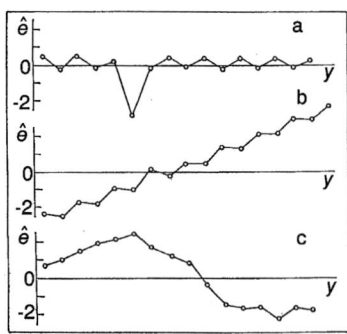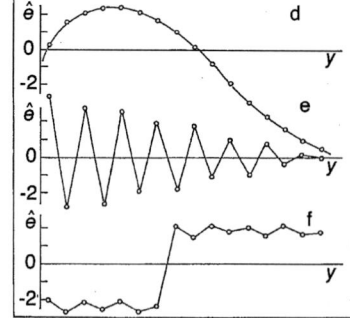

Figure 8.30 Plot II (residuals *vs.* the dependent variable y): (a) detection of an outlier, (b) detection of a trend in residuals, (c) detection of sign changes, (d) detection of a false model, (e) detection of heteroscedasticity, and (f) detection of an abrupt shift in level of the experiment.

8.6.4.2   Statistical Analysis of Residuals

The plots that have been recommended in the previous sections are visual techniques for easy checking of some of the basic assumptions of the least-squares method and of the proposed model. Certain statistics provide a numerical measure for some of the discrepancies previously described.

In many regression programs used in the laboratory, the statistical analysis of residuals represents the main diagnostic tool used to search for the "best" model when more than one are possible or proposed. The goodness-of-fit test analyses the set of residual and examines the following criteria.

(1) The arithmetic mean of residuals known as the *residual bias*, $E(\hat{e})$, should be equal to zero.

(2) The mean of the absolute values of residuals, $|\bar{e}|$, and the square-root of the residual variance (the estimate of the *residual standard deviation*), $s(\hat{e})$, should both be of the same magnitude as the (instrumental) error of the dependent variable (observation, measured quantity y), $s_{inst}(y)$, i.e. $|\bar{e}| \approx s_{inst}(y)$ and $s(\hat{e}) \approx s_{inst}(y)$.

(3) The *residual skewness*, $\hat{g}_1(\hat{e})$, for a Gaussian normal distribution should be equal to zero.

(4) The *residual kurtosis*, $\hat{g}_2(\hat{e})$, for a Gaussian normal distribution should be equal to 3.

(5) The *residual variance* is calculated from the residual sum of squares $\hat{\sigma}^2 = U(\mathbf{b})/(n-m)$.

(6) The *coefficient of determination*, D, is calculated from

$$D = 1 - \frac{U(\mathbf{b})}{\sum_{i=1}^{n}(y_i - \bar{y})^2} \tag{8.179}$$

where $\bar{y} = (1/n)\sum_{i=1}^{n} y_i$. The coefficient of determination is equal to the square of the correlation coefficient, for linear models.

(7) In chemometrics, we often use the Hamilton R-factor:

$$R = \sqrt{\frac{U(\mathbf{b})}{\sum_{i=1}^{n} y_i^2}} \tag{8.180}$$

When $\bar{y} = 0$, $R^2 = 1 - D$, so Eq. (8.180) may be expressed as

$$R = \sqrt{(1-D) - \frac{(1-D)n\bar{y}^2}{\sum\limits_{i=1}^{n} y_i^2}} \qquad (8.181)$$

The Hamilton R-factor illustrates the difference between the two models, $y = f(x, \boldsymbol{\beta})$ and $\bar{y} = 0$. This rule is not correct for models with intercept terms and the values of the Hamilton R-factor are incorrectly low. It should be noted that D and the R-factor are continuous functions of the number of parameters. D is an increasing function of the number of parameters, whereas the R-factor is decreasing function of this number. Therefore, both of these statistics are unsuitable as resolution diagnostics for comparing models with different numbers of parameters.

(8) To distinguish between models, the *Akaike information criterion*, *AIC*, is more suitable

$$AIC = -L(\mathbf{b}) + 2m \qquad (8.182)$$

The "best" model is considered to be that for which this criterion reaches a minimum value. For least-squares models and models that do not belong to the same class, the *AIC* criterion may be expressed as

$$AIC = n \ln\left[\frac{U(\mathbf{b})}{n-m}\right] + 2m \qquad (8.183)$$

It should be noted that the diagnostic use of classical residuals is not rigorous but rather approximate. Classical residuals do not have zero mean, they are biased and they are a linear combination of errors ε. Moreover, they depend on the true values of parameters $\boldsymbol{\beta}^*$, and these are unknown. Therefore, it has been suggested [58] that the projections of residuals are used, because this partly limits all these disadvantages. Lyoness [63] proposed various approximate expressions for the determination of different types of residuals for nonlinear models, for example, Jackknife residuals, recursive residuals and partial residuals, with applications similar to those for linear models.

For more objective examination of residuals, all the statistical regression diagnostics for linear models may also be used for nonlinear models. A difficulty may arise from the distributions of some test criteria, which are affected by the nonlinearity of the model. Some of the test criteria are derived from a general criterion

$$T_{p,q} = \sum_{i=1}^{n} \hat{e}_i^p \left[f(x_i, \mathbf{b}) \right]^q \tag{8.184}$$

As for linear regression models, the following conditions should be met:

(a) The test criterion $T_{1,1}$ should be approximately equal to zero since $\hat{e}^T f(x_i, \mathbf{b}) = 0$. A high value of $T_{1,1}$ indicates a false minimum or a false model.

(b) The test criterion $T_{2,1}$ indicates heteroscedasticity.

(c) The test criterion $T_{1,2}$ indicates that a false model has been proposed.

(d) The test criterion $T_{1,0}$ should be approximately equal to zero. From Eq. (8.154) it follows that the mean of the residuals $E(\hat{e})$ is not equal to zero. Therefore $T_{1,0} = n \sum_{i=1}^{n} \left[d_i - \sum_{k=1}^{q} P_{ik} d_k \right]$. Expressions for the variance and covariance matrix have been proposed [60]. The predictive ability of a proposed model may be examined by the *mean quadratic error of prediction*, defined by

$$MEP = \frac{1}{n} \sum_{i=1}^{n} \left[y_i - f(x_i, \mathbf{b}_{(i)}) \right]^2 \tag{8.185}$$

The one-step approximation $\mathbf{b}_{(i)}^1$, defined by Eq. (8.188), may be used instead of the parameter estimate, $\mathbf{b}_{(i)}$. The lower the values of *MEP*, the better the predictive ability of the proposed model. The correctness of the proposed model can be examined by the White test [67]. The coefficient C, defined as

$$C = (n-1) \sum_{i=1}^{n} \frac{\delta f(x_i, \mathbf{b})}{\delta \beta_j} \frac{\delta f(x_i, \mathbf{b})}{\delta \beta_k} \left[\hat{e}_i - \frac{U(\mathbf{b})}{n} \right] \tag{8.186}$$

should be equal to zero for a correct model only. The test procedure calculates $(m(m + 1)/2 - n)$ variables, and

$$w_s = \frac{\partial f(x_i, \beta)}{\partial \beta_j} \frac{\partial f(x_i, \beta)}{\partial \beta_k} \tag{8.187}$$

for $s = 1, ..., m(m + 1)/2$. The test criterion represents the correlation coefficient of regression of the variable $\hat{e}^2$ on the vector of variables $\mathbf{w}$. In the case of a correct model, this test criterion has the $\chi^2_{1-\alpha}(m(n + 1)/2)$ distribution.

8.6.4.3 Identification of Influential Points

For linear regression models (Chapter 6), all characteristics which aid the identification of influential points are functions of residuals $\hat{e}_i$ and diagonal elements H_{ii} of the projection matrix, $\mathbf{H} = \mathbf{X}(\mathbf{X}^T\mathbf{X})^{-1}\mathbf{X}^T$. Some diagnostics of influential points use estimate $\mathbf{b}_{(i)}$ calculated from all points except the ith one. For linear regression models, the estimate $\mathbf{b}_{(i)}$ is easily obtained from information on the matrix $\mathbf{X}^T\mathbf{X}^{-1}$ and quantities $\hat{e}_i$ and H_{ii}.

For nonlinear regression models, the situation is rather more complicated as the parameter estimates and residuals cannot be expressed simply as a linear combination of experimental data. When a Taylor-type linearization of the original nonlinear model is used, all methods for identification of influential points in linear models can be used. This starts with a one-step approximation of the parameter estimate

$$\mathbf{b}^1_{(i)} = \mathbf{b} - \left(\mathbf{J}^T\mathbf{J}\right)^{-1} \mathbf{J}_i \frac{\hat{e}_i}{1 - P_{ii}} \tag{8.188}$$

where P_{ii} are elements of the projection matrix [Eq. (8.68)]. With the use of Eq. (8.188), the test criterion DFS_{ij} may be written as

$$DFS_{ij} = \frac{b_j - b^1_{j(i)}}{\hat{s}_{(i)} V_{ii}} \tag{8.189}$$

This criterion expresses the influence of the ith point on the estimate of the jth parameter. Quality, $\hat{s}^2_{(i)}$, is the variance estimate calculated when the ith point is omitted, i.e.

$$\hat{s}_{(i)}^2 = \frac{U(\mathbf{b}) - \dfrac{\hat{e}_i^2}{1 - P_{ii}}}{n - m - 1} \tag{8.190}$$

The symbol V_{ii}, in Eq. (8.189), denotes elements of the matrix $\mathbf{V} = (\mathbf{J}^{\mathrm{T}}\mathbf{J})^{-1}$. Applying the DFS_{ij} criterion, the ith point is considered to be influential if $DSF_{ij} > \dfrac{2}{\sqrt{n}}$.

Influential points may be identified readily on the basis of a one-step approximation of the Jackknife residuals $\hat{e}_{Ji}$ expressed as,

$$\hat{e}_{Ji} = \frac{\hat{e}_i}{\hat{s}_{(i)}\sqrt{1 - P_{ii}}} \tag{8.191}$$

To express the influence of individual points on parameter estimates, the quadratic expansion of a regression model may be used. Often an examination of either the changes of the vector of bias with the omission of the ith point, $\mathbf{h}_{(i)}$, or changes of the mean value of the ith residual with the ith point omitted, is suitable [66].

A nonlinear measure of the influence of the ith point on the parameter estimates is represented by the likelihood distance,

$$LD_i = 2\left[\ln L(\mathbf{b}) - \ln L(\mathbf{b}_{(i)})\right] \tag{8.192}$$

In the case of least-squares, the likelihood distance is expressed by

$$LD_i = n \ln\left[\frac{U(\mathbf{b}_{(i)})}{U(\mathbf{b})}\right] \tag{8.193}$$

In Eqs. (8.192) and (8.193), the estimates $\mathbf{b}_{(i)}$ calculated by nonlinear regression, when the ith point was left out, or the one-step approximation $\mathbf{b}_{(i)}^1$ of parameter estimates may be used. When $LD_i > \chi_{1-\alpha}^2(2)$, the ith point is said to be strongly influential. The significance level, α, is usually chosen as 0.05.

Some diagnostics for identification of influential points were compared [66], and the following conclusions were reached.

(a) Influential points affect not only the parameter estimates but also the relative bias h_R, which is rather sensitive to the presence of influential points.

(b) Diagnostics based on linearization or quadratic expansion of a nonlinear model do not always indicate the presence of influential points. They are not suitable for strongly nonlinear models.

(c) The best identification of influential points is given by the likelihood distance LD_i. In some cases, groups of influential points can cause masking effects.

(d) For practical calculations, the approximation of LDS_i is sufficient when the quantity $\mathbf{b}^1_{(i)}$ from Eq. (8.188) is used in Eq. (8.185) instead of $\mathbf{b}_{(i)}$.

Problem 8.22   *Identification of influential points in the dependence of the activity of sea-weed on temperature*

Determine the influential points in the data in Problem 8.15. Calculate the mean values of residuals $E(\hat{e}_i)$, the Jackknife residuals $\hat{e}_{Ji}$ and the likelihood distance LD_i and LDS_i.

Data: from Problem 8.15

Table 8.8 The mean values of residuals and other measures of influential points for the data in Problem 8.15

Point	$E(\hat{e}_i)$	$\hat{e}_{Ji}$	LDS_i	LD_i
1	0.61	0.81	1.05	0.277
2	0.304	0.34	0.82	0.085
3	−0.661	−0.63	1.99	0.376
4	−0.343	−0.35	0.98	0.121
5	0.745	0.89	1.22	4.75
6	−0.419	−0.89	0.93	136
7	0.181	0.415	0.0216	1968

Solution: Diagnostics for the identification of influential points are listed in Table 8.8.

Conclusion: When the diagnostics for identification of influential points are compared, only the nonlinear measure LD_i lead to a conclusion that two points, 6 and 7, are strongly influential, because they control the decreasing part of the curve. The other diagnostics do not indicate obviously influential points. The mean values of residuals $E(\hat{e}_i)$ show their bias directly.

8.7 Procedure for Building and Testing a Nonlinear Model

The quality of a proposed nonlinear model is examined in the same way as for linear models, using the following criteria [100]:

(1) *The quality of parameter estimates*

The quality of parameter estimates obtained is considered according to their confidence intervals or their variances $D(b_j)$. The empirical rule that is often used is that a parameter is considered to be significant when its estimate is greater than 3 standard deviations, i.e. $3\sqrt{D(b_j)} < |b_j|$. High parameter variances are often caused either by termination of the minimization process before a minimum is reached, by inaccuracy of determination of matrix $\mathbf{J}$, or high nonlinearity of the regression model.

(2) *The quality of the curve fitting*

Agreement of the proposed model with the experimental data is examined by the goodness-of-fit test based on the statistical analysis of residuals. The following statistical characteristics for a set of classical residuals are calculated: from the residual square-sum $U(\mathbf{b})$ reached at a minimum, the estimate of residual variance $\hat{\sigma}^2$ and estimates of the determination coefficient D, the regression rabat $100D[\%]$, the arithmetic mean of the residuals $E(\hat{e}_i)$, the mean of the absolute values of the residuals $|\bar{e}|$, the mean of the relative residuals $\bar{e}_{rel}$, the residual standard deviation $\hat{s}(\hat{e})$, the residual skewness $\hat{g}_1(\hat{e})$, the residual kurtosis $\hat{g}_2(\hat{e})$, and the Pearson χ^2-test of normality of the residual distribution are carried out. In addition, the four test criteria $T_{1,1}, T_{2,1}, T_{1,2}$ and $T_{1,0}$ are calculated, for more objective residual analysis.

(3) *The predictive ability of the proposed model*

The predictive ability of a model is classified by the following procedure. Data are divided into two groups, $\mathbf{M}_1$ with indices $i = 1, ..., \text{int}(n/2)$ and $\mathbf{M}_2$ with indices $i = \text{int}(n/2) + 1, ..., n$. Estimates of the parameters are calculated from points in the subgroup $\mathbf{M}_1$ as $\mathbf{b}(\mathbf{M}_1)$. The predictive ability of the model is expressed by

$$K = \frac{U(\mathbf{b})}{\sum_{i \in M_1} [y_i - f(x_i, \mathbf{b}(\mathbf{M}_2))]^2 + \sum_{i \in M_2} [y_i - f(x_i, \mathbf{b}(\mathbf{M}_1))]^2} \qquad (8.194)$$

The predictive ability of the model is higher as the criterion, K, tends towards a value of 1. The mean quadratic error of prediction *MEP* (8.185) is then calculated. The lower the value of *MEP*, the better the predictive ability of the proposed model.

(4) *The quality of the experimental data*

For the examination of the quality of the experimental data, influential points are identified by regression diagnostics. The most important diagnostics are the likelihood distances LD_i and LDS_i. The test criterion DFS_{ij} and the mean value of each residual $E(\hat{e}_i)$ are also useful.

(5) *The correctness of the model proposed*

The White test calculates the coefficient C to prove that proposed model is correct. Some other tests of accuracy of the proposed model have been proposed [68].

(6) *The physical meaning of parameter estimates*

In experimental models, there are often restrictions from the physical meaning of the parameters. For example, concentrations or molar absorptivities must be positive numbers.

8.8 References

[1]   L. Endrenyi, ed., *Kinetic Data Analysis*, p. 47, Plenum Press, New York, 1983.

[2]   R. C. Magel and D. Hertsgaard, *Commun. Stat.*, 1987, 16, 85.

[3]   J. M. Criado, M. Gonzalez, A. Ortega and C. Real, *J. Therm. Anal.*, 1984, 29, 243.

[4]   F. J. Anscombe, *J. Royal Stat. Soc*, 1969, B29, 1.

[5]   A. R. Gallant, *Nonlinear Statistical Models*, Wiley, New York, 1987.

[6]   Y. Bard, *Nonlinear Parameter Estimation*, Academic Press, New York, 1974.

[7]   D. M. Bates and D. G. Watts, *J. Roy. Stat. Soc*, 1980, B24, 1.

[8]   J. C. Nash, *J. Inst. Math. Applies.*, 1977, 19, 321.

[9]   K. Hiebert, *ACM Trans. Math. Software*, 1981, 7, 1.

[10]   J. L. Kuester and J. N. Mize, *Optimization Techniques in FORTRAN*, McGraw Hill, New York, 1973.

[11]   M. A. Wolfe, *Numerical Methods for Unconstrained Optimization*, Van Nostrand, New York, 1978.

[12]   P. E. Gill, W. Murray and M. M. Wright, *Practical Optimization*, Academic Press, London, 1981.

[13] R. Schmidt, *Advances in Nonlinear Parameter Optimization*, Springer, Berlin, 1982.

[14] T. Nakagawa and Y. Oyanagi, *Program System SALS for Nonlinear Least Squares Fitting*, ISE-TR-13, University of Tsukuba, Japan, 1980.

[15] M. M. Rosenbrock and C. Storey, *Computational Techniques for Engineers*, Pergamon, Oxford, 1966.

[16] M. D. J. Powell, *J. Comput.*, 1964, 7, 155.

[17] W. Spendley, G. R. Hext and F. R. Himworth, *Technometrics*, 1962, 4, 441.

[18] J. A. Nelder and R. Mead, *J. Comput.*, 1965, 7, 308.

[19] M. W. Routh, P. A. Swartz and M. B. Denton, *Anal. Chem.*, 1977, 49, 1422.

[20] P. B. Ryan, P. L. Barr and M. D. Tod, *Anal. Chem.*, 1977, 49, 1460.

[21] S. Marsili-Libelli and M. Castelli, *Appl. Mathematics and Comput.*, 1987, 23, 341.

[22] I. A. Volkov, P. I. Grabov and A. B. Potapov, *Zavod. Lab.*, 1985, No. 5, 60.

[23] F. T. Lindstrom, *Am. Statist.*, 1980, 34, 183.

[24] W. Spendley, in *Optimization*, R. Fletcher (ed.), Academic Press, London, 1969.

[25] W. L. Price, *J. Opt. Theor. Appl.*, 1983, 40, 333.

[26] I. O. Bohachevsky, M. E. Johnson and M. L. Stein, *Technometrics*, 1986, 28, 209.

[27] M. V. Henckroth, *AICHE Journal*, 1976, 22, 744.

[28] L. Pronzato, E. Walter, A. Venot and J. F. Lebruchec, *Math. Comput. Simulation*, 1984, 26, 412.

[29] L. G. Sillen and N. Ingri, *Acta Chem. Scand.*, 1962, 16, 173.

[30] M. Meloun and J. Čermák, *Talanta*, 1984, 31, 947.

[31] G. Peckham, *J. Comput.*, 1970, 13, 418.

[32] M. L. Ralston and R. I. Jennrich, *Technometrics*, 1978, 20, 7.

[33] J. E. Dennis, D. M. Gay and R. E. Welsch, *ACM Trans. Math. Software*, 1981, 7, 348.

[34] H. Ramsin and P. Wedin, *BIT*, 1977, 17, 72.

[35] R. I. Jennrich and P. F. Sampson, *Technometrics*, 1968, 10, 63.

[36] R. Schnidt, *Advances in Nonlinear Parameter Optimization*, Springer, Berlin, 1982.

[37] J. E. Dennis and R. E. Welsch, *Commun. Statist.*, 1978, B7, 345.

[38] P. E. Gill and W. Murray, *SI AM J. Num. Anal.*, 1978, 15, 977.

[39] J. M. Chambers, *Biometrika*, 1973, 60, 1.

[40] J. C. Nash, *Compact Numerical Methods for Computers*, Adam Hilger Ltd., Bristol, 1979.

[41] N. Wharton and D. K. Olson, *A Generalized Nonlinear Least-Squares Fitting*, Program Rept.ORNL ITM-6545, Oak Ridge Natl. Lab., 1978.

[42] J. J. More, in *Lecture Notes in Mathematics*, D. Watson (eds.), No. 630, Springer Verlag, Berlin,1978.

[43] S. G. Linquist, *Proc. Conf. COMPSTAT80*, Physica Verlag, Wien, 1980.

[44] R. R. Meyer and D. M. Roth, *J. Inst. Math. Applies*, 1973, 9, 218.

[45] J. E. Dennis and H. H. W. Mei, *J. Opt. Theor. Appl*, 1979, 28, 453.

[46] J. Militký and J. Čáp, *Proc. Conf. CEF 87*, Taormina, Sicilia, May, 1987.

[47] J. V. Beck and K. J. Arnold, *Parameter Estimation in Engineering and Science*, Wiley, New York,1977.

[48] A. R. Gallant, *J. Am. Statist. Assoc*, 1977, 72, 523.

[49] E. Z. Demidenko, *Linejnaja i nelinejnaja regresija*, Finansy i Statistika., Moskva, 1981. [50] G. M. Stanley and R. S. H. Mah, *Chem. Eng. Sci.*, 1981, 36, 259.

[51] V. G. Gorskij, *Zavod. Lab.*, 1987, No. 1, 50.

[52] K. M. Brown and J. E. Dennis, *Num. Math.*, 1972, 18, 289; and A. J. Miller in *Interactive Statistics*,D. McNeil (eds.), North Holland, Amsterdam, 1979.

[53] D. A. Ratkowsky, *Nonlinear Regression Modelling*, Dekker, New York, 1983.

[54] L. Lukšan, *SPONA - Package for Optimization and Nonlinear Approximation*, Resp. Rept. V-4,Central Computing Center of Academy of Sciences, Prague, 1976.

[55] F. James and M. Ross, *Comp. Phys. Commun.*, 1976, 10, 343. [56] D. M. Bates and D. G. Watts, *J. Roy. Stat. Soc*, 1986, B42, 1.

[57] J. R. Donaldson and R. B. Schnabel, *Technometrics*, 1987, 29,

[58] R. D. Cook, E. C.-L. Tsai and B. C. Wei, *Biometrika*, 1986, 73,

[59] M. J. Box, *J. Roy. Stat. Soc*, 1971, B32,

[60] R. Morton, *Biometrika*, 1987, 74, 679.

[61] G. P. Clarke, *J. Am. Statist. Assoc*, 1987, 82, 221.

[62] L. Schwartz, *Anal. Chim. Acta*, 1980, 122, 291.

[63] R. M. Lyoness, *Commun. Statist.*, 1987, 16, 997.

[64] D. M. Himmelblau, *Process Analysis by Statistical Methods*, Wiley, New York, 1970.

[65] M. Meloun, J. Havel and E. Hogfeldt, *Computation of Solution Equilibria*, Ellis Horwood, Chichester,1988.

[66] W. C. Hamilton, *Statistical in Physical Science*, Ronald Press, New York, 1964.

[67] J. Militký, *Proc. 2nd Int. Statist. Conference*, Tampere University Press, 1987.

[68] H. White and I. Dorniwotz, *Econometrica*, 1984, 52, 143.

[69] M. R. Hestena and E. Stiefel, *J. Res. NBS*, 1952, 49, 409.

[70] L. C. W. Dixon, *J. Inst. Math. Appl.*, 1975, 15, 9.

[71] R. Fletcher, *Comput. J.*, 1970, 13, 317.

[72] P. E. Gill and W. Murray, *J. Inst. Maths. Appl*, 1972, 9, 91.

[73] S. S. Oren and D. G. Luenberger, *Management Sci.*, 1974, 20, 845.

[74] S. S. Oren, *Management Sci.*, 1974, 20, 863.

[75] H. Y. Huang, J. P. Chambliss, *J. Optimization Theory Appl.*, 1974, 13, 620.

[76] R. Bass, *Math, of Comp.*, 1972, 26, 129.

[77] D. H. Jacobson and W. Oksma, *J. Math. Anal. Appl.*, 1972, 38, 535.

[78] E. J. Davison and P. Wong, *Automatica*, 1975, 11, 197.

[79] K. Ritter, *Computing*, 1975, 14, 79.

[80] W. C. Davidon, *Math. Programming*, 1975, 9, 1.

[81] W. H. Swann, *Central Instrument Laboratory Research Note*, 1964, 64, 3.

[82] M. J. D. Powell, *J. Comput.*, 1964, 7, 155.

[83] W. I. Zangwill, *J. Comput.*, 1967, 10, 293.

[84] K. W. Brodlie, *J. Inst. Maths. Appl.*, 1975, 15, 385.

[85] L. Nazareth, *Res. Report LBL2692*, University of California 1973.

[86] R. Mifflin, *Math. Programming*, 1975, 9, 100.

[87] R. F. Dennemeyer and E. H. Mookini, *J. Optimization Theory Appl.*, 1975, 16, 67.

[88] P. E. Gill, Murray, *Math. Programming*, 1974, 7, 311.

[89]   W. E. Bosarge and P. L. Fabl, *J. Optimization Theory Appl.*, 1969, 4, 156.

[90]   R. Fletcher, *Res. Report R-6799*, AERE Harwell, 1971.

[91]   M. Meloun and M. Javůrek, *Talanta*, 1985, 32, 973.

[92]   G. P. Box, W. G. Hunter, J. S. Hunter: *Statistics for Experimenters.* J. Wiley, New York, 1978. str. 483-487.

[93]   L. Bennett, L. Swartzendruber, H. Brown: *Superconductivity Magnetization Modeling*, NIST 1994.

[94]   K. Eckerle: *Circular Interference Transmittance Study*, NIST (1975).

[95]   B. Rust: NIST 1996.

[96]   D. Chwirut: *Ultrasonic Reference Block Study*, NIST (1975).

[97]   C. Lanczos: *Applied Analysis.* Englewood Cliffs, NJ., Prentice Hall, 1956, str. 272-280.

[98]   J. J. More, B. S. Garbow, K. E. Hillstrom: *Testing unconstrained optimization software. ACM Transactions on Mathematical Software.* 7(1), (1981), p. 17-41 in book M. R. Osborne: *Some aspects of nonlinear least squares calculations. In Numerical Methods for Nonlinear Optimization*, Lootsma (ed). Academic Press, New York 1972, p. 171-189.

[99]   D. A. Ratkowsky: *Nonlinear Regression Modeling.* Marcel Dekker, New York 1983, p. 61-88.

[100] M. Meloun, J. Militký: *Statistické zpracování experimentálních dat*, Plus Praha 1994 (1st edition), EAST PUBLISHING, Praha 1998 (2nd edition), Academia Praha 2004 (3rd edition).

Supplemented material (Review Questions, Exercises, Results of Exercises) to Chapter 8 is on the enclosed CD.

Index